COOPERATIVE LOCALIZATION AND NAVIGATION

Theory, Research, and Practice

COOPERATIVE LOCALIZATION AND NAVIGATION

Theory, Research, and Practice

Edited by

Chao Gao, Guorong Zhao
Naval Aviation University, Yantai, China

Hassen Fourati
University Grenoble Alpes, Grenoble, France

CRC Press
Taylor & Francis Group
Boca Raton London New York

CRC Press is an imprint of the
Taylor & Francis Group, an **informa** business

MATLAB® is a trademark of the MathWorks, Inc. and is used with permission. The Mathworks does not warrant the accuracy of the text or exercises in this book. This book's use or discussion of MATLAB® software or related products does not constitute endorsement or sponsorship by the MathWorks of a particular pedagogical approach or particular use of the MATLAB® software.

CRC Press
Taylor & Francis Group
6000 Broken Sound Parkway NW, Suite 300
Boca Raton, FL 33487-2742

First issued in paperback 2022

© 2020 by Taylor & Francis Group, LLC
CRC Press is an imprint of Taylor & Francis Group, an Informa business

No claim to original U.S. Government works

ISBN: 978-1-03-240129-4 (pbk)
ISBN: 978-1-138-58061-9 (hbk)
ISBN: 978-0-429-50722-9 (ebk)

DOI: 10.1201/9780429507229

Publisher's Note
The publisher has gone to great lengths to ensure the quality of this reprint but points out that some imperfections in the original copies may be apparent.

**Visit the Taylor & Francis Web site at
http://www.taylorandfrancis.com**

**and the CRC Press Web site at
http://www.crcpress.com**

Dedicated

To my daughter, JiaYi, my wife, Yumiao, and my parents,
without whom I would not be where I am today

Chao Gao

Contents

Section I Preliminaries on Cooperative Navigation: Communications, Coordination and Sensing

Section II Cooperative Strategies for Localization and Navigation

Section III Estimation Methodologies for Cooperative Localization and Navigation

Foreword

Determining the position of objects on the globe has been the subject of great interest since ancient times when people used stars to triangulate their positions. The advent of radar in the 1930s brought a major advance where the direction of targets could be measured by transmitting short radio pulses and measuring the direction of the strongest return. Earlier, Bellini and Tosi had invented a method for finding the direction of a radio transmitter by using two orthogonal loop antennas. Since loop antennas receive both vertical and horizontal polarization, their accuracy was diminished due to skywave noise. This drawback was eliminated by the invention of the Adcock array, which consists of four monopole (or dipole) antennas positioned at the vertices of a square. The Adcock array only receives vertically polarized signals and can measure elevation and azimuth.

The satellite-based global positioning system (GPS) was launched by the US military in the 1970s and became available for civilian use in the 1980s. While GPS remains the most popular positioning system for open outdoor environments, it is inadequate in many real-world scenarios. In indoor environments, in urban canyons, and under forest canopies, the GPS signal is greatly attenuated, rendering positioning systems that rely solely on GPS unreliable, yet there are numerous applications in industrial, commercial, medical, public safety, and military settings that require localization of objects and individuals in these situations.

Apart from GPS, some positioning systems rely on other infrastructures such as cellular network. The impetus behind localization using cellular networks was the 2015 mandate from the US Federal Communication Commission (FCC) that cellular network operators be capable of locating emergency calls from indoors to within 50-meter accuracy. Wireless local area networks based on IEEE 802.11 standards have also been widely employed for indoor localization due to the rapid growth of wireless access points. For some applications, however, a supporting infrastructure does not exist, and localization methods must be devised to operate without an infrastructure.

A new area of localization and navigation for GPS-denied environments has emerged in recent years involving cooperation among clustered nodes. Early examples include localization of nodes in ad-hoc and wireless sensor networks (WSNs). Due to cost and size limitations, sensor nodes are not equipped with GPS modules and are required to obtain their location information after the network is deployed. Many algorithms have been devised for localization in WSNs where the nodes determine their positions in cooperation with other nodes.

Research in cooperative localization and navigation has been accelerating in response to a host of new applications, including fleet navigation of autonomous unmanned ground, underwater, or aerial vehicles, indoor localization, and localization in 5G networks. Localization and navigation of multiple unmanned vehicles moving as a rigid body can be accomplished through cooperation and information exchange with other members of the group. In particular, if a few vehicles can maintain an accurate estimate of their positions and act as reference nodes, other members of the group can evaluate their positions accurately through exchange of information with the reference nodes. The accuracy of indoor positioning systems is limited by the harsh multipath propagation environment. Cooperative localization can not only improve the accuracy of the positioning system but also enable localization in blind spots; namely, areas that may not receive a signal from the reference nodes. In the upcoming 5G cellular networks, localization of mobile terminals (MTs) can be enhanced by adoption of new technologies such as device-centric architectures, millimeter wave (mmWave), and massive multiple-input multiple-output (MIMO) systems. For example, mmWave operates at higher frequencies and provides a larger bandwidth, improving the accuracy of time measurements as well as the resolution of multipath. Massive MIMO provides more accurate directional beams between the base station and MTs, and device-to-device communication provides a large number of line-of-sight links between MTs and enables information exchange to aid in cooperative localization.

While work in the area of localization is rapidly advancing, currently there is no book that assembles the state-of-the-art research into a single reference. This book presents a review of the latest results in cooperative localization in both theory and practice. It provides the reader with a comprehensive view of state estimation techniques for localization and navigation as well as advances in design of algorithms and architectures. Important aspects of cooperative localization and navigation including state estimation techniques, fusion techniques for inertial and vision sensors, computation and communication aspects of cooperative localization, and applications to wireless sensor networks, multi-robot systems, fleet of unmanned ground, underwater, and aerial vehicles are discussed throughout the book.

This book is comprised of four sections and 29 chapters. Section I is dedicated to preliminary issues in cooperative localization and navigation including communication, coordination and sensing. This section includes seven chapters, with Chapter 1 presenting an overview of position location techniques followed by a study of localization in reconfigurable networks, cooperative versus noncooperative localization, indoor positioning, and novel signal processing techniques for localization. Chapter 2 investigates the problem of unique localizability in cooperative localization of wireless ad hoc and sensor networks. Optimization of computational cost and communication overhead in cooperative localization is studied in Chapter 3. Extending radio coverage by positioning radio relays using swarm of UAVs is studied in Chapter 4. Chapter 5 is devoted to the optimization of Euclidean distance matrix (EDM) for sensor network localization. In Chapter 6 the author provides statistical shape analysis and anchor selection in cooperative localization. Finally, in Chapter 7 heterogeneity is exploited as a mechanism to design resilient networks of cooperative robots.

Section II includes eight chapters and is devoted to cooperative strategies for localization and navigation. Chapter 8 presents a preliminary study of cooperative navigation for underwater vehicles utilizing range and range-rate observations. In Chapter 9 the authors review recent cooperative strategies for localization techniques and solutions, which are adapted on mobile-devices via onboard wireless sensor technologies, and compare their performances. Chapter 10 describes a multi-robot navigation system which explores and maps large-scale urban environments while performing visual object recognition. In Chapter 11 the authors study vehicle localization in Global Navigation Satellite System (GNNS)-denied environments. Chapter 12 studies cooperative localization in multi-robot systems based on relative distance measurements. Chapter 13 presents cooperative multi-vehicle navigation for mobile agents and underwater vehicles, using range-based techniques. In Chapter 14 pedestrian positioning is studied using nearby vehicles as anchor nodes. Finally, Chapter 15 considers the problem of network-based mobile phone positioning when multiple providers are available.

Section III, which consists of eight chapters, is dedicated to estimation methods for cooperative localization and navigation. In Chapter 16 the authors investigate centralized and distributed optimization methods for cooperative detection and localization. Chapter 17 considers distributed localization for multi-vehicle networks based on graph Laplacian techniques. In Chapter 18 the authors also consider range-based navigation for marine applications, and in Chapter 19 energy-efficient distributed localization for WSNs is presented. Diffusion Kalman filtering based on covariance intersection is studied in Chapter 20, and in Chapter 21 the authors investigate cooperative localization using GNSS positioning and the positioning based on dedicated short-range communication (DSRC). Chapter 22 considers the problem of coordinating multiple mobile robots to sense and track a moving target. Chapter 23 also considers the motion control of multiple aerial vehicles for mapping and navigation purposes.

Finally, Section IV is dedicated to new directions and current and future challenges in cooperative navigation, which is also the subject of Chapter 24 and discussed in detail in this chapter. In Chapter 25 the authors present a multi-robot localization scheme sharing GNNS corrections using a distributed estimation method. Chapter 26 studies the challenges of navigation of underwater vehicles, and Chapter 27 considers the problem of self-localization and target tracking under sparse communication. Chapter 28 applies deterministic ray-tracing methods for localization of mobile terminals in indoor environments. Finally, Chapter 29 studies cooperative localization of WSNs using kernel Isomaps.

We hope that the readers will be stimulated by the interesting research presented in this book and that it helps spur additional research in the exciting and growing field of cooperative localization and navigation.

Mort Naraghi-Pour, PhD
Michel B. Voorhies Distinguished Professor
School of Electrical Engineering and Computer Science
Louisiana State University
Baton Rouge, Louisiana

Preface

Cooperative navigation, as a novel and promising direction of navigation technique, focuses mainly on the localization and navigation demands of clustered moving bodies based on cooperative bodies or exogenous information sources. Over the past several years, there has been an increasing interest in the critical technique point of this navigation approach. This interest is motivated by the versatility of the diverse areas of application such as indoor localization, aerial and underwater navigation, fleet navigation of unmanned vehicles, localization in 5G networks, spacecraft, and multi-robot applications. Cooperative navigation consists of the determination of rigid body location and heading in 3D space (principally in terms of position, velocity, and azimuth). This research area is a multilevel, multifaceted process involving communication basis, cooperative strategies, estimation methodologies, applicable platforms, and extension directions of cooperative navigation from several sources. Another area of interest is in the fact that the results of this book can provide an alternative navigation technique under the GNSS-denied environment by means of information cooperation.

Cooperative navigation for clustered rigid bodies is a research area that borrows ideas from diverse fields, such as wireless communication, signal processing, multi-agent systems, and estimation theory, where enhancements are involved in point-of-view data authenticity or availability. Cooperative navigation for clustered rigid bodies is motivated by several issues and problems, such as communication architecture, cooperative strategies, estimation methodologies, and processing frameworks. As a majority of these problems have been identified and heavily investigated, no single data fusion algorithm is capable of addressing all of these challenges. Consequently, a variety of methods in the literature focuses on a subset of these issues.

The idea of this book comes as a response to the immense interest and strong activities in the field of cooperative localization and navigation during the past few years, both in theoretical and practical aspects. This book is targeted toward researchers, academics, engineers, and graduate students working in the fields of sensor fusion, filtering, and signal processing for localization and navigation. This book, *Cooperative Localization and Navigation: Theory, Research and Practice*, captures the latest results and techniques for cooperative navigation drawn from a broad array of disciplines. It is intended to provide the reader with a generic and comprehensive view of contemporary state estimation methodologies for localization and navigation, as well as the most recent research and novel advances on cooperative localization and navigation tasks exploring the design of algorithms and architectures, benefits, and challenging aspects, as well as a potential broad array of disciplines, including wireless communication, indoor localization, robotics, and emergency rescue. These issues arise from the imperfection and diversity of cooperative sources, the contention and collision of communication channels, the selection and fusion of cooperative data, and the nature of the application environment. The issues that make cooperative-based navigational state estimation a challenging task, and which are discussed throughout the book, are related to (1) the nature and model of sensors and cooperative sources (e.g., range-based sensor, angle-based sensor, inertial sensor, and vision sensor), (2) the communication medium and cooperative strategies, (3) the theoretical developments of state estimation and data fusion, and (4) the applicable platforms.

With contributions from leading navigation researchers and academicians, this book consists of 29 chapters, divided into four sections. Each chapter is complete in itself and can be read in isolation or in conjunction with other chapters. The book is introduced with a brief foreword on cooperative localization and navigation problems. In Section 1, Chapters 1 through 7 are devoted to some preliminaries on cooperative platform representations, communication requirements, formation coordination, and the feature of cooperative sources. In Section 2, Chapters 8 through 15 focus mostly on cooperative strategies applicable for various navigational scenarios, such as unknown, dynamic, and troublesome scenarios, cooperative underwater navigation, cooperative multi-robot navigation, vehicle localization in GNSS-denied environments, anchor-based cooperative positioning, and network-based localization. In Section 3, Chapters 16 through 23 present the main theories and novel advances in estimation methodologies for cooperative navigation, including centralized and distributed optimization methods, graph Laplacian techniques, range-based navigation algorithms, energy-efficient network localization, diffusion Kalman filtering, and DSRC-enhanced approach. In Section 4, Chapters 24 through 29 focus on the applicable platforms and directions, with several experimental tests, and showcase advancements such as indoor localization, UAV navigation, multiple autonomous underwater vehicles, collaborative self-localization and target tracking, and cooperative localization in wireless sensor networks.

Chao Gao, Guorong Zhao
Naval Aviation University, Yantai, P.R. China

Hassen Fourati
University Grenoble Alpes, Grenoble, France

MATLAB® is a registered trademark of The MathWorks, Inc. For product information, please contact:

The MathWorks, Inc.
3 Apple Hill Drive
Natick, MA, 01760-2098 USA
Tel: 508-647-7000
Fax: 508-647-7001
E-mail: info@mathworks.com
Web: www.mathworks.com

Editors

Chao Gao, PhD, is an assistant professor of control science and engineering at the Naval Aviation University (NAU), Yantai, China, and also a postdoctoral researcher and one of the founders of Aircraft Networked Navigation Technology Laboratory (ANTL), affiliated with NAU. He earned his bachelor of engineering degree in electrical engineering and automation at the Naval Aeronautical Engineering Institute, China, and his master's and PhD degrees in control science and engineering at the Naval Aeronautical and Astronautical University, China, in 2007, 2009, and 2017, respectively. His current research interests include networked navigation technology, mobile wireless networks, cooperative communication, networked control systems, and navigation technology. He holds nine inventive patents, and has published about 30 research papers in scientific journals, international conferences, and book chapters.

Guorong Zhao, PhD, is a full professor of control science and engineering at Naval Aviation University (NAU), Yantai, China, and leader of the Aircraft Networked Navigation Technology Laboratory (ANTL), affiliated with NAU. He earned his bachelor of engineering degree in engineering science from Naval Aeronautical Engineering Institute (NAEI), Yantai, China in 1984, and his master's and PhD degrees in control science and engineering from Harbin Institute of Technology at Harbin, China in 1992 and 1996, respectively. He joined the faculty of the Department of Aircraft Engineering at NAEI, China, as an assistant professor in 1998. He has been the Chair of the Department of Control Engineering, NAEI, since August 2012. He is currently a full professor of control science and engineering at NAU. His research interests include vehicular ad hoc networks, ad hoc wireless networks, distributed systems, navigation, guidance and control.

Hassen Fourati, PhD, is an associate professor of electrical engineering and computer science at the University of Grenoble Alpes, Grenoble, France, and a member of the Networked Controlled Systems Team (NeCS), affiliated with the Automatic Control Department of the GIPSA-Lab. He earned his bachelor of engineering degree in electrical engineering at the National Engineering School of Sfax (ENIS), Tunisia, master's degree in automated systems and control at the University of Claude Bernard (UCBL), Lyon, France, and PhD degree in automatic control at the University of Strasbourg, France, in 2006, 2007, and 2010, respectively. His research interests include nonlinear filtering and estimation and multisensor fusion with applications in navigation, motion analysis, and traffic management. He has published several research papers in scientific journals, international conferences, book chapters, and books.

Contributors

Naseer Al-Jawad
School of Science and Technology
University of Northampton
Northampton, United Kingdom

Ali Al-Sherbaz
Applied Computing Department
University of Buckingham
Buckingham, United Kingdom

Nuha A.S. Alwan
Department of Electrical Engineering
College of Engineering
University of Baghdad
Baghdad, Iraq

Joaquín Aranda
Department of Computer Science
    and Automatic Control
National Distance Education
    University
Madrid, Spain

Miguel Aranda
SIGMA Clermont Engineering
    School
Institut Pascal
Clermont-Ferrand, France

Ian Bajaj
School of Electrical and Electronic
    Engineering
Nanyang Technological University
Singapore

Floriana Benedetti
Department of
    Informatics, Bioengineering,
    Robotics and Systems Engineering
University of Genoa
Genoa, Italy

Philippe Bonnifait
Department of Computer Science
CNRS UMR 7253 Heudiasyc
Sorbonne Universities
University of Technology of Compiègne
Compiègne, France

Thomas Bräunl
Department of Electrical, Electronic and
    Computer Engineering
The University of Western Australia
Perth, Australia

Kevin M. Brink
Munitions Directorate
Air Force Research Lab
Eglin Air Force Base
Florida

Baigen Cai
School of Computer and Information
    Technology
Beijing Jiaotong University
Beijing, China

Alessio Capitanelli
Department of Informatics,
    Bioengineering, Robotics and
    Systems Engineering
University of Genoa
Genoa, Italy

Anusna Chakraborty
Aerospace Engineering and
    Engineering Mechanics Department
University of Cincinnati
Cincinnati, Ohio

Naveen Crasta
Laboratory of Robotics and
    Engineering Systems
ISR/Instituto Superior Tecnico
University of Lisbon
Lisbon, Portugal

Patricio J. Cruz
Department of Automation and Industrial
Control
Escuela Politécnica Nacional
Quito, Ecuador

Daniela De Palma
Department of Innovation Engineering
University of Salento
Lecce, Italy

Antonio del Corte Valiente
Department of Computer Engineering
University of Alcalá
Alcalá de Henares, Spain

Tolga Eren
Department of Electrical and Electronics
Engineering
Kirikkale University
Kirikkale, Turkey

Shih-Hau Fang
Department of Electrical Engineering
Yuan Ze University
MOST Joint Research Center
for AI Technology and All
Vista Healthcare
Taiwan, China

Rafael Fierro
Department of Electrical and Computer
Engineering
University of New Mexico
Albuquerque, New Mexico

Jörg Fliege
School of Mathematics
University of Southampton
Southampton, United Kingdom

Kayhan Zrar Ghafoor
Department of Computer Science
Faculty of Science
Cihan University-Erbil
Kurdistan, Iraq

Seyed Ali Ghorashi
Department of Telecommunications
Faculty of Electrical Engineering
Shahid Beheshti University
Tehran, Iran

Josefa Gómez Pérez
Department of Computer Science
University of Alcalá
Alcalá de Henares, Spain

José Manuel Gómez Pulido
Department of Computer Science
University of Alcalá
Alcalá de Henares, Spain

Francesca Guerriero
Department of Mechanical Energy and
Management Engineering
University of Calabria
Calabria, Italy

Oscar Gutiérrez Blanco
Department of Computer Science
University of Alcalá
Alcalá de Henares, Spain

Zachary J. Harris
Department of Mechanical
Engineering
Johns Hopkins University
Baltimore, Maryland

Jinwen Hu
Northwestern Polytechnical University
Xi'an, China

Zahir M. Hussain
Faculty of Computer Science and
Mathematics
University of Kufa
Najaf, Iraq

and

School of Engineering
Edith Cowan University
Joondalup, Australia

Giovanni Indiveri
Department of Innovation Engineering
University of Salento
Lecce, Italy

Jyoti Kashniyal
Department of Information Technology
Indian Institute of Information
    Technology
Allahabad, India

Linghe Kong
Department of Computer Science and
    Engineering
Shanghai Jiao Tong University
Shanghai, China

Khaoula Lassoued
Department of Computer Science
CNRS UMR 7253 Heudiasyc
Sorbonne Universities
University of Technology of Compiègne
Compiègne, France

Kai Li Lim
Department of Electrical, Electronic and
    Computer Engineering
The University of Western Australia
Perth, Australia

Zhiyun Lin
School of Automation
Hangzhou Dianzi University
Hangzhou, China

Jiang Liu
School of Electronic and Information
    Engineering
Beijing Jiaotong University
Beijing, China

Yang Lyu
School of Automation
Northwestern Polytechnical University
Xi'an, China

Halgurd Sarhang Maghdid
Department of Software
    Engineering
Faculty of Engineering
Koya University
Koysinjaq, Iraq

and

Department of Computer Science
Faculty of Science
Cihan University-Erbil
Erbil, Iraq

Fulvio Mastrogiovanni
Department of Informatics,
    Bioengineering, Robotics and
    Systems Engineering
University of Genoa
Genoa, Italy

Youcef Mezouar
SIGMA Clermont Engineering
    School
Institut Pascal
Clermont-Ferrand, France

Sohum Misra
Aerospace Engineering and
    Engineering Mechanics
    Department
University of Cincinnati
Cincinnati, Ohio

David Moreno-Salinas
Department of Computer Science and
    Automatic Control
National Distance Education University
    (UNED)
Madrid, Spain

Mort Naraghi-Pour
School of Electrical Engineering and
    Computer Science
Louisiana State University
Baton Rouge, Louisiana

Marcelo Borges Nogueira
Science and Technology School
Federal University of Rio Grande do Norte
Natal, Brazil

Sadao Obana
Graduate School of Informatics and
    Engineering
The University of Electro-Communications
Chofu, Japan

Panagiotis Agis Oikonomou-Filandras
True AI LTD
London, United Kingdom

Quan Pan
School of Automation
Northwestern Polytechnical University
Xi'an, China

Gianfranco Parlangeli
Department of Innovation Engineering
University of Salento
Lecce, Italy

António M. Pascoal
Laboratory of Robotics and Engineering
    Systems
ISR/Instituto Superior Tecnico
University of Lisbon
Lisbon, Portugal

Fernando Lobo Pereira
SYSTEC
Electrical Computer Engineering
    Department
Faculty of Engineering
Porto University
Porto, Portugal

Hou-Duo Qi
School of Mathematics
University of Southampton
Southampton, United Kingdom

Ramtin Rabiee
Department of Applied Physics and
    Electronics
Umeå University
Umeå, Sweden

Robert G. Reid
Department of Electrical, Electronic and
    Computer Engineering
The University of Western Australia
Perth, Australia

Francisco Saez de Adana
Department of Computer Science
University of Alcalá
Alcalá de Henares, Spain

Reza Shahbazian
Department of Mathematical and
    Computer Science
University of Calabria
Calabria, Italy

Rajnikant Sharma
Aerospace Engineering and
    Engineering Mechanics
University of Cincinnati
Cincinnati, Ohio

Krishna Pratap Singh
Department of Information Technology
Indian Institute of Information
    Technology
Allahabad, India

Suhua Tang
Graduate School of Informatics and
    Engineering
The University of
    Electro-Communications
Chofu, Japan

Wee Peng Tay
School of Electrical and Electronic
    Engineering
Nanyang Technological University
Singapore

Abdelhamid Tayebi
Department of Computer Science
University of Alcalá
Alcalá de Henares, Spain

Clark N. Taylor
ECE Department
Air Force Institute of Technology
Wright-Patterson Air Force Base
Ohio

Jackeline Abad Torres
Department of Automation and Industrial
    Control
Escuela Politécnica Nacional
Quito, Ecuador

Cesar Vargas-Rosales
Tecnologico de Monterrey
Escuela de Ingenieria y Ciencias
Monterrey, Mexico

Shekhar Verma
Department of Information Technology
Indian Institute of Information Technology
Allahabad, India

Gianni Viardo Vercelli
Department of Informatics,
    Bioengineering, Robotics and
    Systems Engineering
University of Genoa
Genoa, Italy

Rafaela Villalpando-Hernandez
Tecnologico de Monterrey
Escuela de Ingenieria y Ciencias
Monterrey, Mexico

Renato Vizuete
GIPSA-lab
Université Grenoble Alpes
Grenoble, France

Leigang Wang
Luoyang Electronic Equipment Test
    Center of China
Luoyang, China

Louis L. Whitcomb
Department of Mechanical Engineering
Johns Hopkins University
Baltimore, Maryland

Kai-Kit Wong
Wireless Communications
Department of Electronic and
    Electrical Engineering
University College
London, United Kingdom

Naihua Xiu
Department of Applied Mathematics
Beijing Jiaotong University
Beijing, China

Zhao Xu
School of Electronics and Information
Northwestern Polytechnical University
Xi'an, China

LiChuan Zhang
School of Marine Science and Technology
Northwestern Polytechnical University
Xi'an, China

Ping Zhang
School of Computer and Information
Anhui Polytechnic University
Wuhu, China

Tao Zhang
Department of Automation
School of Information Science and
    Technology
Tsinghua University
Beijing, China

Chunhui Zhao
School of Automation
Northwestern Polytechnical University
Xi'an, China

Section I

Preliminaries on Cooperative Navigation

Communications, Coordination and Sensing

1

Advances in Radio Localization Techniques

Cesar Vargas-Rosales, Rafaela Villalpando-Hernandez,
and Mort Naraghi-Pour

CONTENTS

1.1 Introduction

In the last few decades, the demand for accurate positioning systems has been steadily on the rise. During this period, a multitude of positioning systems have been developed, and research and development in this area continues unabated. Positioning systems were originally motivated by military applications and remain important in this arena. However, in the last three decades they have also become important in law enforcement, emergency and rescue missions, and a host of other civilian applications including safety and security, situational awareness, tracking people, vehicles, and assets, and ad-hoc and wireless sensor networks (WSN).

Global positioning system (GPS) has had a tremendous success for localizing and tracking people and objects on the entire globe. However, GPS may not be available or reliable in many situations, including indoors, under forest canopies, and in urban canyons. GPS is also susceptible to jamming and spoofing [65,66]. Local positioning systems, on the other hand, try to locate a target relative to a local coordinate system. For example, several reference receivers, referred to as anchor nodes, whose location is known may collaborate to locate a (mobile) target node. The position of the target node is then computed relative to the location of the anchors. If the position of anchor nodes is known globally, then the global position of target node can also be computed accordingly.

Localization has always been in demand in cellular networks [20,49]. Location-based services have been a part of every generation, combining satellite navigation through GPS with cellular-based services due to the performance degradation of GPS services in dense urban and indoor environments. Localization has also been a fundamental part of every emergency service since the 1990s. Regulation is not the same in every part of the world. Thus, in some areas location services have been a legal mandate, while in others, service providers are asked to provide this service voluntarily. On the commercial side, location information provides opportunities to generate location-based services that can enable market-oriented applications and generate revenue for businesses and application developers. Network optimization will also benefit if position location information is available, adding features to network tasks such as load balancing, capacity assignment, maximizing carried traffic, etc.

In the US, location requirements in cellular networks, as defined by the Federal Communications Commission (FCC), have evolved since 1996, with error tolerances of 125 m in 67% of cases, to a 2015 policy for 3D scenarios to be fulfilled in 2019 with accuracy of 50 m in the 2D location (horizontal) and a vertical accuracy to be determined to comply with the 3D policy [20].

In this chapter, we present an overview of the recent advances in radio localization techniques. In Section 1.2 we discuss the traditional methods the anchor nodes use to evaluate their separation distance from the target node. In particular, we discuss methods of evaluating time of arrival (TOA), time difference of arrival (TDOA), direction of arrival (DOA), and received signal strength (RSS) and how they are used to estimate separation distance and location coordinates.

The advent of new technologies, applications, and services in wireless communication has spurred demand for novel localization techniques. In recent years wireless reconfigurable networks (WRNs) have seen a tremendous growth due to the introduction of Internet of Things (IOT) [83], integration of drones into new applications [25,77], and the development of WSNs [82]. Localization in WRNs is presented in Section 1.3, where positioning techniques for two- and three-dimensional space are presented for single- and multi-hop networks.

Localization of a target node may be classified as cooperative or non-cooperative. In cooperative localization, a target node collaborates in the localization process, say by transmitting a known signal and synchronizing with the anchor nodes when necessary. Moreover, the anchor nodes have a template of the target node's signal and full knowledge of its signal characteristics [102]. On the other hand, in non-cooperative localization, the target node, either intentionally or unintentionally, does not participate in the localization process. Examples of this include law enforcement, intruder detection, patient monitoring, and location-aware services [2,98]. Non-cooperative localization may also include cases where the anchor nodes do not have access to the target node's signal characteristics [51]. For example, when the target node's bandwidth allocation is dynamic, and the anchor nodes do not have access to the bandwidth allocation scheme. Non-cooperative localization techniques are presented in Section 1.4.

Indoor localization is a remarkably challenging problem. Closely positioned walls and partitions, ceiling structures, the high density of objects, furniture, and people create a highly changing environment resulting in strong multipath effects which are difficult to model. The line-of-sight (LOS) component is often obstructed, limiting the reliability of TOA, TDOA, or DOA. In addition, due to small distances involved, accurate positioning requires highly precise measurements. RSS-based positioning is attractive as it puts low demand on hardware and software complexity of the positioning system. However, in an indoor environment, distance estimation based on RSS measurements alone is unreliable due to large fluctuations in RSS resulting from small-scale fading. Consequently, RSS-base indoor localization is enhanced using fingerprinting [12,42], ray tracing [89], or through a combination of RSS and time-based measurements. Challenges in indoor positioning along with fingerprinting and ray-tracing schemes are presented in Section 1.5.

Traditionally, estimation techniques such as least squares (LS), maximum likelihood (ML), filtering (Kalman, Wiener, particle), and subspace methods MUltiple SIgnal Classification (MUSIC) and Estimation of Signal Parameters via Rotational Invariance Techniques (ESPRIT) have been used in measurement of parameters (e.g., TOA, TDOA, etc.), estimation of the separation distance from measurement parameters, and finally in position estimation. Other techniques may be combined with these classical methods to enhance the performance of the positioning system. Three such methods that have recently been considered for localization, namely Euclidean distance matrices (EDM), belief propagation, and consensus algorithms, are discussed in Section 1.6. Finally, Section 1.7 ends this chapter with conclusions and a discussion of future challenges.

1.2 An Overview of Position Location Techniques

In this section, we introduce the traditional methods to estimate separation distance. These include TOA, TDOA, DOA, and RSS. The geometric methods to obtain the location in the noise-free case are also briefly discussed.

1.2.1 Time of Arrival Estimation

A frequently used technique for distance measurement is based on TOA estimation [102]. Here multiple receivers referred to as anchor nodes collaborate to localize a (mobile) target node. The position of the anchor nodes is assumed to be known. In the coplanar case, for a measured TOA τ, the target mobile is located on a circle (sphere in the non-coplanar case) of radius $R = c\tau$ centered at the anchor node, where c is the speed of the signal in the medium of interest. It is evident that for noise-free measurements, in the coplanar case at least three measurements are needed to locate the target (four in the non-coplanar case).

To measure the TOA, the anchors must store a copy of the signal which is transmitted by the target node. Having received the signal from the target, they can estimate the TOA by comparing the received signal with their own template signal. This also requires that the transmitted signal be tagged with a time stamp and that the anchor nodes and the target be synchronized. Maintaining clock synchronization across geographically dispersed devices is difficult and is one of the main drawbacks of this approach.

TOA estimation can be performed using a correlation (or matched filter) receiver [64]. The peak of the cross-correlation is used to estimate the TOA. In a multipath environment, correlation-based TOA estimation is very challenging, as multiple peaks may appear in the correlator output. Since LOS is the first multipath component to arrive, the first peak of the cross-correlation is usually used to estimate the TOA. However, all the other multipath components arriving after the LOS component cause an interference which degrades the effective signal-to-noise ratio (SNR) and deteriorates the performance of this method. Moreover, the fact that LOS may not be the strongest multipath component makes finding the first peak more challenging. Consequently, this approach remains very sensitive to multipath effects and noise.

Other methods for TOA estimation include deconvolution, ML estimation, and subspace methods. Deconvolution is similar to equalization methods where the Fourier transform of the received signal is divided by the channel frequency response to obtain the Fourier transform of the transmitted signal [35]. It is well known that this approach can enhance the noise. To mitigate this effect, constrained LS methods have been developed [91,47]. The ML estimation method attempts to resolve the multipath components by estimating their delays and the channel gains [38,71]. Due to the large number of paths and the optimization algorithm involved in ML estimation, this approach, while optimal, has a very high computational complexity. Subspace methods such as MUSIC and ESPRIT have also been studied for time delay estimation [40,45,97] and are more computationally efficient than ML estimation.

1.2.2 Time Difference of Arrival Estimation

TDOA is another time-based method for localization which measures the difference in time between the signals arriving at two different receivers [13,24,26,63]. As such, this approach has several advantages over the TOA estimation method. Since it only measures the time difference at the anchor nodes, only the clocks between these receivers must be synchronized. In particular, it is not required that these clocks be synchronized with the clock at the target node. Clock synchronization among the anchor nodes is often easy to achieve since these nodes are part of a network infrastructure and may even be synchronized using a reference receiver. Moreover, since only signals at the receivers are compared, the transmitted signal from the target does not need to be time-stamped. These aspects make the TDOA-based localization more suitable in scenarios such as rescue missions, where the target mobile cannot synchronize its clock with the anchor receivers. As in the case of TOA, the anchors' positions must be known. TDOA computation, however, requires more coordination among the anchor nodes. In particular, to compute the TDOA between two anchor nodes, their received signals must be available in the same location, requiring that at least one of the signals be transmitted again. Recently, a new approach to TDOA has been proposed which does not require clock synchronization among the anchors [100].

Geometrically, each noise-free TDOA measurement indicates that the target node is located on a hyperbola with the two anchors located at its foci. Therefore, in the coplanar case, three anchor nodes and two TDOA measurements are needed to locate the target, which will be at the intersection of the two hyperbolas.

1.2.3 Direction of Arrival

Rotating a narrow beam antenna in azimuth and measuring the angle of the antenna with the highest output power is an old scheme for finding the direction of arrival of a signal. Today DOA estimation is obtained by anchor nodes equipped with an antenna array and where each antenna has its own dedicated RF front end [76,87,30]. Clearly the hardware complexity of this approach is high. As in the case of TOA and TDOA, the location of the anchors must also be known. However, in the coplanar case only two anchors will be sufficient to localize the target using triangulation, and for the non-coplanar case, only three anchors are needed.

DOA estimation requires that the antennas and the RF chains be carefully calibrated. This is due to the fact that DOA estimation relies on the phase difference between the signal arriving at each antenna element in the array. Therefore, an unknown phase inserted into the signal by an antenna element or RF chain causes errors in the DOA estimation. DOA-based localization is also very sensitive to multipath effects, and for accurate estimation it requires a strong LOS component. Another important issue is that DOA-based localization is good for near-field applications. As the separation between anchors and the target increases, small errors in DOA estimation translate into large position errors for the target node. The most prevalent methods for DOA estimation employ subspace techniques such as MUSIC and ESPRIT methods [76,68].

1.2.4 Received Signal Strength

As the signal propagates from the target node, it undergoes attenuation which is related to the distance traveled. Therefore, by measuring the RSS, the distance the signal has traveled can be estimated. In modern communication devices RSS measurements are already available for other purposes such as adaptive modulation and coding [73,84], channel sensing [29,103,58], channel quality assessment, or resource allocation [27]. Therefore, RSS measurement does not involve any additional hardware or processing cost. As in the previous cases, in RSS-based localization the location of the anchor nodes which measure the RSS values must be known. Furthermore, as in the case of TOA, in the coplanar case three, and in the non-coplanar case at least four anchors are needed.

As discussed previously, TOA, TDOA, and DOA estimations all require a strong LOS signal. RSS-based localization, however, can also operate in non-line-of-sight (NLOS) environments where a LOS signal may not be present. Three main approaches to RSS-based localization are range-based localization, fingerprinting, and proximity sensing.

To estimate the range from RSS measurements, an accurate model of the multipath channel is needed. However, even when this model is available, large variations in RSS value are expected. Therefore range-based localization in environments where strong multipath effects are present (e.g., indoor environments) is not very accurate. To address the issue of high variability of RSS measurements, RF fingerprinting creates a radio map of the area of interest. In particular, measurements of RSS values are made (on a grid) in the area of interest. This step is performed offline and the data is maintained in a server. During the localization process the anchor nodes measure the RSS values and transmit them to the server, which compares these values to the radio map in a pattern matching scheme. The location of the mobile is then obtained from this pattern matching. The difficulty in RF fingerprinting techniques is that an accurate radio map must be prepared beforehand. This involves considerable offline effort as well as the need to have full access to the area of interest beforehand. Moreover, since the wireless channel is time-varying, the radio map must be updated regularly to account for these variations.

1.2.5 Range-Free Localization

In ad-hoc wireless applications such as WSNs, GPS may not be a viable option for localization. In many deployment scenarios the GPS signal may be too weak or not even available. In addition, the sensors are intended to be inexpensive devices, and for a large-scale network, fitting each sensor with a GPS device may be too costly. Therefore the sensor nodes try to localize themselves relative to the position of their neighbors (i.e., those sensors within their radio range). The absolute positions are then obtained with the aid of a few anchor nodes which are aware of their global positions, say through GPS or manual configuration.

The sensor nodes can obtain the distances to their neighbors through RSS measurements. The range-based localization can be implemented in a centralized or distributed fashion. The former requires a great deal of communication between the sensors and a central processor and therefore is not suitable for resource-constrained WSNs. In contrast, in distributed algorithms, each node computes its position based on information available from its neighbors. Several distributed algorithms have been developed in recent years for distributed localization [55,54,85].

Due to the complexity of range-based localization, recently range-free localization methods have become very attractive for WSNs [88,79]. One approach, proximity-based localization, relies on the proximity or connectivity of the anchor and non-anchor nodes [9,44]. The position of the sensor nodes is computed using one of several methods including triangulation, multilateration, bounding box [9], or geometric methods [79,96].

Another approach for range-free localization is the hop-based method [88], which is reminiscent of some routing algorithms in ad-hoc networks. The minimum number of hops of a non-anchor from the anchor nodes is computed and along with the average hop-size is flooded to all the nodes in the network. The non-anchors then use this information to compute their distances from the anchors [34].

1.3 Reconfigurable Networks

In recent years, wireless reconfigurable networks (WRNs) have experienced an incredible growth due to the birth of the IOT and to the integration of drones into new applications, such as surveillance, product delivery, and rescue missions. As expected, with the emergence of the IOT the number of wireless devices worldwide has increased enormously in a short period of time. Furthermore, new applications require that WRN protocols adapt rapidly to the new challenging and dynamic conditions. Topics like power savings, 3D scenarios, changing user demands, propagation conditions, and reliable high data rate must be taken into consideration in the design and development of new WRN protocols. The availability of big data centers can be used to improve the performance of WRN protocols by the convenient use of data resources [23]. In Figure 1.1, we present an illustration of a WRN, where point-to-point communication, heterogeneous devices, clustering, and mobility are illustrated.

Knowledge of position location (PL) information in WRN is now more than ever crucial for several applications such as device monitoring, medical assistance, rescue operations, sensor battery replacement, and recently for drone trajectory control and location-based services for smart-phone users. In addition, PL Information (PLI) has been used to develop routing and handoff protocols for WRN.

Thus, modern PL techniques designed for WRN have to take into consideration the above challenges to be able to provide a robust and accurate solution in network scenarios composed of a large number of heterogeneous wireless devices (sensors, smartphones, laptops, etc.) with diverse capacities and demands, under highly changing topology

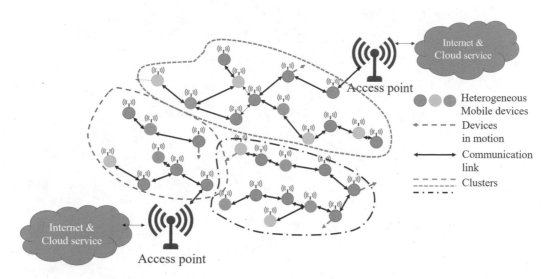

FIGURE 1.1
Reconfigurable network.

conditions and where GPS may not be available. In the following subsections, examples of collaborative PL algorithms applied to WRN (ad-hoc and sensor) are presented in the case of 2D and 3D localization techniques such as those for mobile robots and drones. Single-hop and multi-hop algorithms are also discussed, as well as relational and heuristic techniques.

1.3.1 2D Position Location Techniques

In this section, several PL techniques developed for planar WRN in single and multi-hop environments are discussed.

In [93], the authors present a PL technique called Vertex-Projection Algorithm (VPA) for location estimation based on a pyramidal structure. This algorithm was designed for single- and multi-hop 2D WRN. The performance of the algorithm depends on the deployment of the anchor nodes (also referred to as access points [APs]), node transmission range, and noise. They propose a positioning system based on at least three anchors (**A**, **B**, **C**) placed at known locations. A pyramid is defined based on anchors' locations and on the estimated distances (δ_A, δ_B, δ_C) from the anchors to the target node of interest (NOI), as in Figure 1.2. Distance is estimated from the anchor to the NOI based on TDOA and the concatenation of the estimated distances between nodes in the connecting route (according to Dijkstra's routing algorithm). Then, through an algebraic formulation, the position of the NOI **P** is estimated, and the final location estimation is given by the normal projection of **P** onto the plane of the access points; that is, **P***, as in Figure 1.2. Simulations are performed in noisy reconfigurable environments with different node densities and reachability radii, as well as for single- and multi-hop scenarios.

In [95] the authors present a PL technique for multi-hop planar WRN based on the convenient deployment of three anchor nodes (with known locations) and on the Manhattanization of the network area. They model the PL problem as a linear constrained optimization search in a discrete coordinate system. They discretize the working area defined by anchor coordinates and assume that every node in the network can be located

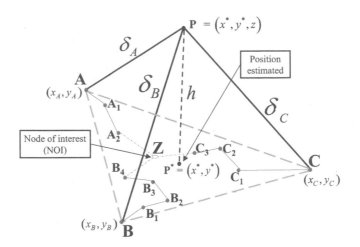

FIGURE 1.2
Pyramidal PL for planar scenarios.

at the corners of the Manhattan grid. Based on this scenario, they define a set of Manhattan distances from anchor node j to node $\mathbf{n}_0$ (δ_j, $j = 1, 2, 3$), then a set of coordinate linear equations is formulated for a node in R_1 and R_2 (see Figure 1.3). After that, they propose a set of linear Manhattan distance constraints (relative to R_1 or R_2) and finally, an optimization search to minimize the difference between estimated (based on hop count criterion) distances

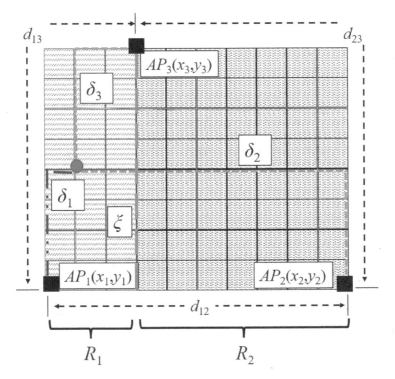

FIGURE 1.3
Positioning system for the linear constrained search PL method.

from anchor node j to node n_0, and real Manhattan distances ($\delta_j, j = 1, 2, 3$) is performed. Feasibility of the PLI algorithm is shown through simulations for different environments.

1.3.2 3D Position Location Techniques

In this section, we present several examples of PL techniques designed for 3D WRN. Also, single-hop and multi-hop scenarios are considered.

In [14], the authors present a simple PL method for a single-hop 3D WRN based on a centroid calculation, where each unknown node randomly selects four anchor nodes (with known coordinates) in range to form a series of tetrahedrons. Basically, each node with unknown location collects beacon signals from various anchor nodes. Then, all these nodes calculate their own coordinates by a centroid determination from all positions of the anchor nodes in range.

In [60], the authors present a single-hop PL method based on the traditional trilateration and TDOA data from several anchor nodes buried in a wheat silo sending ultra-wide band (UWB) pulses to other sensors in order to construct a signal map. They propose to store the pulse propagation delays at several reference points deployed conveniently in the three axes (x, y, z). The authors propose a system that uses pulse generators buried in the grain, x, and z axes booms on top of the silo along with a receive antenna array (see Figure 1.4). Generated pulses are approximately Gaussian and propagate through the grain. The received signal at each array position contains both LOS and NLOS components due to multiple reflections. Clear identification and isolation of the leading pulse edge not contaminated by multi-path interference is required. The received signals at the antenna array positions are recorded using a digital oscilloscope. The resulting digital signals are then processed using a cubic spline interpolation to improve signal representation for position location calculation. Then, they formulate a Euclidean distance equation for every receiver based on known receiver coordinates and the TDOA estimated range. Finally, they calculate the intersection for the sphere set defined by the Euclidean distances to provide an estimation of the sensor location (x_0, y_0, z_0).

In [94], the authors present a PL method suitable for 3D scenarios based on the convenient deployment of several anchor nodes and on the Manhattanization of the network space. They consider a 3D reconfigurable ad-hoc network, where every node communicates

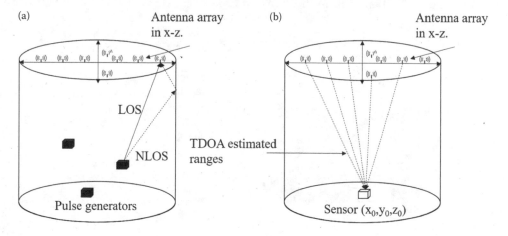

FIGURE 1.4
3D localization in a wheat silo, (a) Calibration phase for the spherical PL method. (b) Illustration of sensor location process.

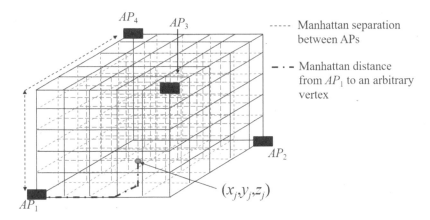

FIGURE 1.5
Manhattanized 3D network scenario.

through multi-hop routes with a set of anchor nodes (at least three). They assume that the anchor nodes can be located at the non-adjacent corners of a cube, as Figure 1.5 illustrates. This positioning system also assumes that every node is located at the vertices of the Manhattanized cube. Next, they associate to every vertex a Manhattan distance δ_i set relative to anchor node i. Later, they formulate equations for coordinates (x_j, y_j, z_j) of every vertex based on Manhattan distances from every anchor to the node of interest, and on the separation between the anchors. Distance estimates from each anchor node to the NOI is based on hop count, and the set of equations is solved to estimate the coordinates (x_j^*, y_j^*, z_j^*). Two sets of equations are formulated for the vertex coordinates when four anchors are considered, which allows error reduction in noisy environments.

In the same reference 94, several improvement algorithms like finer discretization (smaller cubes) and maximum a posteriori (MAP) probability criterion for Manhattan distance assignment are explored. In the MAP criterion, they propose to assign 0, 1, 2, or 3 Manhattan steps instead of always 1, based on the estimated range, on the region definition of the Manhattan cube (see Figure 1.6), and on a MAP probability criterion. They show feasibility of the 3D PL algorithm through simulations for different ad hoc scenarios.

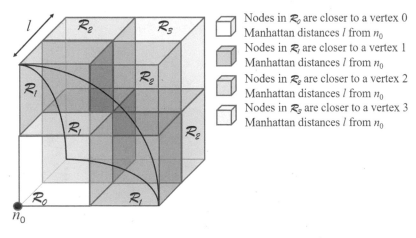

FIGURE 1.6
Region assignment in the Manhattan cube relative to node location.

1.4 Cooperative vs. Non-Cooperative Techniques

Scenarios we have discussed up to now may be considered to be cooperative in nature. In such cases the target node to be localized emits an RF signal which can be used for localization, and often the target node cooperates with the reference receivers (anchor nodes) in the localization process. For example, in self-localization, devices communicate with other devices in order to establish a framework to determine their position. Furthermore, the localizing system is assumed to have full knowledge of the carrier frequency and the bandwidth of the transmitted signal. While some nuisance parameters of the signal and noise (e.g., transmit power, noise power) can be estimated along with the position of the source, the spectral occupancy of the source signal is assumed to be known to the localizing system.

In some real-world applications, we may be interested in locating a subject who does not actively participate in the localization process. For example, the subject may be intentionally avoiding RF transmission so as not to be located. Examples include intruder detection, law enforcement, and counter-terrorism where the subject maintains a low radio profile and will be non-cooperative in the localization process. In other scenarios, the subject may be unable or not care to carry a transmitter. Examples of this include emergency services, patient monitoring, and location services.

The ability to locate and track subjects in buildings significantly improves the chances of operational success for law enforcement (e.g., in hostage situations) and first responders (e.g., building fires or other emergencies). In patient monitoring situations, the patient may be unable or unwilling to bear the extra burden of carrying a radio device to help in the localization. Finally, in location services a subject may not care whether they are located or not. In other scenarios localization needs to be performed by a system that is not a part of the network infrastructure where the device to be localized is operating. Consequently, complete knowledge of the RF operating parameters of the device of interest (e.g., signal bandwidth, operating frequencies, etc.) are not available. In this section we review several non-cooperative localization techniques where the target node does not actively participate in the localization process or when the target node's signal parameters are not available.

1.4.1 Radio Tomography

Radio tomographic imaging (RTI) is a localization technique that does not rely on any signal transmission by the subject and in that sense may be considered a fully passive localization method. RTI exploits the radio shadows created by the subject of localization [2,36,98]. This method assumes that the subject is located in a dense WSN. Each sensor periodically sends packets to other sensors. When the subject physically obstructs one or more radio links, the RSS in those links drops below a nominal pre-calibrated value (see Figure 1.7). By observing the deviation of RSS in those links, the subject can be located and tracked. In this approach, when a line of sight between the sensors is available that can only be shadowed by the subject, the shadowing loss can be accurately predicted. However, when other obstructions are present, many channel measurements are needed in order to obtain an accurate model of the channel and to perform the necessary pre-calibrations.

Localization and tracking based on RTI was first studied in several papers by Patwari and his students [98,99]. In [36], the authors develop an RTI testbed and conduct field test measurements. An RTI network is developed for vehicle identification and tracking in [2].

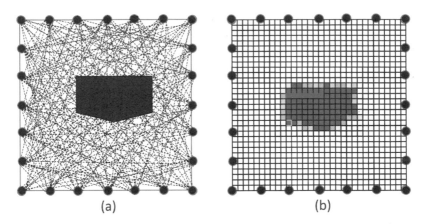

FIGURE 1.7

RSS shadowing example. (a) Signal strength measurements resulting from shadowing sue to the object. (b) Tomographic construction of the object from RSS measurements.

Indoor localization using RTI for applications such as assisted living has been reported in [61,62] where a sensor network can both monitor and locate breathing in a home.

1.4.2 RFID-Based Localization

Radio frequency identification (RFID) tag localization and tracking is another approach which does not require active participation of the subject in the localization process and has received a great of attention in recent years [46,1,10,8,104,101]. An RFID system consists of two main components: a transponder or RFID tag and a detector or reader. The tag may be active or passive. An active tag is powered by an internal battery which can increase the transmission range of the tag, albeit with a larger size and higher cost. Active tags can continuously broadcast a beacon that can be used for accurate localization of the tag. A passive tag, on the other hand, does not have any power source. It is powered by the electromagnetic energy transmitted by the reader using a coupling element. As such, passive tags do not transmit a signal on their own but rather only respond to the probes from the reader.

RFID-based localization schemes have been developed for both tag and reader localization. For tag localization, the algorithms assume that the target nodes to be tracked are attached with RFID tags and they move in an area covered by a grid of RFID readers whose antennas (anchor nodes) have known locations (see Figure 1.8a). The RFID readers are assumed to provide full coverage for the area of interest. Clearly, the cost of such a system is high due to the need for a large number of readers. The information from the readers is sent to a server, which fuses the data and performs the localization algorithm. For localization of the reader, as shown in Figure 1.8b, each target node to be located must carry an RFID reader, while the RFID tags act as the anchor nodes and are deployed in the area with their location known. A combination of these two methods has also been studied [46].

RFID-based localization and tracking methods vary depending on the type of sensor information measured, the assumed RFID signal model, and the inference algorithm. Regarding sensor information, methods based on RSS measurements [41], tag detection events [90], and phase measurement of the received signals have been studied [74]. Methods based on RSS measurements often require active RFID tags and suffer from inaccuracies of RSS measurements due to multipath propagation and channel dynamics. However,

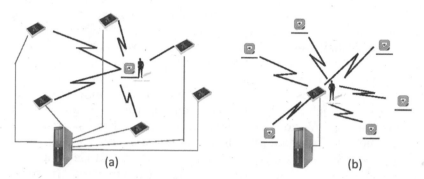

FIGURE 1.8
RFID-based localization. (a) Tag localization. (b) Reader localization.

range-free RFID-based localization has also been studied [46]. Inference algorithms developed for tag localization include multi-lateration [7], Bayesian inference [74], k-nearest neighbor [56], Kalman filtering [7], and particle filtering [28].

1.4.3 UWB Multistatic Radar

Radar systems (in monostatic configuration) have long been used for localization and ranging. Multistatic radar employs at least three non-co-located antennas for transmission and reception (often a single transmitter and a multitude of receivers). Multistatic radar offers improved performance over the monostatic configuration due to its larger coverage area and spatial diversity [15]. UWB is a very attractive technology for multistatic radar [67]. The impulse radio version of UWB transmits pulses of a few nanoseconds' duration, resulting in a very wide band signal. Consequently, this version of UWB provides remarkable resolution and localization accuracy.

In recent works, target TOA is estimated at three or more receivers and the target location is then obtained from multilateration [69,70]. In [16], a pixel-base method is used where the surveillance area is divided into small pixels and at each receiver the correlation of the received signal and the reference signal is computed for each pixel. A fusion algorithm combines the values from all the receivers for each pixel, and the pixel with the maximum combined correlation is estimated as the location of the target. The pixel-based approach is also adopted in [80] for multiple target detection and localization where the energy of the sample received by each receiver for each pixel is measured. A threshold is applied to the energy of each pixel. This approach is reminiscent of the energy detectors for signal detection and results in a constant false alarm rate (CFAR) detector [50]. The position of the target is then estimated to be the pixel for which the energy at most of the receivers exceeds the threshold. Tracking of targets using UWB multistatic radar is developed in [81], where the authors estimate the target's velocity using low-complexity particle filtering. The particle filter is also specifically modified to address the problem of blind zones associated with UWB multistatic radar. Tracking of humans using UWB radar is also studied in [37].

1.4.4 Non-Cooperative Localization Due to Signal Uncertainties

In most localization scenarios, a number of parameters must be unknown to the localizing algorithm. For example, in practical situations the transmit power of the target nodes to be localized is often unknown. Other unknown parameters include the power of the noise in the anchor receivers and the channel fading coefficients. Algorithms have been designed to estimate these nuisance parameters along with the position of the target sources [51]. In these cases, the localizing system has access to the network in which the target nodes operate and is able to obtain information regarding the communication system of the target node. However, in some real-world scenarios this is not feasible. More specifically, when the localizing system operates outside of the network infrastructure in which the target node is communicating, it does not have access to all the communication parameters of the subject. In such cases novel algorithms are required which can obtain the missing information from the targets' received signal without the collaboration of the targets.

Today, most advanced communication standards such as 3rd Generation Partnership Project (3GPP) Long-Term Evolution (LTE), IEEE 802.11a/g wireless LAN, and IEEE 802.16 Worldwide Interoperability for Microwave Access (WiMAX) standards employ multicarrier communication, most notably orthogonal frequency division multiplexing (OFDM). In these standards, the users share the OFDM subcarriers in a multiple access scheme referred to as orthogonal frequency-division multiple access (OFDMA). At the time of connection setup, each user, depending on its bandwidth demands, is allocated a number of subcarriers. In LTE, a resource block consisting of 180 KHz wide in frequency and a slot long in time is the smallest unit of resource that can be allocated to a user. In frequency, the resource block consists of 12 subcarriers, each 15 KHz wide. For example, for the standard-defined bandwidth of 5 MHz, 25 resource blocks will be allocated (with the occupied bandwidth of 4.5 MHz). Moreover, while the resource blocks must be in the same time slot, they do not have to be contiguous in the frequency domain [43,48,86].

To locate a user whose wireless system employs a standard based on OFDMA, exact knowledge of the user's subcarrier allocation is essential in order to achieve acceptable performance. If the localization system has access to the multiple access layer, then this information may be readily available.

However, for non-cooperative geolocation applications operating outside of the network infrastructure, this information may be difficult to obtain. Therefore, in such cases the localization algorithm must not only locate the target nodes in the spatial domain, but also in the spectral domain. This problem is investigated in three recent papers [51–53]. An algorithm based on clustering is proposed in [52] for joint spatial and spectral localization of the OFDMA users. The algorithm is composed of three steps. The proposed method first employs the MUSIC algorithm [76] to obtain the DOA estimates in each subcarrier. Next, multiangulation is used to form a target location estimate in the Cartesian space for each subcarrier. Finally, in a clustering step, the subcarrier-wise estimates are combined into aggregate position estimates while also identifying the subcarrier occupancy of the targets. The clustering algorithm also identifies the subcarriers that are not occupied by any target. This is an efficient, albeit suboptimum method for estimating the locations and spectral occupancy of the target nodes. In addition, a novel algorithm based on expectation maximization (EM) is developed for this problem in [51,53]. The subcarrier occupancy of the target nodes is detected via a likelihood ratio test nested in every EM iteration. Numerical results show that while both algorithms are able to detect the spectral occupancy of the targets and estimate their location, the latter has better performance. However, the former algorithm is more efficient and has a much lower computational complexity. As is well known, while EM algorithm can achieve the ML solution, EM has several drawbacks, including its high computational complexity, slow convergence, and sensitivity to the initial conditions. As suggested in [52], the results from [52] can be used to initialize the EM algorithm in [51,53].

1.5 Indoor Propagation Issues and Their Mitigation Techniques

Indoor positioning has become relevant for many applications in diverse areas such as emergency, safety, health, commercial, advertisement, etc. The fundamental part to deploy a positioning system for localization in indoor environments is the set of sensors and communication technologies [11]. It is well known that communications solutions in indoor environments are related to IEEE 802.11X standards [3]. Although cellular signals can be received inside buildings, the accuracy for indoor localization would be compromised. Even if GPS systems are used, the accuracy in indoor environments is limited. Some of the problems encountered by localization algorithms in indoor environments are high density of objects, multiple materials used in the structures, highly changing obstruction conditions, small spaces, and the conditions changing at different times during the day.

The highly changing obstruction conditions will create strong shadowing and fading effects for signals trying to traverse walls. High attenuation will be faced for all signals that need to reach areas hidden in the back of obstacles. A good planning of the infrastructure for localization would be needed to deploy the needed antennas that guarantee coverage of the signals. This will consequently have an effect in the multilateration algorithms, since propagation in some cases would be due to NLOS paths, and distance estimation would need to take this into account to establish coordinates with respect to fixed references or a coordinate framework.

Small spaces will have an impact on accuracy and precision of measurements, being critical for those algorithms that use time, for example, TOA and TDOA. Due to the short paths that signals will traverse, times that need to be measured are also very small, in the order of tenths or hundredths of a microsecond, hence high precision equipment and measurements would need to be available to capture such time changes in order to have accurate distance estimations.

The multiple materials that signals can encounter as they propagate through the medium is of relevance to model reflections and scattering. It is common to find combinations of carpets on floors and sometimes floors covered with wood or other materials. Also, walls might be of concrete or plaster or might be covered with different fabrics for noise cancelation. All these materials will have an impact on the way signal propagates. Ray tracing techniques model such surfaces and consider the reflexivity of the surfaces as a function of the material to determine how the signal is reflected and propagated.

Indoor environments are highly dynamic at different times during the day; for example, an office during office hours presents challenges that are different from those at other times. Also, an office will be different from a home, a warehouse, a supermarket, a library, etc. Every one of those indoor scenarios needs to be evaluated in order to obtain an accurate characterization of the environment that will allow to planning to be carried out for the deployment of the technology to be used for localization.

There are several things that one might do to design and deploy a localization system in indoor environments. One is to have measurement equipment with high accuracy to obtain channel models and determine how the signal propagates. Another is to simulate the environment using software tools that allow one to study the space using ray tracing. Yet, another approach is to use measurements to generate radio maps of the space with accurate measurements of signal variables such as power. As mentioned previously, this is referred to as fingerprinting.

1.5.1 Fingerprinting

Fingerprinting is the method used to extract information from the context of the environment where it is performed. It helps create maps of the area based on different parameters, and once those maps have been created, real-time measurements of those parameters taken by a device could be compared to the generated map so that location can be estimated. The map would consist of measurements of those parameters with their specific location in the area of interest, stored in a database. Fingerprinting becomes more difficult as one increases the granularity of the scenario (the number of measurements), because it creates large amounts of data; also, trying to get a 3D image of the entire space is not easy given that the environment changes with time.

There are several parameters that can be used to generate a fingerprint of the environment. One can use measurements of power received at different locations, or measurements of ambient light. Another parameter could be the Earth's magnetic field, or if a controlled environment, temperature. All these options need a highly accurate set of sensors and a system capable of storing and processing the data in real time [11,19]. Most of the parameters used would need to be compensated for impairments as measurements for the map were taken. For example, for measurements of power, compensation would be for propagation impairments and additive noise that would cause rapid changes in the values of power at small changes in position. Some solutions for these impairments are based on smoothing performed by digital filters such as Kalman and particle filters.

Figure 1.9a shows an indoor environment where sensors where deployed and measurements of power and TOA were taken to generate fingerprints of the area. Figure 1.9b shows the measurements taken of TOA and power in dB for the indoor environment. One can see how the values obtained suddenly vary with changes in distance.

1.5.2 Ray Tracing

Ray tracing techniques have been used for deterministic propagation prediction through simulations of diverse indoor and outdoor environments. The more precise the description of the scenarios (physical conditions, materials, and signal characteristics), the more accurate the results that are obtained. The scenarios evaluated are divided into 3D cubes where different rays from the transmitted signal are captured and analyzed. In general, from such information one can obtain large- and small-scale propagation parameters and be able to characterize the space according to propagation models and statistical channel models for prediction. Also, diffraction, scattering, and shadowing phenomena can be characterized to be included in the simulations. Ray tracing has been used in vehicular scenarios [6], as well as in indoor environments in the transportation industry such as inside trains [4]. Also, the adaptation of the cube sizes has been analyzed for convergence of the ray-tracing algorithms in [5].

In [89], an algorithm for indoor localization is proposed, where ray tracing is used to determine how the signal propagates in the environment given its physical characteristics. Since in indoor environments multipath effects are common and the absence of LOS components are highly probable, then the proposed method includes fingerprinting using time and power in the measurements in the simulator.

When dealing with communication impairments to obtain measurement of parameters for localization purposes, one has to compensate using signal processing techniques. The combination of localization techniques that combine parameters to estimate separation distance and optimal algorithms to generate estimates of positions can help to reduce errors

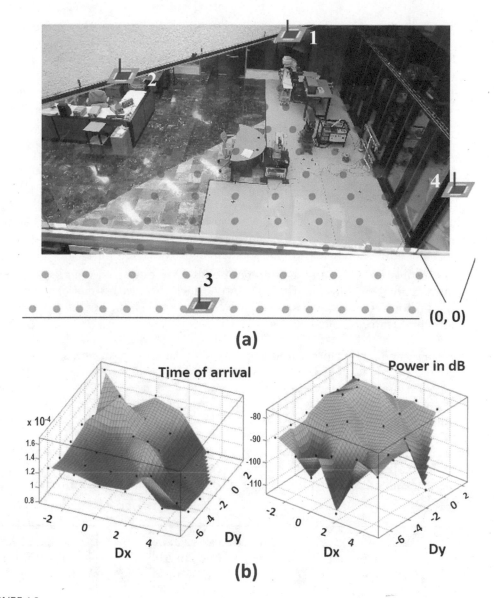

FIGURE 1.9
Measurement set up and maps of the environment.

in the estimation. Digital filters for smoothing, the use of DOA or TOA in combination with RSS from ray tracing techniques and fingerprinting will result in performance improvements for localization algorithms in indoor environments.

1.6 Signal Processing for Localization

Localization techniques are based on the estimation of separation distance between a reference node or anchor and a target node that needs to be located. In order to estimate such distance, parameters such as TOA, DOA, RSS, or TDOA have been used as measurement data for the estimation [49]. Considering these parameters also defines different geometries with circles, hyperbolas, or ellipses [49], and the combination of the distance estimates from different anchor nodes is important, since it determines in part the accuracy of the method. The derivation of a system of equations that defines such geometries is one form of combining such estimates to calculate the estimated position of the node of interest [75].

Every localization algorithm consists of three stages (see Figure 1.10):

1. *Measurement or Evidence.* This stage is where the reference nodes or anchors with known positions obtain or measure the parameter that will help estimate separation distance to the node to be located. Some common parameters used are RSS, TOA, DOA, and TDOA.

2. *Distance Estimation.* In the second stage, the parameter or evidence just obtained in the previous stage is used to estimate separation distance. This stage is executed for each anchor node involved in the localization algorithm.

3. *Position Estimation.* Once all the reference nodes have estimated their separation distance, then algorithms are used to combine such distance estimations and the known positions of the anchors in order to calculate an estimate of the position of the target node to be located.

At every stage, one can perform improvements with different algorithms; for example, new initiatives to estimate TOA in stage 1, or more accurate propagation models that use RSS to estimate separation distance in stage 2, and different ways to combine the individual distance estimations to calculate a more accurate position of the node in stage 3. In [75], several combination techniques using TOA and DOA are presented where results are based on LS solutions.

Classical signal processing algorithms such as LS, ML, MAP, filtering (Kalman, Wiener, particle), splines, MUSIC, and ESPRIT have been used in many localization techniques. In this section, some algorithms that improve estimations along the stages introduced above

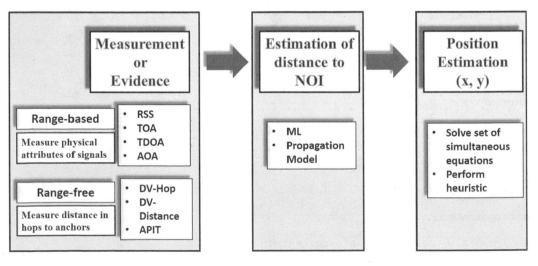

FIGURE 1.10
Localization stages.

are discussed. In particular, three signal processing techniques that have been successfully applied in localization, namely EDMs, consensus algorithm, and belief propagation are presented.

1.6.1 Euclidean Distance Matrices

EDMs have been applied to several areas, one of which is localization (see [18,22,31]). Within the context of sensor networks, EDMs have been applied together with multilateration techniques, for localization purposes [21,22].

Given a point set, an EDM is a square matrix that contains in its elements the distances (or the square of the distances) between any pair of points. Assume that you have a sensor network where sensor nodes have been deployed in an area, and now you would like all sensors to be self-localized. You can consider that each sensor is at a point in space and the set of all sensors will generate the point set. You can collect information about relative distances throughout relational algorithms where neighboring nodes can interact and identify other nodes, but at the end you need the estimated distances among these nodes and eventually a map of the site where you can locate the nodes. One application of EDMs is to reconstruct the point set, that is, the sensor network with localized sensors. The problem is that distances that have been estimated in general have been corrupted by noise, some distances might not be available, or information might be incorrect. Within the context of distance geometry [18], one problem is if there exists a point configuration so that the distances collected throughout the network can be generated. Due to the symmetry of EDMs, eigenvalue decomposition (EVD) is chosen over singular value decomposition (SVD) [22] to reconstruct the original point set.

Multidimensional scaling is used with EDMs to obtain the point set when the distance between points in the set is corrupted; that is, when the distance between nodes i and j, d_{ij} is noisy. Also, semidefinite programming is used to determine if a matrix is EDM. An important result of EDMs is that of its rank, because it depends only on the dimension of the space where the point set is defined, and not on the number of points that generate the matrix, [22].

A problem of interest in localization is that of *unlabeled distances*, which arises when we have distances but we do not know the points that generate them. From the propagation point of view, we could have a distance estimation based on RSS and a propagation model, but in general we will have multiple paths that belong to the same source arriving at different times and with different features that could be considered unlabeled. To obtain the point set in the unlabeled distances problem, one has to test basically all the possibilities with an exhaustive search [22].

1.6.2 Belief Propagation

The mathematical theory of reasoning is focused on the combination of evidence, where the theory assigns a number to a body of evidence to support it. It is in general contrasted to the theory of partial belief also known as Bayesian theory [17,33,78]. In [78], it is shown that the Bayesian theory is a special case of the theory of evidence, where Bayesian belief functions, including Bayes rule, are special cases of belief functions in the theory of evidence which determine the degrees of belief which are the numbers assigned to the evidences.

Consider a sensor network where individual sensors deliver information related to localization of a node of interest indexed in time. Then given such information, we would like to calculate what is the chance that the node of interest is at a certain position. This conditional probability would be the belief function, which then maximized would be the MAP probability position estimation. In general, in this scenario, the evidence (sensors' information) is considered to be Markovian, hence the conditioning set of random variables is reduced by considering only those at the present time.

Bayes theory has been used for location estimation in sensor networks in the form of filters such as Kalman, Extended Kalman, Unscented Kalman, and Particle, among others. In the Kalman filter use for localization, Gaussian distributions are used to represent beliefs, with the mean being associated to the position of the node, and the variance being the error in the estimation.

In [92], a distributed localization algorithm for sensor networks is proposed. The algorithm uses 3D ray launching techniques with IEEE 802.11 WiFi networks. RSS is obtained, and cooperation among sensors is established by fusing RSS and ranging information from neighboring sensors and anchor nodes. The self-localization estimation is based on the calculation of the conditional probability of each sensor's position conditioned on a set of RSS measurements, a set of range measurements, and a training data set that they called calibration data. The authors achieve between 70% and 95% of the time location errors below 2 m for an indoor scenario.

In [39], the self-localization problem is addressed by generating graphs that contain the information of the topology of the network which is built with local information using distance measurements between pairs of sensors. Nonparametric belief propagation (NBP) is used and shown to perform better than belief propagation alone. Considering location errors and the estimation of sensor locations, NBP is a generalization of particle filtering. Authors use graphical models where inference algorithms are applied, where belief propagation and NBP are two of those algorithms. With the definition of the graphical model for localization, the inference algorithms can be iterative processes where sensors generate a belief to be shared with its neighbors, and then with information from the neighbors, the belief is modified and shared once again, repeating until convergence.

1.6.3 Consensus Algorithm

The fundamental idea of consensus, especially for networks, is for all the nodes to agree in the value of a parameter of interest that depends on the state of all the nodes in the network. The consensus algorithm specifies the rules for the exchange of information from one node to all of its neighbors. The applications of consensus algorithms are diverse, but for localization two are related—sensor fusion in networks and rendezvous in space [57]. The former is carried out by using Kalman filters or LS estimators, while the latter considers the network topology to be position dependent through the use of proximity. Performance of consensus algorithms is primarily measured by the speed to reach agreement.

In [32], a distributed spectral clustering algorithm for wireless sensor networks is introduced. The topology of the network is organized in clusters that are determined based on the position of the sensors. The idea is to not have a fusion center in the network, since it produces congestion and is vulnerable to attacks. The algorithm uses *K*-means algorithm in combination with a distributed calculation of eigenvectors of the connectivity of the network represented by the Laplacian. Consensus is included to reach steady state when all the sensors converge to the value of the eigenvector of the Laplacian. Once the eigenvector is obtained using consensus, clustering is carried out using the *K*-means algorithm on the eigenvector, where position of the sensors is used, but not exchanged. The algorithm uses distributed average consensus to compute the eigenvector of the Laplacian where the state of each sensor is its position, and the sample mean of the states is calculated in a distributed manner. It is shown that because distributed spectral clustering uses connectivity information, it provides better results compared to the use of *K*-means.

1.7 Conclusions and Future Challenges

In this chapter, we have presented an overview of localization techniques along with a discussion of recent developments and advances addressing new applications in localization. Localization techniques have evolved from traditional methods using multilateration and triangulation in infrastructure-based networks to algorithms in reconfigurable networks where techniques need to adapt depending on scenarios where cooperation can be compromised, and the estimation of the position needs to be obtained through a combination of alternative techniques. In other scenarios, target nodes may not be actively participating in the localization process or may be intentionally maintaining a low radio profile so as to not be located. Localization of such passive non-cooperative target nodes arises in many civilian and military applications. Another non-cooperative scenario is when knowledge of the target node's signal, such as its spectral occupancy, is not available at the anchor nodes. This case arises, for example, when the target node is subject to dynamic resource allocation and operates outside of the anchor nodes' infrastructure.

In this chapter, we have introduced the position location problem, and how traditional methods such as TOA, TDOA, DOA, and RSS are evaluated and how they are used to estimate separation distance and location coordinates. We have discussed several localization methods in reconfigurable networks. These networks have seen tremendous growth due to the introduction of the IOT and drones into new applications. Next, we have discussed the recent developments in non-cooperative localization of passive target nodes as well as localization in the absence of information on target node's bandwidth occupancy. Also, we have presented ideas where alternative methods such as fingerprinting and ray launching can be combined to improve location estimation accuracy in indoor environments. Finally, recent developments in signal processing for localization such as EDMs, inference algorithms such as belief propagation, and consensus algorithm have been presented.

The next generation of cellular networks, 5G, will usher in a number of innovative technologies including massive multi-input multi-output (MIMO) system and millimeter wave. For 5G, different features will be exploited to perform localization. The use of higher frequencies in millimeter waves along with massive MIMO will allow the generation of highly directional antenna beams that will enable location accuracies up to 1 m. Another feature will be cooperation where direct communications among devices will enable the use of cooperative algorithms for localization as those used in reconfigurable networks. The different features of 5G such as scenarios for device-to-device (D2D) and machine-to-machine (M2M) communications require that the position location algorithms have the flexibility for both infrastructure and infrastructureless networks. Cooperation will be fundamental to achieve the accuracy objectives, and the benefits of localization techniques for ad-hoc and sensor networks will be a base on which new algorithms for the next generation wireless communication standards will be generated.

Localization is becoming essential for 5G networks that are envisioned to enable IOT. The authors in [72] introduce a distributed localization algorithm of low power devices that performs self-localization where those devices do not necessarily have direct access to reference or anchor nodes. The solution is a cooperative algorithm where local measurements are obtained, and a linear distributed iterative process is carried out instead of the nonlinear system of equations given by the geometry.

Today we are witnessing technological developments where intelligence is an important ingredient to achieve quality and performance. With the IOT and the smart *everything* trends, localization techniques will need to evolve to adapt and achieve their goals [59]. The evolution of location-based services will integrate solutions that consider the context of the area, the device heterogeneity, the current and new applications that can be improved by including PLI, and demographics to determine the target groups for such applications and collecting information to adapt and improve such services. Among the fundamental parts of a system for localization purposes in the new smart world, we have sensors, localization and tracking algorithms, classification or discrimination algorithms, and statistical signal processing techniques to extract and collect information. Applications of these smart algorithms will be found in health, retail, transportation, city infrastructures, smart homes/ offices, wearables, etc.

Future trends for localization include the flexible characteristic that algorithms need to integrate heterogeneous devices, to obtain location estimates for 3D spaces, and to work in harsh environments such as highly populated areas or hostile environments such as emergency and disaster recovery scenarios. Mobile devices such as robots and drones must also be part of this smart environment, and user mobility will also be a challenge to address.

References

1. A. Almaaitah, K. Ali, H.S. Hassanein, and M. Ibnkahla. 3D passive tag localization schemes for indoor RFID applications. In *2010 IEEE International Conference on Communications*, pages 1–5, May 2010.

2. C.R. Anderson, R.K. Martin, T.O. Walker, and R.W. Thomas. Radio tomography for roadside surveillance. *IEEE Journal of Selected Topics in Signal Processing*, 8(1):66–79, Feb 2014.

3. A. Aragon-Zavala. *Indoor Wireless Communications. From Theory to Implementation*. Wiley-IEEE Press, New York, NY, USA, 2017.

4. L. Azpilicueta, J.J. Astrain, P. Lopez-Iturri, F. Granda-Gutierrez, C. Vargas-Rosales, J. Villadangos, A. Perallos, A. Bahillo, and F. Falcone. Optimization and design of wireless systems for the implementation of context aware scenarios in railway passenger vehicles. *IEEE Transactions on Intelligent Transportation Systems*, 18(10):2838–2850, Oct 2017.

5. L. Azpilicueta, P. Lopez-Iturri, E. Aguirre, C. Vargas-Rosales, A. Leon, and F. Falcone. Influence of meshing adaption in convergence performance of deterministic ray launching estimation in indoor scenarios. *Journal of Electromagnetic Waves and Applications*, 31(5):544–559, Mar 2017.

6. L. Azpilicueta, C. Vargas-Rosales, and F. Falcone. Intelligent vehicle communication: Deterministic propagation prediction in transportation systems. *IEEE Vehicular Technology Magazine*, 11(3):29–37, Sept 2016.

7. A. Bekkali, H. Sanson, and M. Matsumoto. RFID indoor positioning based on probabilistic RFID map and Kalman filtering. In *Third IEEE International Conference on Wireless and Mobile Computing, Networking and Communications (WiMob 2007)*, pages 21–21, Oct 2007.

8. D. Boontrai, T. Jingwangsa, and P. Cherntanomwong. Indoor localization technique using passive RFID tags. In *2009 9th International Symposium on Communications and Information Technology*, pages 922–926, Sept 2009.

9. A. Boukerche, H.A.B.F. Oliveira, E.F. Nakamura, and A.A.F. Loureiro. Localization systems for wireless sensor networks. *IEEE Wireless Communications*, 14(6):6–12, Dec 2007.

10. J.L. Brchan, L. Zhao, J. Wu, R.E. Williams, and L.C. Prez. A realtime RFID localization experiment using propagation models. In *2012 IEEE International Conference on RFID (RFID)*, pages 141–148, Apr 2012.

11. R. F. Brena, J. P. Garcia-Vazquez, C. E. Galvan-Tejada, D. Munoz-Rodriguez, C. Vargas-Rosales, and J. Fangmeyer Jr. Evolution of indoor positioning technologies: A survey. *Journal of Sensors*, 2017, Article ID 2630413, 2017.

12. M. Bshara, U. Orguner, F. Gustafsson, and L. Van Biesen. Fingerprinting localization in wireless networks based on received-signal-strength measurements: A case study on WiMAX networks. *IEEE Transactions on Vehicular Technology*, 59(1):283–294, Jan 2010.

13. Y.T. Chan and K.C. Ho. A simple and efficient estimator for hyperbolic location. *IEEE Transactions on Signal Processing*, 42(8):1905–1915, Aug 1994.

14. H. Chen, P. Huang, M. Martins, H.C. So, and K. Sezaki. Novel centroid localization algorithm for three-dimensional wireless sensor networks. In *Proceedings of the International Conference on Wireless Communications, Networking and Mobile Computing (WiCOM '08)*, 2008.

15. V.S. Chernyak. *Fundamentals of Multisite Radar Systems*. Gordon and Breach Science Publisher, London, UK, 1998.

16. M. Chiani, A. Giorgetti, M. Mazzotti, R. Minutolo, and E. Paolini. Target detection metrics and tracking for UWB radar sensor networks. In *2009 IEEE International Conference on Ultra-Wideband*, pages 469–474, Sept 2009.

17. A. Darwiche. *Modeling and Reasoning with Bayesian Networks*. Cambridge University Press, 2009.

18. J. Dattorro. *Convex Optimization & Euclidean Distance Geometry*. Meboo Publishing USA, 2005.

19. P. Davidson and R. Piche. A survey of selected indoor positioning methods for smartphones. *IEEE Communications Surveys & Tutorials*, 19(2): 1347–1370, Second quarter 2017.

20. J.A. del Peral-Rosado, R. Raulefs, J.A. Lopez-Salcedo, and G. Seco-Granados. Survey of cellular mobile radio localization methods: From 1G to 5G. *IEEE Communications Surveys & Tutorials*, 20(2):1124–1148, Second Quarter 2018.

21. L. Doherty, K. Pister, and L. El Ghaoui. Convex position estimation in wireless sensor networks. In *Proceedings of the IEEE Conference on Computer Communications (INFOCOM)*, pages 1655–1663. IEEE, 2001.

22. I. Dokmanic, R. Parhizkar, J. Ranieri, and M. Vetterli. Euclidean distance matrices. *IEEE Signal Processing Magazine*, 32(6):12–30, Nov 2015.

23. A. El-Mougy, M. Ibnkahla, G. Hattab, and W. Ejaz. Reconfigurable wireless networks. *Proceedings of the IEEE*, 103(7):1125–1158, 2015.

24. B.T. Fang. Simple solutions for hyperbolic and related position fixes. *IEEE Transactions on Aerospace and Electronic Systems*, 26(5):748–753, Sept 1990.

25. A. Fotouhi, M. Ding, and M. Hassan. Flying drone base stations for macro hotspots. *IEEE Access*, 6:19530–19539, 2018.

26. W.H. Foy. Position-location solutions by Taylor-series estimation. *IEEE Transactions on Aerospace and Electronic Systems*, AES12(2):187–194, March 1976.

27. X. Gao and M. Naraghi-Pour. Computationally efficient resource allocation for multiuser OFDM systems. In *IEEE Wireless Communications and Networking Conference, 2006. WCNC 2006*, volume 2, pages 804–809, April 2006.

28. L. Geng, M.F. Bugallo, A. Athalye, and P.M. Djuri. Indoor tracking with RFID systems. *IEEE Journal of Selected Topics in Signal Processing*, 8(1):96–105, Feb 2014.

29. S.G. Glisic. 1-persistent carrier sense multiple access in radio channels with imperfect carrier sensing. *IEEE Transactions on Communications*, 39(3):458–464, Mar 1991.

30. L.C. Godara. Application of antenna arrays to mobile communications. II. Beam-forming and direction-of-arrival considerations. *Proceedings of the IEEE*, 85(8):1195–1245, Aug 1997.

31. J.C. Gower. Euclidean distance geometry. *Mathematical Scientist*, 7:1–14, 1982.

32. M. Gowtham, S. Zhang, C. Tepedelenlioglu, M.K. Banavar, A. Spanias, C. Vargas-Rosales, and R. Villalpando-Hernandez. Location based distributed spectral clustering for wireless sensor networks. In *Proceedings of the Seventh Conference on Sensor Signal Processing for Defence (SSPD2017)*, 2017.

33. U. Grenander and M. Miller. *Pattern Theory*. Oxford University Press, 2007.
34. M. Guadane, W. Bchimi, A. Samet, and S. Affes. Enhanced rangefree localization in wireless sensor networks using a new weighted hopsize estimation technique. In *2017 IEEE 28th Annual International Symposium on Personal, Indoor, and Mobile Radio Communications (PIMRC)*, pages 1–5, Oct 2017.
35. M.D. Hahm, Z.I. Mitrovski, and E.L. Titlebaum. Deconvolution in the presence of doppler with application to specular multipath parameter estimation. *IEEE Transactions on Signal Processing*, 45(9):2203–2219, Sept 1997.
36. B.R. Hamilton, X. Ma, R.J. Baxley, and S.M. Matechik. Propagation modeling for radio frequency tomography in wireless networks. *IEEE Journal of Selected Topics in Signal Processing*, 8(1):55–65, Feb 2014.
37. Y. He, P. Aubry, and F. Le Chevalier. Ultra-wideband multistatic tracking of human targets. In *IET International Radar Conference 2013*, pages 1–5, Apr 2013.
38. J. Ianniello. Large and small error performance limits for multipath time delay estimation. *IEEE Transactions on Acoustics, Speech, and Signal Processing*, 34(2):245–251, Apr 1986.
39. A.T. Ihler, J.W. Fisher, R.L. Moses, and A.S. Willsky. Nonparametric belief propagation for self-localization of sensor networks. *IEEE Journal on Selected Areas in Communications*, 23(4):3058–3067, Apr 2005.
40. A. Jakobsson, A.L. Swindlehurst, and P. Stoica. Subspace-based estimation of time delays and doppler shifts. *IEEE Transactions on Signal Processing*, 46(9):2472–2483, Sept 1998.
41. D. Joho, C. Plagemann, and W. Burgard. Modeling RFID signal strength and tag detection for localization and mapping. In *2009 IEEE International Conference on Robotics and Automation*, pages 3160–3165, May 2009.
42. C. Koweerawong, K. Wipusitwarakun, and K. Kaemarungsi. Indoor localization improvement via adaptive RSS fingerprinting database. In *The International Conference on Information Networking 2013 (ICOIN)*, pages 412–416, Jan 2013.
43. A. Lackpour, C. Hamilton, M. Jacovic, I. Rasheed, X.R. Rey, and K.R. Dandekar. Enhanced 5G spectrum sharing using a new adaptive NC-OFDM waveform with reconfigurable antennas. In *Proceedings of the IEEE International Symposium o. Dynamic Spectrum Access Networks*, Mar 2017.
44. S. Li, L. Huang, J. Wu, H. Xu, and J. Wang. NBLS: Neighbor-information-based localization system for wireless sensor networks. In *2008 3rd International Conference on Communication Systems Software and Middleware and Workshops (COMSWARE '08)*, pages 91–94, Jan 2008.
45. X. Li and K. Pahlavan. Super-resolution TOA estimation with diversity for indoor geolocation. *IEEE Transactions on Wireless Communications*, 3(1):224–234, Jan 2004.
46. J. Maneesilp, C. Wang, H. Wu, and N. Tzeng. RFID support for accurate 3D localization. *IEEE Transactions on Computers*, 62(7):1447–1459, Jul 2013.
47. T.G. Manickam, R.J. Vaccaro, and D.W. Tufts. A least-squares algorithm for multipath time-delay estimation. *IEEE Transactions on Signal Processing*, 42(11):3229–3233, Nov 1994.
48. H. Minn and A. Khansefid. Massive MIMO systems in noncontiguous bands with asymmetric traffics. *IEEE Transactions on Wireless Communication*, 15(7):4689–4702, Jul 2016.
49. D. Munoz, F. Bouchereau, C. Vargas, and R. Enriquez-Caldera. *Position Location Techniques and Applications*. Academic Press/Elsevier, 2007.
50. M. Naraghi-Pour and T. Ikuma. Autocorrelation-based spectrum sensing for cognitive radios. *IEEE Transactions on Vehicular Technology*, 59(2):718–733, Feb 2010.
51. M. Naraghi-Pour and T. Ikuma. EM-based localization of noncooperative multicarrier communication sources with noncoherent subarrays. *IEEE Transactions on Wireless Communications*, 17(9): 6149–6159, Sept 2018.
52. M. Naraghi-Pour and T. Ikuma. Joint spatial and spectral localization of OFDM sources with noncoherent arrays. In *2018 IEEE Global Communication Conference (Globecom)*, pages 1–6, Dec 2018.
53. M. Naraghi-Pour and T. Ikuma. Localization of non-cooperative OFDM sources with noncoherent snapshots. In *2018 IEEE Wireless Communications and Networking Conference (WCNC)*, pages 1–6, Apr 2018.

54. M. Naraghi-Pour and G.C. Rojas. Sensor network localization via distributed randomized gradient descent. In *MILCOM 2013—2013 IEEE Military Communications Conference*, pages 1714–1719, Nov 2013.

55. M. Naraghi-Pour and G.C. Rojas. A novel algorithm for distributed localization in wireless sensor networks. *ACM Transaction on Sensor Network*, 11(1):1:1–1:25, Sept 2014.

56. L.M. Ni, Y. Liu, Y.C. Lau, and A.P. Patil. Landmarc: Indoor location sensing using active RFID. In *Proceedings of the First IEEE International Conference on Pervasive Computing and Communications, 2003. (PerCom 2003)*, pages 407–415, Mar 2003.

57. R. Olfati-Saber, J.A. Fax, and R.M. Murray. Consensus and cooperation in networked multi-agent systems. *Proceedings of the IEEE*, 95(1):215–233, Jan 2007.

58. M. Orooji, E. Soltanmohammadi, and M. Naraghi-Pour. Improving detection delay in cognitive radios using secondary-user receiver statistics. *IEEE Transactions on Vehicular Technology*, 64(9):4041–4055, Sept 2015.

59. K. Pahlavan, P. Krishnamurthy, and Y. Geng. Localization challenges for the emergence of the smart world. *IEEE Access*, 3:3058–3067, 2015.

60. G. Parkinson, M. Boon, J.G. Davis, and R. Sloan. 3D positioning using spherical location algorithms for networked wireless sensors deployed in grain. In *Proceedings of the IEEE International Microwave Symposium Digest (MTT)*, pages 1417–1420, 2009.

61. N. Patwari, L. Brewer, Q. Tate, O. Kaltiokallio, and M. Bocca. Breathfinding: A wireless network that monitors and locates breathing in a home. *IEEE Journal of Selected Topics in Signal Processing*, 8(1):30–42, Feb 2014.

62. N. Patwari, J. Wilson, S. Ananthanarayanan, S.K. Kasera, and D.R. Westenskow. Monitoring breathing via signal strength in wireless networks. *IEEE Transactions on Mobile Computing*, 13(8):1774–1786, Aug 2014.

63. N.B. Priyantha, A. Chakraborty, and H. Balakrishnan. The Cricket Location-Support System. In *6th ACM MOBICOM*, Boston, MA, Aug 2000.

64. J.G. Proakis and M. Salehi. *Digital Communications*. McGraw-Hill, 2008.

65. M.L. Psiaki and T.E. Humphreys. GNSS spoofing and detection. *Proceedings of the IEEE*, 104(6):1258–1270, Jun 2016.

66. M.L. Psiaki, B.W. O'Hanlon, J.A. Bhatti, D.P. Shepard, and T.E. Humphreys. GPS spoofing detection via dual-receiver correlation of military signals. *IEEE Transactions on Aerospace and Electronic Systems*, 49(4):2250–2267, Oct 2013.

67. R. Zekavat and R. Buehrer. *Wireless Localization Using Ultra-Wideband Signals*. Wiley-IEEE Press, New York, NY, USA, 2012.

68. B.D. Rao and K.V.S. Hari. Performance analysis of esprit and tam in determining the direction of arrival of plane waves in noise. *IEEE Transactions on Acoustics, Speech, and Signal Processing*, 37(12):1990–1995, Dec 1989.

69. J. Rovakova and D. Kocur. TOA association for handheld UWB radar. In *11th International Radar Symposium*, pages 1–4, Jun 2010.

70. J. Rovkov and D. Kocur. UWB radar signal processing for through wall tracking of multiple moving targets. In *The 7th European Radar Conference*, pages 372–375, Sept 2010.

71. H. Saarnisaari. Ml time delay estimation in a multipath channel. In *Proceedings of ISSSTA'95 International Symposium on Spread Spectrum Techniques and Applications*, volume 3, pages 1007–1011, vol. 3, Sept 1996.

72. S. Safavi, U. Khan, S. Kar, and J.M.F. Moura. Distributed localization: A linear theory. *Proceedings of the IEEE*, 106(7):1204–1223, Jul 2018.

73. S. Sampei and H. Harada. System design issues and performance evaluations for adaptive modulation in new wireless access systems. *Proceedings of the IEEE*, 95(12):2456–2471, Dec 2007.

74. S. Sarkka, V.V. Viikari, M. Huusko, and K. Jaakkola. Phase-based UHF RFID tracking with nonlinear Kalman filtering and smoothing. *IEEE Sensors Journal*, 12(5):904–910, May 2012.

75. A.H. Sayed, A. Tarighat, and N. Khajehnouri. Network-based wireless localization. *IEEE Signal Processing Magazine*, 22(4):24–40, May 2005.

76. R. Schmidt. Multiple emitter location and signal parameter estimation. *IEEE Transactions on Antennas and Propagation*, 34(3):276–280, Mar 1986.
77. H. Seliem, R. Shahidi, M.H. Ahmed, and M.S. Shehata. Drone-based highway-VANET and DAS service. *IEEE Access*, 6:20125–20137, 2018.
78. G. Shafer. *A mathematical theory of evidence*. Princeton University Press, 1976.
79. M. Singh, S.K. Bhoi, and P.M. Khilar. Geometric constraint-based range-free localization scheme for wireless sensor networks. *IEEE Sensors Journal*, 17(16):5350–5366, Aug 2017.
80. B. Sobhani, M. Mazzotti, E. Paolini, A. Giorgetti, and M. Chiani. Multiple target detection and localization in UWB multistatic radars. In *2014 IEEE International Conference on Ultra-WideBand (ICUWB)*, pages 135–140, Sept 2014.
81. B. Sobhani, E. Paolini, A. Giorgetti, M. Mazzotti, and M. Chiani. Target tracking for UWB multistatic radar sensor networks. *IEEE Journal of Selected Topics in Signal Processing*, 8(1):125–136, Feb 2014.
82. K. Sohraby, D. Minoli, and T. Znati. *Applications of Wireless Sensor Networks*. Wiley, 2007.
83. E. Soltanmohammadi, K. Ghavami, and M. Naraghi-Pour. A survey of traffic issues in machine-to-machine communications over LTE. *IEEE Internet of Things Journal*, 3(6):865–884, Dec 2016.
84. E. Soltanmohammadi, M. Orooji, and M. Naraghi-Pour. Improving the sensing throughput tradeoff for cognitive radios in Rayleigh fading channels. *IEEE Transactions on Vehicular Technology*, 62(5):2118–2130, Jun 2013.
85. S. Srirangarajan, A. Tewfik, and Z.-Q. Luo. Distributed sensor network localization using SOCP relaxation. *Wireless Communications, IEEE Transactions on*, 7(12):4886–4895, Dec 2008.
86. S. Stefanatos and F. Foukalas. A filter-bank transceiver architecture for massive non-contiguous carrier aggregation. *IEEE Journal on Selected Areas in Communications*, 35(1):215–227, Jan 2017.
87. P. Stoica and A. Nehorai. Performance study of conditional and unconditional direction-of-arrival estimation. *IEEE Transactions on Acoustics, Speech, and Signal Processing*, 38(10):1783–1795, Oct 1990.
88. H. Suo, J. Wan, L. Huang, and C. Zou. Issues and challenges of wireless sensor networks localization in emerging applications. In *2012 International Conference on Computer Science and Electronics Engineering*, volume 3, pages 447–451, Mar 2012.
89. A. Tayebi, J. Gomez, F. Saez de Adana, and O. Gutierrez. The application of ray tracing to mobile localization using the direction of arrival and received signal strength in multipath indoor environments. *Proceedings in Electromagnetics Research*, 91, 2009.
90. T. Tran, C. Sutton, R. Cocci, Y. Nie, Y. Diao, and P. Shenoy. Probabilistic inference over RFID streams in mobile environments. In *2009 IEEE 25th International Conference on Data Engineering*, pages 1096–1107, Mar 2009.
91. R.J. Vaccaro, T. Manickam, C.S. Ramalingam, R. Kumaresan, and D.W. Tufts. Least-squares time-of-arrival estimation for transient signals in a multipath environment. In *[1991] Conference Record of the Twenty-Fifth Asilomar Conference on Signals, Systems Computers*, volume 2, pages 1098–1102, Nov 1991.
92. S. Van de Velde, G. Arora, L. Vallozi, H. Rogier, and H. Steendam. Cooperative hybrid localization using gaussian processes and belief propagation. In *Proceedings of the IEEE Conference on Communications (ICC) Workshop on advances in Network Localization and Navigation*, pages 790–195. IEEE, 2015.
93. C. Vargas-Rosales, D. Munoz-Rodriguez, R. Torres-Villegas, and E. Sanchez-Mendoza. Vertex projection and maximum likelihood position location in reconfigurable networks. *Wireless Personal Communications*, 96(1):1245–1263, 2017.
94. R. Villalpando-Hernandez, D. Munoz-Rodriguez, C. Vargas-Rosales, and L. Rizo-Dominguez. 3-D position location in ad hoc networks: A Manhattanized space. *IEEE Communications Letters*, 21(1):124–127, 2017.
95. R. Villalpando-Hernandez, D. Munoz-Rodriguez, C. Vargas-Rosales, and J.R. Rodriguez-Cruz. Position location in ad-hoc/sensor networks: A linear constrained search. *IEEE Communications Letters*, 15(6):605–607, 2011.

96. C. Wang, K. Liu, and N. Xiao. A range free localization algorithm based on restricted-area for wireless sensor networks. In *2008 The Third International Multi-Conference on Computing in the Global Information Technology (ICCGI 2008)*, pages 97–101, Jul 2008.

97. Y.-Y. Wang and W.-H. Fang. TST-MUSIC for DOA-delay joint estimation. In *2000 IEEE International Conference on Acoustics, Speech, and Signal Processing. Proceedings (Cat. No.00CH37100)*, volume 5, pages 2565–2568, vol. 5, Jun 2000.

98. J. Wilson and N. Patwari. Radio tomographic imaging with wireless networks. *IEEE Transactions on Mobile Computing*, 9(5):621–632, May 2010.

99. J. Wilson and N. Patwari. See-through walls: Motion tracking using variance-based radio tomography networks. *IEEE Transactions on Mobile Computing*, 10(5):612–621, May 2011.

100. B. Xu, G. Sun, R. Yu, and Z. Yang. High-accuracy TDOA-based localization without time synchronization. *IEEE Transactions on Parallel and Distributed Systems*, 24(8):1567–1576, Aug 2013.

101. Q. Yang, D.G. Taylor, and G.D. Durgin. Kalman filter based localization and tracking estimation for HIMR RFID systems. In *2018 IEEE International Conference on RFID (RFID)*, pages 1–5, Apr 2018.

102. R. Zekavat and R. Michael Buehrer. *Handbook of Position Location: Theory, Practice and Advances*. Wiley-IEEE Press, 1st edition, 2011.

103. Q. Zhao, S. Geirhofer, L. Tong, and B.M. Sadler. Opportunistic spectrum access via periodic channel sensing. *IEEE Transactions on Signal Processing*, 56(2):785–796, Feb 2008.

104. W. Zhu, J. Cao, Y. Xu, L. Yang, and J. Kong. Fault-tolerant RFID reader localization based on passive RFID tags. *IEEE Transactions on Parallel and Distributed Systems*, 25(8):2065–2076, Aug 2014.

2

Conditions for Unique Localizability in Cooperative Localization of Wireless ad hoc and Sensor Networks

Tolga Eren

CONTENTS

2.1 Introduction

Locations of nodes are essential in several applications of wireless ad hoc and sensor networks, because the information gathered by sensor nodes are often meaningful and helpful when the locations of received information are available (Akyildiz et al. 2002). GPS receivers may be installed on all sensor nodes, but this is an expensive solution. Manually configuring each sensor node is also impractical when the number of nodes is large. In order to solve this problem, a few nodes are selected as reference nodes in the network. These reference nodes are sometimes called beacons or anchors. The rest of the nodes are called ordinary nodes. Reference nodes know their locations by means of GPS or manual configuration. However, ordinary nodes, which are large in number, do not know their locations. While reference nodes transmit their positions, ordinary nodes make measurement with reference nodes. An ordinary node, which can make measurements to three reference nodes (in the plane), can find out its location. However, several ordinary nodes cannot make measurements with reference nodes because nodes have power limitations or the signals coming from reference nodes are blocked. In order to solve this problem, ordinary nodes not only make measurements with reference nodes, but they also make measurements with other ordinary nodes. They then combine measurements to determine their locations. This method is called "cooperative localization" (Patwari et al. 2005, Wymeersch et al. 2009).

In cooperative localization, the issue of localizability arises once we know the positions of reference nodes and the measurements obtained by ordinary nodes. If there is a unique set of solutions for the positions of ordinary nodes, satisfying the conditions imposed by the positions of reference nodes and the measurements obtained by ordinary nodes, then the network is uniquely localizable. We use graphs from graph theory in mathematics in order to model wireless networks. Localization of networks is related to the problem of graph embedding. Unique localizability of networks is associated with graph rigidity, studied in graph theory. In this chapter, we explore the conditions of unique localizability of networks, in which cooperative localization is implemented.

Various applications of rigidity and global rigidity exist in the literature (Eren et al. 2004, Aspnes et al. 2006, Zhang and Liu et al. 2012, Zhang and Cui et al. 2012, Zhou et al. 2013, Zhu and Hu 2014). For example, unique network localizability has been used in robotics (Zelazo et al. 2012, 2015) and in sensor networks (Eren et al. 2004, Aspnes et al. 2006, Zhang and Cui et al. 2012, Yang et al. 2013, Zhou et al. 2013, Williams et al. 2015, Eren 2016). Moreover, rigidity theory has been used to control robot formations by applying to network topologies (Zelazo et al. 2012, He et al. 2013, Williams et al. 2014). The advantage of rigidity theory is that it helps to keep formations by using distance measurements rather than position measurements. In addition, rigidity helps to find locations from the measurements of distances.

Localizability has been used in the localization process of sensor networks in real applications. For example, Chen et al. (2014) develop a method for localization called the localizability-aided localization (LAL) method, in which the process makes use of the localizability information of the network. They implement real-world experiments and a large number of simulations. Before the LAL method starts the localization scheme, it has an adjustment phase that has the stages of node localizability testing and structure analysis. The method makes use of the node localizability testing in order to make adjustments in the network in a purposeful direction. The real-world data comes from a wireless sensor network system built to collect ecological information in a forest. In such a system, energy plays a central role for the success of the entire system. The method makes use of the node localizability testing information in order to determine affected nodes and inserted edges. They conclude that the information of the node localizability saves time and cost by avoiding unnecessary adjustments.

2.2 Rigidity

2.2.1 Rigid Frameworks

As noted earlier, graphs from graph theory are used to model networks. We denote a graph by $G = (V, E)$, where V is the set of vertices and E is the set of edges. The set of vertices is used to model the set of nodes in the network, whereas the set of edges is used to model the set of sensing and communication links in the network. A *framework* associated with the graph G is denoted by $G(p)$, and it is an embedding of the graph G, where p is a plane configuration, that is, $p : V \to R^2$. Let $v_i v_j$ denote an edge in E and $||.||$ denote the distance. We say that frameworks $G(p)$ and $G(q)$ are *equivalent* if $||p(v_i) - p(v_j)|| = ||q(v_i) - q(v_j)||$, for all

$v_i v_j$ in E. Let v_i, v_j denote vertices in V. We say that frameworks $G(p)$ and $G(q)$ are *congruent* if $||p(v_i)-p(v_j)|| = ||q(v_i)-q(v_j)||$, for any two vertices v_i, v_j in V. If q is in U_p (where U_p denotes a neighborhood of p) and the equivalency of $G(p)$ and $G(q)$ implies the congruency of $G(p)$ and $G(q)$, then the framework $G(p)$ in R^2 is called *rigid*. Related references include (Whiteley 1996, Berg and Jordan 2003).

Rigidity is also studied in civil engineering under the rigidity of bar-joint frameworks. In this approach, bars and joints are represented by edges and vertices, respectively. A bar-joint framework is called rigid if it cannot be deformed by applying forces on bars or joints. For example, a flexible framework is shown in Figure 2.1. Here, the vertices v_1 and v_4 can move to v_1' and v_4'. On the other hand, a rigid framework is shown in Figure 2.2. In the framework, continuous deformation is not possible.

The *rigidity matrix* $R(G,p)$ is used in algebraic study of rigidity. In this matrix, there are $|E|$ rows and $2|V|$ columns, where $|.|$ denotes the number of elements in a set. If the edge

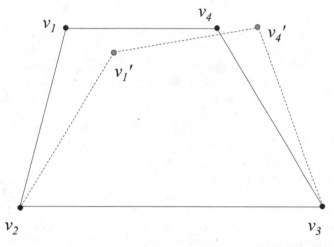

FIGURE 2.1
A non-rigid (flexible) framework. The vertices v_1 and v_4 can move to v_1' and v_4'.

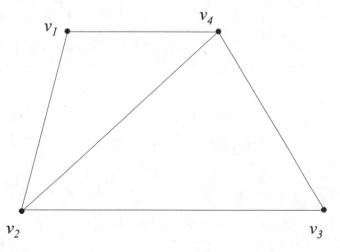

FIGURE 2.2
A rigid framework is shown here. Continuous deformation of the framework is not possible.

$v_i v_j$ is an edge in E, then in the row corresponding to this edge there are only two non-zero entries. One is in column v_i, where the matrix entry is $p(v_i) - p(v_j)$. The other non-zero entry is in column v_j, where the matrix entry is $p(v_j) - p(v_i)$. For example, $R(G,p)$ of the framework in Figure 2.2 is shown below. Since there are 5 edges in the framework, there are 5 rows in the rigidity matrix. The number of columns is 8, which is twice the number of vertices in the framework.

$R(G,p)$	v_1		v_2		v_3		v_4	
(1,2)	x_1-x_2	y_1-y_2	x_2-x_1	y_2-y_1	0	0	0	0
(1,4)	x_1-x_4	y_1-y_4	0	0	0	0	x_4-x_1	y_4-y_1
(2,3)	0	0	x_2-x_3	y_2-y_3	x_3-x_2	y_3-y_2	0	0
(2,4)	0	0	x_2-x_4	y_2-y_4	0	0	x_4-x_2	y_4-y_2
(3,4)	0	0	0	0	x_3-x_4	y_3-y_4	x_4-x_3	y_4-y_3

If $\text{rank}(R(G,p)) = 2|V| - 3$, then the framework (G,p) is called an *infinitesimally rigid* framework. It turns out that any infinitesimally rigid framework is also rigid. The converse is also true provided that the configuration p is generic. A configuration p is called *generic* if any non-zero polynomial equation with integer coefficients does not hold with the coordinates of p.

2.2.2 Graph Rigidity

Generic configurations are an open connected dense subset of $\mathbb{R}^{2n}$. This means that if $G(p)$ is rigid for a configuration p, then it is rigid for almost all p. Thus the rigidity of a framework can be studied by considering the underlying topological structure of the network. We note that the graph component G of $G(p)$ gives the topological structure information of the network. On the other hand, the configuration component p of $G(p)$ informs us about spatial positions of each node in the network. A graph $G = (V,E)$ is called *rigid* if $G(p)$ is rigid for any generic configuration p. A graph $G = (V,E)$ is called *minimally rigid*, if G is rigid and it remains rigid when we remove any one of the edges in E. For example, the graph G in Figure 2.3 is minimally rigid. If we remove any of the edges in G, it becomes flexible. On the hand, the graph in Figure 2.4 is not minimally rigid, because $G = (V,E)$ in Figure 2.4 is rigid, and if we remove the edge $v_1 v_4$, it still remains rigid.

A combinatorial characterization of graph rigidity is as follows. A graph $G = (V,E)$ is rigid if and only if there exists a subgraph $G' = (V,E')$ where $|E'| = 2|V| - 3$, satisfying the condition

$$|E''| \leq 2|V''| - 3, \quad \text{for all } E'' \subseteq E' \tag{2.1}$$

where $G'' = (V'',E'')$ denote the subgraph of G' induced by E''. See (Whiteley 1996) for more details.

2.2.3 Redundant Rigidity

Redundant rigidity is a stronger type of rigidity than (plain) rigidity. A rigid graph $G = (V,E)$ is *redundantly rigid* if $G' = (V,E')$, where $E' = E \backslash \{e\}$ remains rigid for every edge e in E. For example, the graph in Figure 2.4 is not redundantly rigid, because if we remove the edge $v_2 v_3$, it becomes flexible. On the other hand, the graph in Figure 2.5 is redundantly rigid; that

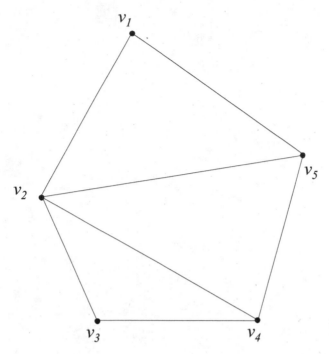

FIGURE 2.3
Minimally rigid graph $G=(V,E)$ is shown. If we remove any one of the edges from E, then the graph G becomes non-rigid.

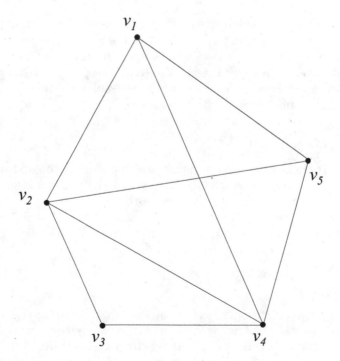

FIGURE 2.4
A rigid graph $G=(V,E)$ that is not minimally rigid is shown here. G is rigid, and if we remove the edge v_1v_4, it still remains rigid.

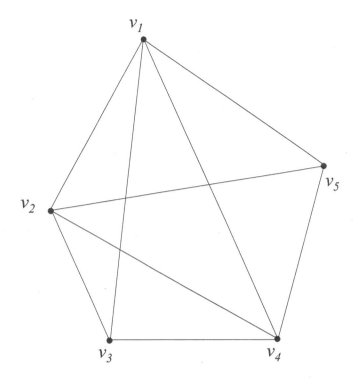

FIGURE 2.5
A redundantly rigid graph is shown here. If we remove any one of the edges in this graph, it still remains rigid.

is, if we remove any one of the edges in this graph, it still remains rigid. We note that a redundantly rigid graph does not need to have all the possible edges between the set of given vertices. For example, the edge v_3v_5 is missing in the graph in Figure 2.5. A graph that has edges between every pair of vertices is called a *complete graph*. An example of a complete graph is shown in Figure 2.6. We note that a complete graph $G = (V,E)$ with $|V| > 3$ is always a redundantly rigid graph.

2.2.4 Graph 3-Connectivity

Because 3-connectivity is related to the rigidity concepts of frameworks, we provide some definitions of 3-connectivity. A graph $G = (V,E)$, where $|V| > 4$, is *3-connected* if it remains connected with the removal of any two vertices in V (Diestel 2000). An example for a 3-connected graph is shown in Figure 2.7. An equivalent definition of 3-connectivity is as follows, from Menger's theorem. A graph $G = (V,E)$, where $|V| > 4$, is 3-connected if there are at least three vertex disjoint paths between any pair of vertices.

2.2.4.1 Global Rigidity

Global rigidity is an even stronger type of rigidity than (plain) rigidity and redundant rigidity. If we reconsider bar-joint frameworks, rigidity is concerned with continuous deformation of the framework. In global rigidity, we consider not only continuous flexing, but also discontinuous deformation and partial reflections of the framework. A discontinuous deformation of a framework is shown in Figure 2.8. The given framework is rigid. Now let us temporarily remove the edge v_3v_6. Now the vertices v_4, v_5, v_6 can rotate,

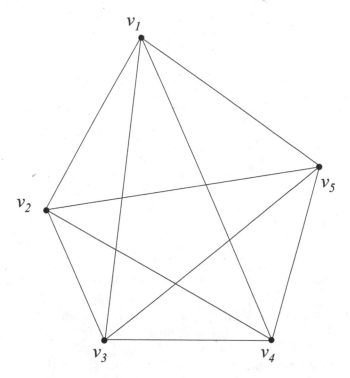

FIGURE 2.6

A complete graph is shown here. If we remove any one of the edges in this graph, it still remains rigid. It has edges between every pair of vertices.

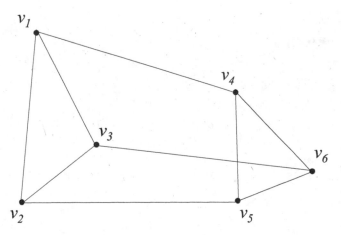

FIGURE 2.7

An example for a 3-connected graph.

and we can obtain a second realization of the framework, in which all the edge length constraints are satisfied, as shown in Figure 2.9. Once we obtain this second realization, we insert the edge v_3v_6 back to its place. From this example, originally given in (Hendrickson 1992), we learn that (plain) rigidity is not sufficient for unique localizability.

An example for partial reflection is shown in Figure 2.10. Here it is possible to flip the vertex v_1 over the edges v_1v_2 and v_1v_4 to its new place, denoted with v_1'. Although the framework

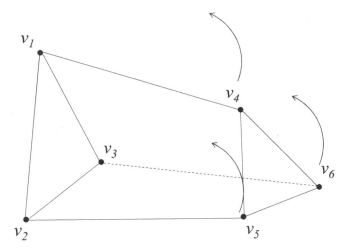

FIGURE 2.8
Discontinuous flexing of a framework. We temporarily remove the edge v_3v_6. Now the vertices v_4, v_5, v_6 can rotate.

is rigid, as observed here, it is still possible to obtain a second realization by a partial reflection. In order to avoid partial reflections, the graph has to be 3-connected. For example, the framework in Figure 2.7 is 3-connected and partial reflections are disallowed in this framework. We note that although the framework in Figure 2.7 is both rigid and 3-connected, it is not uniquely localizable because of discontinuous flexing as explained above.

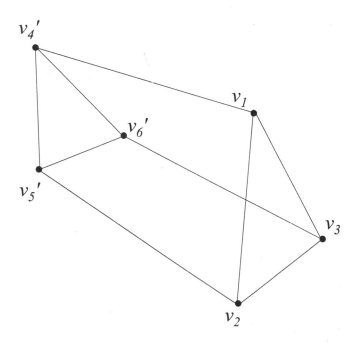

FIGURE 2.9
We can obtain a second realization of the framework in which all the edge length constraints are satisfied. Once we obtain this second realization, we insert the edge v_3v_6 back to its place.

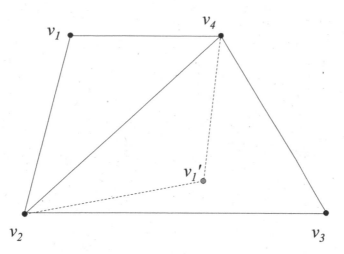

FIGURE 2.10

An example for partial reflection. It is possible to flip the vertex v_1 over the edges v_1v_2 and v_1v_4 to its new place denoted with v_1'.

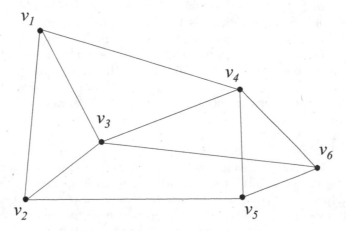

FIGURE 2.11

An example of a globally rigid graph.

A given framework $G(p)$ is *globally rigid*, if every framework $G(q)$ which is equivalent to $G(p)$, is also congruent to $G(p)$. Combinatorial characterization of global rigidity makes use of the previous concepts, namely redundant rigidity and 3-connectivity. A graph $G = (V,E)$ is globally rigid if and only if it is redundantly rigid and 3-connected (Hendrickson 1992, Jackson and Jordan 2005). An example of a globally rigid graph is shown in Figure 2.11.

2.3 Localizability and Rigidity

The connection between localizability of networks and the concept of graph rigidity was provided in (Eren et al. 2004). Specifically, first a grounded graph is created by the network links between ordinary nodes with other ordinary nodes and reference nodes, together

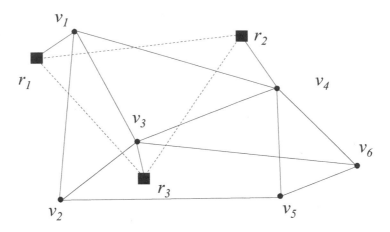

FIGURE 2.12
An example of a grounded graph to determine localizability of a network.

with the insertion of additional links between every pair of reference nodes. It was shown in (Eren et al. 2004) that network is localizable if the associated grounded graph is globally rigid. An example of a grounded graph is shown in Figure 2.12. Here r_1, r_2, r_3 denote the reference nodes. The edges between these reference nodes are implicitly inserted to the existing graph to construct the grounded graph. These implicit edges do exist because we know the distances between reference nodes since their location are already known.

Trilateration graphs are proposed as inductive constructions of globally rigid graphs in (Eren et al. 2004). In such graphs, if a node is inserted with 3 edges to a globally rigid graph in the plane, the resulting graph is also globally rigid. This provides us a sequential method to localize nodes in a network.

Merging globally rigid graphs are also considered in (Eren et al. 2004). If two globally rigid graphs (G_1 and G_2) are connected by a set of edges E_m, the resulting graph is globally rigid if the following two conditions hold: (1) The end vertices of E_m has at least three vertices in G_1 and has at least three vertices in G_2. (2) There are at least four edges links in E_m.

Another result about merging globally rigid graphs is as follows. If two globally rigid graphs, G_1 and G_2, share at least three vertices, the union of these two graphs is globally rigid (Eren et al. 2004).

2.4 Measures of Rigidity, Redundancy, and 3-Connectivity

As explained in previous sections, for unique localizability, rigidity is necessary but not sufficient. The underlying graph has to be redundantly rigid and 3-connected for unique localizability. Let us consider a set of nodes distributed in an area. Let sensor nodes increase their sensing radius gradually in a such a way that the underlying graph of the network becomes rigid and then globally rigid. We wonder whether it is hard to achieve global rigidity once the graph becomes rigid. We may explore this problem by investigating the changes in the rigidity, redundant rigidity, and 3-connectivity properties of the graph as we increase the radii of the nodes. To do that we make use of the rigidity index, redundancy index, and 3-connectivity index as explained below. These indices are introduced in (Eren 2016, 2017).

2.4.1 Rigidity Index

Let us consider the graph $G = (V, E)$. Let us consider a family of subsets of E and call it I. If I satisfies the following three matroid axioms, then it forms the independent sets of a matroid $M(G)$:

Axiom 1: Empty set is an element of I.

Axiom 2: If E' is in I and E'' is a subset of E', then E'' is in I.

Axiom 3: If E_1 and E_2 are in I, and the cardinality of E_1 is less than the cardinality of E_2, then there is an edge e in the difference set between E_2 and E_1 such that the union of E_1 and $\{e\}$ is in I. We refer the reader to Oxley (1992) for more information on matroid theory.

For a given framework (G, p), we define the rigidity matroid by using the linear independence of the rows of the rigidity matrix $R(G, p)$. Now let us consider a graph $G = (V, E)$. Let E' be a non-empty subset of E and let V' be the set of vertices incident with E'. Then the edge set E' is called *independent* if

$$|E''| \leq 2|V'''| - 3 \text{ for all } E''' \subseteq E'. \tag{2.2}$$

Let $G = (V, E)$ be a graph. Let us consider the independent set of edges E', which are subsets of E, and denote the collection of these sets by S. The *rigidity index*, shown by $K_r(G)$, is defined as follows:

$$K_r(G) := \frac{\max_{E' \varepsilon S} |E'|}{2|V| - 3} \tag{2.3}$$

The rigidity index essentially denotes the proportion of the size of the set of independent edges within $2|V| - 3$, which is the maximal number of independent edges. If there are no edges in the graph, the rigidity index takes the value of zero. If there are $2|V| - 3$ independent edges, that is, the graph is rigid, the rigidity index becomes one. So the value of the rigidity index is bigger than or equal to zero, and less than or equal to one. If the graph is getting close to rigidity, the value of K_r gets larger.

2.4.1.1 Redundancy Index

Recall that an edge is called redundant if the graph remains rigid after the removal of this edge. The definition of redundancy can be extended to include non-rigid graphs. Given a graph $G = (V, E)$, which is not necessarily rigid, if the rigidity index does not change after the removal of an edge e, then we call this edge a *generalized redundant edge*. In other words, for a given graph $G = (V, E)$ with a rigidity index $K_r(G)$, if $K_r(G) = K_r(G - \{e\})$, then the edge e is a generalized redundant edge. Now let us consider the set of redundant edges for a given graph, denoted by E_u as follows: $E_u = \{e : K_r(G - \{e\}) = K_r(G)\}$. Given this set of edges, we define the redundancy index, denoted as $K_u(G)$, as follows:

$$K_u(G) := \frac{|E_u(G)|}{|E|} \tag{2.4}$$

If there are no redundant edges in G, then the redundancy index is zero. If all the edges in G are redundant, then the redundancy index becomes one. If there are no edges in *G*, then the redundancy index is defined to be zero. As the redundancy index increases, this means that the graph becomes more robust to keep its rigidity index.

Now let us examine the graph in Figure 2.13. There are 15 edges and all of them are independent. Therefore $K_r = 1$, and $K_u = 0$, which means that it is a minimally rigid graph.

The graph in Figure 2.14 is a non-minimally rigid graph, that is, it is rigid and there are redundant edges. Therefore $K_r = 1$. Now the redundant edges in the graph are as follows: $E_u = \{v_1v_2,\ v_1v_5,\ v_2v_4,\ v_4v_5,\ v_1v_4,\ v_2v_5\}$. There are 6 redundant edges and, in total, 16 edges. Therefore $K_u = 6/16$.

The graph in Figure 2.15 shows a flexible (non-rigid) graph. All of its edges are independent edges, that is, there are no redundancies in this graph. Therefore the redundancy index $K_u = 0$. There are 13 edges and all of them independent. The maximal number of independent edges for this graph is $2|V| - 3 = 15$ edges, therefore the rigidity index $K_r = 13/15$.

The graph in Figure 2.16 is non-rigid. The maximal number of independent edges for this graph is $2|V| - 3 = 15$ edges, since there are 9 vertices. There are 14 independent edges in this graph, so the rigidity index $K_r = 14/15$. There are 6 generalized redundant edges here, $E_u = \{v_1v_2,\ v_1v_5,\ v_2v_4,\ v_4v_5,\ v_1v_4,\ v_2v_5\}$. There are 15 edges here, where some of them are dependent. So the redundancy index $K_u = 6/15$.

2.4.2 3-Connectivity Index

Let a graph $G = (V,E)$ be given. Let V_2 denote the set of all vertex pairs in G. Let V_c denote the vertex pairs such that the graph remains connected when we remove the vertices in V_c. Then we define the 3-connectivity index as follows.

$$K_c(G) = \frac{|V_c|}{|V_2|} \tag{2.5}$$

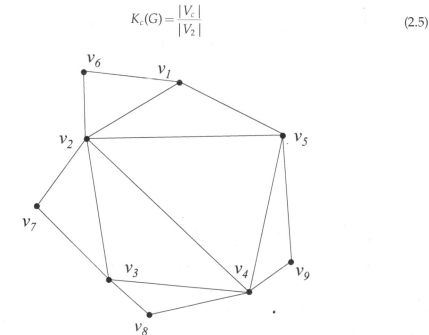

FIGURE 2.13
There are 15 edges and all of them are independent. $K_r = 1$ and $K_u = 0$ in this minimally rigid graph.

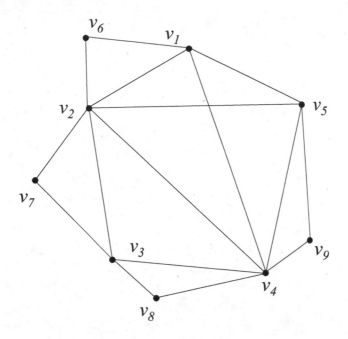

FIGURE 2.14
In this non-minimal rigid graph, the redundant edges are as follows: $E_u = \{v_1 v_2, v_1 v_5, v_2 v_4, v_4 v_5, v_1 v_4, v_2 v_5\}$. So the number of redundant edges is 6, and there are 16 edges in total. Therefore $K_u = 6/16$. Since the graph is rigid, $K_r = 1$.

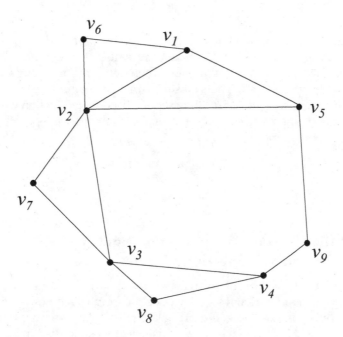

FIGURE 2.15
A non-rigid graph is shown here. All the edges here are independent. This means that there are no redundant edges. So the redundancy index is zero. The maximal number of independent edges for this graph is $2|V| - 3 = 15$ edges, since there are 9 vertices. On the other hand, there are only 13 independent edges. Hence the rigidity index $K_r = 13/15$.

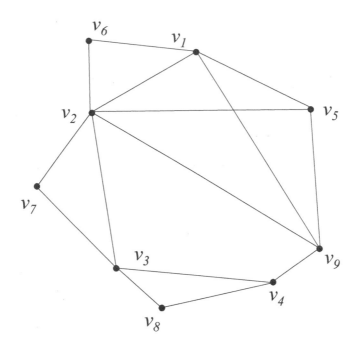

FIGURE 2.16

A non-rigid graph with generalized redundant edges. There are 14 independent edges in this graph. The maximal number of independent edges for this graph is $2|V|-3 = 15$ edges, since there are 9 vertices, so the rigidity index $K_r = 14/15$. There are 15 edges here, in which some of them are dependent. There are 6 generalized redundant edges here, $E_u = \{v_1v_2, v_1v_5, v_2v_4, v_4v_5, v_1v_4, v_2v_5\}$. So the redundancy index $K_u = 6/15$.

$K_c(G)$ takes the value of zero, when there is no vertex pair such that when we remove it, the graph remains connected. $K_c(G)$ takes the value of one, when it is 3-connected. So the 3-connectivity index takes a value between zero and one, and as it gets closer to 3-connectedness, the value of $K_c(G)$ gets closer to one. For example, the graph in Figure 2.2 has a 3-connectivity index value of 5/6, since it gets disconnected when we remove v_2v_4 and it remains connected when we remove the other pairs (there are 6 possible pairs in total). On the other hand, the graph in Figure 2.7 has a 3-connectivity index value of one, that is, it is 3-connected.

2.5 Localizability and the Comparison of Rigidity, Redundancy, and 3-Connectivity Indices

Now let us turn back to our problem that we stated earlier in this chapter. We wonder whether it is hard to achieve global rigidity once the graph becomes rigid. We explore the answer of this problem in random geometric graphs. The reason we explore these graphs is as follows. Localizability of graphs has been historically considered in graphs in a general setting, in which there is no restriction on the existence of edges between any two vertices in the graph. This approach essentially stems from the studies of global rigidity in graph theory, where the graphs are considered in a general setting. However, from a network point of view we have additional restrictions. For example, in a sensor network, a node has a sensing/communication link with the other nodes in its sensing/communication range. In such a setting, we obtain disc graphs to represent the associated network. A related concept to disc graphs is a random geometric graph, which is widely used to represent wireless ad hoc and sensor networks (Santi 2005, Penrose 2003).

Let us model our network with n vertices distributed uniformly and independently on a d by d square area, where d is the side length. In disc graphs, there is an edge between two vertices if the distance between these two vertices is less than or equal to the sensing radius r_s of sensor nodes. Let us consider a set of nodes distributed in an area (with $d = 30$ units) as shown in Figure 2.17. In order to implement localization, we want to achieve global rigidity of the network. So let us consider the situation that sensor nodes increase their sensing radius gradually. At some point in that process, the underlying graph of the network looks like the one shown in Figure 2.18.

Note that the underlying graph of the network becomes rigid as we raise the sensing radius. As we continue to increase r_s, the underlying graph becomes globally rigid. We explore how much additional increase in sensing radius is necessary to reach from rigidity to global rigidity. We investigate the necessary changes in radii by comparing the rigidity, redundant rigidity, and 3-connectivity properties of the network, for which we employ the rigidity index $K_r(G)$, redundancy index $K_u(G)$, and 3-connectivity index $K_c(G)$. For the node distribution in Figure 2.17, plots of the rigidity index, redundancy index, and 3-connectivity index against r_s/d are shown in Figure 2.19.

We repeat this computation of indices for different node distributions. For example, in (Eren 2017), 50 different node distributions are used with the same number of nodes in the given area, and then rigidity, redundancy, and 3-connectivity indices as a function of the ratio between the sensing radius and the side length of the area are computed for every distribution. Then it is possible to compute the sensing radius values at which rigidity index, redundancy index, and 3-connectivity index reach the value of one. The condition that the rigidity index reaches the value of one, when K_u attains the value of one, is also imposed

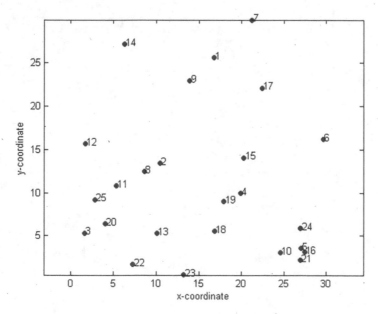

FIGURE 2.17
A set of nodes distributed in an area is shown here.

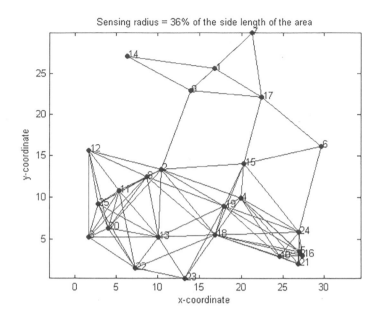

FIGURE 2.18
Underlying graph of the network of the node distribution shown in Figure 2.17.

to make sure that we determine the r_s value for redundant rigidity, since the redundancy index may attain the value of one before the graph becomes rigid. It is observed in (Eren 2017) that in none of the simulations in random geometric graphs does the 3-connectivity index attain the value of one before the redundancy index reaches this value.

Note that the nodes may be distributed uniformly in a random geometric graph. However, there are applications where some sensor nodes are distributed closely, forming clustered

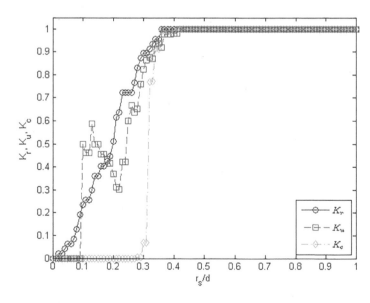

FIGURE 2.19
The changes in the rigidity index, redundancy index and the 3-connectivity index as the sensing radius is increasing.

graphs. When the process is repeated in clustered graph simulations that we did for random geometric graphs, we almost get similar results, with one noteworthy difference. For a few clustered node distributions the graph becomes 3-connected before it becomes redundantly rigid. Therefore, for a few clustered graphs, first the rigidity index, then the 3-connectivity index, and finally the redundancy index reaches the value of one.

2.6 Concluding Remarks

Rigidity, and in particular global rigidity, is associated with unique localizability of wireless ad hoc and sensor networks. Historically, global rigidity in graph theory mostly dealt with graphs in general settings, where there are no geometrical restrictions, that is, it dealt with graphs in which there is no restriction on the existence of edges between any two vertices in the graph. On the other hand, there are geometrical constraints in network-based applications.

For instance, a sensor node has a sensing/communication link, naturally, with other sensor nodes in its sensing/communication range. This results in disc graphs to model the associated network. A closely related graph structure to disc graphs is a random geometric graph, and it is used to model sensor networks. There are also applications in which some sensor nodes are distributed closely, forming clustered graphs.

The effects of graph structures (random geometric graph structure and clustered graph structure) on the conditions of unique localizability are investigated by making use of the rigidity, redundancy, and 3-connectivity indices.

It is shown in (Eren 2016) that considerably less effort is necessary to reach redundant rigidity once the network becomes rigid. This is true for graphs in random geometric graphs and in clustered graphs. When we compare the cases to reach 3-connectivity, as in (Eren 2017), we get a striking difference. It is significantly easy to obtain 3-connectivity once we attain redundant rigidity in random geometric graphs. We note that it is not likely to observe a graph in which 3-connectivity is attained before the graph obtains redundant rigidity in random geometric graphs with uniform distribution. So, in random geometric graphs, testing only 3-connectivity is more likely sufficient for unique localizability. It is discussed in (Eren 2017) that the lack of occurrence of discontinuous flexing is the reason behind this in random geometric graphs.

On the other hand, in clustered graphs, the graph may become 3-connected before it becomes redundantly rigid (Eren 2017). The reason is that there may be rigid components connected by inter-cluster edges, and that may cause discontinuous flexing. Hence, both redundant rigidity and 3-connectivity have to be checked in clustered graphs, for unique localizability.

References

Akyildiz IF, Su W, Sankarasubramaniam Y et al. Wireless sensor networks: a survey. *Computer Networks* 2002; 38(4): 393–422.

Aspnes J, Eren T, Goldenberg D et al. A theory of network localization. *IEEE Transactions on Mobile Computing* 2006; 5(12): 1663–1678.

Berg AR, Jordan T. Algorithms for graph rigidity and scene analysis. In: G Di Battista, U Zwick (Eds.). *Algorithms-ESA 2003. Lecture Notes in Computer Science*, vol. 2832. Springer, Berlin, Heidelberg, 2003, pp. 78–89.

Chen T, Yang Z, Liu Y et al. Localization-oriented network adjustment in wireless ad hoc and sensor networks. *IEEE Transactions on Parallel Distribution Systems* 2014; 25(1): 146–155.

Diestel R. *Graph Theory (Graduate Texts in Mathematics)*. Vol 173. Springer-Verlag, Berlin Heidelberg, 2000.

Eren T. Graph invariants for unique localizability in cooperative localization of wireless sensor networks: rigidity index and redundancy index. *Ad Hoc Networks* 2016; 44: 32–45.

Eren T. The effects of random geometric graph structure and clustering on localizability of sensor networks. *International Journal of Distributed Sensor Networks* 2017; 13(12): 1–14.

Eren T, Goldenberg DK, Whiteley W et al. Rigidity, computation, and randomization in network localization. In *Proceedings of the 2004 International Annual Joint Conference of the IEEE Computer and Communications Societies (INFOCOM 2004)*. Hong Kong, pp. 2673–2684.

Eren T, Anderson BDO, Whiteley W et al. Merging globally rigid formations. In *Proceedings of AAMAS (the Third International Joint Conference on Autonomous Agents & Multi Agent Systems)*. New York, July 2004, pp. 1258–1259.

He F, Wang Y, Yao Y et al. Distributed formation control of mobile autonomous agents using relative position measurements. *IET Control Theory & Applications* 2013 bba; 7(11): 1540–1552.

Hendrickson B. Conditions for unique graph realizations. *SIAM Journal on Computing* 1992; 21(1): 65–84.

Jackson B, Jordan T. Connected rigidity matroids and unique realizations of graphs. *Journal of Combinatorial Theory B* 2005; 94: 1–29.

Oxley JG. *Matroid Theory*. Oxford University Press, New York, 1992.

Patwari N, Ash JN, Kyperountas S et al. Locating the nodes: cooperative localization in wireless sensor networks. *IEEE Signal Processing Magazine July* 2005; 22(4): 54–69.

Penrose M. *Random Geometric Graphs*. Oxford University Press, 2003.

Santi P. Topology control in wireless ad hoc and sensor networks. *ACM Computing Surveys* 2005; 37(2): 164–194.

Whiteley W. Matroids from discrete geometry. In: JE Bonin, JG Oxley, B Servatius (Eds.). *Matroid Theory*. Vol. 197. American Mathematical Society, Contemporary Mathematics, 1996, pp. 171–313.

Williams RK, Gasparri A, Priolo A et al. Evaluating network rigidity in realistic systems: decentralization, asynchronicity, and parallelization. *IEEE Transactions on Robotics* 2014; 30(4): 950–965.

Williams RK, Gasparri A, Soffietti M et al. Redundantly rigid topologies in decentralized multi-agent networks. In *IEEE 54th Annual Conference on Decision and Control (CDC)* 2015, pp. 6101–6108.

Wymeersch H, Lien J and Win M. Cooperative localization in wireless networks. *Proceedings of the IEEE* 2009; 97(2):427–450.

Yang Z, Wu C, Chen T et al. Detecting outlier measurements based on graph rigidity for wireless sensor network localization. *IEEE Transactions on Vehicular Technology* 2013; 62(1): 374–383.

Zelazo D, Franchi A, Allgower F et al. Rigidity maintenance control for multi-robot systems. In: 2012 *Robotics: Science and Systems Conference*, Sydney, Australia, pp. 473–480.

Zelazo D, Franchi A, Bulthoff HH et al. Decentralized rigidity maintenance control with range measurements for multi-robot systems. *International Journal of Robotics Research* 2015; 34(1): 105–128.

Zhang X, Cui Q, Shi Y et al. Cooperative group localization based on factor graph for next-generation networks. *International Journal of Distributed Sensor Networks* 2012; 8(10): 1–15.

Zhang Y, Liu S, Zhao X et al. Theoretic analysis of unique localization for wireless sensor networks. *Ad Hoc Networks* 2012; 10(3): 623–634.

Zhou X, Guo D, Chen T et al. Achieving network localizability in nonlocalizable WSN with moving passive event. *International Journal of Distributed Sensor Networks* 2013; 9(10): 1–12.

Zhu G and Hu J. A distributed continuous-time algorithm for network localization using angle-of-arrival information. *Automatica* 2014; 50(1): 53–63.

3

Optimizing Computational Cost and Communication Overhead in Cooperative Localization

Panagiotis Agis Oikonomou-Filandras and Kai-Kit Wong

CONTENTS

3.1 Introduction

Position location, cf. [51], is a significant technology with a myriad of applications, from industrial and military use to consumer and personal use. From tracking the positions of ships and airplanes to helping tourists navigate inside museums, the ability to localize permeates every aspect of everyday society. The best-known example is satellite-based localization, such as the global positioning system (GPS) [28]. These types of systems use multilateration between the receiver and a set of satellites that have line of sight (LoS) with the receiver and are dependent on coverage, and power consumption requirements. Unfortunately, due to the LoS requirement, these solutions cannot be used inside buildings. To assist the satellite-based system, other types of available signals were repurposed, including cellular and WiFi, for example, [3,13].

The current long-term evolution (LTE) systems can achieve accuracies ranging from 10 to 200 m, [6]. WiFi- or Bluetooth-based localization can be used to enhance the system, to improve accuracy, but they have their own sets of drawbacks. The requirement to create a signal map of the signal, that is, fingerprint the area, and keep it updated, adds an extra layer of complexity in deploying such as solutions, cf. [13,27,41]. A way around the issue is the use of cooperative localization, cf. [49]. The method of ultra-wideband (UWB) has been used a lot for dedicated localization solutions, cf. [47], due to its inherent advantages from the sizeable available bandwidth. In cooperative localization, nodes directly measure distances between them and exchange information, achieving higher accuracy with fewer infrastructure requirements.

An important aspect of localization is the ability to measure the distance between two nodes, using the properties of the radio signal communicating between them. Various schemes have been proposed, including time-of-arrival (ToA), time-difference-of-arrival (TDoA), two-way ToA (TW-ToA), as well as angle-of-arrival (AoA) and received signal strength (RSS), including variations and combinations of the above, cf. [51] for details.

Despite the vast amount of research produced in the field, including the fundamental theory of cooperative localization [47], as well as robust algorithms to be used—see surveys in [6,24]—there have been few if any actual real-life deployments of the technology. There are many reasons, which have hindered the adoption of cooperative localization, given its theoretical advantages. One of them is the fundamental issue of communicating efficiently in ad hoc networks. Time-based distance measurements can have significant errors even from minimal delays, and if the algorithms do not converge fast enough or have energy requirements that are too high, they will be unsuitable for everyday mass scale use. Also, dedicated indoor positioning hardware, for example, using UWB signals, can only be used in niche applications, as they do not scale and are not financially viable for mass deployments. With the advent of 5G networks, low latency communications that allow for large bandwidths are expected, allowing for precise ranging and even using large arrays accurate angle estimation. 5G, as well as the Internet-of-Things (IoT) technologies where a large number of small devices will be required to communicate and possibly localize, cf. [30], strengthen the case for cooperative localization, cf. [6].

The above issues make it very important to focus on optimizing computational cost and communication overhead if there is any possibility of achieving widespread use of the technologies.

As a concrete example of the chapter and to better showcase the practical aspects of cooperative localization, we will consider the scenario where pedestrians aim to localize inside a large department store. The assumption will be that every node in the network is a real pedestrian with a transceiver device on them that allows for cooperative localization algorithms to run, including being able to communicate and measure distances between themselves and other nodes in the network. In practice, we need to consider a lot of practical issues that will arise in this scenario. The first one is the nature of the transceivers the nodes have. If we consider the use of mobile devices, we need to account for the different hardware and software specifications and capabilities. If, on the other hand, the transceivers are provided to the pedestrians from the store, ensuring homogeneity between the devices—for example, attached to the department store's trolleys—then issues of logistics arise, including the costs of provision, maintenance, and support of those devices. We will not try to tackle those in this chapter, but they are worth mentioning, as these costs will mostly drive any real deployment of a positioning system.

3.2 Problem Formulation

In this section, we formulate the network localization problem, briefly discuss different families of algorithms that aim to solve the problem, and provide the mathematical framework for all subsequent algorithms discussed.

We consider a network of nodes in an indoor environment. We assume that N nodes from the network, called the agents, aim to identify their location, while M nodes, called anchors, already have knowledge of their location. We assume that $N \ll M$ and that there are at least three anchors available in the networks. The neighboring agents to the anchors will propagate their information through the network, helping every agent gain some estimate of their position. The nodes are assumed to be able to communicate with each other as well as measure the distance between themselves and nodes within their communication range, but there is no central entity to manage calculations or coordinate the nodes. Hence the cooperative regime is considered decentralized. Let $\theta = [\theta_1, \dots, \theta_i, \dots, \theta_{N+M}]$ be the coordinates of all nodes, with θ_i the coordinates of node i.

$$r = \underbrace{f(x, p)}_{d} + \eta, \tag{3.1}$$

where r is the distance measurements vector, θ is the agent coordinates that we are looking for, f is a nonlinear function that associates θ with d, given some noise variable b, for example, added bias due to NLoS, and η represents the measurement noise. There are many ways to measure the distance between two nodes, depending on the technology used and the information, considered or available. For example, some devices might not provide access to raw waveform data from their sensors. Different ways to associate θ and d from sensor measurements are ToA, TDoA, AoA, and RSS [26]. Another important issue is node synchronization, cf. [10], but in this case we will assume the nodes have perfectly synchronized clocks.

A network of such can be viewed as a graph. The wireless nodes are represented by the set $\mathcal{V}$ of vertices of the graphical model. If two nodes, say node i and node j, are within range, there will be an edge $e_{ji} \in \mathcal{E}$, connecting the two nodes. The set of all nodes j with edges e_{ji} to node i is denoted as the neighborhood $\mathcal{N}_i$. A simple network graph example with agents, represented as "a" labeled circles, and anchors, represented as "A" labeled circles, is depicted

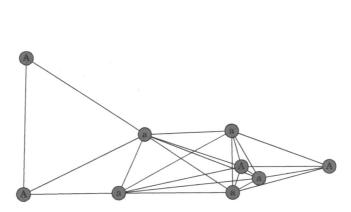

FIGURE 3.1
An example of a network where agents are represented by lowercase "a", anchors are represented by capital "A" and branches (or edges) correspond to communication between them.

in Figure 3.1 After a node has received the coordinates and the corresponding measurements from its neighbors it needs to find an optimal coordinates estimate θ^* for Equation 3.1. Finding an optimal solution can be problematic. First, the equations are nonlinear, resulting in non-convex multi-modal solution spaces, where it is easy to converge to a local optimum. Also, the solution spaces can be quite vast, making an exhaustive search computationally intractable. Other issues are high noise levels, NLoS, or even not enough neighbors, which might create ambiguities, divergent or oscillating behaviors in the optimization, cf. [22]. All the above make position location an exciting and challenging problem, which has stimulated a large and varied number of proposed techniques and algorithms in the literature [6].

Broadly speaking, some methods use linear approximations of Equation 3.1, while others face straight on the nonlinear optimization problem. In the latter category, the problem can be formulated either as an optimization problem or as a probabilistic problem. Each path has its own advantages and disadvantages. Linear approximation optimizations always converge to a global solution, are computationally cheap and algorithmically simple, yet depending on the scenario can provide very low accuracy. By contrast, nonlinear optimization methods, as well as probabilistic modeling, are much more expensive computationally and require more complex algorithms without necessary guaranteeing a global optimum solution, yet tend to provide higher accuracy results. In the nonlinear optimization regime falls a large and varied family of algorithms that can be grossly separated from the probabilistic approach in that they use a point estimate of the position. This lowers the complexity cost relative to the probabilistic approaches but often fails to capture the ambiguity that can exist in cooperative localization, resulting in lower accuracy. Examples of such algorithms include [5,36].

From here on we focus on the probabilistic approach and techniques to manage the extra complexity and communication costs inherent of this family of algorithms.

3.2.1 Probabilistic Graphical Model

Let $p^{(t)}(\Theta_i)$ be the probability density function (pdf), that is, the belief that node i has about its location at time t, where capital symbols, for example, Θ, denote random variables.

Initially, the belief for the agents can be an information-less uniform pdf over the grid, while the anchors' pdfs would have most of its mass at their real position.

Nodes can measure a corresponding distance estimate via ranging. For example, node i receiving a message from node j at time slot t can derive a noisy $r_{ji}^{(t)}$ of the distance between them. It is assumed that the measurement taken from node i receiving a message from node j and that from node j receiving a message from node i are the same, and as a result, the direction of the message plays no role, that is, $r_{j\to i}^{(t)} = r_{i\to j}^{(t)} = r_{ji}^{(t)}$. In practice, distance measurements will differ, and the nodes can share their measurements and use the average. We assume that the nodes use ToA distance measurements, as in [5], and we define the random variable r_{ji} as

$$r_{ji} = d_{ji} + \eta_{ji} \equiv \|\theta_i - \theta_j\| + \eta_{ji}, \tag{3.2}$$

where η_{ji} is a noise factor following a Gaussian distribution with variance $K_e \|\theta_i - \theta_j\|^{\beta_{ji}}$, in which K_e is a proportionality constant capturing the combined physical layer and receiver effect [37], and β_{ji} denotes the path loss exponent. In the case of LoS, η_{ji} is assumed to have zero mean, and $\beta_{ji} = 2$, that is, $\eta_{ji} \sim \mathcal{N}\left(0, K_e \|\theta_i - \theta_j\|^2\right)$. Alternatively, in the case of NLoS, the Gaussian random variable has a positive mean $b_{ji} \gg s_{ji}^2$, where $b_{ji} \sim u\left(\frac{d_{ji}}{3}, \frac{2d_{ji}}{3}\right)$ (i.e., a uniform distribution) and $\beta_{ji} = 3$, that is, $\eta_{ji} \sim \mathcal{N}\left(b_{ji}, K_e \|\theta_i - \theta_j\|^3\right)$. Let $p^{(t)}\left(R_{ji} = r_{ji}^{(t)} \mid \Theta_i, \Theta_j\right)$ be the conditional pdf (cpdf) of observing distance $r_{ji}^{(t)}$ at time slot t, given the location beliefs $p^{(t)}(\Theta_i)$ and $p^{(t)}(\Theta_j)$ for node i and node j, respectively. Assuming statistically independent noise and statistically independent priors between nodes, the joint pdf of the whole probabilistic model with $\vartheta \triangleq \left(\{\forall \Theta_i \in \mathcal{V}\}\right)$ and $R \triangleq (\{R_{ji}\}_{\forall i,j \in \varepsilon})$, for a given time slot t, is found as

$$
\begin{aligned}
p^{(t)}(\vartheta, R) &= p^{(t)}(R \mid \vartheta) p^{(t)}(\vartheta) \\
&= \prod_{i,j \in \varepsilon} p^{(t)}(R_{ji} = r_{ji} \mid \Theta_i, \Theta_j) \prod_{\Theta_i \in \mathcal{V}} p^{(t)}(\Theta_i).
\end{aligned}
\tag{3.3}
$$

Our aim is to find the most probable position for every node given the observed positions and prior information, that is, find the maximum a posteriori (MAP):

$$\hat{\vartheta} = \arg\max_{\vartheta} p^{(t)}(\vartheta \mid R = r), \tag{3.4}$$

where r is a vector representing all the observed distances. To accomplish this, we employ a loopy belief message passing algorithm [23], in which information from the graph can be summarized in local edge information, allowing for an efficient distributed algorithm, despite its lack of guaranteed optimal solution or even convergence for the given random graph geometry.

3.2.2 Belief Message Passing

The network can be modeled as a probabilistic graphical model, namely a cluster graph. We choose the Bethe cluster graph [4] which is composed of two types of factors. The lower factors, which represent the node beliefs, are composed of univariate potentials $\psi(\Theta_i)$, whereas the upper region, which represents the interactions between the node variables,

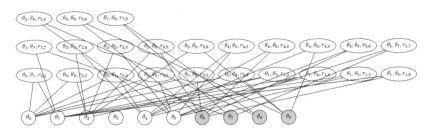

FIGURE 3.2
A Bethe cluster graph of the network in Figure 3.1.

is composed of "large" factors equal to $\psi(\Theta_i, \Theta_j, r_{ji})$. An example of the network in Figure 3.1 is shown in Figure 3.2. The lower factors are set to the initial node beliefs for the given time slot (t), and the upper factors to the corresponding cpdfs, that is,

$$\psi(\Theta_i) = p^{(t)}(\Theta_i), \tag{3.5}$$

$$\psi\left(\Theta_i, \Theta_j, r_{ji} = r_{ji}^{(t)}\right) = p^{(t)}\left(r_{ji} = r_{ji}^{(t)} \mid \Theta_i, \Theta_j\right). \tag{3.6}$$

Messages are then passed between nodes for multiple iterations until the node beliefs have converged or a predetermined number of iterations has passed. The message from node j to node i at BP (belief propagation) iteration ($s + 1$) is found by

$$\delta_{j\rightarrow i}^{(s+1)}(\Theta_i) = \int \psi\left(\Theta_i, \Theta_j, R_{ji} = r_{ji}^{(t)}\right) \frac{b_{j\rightarrow i}^{(s)}(\Theta_j)}{\delta_{i\rightarrow j}^{(s)}(\Theta_j)} d\Theta_j, \tag{3.7}$$

where $r_{ji}^{(t)}$ is the observed value of the distance between the nodes, at time slot t. Intuitively, a message $\delta_{j\rightarrow i}^{(s+1)}(\Theta_i)$ is the belief that node j has about the location of node i and is a function of Θ_i. After it calculates all the incoming messages, each node updates its belief by

$$b_i^{(s+1)}(\Theta_i) = \lambda\psi(\Theta_i)\prod_{k\in\mathcal{N}_i}\delta_{k\rightarrow i}^{(s+1)}(\Theta_i) + (1-\lambda)b_i^{(s)}(\Theta_i), \tag{3.8}$$

in which $\lambda \in [0,1]$ is a dampening factor used to facilitate convergence, c.f. [23].[*] The algorithm continues until convergence or a set number of iterations $\mathcal{S}_{\max}$ has elapsed. The belief that represents an approximation to the true marginal for each node is given by Equation 3.8, that is,

$$p^{(t+1)}(\Theta_i) = b_i^s(\Theta_i). \tag{3.9}$$

This can be thought of as a process of merging every node's belief about a specific node's location to get the best estimate. This message passing analysis naturally leads to a distributed cooperative system because each node is only required to do calculation concerning its local factors and messages.

---

[*] Dampening is not required for 2D localization, but simulations show that it helps in 3D localization where the increased ambiguity results in oscillating behaviors more often.

3.2.3 Computational Issues and Communication Overhead

A major issue in calculating the messages 3.7 and 3.8 is the computational cost. If Θ_i can take one of D discrete values, then the messages and marginals will be represented by dimensional vectors of cardinality D. The integral in Equation 3.7 becomes a matrix-vector product, and this in general requires $\mathcal{O}(D^2)$ operations [25], making the problem intractable because the dimensions of the grid and the resolution of the required localization will increase quadratically. This makes the use of approximation methods necessary.

In the literature, various methods have been proposed, such as particle filtering methods [21,29], Monte-Carlo methods [44], and the more general NBP, for example [17,18,40]. They try to approximate the cpdf by a Gaussian mixture of particles, achieving a complexity of $\mathcal{O}(L^2)$, where L denotes the number of particles and $L \ll D$. Due to the curse of dimensionality [12] this issue becomes even more pronounced in the 3D case, cf. [31]. For example, if the density of the number of particles is to remain approximately the same in the 3D case as in 2D, then the number of particles will increase by a factor of $2d_{ji}$ (d_{ji} being the exact distance). This calculation would require finding the product of large Gaussian mixtures, which is already an expensive operation and can quickly become prohibitively expensive. Especially considering that these algorithms must run on mobile devices that have severe hardware and energy constraints, this issue becomes even more important. It is critical for the algorithms to have as low a complexity cost and energy fingerprint on the device as possible for them to be viable in any real production scenario. The other side of the story is the communication overhead of the algorithms. The information propagated through the network for localization should be as compressed as possible, both to minimize the load of the communications and to decrease the localization time due to lengthy communication overhead.

The above becomes more critical in the case of mobile nodes, where the actual positions change regularly; for example, consumers walking around a shopping mall. Approximating the multimodal $p^{(t)}(\Theta_i)$ position pdf with particles can very quickly become a costly communication overhead, and we need to consider ways to summarize the information. Network operations have been considered in [10], as to when to deploy, activate, and how to prioritize nodes. It is clear that the commutation strategy of the network is of great importance to the lifetime of the network as well as the localization accuracy.

In the next sections, we present a variety of algorithms that can be used for probabilistic cooperative localization. The aim is to highlight techniques that can be used to decrease the complexity of cooperative localization as well as to improve the communication overhead of the algorithms and optimize the communication strategies of the nodes. First, we propose a solution for cooperative localization that dramatically reduces the complexity of the position pdf by taking advantage of the geometry of the localization area. Then we extend the algorithm to the mobile case, and finally suggest a distributed message scheduling strategy that improves the communication overhead of the network.

3.3 Grid-BP (Static Case)

An often-unexplored issue in localization is how the coordinate system itself works. Localization occurs relative to some commonly accepted reference points. These are called geodetic datum. Two of the more commonly known ones are the WGS84 and the NAD83, cf. [1]. GPS uses a polar system, and the origin of the coordinate system is the center of the planet. In cooperative localization systems, though, typically a Cartesian system is assumed, and the nodes localize relative to themselves. Hence, all nodes localize with respect to a *local coordinate system (LCS)*. As a result, an implicit assumption is hidden in every model: *All nodes know the precise location of the origin point and their relative position to it.* This assumption means that even though the nodes use their LCS to localize, a global coordinate system (GCS) is assumed to be known by everyone so that every node can convert their LCS to the GCS coordinates. Otherwise, we could either arbitrarily rotate the whole coordinate system or move the origin arbitrarily around and the solutions of the node coordinates in the LCS would still be valid. The problem has been identified in [17], and a solution assuming the anchors can provide knowledge of their orientation is suggested in [6]. Grid-BP as a general solution using the NATO Military Grid Reference System (MGRS) has been suggested in [32]. In practice, besides the MGRS, any arbitrary coordinate system can be easily used as long as it is fixed and shared across all nodes, and uniquely identifies every grid "box" in the area of interest. For example, a plan map of an airport, split into grids.

Using a grid system not only solves the issues of converting the localized coordinates from an LCS to the GCS, but using the grid intelligently also allows considerable savings in complexity, as explained below. Areas can be split in a more coarse or fine-grained way depending on the requirements of the implementation, and also any areas in the grid where the nodes cannot possibly go—for example, walls inside a building—can be removed from the grid, substantially minimizing the number of operations required to calculate the position pdf.

3.3.1 Problem Formulation

Let $X = [X_1, ..., X_i, ..., X_{N+M}]$ be the random variable vector for the locations of all nodes, with X_i representing the random variable of the grid unique identifier (UID) of node i and $X_i \in \{x_1, ... x_k, ... x_K\}$, where K are the total UIDs in the area. We assume a trivial exampling of splitting a map into a grid as follows. The map is split into 1 m^2 squares split with the top left corner square having as ID the tuple (0 0), with the left number increasing southward and the right number increasing eastward, a pretty standard convention in most graphical user interface canvases that can be used directly to identify the position of a node in a map of a building. Also let θ denote the coordinates of all nodes, with θ_i representing the coordinates of node i, and the domain of θ_i is $\Re^2$. As before, the use of upper case Θ_i and Θ represent the respective random variable and vector random variable of the coordinates of all nodes. It should be emphasized that the nodes will not use the coordinates in their calculations at all. Instead they will be used as labels of the different categories of the multinomial distribution. However, it is still necessary to define them in the problem formulation in order to define the distance between neighboring nodes. The nodes communicate wirelessly, and it is assumed that the maximum communication range for each node is $R_{\max}$. Time is slotted, and time slots are denoted by the time index superscript (t) for $t = 1, 2, ..., \infty$.

We represent the problem as a joint probability distribution. Let $\Pr^{(t)}(X_i)$ be the pdf, that is, the belief that node i has about its location at time t. We model $\Pr^{(t)}(X_i)$ as a multinomial distribution with parameters $z_1, ..., z_k$ where $\Pr(X_i = x_k) = z_k$ is the probability of node i being in UID x_k and $\sum_k z_k = 1$. Also let the set of all nodes j within range of node i be denoted as the neighborhood $ne(i)$.

Initially, the belief for the agents can be a non-informative uniform pdf over the grid, while the anchors' pdfs are focused in the IDs close to the real position, that is, within 10 m.

Nodes obtain distance estimate via ranging, as discussed previously. Consequently, for two nodes i and j we define the random variable R_{ji} with values r_{ji} as

$$r_{ji} = \left\| \theta_i - \theta_j \right\| + \eta_{ji}, r_{ji} = \left\| \theta_i - \theta_j \right\| + \eta_{ji}, \tag{3.10}$$

where η_{ji} is a noise factor following a Gaussian distribution with variance $s_{ji}^2 = K_e \left\| \theta_i - \theta_j \right\|^{\beta_{ji}} s_{ji}^2 = K_e \left\| \theta_i - \theta_j \right\|^{\beta_{ji}}$ in which K_e is a proportionality constant capturing the combined physical layer and receiver effect, and β_{ji} denotes the path loss exponent, as discussed in the previous chapters.

We define the likelihood of node i and node j measuring distance $R_{ji} = r_{ji}$ between them at time t, given X_i, X_j as

$$\overset{(t)}{\Pr}(R_{ji} = r_{ji} \mid X_i, X_j) \propto \exp\left(-\left(\frac{r_{ji} - \left\| C_i - C_j \right\|_2}{h}\right)^2\right), \tag{3.11}$$

where h controls steepness, C_i and C_j are the coordinates of the centers of the grids' squares X_i and X_j, respectively. Therefore, our objective is to find the MAP, that is, the values that maximize $\Pr(X \mid R)$ given distance measurements $R = [R_{ji}]$. For a specific node i, we have

$$\hat{X}_i = \arg\max_{X_i} \overset{(t)}{\Pr}(X_i \mid R_i). \tag{3.12}$$

Thus, $\Pr^{(t)}(X_i \mid R_i)$ can be evaluated using the Bayes' rule as

$$\begin{aligned}
\overset{(t)}{\Pr}(X_i \mid R_i) &\propto \overset{(t)}{\Pr}(X_i) \prod_{j \in \text{ne}(i)} \overset{(t)}{\Pr}(R_{ji} \mid X_i) \\
&\propto \overset{(t)}{\Pr}(X_i) \prod_{j \in \text{ne}(i)} \int \overset{(t)}{\Pr}(R_{ji} \mid X_i, X_j) \overset{(t)}{\Pr}(X_j) dX_j,
\end{aligned} \tag{3.13}$$

in which the sign "$\propto$" means "is proportional to," and normalization should be done to obtain the pdf.

First, we will formulate a valid cluster graph and the respective message passing equations, for the X random variables as discussed in Section 3.2.2. Then efficient approximation for the marginalization and the product operation will be given.

3.3.2 Belief Message Passing

We will model the network as a Bethe cluster graph. The lower factors are composed of univariate potentials $\psi(X_i)$, while the upper region is composed of "large" clusters with one cluster for each factor $\psi(X_i, X_j, R_{ji})$. An example can be seen in Figure 3.3.

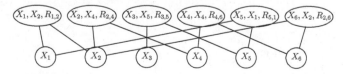

FIGURE 3.3
The cluster graph for the Grid-BP PGM.

The lower factors are set to the initial beliefs for the given time slot (t), and the upper factors to the corresponding cpdfs

$$\psi(X_i) = \overset{(t)}{\mathrm{Pr}}(X_i), \tag{3.14}$$

$$\psi\left(X_i, X_j, R_{ji} = r_{ji}^{(t)}\right) = \overset{(t)}{\mathrm{Pr}}(R_{ji} = r_{ji}^{(t)} \mid X_i, X_j). \tag{3.15}$$

Messages are then passed between nodes for multiple iterations until the node beliefs have converged. The message from node j to node i, at BP iteration $(s + 1)$ is calculated by

$$\mu_{j \to i}^{(s+1)}(X_i) = \int \psi\left(X_i, X_j, R_{ji} = r_{ji}^{(t)}\right) \frac{b_{j \to i}^{(s)}(X_j)}{\mu_{i \to j}^{(s)}(X_j)} dX_j, \tag{3.16}$$

where intuitively, a message (3.16) is the belief that node j has about the location of node i and $r_{ji}^{(t)}$ is the observed value of the distance between the nodes, at time slot t.

Then the belief of node i is updated as

$$b_i^{(s+1)}(X_i) = \lambda \psi(X_i) \prod_{k \in \mathrm{ne}(i)} \mu_{k \to i}^{(s+1)}(X_i) + (1 - \lambda) b_i^{(s)}(X_i), \tag{3.17}$$

where λ is a dampening factor used to facilitate convergence.

BP continues until convergence, or if convergence is not guaranteed, $\mathcal{S}$ reaches a maximum number of iterations $I_{\max}$. Then the beliefs, representing approximations to the true marginals, are found by Equation 3.17, that is, $\mathrm{Pr}^{(t+1)}(X_i) = b_i^{(S+1)}(X_i)$, for each node. The proposed Grid-BP is given as Algorithm 3.1. Each node needs to perform a marginalization operation (3.16) and a product operation 3.17. Approximations are required for both complex operations. In Grid-BP, we take advantage of the multinomial parametric form, which we discuss next.

Algorithm 3.1 Grid-BP

1: Initialize beliefs $p^{(0)}(X_i) \; \forall i \in$ nodes
2: **for** $t = 0$ to T **do**
3:   **for all** $i \in$ Nodes **do**
4:     Broadcast current belief $p^{(0)}(X_i)$
5:     **for all** $j \in \mathrm{ne}(i)$ **do**
6:       Collect distance estimates $r_{ji}^{(t)}$
7:     **end for**
8:   **end for**
9:   Initialize $\psi(X_i) = \mathrm{Pr}^{(t)}(X_i)$
10:   Initialize $\psi(X_i, X_j, R_{ij}) = \mathrm{Pr}^{(t)}(R_{ij} = r_{ji}^{(t)} \mid X_i, X_j)$
11:   **repeat**
12:     **for all** $i \in$ Nodes **do**
13:       **for all** $j \in \mathrm{ne}\,(i)$ **do**
14:         Receive $b_j^{(s)}(X_j)$
15:         Calculate $\mu_{j \to i}^{(s+1)}(X_i)$, using Equation 3.16 using HSM Gibbs sampling
          (i.e., Algorithm 3.2)

16:        **end for**
17:        Calculate $b_i^{(s+1)}(X_i)$, using Equation 3.17.
18:        Check for convergence
19:        Send $b_i^{(s+1)}(X_i)$
20:    **end for**
21: **until** convergence or s reaches I_{max}
22: Update belief $\Pr^{(t+1)}(X_i)$, using Equation 3.17.
23: **end for**

3.3.3 Marginalization Operation

The calculation of Equation 3.16 essentially gives the belief node j has about node i. Since the marginalization is too costly, we will use a Gibbs sampler to approximate the message. To understand the effect of R_{ji}, we consider the case that in a specific time slot t', $b_j^{(S)}(X_j)$ has only one ID, that is, x_j. Then the high probability grid squares of Equation 3.16 will be approximated by IDs of grid buckets that approximately form a circle of radius $r_{ji}^{(t')}$ and center the x_j. With that in mind, first we sample L particles $x_j^{(l)} \sim \mu_{j \to i}(X_j)$. Then we sample L samples $\phi^{(l)} \sim \mathcal{U}(0, 2\pi)$ and finally L samples from $\hat{r}_{ji}^{(l)} \sim \mathcal{N}\left(r_{ji}^{(t)}, h\right)$. The Gibbs sampling algorithm is provided as Algorithm 3.2.

Algorithm 3.2 Gibbs Sampling

1: Set $\mathcal{D}_{X_i}$ to empty
2: **for all** $j \in \text{ne}(i)$ **do**
3: Sample $x_j^{(l)} \sim \mu_{j \to i}(X_j)$ which is a multinomial pdf
4: Sample $\phi^{(l)} \sim \mathcal{U}[0, 2\pi]$
5: Sample $\hat{r}_{ji}^{(l)} \sim \mathcal{N}\left(r_{ji}^{(t)}, h\right)$
6: $x_i^{(l)} = \text{MAP-DMtoID}\left(x_j^{(l)}, r_{ji}^{(t)}, \phi^{(l)}\right)$ which maps the distance metric to IDs
7: Add $\left\{x_i^{(l)}\right\}_{l=1}^{L}$ to $\mathcal{D}_{X_i}$
8: **end for**
9: **return** $\mathcal{D}_{X_i}$

Sampling is repeated for all incoming messages, and then for all $\left\{x_j^{(l)}, \hat{r}_{ji}^{(t)}, \phi^{(l)}\right\}_{l=1, j=1}^{L, |\text{ne}(i)|}$ we use

$$x_i^{(l)} = MAP - DMtoID\left(x_j^{(l)}, \hat{r}_{ji}^{(t)}, \phi^{(l)}\right) \tag{3.18}$$

to get the set $\mathcal{D}_i = \left\{x_i^{(l)}\right\}$. Intuitively we do this by counting for each sampled ID $x_j^{(l)}$ the number of IDs to the east and to the north, node i will be, given the measured distance $r_{ji}^{(t)}$ normalized by D as follows

$$\begin{bmatrix} d_H \\ d_V \end{bmatrix} = \text{int}\left(\frac{\hat{r}_{ji}^{(t)}}{D}\begin{bmatrix} \cos(\phi^{(l)}) \\ \sin(\phi^{(l)}) \end{bmatrix}\right). \tag{3.19}$$

Then, the displacement d_H, d_V is translated to a new ID and return it as $x_j^{(l)}$. The displacement calculation depends on the nature if UIDs chosen. The distance to ID mapping function is given as Algorithm 3.3. It should be noted that no reverse mapping is required in the case of MGRS.

Finally, as will be shown in the sequel, the set $\mathcal{D}_{X_i}$ of all samples obtained from all incoming messages is used to find (3.17).

Algorithm 3.3 MAP-DMtoID

1: Calculate horizontal and vertical steps using Equation 3.19
2: Map horizontal ID $x_j^{(l)} \rightarrow b_j^{(l)}$
3: $b_h^{(l)} = b_j^{(l)} + d_H$
4: Inverse horizontal mapping $b_h^{(l)} \rightarrow x_h^{(l)}$
5: Map vertical ID $x_h^{(l)} \rightarrow b_v^{(l)}$
6: $b_i^{(l)} = b_v^{(l)} + d_V$
7: Inverse vertical mapping $b_i^{(l)} \rightarrow x_i^{(l)}$
8: **return** $x_i^{(l)}$

3.3.4 Product Operation

To obtain (3.17), first we create parametric forms of the incoming messages (3.16) by using the particles in $\mathcal{D}_{X_i}$. We assume that the parameters of the multinomial are random variables $\mathbf{Z}_i$ with a uniform Dirichlet prior with parameters α_k, where $k \in \{1, \ldots, K\}$ and K is the number of unique IDs in the multinomial. We use the samples from each incoming message as observations and get the MAP estimate $\hat{z}_i$ of the parameters $\mathbf{Z}_i$ of each (3.16). We also consider that all incoming messages have the same prior distribution. Based on the above we calculate the parameters for each multinomial as follows

$$\hat{z}_i = \mathbb{E}\big[\Pr(Z_i \mid \mathcal{D}_{X_i})\big] = \mathbb{E}\big[\Pr(\mathcal{D}_{X_i} \mid Z_i)\Pr(Z_i)\big], \tag{3.20}$$

which gives

$$\hat{z}_{i,k} = \frac{M_k + |\text{ne}(i)| \alpha_k}{|\text{ne}(i)| \sum_k (M_k + \alpha_k)}, \tag{3.21}$$

where M_k is number of particles x_k in $\mathcal{D}_{X_i}$. The algorithm is presented in Algorithm 3.4.

Algorithm 3.4 MAP Parameter Estimation

1: Let $|X_i|$ be the number of unique IDs in $\Pr(X_i|\mathbf{Z}_i)$
2: Calculate $M_k = count(x_k, \mathcal{D}_{X_i}) \forall k \in |X_i|$
3: **for all** $k \in |X_i|$ **do**
4:     Calculate $\hat{z}_{i,k}$, with (3.21)
5: **end for**
6: **return** $\hat{z}_i$

For clarity, the quantities of each ID in the samples are shown as being found by a count function but in practice it can be done during the Gibbs sampling step allowing for a more efficient algorithm.

After obtaining the pdfs involved in the calculation of Equation 3.17, the resulting pdf will simply be the dot product of the z_i parameters of each incoming message. The parameters of Equation 3.17 can be directly calculated by adding the logs of the corresponding $\hat{z}_i$'s.

3.3.5 Message Filtering

As it makes no sense to keep all the UIDs on the map, we can assume that each node constructs its belief pdf with the UIDs within a specific range of the UIDs it received in the first iteration, for example, within 80 m. To reduce the IDs further, we propose a simple filter that only keeps the most probable UIDs summing up to an energy threshold of the respective cpdf. The rest are assumed to share the remainder of the pdf energy uniformly. After Monte Carlo simulations, it was evident that by keeping $\sim$80% of the total energy of the pdf, the size of the messages transmitted is decreased by $\sim$90% with barely any increase in localization error. Thus, assuming that each message covers a 100×100 m^2 grid, the total number of IDs used without the filter would be 10^4. After the filter, however, only $\sim$10^2 IDs will be transmitted. In practice, we can decrease the number of UIDs even further by zeroing out UIDs where it is impossible for a node to be, for example, inside walls or other obstacles. Depending on the geometry and architecture of the space—for example, department stores have a lot of "dead" space and areas where consumers do not have access—the number of actual possible UIDs can drop drastically.

3.3.6 Convergence

A cluster graph is not guaranteed to determine the optimal solution. However, if the running intersection property and the family preservation property are held, typically most clusters will converge to a local optimum with only a few clusters oscillating unable to converge [23]. To facilitate convergence, a dampened BP message passing is used, which empirically helps, and even the oscillating nodes converge to a local optimum. This is accomplished by setting

$$\delta_{i \to \text{all } j \in \text{ne}(i)}^{\text{lower} \to \text{upper}}(X_i) = \lambda \psi(X_i) \prod_{k \in \text{ne}(i)} \delta_{k \to i}^{\text{upper} \to \text{lower}}(X_i) + (1 - \lambda) \delta_{i \to \text{all } j \in \text{ne}(i)}^{(\text{old})\text{lower} \to \text{upper}}(X_i), \tag{3.22}$$

where λ is the dampening factor. The Kullback-Leibler (KL) divergence between the current and previous iterations message is calculated and if it falls within a certain threshold the node will consider that it has converged to a solution.

3.3.7 Complexity

The complexity for the marginalization filtering i. $\mathcal{O}(L|\text{ne}(i)|)$ and the computational cost for the product operation is also $\mathcal{O}(L|\text{ne}(i)|)$, which corresponds to the complexities suggested in [25] for parametric techniques. For completeness, the different complexities for discretized, non-parametric, and parametric techniques are shown in Table 3.1 as shown in [25].

TABLE 3.1

Comparison of Complexity Costs for Discretized, Non-Parametric, and Parametric Techniques where L is the Number of Particles and M is the Number of Messages Involved in the Operations

Approach	Operation	Complexity	Value of L
Descretized	Marginalization	$\mathcal{O}(L^2)$	Large
Descretized	Product	$\mathcal{O}(LM)$	Large
Non-Parametric	Marginalization	$\mathcal{O}(L)$	Small
Non-Parametric	Product	$\mathcal{O}(L^2M)$	Small
Parametric	Marginalization	$\mathcal{O}(L)$	Small
Parametric	Product	$\mathcal{O}(LM)$	Small

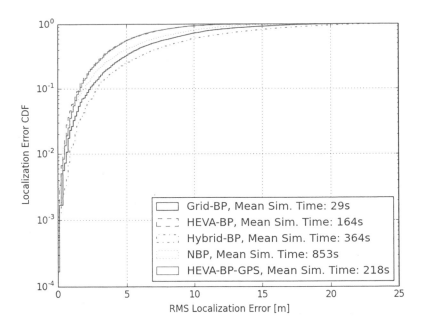

FIGURE 3.4
Comparison of localization error cpdf with Grid-BP.

3.3.8 Simulations

To evaluate the performance, 100 Monte Carlo simulations were conducted and the localization error cpdf was calculated. In each simulation 100 nodes with 20 anchors are placed randomly in a 100 m × 100 m area and the communication range is limited to 12 m and $|ne(i)|_{avrg} = 4.03$. Anchor locations are modeled as multivariate Gaussian pdfs with an identity variance matrix. The grid resolution for Grid-BP is $D = 1$ m. We compare Grid-BP with the NBP [17] and Hybrid-BP [7].* We also compare Grid-BP with HEVA-BP [34], a computationally cheaper variation of NBP, [16]. In addition, $I_{max} = 15$ and 800 particles were sampled. Finally, the noise factor used was $K_e = 0.3$. A variant of HEVA that uses GPS coordinates as a reference system was also provided. In HEVA-GPS, messages contain GPS coordinates that every node converts to an LCS before calculating (3.16) and (3.17).

Figure 3.4 shows the localization error cpdf of all the algorithms. Results illustrate that HEVA-BP, NBP, and Grid-BP achieve similar results, with Grid-BP having slightly better results than Hybrid-BP, which is the only other parametric algorithm. Note that all three algorithms, HEVA, NBP, and Hybrid-BP have the strong assumption of sharing knowledge of the GCS origin, while Grid-BP and HEVA-GPS do not (the realistic scenario). Even though HEVABP-GPS achieves as good results as HEVA-BP, the 25% increase in computational cost with HEVA-BP, due to the mapping of the GPS coordinates to a local Cartesian reference frame, is evident in the mean simulation time. The relative computational efficiency of Grid-BP can also be seen. This is due to the very efficient calculations of both message passing operations compared to the other algorithms. Note that all simulations were run on an Intel i7 2.6 GHz, using Python for scientific computing [35].

---

* For Hybrid-BP, we did not use information given from satellites as in [7].

3.4 PHIMTA (Mobility Case)

Grid-BP provides an elegantly simple and computationally efficient probabilistic algorithm for cooperative localization. It does not, though, take into account the fact that the nodes of the localization network will probably be mobile and change locations. Coping with mobility has been vital in localization research. The typical scenario is to complement GPS with information from inertial measurement units (IMUs) and odometers to provide uninterrupted navigation solutions during GPS outages. The typical way to achieve this integration is to use a Kalman filter (KF) [39] or some variant; for example, the extended KF [51]. Unfortunately, the significant errors of micro-mechanical systems (MEMS) inertial sensors, as well as the time-varying models, cannot be modeled accurately by the KF linearized models. As an alternative to better capture the nonlinearities, the use of a particle filter has been proposed [11]. Even though the traditional context has been vehicles, due to the popularization of smartphones there has been continuing development of low-cost MEMS-based IMUs that can integrate multiple sensors, including orthogonal gyroscopes and accelerometers, in a compact inertial sensor module. Increasingly, new models are designed with better sensor sensitivity and lower errors, and nowadays it is reasonably safe to assume that a man-held device will have integrated gyroscope, accelerometer, and magnetometer sensors.

The information from the sensors provides all the necessary information required for tracking of human movement. There has been much research on pedestrian dead reckoning (PDR) and its applications. In PDR, the frequency of the pedestrian's steps is extracted from the sensor information, and assuming some statistical model for the length and course of the steps, the pedestrian's direction and distance traveled are calculated by summing up all the steps; for example, see [20]. There has also been extensive research in the classification and modeling of sensor outputs with different human movement, for example, see [2,43], but the reality is that erratic movement—for example, walking on slopes, or abrupt movements—will cause the error to multiply.

Alternatively, pedestrian tracking can be treated as an application of a strap-down inertial navigation system (INS). In this case, the orientation of a sensor module is tracked by integrating the angular velocities, which are subsequently used to determine the acceleration components in the GCS. Then the gravity acceleration is subtracted, and the remaining acceleration is integrated over time to find the sensors' displacement. Unfortunately, low-cost MEMS-IMUs are susceptible to errors, such as misalignment errors, scale factor, bias turn-on error, bias drift error, etc. Though we can remove deterministic errors typically via calibration, stochastic errors cannot be removed and can increase quickly. Analysis and modeling of the MEMS-IMU errors can be found in [8,11,38].

A solution proposed in [15] has been to provide a synergism between PDR and INS. Mostly the movement of the pedestrian will be calculated by finding the orientation and number of steps as in PDR, but we will derive the characteristics from the IMU measurements instead of using a statistical model.

The combination of PDR localization and cooperative localization for GPS-denied environments, however, has not been well investigated. The SPAWN framework in [49] considered mobility, but it was demonstrated in [14] that it is too computationally expensive for real-world hand-helds, and a heuristic cooperative localization algorithm, called stop-and-go (SnG), was proposed as an alternative. SnG keeps the computational cost low for mobile devices while synergizing with PDR. Still, due to the heuristics in SnG, the network is highly susceptible to node placement, and if the placement is not uniform enough, then the whole network localization will collapse.

As an example of an algorithm that combines PDR with effective computationally efficient cooperative localization we will present PHIMTA, cf. [34]. PHIMTA extends the static cluster graph of Grid-BP, presented in the previous section, to a dynamic Bayesian network, i.e. a probabilistic temporal model allowing for mobile pedestrian tracking with low computational cost, suitable for our department store scenario.

We extend the previous formulation, Equations 3.14, 3.15, by considering two types of time slots based on how the nodes behave. First, the nodes might move and use IMU information to update their information, namely IMU time slots, or they might stay idle and cooperate with their neighbors to update their location estimate, namely belief propagation (BP) time slots. The nodes use cooperative localization every n time slots, while in between they have a probability $p(W^{(t)})$ at each time slot to wait or to move. If a node is moving during a BP time slot, then it will not participate in the message passing algorithm. Nodes use the SHOE filter, cf. Section 3.4.1, to discriminate between being idle or not.

Let $p\left(X_i^{(t)}\right)$ be the pdf, that is, the belief that node i has about its location at time t. We model $p\left(X_i^{(t)}\right)$ as a multinomial distribution with parameters θ_k, where θ_k is the probability of node i being in ID x_k and $\sum_k \theta_k = 1$. Let $p(X^{(t)})$ denote the state of the system at time slot (t). We assume that the system is Markovian and represent it as a pdf. Then we have

$$P\left(X^{(0)}, X^{(1)}, \ldots, X^{(t)}\right) = p\left(X^{(0)}\right) \prod_{\tau=0}^{t-1} p(X^{(\tau+1)} \mid X^{(\tau)}),$$
(3.23)

where $p(X^{(0)})$ is the initial system state and $p(X^{(\tau+1)} \mid X^{(\tau)})$ is the transition probability. Depending on the type of time slot the transition probability will change. Thus, we have

$$p(X^{(\tau+1)} \mid X^{(\tau)}) = \begin{cases} p(X^{(\tau+1)} \mid X^{(\tau)}, R = r), & \text{for BP time slots,} \\ p(X^{(\tau+1)} \mid X^{(\tau)}, O = o), & \text{for IMU time slots,} \end{cases}$$
(3.24)

in which r denotes a vector with all the distance measurements between nodes, and o is a vector with the IMU observations. The above can be described graphically in Figure 3.5.

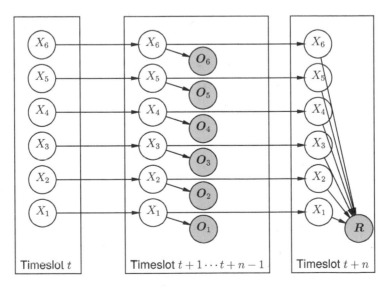

FIGURE 3.5
A dynamic Bayesian network model.

During the BP time slots, the Grid-BP is run as previously suggested on all nodes that are static. BP timeslots occur at constant intervals and are initiated by the anchors, which are typically controlled by the provider of the localization solution—for example, the department store itself in our scenario. Static nodes exchange information with their neighbors and at the end of the BP timeslot the nodes will have new beliefs, that is, $p(X_i^{(t+1)} \mid R_i^{(t+1)}, X_i^{(t)})$, *using* (3.13).

Similarly, during the IMU time slots, we have

$$p(X_i^{(t+1)} \mid O_i^{(t+1)}, X_i^{(t)}) \propto p\left(X_i^{(t)}\right) p(O_i^{(t+1)} \mid X_i^{(t+1)}), \tag{3.25}$$

The model is summarized in Algorithm 3.5.

Algorithm 3.5 Dynamic Bayesian Network

1: **for all** $i \in N$ **do**
2:   Initialize $p\left(X_i^{(0)}\right), v_i^{(0)}, \omega_i^{(0)}, a_i^{(0)}, \mu_i^{(0)}$
3:   **for all** $t \in$ Time slots **do**
4:     **if** time slot is *BP* **then**
5:       calculate Equation 3.24 using GBP, Algorithm 3.1
6:     **else**
7:       calculate Equation 3.24 using PHIMTA, Algorithm 3.6
8:     **end if**
9:   **end for**
10: **end for**

In the next sections, we will describe how PHIMTA solves Equation 3.25 during the IMU time slots.

3.4.1 IMU Time Slots

During the IMU time slots, our is aim is to approximate the transition probability in Equation 3.24. This is accomplished by using a non-parametric particle representation [23]. Assuming at time slot (t) the location pdf for node i is $p\left(X_i^{(t)}\right)$, we represent it by a set of L random samples, or particles $\mathcal{S}_i^{(t)} = \left\{s_i^{(t,l)}\right\}_{l=1}^{L}$, sampled from $p\left(X_i^{(t)}\right)$. The i-th sample is denoted as $s_i^{(t,l)} = \left(x_i^{(t,l)}, \pi_i^{(t,l)}\right)$, where $x_i^{(t,l)}$ is the value of the node state and $\pi_i^{(t,l)} = (1/L)$ is the respective weight. Then the motion model is applied to each sample $s_i^{(t,l)}$ and we obtain a new sample

$$s_i^{(t+1,l)} = \left(x_i^{(t+1,l)}, \frac{1}{L}\right), \tag{3.26}$$

where

$$x_i^{(t+1,l)} \sim p(X^{(t+1)} \mid X_i^{(t)} = x_i^{(t,l)}, O_i^{(t)} = o_i^{(t)}), \tag{3.27}$$

where $o_i^{(t)}$ are the IMU measurements at time (t). Finally, the same parameter estimation Algorithm 3.4 is used to obtain a multinomial parametric form of Equation 3.24 from the samples.

3.4.2 PHIMTA

Our aim is to derive an updated location particle given a location particle $x_i^{(t+1,l)}$ and the IMU observation $o_i^{(t)}$. We will derive the displacement, speed, and attitude vectors from the IMU sensors. We assume a typical MEMS sensor, consisting of an accelerometer, a magnetometer, and a gyroscope. The measurements provided by the IMU and the magnetometer comprise the control input

$$o^{(t)} = \left[a^{(t)}, \omega^{(t)}, \mu_x^{(t)} \right],$$

and we denote their respective noise vector as

$$w^{(t)} = \left[n_a^{(t)}, n_w^{(t)}, n_\mu^{(t)} \right].$$

We assume that the input signal vectors from each sensor have length Y. Also, the noise is assumed to be white Gaussian noise and independent of previous states. A rotation matrix that maps the LCS of the sensors to GCS of the node is required, as the sensor axes may not match the nodes. Then the accelerometer observations will be mapped to the nodes coordinate system and used to calculate the displacement and speed of the node.

To alleviate the noise, a number of schemes will be used. First is the fact that pedestrian walking is cyclical and significantly consistent. Each stride can be split into two phases. The stance phase, when the foot or part of the foot is placed on the ground, and the swing phase, when the foot is mid-air. Both the velocity and the angular velocity can be reset to zero at each stance phase, thus reducing the drift error accumulation. As the gyroscopes cannot be used in the static phase, signals from the accelerometer and magnetometer have to be used to calculate the orientations of the sensor module. To overcome tilt errors, the algorithm presented in [15] is used. The stance phase can be easily detected using peak detection, taking into consideration of the existence of zero crossings, for example, see [43]. In this work, we use the SHOE algorithm [42].

Hence, the system iterates through the following steps:

- Stance phase
    - Reset angular velocity to zero
    - Reset velocity to zero
    - Use magnetometer and accelerometer data to calculate rotation matrix
- Swing phase

$$R^{(t)} = \begin{pmatrix} \cos(p)\cos(a) & -\cos(r)\sin(a)+\sin(r)\sin(p)\cos(a) & \sin(r)\sin(a)+\cos(r)\sin(p)\cos(a) \\ \cos(r)\sin(a) & \cos(r)\cos(a)+\sin(r)\sin(p)\sin(a) & -\sin(r)\cos(a)+\cos(r)\sin(p)\sin(a) \\ -\sin(p) & \sin(r)\cos(p) & \cos(r)\cos(p), \end{pmatrix}$$

(3.28)

    - Use gyroscope data to calculate the rotation matrix
    - Calculate the velocity and displacement using accelerometer data

Our derivation follows closely the work in [15]. In subsequent sections, as everything involves internal calculations at each agent, the node subscript is dropped for simplicity.

3.4.3 Coordinate Systems and Transformation Matrix

The global cartesian coordinate system used is the north-east-down (NED) frame (x^n, y^e, z^d). Consequently, the rotation matrix derived by using direction cosine representations is given by Equation 3.28, where p, r, and a correspond to the pitch, roll, and attitude, respectively, and the time slot superscript has been dropped for simplicity.

3.4.4 Swing Phase

During the swing phase, the orientation of a moving object is tracked by integrating the angular velocity vector $\omega^{(t)} = \left[\omega_x^{(t)} - n_{\omega x}^{(t)}, \omega_y^{(t)} - n_{\omega y}^{(t)}, \omega_z^{(t)} - n_{\omega z}^{(t)} \right]$, obtained from the gyroscope after we correct for noise. Let the sampling period δt be short and $\delta \Psi = [\delta a, \delta p, \delta r]$ be the rotated angle vector of the sensors. Then $\delta \Psi = \omega \delta t$. Assuming a small δt the rotation matrix for a period can be approximated by

$$C^{(t)} = \begin{pmatrix} 1 & -\delta a & \delta p \\ \delta a & 1 & -\delta r \\ -\delta p & \delta r & 1 \end{pmatrix} = I + \Omega^{(t)} \delta t, \tag{3.29}$$

where

$$\Omega^{(t)} = \begin{pmatrix} 0 & -\omega_z & \omega_y \\ \omega_z & 0 & -\omega_x \\ -\omega_y & \omega_x & 0 \end{pmatrix}. \tag{3.30}$$

This allows us to relate the rotation matrix $R^{(t)}$ with the rotation matrix of the next sampling period $R^{(t+\delta t)}$. Then

$$R^{(t+\delta t)} = R^{(t)} \times C^{(t)}, \tag{3.31}$$

where we have overloaded the superscript to mean the current sampling period besides the time slot. This gives

$$\frac{dR^{(t)}}{dt} = R^{(t)} \times \Omega, \tag{3.32}$$

$$R^{(t+\delta t)} = R^{(t)} \times \exp \left(\int_t^{t+\delta t} \Omega dt \right). \tag{3.33}$$

The DCM update equation is obtained as each new angular velocity samples comes by

$$R^{(t+\delta t)} = R^{(t)} \left[I + \frac{\sin(\|\omega\| \delta t)}{\|\omega\|} \Omega + \frac{1 - \cos(\|\omega\| \delta t)}{\|\omega\|^2} \Omega^2 \right]. \tag{3.34}$$

With the DCM updated at each sample, the accelerometer data can easily be mapped from LCS to GCS by

$$a^{(G,t)} = R^{(t)} \cdot a^{(t)} \tag{3.35}$$

where G specifies that the vector is the GCS. Finally, the updated velocity vector is given by

$$\begin{bmatrix} v_n^{(t+1)} \\ v_e^{(t+1)} \\ v_d^{(t+1)} \end{bmatrix} = \begin{bmatrix} v_n^{(t)} \\ v_e^{(t)} \\ v_d^{(t)} \end{bmatrix} + \begin{bmatrix} a_x^{(G,t)} - n_{ax}^{(t)} \\ a_y^{(G,t)} - n_{ay}^{(t)} \\ a_z^{(G,t)} - n_{az}^{(t)} - g \end{bmatrix} \delta t \tag{3.36}$$

and the corresponding displacement vector

$$\begin{bmatrix} d_x^{(t+1)} \\ d_y^{(t+1)} \\ d_z^{(t+1)} \end{bmatrix} = \begin{bmatrix} v_n^{(t)} \delta t \\ v_e^{(t)} \delta t \\ -v_d^{(t)} \delta t \end{bmatrix} \tag{3.37}$$

is used in Algorithm 3.3 to obtain the particle $x_i^{(t+1,l)}$.

3.4.5 Static Phase

During the static phase data from the accelerometer and the magnetometer are used to derive the pitch, roll, and attitude required for the rotation matrix, using Equation 3.28. To compensate the tilt errors, the following algorithm is used as presented in [15].

First, a linear-phase finite impulse response (FIR) low-pass filter (LPF) is used to filter the accelerometer signal. The LPF is designed with a cutoff frequency of less than 1 Hz, as a typical human stride takes ≈ 1 s. The filtered acceleration $g^{(L)}$ is then normalized and redefined as a gravity vector in LCS.

The normalized GCS gravity vector is then given by

$$g^{(G)} = R \cdot g^{(L)}, \tag{3.38}$$

where $g^{(G)} = [0, 0, 1]^T$.

Solving the above equation for roll and pitch gives

$$p^{(t)} = \text{atan2}\left(g_x^{(t)}, \sqrt{\left(g_y^{(t)}\right)^2 + \left(g_z^{(t)}\right)^2} \right), \tag{3.39}$$

$$r^{(t)} = \text{atan2}\left(g_y^{(t)} \text{sign}\left(\cos(p^{(t)})\right), g_z^{(t)} \text{sign}\left(\cos(p^{(t)})\right) \right). \tag{3.40}$$

After both pitch and roll have been found from the acceleration data, the attitude can be calculated from the magnetic field data. Let $\mu^{(L,t)} = \left[\mu_x^{(t)}, \mu_y^{(t)}, \mu_z^{(t)}\right]$ be the LCS magnetometer readings. Then the compensated magnetic field can be calculated as

$$h_x^{(t)} = \mu_x^{(t)} \cos(p^{(t)}) + \mu_y^{(t)} \sin(p^{(t)}) \sin(r^{(t)})$$
$$+ \mu_z^{(t)} \sin(p^{(t)}) \cos(r^{(t)}) \tag{3.41}$$

$$h_y^{(t)} = \mu_y^{(t)} \cos(r^{(t)}) - \mu_z^{(t)} \sin(r^{(t)}) \tag{3.42}$$

$$a^{(t)} = \text{atan2}\left(-h_y^{(t)}, h_x^{(t)}\right) - \text{D}, \tag{3.43}$$

where D is the magnetic declination, or the difference between the magnetic north and the true north, caused by the tilt of the earth magnetic field generator relative to the earth spin axis.

The algorithm is summarized as Algorithm 3.6.

Algorithm 3.6 PHIMTA

1: Sample $\left\{\pi_i^{(t,l)}, x_i^{(t,l)}\right\}_{l=1}^{L} \sim p\left(X_i^{(t)}\right)$
2: Detect Stride Phase using SHOE Algorithm
3: **if** Stride Phase is *Stance* **then**
4:   Set $\omega^{(t)} = 0$
5:   Set $v^{(t)} = 0$
6:   Extract g from a using LPF
7:   Calculate p, r, a using Equations 3.39–3.43
8:   Calculate Rotation Matrix $R^{(t)}$ using Equation 3.28
9: **else**
10:   sample $\left\{n_\omega^{(l,t)}\right\}_{l=1}^{L} \sim \mathcal{N}(n_\omega)$
11:   For each sample calculate $R^{(t)}$ using Equation 3.34
12: **end if**
13: Sample $\left\{n_a^{(l,t)}\right\}_{l=1}^{L} \sim \mathcal{N}(n_a)$
14: Sample $\left\{n_\mu^{(l,t)}\right\}_{l=1}^{L} \sim \mathcal{N}(n_\mu)$
15: for each sample calculate $\left\{a^{(l,G,t)}\right\}_{l=1}^{L}$ using Equation 3.35
16: Calculate $\left\{x^{(l,t+1)}\right\}_{l=1}^{L}$ using Equation 3.37) and Algorithm 3.3
17: Convert to parametric form using Algorithm 3.4
18: Update belief $p\left(X_i^{(t+1)}\right)$

3.4.6 Complexity

For the BP time slots, the complexity is due to the message passing algorithm used. In our case, as GBP is a parametric form message passing algorithm, the computational cost is $\mathcal{O}(\mathcal{N}_i L)$, cf. [32]. This makes the algorithm an order of magnitude faster than non-parametric BP algorithms, for example, SPAWN [49]. We also compare GBP with the SnG algorithm [14], which has a complexity of $\mathcal{O}\left(\bar{\mathcal{N}}_i L\right)$, where $\bar{\mathcal{N}}_i$ symbolizes the average pseudo-anchors of node i. The number of particles used in both algorithms is approximately $L = 100$, while the number of average pseudo-anchors will be less than or equal to the average number of neighbors. As such, the algorithms tend to have similar complexity, with SnG being slightly faster. Even though the two algorithms seem to have the same computational cost, a step-by-step comparison in Table 3.2 clearly shows that GBP is faster, as it has fewer steps and there is no need to optimize the objective function at every iteration.

We assume that both algorithms use approximately the same number of particles, and |ID| is the cardinality of relevant IDs in GBP, while U and $U_{\max}$ are the number of candidate points and number of highest likelihood candidate points, respectively. Finally, I_{IWLS} is the number of iterations IWLS can run.

TABLE 3.2

Complexity of GBP vs SnG

GBP		SnG [14]	
Step	**Complexity**	**Step**	**Complexity**
Sample incoming message	L	Sample incoming message	L
Count sample IDS	$|ID|$	Count-sort likelihoods	$\bar{\mathcal{N}}_i L$
Multiply multinomial pdfs	$N_i |ID|$	Sort candidate points	$U \log(U)$
Filter IDs	$|ID| \log(|ID|)$	Get centroid	$U_{max} \log(U_{max})$
—	—	IWLS	$\mathcal{N}_i \, \Pi_{WLS}$

TABLE 3.3

Complexity of PHIMTA

Step	**Complexity**
Sample $p\left(X_i^{(t)}\right)$	L
SHOE Algorithm	Y
Stride phase	$Y L$
Quasi-static phase	$Y L$
Update position	L
Convert to parametric form	L

TABLE 3.4

Complexity Costs of GBP/PHIMTA vs SnG/PDR

Algorithm	**Complexity**
Grid-BP	$\mathcal{O}(\mathcal{N}_i L_{GBP})$
PHIMTA	$\mathcal{O}(Y \, L_{PHIMTA})$
SnG	$\mathcal{O}(\mathcal{N}_i L_{SNG})$
PDR	$\mathcal{O}(Y \, L_{PDR})$

Note that $L_{GBP} = L_{PHIMTA} = L_{SNG} = L_{PDR} \simeq 100$.

For the IMU time slots, the complexity is due to PHIMTA. All the steps are proportional to either the number of signals Y obtained from MEMS or the number of particles used in the calculations L. Each step is given with the corresponding cost in Table 3.3. Hence the complexity is $\mathcal{O}(Y \, L)$. We compare PHIMTA with the PDR algorithm in [19]. Even if the complexity scales in the same way, PDR has fewer computations per iteration than the hybrid algorithm. Essentially, it is a compromise between computational cost and accuracy, as will be seen in the sequel. By using GBP, though, the computational increase from PHIMTA can be easily compensated. The complexity cost comparison between GBP, PHIMTA, SnG and PDR can be seen in Table 3.4.

3.4.7 Simulation Results

Our proposed algorithm was evaluated using Monte Carlo simulations. We considered a 2D grid 20×20 m with four anchors at the corners of the grid. Ten nodes are randomly placed

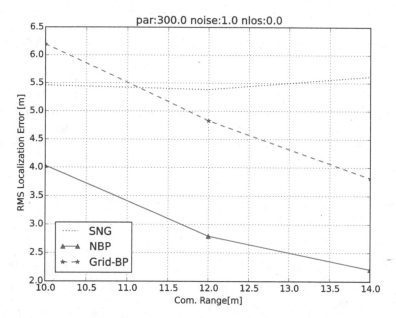

FIGURE 3.6
The RMS positioning error versus the communication range.

inside the grid and are trying to localize. We first consider a static scenario in which the root-mean-square (RMS) localization error is compared between GBP and SnG and NBP [17] as an implementation of the indoors positioning framework.

In Figure 3.6, we illustrate the average RMS error for various communication ranges. We assume that the number of particles used is 300 and $K_e = 0.001$. As expected, NBP outperforms both SnG and GBP, but at a greater computational cost. While GBP provides similar RMS to SnG for lower communication range, it is interesting to note that both NBP and GBP take advantage of the availability of more neighbors, while SnG seems to keep a constant RMS error. We should also mention that SnG would collapse if the average number of neighbors is too low. Finally, for the SnG simulations, we initially ran cooperative least-square (LS), cf. [49], as it requires initial estimates to run, which is not required for NBP and GBP.

3.5 SR-BP (Message Passing Scheduling)

In the last subsection, we consider how we can modify the message schedule in order to decrease communication overhead as well as improve the algorithm convergence rate. As we have already discussed, loopy belief propagation has no guarantees on convergence. In [23], it was shown that using asynchronous message passing schedules can significantly improve the chance and speed of convergence. The problem is that any complicated schedule scheme requires a scheduler to find the solution. Proposed schedulers have been suggested in [46]. Wholly distributed scheduling solutions found in the literature are the Uniform Tree Reweighting (UTR) BP algorithm that was proposed in [48,50], and the Stochastic-Residual Belief propagation (SR-BP) proposed in [33]. We will derive SR-BP and explain why it works, as an example of optimizing the message passing scheduling in the distributed scenario of cooperative localization.

3.5.1 Residual Belief Propagation

Eliden et al. in [9] described a different way of thinking about belief propagation from the one presented up to now. Each message can be viewed as residing in some message space ($\mathcal{R} \subset \mathbb{R}^+)^d$, where d is the dimension of the messages.* Hence, the set of messages $\mathcal{M}$, in a cluster graph, is a subset of $\mathcal{R}^{|\mathcal{M}|}$, $|\mathcal{M}| = 2|\mathcal{E}|$. Let m denote the index of individual messages, v_m denote the mth message, and $v \in \mathcal{R}^{|\mathcal{M}|}$ denote a joint assignment of messages. The updated equation (3.16) can therefore be seen as a mapping function $f_m : \mathcal{R}^{|\mathcal{M}|} \to \mathcal{R}$ that defines the mth message as a function of a subset of the messages in $\mathcal{R}^{|\mathcal{M}|}$. Then we can define the iterative method

$$v_m^{(t+1)} = f_m(v^{(t)}). \tag{3.44}$$

Assuming convergence, we have the fixed point

$$f_m(v^*) = v_m^*. \tag{3.45}$$

Finally, the *global* update functions can be defined for both the synchronous and asynchronous cases, respectively, as

$$F^s\left(v_1, \ldots, v_{|\mathcal{M}|}\right) = \left(f_1(v), \ldots, f_{|\mathcal{M}|}(v)\right), \tag{3.46}$$

$$F_m^a\left(v_1, \ldots, v_{|\mathcal{M}|}\right) = \left(v_1, \ldots, f_m(v), \ldots, v_{|\mathcal{M}|}\right). \tag{3.47}$$

In the asynchronous case, we assume that there is a set of times $T = \{0, 1, 2, \ldots\}$ at which one or more components v_m are updated. Also, let T^m be the set of times v_m is updated. Then for the asynchronous case we adopt Assumption 3.1 of [9].

For every message m, there is a finite time T_m so that for any time $t \geq 0$, the update $v := f_m^a(v)$ is executed at least once in the time interval $[t, t + T_m]$. Essentially this means that all messages will be updated infinitely until convergence.

3.5.2 Convergence Analysis

One of the main tools in convergence analysis is the contraction. Assuming a finite dimensional vector space V that has a vector norm $\| \cdot \|$, we define a mapping $F: V \to V$ to be a $\| \cdot \|$-contraction if

$$\|F(v - F(w)\| \leq a \|v - w\|, \tag{3.48}$$

for some $0 \leq a < 1$, for all $v, w \in V$. Respectively, if $F(\cdot)$ is a $\| \cdot \|$-contraction then a unique fixed point v^* is guaranteed to exist, and applying $F(\cdot)$ iteratively we have that

$$v^{(t+1)} = F(v^{(t)}) \tag{3.49}$$

is guaranteed to converge to v^*, for all possible initial vectors $v^{(0)} \in V$. In the message space R we define a message norm $\|v_m - w_m\|_m$ that measures distances between individual messages

---

* For convenience, all messages are assumed to have the same dimension d.

and a global norm that measures distances between points in $R|M|$. Following the analysis in [9], we also use the max norm $\|\cdot\|_\infty$ for the global norm defined as

$$\|v - w\|_\infty = \max_{m \in \mathcal{M}} \|v_m - w_m\|_m. \tag{3.50}$$

Assuming convergence is guaranteed for the synchronous BP, that is, F^s is a $\|\cdot\|_\infty$-contraction, Elidan et al. showed that the asynchronous BP will also converge if there is a propagation schedule that guarantees that every message will be updated until convergence; that is, Assumption 3.1. Also, they suggested the intuition that in an asynchronous message passing scheme, messages that "carry" more information should be propagated first as they will help the algorithm converge faster, and defined the residual of a message as

$$r_m(v) = \|f_m(v) - v_m\|_m. \tag{3.51}$$

This led to the proposal of the R-BP message passing scheme, where at each iteration (t) of BP, all residuals are calculated and the message with the largest residual is chosen to be propagated, namely

$$m^{(t+1)} = \arg\max_m r_m(v^{(t)}). \tag{3.52}$$

Empirically, Elidan et al. also showed that even in complicated real-life application PGMs, R-BP will converge more often and with less messages than both synchronous and asynchronous BP propagation schemes, which led to a variety of R-BP variants, especially in the LDPC decoding. Unfortunately, R-BP and its variants require a centralized entity to compare all the residuals, find the largest, and propagate the corresponding message, making it unsuitable for BP in distributed networks. In the sequel we present stochastic R-BP (SR-BP), which overcomes the requirement of a centralized entity by using a stochastic propagation scheme, suitable for distributed networks.

3.6 SR-BP

First, we reformulate BP so that each node propagates its belief at each iteration instead of a distinct message for each neighbor. The belief, that is an approximation to the true marginal [45], at each timeslot is calculated as

$$b_s^{(t)}(X_s) = \Phi_s(X_s) \prod_{r \in N_s} \mu_{r \to s}^{(t)}(X_s) \tag{3.53}$$

and the messages calculated from the incoming beliefs $\mu_{r \to s}^{(t)}(X_{r,s})$ are calculated as

$$\mu_{r \to s}^{(t)}(X_s) \propto \sum_{X_r} \Phi_{r,s}(X_r, X_s) \prod_{t \in N_{r \to s}} \mu_{t \to r}^{(t-1)}(X_r)$$

$$\propto \sum_{X_r} \Phi_{r,s}(X_r, X_s) \frac{b_r^{(t-1)}(X_r)}{\mu_{s \to r}^{(t-1)}(X_r)}. \tag{3.50}$$

Consequently, $|\mathcal{M}|$ is equal to the number of nodes. In the distributed case there is no centralized entity to compare all residuals and decide on a message schedule. As a result, each node will have to decide on itself if its message is "important" enough to transmit. Following the intuition behind residuals, we propose a stochastic message passing schedule, where each node transmits its belief at timeslot (t) with probability $\Pr(r_m^{(t)})$. The probability is calculated as

$$\Pr\left(r_m^{(t)}\right) = S\left(r_m^{(t)}\right) = \frac{1}{1 + \exp\left(-r_m^{(t)}\right)},\tag{3.54}$$

where $S(\cdot)$ is the sigmoid function. We now analyze the convergence of SR-BP. To do so, we use Theorem 3.2 of [9] propagation that states If F^s is a max-norm contraction, then any asynchronous propagation schedule that satisfies Assumption 3.1 will converge to a unique fixed point. Assuming that a PGM already satisfies the max-norm contraction condition for the synchronous BP case, we only need to prove that SR-BP satisfies Assumption 3.1.

Assumption 3.1

For every message m, there is a finite time T_m so that for any time $t \geq 0$, the update $v := f_m^a(v)$ is executed at least once in the time interval $[t, t + T_m]$.

By definition, the residual $r_m^{(t)} \geq 0$, $\forall m \in \mathcal{M}$. Then by construction we have

$$r_m^{(t)} \geq 0 \Rightarrow \frac{1}{1 + \exp\left(-r_m^{(t)}\right)} \geq 0.5.\tag{3.55}$$

Consequently, there will always be a positive probability that message m will be transmitted, hence there will be a T_m, so that for any time $t \geq 0$, message m will be updated and Assumption 3.1 is satisfied. Therefore, SR-BP will converge to a fixed point. It is difficult to analyze the convergence rate of SR-BP but intuitively it should be somewhere between the synchronous case, and the centralized residual case. In the sequel we will provide experimental results that showcase this intuition.

3.6.1 Experimental Results

Monte Carlo simulations were carried out to analyze the convergence rate, message overhead, and quality of the marginals of SR-BP. Comparisons were made with synchronous BP, that is, BP, asynchronous BP (ABP). Also, we compare it with centralized R-BP that of course could not be used in practice in a distributed scenario. Finally, we also use the aforementioned message scheduling schemes with UTR-BP [48]. It should be noted that all algorithms use respectively the same codebase, and only the message schedule differs.

3.6.1.1 *Ising Model*

As a simple example, we will showcase the algorithms considering random g$\mathcal{U}$rids parameterized by the Ising model [4]. These allow for a systematic way to analyze an algorithm, as both the difficulty and the scale of the inference task can be readily controlled. Random grids with 11 nodes were created, with univariate potentials $\Phi_i(X_i)$ sampled from $\mathcal{U}[0, 1]$ for each variable, and pairwise potential given by

$$\Phi_{i,j}(X_i, X_j) = \exp(\lambda C * (2 \mathbb{1}(X_i = X_j) - 1)),\tag{3.56}$$

TABLE 3.5

Results for the Ising Model

	Conv. %	Avg. Conv. Iterations	Avg. Messages	KLD
BP	83%	78.5	8634	0.255
ABP	88%	63.2	6905	0.257
R-BP	100%	21.5	315	0.369
SR-BP	94.5%	28.4	393	0.365

where λ is sampled uniformly from $[-0.5, 0.5]$ randomly allowing some nodes to agree and some to disagree with each other. In addition, $\mathbf{1}(X_i = X_j)$ is the indicator function. Finally, C is an agreement factor, where higher values impose stronger constraints on the "negotiations" between nodes, making convergence harder. We run 200 MC simulations for a network with 11 nodes, where $C = 10$ and the algorithms were allowed to run until convergence, or 300 iterations had passed. The results are summarized in Table 3.5.

There are a few interesting points to be made here. First, as expected, R-BP achieves convergence every time in the given scenario, requiring 73% fewer iterations on average and almost 96% fewer messages. Furthermore, SR-BP is pretty close to R-BP, achieving only slightly worse convergence rate and requiring 64% fewer iterations than BP and around 95% fewer messages, with the corresponding decrease in computational cost and overhead, of course. The other thing of interest is the significant increase in convergence rate when UTR-BP is used to compute the messages, as well as the decrease in required iterations and messages to reach it, for BP and ABP. UTR-R-BP seems to work slightly better than normal R-BP, and UTR-SR-BP seems to converge a bit slower and worse than before. Nevertheless, it achieves a better approximation to the real marginals, and the number of messages required continues to be quite small.

Figures 3.7 and 3.8 show the cumulative percentage of convergence of the different algorithms as a function of iterations passed (essentially computational time given to the algorithm). Again, it is worthy to note that in both cases, even though SR-BP converges less than centralized R-BP, when it will converge it converges much faster.

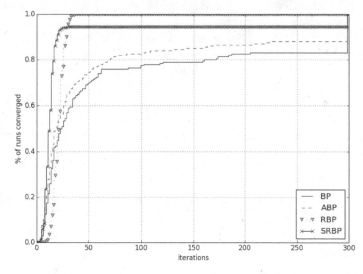

FIGURE 3.7
Percentage of convergence runs using Ising model.

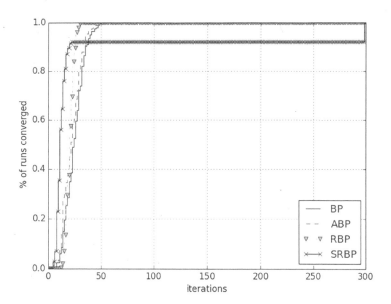

FIGURE 3.8
Percentage of convergence runs using UTR Ising model.

3.6.1.2 Cooperative Spectrum Sensing Model

Next, we study cooperative spectrum sensing, using the same model presented in [48], as a real-life wireless distributed application. We assume that a primary user (PU) is transmitting and secondary users (SU) are collaboratively sensing the spectrum trying to decide if the channel is busy or not. SU that are close by are more likely to sense the same channel state. Let $X_i \in \{0, 1\}$ be binary variable representing the state of the channel close to SU i, and let $Y_i = y_i$ represent the observation made by SU i. Then we have univariate variables $\Phi_i(X_i) = \Pr(y_i|x_i)$ and pairwise potentials $\Phi_{i,j}(X_i, X_j) = \exp(\lambda_{i,j}\mathbf{1}(X_i = X_j))$, where $\lambda_{i,j}$ is the correlation factor between nodes i and j and in this model is sampled uniformly between [0.2, 4]. The observations Y_i are sampled from complex Gaussian distributions with noise variance $\sigma_n^2 \mathbf{I}_s$ and signal variance $\sigma_s^2 \mathbf{I}_s$ for $X_i = 0$ and $X_i = 1$, respectively. $\mathbf{I}_s$ is the $s \times s$ identity matrix, and s is the number of i.i.d signal received samples used in the channel detection. Nodes are deployed randomly in a circular area with unit diameter, and $R = 0.7$ is set as the maximum communication range. One hundred Monte Carlo simulations were run, for 100 transmission timeslots each, with a maximum iteration number of 300. The results are summarized in Table 3.6. It is quite evident that convergence is much more difficult to achieve due to a large number of interconnections cause by the high communication range and the relatively significant correlation factors $\lambda_{i,j}$. Still, SR-BP manages to double the convergence percentage, using approximately 29% of the message

TABLE 3.6

Results for the Cooperative Spectrum Sensing Model

	Conv. %	Avg. Conv. Iterations	Avg. Messages
BP	9.8%	256.1	2817.26
ABP	13.4%	259.1	2849.6
R-BP	19.5%	269.0	279.0
SR-BP	18.5%	267.9	804.7

propagations required by BP. In summary, in a real-life application, SR-BP achieves better convergence, with a 71% decrease in required transmitted messages, while at the same time improving the RoC curve of the spectrum sensing nodes slightly.

It is hard to compute exact marginals for the spectrum sensing case. Instead we compute the ROC curves for the various algorithms, as can be seen in Figures 3.9 and 3.10. As can be seen in both cases, SR-BP achieves a better ROC curve than BP and ABP, achieving a curve that almost matches the one by centralized R-BP.

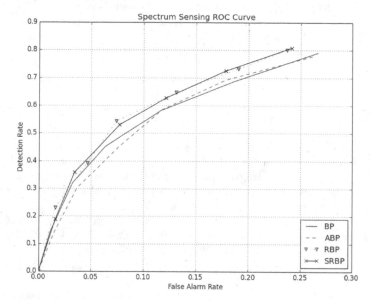

FIGURE 3.9
Spectrum sensing ROC curve.

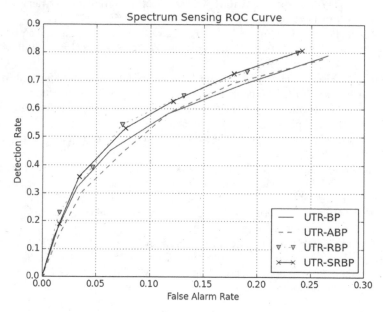

FIGURE 3.10
Spectrum sensing UTR ROC curve.

3.7 Concluding Remarks

This chapter presented techniques that reduce complexity and communication overhead when using message passing algorithms for cooperative localization. The advantage of using a parametric representation of the messages using Grid-BP was successfully shown to decrease computational cost drastically. Then the technique was extended to the mobile case. Finally, message passing scheduling was suggested as a technique to reduce the number of messages required for the algorithm to converge, and SR-BP was presented as a suitable algorithm.

References

1. Hager, John W. et al. Datums, ellipsoids, grids, and grid reference systems. No. DMA-TM-8358.1. Defense Mapping Agency Hydrographic/Topographic Center Washington DC, 1992.
2. K. Altun and B. Barshan. Pedestrian dead reckoning employing simultaneous activity recognition cues. *Measurement Science and Technology*, 23(2):025103, 2012.
3. M.M. Atia, M. Korenberg, and A. Noureldin. A consistent zero-configuration GPS-like indoor positioning system based on signal strength in IEEE 802.11 networks. *2012 IEEE/ION Position Location and Navigation Symposium (PLANS)*, pp. 1068–1073, 2012.
4. D. Barber. *Bayesian reasoning and machine learning*. Cambridge University Press, Cambridge, NY, 2011.
5. R.M. Buehrer, T. Jia, and B. Thompson. Cooperative indoor position location using the parallel projection method. *2010 International Conference on Indoor Positioning and Indoor Navigation, IPIN 2010 Conference Proceedings*, pp. 1–10. IEEE, 2010.
6. R.M. Buehrer, H. Wymeersch, and R.M. Vaghefi. Collaborative sensor network localization: Algorithms and practical issues. *Proceedings of the IEEE*, 106(6):1089–1114, 2018.
7. M.A. Caceres, F. Penna, H. Wymeersch, and R. Garello. Hybrid cooperative positioning based on distributed belief propagation. *IEEE Journal on Selected Areas in Communications*, 29(10):1948–1958, 2011.
8. N. El-Sheimy, H. Hou, H. Hou, X. Niu, and X. Niu. Analysis and modeling of inertial sensors using Allan variance. *IEEE Transactions on Instrumentation & Measurement*, 57(1):140–149, 2008.
9. G. Elidan, I. McGraw, and D. Koller. Residual belief propagation: Informed scheduling for asynchronous message passing. arXiv.org, p. 6837, June 2012.
10. B. Etzlinger and H. Wymeersch. *Synchronization and localization in wireless networks*. Now Publishers, 2018.
11. J. Georgy, A. Noureldin, and M. Bayoumi. Mixture particle filter for low cost ins/odometer/GPS integration in land vehicles. *IEEE 69th Vehicular Technology Conference, 2009. VTC Spring 2009*, pp. 1–5, April 2009.
12. T. Hastie, R. Tibshirani, J. Friedman, T. Hastie, J. Friedman, and R. Tibshirani. *The elements of statistical learning*. Springer, 2009.
13. S. He and S.H.G. Chan. Wi-Fi fingerprint-based indoor positioning: Recent advances and comparisons. *IEEE Communications Surveys and Tutorials*, 18(1):466–490, 2016.
14. T. Higuchi, S. Fujii, H. Yamaguchi, and T. Higashino. Mobile node localization focusing on stop-and-go behavior of indoor pedestrians. *IEEE Transactions on Mobile Computing*, 13(7):1564–1578, July 2014.
15. C. Huang, Z. Liao, and L. Zhao. Synergism of INS and PDR in self-contained pedestrian tracking with a miniature sensor module. *IEEE Sensors Journal*, 10(8):1349–1359, August 2010.

16. A.T. Ihler, J.W. Fisher, R.L. Moses, and A.S. Willsky. Nonparametric belief propagation for self-localization of sensor networks. *IEEE Journal on Selected Areas in Communications*, 23(4):809–819, 2005.

17. A.T. Ihler, J.W. Fisher III, R.L. Moses, A.S. Willsky, and J.W. Fisher. Nonparametric belief propagation for self-calibration in sensor networks. *IPSN 2004 Third International Symposium on Information Processing in Sensor Networks, 2004*, 23(4), pp. 225–233, 2004.

18. A.T. Ihler, J.W. Fisher, R.L. Moses, and A.S. Willsky. Nonparametric belief propagation for self-calibration in sensor networks. *IPSN 2004 Third International Symposium on Information Processing in Sensor Networks, 2004*, pp. 225–233, April 2004.

19. Y. Jin, H.-S. Toh, W.-S. Soh, and W.-C. Wong. A robust dead-reckoning pedestrian tracking system with low cost sensors. *2011 IEEE International Conference on Pervasive Computing and Communications (PerCom)*, pp. 222–230, March 2011.

20. W. Kang, S. Nam, Y. Han, and S. Lee. Improved heading estimation for smartphone-based indoor positioning systems. *2012 IEEE 23rd International Symposium on Personal Indoor and Mobile Radio Communications (PIMRC)*, 2012, Sydney, Australia.

21. N. Kantas, S.S. Singh, and A. Doucet. Distributed self localisation of sensor networks using particle methods. *2006 IEEE Nonlinear Statistical Signal Processing Workshop*, pp. 164–167, September 2006.

22. J. Khodjaev, Y. Park, and A.S. Malik. Survey of NLOS identification and error mitigation problems in UWB-based positioning algorithms for dense environments. *Annals of Telecommunications*, 65(5–6):301–311, 2010.

23. D. Koller and N. Friedman. *Probabilistic graphical models: Principles and techniques*. MIT Press, 2009.

24. H. Koyuncu and S.H. Yang. A survey of indoor positioning and object locating systems. *International Journal of Computer Science and Network Security (IJCSNS '10)*, 10(5):121–128, 2010.

25. J. Lien, U.J. Ferner, W. Srichavengsup, H. Wymeersch, and M.Z. Win. A comparison of parametric and sample-based message representation in cooperative localization. *International Journal of Navigation and Observation*, 2012, 2012.

26. Y. Liu, Z. Yang, X. Wang, and L. Jian. Location, localization, and localizability. *Journal of Computer Science and Technology*, 25(2):274–297, 2010.

27. A.K.M. Mahtab Hossain, Y. Jin, W.S. Soh, and H. Nguyen Van. SSD: A robust RF location fingerprint addressing mobile devices' heterogeneity. *IEEE Transactions on Mobile Computing*, 12(1):65–77, 2013.

28. P. Misra and P. Enge. *Global positioning system: Signals, measurements and performance*, second edition. Massachusetts: Ganga-Jamuna Press, 2006.

29. S. Movaghati and M. Ardakani. Particle-based message passing algorithm for inference problems in wireless sensor networks. *IEEE Sensors Journal*, 11(3):745–754, March 2011.

30. C.L. Nguyen, O. Georgiou, Y. Yonezawa, and Y. Doi. The wireless localization matching problem. *IEEE Internet of Things Journal*, 4(5):1312–1326, 2017.

31. P.-A. Oikonomou-Filandras and K.-K. Wong. HEVA: Cooperative localization using a combined non-parametric belief propagation and variational message passing approach. *Journal of Communications and Networks*, 18(3), 2016.

32. P.-A. Oikonomou-Filandras, K.-K. Wong, and Y. Zhang. Grid-based belief propagation for cooperative localization. arXiv.org, 2015.

33. P.-A. Oikonomou-Filandras, K.-K. Wong, and Y. Zhang. Informed scheduling by stochastic residual belief propagation in distributed wireless networks. *IEEE Wireless Communications Letters*, 4(1):90–93, February 2015.

34. P.-A. Oikonomou-Filandras, K.-K. Wong, Z. Zheng, and Y. Zhang. Cooperative localisation with hybrid inertial navigation system/pedestrian dead reckoning tracking for GPS-denied environments. *Proceedings of the 31st Annual ACM Symposium on Applied Computing*. ACM, 2016.

35. T.E. Oliphant. Python for scientific computing. *Computing in Science Engineering*, 9(3):10–20, May 2007.

36. R.W. Ouyang, A.K.S. Wong, and C.T. Lea. Received signal strength-based wireless localization via semidefinite programming: Noncooperative and cooperative schemes. *IEEE Transactions on Vehicular Technology*, 59(3):1307–1318, 2010.

37. Y. Qi and H. Kobayashi. On relation among time delay and signal strength based geolocation methods. *Global Telecommunications Conference, 2003. GLOBECOM'03*. IEEE, vol. 7, pp. 4079–4083, December 2003.

38. A.G. Quinchia, C. Ferrer, G. Falco, E. Falletti, and F. Dovis. Analysis and modelling of MEMS inertial measurement unit. *2012 International Conference on Localization and GNSS (ICLGNSS)*, 2012.

39. S. Rezaei and R. Sengupta. Kalman filter-based integration of DGPS and vehicle sensors for localization. *IEEE Trans. Control Syst. Technol.*, 15(6):1080–1088, November 2007.

40. V. Savic and S. Zazo. Nonparametric boxed belief propagation for localization in wireless sensor networks. *Third International Conference on Sensor Technologies and Applications, 2009. SENSORCOMM '09*, pp. 520–525, June 2009.

41. S.M.H. Sharhan and S. Zickau. Indoor mapping for location based policy tooling using Bluetooth low energy beacons. *2015 IEEE 11th International Conference on Wireless and Mobile Computing, Networking and Communications, WiMob 2015*, pp. 28–36, 2015.

42. I. Skog, P. Handel, J.O. Nilsson, and J. Rantakokko. Zero-velocity detection, an algorithm evaluation. *IEEE Transactions on Biomedical Engineering*, 57(11):2657–2666, 2010.

43. Z. Sun, Z. Sun, X. Mao, X. Mao, W. Tian, W. Tian, X. Zhang, and X. Zhang. Activity classification and dead reckoning for pedestrian navigation with wearable sensors. *Measurement Science and Technology*, 20(1):015203, 2009.

44. S. Thrun, D. Fox, W. Burgard, and F. Dellaert. Robust Monte Carlo localization for mobile robots. *Artificial Intelligence*, 128(1):99–141, 2001.

45. M.J. Wainwright, T.S. Jaakkola, and A.S. Willsky. Treebased reparameterization framework for analysis of sum-product and related algorithms. *IEEE Transactions on Information Theory*, 49(5):1120–1146, 2003.

46. T. Wang, Y. Shen, S. Mazuelas, and M.Z. Win. Distributed scheduling for cooperative localization based on information evolution. *2012 IEEE International Conference on Communications (ICC)*, pp. 1–6, 2012.

47. M.Z. Win, Y. Shen, and W. Dai. A theoretical foundation of network localization and navigation. *Proceedings of the IEEE*, 106(7):1136–1165, 2018.

48. H. Wymeersch, F. Penna, and V. Savic. Uniformly reweighted belief propagation for estimation and detection in wireless networks. *IEEE Transactions on Wireless Communication*, 11(4):1587–1595, April 2012.

49. H. Wymeersch, J. Lien, and M.Z. Win. Cooperative localization in wireless networks. *Proceedings of the IEEE*, vol. 97, pp. 427–450, 2009.

50. H. Wymeersch, F. Penna, and V. Savíc. Uniformly reweighted belief propagation: A factor graph approach. *2011 IEEE International Symposium on Information Theory Proceedings (ISIT)*, pp. 2000–2004. IEEE, 2011.

51. R. Zekavat and R.M. Buehrer. *Handbook of position location: Theory, practice and advances*, vol. 27. John Wiley & Sons, 2011.

4

Long-Term Area Coverage and Radio Relay Positioning Using Swarms of UAVs

Floriana Benedetti, Alessio Capitanelli, Fulvio Mastrogiovanni, and Gianni Viardo Vercelli

CONTENTS

4.1 Introduction: Problem and Background

Unmanned Aerial Vehicles (UAVs) are becoming increasingly useful for tasks that require the acquisition of data over large areas. Despite that, many real-world scenarios still offer significant challenges due to communication issues and high deployment costs. The coverage problem—that is, the problem of periodically visiting all sub-regions of an area possibly at a given frequency—is especially interesting because of its practical applications, such as surveillance, agricultural operations, and continuous, long-term monitoring of areas hit by a natural disaster. We focus here on this last scenario, considering that in the case of events such as earthquakes, flooding, or hurricanes, aerial coverage could help search and rescue teams to locate victims and assess the evolving state of the surrounding buildings and infrastructures.

On the one hand, it is important to highlight the specific issues that set first response apart from other coverage and exploration applications and why this is a challenging problem given the current state of the art. First, in this application the main interest lies in efficient long-term patrolling of a given area (*Requirement* R1), whose map is supposed to be at least roughly available. This contrasts with the more popular exploration problem, where the focus is exactly on obtaining such a map as quickly as possible. On the other hand, with respect to most commercial aerial coverage applications in the field of agriculture and surveillance, a hypothetical swarm of UAVs should be able to communicate back to the base station the gathered data in real time, without relying on pre-existing infrastructure (R2). Furthermore, it should be robust to changes in the environment (R3) and to other technical difficulties that naturally arise in long-term operations and cannot be reasonably foreseen before the mission starts, such as a UAV needing to recharge or experiencing a fault (R4).

In such a scenario, R1 is implicitly solved by employing a coverage algorithm with one or more UAVs, while it is possible to satisfy R2 by adopting several supporting drones acting together as a signal relay chain to the base station. It is noteworthy that relay positioning has already been successfully employed to increase the effective communication range of mobile robots, route the signal around obstacles, and drive down operation costs, as smaller, simpler UAVs can be used rather than more expensive models. Despite this approach being common in exploration scenarios [28,24], to the best of our knowledge it has not been employed in coverage applications so far. Finally, R3 and R4 can be addressed using *online* coverage algorithms able to dynamically react to a changing environment, instead of relying on *offline* methods usually adopted in most coverage applications. Yet, a system of this sort can be defined only as a *naïve* simultaneous coverage and relay positioning system, as it suffers from an evident problem, that is, the growing number of UAVs which are not directly participating to the task at hand but are uniquely useful to support communication, something that effectively limits the overall system's efficiency.

In this Chapter, we propose an approach to greatly mitigate this issue by involving the supporting UAVs directly into the coverage process in such a way that a drone acts in autonomy while the rest of the swarm provides a robust relay chain with the basecamp and simultaneously helps in the coverage operation, leading to *real* simultaneous coverage and relay positioning.

The conceptual framework described in this Chapter can be applied to any type of mobile robot, but we will focus on the UAVs example given their suitability to face natural disasters, as they allow for quickly acquiring highly informative data from a given height, and are able to fly over areas potentially unreachable to ground robots or first responders in the case of a catastrophic event.

Throughout the Chapter, we will assume that a swarm of drones is tasked to repeatedly fly over an area just struck by a catastrophic event in order to collect data about the environment and possibly locate survivors in need of help as they leave the buildings hit by the disaster. An indicative map of the area may be available, either from prior knowledge of the environment or acquired by the same swarm in a preliminary exploration mission. The available map is then discretized into nodes with a suitable resolution, each linked to the reachable neighboring nodes through edges to form a connected graph. The available UAVs will travel from node to node, trying to keep the average number of node visits as homogeneous as possible but also taking into account communication concerns. The system should be able to be robust to any of the UAVs going missing or being unable to operate, because a drone can (i) lose connection with the swarm due to unforeseeable reasons, (ii) need to leave the swarm due to technical difficulties, either temporarily or permanently (e.g., low battery, hardware faults), and (iii) still be part of the swarm but unable to leave the current node, as it is busy with a secondary task (e.g., offering assistance to or communicating with survivors, gathering specific data). In these pages we present an algorithm to deal with such events, develop an architecture employing it, and simulate its behavior in a Gazebo simulation with real UAV software in the loop. Finally, we compare our results to the case where a single drone is tasked to cover the area while the others only offer communication support (i.e., the naïve case). As we will see in the Conclusions, the proposed approach is able to provide a sensible improvement in coverage time compared to the reference method, with just a small hit on the average communication cost and barely affecting the worst-case communication cost. We conclude that such a system can be deployed to swarms of UAVs already operating in the aforementioned mode in order to improve its performance.

4.2 State of the Art

As discussed above, the main idea underlying this Chapter is to jointly treat two distinct problems in the literature, namely *area coverage* and *relay positioning*. While both problems have been widely reviewed in the literature singularly, no technique has been formalized to achieve a functional tradeoff between the two, similarly to what has been previously done with simultaneous exploration and relay positioning [3]. In this Section we discuss the current advancements in both fields and present tools that can be used to address the issue at hand.

4.2.1 Coverage

The problem of coverage is defined as the problem of sweeping a given area with the highest possible efficiency, that is, being able to visit each point in that area with a possibly constant frequency that is as high and as homogeneous among all the nodes as possible. It is clear that this is very different from the more common shortest-path problem, where one is mainly interested in finding the minimum cost trajectory between any two points. Despite that, coverage also has a large number of important applications, ranging from harvesting [15] to mine hunting [26], and this is why a multitude of algorithms have been proposed in the past. Historically, these algorithms are categorized in two ways: *heuristic* vs. *complete* [18] algorithms, and *online* vs. *offline* [14] algorithms. According to such classification, we define as complete only the algorithms which can be mathematically proven to obtain complete space coverage [21], while we define as heuristic the others, as in [6]. The second classification instead is more complex than it appears at first sight, due to its practical implications in certain applications:

- *Offline methods* precompute an (often optimal) coverage strategy for a given area before runtime operations occur, using static *a priori* information. Notable examples are Space Decomposition [13] and Spanning Tree Coverage [20].

- *Online methods* do not perform calculations offline based on *a priori* knowledge; instead, they periodically determine at runtime the next best action to achieve coverage based on current environmental knowledge and sensory input. The most famous example is given by Real Time Search [17].

Historically, offline methods have been the preferred choice in most cases, as the majority of the coverage applications happen to deal with mostly static environments—for example, large fields in either agricultural or security applications [30]—as well as with very structured environments, like perfectly known surfaces in painting applications [31]. In the case presented here, though, online methods are the preferred choice as they overcome some limitations of offline methods, which are clearly in contrast with the requirements of the proposed scenario:

- It is not possible to assume perfect *a priori* knowledge of the environment in the event of a disaster.

- It is computationally expensive to precompute or recompute a strategy for large areas under optimality constraints.

- Offline algorithms exhibit very low tolerance to unpredictable event, as they cannot be taken into account when the strategy is computed offline.

As a notable consequence to the last point, offline methods strongly depend on every single UAV part of the swarm, as they assign to each UAV a list of nodes to visit. If that UAV is busy, missing, or otherwise unable to act, that would cause a major system failure.

Online coverage algorithms obviously pose a greater challenge to ensure efficient operations with multiple UAVs compared to offline ones. As an example, it is often necessary to make a number of assumptions about which knowledge can be shared among UAVs, something that is not necessary if the whole strategy is computed offline. Examples of algorithms employed in the online multi-agent case are Node Count [23], LRTA [23], Edge Counting [22], and PatrolGRAPH [7]. While it is not surprising to know that increasing the number of UAVs generally leads to better coverage performance, it should be noted that more complex algorithms that may expand the single-agent case can be overtaken by simpler algorithms in the multi-agent case [25]. As an example, Node Count [9] is the simplest algorithm among the cited ones but also the one that performs better in comparative benchmarks [25]. Fundamentally, it consists in maintaining a database of the number of times each node has been visited, shares it among the agents, and lets each agent move toward the neighboring reachable node with the lowest score. In the following Sections we show how this idea can be used to achieve simultaneous relay positioning and area coverage.

4.2.2 Relay Positioning

Optimal relay positioning is a widely explored problem in the telecommunication field, where it is used to place antennas in a given area to provide robust wireless connection [29,32]. With mobile robots, though, a new problem arises, that is, how we can create robust *dynamical* relay chains able to rearrange themselves online [12,19]. Keeping in mind that it is generally desirable to relay a signal from one point to another, this translates in our case into relaying data from one far UAV back to the base station, through a number of intermediate UAVs. This implies that only the leading UAV, i.e., the master, benefits from complete freedom of action, while the other UAVs, which must primarily provide communication support, are forced to assume the positions imposed to them by the relaying algorithm. Furthermore, it is noteworthy that this kind of algorithm not only provides a viable relay chain, but also strives to achieve one of the following secondary goals, or a balance of the two: (i) maximizing the quality of the resulting chain in terms of inter-agent and master-base communication, and (ii) minimizing resource requirements, that is the number of agents employed in the chain [10,27]. In Section 4.2 we detail how one can dynamically generate chains of optimal communication cost with the available resources, and how this can be relaxed to guarantee supporting UAVs with a certain degree of movement margin in the coverage process.

4.3 Methodology

In this Section we show how it is possible to combine relay positioning with coverage algorithms. Such an integration can be achieved by letting a single UAV, i.e., the master, act freely, either directly controlled by an operator or automatically by a coverage algorithm, while the rest of the swarm offers communication support in the form of a relay chain. The main difference with previous approaches in the literature is that we relax the communication optimality criteria of the chain to allow the supporting drones to increase the rate at which the area is covered. To do so, we first formalize the problem of dynamical single-target relay positioning, then we proceed by illustrating a typical solution in the case in which one wants to achieve the best optimal chain given a limited number of UAVs. We conclude the Section introducing a new cost term penalizing support drones entering areas which have already been covered multiple times, effectively improving the efficiency of the coverage task.

4.3.1 Dynamical Single-Target Relay Positioning

Let us provide a more formal description of the dynamical single-target relay positioning problem as described in [10]. Let $x_0, x_n \in \mathcal{U}$, where x_0 is the position of the base station, x_n is the position of the UAV we want to relay to the base, and $\mathcal{U}$ is the space (bi- or tri-dimensional) of positions achievable by the UAVs. Given any two points x, x' in $\mathcal{U}$, we define the *communication reachability function* $f_{comm}(x, x')$ as a function that yields 1 if the communication between the two points is feasible, and 0 otherwise. It is now possible to define a relay chain rc_j as an ordered set $rc_j = (x_0, x_1, \ldots, x_n)$, such that $x_0, x_1, \ldots, x_n \in \mathcal{U}$ and n is the *length* of the chain. Such chain is said to be *valid* if and only if $f_{comm}(x_i, x_{i+1}) = 1$ for each $i = 1, \ldots, n-1$. Finally, let us introduce the *communication cost function* $c_{comm}(x, x')$, which models the cost of transmitting information from an agent placed in x to another one placed in x'. The total communication cost C_{comm} of a given chain can then be given by the relation:

$$C_{comm}(rc_j) = \sum_{i=0}^{n-1} c_{comm}(x_i, x_{i+1}). \tag{4.1}$$

Additional costs may be taken into account to model specific desired properties of the chain (e.g., the distance of the last element of the chain from a specific target); otherwise, one can assume the total communication cost as equivalent to the total cost of the chain C_{tot}. It is noteworthy that we will not give a specific definition for $c_{comm}(x, x')$, as this cost depends on the actual implementation and the way one desires to model the quality of data transmission between any two points (e.g., strength of the signal, number of packets lost).

Once the definition of total cost of the chain has been given, one can generate different kinds of chains between any two points by solving different sub-problems, such as finding the chain of minimum length with minimum cost, or conversely, finding the chain of minimum cost with minimum length. In most practical applications with UAVs, though, the focus is on finding a minimal cost chain that employs at most the number of UAVs available for the specific application at hand.

4.3.2 Determination of the Limited Length Chain of Minimal Cost

In this Section we present how to compute the chain of minimal cost given a limited number of UAVs available for the relay chain based on the methods described in [10,11], following the definition of the *hop-constrained shortest path problem* [16]. This approach is based on a combination of the *Dual Ascent method* [8] and the *Dijkstra's algorithm* [10]. Please note that in our scenario the Dijkstra's algorithm is not used to generate a path, but a *chain*, that is, the resulting series of nodes is not meant to be traversed by a single UAV but is rather a list of positions which must be occupied by the available UAVs to form a valid relay chain. The basic idea is to use the Djikstra's theorem to get a solution as a series of nodes from the environment map going from the base station to the last, independent, UAV of the chain, without regard for the resulting series' length. If the solution initially obtained satisfies the limit on the number of UAVs, it is immediately accepted; otherwise the Dual Ascent algorithm is used to refine the costs used by the Djikstra's algorithm in a way to favor shorter-length solutions. The two algorithms are then iterated until a valid solution is found or no further refinement is possible. In particular, our use of the dual ascent algorithm is detailed in Algorithm 4.1.

More formally, let N be the set of all nodes n which the agents can occupy, each one corresponding to a given position x in the map, and let n_0 be the node corresponding to the first fixed element of the chain (i.e., the base station). Furthermore, any of such nodes can be connected to several peers by directed edges (n, n'), which form the set E. Together, N and E define the navigation graph G, that is, a connected, non-oriented graph of arbitrary order, which may include cycles. Considering that each node n is associated with a position x, the cost definitions given in the previous Section hold equally for both positions and nodes, as long as the cost of a node n is computed on its assigned position x. In particular, we remark that $c_{comm}(n, n') \equiv c_{comm}(x, x')$, allowing us to compute the tree of minimum length minimum cost chains (MLMC) on the navigation graph G, that is the tree of the optimal chains in terms of cost and length (i.e., the number of employed UAVs), from n_0 to all the other nodes in N. This can be achieved by using a modified version of the Djikstra's algorithm, in such a way that: (i) given a chain π, the cost of a chain is actually a compound of the form $(len(\pi), cost(\pi))$, and a chain ordering is induced as $(len_1, cost_1) < (len_2, cost_2)$ if and only if $(cost_1 < cost_2) \vee (cost_1 = cost_2 \wedge len_1 < len_2)$, that is, cost has higher priority than length; (ii) the algorithm returns the whole tree it has built, instead of just the minimum cost chain between any two given nodes.

Algorithm 4.1 Dual Ascent algorithm
___

1:   $\alpha \leftarrow \alpha_0$

2:   **loop**

3:      Calculate MLMC tree from n_0 using $c'_{comm}(n, n') = c_{comm}(n, n') + \alpha$

4:      Obtain from MLMC $y_n, q_n \; \forall n \in N$ and π from n_0 to n_i

5:      **if** $len(\pi) \leq n_{UAVs} + 1$ **then**

6:          **return** π

7:      $S \leftarrow \{(n, n') \in E : q_{n'} > q_n + 1\}$

8:      **if** $S = \infty$ **then**

9:          **fail**

10:     $\in_{n,n'} \leftarrow \dfrac{(y_n + c'_{comm}(n, n')) - y_{n'}}{q_{n'} - (q_n + 1)} \quad \forall (n, n') \in S$

11:     $\epsilon \leftarrow \min \epsilon_{n,n'}$

12:     $\alpha \leftarrow \alpha + \epsilon$

13: **end loop**
___

Before proceeding with the Dual Ascent algorithm, let us define a few more important quantities: (i) n_{UAV}, being the maximum number of UAVs available to build a valid chain; (ii) $q_n(n)$, being the depth of a node n, that is its distance from n_0 in terms of hops in the MLMC tree; and (iii) $y_n(n)$, being the current *fd cost* of the chain from n_0 to n, that is the cost of the chain obtained by adding to each edge in the chain an additional component α, usually initialized to 0 and updated by the Dual Ascent algorithm at every iteration to increasingly favor shorter paths in terms of hops.

Let us now have a look at Algorithm 4.1, which starts by computing the MLMC tree (line 3) with each edge cost increased by α. It is noteworthy that this implies that also $q_n(n)$ and $y_n(n)$ for every n in N are known after this step (line 4). If the MLMC tree contains a chain π between n_0 and the node currently occupied by the chain's master n_μ using at most n_{UAV} drones (line 5), then a valid solution has been found (line 6); otherwise, the MLMC tree may be refined to favor shorter solutions. This is achieved by computing ϵ, which is the amount by which α should be increased to ensure that next time the Djikstra's algorithm will be called, the resulting tree will yield at least one shorter chain (lines 7–12). It is important, though, that the choice of ϵ ensures that no solution can be missed. To this purpose, given any node n', let us consider all edges $(n, n') \in E$ which are not part of the current MLMC tree. One of such edges constitutes a strict improvement if added to the MLMC tree, if and only if the relation $q'_n > q_n + 1$ holds; we refer to the set of such edges as S. Hence, by computing the value $\epsilon_{n,n'}$ by which all edge costs should be increased for an edge n' in S to be included in the MLMC tree, and taking the minimum among all those computed, we obtain an ϵ with the desired characteristics. Following from such statements and without delving into the mathematical details, the value $\epsilon_{n,n'}$ can be obtained by applying the following equation:

$$\epsilon_{n,n'} = \frac{(y_n + c'_{comm}(n, n')) - y_{n'}}{q_{n'} - (q_n + 1)} \quad \forall (n, n') \in S, \tag{4.2}$$

where $c_{comm}(n, n')$ is the communication cost between two nodes, and $c'_{comm}(n, n'$ is the current modified cost, such that:

$$c'_{comm}(n, n') = c_{comm}(n, n') + \alpha. \tag{4.3}$$

Finally, if S is empty, then no chain can possibly be shortened and Algorithm 4.1 fails (lines 8–9). It is noteworthy that while the algorithm may seem computationally complex, it is actually possible to optimize the process, for example by repairing the MLMC tree instead of computing a new one from scratch every time, as described in [10].

4.3.3 Reformulation of the Chain Cost Function

We now want to modify the cost function defined in Equation 4.1 to favor nodes that have been visited fewer times, by introducing a term for the *coverage cost* between two nodes $c_{cov}(n, n')$, which will behave similarly to the node count algorithm. To that aim, let us highlight the differences between *node count* and *edge count*: on the one hand, the former counts the number of times a node has been occupied at each iteration, and an UAV would move from the current node to the neighboring connected node that has been previously visited less; on the other hand, the latter counts the number of times any edge has been traversed and strives to make sure that all edges of a given node are traversed with the same relative frequency. While node count gives the best result in the multi-agent case, edge count can be more naturally integrated with the Djikstra's algorithm and the definition of $c_{comm}(x, x')$ given in Section 4.1, as it assigns costs to edges, not nodes. Hence, we can slightly modify the common structure of node count by working in terms of edges like in edge count and assigning to all inbound edges to a node the same count value, that is the number of times the destination node has been visited. More formally, from the standard Node Count procedure in Algorithm 4.2, we define an operator $count(n_i)$ that measures the amount of visits to a node. At each instant, the agent will select the less visited node from those in the *neighborhood* of its current location n_a, that is, the set of all nodes n_i for which there exists an edge (n_a, n_i). We expand this basic definition stating that there exists a similar operator for edge $count(n_a, n_i)$, such that $count(n_a, n_i) = count(n_i)$. Hence, the agent will select the edge leading to the less visited node in its neighborhood. From this definition, it follows that edges must be directed, meaning that the edge $(n, n') \neq (n', n)$, and consequently the associated costs may be different as well. This may not be evident while dealing with communication costs only, as communicating from one point to another is just as difficult as the opposite, and hence the costs is the same, but it does matter to the purpose of computing the coverage cost.

Algorithm 4.2 Node Count algorithm

Require: The connected graph G, and an agent's current location node n_a
1: **loop**
2:      $n_a \leftarrow$ *one of*$(argmin_{n_i \in neighboroood(G, n_a)} count(n_i))$
3:      $count(n_i) = count(n_i) + 1$
4: **end loop**

We can now provide a definition for the term $c_{cov}(n, n')$ representing the coverage cost between any two nodes n and n'. To this purpose, multiple candidates are equally legitimate as long as the term is directly proportional to the count associated with the edge (n, n'). Following the discussion above, the simplest candidate is then given by:

$$c_{cov}(n, n') = \beta count(n, n'), \tag{4.4}$$

where β is a scaling factor whose value is determined based on the order of magnitude of the communication cost and how much the designer wants to favor coverage performance over communication stability, and vice-versa. It follows that the total cost of an edge becomes:

$$c_{tot}(n, n') = c_{comm}(n, n') + c_{cov}(n, n'), \tag{4.5}$$

while the total cost of a valid chain becomes:

$$C_{tot}([n_0, \ldots, n_t]) = \sum_{i=0}^{t-1} c_{tot}(n_i, n_{i+1}). \tag{4.6}$$

In this regard, the reader may notice that, in long-term operation contexts, the coverage component may grow unbounded, leading to chains which increasingly favor coverage over communication. It is possible to address this issue in several ways, for example by normalizing the count value for every edge across the graph, and/or computing $c_{cov}(n, n')$ over a sliding time window.

In conclusion, by employing the cost in Equation 4.6 to compute a modified MLMC tree and by substituting c_{tot} to c_{comm} in Equations 4.2 and 4.3, we obtain a solution to the relay positioning problem which also takes into account the necessity to fly over under-visited nodes. No changes to the Algorithm 4.2 should be made to ensure the correct operation of the system.

4.4 System Architecture

Following the methods presented in the previous Section, we will now proceed to illustrate a sample architecture that can implement the ideal system outlined in Section 4.1. Figure 4.1 depicts the architecture structure and its main components. The main idea of the architecture is to act in *interleaved rounds*, which means that initially the master UAV acts, or at least takes a decision on its next action, then the swarm reacts, by computing a valid hybrid relay-coverage chain and taking the corresponding configuration. Once all the UAVs have reached their designed destinations, the process starts over. It is noteworthy that the master UAV does not necessarily have to follow a coverage algorithm as long as the nodes it visits are recorded together with the ones visited by the rest of the swarm. Obviously enough, letting the master UAV acts on pure *Node Count* and share the map with the swarm's simultaneous relay chain and coverage algorithm is the most natural solution, and the one we explore here, but also manual control by an operator is a viable alternative.

4.4.1 Communication Cost and Map Discretization

We define $c_{comm}(n,n')$ as a positive monotonic function directly proportional to the shortest distance between the nodes n and n', and ranging from 0 in the ideal case where the two nodes coincide to $c_{comm_{max}}$ when the distance between the two nodes is greater than a given threshold distance $d_{comm_{max}}$. Similarly, we define $f_{comm}(n,n') = 0$ if the distance between the two nodes is greater than $d_{comm_{max}}$, and 1 otherwise. Then, we

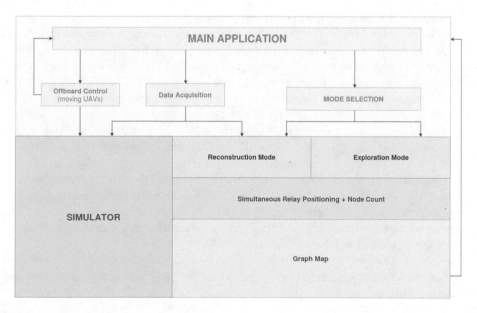

FIGURE 4.1

The proposed architecture. The main application acquires the UAVs' positions and checks whether a new target has been found. If this is the case, it changes the master UAV to the one closest to the target. Then, after determining the new master position, computes the hybrid relay chain and dispatches the target positions to all available UAVs.

arbitrarily discretize the map in a finite number of nodes and define edges between all pairs of nodes for which $f_{comm}(n, n') = 1$ holds true. In this way, we ensure that when we generate a chain, communication is always possible between any two nodes connected by an edge while also providing a high number of connections, hence obtaining a valid discretized map rich of alternative paths where to compute MLMC trees. Finally, edges traversing large obstacles are pruned from the tree, either systematically or as a direct consequence of $c_{comm}(n, n')$, if the communication cost is already taking obstacles into account.

4.4.2 Robustness to Secondary Tasks and Missing UAVs

As described in Section 4.1, one or more UAVs may be busy with a secondary task, which requires them to fly over a given node for a prolonged amount of time. A robust solution to this issue as long as at most one UAV is involved in other tasks at each time instant is to promote such UAV as the master, and let the rest of the swarm rearrange according to the simultaneous coverage and relay positioning algorithm. Since the master UAV is not required to have any specific capability, i.e., the distinction from the rest of the swarm is only logical, the master UAV can be changed on the fly without consequences. To reinforce this point, it is noteworthy that what we get from the algorithm is just a chain of nodes without any reference to a specific UAV. Clearly, while the new master UAV is busy in the same node over several rounds, the rest of the swarm will still be providing assistance with the coverage task and update the node counts accordingly. Once the secondary task is complete, the master UAV privileges can be returned to the original UAV or the execution can proceed with the new one. If more than one agent is expected to be busy with a secondary task at any time, then a more complex solution is necessary, for example, implementing a queue and/or modifying the tree generation algorithm to enforce solutions including the nodes corresponding to the desired task positions.

Another issue raised in Section 4.1 is how to provide robustness to the absence of any single UAV, either when this is planned, for example, in the case of UAVs requiring a recharge, or an unexpected event, for example, when an UAV is gone missing. In this case, a distinction should be made between the master UAV and rest of the swarm. If the master UAV leaves the swarm, either suddenly or in a regulated manner, there would be only minor consequences, as the chain would not be damaged and the only action required would be selecting a new master UAV and reducing the number of available UAVs for the simultaneous coverage and relay positioning algorithm (unless a substitute is readily available). On the contrary, the same does not apply for members of the relay chain. If the UAV is scheduled to leave the chain, preliminary measures can be taken before removing it and reducing the number of UAVs available for the algorithm. Otherwise, the only way to ensure that a chain would remain valid even if one or more UAVs were missing is to use an overly restrictive communication cost policy in order to ensure (or at least promote) two nonconsecutive UAVs to still be able to communicate.

Any missing UAVs returning to the swarm can be easily reintegrated by increasing again the number of agents available to the algorithm. UAVs that have gotten detached from the swarm do not have access to the updated number of visits for each node, but assuming they have a reasonable amount of computing power onboard, they can still fall back to pure node count. As they approach less visited areas of the map, they have a chance to again join the swarm and synchronize their respective knowledge about visited nodes. If this is not possible or the UAV cannot regroup with the swarm, it would still be able to fly back to the base station.

4.5 Experimental Results

In this Section, we present the experiments used to assess the performance of the simultaneous coverage and relay-positioning algorithm and the related architecture, as presented in Sections 4.3 and 4.4, respectively. Experiments have been run in simulation, as we are not interested in the dynamics of the UAVs in this context. However, we run UAV software in the loop to make sure the proposed algorithm is compatible with real-world UAVs. The UAVs we simulate are 3DR IRIS [1], while the simulation environment is Gazebo [4], running on a machine with Intel I7 4500U@3.00 GHz processor and 8 GB of RAM. The proposed architecture has been implemented in the Robot Operating System (ROS) [5] as a single module acting as a base station and sending commands to multiple instances of the UAV software, which run in parallel to simulate both the master UAV and the fellow UAVs available for the hybrid relay-coverage chain. Communication between the base station and the simulated UAVs is implemented using MAVLink [3] through the MAVROS bridge [2]. A total of five UAVs have been simulated in a squared environment measuring 70×70 m, discretized in 196 nodes, each 5 m apart from the others. Several tests were run with these settings with a number of randomly positioned obstacles and targets. When a UAV enters a node containing a target, a secondary task as the one described in Section 4.4.2 begins. In this particular case, the UAV flies around the target and gathers point cloud data until a complete 3D model of the target is acquired. A sample simulation scenario is reported in Figure 4.2.

All the tests have been conducted using a formulation for $c_{comm}(n, n')$ which increases with the distance between n and n', plus a penalty for the surrounding volume occupied by obstacles to model communication difficulties in cluttered environments. For $c_{cov}(n, n')$ we employed the simple definition given in Equation 4.4 and repeated the test several times for different values of the parameter β. Two kinds of tests were run, one where all nodes are to be visited at least once and one where they must be visited at least 10 times. In all tests the following metrics have been measured: (i) the number of iterations needed to achieve the results; (ii) the number of times each node has been visited; (iii) minimum, maximum, and average communication costs among all generated chains.

Let us focus on the impact of the parameter β. Figures 4.3 through 4.5 represent the same environment with the same obstacle locations but different values for β. Nodes with

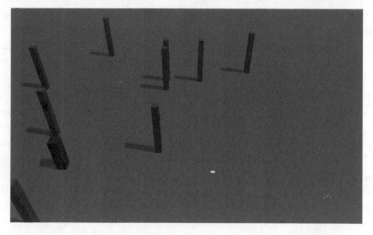

FIGURE 4.2
An example of the performed simulations. Brown boxes corresponds to obstacles, while the blue one is a target.

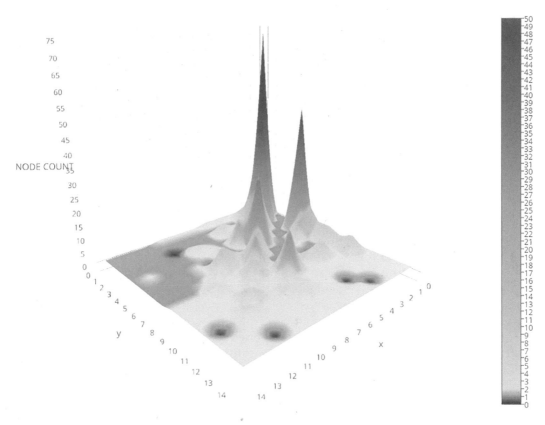

FIGURE 4.3
Number of visits for each node after they have all been visited at least once for β = 0.

a visit value of 0 are occupied either by obstacles or because of obstacles themselves. We invite particular attention to the case in Figures 4.3 where $\beta = 0$, that is, no coverage cost is taken into account and the problem reduces to the naïve case. In such a scenario, we observe large peaks of visits, especially in the nodes close to the base station and where the obstacles form corridors. This is expected, as it is always necessary to keep a valid relay chain, and these locations are fundamental to that purpose. For example, considering the case where the UAVs are tasked to visit all nodes at least once, the highest peaks occur exactly in those nodes and reach 77 number of visits. It took exactly 206 iterations to reach the desired goal, and the average number of visits per node is 3.68. As we increase the value of 1 to 0.5 (Figures 4.4 and 4.5), we notice that the peaky areas get smaller and so do the peak heights. In particular, the maximum visit value is lowered to 11 and 4, respectively, while the number of iterations is reduced to 159 and 74. Furthermore, the average values are lowered to 3.14 and 1.57. Figures 4.6 through 4.8 report the results for the case in which each node must be visited at least 10 times, again with the values of β set to 0, 0.5, and 1, respectively. In this case, we observe even larger peaks in these areas, but the same downward trend as β increases, since the maximum number of visits goes from 673 to 47 and 36. Back to the single-visit case, let us consider the impact of β at every iteration on the communication costs reported in Figures 4.9 through 4.11. The mean and maximum costs are equal to 28.22 and 43.00 for $\beta = 0$, 30.22 and 43.00 for $\beta = 0.5$, and 33.97 and 43.28 for $\beta = 1$. As expected, the average cost slowly rises when increasing β, but the

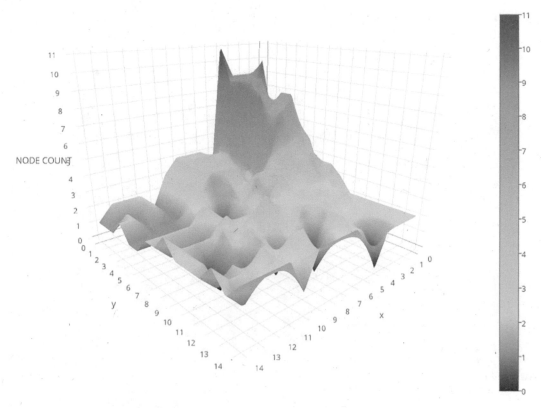

FIGURE 4.4
Number of visits for each node after they have all been visited at least once for $\beta = 0.5$.

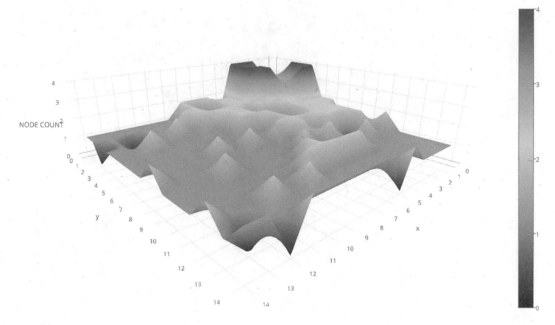

FIGURE 4.5
Number of visits for each node after they have all been visited at least once for $\beta = 1$.

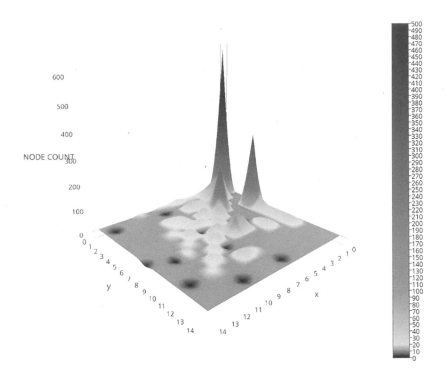

FIGURE 4.6
Number of visits for each node after they have all been visited at least 10 times for $\beta = 0$.

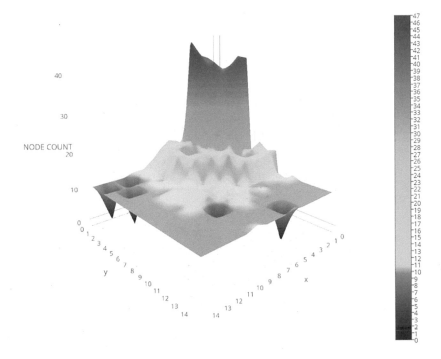

FIGURE 4.7
Number of visits for each node after they have all been visited at least 10 times for $\beta = 0.5$.

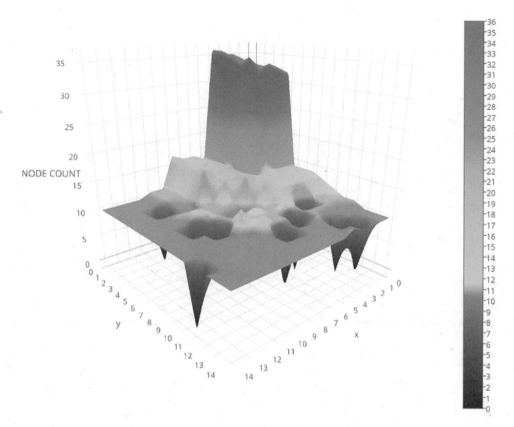

FIGURE 4.8
Number of visits for each node after they have all been visited at least 10 times for $\beta = 1$.

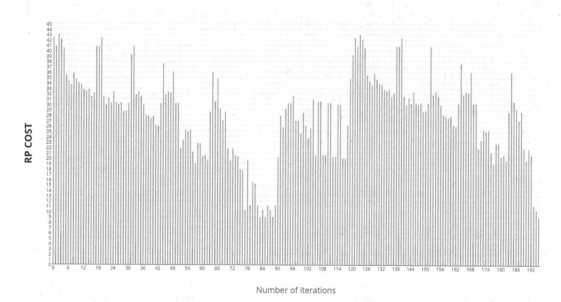

FIGURE 4.9
Total cost of the chain at each iteration until every node has been visited at least once for $\beta = 0$.

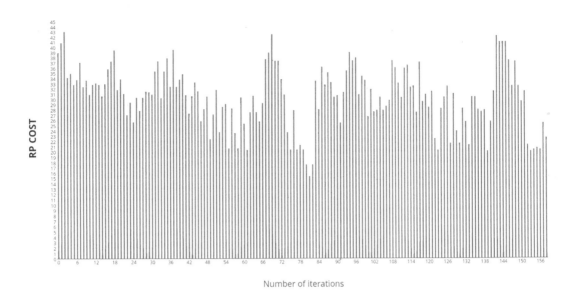

FIGURE 4.10
Total cost of the chain at each iteration until every node has been visited at least once for $\beta = 0.5$.

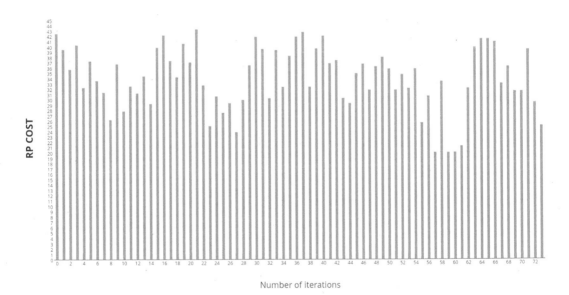

FIGURE 4.11
Total cost of the chain at each iteration until every node has been visited at least once for $\beta = 1$.

worst-case maximum communication cost remains steady. Comparing the results for $\beta = 0$ with $\beta = 1$ in the single-visit case, we observe 64% shorter time to reach the goal in terms of iterations, but only 17% worse average communication cost. The worst-case maximum communication cost remains constant within a reasonable margin of error. Similar results have been achieved on long-term operation with a higher target visits value.

4.6 Conclusions

In this Chapter we propose a novel approach to improve the performance of a UAVs swarm tasked to cover a given area in difficult contexts such as natural disasters. This is achieved by providing a relay chain to the base station so that UAVs do not need to rely on preexisting communication infrastructures. A number of supporting UAVs are employed to build the relay chain, but the proposed algorithm also allows them to provide assistance in the coverage task, decreasing the time required to completely cover the area without affecting the communication reliability too much. This is because the proposed algorithm does not seem to significantly affect the worst-case communication cost of the chain, as highlighted by the reported experimental results. Moreover, the system allows UAVs to be momentarily busy with secondary tasks, such as target data acquisition, without necessarily leaving the swarm. On the robustness side, the architecture is robust to a missing master UAV despite the fact that it presents a master-slave paradigm and can also be extended to be more resilient to interruptions in the relay chain. Finally, the architecture allows for an easy and dynamic modification in the number of UAVs in the chain. This is useful both in the case of a scheduled task, as an UAV needing to recharge batteries, as well as in the case of one UAV getting separated from the swarm and needing to regroup.

Future developments will deal with further increasing the robustness of the proposed architecture by exploring in detail the best strategies to adopt when multiple UAVs are busy with secondary tasks or to avoid the chain to be interrupted. On the performance side, improved formulations of the coverage cost will be formalized. Testing in real-world situations is also in the works.

References

1. 3DR Iris+. https://3dr.com/support/iris/. Last accessed on September 13, 2018.
2. MAVLINK for ROS (MAVROS). http://erlerobotics.com/blog/ros-mavros/. Last accessed on September 13, 2018.
3. Micro Air Vehicle Communication Protocol (MAVLINK). http://qgroundcontrol.org/mavlink/start. Last accessed on September 13, 2018.
4. The Gazebo simulator. http://gazebosim.org/. Last accessed on September 13, 2018.
5. The Robot Operating System (ROS). http://www.ros.org/. Last accessed on September 13, 2018.
6. M. Alinaghian and A. Goli. Two new heuristic algorithms for covering tour problem. *Journal of Industrial and Systems Engineering*, 8(3):24–41, 2015.
7. M. Baglietto, G. Cannata, F. Capezio, A. Grosso, A. Sgorbissa, and R. Zaccaria. Patrolgraph: A distributed algorithm for multi-robot patrolling. In *IAS10 - The 10th International Conference on Intelligent Autonomous Systems*, Baden Baden, Germany, pages 415–424, 2008.
8. A. Balakrishnan and K. Altinkemer. Using a hop-constrained model to generate alternative communication network design. *ORSA Journal on Computing*, 4(2):192–205, 1992.
9. T. Balch and R. Arkin. Avoiding the past: A simple but effective strategy for reactive navigation. In *ICRA (1)*, pages 678–685, 1993.
10. O. Burdakov, P. Doherty, K. Holmberg, J. Kvarnström, and P-M. Olsson. Relay positioning for unmanned aerial vehicle surveillance. *The International Journal of Robotics Research*, 29(8):1069–1087, 2010.

11. K. Cesare, R. Skeele, S-H. Yoo, Y. Zhang, and G. Hollinger. Multi-UAV exploration with limited communication and battery. In *Robotics and Automation (ICRA), 2015 IEEE International Conference on, pages 2230–2235*. IEEE, 2015.

12. J. Chen and D. Gesbert. Optimal positioning of flying relays for wireless networks: A LOS map approach. In *Communications (ICC), 2017 IEEE International Conference on*, pages 1–6. IEEE, 2017.

13. H. Choset. Coverage of known spaces: The boustrophedon cellular decomposition. *Autonomous Robots*, 9(3):247–253, 2000.

14. H. Choset. Coverage for robotics–a survey of recent results. *Annals of Mathematics and Artificial Intelligence*, 31(1–4):113–126, 2001.

15. L.D. Couto, P.W.V. Tran-Jørgensen, and G.T.C. Edwards. Model-based development of a multi-algorithm harvest planning system. In *International Conference on Simulation and Modeling Methodologies, Technologies and Applications*, pages 19–33. Springer, 2016.

16. G. Dahl and L. Gouveia. On the directed hop-constrained shortest path problem. *Operations Research Letters*, 32(1):15–22, 2004.

17. T.M.C.E.P.R. Ferreira. Terrain coverage with UAVs: Real-time search and geometric approaches applied to an abstract model of random events. In *2016 XIII Latin American Robotics Symposium and IV Brazilian Robotics Symposium (LARS/SBR)*, pages 151–156, Oct 2016.

18. E. Galceran and M. Carreras. A survey on coverage path planning for robotics. *Robotics and Autonomous Systems*, 61(12):1258–1276, 2013.

19. F. Jiang and A.L. Swindlehurst. Dynamic UAV relay positioning for the ground-to-air uplink. In *GLOBECOM Workshops (GC Wkshps), 2010 IEEE*, pages 1766–1770. IEEE, 2010.

20. P.J. Jones. Cooperative area surveillance strategies using multiple unmanned systems. PhD thesis, Georgia Institute of Technology, 2009.

21. A. Khan, I. Noreen, and Z. Habib. On complete coverage path planning algorithms for non-holonomic mobile robots: Survey and challenges. *Journal of Information Science and Engineering*, 33(1):101–121, 2017.

22. S. Koenig and R.G. Simmons. Easy and hard testbeds for real-time search algorithms. In *AAAI/IAAI*, Vol. 1, pages 279–285. Citeseer, 1996.

23. R.E. Korf. Real-time heuristic search. *Artificial Intelligence*, 42(2–3):189–211, 1990.

24. P. Mukhija, K.M. Krishna, and V. Krishna. A two phase recursive tree propagation based multi-robotic exploration framework with fixed base station constraint. In *Intelligent Robots and Systems (IROS), 2010 IEEE/RSJ International Conference*, pages 4806–4811. IEEE, 2010.

25. C. Nattero, C.T. Recchiuto, A. Sgorbissa, and F. Wanderlingh. Coverage algorithms for search and rescue with UAV drones. In *Workshop of the XIII AIIA Symposium on Artificial Intelligence*, Pisa, 2014.

26. J-D. Nicoud and M.K. Habib. The Pemex-B autonomous demining robot: perception and navigation strategies. In *Intelligent Robots and Systems 95.'Human Robot Interaction and Cooperative Robots, Proceedings. 1995 IEEE/RSJ International Conference*, volume 1, pages 419–424. IEEE, 1995.

27. P-M. Olsson. Positioning algorithms for surveillance using unmanned aerial vehicles. PhD thesis, Linköping University Electronic Press, 2011.

28. N. M. Rooker and A. Birk. Multi-robot exploration under the constraints of wireless networking. *Control Engineering Practice*, 15(4):435–445, 2007.

29. C. Shen and D. Pesch. A heuristic relay positioning algorithm for heterogeneous wireless networks. In *Vehicular Technology Conference, 2009. VTC Spring 2009. IEEE 69th*, pages 1–5. IEEE, 2009.

30. J. Valente, A.B. Cruz, J. del Cerro Giner, and D.S. Muñoz. A waypoint-based mission planner for a farmland coverage with an aerial robot - a precision farming tool. In *Proceedings of Precision Agriculture 2011*, pages 427–436, 2011.

31. A.S. Vempati, M. Kamel, N. Stilinovic, Q. Zhang, D. Reusser, I. Sa, J. Nieto, R. Siegwart, and P. Beardsley. Paintcopter: An autonomous UAV for spray painting on three-dimensional surfaces. *IEEE Robotics and Automation Letters*, 3(4):2862–2869, 2018.

32. M. Younis and K. Akkaya. Strategies and techniques for node placement in wireless sensor networks: A survey. *Ad Hoc Networks*, 6(4):621–655, 2008.

5

Euclidean Distance Matrix Optimization for Sensor Network Localization*

Jörg Fliege, Hou-Duo Qi, and Naihua Xiu

CONTENTS

* The paper is supported by the Royal Society's International Exchange Scheme IE150919 and the National Natural Science Foundation of China (11728101).

5.1 Introduction

The sensor network localization (SNL), rather than being treated as an individual localization problem, will be employed as a general framework to model various types of localization problems under often partially given distance information. The global positioning system (GPS) is one such problem. The essential information that GPS uses is the distances between the receiver and a few satellites, where distances are measured in terms of the amount of time it takes to receive a transmitted signal. The GPS then calculates the location of the user by the trilateration method based on the obtained distances. Another example where GPS does not work is the indoor facility localization [30]. Another significantly different localization problem is geometric graph embedding [27]: a typical example is to place all nodes of a graph on the surface of the smallest sphere so that the distances of all neighboring nodes are of unit distance. We refer to [26] for a description of many such problems and note that the SNL framework covers many instances of that. In the following, we describe the SNL framework, its four widely used optimization formulations, and the corresponding computational challenges.

5.1.1 Sensor Network Localization

Suppose there are m anchors, denoted by column vectors $x_i = \mathbf{a}_i \in \mathbb{R}^r, i = 1, \ldots, m$, and $(n - m)$ unknown sensor location $x_i \in \mathbb{R}^r$, $i = m+1, \ldots, n$. We are given some (noisy) distances (denoted as δ_{ij}) between sensors and anchors and between sensors and sensors:

$$\begin{cases} \delta_{ij} \approx \|\mathbf{x}_j - \mathbf{a}_i\|, & \text{for some } j \in \{m+1, \ldots, n\} \\ \delta_{ij} \approx \|\mathbf{x}_i - \mathbf{x}_j\|, & \text{for some } i, j \in \{m+1, \ldots, n\}, \end{cases} \tag{5.1}$$

where $\|\cdot\|$ is the Euclidean norm. The purpose is to locate those unknown sensors. This is often done by optimizing certain loss functions. The most often used loss function is Kruskal's stress function [25], which is a direct application of the least-squares criterion to (5.1):

$$\min \quad f_1(X) := \sum_{i,j=1}^n w_{ij}(\|\mathbf{x}_i - \mathbf{x}_j\| - \delta_{ij})^2$$
$$\text{s.t.} \quad x_i = \mathbf{a}_i, i = 1, \ldots, m, \tag{5.2}$$

where ":=" means "define," $X := [\mathbf{x}_1, \ldots, \mathbf{x}_n]$, and the weights w_{ij} are given by

$$w_{ij} := \begin{cases} \text{positive constant} & \text{if } \delta_{ij} \text{ is given} \\ 0 & \text{otherwise.} \end{cases}$$

When both $\mathbf{x}_i$ and $\mathbf{x}_j$ are anchors, $\delta_{ij} := \|\mathbf{a}_i - \mathbf{a}_j\|$. We note that $f_1(X)$ is nonconvex and nondifferentiable. If the least-squares criterion is applied to the squared distances, we end up with a differentiable objective known as the squared stress function [8,10]:

$$\min \quad f_2(X) := \sum_{i,j=1}^n w_{ij}\left(\|\mathbf{x}_i - \mathbf{x}_j\|^2 - \delta_{ij}^2\right)^2$$
$$\text{s.t.} \quad x_i = \mathbf{a}_i, \quad i = 1, \ldots, m. \tag{5.3}$$

It has long been known that the absolute value loss function is more robust than the least-squares function. Therefore, one can consider the following robust variant of (5.2) and (5.3) [9,20], respectively

$$\min \quad f_3(X) := \sum_{i,j=1}^{n} w_{ij} \big| \|\mathbf{x}_i - \mathbf{x}_j\| - \delta_{ij} \big|$$
$$\text{s.t.} \quad \mathbf{x}_i = \mathbf{a}_i, \quad i = 1, \ldots, m, \tag{5.4}$$

and

$$\min \quad f_4(X) := \sum_{i,j=1}^{n} w_{ij} \big| \|\mathbf{x}_i - \mathbf{x}_j\|^2 - \delta_{ij}^2 \big|$$
$$\text{s.t.} \quad \mathbf{x}_i = \mathbf{a}_i, \quad i = 1, \ldots, m. \tag{5.5}$$

The question now boils down how to efficiently solve those optimization problems.

5.1.2 Computational Challenges

We will now explain why those optimization problems are computationally challenging, though some may be more difficult to solve than the others. There are essentially three intrinsic difficulties.

The first stems from the fact that embedding space $\mathbb{R}^r$ is usually of low dimension. If r is allowed to be no less than $(n-1)$, those problems can be reformulated as convex optimization and hence would become significantly easier to solve. We will see below that low-dimensionality of the embedding is equivalent to a rank constraint when the problem is reformulated as a semi-definite programming (SDP) or Euclidean distance matrix (EDM) optimization problem, and the rank constraint is notoriously known to be hard to tackle in optimization.

The second challenge is from possibly a large number of lower and upper bounds that some embedding points should obey. These correspond to additional constraints of the form

$$\ell_{ij} \le \|\mathbf{x}_i - \mathbf{x}_j\| \le u_{ij}, \tag{5.6}$$

where $0 \le \ell_{ij} \le u_{ij}$ are given lower and upper bounds. Such constraints are known as *interval distance constraints* (see [18]). Such constraints appear in molecular conformation [5] and graph embedding [27], and the number of them could be as high as $O(n^2)$. As an example, the number of constraints could be in millions if the number of atoms considered in a molecular conformation problem is just a couple of thousand. Adding those constraints to any of the four optimization problems would make them extremely difficult to solve.

The third challenge is the non-differentiability of some of the loss functions above. In particular, it is more difficult to characterize a stationary point and it is hard to analyze the convergence of corresponding algorithms. This last challenge is also coupled with possibly infinitely many local minima. This can be clearly seen when there are no anchors (i.e., $m = 0$). If $\{\mathbf{x}_i^*\}$ is a local solution, then any of the $\{Q\mathbf{x}_i^* + \mathbf{b}\}$ would also be a local solution, where Q is a $r \times r$ rotation matrix and $\mathbf{b} \in \mathbb{R}^r$ is any given (translation) vector.

5.1.3 Organization of This Paper

There exist many algorithms for sensor network localization, each developed for one of the particular four problems from above. The purpose of this paper is to introduce a recently developed EDM optimization approach that is suitable for *all* of the four problems and always follows a differentiable path regardless of the loss function used. The rest of this paper is organized as follows. We give a short review on relevant research in next section, including the coordinate minimization, the widely used SDP approach, and more recent EDM optimization. In Section 5.3, we provide some essential mathematical background for EDM optimization with particular attention on why EDM optimization has no easy solution, by focusing on the notion of the EDM cone. In Section 5.4 we review three main algorithms for EDM optimization: the method of alternating projections, Newton's method, and a penalty approach. The former two can be regarded as convex relaxation methods. In Section 5.5 we study the role of regularization and investigate two types of regularization. We show that each regularization can be easily incorporated into the penalty approach. Numerical results are reported in Section 5.6, and we conclude the paper in Section 5.7.

5.2 Literature Review

In our review, we focus on three groups of approaches/algorithms because these represent the overall spectrum of solution approaches well, and they facilitate a discussion on the difficulties faced and favorable properties enjoyed by each of the problems (5.2–5.5). The first group is the direct approach, which solves those problems directly without having to reformulate them to other forms. The second group is the convex relaxation approach via SDP. The last group is the focus of the chapter: EDM optimization.

5.2.1 The Direct Approach

By a direct approach, we mean any algorithm that solves one of the four problems without reformulating it into another form. That is, $\{\mathbf{x}_i\}$ are its main variables. Let us use the problem (5.2) to illustrate this approach. One efficient way to solve it is by a *majorization and minimization* (MM) scheme. Consider an optimization problem

$$\min \quad f(\mathbf{x}), \quad \text{s.t.} \quad \mathbf{x} \in \mathcal{B},$$

where $f(\cdot) : \mathbb{R}^p \mapsto \mathbb{R}$ is a loss function and $\mathcal{B} \subseteq \mathbb{R}^p$ represents constraints. More often than not, $f(\mathbf{x})$ is difficult to minimize. Suppose we have another function: $f_m(\cdot, \cdot) : \mathbb{R}^p \times \mathbb{R}^p \mapsto \mathbb{R}$ satisfying the following property:

$$f_m(\mathbf{x}, \mathbf{y}) \geq f(\mathbf{x}) \quad \text{and} \quad f_m(\mathbf{y}, \mathbf{y}) = f(\mathbf{y}), \quad \forall\, \mathbf{x}, \mathbf{y} \in \mathbb{R}^p. \tag{5.7}$$

In other words, for any fixed point $\mathbf{y} \in \mathbb{R}^p$, the graph of $f_m(\cdot, \mathbf{y})$ is always above that of $f(\cdot)$ and they meet at $\mathbf{x} = \mathbf{y}$. Such a function $f_m(\cdot, \cdot)$ is called a majorization of $f(\cdot)$. The main purpose of constructing a majorization is to minimize it instead of minimizing $f(\cdot)$.

Starting from a given point $\mathbf{x}^k \in \mathbb{R}^p$, compute

$$\mathbf{x}^{k+1} = \arg\min \quad f_m(\mathbf{x}, \mathbf{x}^k), \quad \text{s.t.} \quad \mathbf{x} \in \mathcal{B}. \tag{5.8}$$

We then have

$$f(\mathbf{x}^{k+1}) \overset{(5.7)}{\leq} f_m(\mathbf{x}^{k+1}, \mathbf{x}^k) \overset{(5.8)}{\leq} f_m(\mathbf{x}^k, \mathbf{x}^k) \overset{(5.7)}{=} f(\mathbf{x}^k).$$

Therefore, the MM scheme (5.7) and (5.8) generates a sequence $\{x^k\}$ with non-increasing function values. Consequently, $\{f(x^k)\}$ will converge provided that $f(x)$ is bounded from below on $\mathcal{B}$. The whole point of (5.8) is of course that the majorization function $f_m(\cdot, x^k)$ should be easier to minimize than f. As a general rule of thumb, f_m should be constructed in such a way that (5.8) has a closed form solution. This is exactly what can be achieved for the stress function (5.2). Indeed, from the Cauchy-Schwartz inequality,

$$\langle \mathbf{x}_i - \mathbf{x}_j, \mathbf{y}_i - \mathbf{y}_j \rangle \leq \|\mathbf{x}_i - \mathbf{x}_j\| \, \|\mathbf{y}_i - \mathbf{y}_j\|$$

it follows that

$$\phi(\mathbf{x}_i, \mathbf{x}_j) := -\|\mathbf{x}_i - \mathbf{x}_j\| \leq \phi_m(\mathbf{x}_i, \mathbf{x}_j, \mathbf{y}_i, \mathbf{y}_j) := \begin{cases} -\dfrac{\langle \mathbf{x}_i - \mathbf{x}_j, \mathbf{y}_i - \mathbf{y}_j \rangle}{\|\mathbf{y}_i - \mathbf{y}_j\|} & \text{if } \mathbf{y}_i \neq \mathbf{y}_j \\ 0 & \text{otherwise.} \end{cases}$$

Therefore,

$$f_1(X) = \sum_{i,j=1}^{n} w_{ij}(\|\mathbf{x}_i - \mathbf{x}_j\| - \delta_{ij})^2$$

$$\leq \sum_{i,j=1}^{n} \left(w_{ij}\|\mathbf{x}_i - \mathbf{x}_j\|^2 + 2w_{ij}\delta_{ij}\phi_m(\mathbf{x}_i, \mathbf{x}_j, \mathbf{y}_i, \mathbf{y}_j) + \delta_{ij}^2 \right)$$

$$=: f_1^m(X, Y).$$

We note that the majorization function $f_1^m(X, Y)$ is quadratic in X and is easy to minimize when $\mathcal{B} = \mathbb{R}^p$ (unconstrained). The resulting algorithm is the famous SMACOF (Scaling by MAjorizing a COmplicated Function) algorithm [12]. SMACOF is widely used in the field of multi-dimensional scaling (MDS) [8]. A major weakness of SMACOF is that (5.8) does not have a closed form anymore when the original problem is constrained (e.g., $\mathcal{B}$ contains the lower and upper bounds in Equation (5.6)), severely limiting its applications in many other practical problems. It is worth pointing out that the MM scheme is very popular in the direct approach. We will see that it also plays a key role in our EDM optimization approach.

5.2.2 SDP Convex Relaxation

As emphasized earlier, none of the four problems under consideration is convex. The main idea in convex relaxation is to construct a close convex approximation to those nonconvex problems in the hope that the convex optimization will yield a good approximate solution to the true solution sought. Let us use (5.3) to illustrate this approach. We define some notation first. Let $\mathcal{S}^n$ be the space of $n \times n$ symmetric matrices, endowed with the standard trace inner product $\langle \cdot, \cdot \rangle$ and the induced Frobenius norm $\|\cdot\|$. Let $\mathcal{S}_+^n$ be the cone of all positive semidefinite matrices in $\mathcal{S}^n$. Note that

$$\|\mathbf{x}_i - \mathbf{x}_j\|^2 = \|\mathbf{x}_i\|^2 + \|\mathbf{x}_j\|^2 - 2\langle \mathbf{x}_i, \mathbf{x}_j \rangle = Y_{ii} + Y_{jj} - 2Y_{ij},$$

where the Gram matrix Y is defined by

$$Y := X^T X = \left(\langle \mathbf{x}_i, \mathbf{x}_j \rangle \right)_{i,j=1}^n.$$

Therefore, the squared stress function can be represented as

$$f_2(X) = \sum_{i,j=1}^n w_{ij} \left(Y_{ii} + Y_{jj} - 2Y_{ij} - \delta_{ij}^2 \right),$$

which is a linear function of Y. The constraint on Y is that $Y \in \mathcal{S}_+^n$ and that the matrix has rank r (the embedding dimension). This approach was initially used by Biswas and Ye [6] to derive an SDP for SNL. By dropping the rank constraint, one obtains a convex SDP relaxation for the problem (5.3). The approach in [6] has been extensively investigated in several subsequent studies; see for example [7,23] and the references therein. A different line of development, still employing SDP, can be found in [21,22], where an efficient proximal point algorithm with a semismooth Newton step is developed. When there are many constraints in Equation (5.6), those SDP-based methods become computationally expensive. We would also like to point out that the "plain" distance $\| \mathbf{x}_i - \mathbf{x}_j \|$ does not have a straightforward SDP representation; see [29] for such an attempt. Therefore, this approach becomes more involved when it tries to approximate the problem (5.4). We also refer to the thesis [13] for further applications of SDP to Euclidean distance geometry problems.

5.2.3 The Origin of EDM Optimization

5.2.3.1 Classical Euclidean Geometry: From Distances to Embedding Points

We will now explain why EDM optimization is a natural reformulation for the SNL problems. Suppose we have n points $\mathbf{x}_i \in \mathbb{R}^r$, $i = 1, \ldots, n$. We can easily calculate the pairwise (squared) Euclidean distances

$$D_{ij} = d_{ij}^2 = \| \mathbf{x}_i - \mathbf{x}_j \|^2, \quad i, j = 1, \ldots, n. \tag{5.9}$$

Now suppose we only know the matrix D, can we generate a set of points $\{\mathbf{x}_i\}$ that satisfy (5.9)? There is a classical way to tackle this problem.

Consider the *double centering matrix*

$$B := -\frac{1}{2} JDJ \quad \text{with } J := I - \frac{1}{n} \mathbf{1}_n \mathbf{1}_n^T, \tag{5.10}$$

where I is the identity matrix in $\mathcal{S}^n$ and $\mathbf{1}_n$ is the column vector of all ones in $\mathbb{R}^n$. The matrix J is usually called the *centering matrix*. It is known [35,39,37] that B is positive semidefinite with rank $(B) = r$. Hence, it has the following eigenvalue-eigenvector decomposition:

$$B = [\mathbf{p}_1, \ldots, \mathbf{p}_r] \begin{bmatrix} \lambda_1 & & \\ & \ddots & \\ & & \lambda_r \end{bmatrix} \begin{bmatrix} \mathbf{p}_1^T \\ \vdots \\ \mathbf{p}_r^T \end{bmatrix}, \tag{5.11}$$

where $\lambda_1 \geq \cdots \geq \lambda_r > 0$ are the positive eigenvalues of B and $\mathbf{p}_i, i = 1, \ldots, r$ are the corresponding orthonormal eigenvectors. The embedding points are then given by

$$X := [\mathbf{x}_1, \ldots, \mathbf{x}_n] = \left[\sqrt{\lambda_1} \mathbf{p}_1, \ldots, \sqrt{\lambda_r} \mathbf{p}_r \right]^T, \tag{5.12}$$

which must satisfy (5.9).

If we already know that the first m points are anchors, as in the SNL problem, we can always find a mapping such that

$$\mathbf{a}_i = Q\mathbf{x}_i + \mathbf{b}, \quad i = 1, \ldots, m \tag{5.13}$$

where Q is a rotation matrix and $\mathbf{b}$ is a shifting vector [2,34]. We then map the rest of the points $\mathbf{x}_i, i = m+1, \ldots, n$ to the coordinates system defined by the anchors. This part is known as the *Procrustes analysis*, mapping one set of points to another as accurately as possible.

5.2.3.2 Classical Multi-Dimensional Scaling (cMDS)

The computational procedure from (5.10) to (5.12) is fundamental in generating embedding points that preserve the given Euclidean distances. This is based on the assumption that D is a true EDM and has no missing values. However, in practice, the given distances are often noisy and many may be missing. Let Δ represent such given distances. In particular, if the value of δ_{ij} is missing, we simply set $\delta_{ij} = 0$. We can modify the computational procedure in Equations (5.10)–(5.12) to generate a set of embedding points $\mathbf{y}_i, i = 1, \ldots, n$ such that

$$\|\mathbf{y}_i - \mathbf{y}_j\| \approx \delta_{ij} \quad \text{if } \delta_{ij} > 0. \tag{5.14}$$

This is done as follows. Let $\overline{\Delta} := \left(\delta_{ij}^2 \right)$ (the squared dissimilarity matrix). Compute

$$\overline{B} := -\frac{1}{2} J \overline{\Delta} J \tag{5.15}$$

and decompose it as

$$\overline{B} = [\overline{\mathbf{p}}_1, \ldots, \overline{\mathbf{p}}_n] \begin{bmatrix} \overline{\lambda}_1 & & \\ & \ddots & \\ & & \overline{\lambda}_n \end{bmatrix} \begin{bmatrix} \overline{\mathbf{p}}_1 \\ \vdots \\ \overline{\mathbf{p}}_n \end{bmatrix}, \tag{5.16}$$

where $\overline{\lambda}_1 \geq \cdots \geq \overline{\lambda}_n$ are the real eigenvalues of $\overline{B}$ (because the matrix is symmetric). Suppose that $\overline{B}$ has s positive eigenvalues. The embedding points $\{\mathbf{y}_i\}$ can be generated by

$$[\mathbf{y}_1, \ldots, \mathbf{y}_n] = \left[\sqrt{\overline{\lambda}_1} \overline{\mathbf{p}}_1, \ldots, \sqrt{\overline{\lambda}_s} \overline{\mathbf{p}}_s \right]^T. \tag{5.17}$$

The corresponding Euclidean distance matrix is

$$D^{\text{mds}} := \left(\|\mathbf{y}_i - \mathbf{y}_j\|^2 \right)_{i,j=1}^n.$$

The computational procedure from (5.15) to (5.17) is known as *classical multidimensional scaling* (cMDS), originating from the work of Schoenberg [35] and Young and Householder [39], popularized by Torgerson [37] and Gower [19].

A natural question is how good cMDS is. This has been partially answered by Sibson [36] for the case $\delta_{ij} = d_{ij} + \epsilon_{ij}$, with ϵ_{ij} being a small perturbation of the true Euclidean distances d_{ij}. We will not review this analysis and its subsequent development. Instead, we cast cMDS as an optimization problem, which will reveal its sub-optimality in an elegant way. To this purpose, we formally state the definition of the (squared) Euclidean distance matrix.

Definition 5.2.1

A matrix $D \in \mathcal{S}^n$ is called a (squared) Euclidean distance matrix (EDM) if there exist a set of points $\mathbf{x}_i \in \mathbb{R}^p$, $i = 1, \ldots, n$ with p being any positive integer, such that

$$D_{ij} = \left\| \mathbf{x}_i - \mathbf{x}_j \right\|^2, \quad i, j = 1, \ldots, n.$$

The smallest such p is called the embedding dimension of D, denoted as r. It is known that $r = \text{rank}(JDJ)$. We let $\mathcal{D}^n$ denote the set of all $n \times n$ EDM D.

We also define $\|A\|_J := \|JAJ\|$ for $A \in \mathcal{S}^n$. Then $\|A\|_J$ is a semi-norm, but not a true norm as $\|A\|_J = 0$ does not imply $A = 0$. What makes this semi-norm interesting is that D^{mds} is the optimal solution of the following problem:

$$D^{\text{mds}} = \arg\min \left\| D - \overline{\Delta} \right\|_J, \quad \text{s.t.} \quad D \in \mathcal{D}^n.$$

That is, cMDS computes D^{mds} that is nearest to $\overline{\Delta}$ under the semi-norm $\|\cdot\|_J$; see [28] for further details. A more natural measurement of nearness uses the Frobenius norm and thus gives rise to the following optimization problem:

$$D^{\text{edm}} = \arg\min \left\| D - \overline{\Delta} \right\| \quad \text{s.t.} \quad D \in \mathcal{D}^n. \tag{5.18}$$

This is the basic model of EDM optimization. It purely focusses on D as its main variable and leaves aligning the generated embedding points to existing anchors to the Procrustes analysis.

5.2.3.3 Early Research on EDM Optimization

The earliest research that seriously attempted to solve (5.18) includes the work of Gaffke and Mathar [15] and Glunt et al. [16]. Both papers treated D as the main variable. [17] has also successfully applied the method of alternating projections described in [16] to molecular conformation. The methods developed in [15] and [16] are first-order methods. A second-order method (i.e., a variant of Newton's method) enjoying a quadratic convergence rate was later developed by Qi [31]. Alfakih et al. [1] took a different route that relies on SDP solvers. They map $\mathcal{D}^n$ to $\mathcal{S}_+^{n-1}$ through a linear transformation. This approach has been further investigated by Wolkowicz in a series of papers, see [24,14] and the references therein.

Compared to the large number of research papers on convex relaxations on SNL, EDM optimization has not been well explored, mainly because the prevalent belief seems to be that SDP is the best vehicle to solve such problems—a notion warranted by the fact that off-the-shelf SDP solvers are well developed and readily available. However, it is widely known that SDP encounters a serious computational bottleneck when the size of the problem grows beyond a thousand rows and columns of the unknown matrix, unless the problem has some special structure. Moreover, this approach only works for some particular loss functions. We will see in our numerical tests that EDM optimization approaches can solve general SNL problems of size $n = 3000$ in under 3 minutes on a standard laptop. Moreover, EDM optimization has the capability of tackling problems with any of the given four loss functions and more complicated constraints. We start by reformulating the four problems given in our introduction as EDM optimization below.

5.2.4 SNL as EDM Optimization

Let us recall the data that we are going to use. We are given a dissimilarity matrix $\Delta = (\delta_{ij})$, which contains the given distance measurements among anchors and sensors, and the corresponding weight matrix $W = (w_{ij})$, which indicates the quality of the given δ_{ij}. We require $w_{ij} > 0$ when $\delta_{ij} > 0$ and $w_{ij} = 0$ when $\delta_{ij} = 0$ (i.e., when data is missing or not given).

Our main variable is $D \in \mathcal{D}^n$. If there are m anchors $\mathbf{x}_i = \mathbf{a}_i$, $i = 1, \ldots, m$, we may enforce

$$D_{ij} = \|\mathbf{a}_i - \mathbf{a}_j\|^2, \quad i, j = 1, \ldots, m \tag{5.19}$$

by using this equation as a constraint. This condition may be fed into the lower and upper bound constraints in Equation 5.6 by setting

$$\ell_{ij} = u_{ij} = \|\mathbf{a}_i - \mathbf{a}_j\|, \quad i, j = 1, \ldots, m.$$

In general, we define

$$\mathcal{B} := \{D \in \mathcal{S}^n \mid L_{ij} \le D_{ij} \le U_{ij}, \quad i, j = 1, \ldots, n\},$$

where L_{ij} and U_{ij} are given lower and upper bounds for D_{ij}. For example, we have $L_{ii} = U_{ii} = 0$ for all $i = 1, \ldots, n$ because $D \in \mathcal{D}^n$ must have zero diagonal. Furthermore, in the case of having m anchors, we may set $L_{ij} = U_{ij} = \|\mathbf{a}_i - \mathbf{a}_j\|^2$ for $i, j = 1, \ldots, m$. Then any embedding points of D satisfying (5.19) can be mapped to the anchors $\mathbf{a}_i$ through the linear mapping (5.13). Therefore, the anchor information is completely encoded by the box constraint $\mathcal{B}$. In this way, the problem (5.2) can be reformulated as EDM optimization:

$$\min \quad f^{(d,2)}(D) := \frac{1}{2} \left\| \sqrt{W} \circ \left(\sqrt{D} - \Delta \right) \right\|^2 \tag{5.20}$$
$$\text{s.t.} \quad D \in \mathcal{D}^n, \quad D \in \mathcal{B}, \quad \text{and} \quad \text{rank}(JDJ) \le r,$$

where $\sqrt{W}$ is the elementwise square root (hence $\sqrt{D_{ij}} = d_{ij}$), and $\circ$ is the elementwise multiplication known as the Hadamard product. It is worth pointing out that the objective function is convex in D because

$$f^{(d,2)}(D) = \frac{1}{2} \langle W, D \rangle - \langle W \circ \Delta, \sqrt{D} \rangle + \frac{1}{2} \langle W, \overline{\Delta} \rangle$$

The first term in the above is linear in D, the second term is negative of a square root term and hence convex, while the third term is constant. Since the rank of J is $(n-1)$, we have rank$(JDJ) \leq n-1$. If the embedding dimension r is allowed to be no less than $(n-1)$, then the problem (5.20) is convex in D because $\mathcal{D}^n$ is a convex cone (the latter is discussed in detail later).

Similarly, the other three problems can also be phrased as EDM optimization problems. We list them here for easy reference:

$$\min \quad f^{(D,2)}(D) := \frac{1}{2}\left\|\sqrt{W} \circ (D-\Delta)\right\|^2$$
$$\text{s.t.} \quad D \in \mathcal{D}^n, \quad D \in \mathcal{B}, \quad \text{and} \quad \text{rank}(JDJ) \leq \tag{5.21}$$

$$\min \quad f^{(d,1)}(D) := \frac{1}{2}\left\|\sqrt{W} \circ \left(\sqrt{D}-\Delta\right)\right\|_1$$
$$\text{s.t.} \quad D \in \mathcal{D}^n, \quad D \in \mathcal{B}, \quad \text{and} \quad \text{rank}(JDJ) \leq r, \tag{5.22}$$

where $\|A\|_1 := \Sigma_{i,j=1}^n |A_{ij}|$ is the ℓ_1 norm in $\mathcal{S}^n$. The last EDM optimization problem corresponding to (5.5) is

$$\min \quad f^{(D,1)}(D) := \frac{1}{2}\left\|\sqrt{W} \circ (D-\Delta)\right\|_1$$
$$\text{s.t.} \quad D \in \mathcal{D}^n, \quad D \in \mathcal{B}, \quad \text{and} \quad \text{rank}(JDJ) \leq r. \tag{5.23}$$

In the above objective functions we used D to mean that the "squared" distance is used in the objective and d meaning the "plain" distance is used. For example, $f^{(D,2)}$ means that the squared distance is used with least squares, while $f^{(d,1)}$ indicates that the "plain" distance is used with the least absolute-values (i.e., ℓ_1 norm). The remaining task is to show how those EDM optimizations can be efficiently solved even when n is large. Obviously, any efficient algorithm would heavily depend on how efficiently we can handle the essential constraints in EDM optimization: the rank constraint and the cone constraint on $\mathcal{D}^n$. We explain below that they can be dealt with through an efficient computational manner.

5.3 EDM Optimization: Mathematical Background

Let us set up the minimal requirements for what we would consider to fall in the framework of EDM optimization. EDM optimization problems usually take the form

$$\min f(D), \quad \text{s.t.} D \in \mathcal{D}^n, \quad D \in \mathcal{B}, \quad \text{rank}(JDJ) \leq r, \tag{5.24}$$

where $f : \mathcal{S}^n \mapsto \mathbb{R}$. The three constraints here are essential in the sense that most practical problems would require their embedding points to satisfy them. Of course, other variables and constraints can be added to the basic model (5.24). For example, the spherical constraints studied in [4] may be considered. But such extra features are not our concern here.

The two most difficult parts in solving EDM optimization are the constraint on $\mathcal{D}^n$ (a conic constraint) and the rank constraint on rank$(JDJ) \leq r$. In this section, we review their mathematical properties, essential for our computational approach.

5.3.1 EDM Cone $\mathcal{D}^n$

The first issue to address is the membership of $\mathcal{D}^n$, as we must know if a candidate matrix belongs to the set or not, in order to solve the EDM optimization efficiently. One answer to this question is to see if we can calculate the orthogonal projection $\Pi_{\mathcal{D}^n}(A)$ defined by

$$\Pi_{\mathcal{D}^n}(A) := \arg\min \|A - D\| \quad \text{s.t.} \ D \in \mathcal{D}^n. \tag{5.25}$$

We note that $\Pi_{\mathcal{D}^n}(A)$ is well defined because $\mathcal{D}^n$ is a close convex cone, which is a direct consequence of its characterization below due to Schoenberg [35]:

$$D \in \mathcal{D}^n \quad \Leftrightarrow \quad \text{diag}(D) = 0 \quad \text{and} \quad -D \in \mathcal{K}_+^n, \tag{5.26}$$

where $\mathcal{K}_+^n$ is the conditionally positive semidefinite cone defined as follows (also see [32]). We recall that the semidefinite cone $\mathcal{S}_+^n$ can be defined as

$$\mathcal{S}_+^n = \{A \in \mathcal{S}^n \mid \mathbf{v}^T A \mathbf{v} \geq 0 \ \forall \mathbf{v} \in \mathbb{R}^n\}.$$

If we restrict $\mathbf{v}$ to the subspace $\mathbf{1}_n^\perp := \{\mathbf{v} \in \mathbb{R}^n \mid \mathbf{v}^T \mathbf{1} = 0\}$, then we have as $\mathcal{K}_+^n$ the set

$$\mathcal{K}_+^n := \{A \in \mathcal{S}^n \mid \mathbf{v}^T A \mathbf{v} \geq 0 \ \forall \mathbf{v} \in \mathbf{1}_n^\perp\}.$$

Since the centering matrix J is the projection matrix onto $\mathbf{1}_n^\perp$, we arrive at the equivalent representation

$$\mathcal{K}_+^n = \left\{A \in \mathcal{S}^n \mid JAJ \in \mathcal{S}_+^n\right\}. \tag{5.27}$$

It is easy to see that $\mathcal{K}_+^n$ is closed and convex, and so is $\mathcal{D}^n$ by Equation 5.26. It is worth noting that $\mathcal{S}_+^n \subset \mathcal{K}_+^n$ and $\mathcal{K}_+^n$ is strictly larger than $\mathcal{S}_+^n$. Moreover, $\mathcal{K}_+^n$ is not self-dual in the sense that

$$\left(\mathcal{K}_+^n\right)^* \subset \mathcal{K}_+^n \quad \text{and} \quad \left(\mathcal{K}_+^n\right)^* \neq \mathcal{K}_+^n,$$

where for a cone $\mathcal{C}$ in $\mathcal{S}^n$ its dual cone $\mathcal{C}^*$ is defined by

$$\mathcal{C}^* := \left\{Y \in \mathcal{S}^n \mid \langle Y, X \rangle \geq 0 \ \ \forall X \in \mathcal{C}\right\}.$$

But $\mathcal{S}_+^n$ is self-dual. Hence, EDM optimization is significantly different from SDP.

What does $\mathcal{D}^n$ look like? Let us use $\mathcal{D}^3$ as an example to understand its (computational) complexity.

Example 5.3.1

(Projection onto $\mathcal{D}^3$) It follows from Equation 5.26 that

$$\mathcal{D}^3 = \left\{D = \begin{bmatrix} 0 & x & y \\ x & 0 & z \\ y & z & 0 \end{bmatrix} \begin{matrix} 2(x+y) - z \geq 0 \\ 2(x+z) - y \geq 0 \\ 2(y+z) - x \geq 0 \end{matrix} \quad \text{and} \quad x^2 + y^2 + z^2 \leq \frac{1}{2}(x+y+z)^2 \right\}.$$

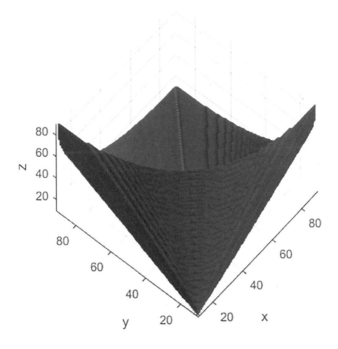

FIGURE 5.1
Shape of $\mathcal{D}^3$.

For a given matrix $A \in \mathcal{S}^3$ specified by (a, b, c), its projection to $\mathcal{D}^3$ is given by

$$\Pi_{\mathcal{D}^3}(A) = \arg \quad \min \left\{ \|(a,b,c)-(x,y,z)\| \left| \begin{array}{l} 2(x+y)-z \geq 0 \\ 2(x+z)-y \geq 0 \\ 2(y+z)-x \geq 0 \end{array} \right. \text{ and } \begin{array}{l} \|(x,y,z)\| \leq \dfrac{1}{2}t \\ t = x+y+z \end{array} \right\}.$$

The constraint $\|(x,y,z)\| \leq 0.5t$ is a second-order cone constraint. Hence, computing $\Pi_{\mathcal{D}^n}(A)$ is a quadratic second order cone optimization problem and in general would need an iterative algorithm for its solution. The "shape" of $\mathcal{D}^3$ can be seen in Figure 5.1.

Although $\Pi_{\mathcal{D}^n}$ is difficult to compute, it is surprisingly easy to compute the projection onto $\mathcal{K}^n_+$ [15]:

$$\Pi_{\mathcal{K}^n_+}(A) = A - \Pi_{\mathcal{S}^n_+}(JAJ). \tag{5.28}$$

This projection formula is fundamental to the alternating projection methods described in [15,16] and the Newton-type method from [11,31].

Another attempt to answer the question of membership of $\mathcal{D}^n$ is by the *triangle inequality test*. Let there be given a pre-distance matrix D satisfying the condition $\text{diag}(D)=0$ and $D_{ij} \geq 0$. We test all the triangle inequalities:

$$\sqrt{D_{ij}} + \sqrt{D_{jk}} \geq \sqrt{D_{ik}} \quad \text{for all triple } (i,j,k).$$

The following example shows that the triangle inequality test is not enough to verify $D \in \mathcal{D}^n$:

$$D = \begin{bmatrix} 0 & 1 & 1 & 1 \\ 1 & 0 & 4 & 4 \\ 1 & 4 & 0 & 4 \\ 1 & 4 & 4 & 0 \end{bmatrix}.$$

For this example, all triangle inequalities hold. However, D is not Euclidean because the matrix $(-JDJ)$ has a negative eigenvalue (-0.5), which implies $(-D) \notin \mathcal{K}_+^n$ and hence not in $\mathcal{D}^n$ due to (5.26).

5.3.2 Two Representations of the Rank Constraint

The second issue we need to address concerns the rank constraint. We describe two representations that will lead to two different types of algorithms for EDM optimization.

5.3.2.1 Representation by Eigenvalue Functions

Recall that the fact $(-D) \in \mathcal{K}_+^n$ means $(-JDJ) \in \mathcal{S}_+^n$. This implies

$$\text{rank}(JDJ) \le r \quad \Leftrightarrow \quad \sum_{i=1}^{n} \lambda_i = \sum_{i,j=1}^{r} \lambda_i, \tag{5.29}$$

where $\lambda_1 \ge \cdots \ge \lambda_n \ge 0$ are the eigenvalues of $(-JDJ)$ in non-increasing order. Define the eigenvalue function:

$$p(D) := \sum_{i=1}^{n} \lambda_i - \sum_{i=1}^{r} \lambda_i = \text{trace}(-JDJ) - \sum_{i,j=1}^{r} \lambda_i \ge 0, \quad \forall -D \in \mathcal{K}_+^n.$$

It is known that the sum of largest eigenvalues is a convex function. Therefore, $p(D)$ is a difference of two convex functions, that is, a so-called *DC function*. The characterization in Equation 5.26 and the equivalence in Equation 5.29 translate to

$$\{D \in \mathcal{D}^n \quad \text{and} \quad \text{rank}(JDJ) \le r\} \quad \Leftrightarrow \quad \{\text{diag}(D) = 0, \quad -D \in \mathcal{K}_+^n, \quad \text{and} \quad p(D) = 0\}. \tag{5.30}$$

5.3.2.2 Representation by Distance Functions

The essential constraints in EDM optimization can be grouped as follows.

$$\{D \in \mathcal{D}^n \quad \text{and} \quad \text{rank}(JDJ) \le r\} \quad \overset{(5.26)}{\Leftrightarrow} \quad \{\text{diag}(D) = 0 \quad \text{and} \quad -D \in \mathcal{K}_+^n(r)\}$$

where $\mathcal{K}_+^n(r)$ is the rank-r cut of $\mathcal{K}_+^n$:

$$\mathcal{K}_+^n(r) := \mathcal{K}_+^n \cap \{A \in \mathcal{S}^n \mid \text{rank}(JAJ) \le r\}.$$

An important point to consider here is that it is computationally rather inexpensive to check the membership of $\mathcal{K}_+^n(r)$ as we see below.

Let us define the Euclidean distance from a given point $A \in \mathcal{S}^n$ to $\mathcal{K}_+^n(r)$ by

$$\text{dist}\left(A, \mathcal{K}_+^n(r)\right) := \min\left\{\|A - Y\| \big| Y \in \mathcal{K}_+^n(r)\right\}$$

and define the function $g(\cdot): \mathcal{S}^n \mapsto \mathbb{R}$ to be the squared distance function from $(-A)$ to $\mathcal{K}_+^n(r)$:

$$g(A) := \frac{1}{2}\text{dist}^2\left(-A, \mathcal{K}_+^n(r)\right). \tag{5.31}$$

It can be seen that

$$-D \in \mathcal{K}_+^n(r) \quad \Leftrightarrow \quad g(D) = 0. \tag{5.32}$$

We further define $h(\cdot): \mathcal{S}^n \mapsto \mathbb{R}$ by

$$h(A) := \frac{1}{2}\| A \|^2 - g(-A). \tag{5.33}$$

It is shown in [33, Proposition 3.4] that $h(\cdot)$ is convex and

$$h(A) = \frac{1}{2}\left\|\Pi_{\mathcal{K}_+^n(r)}(A)\right\|^2 \quad \text{and} \quad \Pi_{\mathcal{K}_+^n(r)}(A) \in \partial h(A), \tag{5.34}$$

where $\partial h(A)$ is the subdifferential of $h(\cdot)$ at A and $\Pi_{\mathcal{K}_+^n(r)}(A)$ is a nearest point in $\mathcal{K}_+^n(r)$ from A:

$$\Pi_{\mathcal{K}_+^n(r)}(A) \in \arg\min\|A - Y\|, \quad \text{s.t.} \quad Y \in \mathcal{K}_+^n(r).$$

We note that there might be many nearest points since $\mathcal{K}_+^n(r)$ is not convex. The characterizations in Equations 5.26 and 5.32 translate to

$$\{D \in \mathcal{D}^n \quad \text{and} \quad \text{rank}(JDJ) \leq r\} \quad \Leftrightarrow \quad \{\text{diag}(D) = 0 \quad \text{and} \quad g(D) = 0\}. \tag{5.35}$$

5.3.2.3 Efficient Computation

Using the representations in Equations 5.30 and 5.35 relies on computing the functions $p(D)$ and $g(D)$. We show that both can be computed efficiently.

Suppose $A \in \mathcal{S}^n$ has the following spectral decomposition:

$$A = \lambda_1 \mathbf{p}_1 \mathbf{p}_1^T + \lambda_2 \mathbf{p}_2 \mathbf{p}_2^T + \cdots + \lambda_n \mathbf{p}_n \mathbf{p}_n^T,$$

where $\lambda_1 \geq \lambda_2 \geq \ldots \geq \lambda_n$ are the eigenvalues of A in non-increasing order, and $\mathbf{p}_i$, $i = 1, \ldots, n$ are the corresponding orthonormal eigenvectors. We define a PCA-style matrix truncated at r:

$$\text{PCA}_r^+(A) := \sum_{i=1}^{r} \max\{0, \lambda_i\} \mathbf{p}_i \mathbf{p}_i^T.$$

It can be shown [33, Proposition 3.3] that one particular $\Pi_{\mathcal{K}_+^n(r)}(A)$ can be computed via

$$\Pi_{\mathcal{K}_+^n(r)}(A) = \mathrm{PCA}_r^+(JAJ) + (A - JAJ).$$

Therefore, both $p(D)$ and $g(D)$ can be computed easily after obtaining $\mathrm{PCA}_r^+(-JDJ)$, which in itself can be obtained by computing just the first r largest eigenvalues of $(-JDJ)$ and their corresponding eigenvectors (e.g., via `eigs.m` in MATLAB®). Please refer to [40, Section II] for more detail.

5.4 EDM Optimization: Algorithms

We recall that the EDM optimization problem is essentially nonconvex mainly due to the rank constraint and the possibly non-convex loss function. We will now review some important convex relaxation methods because of two reasons: (i) they are globally convergent; and (ii) they may act as a subproblem solver in nonconvex approaches. However, our main focus will be on a recently proposed nonconvex approach, for its easy implementation and low computational cost.

5.4.1 Convex Relaxation Methods

If we drop the rank constraint from the least-squares models (5.20) and (5.21), we arrive at a convex problem. Let us use (5.21) as an example. We start by dropping the rank constraint and set all weights to 1 in Equation 5.21:

$$\min \frac{1}{2}\left\|D - \overline{\Delta}\right\|^2, \quad \text{s.t.} \quad D \in \mathcal{S}_h^n \quad \text{and} \quad -D \in \mathcal{K}_+^n, \tag{5.36}$$

where $\mathcal{S}_h^n$ is the hollow subspace

$$\mathcal{S}_h^n := \left\{ A \in \mathcal{S}^n \,\middle|\, \mathrm{diag}(A) = 0 \right\}.$$

5.4.1.1 Method of Alternating Projections

It is easy to see that (5.36) is a projection problem onto $\mathcal{S}_h^n \cap \mathcal{K}_-^n$, the intersection of the subspace $\mathcal{S}_h^n$ and the convex cone $\mathcal{K}_-^n := -\mathcal{K}_+^n$. For this, a method of alternating projections (MAP) was developed in [15] and [16]: start with an initial point $D^0 \in \mathcal{S}^n$ and then compute

$$D^{k+1} = \Pi_{\mathcal{K}_-^n}\left(\Pi_{\mathcal{S}_h^n}(D^k)\right), \quad k = 0, 1, 2\ldots,. \tag{5.37}$$

The two projections are easy to compute. In fact,

$$\Pi_{\mathcal{S}_h^n}(A) = A_0, \quad \forall A \in \mathcal{S}^n,$$

where A_0 is A except its diagonal is being replaced by 0, and by Equation 5.28

$$\Pi_{\mathcal{K}_-^n}(A) = -\Pi_{\mathcal{K}_+^n}(-A) = A + \Pi_{\mathcal{S}_+^n}(-JAJ).$$

MAP can easily be extended to deal with the weighted problem

$$\min \quad f(D) = \frac{1}{2}\left\| \sqrt{W} \circ (D - \overline{\Delta}) \right\|^2, \quad \text{s.t.} \quad D \in \mathcal{S}_h^n \cap \mathcal{K}_-^n. \tag{5.38}$$

For this, let w_i be the largest element in the ith row of $\sqrt{W}$,

$$w_i := \max\left\{ \sqrt{W_{ij}} \mid j = 1, \dots, n \right\},$$

and set $\mathbf{w} := (w_1, \dots, w_n)^T$. Let $\text{Diag}(\mathbf{w})$ be the diagonal matrix formed by $\mathbf{w}$. A majorization function of $f(D)$ is

$$f_m(D, D^k) := f(D^k) + \left\langle \nabla f(D^k), D - D^k \right\rangle + \frac{1}{2}\left\| \text{Diag}\left(\sqrt{\mathbf{w}}\right)\left(D - \overline{\Delta}\right)\text{Diag}\left(\sqrt{\mathbf{w}}\right) \right\|^2,$$

where D^k is the current iterate. Hence, the MM (majorization and minimization) scheme can be applied to $f_m(D, D^k)$ to arrive at

$$D^{k+1} = \arg\min f_m(D, D^k), \quad \text{s.t.} \quad D \in \mathcal{S}_h^n \cap \mathcal{K}_-^n,$$

which is known as the *diagonally weighted projection problem* for which there exists a computational formula for weighted projection onto $\mathcal{K}_-^n$, see [31, Section 4.2].

MAP is guaranteed to converge. However, it is incapable of dealing with the rank constraint. The scheme has to rely on other mechanism to get the embedding dimension right, for example by enforcing many box constraints (5.6); see [17] for such a mechanism.

5.4.1.2 Newton's Method

MAP is a first-order method and hence may take many iterations to terminate. A corresponding Newton-type method was developed in [31] by considering the Lagrange dual problem of (5.36):

$$\min_{\mathbf{y} \in \mathbb{R}^n} \quad \theta(\mathbf{y}) := \frac{1}{2}\left\| \Pi_{\mathcal{K}_+^n}(\overline{\Delta} + \text{Diag}(\mathbf{y})) \right\|^2.$$

Being a Lagrange dual function, $\theta(\mathbf{y})$ is convex and in this case also differentiable. The optimality condition for the dual is

$$F(\mathbf{y}) := \nabla\theta(\mathbf{y}) = \text{diag}\left(\Pi_{\mathcal{K}_+^n}(\overline{\Delta} + \text{Diag}(\mathbf{y})) \right) = 0.$$

This is a system of nonlinear equations. A Newton-type method for this system hence takes the following form:

$$\mathbf{y}^{k+1} = \mathbf{y}^k - V_k^{-1}F(\mathbf{y}^k), \quad k = 0, 1, \dots,$$

where V_k is a generalized Jacobian of $F(\cdot)$ at $\mathbf{y}^k$. Suppose $\{\mathbf{y}^k\}$ converges to $\mathbf{y}^*$ satisfying $F(\mathbf{y}^*) = 0$ ($\mathbf{y}^*$ must be optimal for the dual problem), then the corresponding optimal EDM solution is

$$D^* = -\Pi_{\mathcal{K}_+^n}(\overline{\Delta} + \text{Diag}(\mathbf{y}^*)).$$

It can be shown that this Newton-type method converges quadratically and can be further generalized to deal with the weighted problem (5.38) as well. We refer to [31] for detailed analysis of this method, which is further used to solve the subproblems for the rank constrained problem (5.21) in [33].

5.4.2 Penalty Methods

If we take into account the rank representations in Equations 5.30 and 5.35, a penalty approach appears to be naturally and promising when applied to EDM optimization problems. We describe two of them separately, with the main focus on penalizing (5.35).

5.4.2.1 Penalizing the Eigenfunction

Consider for example problem (5.21). By using (5.30), the problem is equivalent to

$$\min \quad f^{(D,2)}(D) = \left\| \sqrt{W} \circ (D - \overline{\Delta}) \right\|^2$$
$$\text{s.t.} \quad D \in \mathcal{B}, -D \in \mathcal{K}_+^n, \quad p(D) = 0,$$

where the diagonal constraint diag$(D) = 0$ is submerged into the box constraint $\mathcal{B}$. We emphasize that all the constraints except $p(D) = 0$ are convex. The penalized problem is

$$\min \quad f_\rho^{(D,2)}(D) = \left\| \sqrt{W} \circ (D - \overline{\Delta}) \right\|^2 + \rho p(D)$$
$$\text{s.t.} \quad D \in \mathcal{B}, -D \in \mathcal{K}_+^n,$$

where $\rho > 0$ is a penalty parameter. We note that $p(D) \geq 0$ whenever $-D \in \mathcal{K}_+^n$. Therefore, p is a penalty function over the feasible region. This penalty problem was thoroughly investigated in [33], where the Newton method reviewed above was used to solve the corresponding subproblems. The corresponding analysis is quite involved, and it is not possible here to give it a complete review. Instead, we focus now on the other penalty method using (5.35). It will turn out that this method is easy to implement and has a low computational cost per step.

5.4.2.2 Penalizing the Distance Function

To illustrate the power of this approach, we consider the nondifferentiable problem (5.20) as an example. We note that this approach can also be used for other EDM optimization problems. By Equation 5.35, the problem (5.20) is equivalent to

$$\min \quad f_\rho^{(d,2)}(D) = \left\| \sqrt{W} \circ \left(\sqrt{D} - \Delta \right) \right\|^2$$
$$\text{s.t.} \quad D \in \mathcal{B}, g(D) = 0.$$

The penalized problem is

$$\min \quad F(D) := \left\| \sqrt{W} \circ \left(\sqrt{D} - \Delta \right) \right\|^2 + \rho g(D), \quad \text{s.t.} \quad D \in \mathcal{B}. \tag{5.39}$$

The function $F(\cdot)$ is still difficult to minimize. We will construct a majorization that is easy to minimize.

Suppose D^k is the current iterate. We recall from Equation 5.33 that

$$
\begin{aligned}
g(D) &= \frac{1}{2}\|D\|^2 - h(-D) \\
&\leq \frac{1}{2}\|D\|^2 - \underbrace{\left\{ h(-D^k) + \left\langle \Pi_{\mathcal{K}^n_+(r)}(-D^k), -D + D^k \right\rangle \right\}}_{\text{by convexity of } h(\cdot) \text{ and } (5.34)} \\
&= \frac{1}{2}\|D\|^2 - h(-D^k) + \left\langle \Pi_{\mathcal{K}^n_+(r)}(-D^k), D - D^k \right\rangle \\
&=: g_m(D, D^k).
\end{aligned}
$$

That is, $g_m(D, D^k)$ is a majorization of $g(D)$ (one can see this by verifying (5.7)). Consequently, a majorization for $F(D)$ is

$$
F_m(D, D^k) = \left\| \sqrt{W} \circ \left(\sqrt{D} - \Delta \right) \right\|^2 + \rho g_m(D, D^k).
$$

By applying the MM scheme (5.8), we arrive at the following algorithm:

$$
D^{k+1} = \arg\min F_m(D, D^k), \quad \text{s.t.} \quad D \in \mathcal{B}. \tag{5.40}
$$

We see below that the majorization problem has a very nice structure:

$$
\begin{aligned}
D^{k+1} &= \arg\min_{D \in \mathcal{B}} \left\| \sqrt{W} \circ \left(\sqrt{D} - \Delta \right) \right\|^2 + \rho/2 \|D\|^2 + \left\langle \rho \Pi_{\mathcal{K}^n_+(r)}(-D^k), D \right\rangle \\
&= \arg\min_{D \in \mathcal{B}} \frac{\rho}{2}\|D\|^2 + \left\langle W + \rho \Pi_{\mathcal{K}^n_+(r)}(-D^k), D \right\rangle - 2\left\langle W \circ \Delta, \sqrt{D} \right\rangle \\
&= \arg\min_{D \in \mathcal{B}} \frac{1}{2}\left\| D - D_\rho^k \right\|^2 - \frac{2}{\rho}\left\langle W \circ \Delta, \sqrt{D} \right\rangle,
\end{aligned}
$$

where

$$
D_\rho^k := -W/\rho - \Pi_{\mathcal{K}^n_+(r)}(-D^k).
$$

Hence, D^{k+1} can be computed through the following $n(n-1)/2$ one-dimensional optimization problems:

$$
D_{ij}^{k+1} = \arg\min \frac{1}{2}\left(D_{ij} - \left(D_\rho^k \right)_{ij} \right)^2 - 2(w_{ij}\delta_{ij}/\rho)\sqrt{D_{ij}}, \quad \text{s.t.} \quad L_{ij} \leq D_{ij} \leq U_{ij}. \tag{5.41}
$$

This is equivalent to solving the following one-dimensional problem. For given $\omega \in \mathbb{R}$, $\alpha > 0$ and an interval $B := [a, b]$ with $0 \leq a \leq b$, find the optimal solution $\mathcal{S}_B(\omega, \alpha)$ of the problem:

$$
\mathcal{S}_B(\omega, \alpha) := \arg\min \frac{1}{2}(x - \omega)^2 - 2\alpha\sqrt{x}, \quad \text{s.t.} \quad x \in B = [a, b], \tag{5.42}
$$

Define

$$
u := \frac{\alpha}{2}, \quad v := \frac{\omega}{3}, \quad \text{and} \quad \tau := u^2 - v^3.
$$

Let

$$
S(\omega,\alpha) := \begin{cases} \left[(u+\sqrt{\tau})^{1/3} + (u-\sqrt{\tau})^{1/3}\right]^2 & \text{if } \tau \geq 0 \\ 4v\cos^2(\phi/3) & \text{if } \tau < 0, \end{cases}
$$

where the angle ϕ is defined by $\cos(\phi) = u/v^{3/2}$.
The following results are proved in [40, Propositions 3.4 and 3.5]:

Proposition 5.4.1

Suppose we are given $\omega \in \mathbb{R}$, $\alpha > 0$ and $B = [a,b]$ with $0 \leq a \leq b$. Then we have
  (i) *The solution of (5.42) is given by*

$$
S_B(\omega,\alpha) = \min\{b, \quad \max\{a, S(\omega,\alpha)\}\}.
$$

  (ii) *Suppose $b > 0$ and we have two positive constants $\omega_{\max} > 0$ and $\alpha_0 > 0$. Then there exists $\gamma > 0$ such that*

$$
S_B(\omega,\alpha) \geq \gamma
$$

for any ω and α satisfying

$$
|\omega| \leq \omega_{\max} \quad \text{and} \quad \alpha \geq \alpha_0.
$$

Therefore, D^{k+1} in Equation (5.41) can be computed as follows

$$
D_{ij}^{k+1} = \begin{cases} \min\{U_{ij}, \quad \max\{(D_\rho^k)_{ij}, \quad L_{ij}\} & \text{if } \delta_{ij} = 0 \\ S_{B_{ij}}\left((D_\rho^k)_{ij}, \quad w_{ij}\delta_{ij}/\rho\right) & \text{if } \delta_{ij} > 0, \end{cases} \tag{5.43}
$$

where $B_{ij} := [L_{ij}, U_{ij}]$. For simplicity, we denote the update formula in Equation (5.43) by

$$
D^{k+1} = T(D^k), k = 0,1,\dots,. \tag{5.44}
$$

The algorithm just outlined is called SQREDM (square-root model via EDM for (5.20)) in [40]. We summarize it below.

Algorithm 5.1 SQREDM Method

1: **Input data:** Dissimilarity matrix Δ, weight matrix W, penalty parameter $\rho > 0$, lower-bound matrix L, upper-bound matrix U and the initial D^0. Set $k := 0$.
2: **Update:** Compute D^{k+1} by Equation 5.44 via (5.43).
3: **Convergence check:** Set $k := k + 1$ and go to Step 2 until convergence.

We provide a few remarks on the algorithm SQREDM.

(R1) (Differentiable path) Without loss of generality, we may assume that the box constraint $\mathcal{B}$ is bounded. Then the generated sequence $\{D^k\}$ is also bounded because $D^k \in \mathcal{B}$. Moreover,

$$\left\| D_\rho^k \right\| = \left\| W / \rho + \Pi_{\mathcal{K}_+^n(r)}(-D^k) \right\| \leq \frac{1}{\rho} \|W\| + \left\| \Pi_{\mathcal{K}_+^n(r)}(-D^k) \right\| \leq \frac{1}{\rho} \|W\| + \left\| D^k \right\|,$$

where the last inequality used the fact that $0 \in \mathcal{K}_+^n(r)$. The boundedness of $\{D^k\}$ implies the boundedness of $\left\{ D_\rho^k \right\}$. It follows from Proposition 4.1 (ii) that there exists $\gamma > 0$ such that $D_{ij}^k \geq \gamma$ whenever $\delta_{ij} > 0$. If $\delta_{ij} = 0$, then $w_{ij} = 0$ and the term D_{ij} does not enter the objective function $F(D)$. Consequently, $F(D)$ is differentiable on the generated iterates $\{D^k\}$ although it may not be differentiable at other places. Since D_{ij}^k is bounded away from 0 by a constant $\gamma > 0$ (when $\delta_{ij} > 0$), $F(D)$ is also differentiable at any of the accumulation points of $\{D^k\}$. The property of being differentiable along the generated path has greatly simplified the convergence analysis in [40].

(R2) (Computational complexity) The major computation in SQREDM concerns $\Pi_{\mathcal{K}_+^n(r)}(-D^k)$, which has been dealt with in Section 3(c). The overall complexity is $O(Nrn^2)$, where $O(rn^2)$ stems from computing $\Pi_{\mathcal{K}_+^n(r)}(-D^k)$ and the formula in Equation 5.43, and N is the number of updates, that is number of iterations.

(R3) (Generalization) The convergence of SQREDM has been established in [40]. However, the corresponding framework can be generalized to other EDM optimization problems; see, for example [41] for a generalization to the robust model (5.22).

5.5 Regularization

Regularization in SNL seems to have been motivated by a similar strategy for dimension reduction in manifold learning [38], where data sitting on a manifold in a high-dimensional space are mapped to a low-dimensional Euclidean space. Flattening a manifold to a low-dimensional space often causes points to be gathered around the geometric center of those points. The idea considered in [38] is to pull apart the embedding points via enforcing some regularization terms. This regularization approach has not yet been well addressed in EDM optimization. We show below that this can be done without causing too much extra computational burden to the algorithm SQREDM.

Suppose we have n points that we would like to "push apart." We introduce two ways to achieve this purpose. One is to maximize the total Euclidean distances among points

$$\mathcal{R}_1(D) := \sum_{i,j=1}^n \|\mathbf{x}_i - \mathbf{x}_j\| = \left\langle \mathbf{1}_n \mathbf{1}_n^T, \ \sqrt{D} \right\rangle$$

and the other is to maximize the total squared Euclidean distance, as suggested by [7]

$$\mathcal{R}_2(D) := \sum_{i,j=1}^n \|\mathbf{x}_i - \mathbf{x}_j\|^2 = \left\langle \mathbf{1}_n \mathbf{1}_n^T, \ D \right\rangle,$$

where $\mathbf{x}_i$, $i = 1, \ldots, n$ are the embedding points of the EDM D. In particular, $\mathcal{R}_2$ has an interesting interpretation. Suppose the embedding points $\{\mathbf{x}_i\}$ are centered, that is, $\Sigma_{i=1}^n \mathbf{x}_i = 0$. (This is so when they are embedding points form D by MDS.) Then

$$\mathcal{R}_2(D) = \sum_{i,j=1}^n \left(\|\mathbf{x}_i\|^2 - 2\langle \mathbf{x}_i, \mathbf{x}_j \rangle + \|\mathbf{x}_j\|^2 \right) = 2n \sum_{i=1}^n \|\mathbf{x}_i\|^2,$$

which is called the *variance* among the embedding points. Maximizing $\mathcal{R}_2$ is known as maximum variance unfolding in [38].

Adding each term to the penalized problem (5.39) yields

$$\min \quad F^{(i)}(D) := \left\| \sqrt{W} \circ \left(\sqrt{D} - \Delta \right) \right\|^2 + \rho g(D) - \beta \mathcal{R}_i(D), \quad \text{s.t.} \quad D \in \mathcal{B}, \qquad (5.45)$$

where $i \in \{1, 2\}$ and $\beta > 0$ is the regularization parameter. Again, we can use the MM scheme to solve this regularized problem. A natural majorization function is

$$F_m^{(i)}(D, D^k) = F_m(D, D^k) + \beta \mathcal{R}_i(D).$$

We solve

$$D^{k+1} = \arg\min F_m^{(i)}(D, D^k) \quad \text{s.t.} \quad D \in \mathcal{B}.$$

When $i = 1$, it is easy to see

$$D^{k+1} = \arg \min_{D \in \mathcal{B}} F_m^{(1)}(D, D^k)$$

$$= \arg \min_{D \in \mathcal{B}} \frac{1}{2} \left\| D - D_\rho^k \right\|^2 - 2 \left\langle W \circ \Delta / \rho + \beta / (2\rho) \mathbf{1}_n \mathbf{1}_n^T, \ \sqrt{D} \right\rangle.$$

Consequently,

$$D_{ij}^{k+1} = \mathcal{S}_{\mathcal{B}_{ij}} \left((D_\rho^k)_{ij}, \ w_{ij} \delta_{ij} / \rho + \beta / (2\rho) \right).$$

We denote this update by

$$D^{k+1} = \mathcal{T}_1(D^k), \quad k = 0, 1, \ldots, \qquad (5.46)$$

and refer to it as SQREDMR1.

Algorithm 5.2 SQREDMR1 Method

1: **Input data:** Dissimilarity matrix Δ, weight matrix W, penalty parameter $\rho > 0$, lower-bound matrix L, upper-bound matrix U and the initial D^0. Set $k := 0$.
2: **Update:** Compute D^{k+1} by Equation 5.46.
3: **Convergence check:** Set $k := k + 1$ and go to Step 2 until convergence.

When $i = 2$, it is easy to see

$$D^{k+1} = \arg\min_{D \in \mathcal{B}} F_m^{(2)}(D, D^k)$$

$$= \arg\min_{D \in \mathcal{B}} \frac{1}{2} \left\| D - \left(D_\rho^k + (\beta/\rho)\mathbf{1}_n\mathbf{1}_n^T \right) \right\|^2 - 2\left\langle W \circ \Delta/\rho, \sqrt{D} \right\rangle.$$

Consequently,

$$D_{ij}^{k+1} = \begin{cases} \min\{U_{ij}, \max\left\{\left(D_\rho^k\right)_{ij}, L_{ij}\right\} & \text{if } \delta_{ij} = 0 \\ \\ \mathcal{S}_{\mathcal{B}_{ij}}\left(\left(D_\rho^k\right)_{ij} + \beta/\rho, w_{ij}\delta_{ij}/\rho\right) & \text{if } \delta_{ij} > 0, \end{cases}$$

We denote this update by

$$D^{k+1} = \mathcal{T}_2(D^k), \quad k = 0, 1, \ldots,. \tag{5.47}$$

and refer to it as SQREDMR2.

Algorithm 5.3 SQREDMR2 Method

1: **Input data:** Dissimilarity matrix Δ, weight matrix W, penalty parameter $\rho > 0$, lower-bound matrix L, upper-bound matrix U and the initial D^0. Set $k := 0$.
2: **Update:** Compute D^{k+1} by Equation 5.47.
3: **Convergence check:** Set $k := k + 1$ and go to Step 2 until convergence.

Comparing (5.46) and (5.47) to the original algorithm (5.44), we can see that the regularization term does not create any additional computational difficulties. In contrast to this, it would be difficult to include the regularization term $\mathcal{R}_1$ in the SDP model in [7] without causing a significant computational burden.

5.6 Numerical Examples

The purpose of this section is not to provide a full assessment of EDM optimization schemes on a wide range of SNL problems. Instead, we illustrate the role of the regularizations $\mathcal{R}_1(D)$ and $\mathcal{R}_2(D)$. The main finding is that they do improve the localization quality as long as the regularization parameter is small (e.g., less than 1 in our tested cases), though it remains unclear how to choose the best value for this parameter.

5.6.1 Implementation and Test Problems

Details on the implementation, including the stopping criterion for SQREDM, can be found in [40]. We describe our test problems below. There are two cases, depending on the positions of the anchors: (i) outer position and (ii) inner position. The case (ii) was also considered in [40].

Example 5.6.1

This example has been widely tested since its detailed study in [7]. In the square region $[-0.5, 0.5]^2$, four anchors $\mathbf{x}_1 = \mathbf{a}_1, \ldots, \mathbf{x}_4 = \mathbf{a}_4$ ($m = 4$) are placed at

$$\text{Case(i)(Outer position)}: \quad (\pm 0.45, \pm 0.45)$$
$$\text{Case(ii)(Inner position)}: \quad (\pm 0.20, \pm 0.20)$$

The generation of the remaining $(n - m)$ sensors $(\mathbf{x}_{m+1}, \ldots, \mathbf{x}_n)$ follows the uniform distribution over the square region. The noisy Δ is usually generated as follows.

$$\delta_{ij} := \| \mathbf{x}_i - \mathbf{x}_j \| \times \left| 1 + \epsilon_{ij} \times nf \right|, \quad \forall (i, j) \in \mathcal{N}$$
$$\mathcal{N} := \mathcal{N}_x \cup \mathcal{N}_a$$
$$\mathcal{N}_x := \left\{ (i, j) \big| \| \mathbf{x}_i - \mathbf{x}_j \| \leq R, \quad i > j > m \right\}$$
$$\mathcal{N}_a := \left\{ (i, j) \big| \| \mathbf{x}_i - \mathbf{a}_j \| \leq R, \quad i > m, \quad 1 \leq j \leq m \right\},$$

where R is known as the radio range, ϵ_{ij}'s are independent standard normal random variables, and nf is the noise factor (here, $nf = 0.2$ was used in our tests, which corresponds to 20% noise level). In the literature (see, e.g., [7]), this type of perturbation in δ_{ij} is known to be multiplicative and follows the unit-ball rule in defining $\mathcal{N}_x$ and $\mathcal{N}_a$ (see [3, Section 3.1] for more detail). The corresponding weight matrix W and the lower and upper bound matrices L and U are given in the table below. Here, M is a large positive quantity. For example, $M := n \max_{ij} \Delta_{ij}$ is the upper bound of the longest shortest path if the network is viewed as a graph.

(i, j)	W_{ij}	Δ_{ij}	L_{ij}	U_{ij}
$i = j$	0	0	0	0
$i, j \leq m$	0	0	$\| \mathbf{a}_i - \mathbf{a}_j \|^2$	$\| \mathbf{a}_i - \mathbf{a}_j \|^2$
$(i, j) \in \mathcal{N}$	1	δ_{ij}	0	R^2
otherwise	0	0	R^2	M^2

To compare the embedding quality, we use a widely used measure RMSD (root of the mean squared deviation) defined by

$$\text{RMSD} := \left[\frac{1}{n - m} \sum_{i=m+1}^{n} \| \hat{\mathbf{x}}_i - \mathbf{x}_i \|^2 \right]^{1/2},$$

where the $\mathbf{x}_i$ are the true positions of the sensors in our test problems and the $\hat{\mathbf{x}}_i$ are their corresponding estimates. The $\hat{\mathbf{x}}_i$ were obtained by applying CMDS to the final output of the distance matrix, followed by aligning them to the existing anchors through the well-known Procrustes procedure (see [8, Chapter 20] or [34, Proposition 4.1]. All tests were run in MATLAB 2017b.

5.6.2 Numerical Experiments

First, we would like to assess the levels of improvement the two regularizations are able to provide. We set $n = 104$ (100 sensors), $nf = 0.2$ and $R = 0.3$, and set the random seed to be

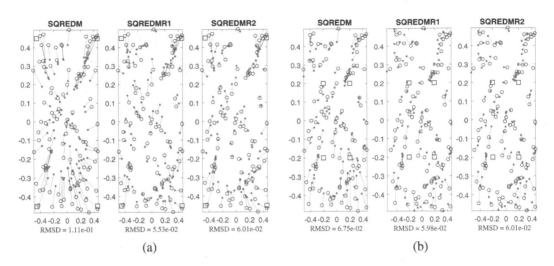

(a)   (b)

FIGURE 5.2
Improving embedding quality by regularization: (a) considers Case (i) and shows ca. 50% reduction in RMSD for both regularizations. In contrast, (b) (Case (ii)) shows a small improvement for both regularizations.

rng(0). We also choose $\beta = 0.8$ for both SQREDMR1 and SQREDMR2. Figure 5.2 contains the recovered sensors for both cases. Figure 5.2 considers Case (i) of outer anchor positions while Figure 5.2 considers for Case (ii) of inner anchor positions. The RMSD is decreased by more than 50% in the first case, while RMSD is only reduced marginally in the second case. This is reasonable as the choice of β is more suitable to Case (i) than to Case (ii).

We now test how sensitive SQREDM is to both regularization terms. As already seen in Figure 5.3, the two terms behave quite similarly at $\beta = 0.8$. Surprisingly, this

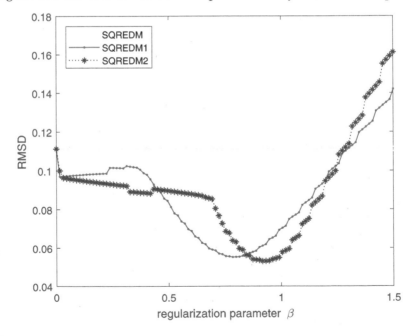

FIGURE 5.3
Sensitivity of regularization parameter β on Case (i). Both SQREDMR1 and SQREDMR2 behave quite similarly and both improved the embedding quality over a wide range of β from 0 to about 1:3.

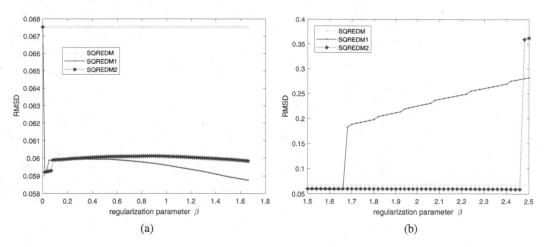

FIGURE 5.4
Sensitivity of regularization parameter β on Case (ii). Both SQREDMR1 and SQREDMR2 behave quite similarly and both improved the embedding quality over the range of β from 0 to about 1:6 (a). Beyond this range, SQREDMR2 continues its steady behavior, while the RMSD for SQREDMR1 soon jump to values than those of SQREDM (b).

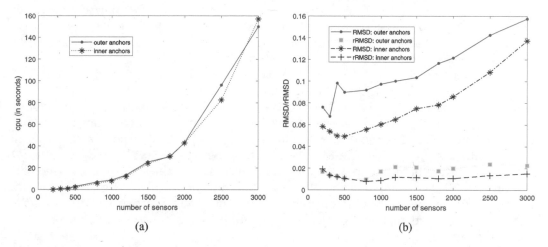

FIGURE 5.5
Speed and quality of SQREDM for both Case (i) and Case (ii). The amount of time consumed by SQREDM is similar for both cases (a). However, the quality for Case (ii) is constantly better than for Case (i) (b, where we also plotted the refined RMSD.)

phenomenon abides over a range of choices. This is clearly shown in Figure 5.4 (Case (i)), where RMSD vs β is plotted. SQREDMR1 and SQREDMR2 follow each other quite closely and they produced better RMSD than SQREDM for all $\beta \in (0, 1.2)$. Since we are testing the same problem with different choices of β, the plot for SQREDM is a straight line. This behavior is also observed for Case (ii) in Figure 5.4, where we see a sudden jump for SQREDMR1 at $\beta \approx 1.66$ and for SQREDMR2 at $\beta \approx 2.45$. It seems that both terms are less sensitive when $\beta \leq 1$, as opposed to the observed behavior for $\beta \geq 1$. Our suggestion is thus to choose β between 0.5 and 1. Of course, an optimal choice of β is highly problem dependent.

Finally, we test how fast SQREDM is for large values of n. We test problems of size ranging from $n = 200$ to $n = 3000$. Figure 5.5 reports the average results over 20 runs on randomly generated instances, using rng("shuffle") for the random number generation. As seen from Figure 5.5b, SQREDM used a similar amount of time for both cases (Case (i) and Case (ii)) and is in general rather competitive. For $n = 2000$, the algorithm only uses about 40 seconds before terminating. This time climbs to about 160 seconds for the $n = 3000$ case. (All experiments have been run on a Dell laptop with CPU 2.50 Ghz and 8 GB RAM.) The quality of the localization is shown in Figure 5.5b, where we also report the refined RMSD (rRMSD) obtained by using the heuristic gradient method of [7], called the refinement step. We see that the quality after the refinement step is very satisfactory, demonstrating that the final embedding provided by SQREDM is a good starting point for the refinement step.

5.7 Conclusion

Sensor network localization has been extensively studied and has a capacity of modeling various practical problems. In this chapter, we reviewed some major computational approaches such as the coordinate minimization, SDP, and EDM optimization, with the latter two being based on matrix optimization. Both SDP and EDM approaches can be seen as centralized approaches, while we note that coordinate minimization can be seen as a case of distributed optimization. We also note that the majorization-minimization technique has played an important role in both coordinates' minimization and EDM optimization.

In particular, we discussed the origin of EDM optimization and reviewed in detail some of its major algorithms, including the convex relaxation and the latest penalty method SQREDM. We addressed the issue of regularization, which has not been studied previously in EDM optimization. We considered two regularization terms, one regarding "plain" distances and the other regarding "squared" distances. We showed that both terms can be naturally incorporated into SQREDM, resulting in SQREDM R1 and SQREDM R2, respectively. Their numerical performance was demonstrated through a widely-used SNL test problem. Both regularizations lead to an improvement in embedding quality, but the level of improvement varies with the problem type. The potential of regularization in EDM optimization is worth further investigations with different loss functions and test problems from other application areas.

References

1. A.Y. Alfakih, A. Khandani and H. Wolkowicz, Solving Euclidean distance matrix completion problems via semidefinite programming, *Comput. Optim. Appl.*, 12, 1999, pp. 13–30.
2. K.S. Arun, T.S. Huang and S.D. Blostein, Least-squares fitting of two 3-D point sets, *IEEE Trans. Pattern Anal. Machine Intell.*, 9, 1987, pp. 698–700.
3. S. Bai and H.-D. Qi, Tackling the flip ambiguity in wireless sensor network localization and beyond, *Digital Signal Process.*, 55, 2016, pp. 85–97.
4. S. Bai, H.-D. Qi and N. Xiu, Constrained best Euclidean distance embedding on a sphere: A matrix optimization approach, *SIAM J. Optim.*, 25, 2015, pp. 439–467.

5. H.M. Berman, J. Westbrook, Z. Feng, G. Gillilan, T.N. Bhat, H. Weissig, I.N. Shindyalov and P.E. Bourne, The protein data bank, *Nucleic Acids Res.*, 28, 2000, pp. 235–242.

6. P. Biswas and Y. Ye, Semidefinite programming for ad hoc wireless sensor network localization, In *Proceedings of the 3rd IPSN*, Berkeley, CA, pp. 46–54, 2004.

7. P. Biswas, T.-C. Liang, K.-C. Toh, T.-C. Wang and Y. Ye, Semidefinite programming approaches for sensor network localization with noisy distance measurements, *IEEE Trans. Auto. Sci. Eng.*, 3, 2006, pp. 360–371.

8. I. Borg and P.J.F. Groenen, *Modern Multidimensional Scaling: Theory and Applications*, 2nd Ed., Springer Series in Statistics, Springer, 2005.

9. L. Cayton and S. Dasgupta, Robust Euclidean embedding, in *Proceedings of the 23rd International Conference on Machine Learning*, Pittsburgh, PA, 2006, pp. 169–176.

10. T.F. Cox and M.A.A. Cox, *Multidimensional Scaling*, 2nd Ed., Chapman and Hall/CRC, 2001.

11. C. Ding and H.-D. Qi, Convex optimization learning of faithful Euclidean distance representations in nonlinear dimensionality reduction, *Math. Program.*, 164, 2017, pp. 341–381.

12. J. de Leeuw, *Applications of Convex Analysis to Multidimensional Scaling*, In J Barra, F Brodeau, G Romier, B van Cutsem (eds.), *Recent Developments in Statistics*, pp. 133–145. North Holland Publishing Company, Amsterdam, The Netherlands, 1977.

13. J. Dattorro, *Convex Optimization and Euclidean Distance Geometry*, Meboo Publishing, USA, 2005.

14. D. Drusvyatskiy, N. Krislock, Y.-L. Voronin and H. Wolkowicz, Noisy Euclidean distance realization: Robust facial reduction and the Pareto frontier, *SIAM J. Optim.*, 27, 2017, pp. 2301–2331.

15. N. Gaffke and R. Mathar, A cyclic projection algorithm via duality, *Metrika*, 36, 1989, pp. 29–54.

16. W. Glunt, T.L. Hayden, S. Hong and J. Wells, An alternating projection algorithm for computing the nearest Euclidean distance matrix, *SIAM J. Matrix Anal. Appl.*, 11, 1990, pp. 589–600.

17. W. Glunt, T.L. Hayden and R. Raydan, Molecular conformations from distance matrices, *J. Comput. Chemistry*, 14, 1993, pp. 114–120.

18. D.S. Goncalves, A. Mucherino, C. Lavor and L. Liberti, Recent advances on the interval distance geometry problem, *J. Glob. Optim.*, 69, 2017, pp. 525–545.

19. J.C. Gower, Some distance properties of latent root and vector methods used in multivariate analysis, *Biometrika*, 53, 1966, pp. 325–338.

20. W.J. Heiser, Multidimensional scaling with least absolute residuals, in *Proceedings of the First Conference of the International Federation of Classification Societies (IFCS)*, Aachen, Germany, June 1987, pp. 455–462.

21. K.F. Jiang, D.F. Sun and K.C. Toh, Solving nuclear norm regularized and semidefinite matrix least squares problems with linear equality constraints, in *Discrete Geometry and Optimization*, Springer International Publishing, pp. 133–162, 2013.

22. K.F. Jiang, D.F. Sun and K.-C. Toh, A partial proximal point algorithm for nuclear norm regularized matrix least squares problems, *Math. Programming Comput.*, 6, 2014, pp. 281–325.

23. S. Kim, M. Kojima, H. Waki and M. Yamashita, Algorithm 920: SFSDP: A sparse version of full semidefinite programming relaxation for sensor network localization problems, *ACM Trans. Math. Softw.*, 38(4), 2012, pp. 27:1–27:19.

24. N. Krislock and H. Wolkowicz, Euclidean distance matrices and applications, In M. Anjos and J. Lasserre (eds.), *Handbook on Semidefinite, Cone and Polynomial Optimization*, Springer, New York, pp. 879–914, 2012.

25. J.B. Kruskal, Nonmetric multidimensional scaling: A numerical method, *Psychometrika*, 29, 1964, pp. 115–129.

26. L. Liberti, C. Lavor, N. Maculan and A. Mucherino, Euclidean distance geometry and applications, *SIAM Rev.*, 56, 2014, pp. 3–69.

27. L. Lovász, Semidefinite programs and combinatorial optimization. Lecture notes, Microsoft Research, Redmond, WA, 1992.

28. K.V. Mardia, Some properties of classical multidimensional scaling, *Comm. Statist. A – Theory Methods*, A7, 1978, pp. 1233–1243.

29. P. Oğuz-Ekim, J.P. Gomes, J. Xavier and P. Oliveira, Robust localization of nodes and time-recursive tracking in sensor networks using noisy range measurements, *IEEE Trans. Signal Process.*, 59, 2011, pp. 3930–3942.

30. N. Patwari, A.O. Hero, M. Perkins, N.S. Correal and R.J. O'Dea, Relative location estimation in wireless sensor networks, *IEEE Trans. Signal Processing*, 51, 2003, pp. 2137–2148.

31. H.-D. Qi, A semismooth Newton method for the nearest Euclidean distance matrix problem, *SIAM J. Matrix Anal. Appl.*, 34, 2013, pp. 67–93.

32. H.-D. Qi, Conditional quadratic semidefinite programming: Examples and methods, *J. Oper. Res. Society of China*, 2, 2014, pp. 143–170.

33. H.-D. Qi and X.M. Yuan, Computing the nearest Euclidean distance matrix with low embedding dimensions, *Math. Prog.*, 147, 2014, pp. 351–389.

34. H.-D. Qi, N.H. Xiu and X.M. Yuan, A Lagrangian dual approach to the single source localization problem, *IEEE Trans. Signal Process.*, 61, 2013, pp. 3815–3826.

35. I.J. Schoenberg, Remarks to Maurice Frechet's article Sur la definition axiomatque d'une classe d'espaces vectoriels distancies applicbles vectoriellement sur l'espace de Hilbet, *Ann. Math.*, 36, 1935, pp. 724–732.

36. R. Sibson, Studies in the robustness of multidimensional scaling: Perturbational analysis of classical scaling, *J. Royal Statistical Society, B*, 41, 1979, 217–219.

37. W.S. Torgerson, Multidimensional scaling: I. Theory and method, *Psychometrika*, 17, 1952, pp. 401–419.

38. K.Q. Weinberger and S.K. Saul, An introduction to nonlinear dimensionality reduction by maximum variance unfolding, *American Association for Artificial Intelligence*, 2006, pp. 1683–1686.

39. G. Young and A.S. Householder, Discussion of a set of points in terms of their mutual distances, *Psychometrika*, 3, 1938, pp. 19–22.

40. S. Zhou, N.H. Xiu and H.-D. Qi, A fast matrix majorization-projection method for penalized stress minimization with box constraints, *IEEE Trans. Signal Process.*, 66, 2018, pp. 4331–4346.

41. S. Zhou, N.H. Xiu and H.-D. Qi, *Robust Euclidean embedding via EDM optimization*, Technical Report, March, 2018.

6

Statistical Shape Analysis in Cooperative Localization

Ping Zhang

CONTENTS

In cooperative localization, internode measurements provide the relative location information of the nodes. This relative location information depicts the node location relative to each other, which is invariant in global translation, rotation/reflection, and in some cases scaling transformation of all nodes [1]. While in statistical shape analysis, when location, scale, and rotational effects are filtered out, the geometrical information remained is called shape, which is quite similar to the relative location. In this chapter, the concept of relative configuration is adopted to describe the shape or its variations of the node network, without considering its absolute location, orientation, and/or scaling. By using this concept, we investigate how to construct the optimal minimally constrained system (MCS) and give its geometric meaning, and discuss the anchor selection problem in cooperative localization.

6.1 Introduction of Statistical Shape Analysis

Shape is all the geometrical information that remains when location, scale, and rotational effects are filtered out from an object [2]. It can be found in the study of biology, medicine, image analysis, computer vision, archaeology, geography, geology, agriculture, and genetics, where shape is usually represented by a set of labeled points. In fact, shape is invariant under the Euclidean similarity transformations of translation, scaling, and rotation.

The Euclidean similarity transformation is represented as below. Let $\mathbf{s}_i = [s_{i,x}, s_{i,y}]^T$, $i = 1$, $2, \ldots, n$ denote the coordinates of n labeled points. The Euclidean similarity transformation of the location vector $s = [s_1^T, s_2^T, \ldots, s_n^T]^T$ (or named configuration) can be represented as [2]

$$T(\mathbf{s}) = \beta\Gamma\mathbf{s} + x\mathbf{1}_x + y\mathbf{1}_y \tag{6.1}$$

where $\mathbf{1}_x = [1, 0, \ldots, 1, 0]^T \in \mathbb{R}^{2n}$, $\mathbf{1}_y = [0, 1, \ldots, 0, 1]^T \in \mathbb{R}^{2n}$, x and y indicate the translation parameters in x and y directions, respectively, the total orthogonal matrix $\Gamma = \text{diag}(\Gamma_0, \Gamma_0, \ldots, \Gamma_0)$ is a $2n$-by-$2n$ block diagonal matrix composed of n rotation matrices Γ_0 in the 2-by-2 orthogonal group $\mathcal{O}(2)$, and the scaling factor $\beta > 0$ controls the scaling of the network.

Besides the definition of shape, there are several variations, such as size-and-shape and reflection shape. Size-and-shape removes location and rotational effects but not scale. It is invariant to rigid-body transformations. Reflection shape extends rotation to orthogonal transformation, which includes reflection. Since reflection is always unidentifiable through internode measurements in cooperative localization, we restrict our attention in this chapter to reflection shape and its varieties, and omit reflection without loss of clarity.

6.2 Statistical Shape Analysis for Relative Configuration

Internode measurements provide only the relative location information of the nodes, where the network absolute location, orientation, or even scaling are missing. For example, internode distance measurements cannot specify the absolute location and orientation of the network, angle of arrival (AoA) cannot specify the absolute location and scaling of the network, and the connectivity measurements can specify neither location, orientation, nor scaling of the network. In this section, statistical shape analysis is performed on the introduced concept relative configuration.

6.2.1 Definition of Relative Configuration

The relative configuration, or named relative map, describes the shape of a network without considering the networks absolute location, orientation, and in some cases scaling [3]. As seen in Figure 6.1, for 3-node networks, A, B, C, and D have the same relative configuration. Compared with A, triangle B involves a rotation, C involves a reflection, and D involves a scaling. E has a different relative configuration, except for exchanging the node labels 2 and 3. The relative configuration of F is different from all the others. Furthermore, when the scaling information is retained in the definition of the relative configuration, D becomes a relative configuration different from A, B, and C.

Mathematically, the relative configuration of a network with location vector $\mathbf{s}$ can be representation as the class

$$[\mathbf{s}] = \left\{ T(\mathbf{s}) = \beta\Gamma\mathbf{s} + x\mathbf{1}_x + y\mathbf{1}_y \mid \beta \in \mathbb{R}^+, \Gamma \in \mathcal{O}(2), x, y \in \mathbb{R} \right\}. \tag{6.2}$$

It depicts the geometrical information that can be specified by the same internode measurement, such as signal travel time (without the propagation speed known). But for

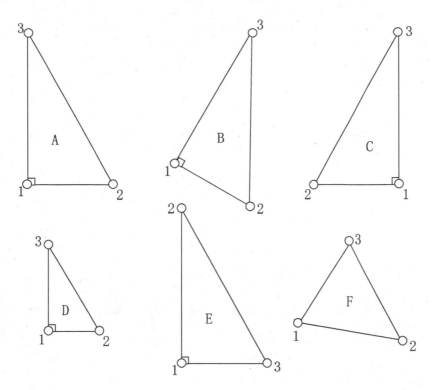

FIGURE 6.1
The relative configuration of 3-node networks.

other measurements, for example, internode distances, we can modify the definition of the relative configuration as

$$[\mathbf{s}] = \left\{ \mathcal{T}(\mathbf{s}) = \Gamma \mathbf{s} + x \mathbf{1}_x + y \mathbf{1}_y \mid, \Gamma \in \mathcal{O}(2), x, y \in \mathbb{R} \right\}. \tag{6.3}$$

In addition, for internode angle measurements, the relative configuration can also be changed as

$$[\mathbf{s}] = \left\{ \mathcal{T}(\mathbf{s}) = \beta \mathbf{s} + x \mathbf{1}_x + y \mathbf{1}_y \mid \beta \in \mathbb{R}^+, x, y \in \mathbb{R} \right\}. \tag{6.4}$$

6.2.2 Coordinate Representation

Any $\mathcal{T}(\mathbf{s})$ can be viewed as a coordinate representation of the relative configuration $[\mathbf{s}]$, which actually specifies a coordinate system implicitly. Here, the *full Procrustes coordinates* is introduced to as a coordinate representation of the relative configuration. Geometrically, the *full Procrustes coordinates* of the relative configuration are derived through superimposing $\mathbf{s}$ onto a $2n$-dimensional reference location vector $\mathbf{r} = [\mathbf{r}_1^T, \mathbf{r}_2^T, \ldots, \mathbf{r}_n^T]^T$ where $\mathbf{r}_i = [r_{i,x}, r_{i,y}]^T, i = 1, 2, \ldots, n$, as [2]

$$\mathbf{s_r} = \arg\min_{\mathcal{T}(\mathbf{s})} \left| \mathcal{T}(\mathbf{s}) - \mathbf{r} \right| \tag{6.5}$$

where $\|\cdot\|$ denotes Euclidean norm.

As an important attribution of the full Procrustes coordinates, the optimization problem (6.5) admits a closed-form solution, named full Procrustes fit [2,4], given by

$$\mathbf{s_r} = \mathcal{T}^*(\mathbf{s}) = \beta^* \mathbf{\Gamma}^* \mathbf{s} + x^* \mathbf{1}_x + y^* \mathbf{1}_y \tag{6.6}$$

where

$$\mathbf{\Gamma}_0^* = \mathbf{V}\mathbf{W}^T \tag{6.7}$$

$$\beta^* = \frac{\mathrm{tr}(\mathbf{\Sigma}_{s,r}\mathbf{\Gamma}_0^*)}{\mathrm{tr}(\mathbf{\Sigma}_{s,s})} \tag{6.8}$$

$$[x^*, y^*]^T = \mu_r - \beta^* \mathbf{\Gamma}_0^* \mu_s \tag{6.9}$$

$$\mathbf{\Gamma}^* = \mathrm{diag}(\mathbf{\Gamma}_0^*, \mathbf{\Gamma}_0^*, \ldots, \mathbf{\Gamma}_0^*). \tag{6.10}$$

Here, $\mathbf{W}\mathbf{D}\mathbf{V}^T$ is the singular value decomposition (SVD) of the covariance matrix.

$$\mathbf{\Sigma}_{s,r} = \frac{1}{n}\sum_{i=1}^{n}(\mathbf{s}_i - \mu_s)(\mathbf{r}_i - \mu_r)^T, \quad \mathbf{\Sigma}_{s,s} = \frac{1}{n}\sum_{i=1}^{n}(\mathbf{s}_i - \mu_s)(\mathbf{s}_i - \mu_s)^T$$

is the covariance matrix of s,

$$\mu_s = \frac{1}{n}\sum_{i=1}^{n}\mathbf{s}_i \quad \text{and} \quad \mu_r = \frac{1}{n}\sum_{i=1}^{n}\mathbf{r}_i$$

are the mean vectors of the node locations $\mathbf{s}_i$ and the reference locations $\mathbf{r}_i$, $i = 1, 2, \ldots, n$, respectively, and $\mathrm{tr}(\cdot)$ denotes trace operation.

The full Procrustes coordinates of the relative configuration depends on the selection of the reference vector $\mathbf{r}$. In fact, if $\mathbf{s}$ is the full Procrustes coordinates at the reference $\mathbf{r}$, it should fulfill the following parameter equation

$$\mathbf{g_r}(\mathbf{s}) = \begin{vmatrix} \mathbf{1}_x^T \mathbf{s} - \mathbf{1}_x^T \mathbf{r} \\ \mathbf{1}_y^T \mathbf{s} - \mathbf{1}_y^T \mathbf{r} \\ \mathbf{v}_r^T \mathbf{s} \\ \mathbf{s}^T \mathbf{s} - \mathbf{s}^T \mathbf{r} \end{vmatrix} = 0 \tag{6.11}$$

where $\mathbf{1}_x$ and $\mathbf{1}_y$ are defined in Equation 6.1, and

$$\mathbf{v_r} = [r_{1,y}, -r_{1,x}, \ldots, r_{n,y}, -r_{n,x}]^T. \tag{6.12}$$

In cooperative localization, a convenient choice of the reference vector is setting $\mathbf{r}$ at the ground truth $\mathbf{s}$. Then, the full Procrustes coordinates of the relative configuration estimate can be represented by $\hat{\mathbf{s}}_s$. Among all choices of the reference $\mathbf{r}$, $\hat{\mathbf{s}}_s$ owns the lowest squared error to the ground truth location $\mathbf{s}$. Also, this choice has a close relationship with the error metric explained in the subsection below.

For internode distance measurements, the scaling information should be retained. As a result, the coordinate of the relative configuration can be defined by *partial Procrustes coordinates/fit* as [2]

$$\mathbf{s_r} = \mathcal{T}^*(\mathbf{s}) = \mathbf{\Gamma}^*\mathbf{s} + x^*\mathbf{1}_x + y^*\mathbf{1}_y \tag{6.13}$$

where $\mathbf{\Gamma}^*$, x^*, and y^* are defined in Equations 6.9 and 6.10 with $\beta^* = 1$. Additionally, if $\mathbf{s}$ is the partial Procrustes coordinates at the reference $\mathbf{r}$, it should fulfill the following parameter equation

$$\mathbf{g_r}(\mathbf{s}) = \begin{bmatrix} \mathbf{1}_x^T\mathbf{s} - \mathbf{1}_x^T\mathbf{r} \\ \mathbf{1}_y^T\mathbf{s} - \mathbf{1}_y^T\mathbf{r} \\ \mathbf{v}_r^T\mathbf{s} \end{bmatrix} = \mathbf{0}. \tag{6.14}$$

6.2.3 Error Metric

To evaluate the error on the estimation of relative configuration, it is confusing that the Euclidean distance of the coordinate representation differs under different choices of the reference vector. To solve this problem, the *relative error* [3] is selected. Specifically, suppose $\hat{\mathbf{s}}$ is an estimation the location vector $\mathbf{s}$, the relative error is defined as the shortest squared distance from $\mathbf{s}$ to (seen in Figure 6.2) as

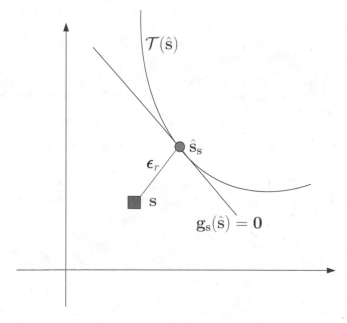

FIGURE 6.2
Illustration of the relative error.

$$\epsilon_r = \epsilon_r(\mathbf{s}) = \min_{\mathcal{T}(\hat{\mathbf{s}})} \| \mathcal{T}(\hat{\mathbf{s}}) - \mathbf{s} \|^2$$

$$= \| \hat{\mathbf{s}}_\mathbf{s} - \mathbf{s} \|^2 . \tag{6.15}$$

Although ϵ_r is not invariant to the scaling operation, it can be proved that it normalized version $\epsilon_r(\mathbf{s})/\mathrm{tr}(\Sigma_{\mathbf{s},\mathbf{s}})$ is similarity transformation independent.

For internode distance measurements, the relative configuration that retains the scaling information and the relative error can be defined as

$$\epsilon_r = \min_{\mathcal{T}(\hat{s})} \left| \mathcal{T}(\hat{\mathbf{s}}) - \mathbf{s} \right|^2 = \left| \hat{\mathbf{s}}_\mathbf{s} - \mathbf{s} \right|^2 \tag{6.16}$$

where $\hat{\mathbf{s}}_\mathbf{s}$ denotes the partial Procrustes coordinates of the relative configuration estimate at the ground truth $\mathbf{s}$. The following sections adopt the relative configuration that retains the scaling information.

6.3 Optimal Minimally Constrained System and Its Geometric Explanation

MCS refers to the minimal number of global (continuous) constraints needed to derive node absolute locations [3,5]. For example, in a two-dimensional rigid network, three constraints are required to specify the central location and rotation of the network. In this section, a geometric approach to construct an optimal MCS is proposed, which produces the lowest localization error.

6.3.1 Minimally Constrained System

Suppose a network composed of n nodes, whose locations are $\mathbf{s}_i = [s_{i,x}, s_{i,y}]^T$, $i = 1, 2, ..., n$. Under Gaussian additive noise assumption, the internode distance measurements can be modeled as

$$d_{i,j} = \| \mathbf{s}_i - \mathbf{s}_j \| + \epsilon_{i,j}, \quad \text{for } (i,j) \in \mathcal{E}, \tag{6.17}$$

where $\|\cdot\|$ denotes Euclidean norm, $\epsilon_{i,j}$'s are independent zero mean Gaussian stochastic variables with variance σ^2, and $\mathcal{E}$ is composed of the index pairs of the connected nodes. In this paper, the internode distance measurements are assumed to be symmetric so that any index pair $(i, j) \in \mathcal{E}$ fulfills $i < j$.

It can be seen in Equation 6.17 that global translation and rotation/reflection do not affect the internode distance measurement, which needs at least three global constraints to fix when sufficient internode distances make the network rigid. Here, we denote the global constraints as $\mathbf{h}(\mathbf{s}) = \mathbf{0}$.

Under the constraints $\mathbf{h}(\mathbf{s}) = \mathbf{0}$, the Cramér-Rao lower bound (CRLB) of the location vector $\mathbf{s}$ can be derived through constrained CRLB theory [6,7] as

$$C_s = \mathbf{U}(\mathbf{U}^T \mathbf{J}_s \mathbf{U})^{-1} \mathbf{U}^T, \tag{6.18}$$

where the columns of $\mathbf{U}$ form an orthonormal basis of the null space of $\partial \mathbf{h}(\mathbf{s})/\partial \mathbf{s}^T$, and $\mathbf{U}^T \mathbf{J}_s \mathbf{U}$ is reversible under mild restrictions that the constraints $\mathbf{h}(\mathbf{s}) = 0$ are non-degenerate. Here, the Fisher information matrix (FIM) is

$$\mathbf{J}_s = -\mathrm{E}\left[\frac{\partial^2 l(\mathbf{s})}{\partial \mathbf{s} \partial \mathbf{s}^T}\right]$$

$$= \sigma^{-2} \mathbf{F}^T \mathbf{F}, \tag{6.19}$$

where $\mathbf{F}$ is obtained by stacking

$$\left[\mathbf{0}_{1\times 2(i-1)}, \frac{\mathbf{s}_i^T - \mathbf{s}_j^T}{\|\mathbf{s}_i - \mathbf{s}_j\|}, \mathbf{0}_{1\times 2(j-i-1)}, \frac{\mathbf{s}_j^T - \mathbf{s}_i^T}{\|\mathbf{s}_i - \mathbf{s}_j\|}, \mathbf{0}_{1\times 2(n-j)}\right] \tag{6.20}$$
$$\text{for } (i,j) \in \mathcal{E}$$

in row. Since the internode measurements cannot specify the network location and orientation, the FIM $\mathbf{J}_s$ is rank deficient. When the internode distances are sufficient to make the network rigid, it can be verified that $\mathbf{J}_s$'s null space is spanned by $\mathbf{1}_x = [1, 0, ..., 1, 0]^T \in \mathbb{R}^{2n}$, $\mathbf{1}_y = [0, 1, ..., 0, 1]^T \in \mathbb{R}^{2n}$, and $\mathbf{v}_s = [s_{1,y}, -s_{1,x}, s_{2,y}, ..., -s_{n,x}]^T \in \mathbb{R}^{2n}$.

6.3.2 Optimality, Uniformity, and Geometric Explanation

The CRLB (6.18) depends on the MCS $\mathbf{h}(\mathbf{s}) = 0$.

In this section, we construct an optimal MCS which minimizes the trace of the CRLB, discuss the practical feasibility of the construction, and provide a geometric approach to construct a practicable MCS.

6.3.2.1 Optimality

The trace of $\mathbf{J}_s^\dagger$ lower bounds the trace of the CRLB (6.18), where, $\mathbf{J}_s^\dagger$ denotes the Moore-Penrose pseudo-inverse of the FIM (6.19). A brief proof of this fact is given below.

Let $\mathbf{J} = \mathbf{U}_s \mathbf{\Lambda}_s \mathbf{U}_s^T$ be the compact SVD [8] of $\mathbf{J}$, where $\mathbf{\Lambda}_s$ is a $(2n - 3)$-by-$(2n - 3)$ diagonal matrix with positive diagonal entities, and $\mathbf{U}_s$ is a $2n$-by-$(2n - 3)$ matrix whose columns are composed of the eigenvectors corresponding to the positive eigenvalues. Then

$$(\mathbf{U}^T \mathbf{J} \mathbf{U})^{-1} = (\mathbf{U}_s^T \mathbf{U})^{-1} \mathbf{\Lambda}_s^{-1} (\mathbf{U}^T \mathbf{U}_s)^{-1} \tag{6.21}$$

$$\mathrm{tr}((\mathbf{U}^T \mathbf{J} \mathbf{U})^{-1}) = \mathrm{tr}(\mathbf{\Lambda}_s^{-1}(\mathbf{U}_s^T \mathbf{U} \mathbf{U}^T \mathbf{U}_s)^{-1}). \tag{6.22}$$

Since

$$\mathbf{U}_s^T \mathbf{U} \mathbf{U}^T \mathbf{U}_s \leq \mathbf{U}_s^T \mathbf{U}_s = \mathbf{I}_{2n-3} \tag{6.23}$$

we have

$$(\mathbf{U}_s^T \mathbf{U} \mathbf{U}^T \mathbf{U}_s)^{-1} \geq \mathbf{I}_{2n-3}. \tag{6.24}$$

Hence

$$\text{tr}(\mathbf{U}(\mathbf{U}^T\mathbf{J}\mathbf{U})^{-1}\mathbf{U}^T) \geq \text{tr}(\mathbf{\Lambda}_s^{-1}) = \text{tr}(\mathbf{J}^\dagger) \tag{6.25}$$

where it is easy to prove that the equality holds if $\mathbf{h(s)} = \mathbf{g_r(s)}$ given in Equation 6.14 with $\mathbf{r} = \mathbf{s}$ as a known parameter.

Besides (6.14), the linear constraints $\mathbf{h(s)} \triangleq \mathbf{As} - \mathbf{b} = \mathbf{0}$ can minimize the CRLB trace by letting the rows of the coefficient matrix $\mathbf{A}$ composed of an orthonormal basis of the null space of the FIM $\mathbf{J_s}$, which are $\mathbf{1}_x^T$, $\mathbf{1}_y^T$, and $\mathbf{v}_s^T$ when the network is rigid. The observation $\mathbf{b} = \mathbf{As}$, which is $[\mathbf{1}_x^T\mathbf{s}, \mathbf{1}_y^T\mathbf{s}, 0]^T$ for rigid networks. Here, both $\mathbf{A}$ and $\mathbf{b}$ are assumed to be known *a priori*.

6.3.2.2 Uniformity

Constructing the coefficient matrix $\mathbf{A}$ and the observation $\mathbf{b}$ needs to know the ground truth locations $\mathbf{s}$ *a priori*. In practice, an optimal MCS is preferred to be constructed by some consistent measurements on the node locations, so that the MCS may keep the optimal uniformly across all possible node locations.

However, *no* such MCS exists. For globally rigid networks, it is proved that there is *no* MCS whose CRLB trace achieves the lower bound $\text{tr}(\mathbf{J}^\dagger)$ for all $\mathbf{s} \in \mathbb{R}^{2n}$, as given below.

Without loss of generality, we assume the network is globally rigid. Let $\mathbf{U}$ be a $2n$-by-$(2n-3)$ matrix whose columns form an orthogonal basis of the null space of $\mathbf{H}^T = \partial \mathbf{h(s)}/\partial \mathbf{s}^T$, then the equality

$$\text{tr}\left(\mathbf{U}(\mathbf{U}^T\mathbf{J}\mathbf{U})^{-1}\mathbf{U}^T\right) = \text{tr}(\mathbf{J}^\dagger) \tag{6.26}$$

holds if and only if the columns of $\mathbf{U}$ span the eigenspace of $\mathbf{J}$ that corresponds to the non-zero eigenvalues. Equivalently, this requires the columns of $\mathbf{H}$ span the null space of $\mathbf{J}$.

Since the null space of $\mathbf{J}$ is spanned by the vectors $\mathbf{1}_x$, $\mathbf{1}_y$, and $\mathbf{v}_s$, we get immediately that there should exist an invertible 3-by-3 matrix $\mathbf{T(s)}$ which makes the equalities hold.

$$\mathbf{T(s)}\frac{\partial \mathbf{h(s)}}{\partial \mathbf{s}^T} = [\mathbf{1}_x, \mathbf{1}_y, \mathbf{v}_s]^T \tag{6.27}$$

Note that

$$\mathbf{T(s)}\frac{\partial \mathbf{h(s)}}{\partial \mathbf{s}^T} = \frac{\partial \mathbf{T(s)}\mathbf{h(s)}}{\partial \mathbf{s}^T}$$

holds because $\mathbf{h(s)} = \mathbf{0}$, there must exist a function $\tilde{h}(\mathbf{s})$ satisfying $\partial \tilde{h}(\mathbf{s})/\partial \mathbf{s} = \mathbf{v}_s$ to make the last equality in Equation 6.27 hold. But such $\tilde{h}(\mathbf{s})$ does not exist since $\partial \tilde{h}(\mathbf{s})/\partial s_{1,x} = s_{1,y}$ and $\partial \tilde{h}(\mathbf{s})/\partial s_{1,y} = -s_{1,x}$ contradictory expressions of $\tilde{h}(\mathbf{s})$ as

$$\tilde{h}(\mathbf{s}) = s_{1,x}s_{1,y} + c_1 = -s_{1,x}s_{1,y} + c_2 \tag{6.28}$$

where c_1 and c_2 are independent of $s_{1,x}$ and $s_{1,y}$, respectively. As a result, there is no MCS $\mathbf{h(s)} = \mathbf{0}$ whose CRLB trace achieves the lower bound $\text{tr}(\mathbf{J}^\dagger)$ for all $\mathbf{s} \in \mathbb{R}^{2n}$.

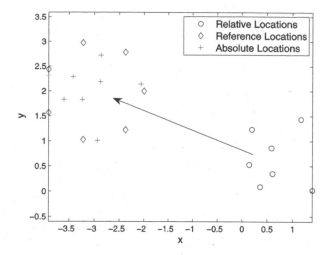

FIGURE 6.3
Superimposing operation: The absolute locations are obtained by "superimposing" the relative locations onto the reference locations, so that the squared distance between the absolute and the reference locations is minimized.

This means that no measurements can behave as an MCS, which keeps the optimal uniformly across all possible node locations. Since no MCS is the optimal for all node locations, we redirect our attention to find some practically available MSCs which maintain the performance under some given node locations.

6.3.2.3 Geometric Approach

The constraint (6.14) is equivalent to superimposing $\hat{s}$ onto r as (6.13), seen in Figure 6.3. But $\hat{s}_r$ is a biased estimate, which approaches s_r, not s, as the internode measurement noise decreases to zero. Note that both $\hat{s}_r$ and s_r fulfill the MCS (6.11); we get a lower bound for the mean squared error of $\hat{s}_r$ through biased CRLB

$$E[(\hat{s}_r - s)(\hat{s}_r - s)^T]$$
$$\geq \Gamma^{*T} U_r (U_r^T J_{s_r} U_r)^{-1} U_r^T \Gamma^* + bb^T, \qquad (6.29)$$

where $\hat{s}_r$ is an unbiased estimate of s_r, $b = s_r - s$ denotes the bias, and the columns of U_r form an orthonormal basis of the null space of $[1_x, 1_y, v_r]^T$.

Peculiarly, when $r = s$, $s = s_s$ is unbiased, and thus

$$E[(\hat{s}_s - s)(\hat{s}_s - s)^T] \geq J_s^\dagger \qquad (6.30)$$

6.4 Anchor Selection in Cooperative Localization

By using anchors, the constraints can be represented as $Hs = b$, where H is a $2n$-by-$2(n - m)$ matrix which is stacked in column by

$$[0_{2 \times 2(j-1)}, I_2, 0_{2 \times 2(n-j)}]^T, \quad j \in \mathcal{U}$$

in column. Here, $\mathcal{U}$ and $\mathcal{A}$ denote the index set of the unknown nodes and the anchor nodes, respectively. Under the constraints $\mathbf{Hs} = \mathbf{b}$, the CRLB of the unknown node locations can be represented as

$$\mathbf{C} = (\mathbf{H}^T \mathbf{J_s} \mathbf{H})^{-1} \tag{6.31}$$

In addition, let $\mathbf{V} = [\mathbf{1}_x, \mathbf{1}_y, \mathbf{v}]$; we can perform a compact singular value decomposition (SVD) of the FIM $\mathbf{J_s}$ as

$$\mathbf{J_s} = \mathbf{U}\boldsymbol{\Lambda}\mathbf{U}^T \tag{6.32}$$

where $\boldsymbol{\Lambda}$ is a diagonal matrix with diagonal elements $\lambda_1 \geq \lambda_2 \geq \cdots \geq \lambda_{2n-3} > 0$, and the columns of $\mathbf{U}$ form an orthonormal basis of the null space of $\mathbf{V}^T$.

6.4.1 Bound Analysis

It is somewhat difficult to derive the perimeter anchor deployment strategy from the analytic expression of Equation 6.31. To cope with this problem, we introduce an approximation $b(\mathcal{A})$ of tr $(\mathbf{C})$, where the difference between tr $(\mathbf{C})$ and $b(\mathcal{A})$ is bounded as below.

The ratio between tr$(\mathbf{C})$ and $b(\mathcal{A})$ is bounded as

$$\lambda_1^{-1} \leq \frac{\mathrm{tr}(\mathbf{C})}{b(\mathcal{A})} \leq \lambda_{2n-3}^{-1}. \tag{6.33}$$

where λ_1 and λ_{2n-3} are the first and the $(2n-3)$th largest eigenvalues of the distance FIM $\mathbf{J_s}$, and

$$b(\mathcal{A}) = \mathrm{tr}((\mathbf{H}^T\mathbf{U}\mathbf{U}^T\mathbf{H})^{-1})$$
$$\frac{n}{m}\left(\frac{(\bar{s}_{A,x} - \bar{s}_x)^2 + (\bar{s}_{A,y} - \bar{s}_y)^2 + \rho^2}{\rho_A^2} + 2\right) + 2(n-m) - 3 \tag{6.34}$$

where

$$\bar{s}_x = \frac{1}{n}\sum_{i=1}^{n}s_{i,x}, \quad \bar{s}_y = \frac{1}{n}\sum_{i=1}^{n}s_{i,y}, \quad \bar{s}_{A,x} = \frac{1}{m}\sum_{j \in \mathcal{A}}s_{j,x},$$

$$\bar{s}_{A,y} = \frac{1}{m}\sum_{j \in \mathcal{A}}s_{j,y}, \quad \rho^2 = \frac{1}{n}\sum_{i=1}^{n}(s_{i,x} - \bar{s}_x)^2 + (s_{i,y} - \bar{s}_y)^2,$$

and

$$\rho_A^2 = \frac{1}{m}\sum_{j \in \mathcal{A}}(s_{j,x} - \bar{s}_{A,x})^2 + (s_{j,y} - \bar{s}_{A,y})^2.$$

The derivation of Equations 6.33 and 6.34 can be found in [9].

Minimizing $b(\mathcal{A})$ leads to the uniform perimeter anchor deployment strategy. In Equation 6.34, $b(\mathcal{A})$ can be viewed as a function of $(\bar{s}_{A,x} - \bar{s}_x)^2 + (\bar{s}_{A,y} - \bar{s}_y)^2$ and $\rho\mathcal{A}$ that are affected by the anchor selection. To minimize $b(\mathcal{A})$, $(\bar{s}_{A,x} - \bar{s}_x)^2 + (\bar{s}_{A,y} - \bar{s}_y)^2$ should be minimized, and $\rho\mathcal{A}$ should be maximized. $(\bar{s}_{A,x} - \bar{s}_x)^2 + (\bar{s}_{A,y} - \bar{s}_y)^2$ evaluates the squared distance between the centroids of the anchors and the nodes. It can be reduced to zero if $\bar{s}_{A,x} = \bar{s}_x$ and $\bar{s}_{A,y} = \bar{s}_y$. $\rho\mathcal{A}$ quantifies the diameter of the network composed of the anchors. To maximize it, the anchors should be deployed on the perimeter of the network. When the nodes are randomly deployed in a two-dimensional plane, deploying the anchors uniformly around the perimeter of the network meets the requirements above, so that it can be viewed as an optimal strategy to reduce $b(\mathcal{A})$.

6.4.2 Isotropic Discriminability

What is $b(\mathcal{A})$? By using relative-transformation decomposition, the location vector $\mathbf{s}$ can be reparameterized as

$$\mathbf{s} = \mathbf{U}\eta + \mathbf{V}\zeta \qquad (6.35)$$

where $\eta \in \mathbb{R}^{2n-3}$ and $\zeta \in \mathbb{R}^3$ refer to the coordinates in the relative and transformation subspace, respectively.

Under the assumption that the deformation of the relative configuration is isotropically discriminable; that is, the FIM $\mathbf{J}_\eta = \alpha\mathbf{I}_{2n-3}$, $\alpha > 0$, and no information on the global transformation is available; that is, the FIM $\mathbf{J}_\zeta = \mathbf{0}$, we have

$$\mathbf{J}_s = \alpha\mathbf{U}\mathbf{U}^T. \qquad (6.36)$$

Throughout this paper, we ignore the factor α without loss of generality.

After setting m anchors, we get the CRLB of $\mathbf{s}$ from Equation 6.18, whose trace is just $b(\mathcal{A})$. Comparing (6.32) and (6.36), we find that the approximation $b(\mathcal{A})$ is equivalent to an isotropic discriminability approximation of the deformation of the relative configuration, where the similarity can be evaluated by the eigenvalue ratio λ_1/λ_{2n-3}.

6.4.3 Performance Analysis

The performance of the approximation (6.34) can be quantified through the CRLB ratio $\mathrm{tr}(\mathbf{C}(\mathcal{A}_b))/\mathrm{tr}(\mathbf{C}(\mathcal{A}_c))$, where the numerator is the trace of the CRLB (6.18) corresponding to the anchor set $\mathcal{A}_b$ obtained by minimizing $b(\mathcal{A})$ and the denominator referring to the anchor set $\mathcal{A}_c$ obtained by minimizing the CRLB trace directly. This performance metric is bounded as below.

The CRLB ratio is bounded as

$$1 \le \frac{\mathrm{tr}(\mathbf{C}(\mathcal{A}_b))}{\mathrm{tr}(\mathbf{C}(\mathcal{A}_c))} \le \frac{\lambda_1}{\lambda_{2n-3}}. \qquad (6.37)$$

The proof is given below.

Let $\mathcal{A}_c$ and $\mathcal{A}_b$ denote the anchor set obtained by minimizing $\mathrm{tr}(\mathbf{C})$ and $b(\mathcal{A})$, respectively. After replacing $\mathbf{C}$ with $\mathbf{C}(\mathcal{A})$ to stress its dependence on the anchor selection, we get

$$\frac{\mathrm{tr}(\mathbf{C}(\mathcal{A}_b))}{\mathrm{tr}(\mathbf{C}(\mathcal{A}_c))} \ge 1 \qquad (6.38)$$

$$\frac{b(\mathcal{A}_b)}{b(\mathcal{A}_c)} \leq 1. \tag{6.39}$$

From Equation 6.33, we have

$$\mathrm{tr}(\mathbf{C}(\mathcal{A}_c)) \geq \lambda_1^{-1} b(\mathcal{A}_c) \tag{6.40}$$

$$\mathrm{tr}(\mathbf{C}(\mathcal{A}_b)) \leq \lambda_{2n-3}^{-1} b(\mathcal{A}_b) \tag{6.41}$$

and thus

$$\frac{\mathrm{tr}(\mathbf{C}(\mathcal{A}_b))}{\mathrm{tr}(\mathbf{C}(\mathcal{A}_c))} \leq \frac{\lambda_{2n-3}^{-1} b(\mathcal{A}_b)}{\lambda_1^{-1} b(\mathcal{A}_c)} \leq \frac{\lambda_1}{\lambda_{2n-3}} \tag{6.42}$$

where the right inequality is obtained by using Equation 6.39.

From Equations 6.38 and 6.42, we get

$$1 \leq \frac{\mathrm{tr}(\mathbf{C}(\mathcal{A}_b))}{\mathrm{tr}(\mathbf{C}(\mathcal{A}_c))} \leq \frac{\lambda_1}{\lambda_{2n-3}}. \tag{6.43}$$

The eigenvalue ratio λ_1/λ_{2n-3} is independent of the anchor selection. When λ_1/λ_{2n-3} approaches 1, the anchor set selected by minimizing $b(\mathcal{A})$ would be close to the optimal one, but there is a negative result.

For fully connected networks, the eigenvalue ratio λ_1/λ_{2n-3} is bounded as

$$\lambda_1/\lambda_{2n-3} \geq 2. \tag{6.44}$$

The proof is given below.

Without loss of generality, we set $\sigma^2 = 1$. Then for a fully connected network, it can be verified that n is an eigenvalue of $\mathbf{J}_s$ with eigenvector proportional to $\mathbf{s} - (1/n)\mathbf{1}_x\mathbf{1}_x^T\mathbf{s} - (1/n)\mathbf{1}_y\mathbf{1}_y^T\mathbf{s}$. Note that the ith 2-by-2 diagonal block of $\mathbf{J}_s$ can be rewritten as $\sum_{j,j \neq i} \tau_{i,j}\tau_{i,j}^T$, and

$$\mathrm{tr}(\tau_{i,j}\tau_{i,j}^T) = \mathrm{tr}(\tau_{i,j}^T\tau_{i,j}) = 1, \quad i \neq j \tag{6.45}$$

we have

$$\mathrm{tr}(\mathbf{J_s}) = \sum_{i=1}^{n} \mathrm{tr}\left(\sum_{j,j \neq i} \tau_{i,j}\tau_{i,j}^T\right) = n(n-1). \tag{6.46}$$

Since $\mathbf{J}_s$ is positive semidefinite with eigenvalues $\lambda_1 \geq \lambda_2 \geq \cdots \geq \lambda_{2n-3} \geq \lambda_{2n-2} = \lambda_{2n-1} = \lambda_{2n} = 0$, we get

$$\lambda_{2n-3} = \min_{i=2,3,\ldots,2n-3} \lambda_i \leq \frac{n(n-1)-n}{2n-4} = \frac{n}{2}. \tag{6.47}$$

Therefore, $\lambda_1/\lambda_{2n-3} \geq 2$ because $\lambda_1 \geq n$.

References

1. P. Zhang and Q. Wang. On using the relative configuration to explore cooperative localization. *IEEE Transactions on Signal Processing*, 62(4):968–980, 2014.
2. I.L. Dryden and K.V. Mardia. *Statistical Shape Analysis*. Wiley, New York, 1998.
3. J.N. Ash and R.L. Moses. On the relative and absolute positioning errors in self-localization systems. *IEEE Transactions on Signal Processing.*, 56(11):5668–5679, Nov. 2008.
4. K.S. Arun, T.S. Huang, and S.D. Blostein. Least-squares fitting of two 3-d point sets. *IEEE Transactions on Pattern Analysis and Machine Intelligence*, 9:698–700, Sep. 1987.
5. P. Zhang, N. Yan, J. Zhang, and C. Yuen. Optimal minimally constrained system in cooperative localization [C]. *Wireless Communications & Signal Processing (WCSP), International Conference on, Nanjing, 2015*: 1–5.
6. J.D. Gorman and A.O. Hero. Lower bounds for parametric estimation with constraints. *IEEE Transactions on Information Theory*, 36(6):1285–1301, Nov. 1990.
7. P. Stoica and B.C. Ng. On the Cramér-Rao bound under parametric constraints. *IEEE Signal Processing Letters*, 5(7):177–179, Jul. 1998.
8. T.S. Shores. *Applied Linear Algebra and Matrix Analysis*, volume 10. Springer Verlag, 2007.
9. P. Zhang, A.L. Cao, and T. Liu. Bound analysis for anchor selection in cooperative localization. In *Industrial IoT Technologies and Applications Second EAI International Conference*, pages 1–10, Mar. 2017.

7

On Resilience and Heterogeneity in Robotic Networks

Jackeline Abad Torres, Patricio J. Cruz, Renato Vizuete, and Rafael Fierro

CONTENTS

Significant research gaps need to be overcome in order to enable teams of collaborative mobile agents to operate in dynamical and possibly hostile environments such as inside collapsed structures or dense urban areas. For example, the operational capability can be limited by the complexity of the physical environment, prior access or knowledge of the area, availability of supporting infrastructure, and presence of peer adversaries. Therefore, future intelligent robotic systems will need to work across large heterogeneous teams of mobile agents and exhibit adaptable levels of autonomy. Fundamental gaps exist in the understanding of heterogeneous resilient cooperative robotic networks. Designing and operating such networks are difficult as the number of agents, the degree heterogeneity, and the agents' adaptability change over time. Although resilience and heterogeneity play a critical role in multi-agent systems, they are not well understood. In this chapter, we introduce a series of definitions, metrics, and theory related to heterogeneity and resilience in the context of multi-robotic systems. A deeper understanding of these characteristics is critical to integrate them in the design and control of such intelligent mobile networks. We also present a case study based on the containment formation control problem for a group of homogeneous and heterogeneous agents.

7.1 Introduction

Cooperative mobile agents have the potential to provide essential support to human teams in time-critical missions such as search-and-rescue and disaster relief operations. This type of intelligent system is expected to have the ability to understand and learn from the environment, adapt to highly complex situations, conduct useful activity, assist in making rapid decisions, and enable new missions that would otherwise be impossible. Furthermore, future intelligent networked systems will need to exhibit collaborative levels of autonomy within well prescribed bounds in order to work across large heterogeneous teams of mobile agents and humans.

Our vision is to exploit the distributed perception abilities of a collaborative heterogeneous team as well as their processing and intelligence capabilities to make rapid joint decisions to adapt missions in the face of unexpected events. This vision is not about developing a singular technology component, but rather how to integrate varying levels of perception and autonomy all under one command and control architecture. This type of collaborative approach can augment the capability of the collective to solve the challenge of operating effectively in complex environments with limited or non *a priori* information.

As an example, a mobile robotic team can consist of unmanned aerial vehicles (UAVs) which can enter the building quickly and record aerial images of the interior. However, their range is limited by their battery capacity and the possibility that there are features such as narrow passages or blocked hallways which prevent UAVs from flying into certain critical areas. The areas inaccessible to aerial robots may be accessed by small, agile unmanned ground vehicles (UGVs), which can go under or around the obstacles and explore the areas beyond. A sketch of a collaborative heterogeneous team of UAVs and UGVs operating in a collapsed environment is shown in Figure 7.1. Therefore, these mobile agents are complementary, and by coordinating them, one can exploit their particular capabilities to optimize and adapt missions in the face of compromised environments, unexpected events, and adversarial actions.

In the last few years, a significant effort has been made to design, implement, and test planning algorithms for a prescribed heterogeneous multi-agent team. Collaborative multi-robot teams have been proposed for applications such as exploration and mapping, package delivery, sensor coverage, and connectivity maintenance [8,10,12,21]. When dealing with these systems, one of the main challenges is the development of control methods that can complement the distinct capabilities of the agents within the system by coordinating actions to accomplish a common goal. Selecting the heterogeneous agents that can execute a given task efficiently is also a critical step. The common solution to this problem is to map agent capabilities to mission requirements and to find the minimum set of agents required for a task [6,9]. Also, the effects of the diversity on the completion of a goal in a multi-agent team have been explored [24].

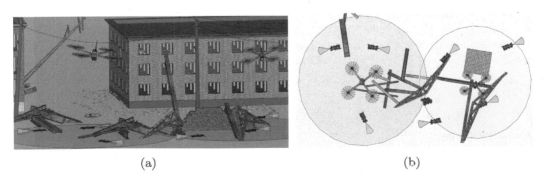

(a)   (b)

FIGURE 7.1
A group of aerial vehicles collaborating with a group of ground mobile robotic sensors in a disaster relief scenario. (a) 3D view and (b) 2D view.

Despite all these efforts, there is no general framework or theory that leads directly to algorithms and implementations for cooperative heterogeneous multi-agent networks. Novel and possibly hybrid control architectures for large heterogeneous teams are needed that may include global and local control, emergent behaviors, and resilience. Heterogeneous agents, which can be physical and/or virtual, may differentiate in various ways including dynamics, size, sensing capacity, computation power, communications, and degree of autonomy. Thus, it is important to study and design how to merge all these diverse capabilities in the context of group control.

Re-teaming and adaptive mission planning are needed when faced with difficult operational conditions in the physical world. Therefore, it is desirable to merge resilient-based control methods for heterogeneous teams. Resilient approaches enable the team to be robust, for instance to networking and information uncertainty or to platform-level malfunctions. The notion of *robustness* relates to the closed loop system's ability to handle uncertainties on the model's parameters and disturbances. This property is considered a pre-event or offline concept since the system is generically designed to satisfy this characteristic [30,35]. On the other hand, *resilience* is the system's ability to self-recover after adversarial attacks or system failure, which is typically considered an online or post-event concept [35]. A team of UAVs or UGVs that are required to work in an adversarial environment should be capable of maintaining the state of awareness and working with acceptable performance under threats and disturbances of an unexpected or malicious nature [26]. The demands of such systems require knowledge of what types of uncertainties or malicious events should be considered for pre-event robustness and post-event resilience, which implies a compromise between control performance and security of the agent's network. Further, both robustness and resilience are studied together, since the resilience of a system relies on its robustness by considering the post-event model features.

In this chapter, we consider that system failures or adversarial attacks affect a robot such that it communicates incorrect information and on a global level prevents achievement of the team's task. For instance, an adversarial attack can manipulate a robot to transmit erroneous or no data; meanwhile, a system failure—for example, battery depletion or defective sensors—can induce the same problem.

7.2 Heterogeneity in Multi-Agent Systems

In this section we discuss how *heterogeneity* has been defined or categorized in the context of multi-agent robotic systems. Heterogeneity is a term generally used to characterize the variations between the members of a system. In the case of multi-robot networks, these variations may be evident as different agent dynamics; for example a group of UGVs and UAVs, or as subtle as the individual mission objectives; for example mobile sensors together with mobile communication relays. The term heterogeneous system can be used to describe both types of systems. However, it does not explain how these variations are combined to solve a global goal.

In order to characterize heterogeneity to better define the problem being addressed, two main categories, *hardware-based* and *objective-based* heterogeneity, are proposed in [10]. The former is given by the disparity in the hardware between the robots within the network; for example, different platform dynamics, sensor footprints, communication ranges, and computational resources. The latter focuses on giving the agents local goals that must be met in order to accomplish a global mission objective. An example of hardware-based heterogeneity is a cooperative team formed by a truck and a quadrotor that are deployed to perform autonomous deliveries in urban environments [21]. On the other hand, a clear example of objective-based heterogeneity is a network of mobile platforms with identical hardware characteristics deployed to fully explore an unknown environment, but a group of them acts as localizers and the rest as mappers [4].

From this brief discussion, the term heterogeneity may not be descriptive enough to characterize the diversity within a multi-agent intelligent system. Because of this, there has been an effort to define a metric for the heterogeneity in this type of system such as in [3,11,14,32]. For details about the following definitions, please refer to these references.

Let $\mathcal{R}$ be a set of N robots with $\mathcal{R} = \{r_1, r_2, \ldots, r_N\}$ which forms a multi-agent system. Let $\mathcal{C} = \{c_1, c_2, \ldots, c_M\}$ be a classification of $\mathcal{R}$ into M possibly overlapping subsets, so c_i is an individual subset of $\mathcal{C}$. Let p_i be the ratio of the robots in the ith subset that is given by

$$p_i = \frac{|c_i|}{\sum_{j=1}^{M} |c_j|}, \tag{7.1}$$

where $|c_i|$ is the cardinality of the ith subset. Notice that $\sum_{i=1}^{N} p_i = 1$.

The variety of a multi-robotic system describes the diverse homogeneous subgroups that are part of the system. Thus, the terms variety and diversity are generally equivalent. For example, consider that we have available the six sets illustrated in Figure 7.2. These are $\mathcal{R}_0$, $\mathcal{R}_1$, $\mathcal{R}_2$, $\mathcal{R}_3$, $\mathcal{R}_4$, and $\mathcal{R}_5$. Within these sets of four robots, in each one we have two types of agents: aerial robotic platforms and ground vehicles. Each set has a different number of subsets, from one in Figure 7.2a to four in Figure 7.2f. Thus, the number of subsets is an important component of the variety of the set of robots. Now focusing on Figure 7.2b and c, which set is more diverse? Both cases have six robots and two different subsets. However, $\mathcal{R}_1$ has a higher number of aerial vehicles than $\mathcal{R}_2$. Therefore, the relative proportion of members in each subset is also important in the meaning of diversity. These examples highlight that the number of subsets in the system and the distribution of the members into those subsets are fundamental in a metric for diversity. Next, we provide a definition of this metric.

Definition 7.1 (Variety/Diversity)

The variety of a set of robots $\mathcal{R}$ whose classification is given by $\mathcal{C} = \{c_1, c_2, \ldots, c_M\}$ can be defined as

$$V(\mathcal{R}) = [p_1, p_2, \ldots, p_M], \tag{7.2}$$

where p_i is given by Equation 7.1.

For instance, the diversity of $\mathcal{R}_1$ shown in Figure 7.2b is $V(\mathcal{R}_1) = [3/4, 1/4]$, while for $\mathcal{R}_2$ shown in Figure 7.2c is $V(\mathcal{R}_2) = [1/2, 1/2]$. Thus, the variety of a system describes how well spread out the members are among the classification subsets. This measure of disorder can be also captured by the *entropy* of the multi-robotic system.

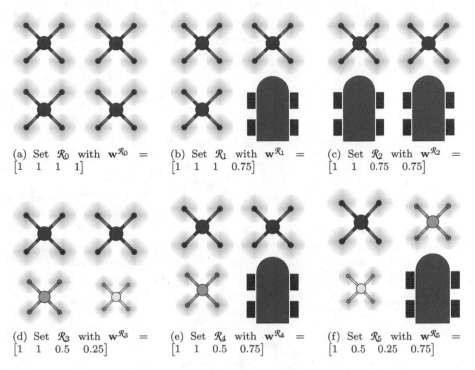

(a) Set $\mathcal{R}_0$ with $\mathbf{w}^{\mathcal{R}_0} = \begin{bmatrix} 1 & 1 & 1 & 1 \end{bmatrix}$

(b) Set $\mathcal{R}_1$ with $\mathbf{w}^{\mathcal{R}_1} = \begin{bmatrix} 1 & 1 & 1 & 0.75 \end{bmatrix}$

(c) Set $\mathcal{R}_2$ with $\mathbf{w}^{\mathcal{R}_2} = \begin{bmatrix} 1 & 1 & 0.75 & 0.75 \end{bmatrix}$

(d) Set $\mathcal{R}_3$ with $\mathbf{w}^{\mathcal{R}_3} = \begin{bmatrix} 1 & 1 & 0.5 & 0.25 \end{bmatrix}$

(e) Set $\mathcal{R}_4$ with $\mathbf{w}^{\mathcal{R}_4} = \begin{bmatrix} 1 & 1 & 0.5 & 0.75 \end{bmatrix}$

(f) Set $\mathcal{R}_5$ with $\mathbf{w}^{\mathcal{R}_5} = \begin{bmatrix} 1 & 0.5 & 0.25 & 0.75 \end{bmatrix}$

FIGURE 7.2
Heterogeneity example. (a) Homogeneous group, results: $V(\mathcal{R}_0) = [1]$, $E(\mathcal{R}_0) = Q(\mathcal{R}_0) = H(\mathcal{R}_0) = 0$.
(b) Heterogeneous group 1, results: $V(\mathcal{R}_1) = [3/4, 1/4]$, $E(\mathcal{R}_1) = 0.244$, $Q(\mathcal{R}_1) = 0.070$, and $H(\mathcal{R}_1) = 0.018$.
(c) Heterogeneous group 2, results: $V(\mathcal{R}_2) = [1/2, 1/2]$. $E(\mathcal{R}_2) = 0.301$, $Q(\mathcal{R}_1) = 0.125$, and $H(\mathcal{R}_2) = 0.038$.
(d) Heterogeneous group 3, results: $V(\mathcal{R}_3) = [1/2, 1/4, 1/4]$, $E(\mathcal{R}_3) = 0.452$, $Q(\mathcal{R}_1) = 0.438$, and $H(\mathcal{R}_3) = 0.198$.
(e) Heterogeneous group 4, results: $V(\mathcal{R}_4) = [1/2, 1/4, 1/4]$, $E(\mathcal{R}_4) = 0.452$, $Q(\mathcal{R}_1) = 0.170$, and $H(\mathcal{R}_4) = 0.077$.
(f) Heterogeneous group 5, results: $V(\mathcal{R}_5) = [1/4, 1/4, 1/4, 1/4]$, $E(\mathcal{R}_5) = 0.602$, $Q(\mathcal{R}_1) = 0.210$, and $H(\mathcal{R}_5) = 0.126$.

Definition 7.2 (Entropy)

The entropy of the system formed by the robots in $\mathcal{R}$ is given by

$$E(\mathcal{R}) = -\sum_{i=1}^{M} p_i \log(p_i). \tag{7.3}$$

Consider the group $\mathcal{R}_0$ depicted in Figure 7.2a. It is clear that this set is a homogeneous multi-robot system, so there is only one subset, that is, $\mathcal{C} = \{c_1\}$. Then, $p_1 = 1$ and the entropy is

$$E(\mathcal{R}_0) = -\sum_{i=1}^{1} p_i \log(p_i) = -p_1 \log(p_1)$$
$$= -\log(1) = 0.$$

Notice that $E(\mathcal{R})$ is minimum when the system is homogeneous. Next the entropy of $\mathcal{R}_3$ in Figure 7.2d is evaluated. This set has three subsets, so $\mathcal{C} = \{c_1, c_2, c_3\}$ with $p_1 = 1/2, p_2 = 1/4$, and $p_3 = 1/4$ Then,

$$E(\mathcal{R}_3) = -\sum_{i=1}^{3} p_i \log(p_i) = -[p_1 \log(p_1) + p_2 \log(p_2) + p_3 \log(p_3)]$$

$$= -\left[\frac{1}{2}\log\left(\frac{1}{2}\right) + \frac{1}{4}\log\left(\frac{1}{4}\right) + \frac{1}{4}\log\left(\frac{1}{4}\right)\right]$$

$$= 0.452$$

Looking now at the sets $\mathcal{R}_3$ and $\mathcal{R}_4$ in Figure 7.2, both have the same values of variety and entropy. However, the members of $\mathcal{R}_3$ are aerial robots of different sizes, while $\mathcal{R}_4$ has a ground vehicle as member. Therefore, the metrics associated with variety are not enough to characterize the heterogeneity of a system. It is then important to quantify the system's disparity, that is, to measure how alike or unlike one agent is with respect to another. Next, we present the definition of this metric.

Definition 7.3 (Disparity)

The disparity of a set of robots $\mathcal{R} = \{r_1, r_2, \ldots, r_N\}$

$$D(r_i) = \sum_{j=1}^{N} d_{ij}^2, \forall i \text{ such that } r_i \in \mathcal{R} \tag{7.4}$$

where d_{ij} is a metric of the difference between $r_i, r_j \in \mathcal{R}$.

This last metric evaluates the disparity of a single agent in $\mathcal{R}$. It is important to determine the disparity of the whole set, so it can be computed using the Rao's quadratic entropy.

Definition 7.4 (Rao's Quadratic Entropy)

Rao's quadratic entropy of the set of robots $\mathcal{R}$ is given by

$$Q(\mathcal{R}) = \sum_{i, r_i \in \mathcal{R}} \sum_{j, r_j \in \mathcal{R}} p_i p_j d_{ij}^2, \tag{7.5}$$

where d_{ij} is the same metric as for Definition 7.3.

Notice that d_{ij} is required to compute 7.4 and 7.5. Next, we provide an example definition of this metric. Let $\omega^{r_i} \in [0, 1]$ be the resource weight which quantifies the amount of a resource available at the ith robot. For the case of the set $\mathcal{R} = \{r_1, \ldots, r_N\}$, then we have that $\mathbf{w}^{\mathcal{R}} = [\omega^{r_1} \ldots \omega^{r_N}]$ is the resource weight vector. Based on this resource weight index, we define the metric of the difference between $r_i, r_j \in \mathcal{R}$ as

$$d_{ij} = \frac{\omega^{r_i} - \omega^{r_j}}{\max(\omega^{r_i}, \omega^{r_j})}, \tag{7.6}$$

where $\max(a, b)$ gives the maximum between a and b.

As an example, we assume that $\mathbf{w}^{\mathcal{R}_5} = [1 \quad 0.5 \quad 0.25 \quad 0.75]$ whose values quantify, for instance, the communication range of each agent in $\mathcal{R}_5$ of Figure 7.2f. In $\mathbf{w}^{\mathcal{R}_5}$, 1 denotes the top range and 0.25 the bottom one. Therefore, the disparity of the 4th agent in $\mathcal{R}_5$ is

$$D(r_4) = \sum_{j=1}^{4} d_{4j}^2$$

$$= \left(\frac{0.75-1}{1}\right)^2 + \left(\frac{0.75-0.5}{0.75}\right)^2 + \left(\frac{0.75-0.25}{0.75}\right)^2 + \left(\frac{0.75-0.75}{0.75}\right)^2$$

$$= 0.618,$$

and the Rao's quadratic entropy of $\mathcal{R}_5$ is

$$Q(\mathcal{R}_5) = \sum_{i=1}^{4}\sum_{j=1}^{4} p_i p_j d_{ij}^2$$

$$= 2\left(\frac{1}{4}\right)\left(\frac{1}{4}\right)\left[\left(\frac{0.5}{1}\right)^2 + \left(\frac{0.75}{1}\right)^2 + \left(\frac{0.25}{1}\right)^2\right.$$

$$\left. + \left(\frac{0.25}{0.5}\right)^2 + \left(\frac{0.25}{0.75}\right)^2 + \left(\frac{0.5}{0.75}\right)^2\right]$$

$$= 0.210$$

Finally, the heterogeneity of a multi-robotic system can be formulated by combining Definitions 7.2 and 7.4.

Definition 7.5 (Heterogeneity)

The heterogeneity of the set of robots $\mathcal{R}$ is given by

$$H(\mathcal{R}) = E(\mathcal{R})Q(\mathcal{R}), \tag{7.7}$$

with $E(\mathcal{R})$ and $Q(\mathcal{R})$ given by Equations 7.3 and 7.5, respectively.

For the set $\mathcal{R}_5$ in Figure 7.2f, the result for its heterogeneity is

$$H(\mathcal{R}_5) = E(\mathcal{R}_5)Q(\mathcal{R}_5)$$

$$= (0.602)(0.210)$$

$$= 0.126$$

The results for the all sets are given in Figure 7.2.

7.3 Resilience in Multi-Agent Systems

Consensus, or scaled consensus [27], allows multi-agent systems to achieve an agreement on the estimate of a variable of interest. However, consensus algorithms do not typically consider the presence of misbehaving agents in the network. Instead, the capacity of a multi-agent system to be connected against loss of communication links, or agents (*structural robustness or tolerance*) [2,5], and the effect on the global dynamics of topological changes due to links, agent failures, or local perturbations (*dynamical robustness*) [20,22,23] have been studied. In any case, the robustness of the system is closely related to the network topology before and after a failure or attack.

Recently, new resilient consensus algorithms have appeared [18,28,29]. The aim of these new algorithms is to indirectly identify untrustworthy agents and estimate the variable of interest (to reach consensus) based only on trustworthy agents. Thus, the consensus of the multi-agent system is assured as long as the network meets some topological conditions such as the number of misbehaving agents.

Definition 7.6 (Connectivity Model)

We consider a robotic team coupled by a communication network modeled as graph $\mathcal{G} = (\mathcal{V}, E : \Gamma)$ where the set of vertices $\mathcal{V} = \{1, 2, \ldots, n\}$ represents the agents in the system, and the set of ordered pairs of vertices E represents the communication links connecting the agents. Further, each directed edge (i, j) in the graph has associated with it a positive weight $\gamma_{i,j}$ as specified in the weight set Γ. Both an edge and its weight in the graph can represent the quality of the communication link between two vertices. Note that a directed edge represents a unilateral communication of information. The communication links can be modeled using a disk model [31] or designed to maximize a metric of the communication network or its graph [17].

To ensure resilient coordinated motion of a robotic team, one can use a control rule based on distributed consensus [25,29]. That is, each robot i aims to estimate a global variable of interest, w_i, using local information. That is,

$$w_i[t+1] = g(w_i[t], \{w_j[t] \mid j \in \mathcal{N}_i\}), \tag{7.8}$$

where $\mathcal{N}_i$ is a set of neighbors of i. Provided that the graph $\mathcal{G}$ is connected, the variable of interest reaches consensus or scaled consensus on the weighted average of the initial values [15,27]. However, consensus is achieved if all the robots communicate their truthful value. This idea leads to definition of the following thread model [29].

Definition 7.7 (Non-Cooperative Robot)

A robot is cooperative if it applies the consensus updating rule (7.8) at each time-step t and shares the results with its neighbors. It is called ***non-cooperative*** otherwise.

Definition 7.8 (Resilient Consensus)

A group of robots is said to reach resilient consensus if all the cooperative robots achieve consensus with a value that lies between the minimum and maximum initial values of the cooperative robots, even in the presence of F non-cooperative robots.

Usually, a non-cooperative robot is removed or isolated from the network [2]. Thus, the robustness or resilience of the system is expressed in terms of the graph connectivity.

Definition 7.9 (*r*-Connected Graph)

A graph is *r*-connected if there is not a set of $r - 1$ vertices whose removal disconnects the graph.

With this attack model, the non-cooperative robot and all the information that it can transmit are not considered for consensus. However, the remaining robots in the team reach consensus.

Recently, the weighted-mean-subsequence-reduced (W-MSR algorithm has proven to achieve consensus on a weighted average of cooperative vertices' values under certain topological conditions. To explain the topological conditions, we introduce the following definitions:

Definition 7.10 (*r*-Reachable Graph)

A non-empty set $\mathcal{S} \in \mathcal{V}$ is *r*-reachable if there exists a node $i \in \mathcal{S}$ such that $|\mathcal{N}_i \setminus \mathcal{S}| \geq r$, where $\mathcal{N}_i$ are the neighbors of i.

Definition 7.11 (*r*-Robust Graph)

A non-trivial graph is *r*-robust if for each pair of disjoint non-empty subsets of $\mathcal{V}$ *at least one is r-reachable.*

A robotic network modeled using a graph $\mathcal{G} = (\mathcal{V}, E)$, where each cooperative robot updates its estimated variable w_i using the W-MSR algorithm (described below) achieves resilient consensus if it has at most F non-cooperative neighbors, that is, graph $\mathcal{G}$ is $(2F + 1)$-robust.

The W-MSR algorithm can be summarized in three steps:

1. **(Sort)** At t, each robot i creates a sorted list of the estimated variable $w_j[t]$ values received from its neighbors $j \in \mathcal{N}_i$.

2. **(Identify)** Each robot compares the shorted list with its own value $w_i[t]$. If there are F or less values strictly greater than $w_i[t]$, they are removed from the list; otherwise, it removes precisely the F largest values. The robot applies a similar process to the values less than $w_i[t]$.

3. **(Update)** Consider $\mathcal{R}_i$ the neighbors of i whose values remain in the sorted list after applying step 2, the robot i applies an updating rule (weighted average) using only the robots in $\mathcal{R}_i$:

$$w_i[t+1] = \gamma_{i,i}[t]w_j[t] + \sum_{j \in \mathcal{R}_i} \gamma_{i,j}[t]w_j[t], \qquad (7.9)$$

where $\gamma_{i,j}[t] \geq 0$, and $\Sigma_{i \in \mathcal{R}_i} \gamma_{i,i}[t] = 1$.

Since $(2F + 1)$-robust property may be difficult to satisfy, [28] proposes a sliding window approach that uses old w_j values in order to achieve consensus on a time-varying communication network. This parallel approach requires (T, r)-robustness of the connectivity graph.

Definition 7.12 ((*T, r*)-Robust Graph)

The dynamic graph $\mathcal{G}[t]$ is (T, r)-robust if $\mathcal{G}^T[t] = \bigcup_{\tau=0}^{T} \mathcal{G}[t - \tau]$ satisfy the conditions of an *r*-robust graph for every $t \geq T$.

The above definition is a relaxed robustness property since the graph is not required to be *r*-robust at each instant of time. Instead, the robustness is seen as a joint property over a bound interval of time *T*.

The sliding weighted mean-subsequence-reduced (SW-MSR) algorithm extends the classical W-MSR algorithm by including a sliding window with the duration of *T* steps. The algorithm is summarized as follows:

1. **(Sort and Store)** At *t*, each robot *i* creates a sorted list of the estimated values of $w_j[t - \tau_{i,j}]$ received from its neighbors $j \in \mathcal{N}_i^T = \bigcup_{\tau=0}^{T} \mathcal{N}_i[t - \tau]$, where $\tau_{i,j} = \max(\{\tau \in [0, T] \mid j \in \mathcal{N}_i[t - \tau]\}, \forall j \in \mathcal{N}_i^T)$.

2. **(Identify)** Each robot compares the sorted list with its own value $w_i[t]$ and removes *F* values strictly greater than $w_i[t]$ and *F* values strictly less than $w_i[t]$ analogously to the W-MSR algorithm.

3. **(Update)** Consider $\mathcal{R}_i$ the neighbors of *i* over the interval $[t - \tau, t]$ whose values remain in the sorted list after applying step 2; the robot *i* applies an updating rule (weighted average):

$$w_i[t + 1] = \gamma_{i,i}[t] w_i[t] + \sum_{j \in \mathcal{R}_i} \gamma_{i,j}[t] w_j[t - \tau_{i,j}], \tag{7.10}$$

where $\gamma_{i,j}[t] \geq 0$ and $\sum_{j \in \mathcal{R}_i} \gamma_{i,j}[t] = 1$.

7.4 Case Study

In this case study, we focus on a set $\mathcal{R}$ of mobile robots with hardware heterogeneity, whose task is trajectory tracking. Additionally, the formation of robots is given by a convex hull defined by a set of virtual robots [13,19,34], and the communication network among the agents is represented by a graph. Thus, $\mathcal{R} = \mathcal{R}_0 \cup \mathcal{R}_1 \cup \mathcal{R}_2$, where $\mathcal{R}_0 = \{1, \ldots, m_1\}$ represents the set of *virtual* agents and $\mathcal{R}_1 = \{m_1 + 1, \ldots, m_1 + m_2\}$ and $\mathcal{R}_2 = \{m_1 + m_2 + 1, \ldots, n\}$ represent *real* robots.

7.4.1 Network Connectivity

We use the communication network of the robots' team defined in Section 7.3. For simplicity, we use the same notation for the robots and the vertices in the graph. We consider a directed graph since the receiver noise may be different for the connected agents. In this case, an agent will hear the other but not be able to reply. For instance, virtual robots are able to transfer information but not receive it, and hence there is an edge between them and a real robot; for example, the vertices $\mathcal{R}_0 = \{1, \ldots, 9\}$ represent the set of virtual robots (see Figure 7.3a).

Although the connectivity model defines the edge's weight (see Section 7.3), we consider that two agents are either connected or not. Further, a non-cooperative agent is removed from the network, and then the communication network is reconfigured so that consensus is achieved.

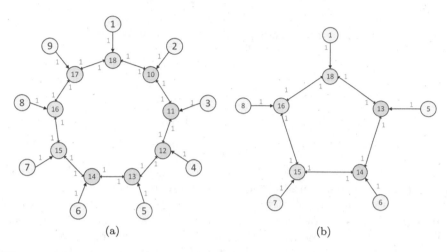

FIGURE 7.3
Connectivity graph.

7.4.2 Motion Dynamics

For every agent $q \in \mathcal{R}_1$, we define a state vector $[x_q \quad y_q \quad \psi_q]^T$, where (x_q, y_q), is the robot's position and ψ_q is the orientation. Considering a point located near the wheel axle for the motion control, the dynamics is given by

$$\begin{bmatrix} \dot{x}_q \\ \dot{y}_q \\ \dot{\psi}_q \end{bmatrix} = \begin{bmatrix} \cos\psi_q & -b\sin\psi_q \\ \sin\psi_q & b\cos\psi_q \\ 0 & 1 \end{bmatrix} \begin{bmatrix} v_q \\ \omega_q \end{bmatrix}, \qquad (7.11)$$

where v_q and ω_q are the linear and rotational velocities, and b is a constant.

We define $\mathbf{u}_q = [v_q \quad \omega_q]^T$ as the control input of the qth mobile robot.

For every agent $p \in \mathcal{R}_2$, we define a state vector $[x_p \quad v_{x_p} \quad y_p \quad v_{y_p}]^T$ where (x_p, y_p) and (v_{x_p}, v_{y_p}) are the robot's position and velocity, respectively. The decoupled motion dynamics of an agent in $\mathcal{R}_2$ at x-axis is given by

$$\begin{bmatrix} \dot{x}_p \\ \dot{v}_{x_p} \end{bmatrix} = A \begin{bmatrix} x_p \\ v_{x_p} \end{bmatrix} + \begin{bmatrix} 0 \\ 1 \end{bmatrix} a_{xp}, \qquad (7.12)$$

where A is the state matrix and a_{xp} (or a_{yp}) is the acceleration (control input). We define $\mathbf{u}_p = [a_{xp} \quad a_{yp}]^T$ as the control input of the pth mobile robot.

Finally, an agent $c \in \mathcal{R}_0$ has the state vector $[x_c \quad \dot{x}_c \quad y_c \quad \dot{y}_c]^T$, where (x_p, y_p) and $(\dot{x}_p, \dot{y}_p)$ are the robot's position and velocity, respectively. The motion dynamics of the virtual agent at each axis is represented by a double integrator,

$$\begin{bmatrix} \ddot{x}_c \\ \ddot{y}_c \end{bmatrix} = \mathbf{u}_c, \qquad (7.13)$$

where $\mathbf{u}_c$ is an acceleration vector defined by the desired trajectory.

We define the acceleration for a robot $p \in \mathcal{R}_1 \cup \mathcal{R}_2$ at the x-axis as:

$$a_{xp} = -\sum_{j \in \mathcal{R}} \gamma_{p,j} \left[(x_p - x_j) + \alpha_{\mathcal{R}_{1,2}} \left(v_{x_p} - v_{x_j} \right) \right]$$

$$\qquad - \beta_{\mathcal{R}_{1,2}} \mathrm{sgm} \left(\sum_{j \in R} \gamma_{p,j} \left[(x_p - x_j) + \xi_{\mathcal{R}_{1,2}} \left(v_{x_p} - v_{x_j} \right) \right] \right), \tag{7.14}$$

where $\alpha \mathcal{R}_{1,2}$, $\beta \mathcal{R}_{1,2}$, and $\xi \mathcal{R}_{1,2}$ are positive constants defined for the controller, and $\mathrm{sigm}(x) = x / (|x| + \epsilon)$ is a sigmoid function. A similar controller is used for the y-axis. We note that the same kind of controller can be applied for mobile agents in $\mathcal{R}_1 \cup \mathcal{R}_2$; however, pre-processing is necessary for agents in $\mathcal{R}_1$. The reader is referred to [33] for additional details. To prove stability of the closed loop multi-agent system, one can follow a similar approach as in [7] and [33]. Note that in Equation 7.14, there are two variables of interest: position and velocity. Further, the updating rule 7.8 is implicitly used to compute the control input.

7.4.3 Simulation Results

We consider a group of heterogeneous robots formed by five mobile agents of type $\mathcal{R}_1$, four $\mathcal{R}_2$, and nine *virtual agents*, whose communication network is represented by the graph in Figure 7.3a. The vertices $\{1, \dots, 9\}$, $\{10, \dots, 14\}$, and $\{15, \dots, 18\}$ in $\mathcal{G}$ represent the sets $\mathcal{R}_0$, $\mathcal{R}_1$, and $\mathcal{R}_2$, respectively (Figure 7.3).

The initial position of the agents in $\mathcal{R}_1 \cup \mathcal{R}_2$ is randomly generated, and the nine virtual agents define the convex hull where the robots must converge (dashed line in Figure 7.4a). This problem is known in the literature as the *containment formation problem* [7].

The robots' parameters for Equations 7.11 and 7.12 are: $b = 0.2$ m, $A = \begin{bmatrix} 0 & 1 \\ -0.4 & -0.01 \end{bmatrix}$. Further, the linear and rotational velocities of the mobile

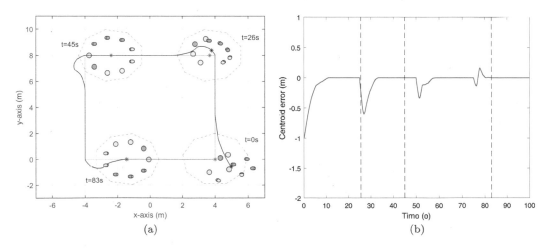

FIGURE 7.4
(a) Team's trajectory. (b) Centroid error.

platform are bounded according to the limits of this robot specified in [1] and [16] $-1.2 \leq u \leq 1.2$ m/s, $-5.24 \leq \omega \leq 5.24$ rad/s, $-1.5 \leq a_x \leq 1.5$ m/s^2 and $-1.5 \leq a_y \leq 1.5$ m/s^2. Finally, the controller parameters are as follows: $\alpha_{\mathcal{R}_{1,2}} = 10$, $\beta_{\mathcal{R}_{1,2}} = 2.4$, $\xi_{\mathcal{R}_{1,2}} = 30$, and $\varepsilon = 0.25$.

When all the robots are cooperative, the connectivity network has the form shown in Figure 7.3a. We consider that at time τ, the robots {10, 11, 12, 17} become non-cooperative due to the agents' failures or an attack. Thus, the remaining robots in the team stop receiving information from them and the virtual robots {2, 3, 4, 19} because of the network topology. Figure 7.3b shows the cooperative graph used after a failure or attack at time τ. Consequently, control law 7.14 changes accordingly since the implicit updating rule uses only the information of the remaining robots.

The first simulation provides a general idea of how the heterogeneous team is working when all the robots are cooperative. Figure 7.4a shows the team's trajectory as well as the centroid trajectory (solid line), and the convex hull built upon the agents in $\mathcal{R}_0$ (dashed line). Figure 7.4b shows the centroid error whose maximum is at the beginning of the simulation and each time that the reference changes abruptly. Figure 7.5a and b shows the control effort required to keep the desired trajectory.

Before presenting the resilience of the heterogeneous team, let us show the effect of losing cooperative agents in a *quasi-homogeneous* network. In his case, agents of the type $\mathcal{R}_0$ and $\mathcal{R}_1$ form the quasi-homogeneous group. Figure 7.6a shows the convex hull (dashed line) and the centroid (solid line) before and after the agents $\mathcal{N} = \{2, 3, 4, 9, 10, 11, 12, 17\}$ become non-cooperative at $\tau = 65$ s. The weights $\gamma_{i,j}$ in the control law 7.14, or equivalently in the graph $\mathcal{G}$, change to fit the new connectivity graph (Figure 7.3b). Because of this change, the convex hull also changes (dashed line in Figure 7.6a). Note that the maximum centroid error is caused by the agents' initial position ($t = 0$ s) and the failure of the agents at $t = \tau$ (Figure 7.6b). Further, control input peaks at the linear and rotational velocity also occur in that situation (Figure 7.7).

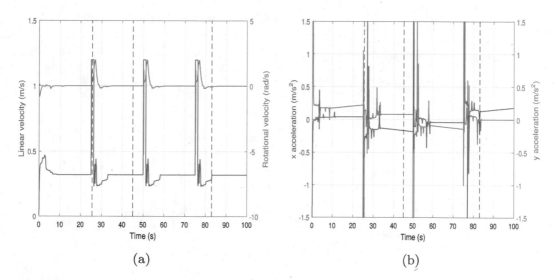

(a) (b)

FIGURE 7.5

(a) Control inputs (linear and rotational velocity) for an agent in $\mathcal{R}_1$. (b) Control inputs (x- and y- acceleration) for an agent in $\mathcal{R}_2$.

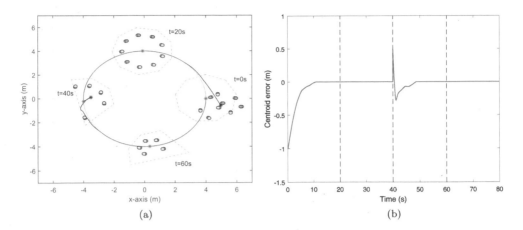

FIGURE 7.6
(a) Team's trajectory. (b) Centroid error.

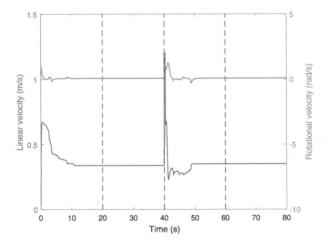

FIGURE 7.7
Control inputs (linear and rotational velocities) for agents in $\mathcal{R}_1$.

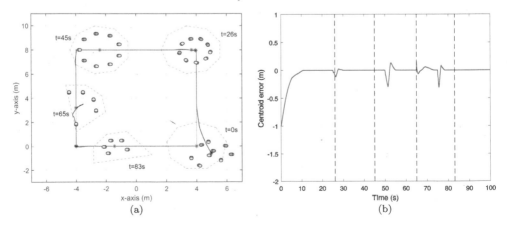

FIGURE 7.8
(a) Team's trajectory. (b) Centroid error.

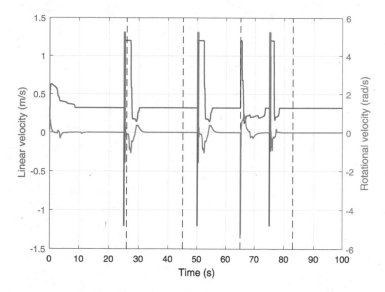

FIGURE 7.9
Control inputs (linear and rotational velocities) for agents in $\mathcal{R}_1$.

We note that the effect on the control input of the agents and the centroid's formation, after the robots in $\mathcal{N}$ become non-cooperative, are similar to an abrupt change in the reference (see Figures 7.8a,b, and 7.9).

The following simulation illustrates the team behavior when the agents $\mathcal{N} = \{2, 3, 4, 9, 10, 11, 12, 17\}$ become non-cooperative in a heterogeneous group. At time $\tau = 65$ s, the agents in $\mathcal{N}$ stop sharing information. Thus, the weights $\gamma_{i,j}$ in the control law 7.14 change to fit the new connectivity graph (Figure 7.3b) and, consequently there is a modification in the convex hull depicted in Figure 7.10a.

Due to the abrupt change of the team's configuration, we observe that centroid error occurs at time $\tau = 65$ s (Figure 7.10b), which demands more energy, as seen in Figure 7.11a and b.

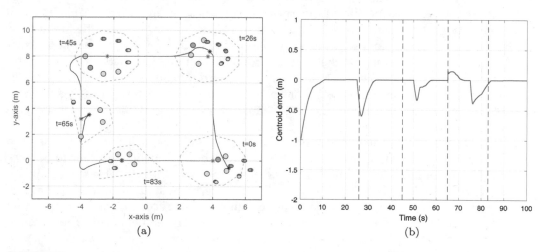

(a)   (b)

FIGURE 7.10
(a) Team's trajectory. (b) Centroid error.

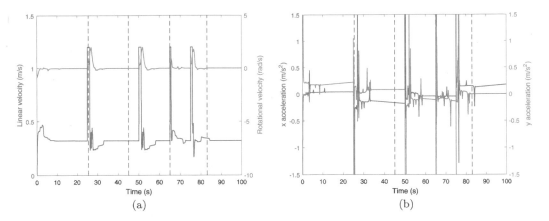

FIGURE 7.11
(a) Control inputs (linear and rotational velocity) for an agent in $\mathcal{R}_1$. (b) Control inputs (x- and y- acceleration) for an agent in $\mathcal{R}_2$.

7.5 Conclusions

The variety and disparity capabilities of heterogeneous agents greatly increase their resilience and ability to successfully accomplish complex tasks. However, coordinating their actions is a challenging problem, especially since the agents may all possess different resources and constraints. Consequently, it is critical to model, measure, and analyze the heterogeneity and adaptability/resilience of a cooperative multi-robotic system.

In general, there is no agreed-upon definition of these two fundamental characteristics, heterogeneity and resilience, across scientific disciplines. Moreover, the effects of incorporating heterogeneity and/or resilience as design variables of a multi-agent system have been mostly unexplored in the literature. We believe that only after a universal definition of heterogeneity and resilience have been established can they be fully exploited in the design and application of collaborative multi-robotic systems.

In this chapter, we summarized a series of definitions and metrics related with heterogeneity of a multi-robotic system. We also provided concepts and theory related with resilient consensus for cooperative robots. A case study of a resilient heterogeneous multi-agent mobile system for the case of group formation was also introduced and analyzed. The group behavior results of the heterogeneous case showed a better performance than its homogeneous counterpart. In addition, the heterogeneous group was capable of reconfiguring and maintaining the formation after a number of robots became non-cooperative.

Future work will be centered around studying how heterogeneity and resilience can be employed as part of the design of a cooperative robotic network in order to balance the complexity, modularity, applicability, and adaptability of the overall system. A better understanding of how heterogeneity can be analyzed can help to find the range of missions that an intelligent multi-agent system can or cannot accomplish. Morphing and reconfigurable platforms and system goals are ways to increase resiliency. Thus, theory, methods, and algorithms are needed to leverage heterogeneity and mechanical/goal adaptability for enhanced system resilience.

Acknowledgments

The authors gratefully acknowledge the financial support provided by the Escuela Politécnica Nacional for the development of the project PIS-17–02 "Estudio, coordinación y consenso de sistemas multi-agentes heterogéneos con aplicaciones en robótica móvil." The work of Rafael Fierro was supported in part by Air Force Research Laboratory (AFRL) under agreement number FA9453-18-2-0022. We would like to thank LabEx PERSYVAL-Lab (ANR-11-LABX-0025-01) funded by the French program Investissement d'avenir, for the scholarship providing the financial support to Renato Vizuete.

References

1. Adept Technology, Inc. *Pioneer 3-DX Datasheet*, 2011. Rev. A.
2. R. Albert, H. Jeong, and A.-L. Barabási. Error and attack tolerance of complex networks. *Nature*, 406(6794):378, 2000.
3. T. Balch. Hierarchic social entropy: An information theoretic measure of robot group diversity. *Autonomous Robots*, 8(3):209–238, 2000.
4. B. Benjamin, G. Erinc, and S. Carpin. Real-time wifi localization of heterogeneous robot teams using an online random forest. *Autonomous Robots*, 39(2):155–167, Aug 2015.
5. B. Bollobás and O. Riordan. Robustness and vulnerability of scale-free random graphs. *Internet Mathematics*, 1(1):1–35, 2004.
6. A. Campbell and A.S. Wu. Multi-agent role allocation: Issues, approaches, and multiple perspectives. *Autonomous Agents and Multi-Agent Systems*, 22(2):317–355, Mar 2011.
7. Y. Cao, D. Stuart, W. Ren, and Z. Meng. Distributed containment control for double-integrator dynamics: Algorithms and experiments. In *American Control Conference (ACC), 2010*, pages 3830–3835. IEEE, 2010.
8. P. Chand and D.A. Carnegie. Mapping and exploration in a hierarchical heterogeneous multi-robot system using limited capability robots. *Robotics and Autonomous Systems*, 61(6):565–579, 2013.
9. J. Chen and D. Sun. Coalition-based approach to task allocation of multiple robots with resource constraints. *IEEE Transactions on Automation Science and Engineering*, 9(3):516–528, July 2012.
10. R.A. Cortez, R. Fierro, and J. Wood. Connectivity maintenance of a heterogeneous sensor network. In: A. Martinoli, F. Mondada, N. Correll, G. Mermoud, M. Egerstedt, M. Ani Hsieh, L.E. Parker, and K. Støy, editors, *Distributed Autonomous Robotic Systems: The 10th International Symposium*, pages 33–46. Springer, Berlin Heidelberg, 2013.
11. P.J. Cruz and R. Fierro. Building coalitions of heterogeneous agents using weighted bipartite graphs. In *54th IEEE Conference on Decision and Control (CDC)*, pages 2822–2828, December 2015.
12. P.J. Cruz, R. Fierro, and C.T. Abdallah. Cooperative learning for robust connectivity in multirobot heterogeneous networks. In K.G. Vamvoudakis and S. Jagannathan, editors, *Control of Complex Systems*, pages 451–473. Butterworth-Heinemann, 2016.
13. G. Droge. Distributed virtual leader moving formation control using behavior-based MPC. In *American Control Conference (ACC), 2015*, pages 2323–2328. IEEE, 2015.
14. D.A. Harrison and K.J. Klein. What's the difference? diversity constructs as separation, variety, or disparity in organizations. *Academy of Management Review*, 32(4):1199–1228, 2007.
15. A. Jadbabaie, J. Lin, and A Stephen Morse. Coordination of groups of mobile autonomous agents using nearest neighbor rules. *IEEE Transactions on Automatic Control*, 48(6):988–1001, 2003.

16. J.D. Jeon and B.H. Lee. Wheel velocity obstacles for differential drive robot navigation. *Journal of Automation and Control Engineering*, 3(5), 2015.

17. Y. Kim and M. Mesbahi. On maximizing the second smallest eigenvalue of a state-dependent graph laplacian. In *American Control Conference, 2005. Proceedings of the 2005*, pages 99–103. IEEE, 2005.

18. H.J. LeBlanc, H. Zhang, X. Koutsoukos, and S. Sundaram. Resilient asymptotic consensus in robust networks. *IEEE Journal on Selected Areas in Communications*, 31(4):766–781, 2013.

19. Z. Liu, W. Chen, J. Lu, H. Wang, and J. Wang. Formation control of mobile robots using distributed controller with sampled-data and communication delays. *IEEE Transactions on Control Systems Technology*, 24(6):2125–2132, 2016.

20. J. Lu, X. Yu, G. Chen, and D. Cheng. Characterizing the synchronizability of small-world dynamical networks. IEEE Transactions on Circuits and Systems Part 1. *Regular Papers*, 51(4):787–796, 2004.

21. N. Mathew, S.L. Smith, and S.L. Waslander. Planning paths for package delivery in heterogeneous multirobot teams. *IEEE Transactions on Automation Science and Engineering*, 12(4):1298–1308, Oct 2015.

22. A.E. Motter and Y-C. Lai. Cascade-based attacks on complex networks. *Physical Review E*, 66(6):065102, 2002.

23. F. Pasqualetti, A. Bicchi, and F. Bullo. Consensus computation in unreliable networks: A system theoretic approach. *IEEE Transactions on Automatic Control*, 57(1):90–104, 2012.

24. A. Prorok, M.A. Hsieh, and V. Kumar. Formalizing the impact of diversity on performance in a heterogeneous swarm of robots. In *IEEE International Conference on Robotics and Automation (ICRA)*, pages 5364–5371, May 2016.

25. W. Ren and R.W. Beard. *Distributed consensus in multivehicle cooperative control*. Springer, 2008.

26. C.G. Rieger, D.I. Gertman, and M.A. McQueen. Resilient control systems: Next generation design research. In *Human System Interactions, 2009. HSI'09. 2nd Conference on*, pages 632–636. IEEE, 2009.

27. S. Roy. Scaled consensus. *Automatica*, 51:259–262, 2015.

28. D. Saldana, A. Prorok, S. Sundaram, M.F.M. Campos, and V. Kumar. Resilient consensus for time-varying networks of dynamic agents. In *American Control Conference (ACC), 2017*, pages 252–258. IEEE, 2017.

29. K. Saulnier, D. Saldana, A. Prorok, G.J. Pappas, and V. Kumar. Resilient flocking for mobile robot teams. *IEEE Robotics and Automation Letters*, 2(2):1039–1046, 2017.

30. S. Skogestad and I. Postlethwaite. *Multivariable feedback control: Analysis and design*, volume 2. Wiley New York, 2007.

31. E. Stump, A. Jadbabaie, and V. Kumar. Connectivity management in mobile robot teams. In *Robotics and Automation, 2008. ICRA 2008. IEEE International Conference on*, pages 1525–1530. IEEE, 2008.

32. P. Twu, Y. Mostofi, and M. Egerstedt. A measure of heterogeneity in multi-agent systems. In *American Control Conference (ACC)*, pages 3972–3977, June 2014.

33. R. Vizuete, J.A. Torres, and P. Leica. Trajectory tracking based on containment algorithm applied to a formation of mobile manipulators. In *ICINCO 2017—Proceedings of the 14th International Conference on Informatics in Control, Automation and Robotics*, pages 122–131, 2017.

34. Z. Yan, X. Liu, A. Jiang, and L. Wang. Formation control of multiple UUVs based on virtual leader. In *Control Conference (CCC), 2016 35th Chinese*, pages 4621–4626. IEEE, 2016.

35. Q. Zhu and T. Basar. Game-theoretic methods for robustness, security, and resilience of cyberphysical control systems: Games-in-games principle for optimal cross-layer resilient control systems. *IEEE Control Systems*, 35(1):46–65, 2015.

Section II

Cooperative Strategies for Localization and Navigation

8

Preliminary Study of Cooperative Navigation of Underwater Vehicles without a DVL Utilizing Range and Range-Rate Observations

Zachary J. Harris and Louis L. Whitcomb

CONTENTS

8.1 Introduction

This paper addresses the navigation problem arising in underwater vehicles equipped with an acoustic modem, attitude and depth sensors, but lacking a Doppler velocity log (DVL), and a surface vehicle equipped with an acoustic modem and aglobal positioning system (GPS) unit. The case of underwater vehicle navigation without a DVL sensor is relevant to low-cost vehicles for which the cost of a DVL may be prohibitive, and for missions in which the vehicle's altitude above the sea floor (or depth beneath overhead ice) exceeds its DVL bottom-lock range—the most commonly used DVLs operate at frequencies of 300–1200 kHz with maximum bottom-lock range in sea water of about 25–200 m, respectively.

Navigation methods for underwater vehicles utilizing velocity signals (for example, dead-reckoning navigation, Doppler sonar navigation) or acceleration signals (for example, inertial navigation) all accumulate errors due to signal bias, signal noise, and a variety of other sensor-calibration issues, and thus require regular external corrections. The corrections are independent observations of absolute position or velocity from a separate sensor or system that are used to correct the drift accumulated in the navigation estimate [13,62,46].

For land and air vehicle navigation, GPS provides an ideal independent source of position corrections for acceleration and velocity-based navigation systems [13], but GPS is unavailable to submerged underwater vehicles. Bottom-lock Doppler-sonar velocity measurements are an excellent correction source for velocity estimates. Pressure depth sensor measurements are an excellent correction source for the Z-dimension (only) of velocity-based and acceleration-based position estimates.

For submerged underwater vehicles (for which GPS is unavailable), few methods currently exist for absolute X-Y position corrections. The most common X-Y position correction methods are time-of-flight acoustic navigation systems, such as long-baseline (LBL) and ultra-short baseline (USBL) acoustic navigation [36,50,31].

Range-only one-way travel-time (OWTT) cooperative underwater navigation uses ranges estimated from the acoustic time-of-flight between subsea nodes, for example, between two vehicles or between a client vehicle and a server reference beacon of known (fixed or moving) location. When all vehicles and beacons (nodes) are equipped with precision clocks, each node's acoustic data transmission can be received by multiple receiving nodes—enabling all nodes within acoustic range to simultaneously (a) measure range to the transmitting node from the measured time-of-flight and (b) decode the data encoded in the acoustic data packet. This method provides both bounded-error position estimates and long-range capabilities with reduced need for multiple costly fixed beacons, as is the case with most LBL systems. OWTT navigation also provides scalability by allowing all vehicles within acoustic range to simultaneously use the same acoustic data packet broadcast independent of the number of vehicles.

This paper examines the feasibility of employing cooperative navigation with underwater acoustic networks to provide both position and velocity corrections for low-cost underwater vehicles not employing a Doppler sonar for velocity measurements. The present study complements our previous results reported in [29], in which we reported that the addition of acoustic range-rate observations does not appear to offer significant advantages for underwater vehicle navigation when the vehicle employs high-accuracy Doppler sonar navigation in addition to heading, attitude, and depth sensors with reasonable range and range-rate noise statistics. This previous study motivated the present study, which investigates the performance of cooperative navigation when high-accuracy Doppler sonar signals are unavailable. The examination of range-rate observations in addition to range observations is motivated by the fact that most underwater acoustic modems compute a range rate as part of the acoustic receptions processing of each incoming acoustic data packet.

8.2 Literature Review

Radio-frequency (RF) telemetry is the preferred telemetry method for land, air, and space, but is not useful for underwater telemetry because the ocean is opaque to high-frequency RF telemetry. Even extremely low-frequency RF telemetry can penetrate seawater only to tens or hundreds of feet [51,2], The development of underwater acoustic modems, however, has enabled underwater data telemetry at ranges up to tens of kilometers [12,57,35,52,14]. The propagation delay of acoustic telemetry in seawater is about 1.5 km per second, and the data bandwidth varies with range, carrier frequency, and encoding (modulation) method.

Previous results by the authors and others [67,69,7,64,53] have shown the effectiveness of position corrections for Doppler and inertial navigation with range-only OWTT underwater navigation using ranges estimated from the acoustic time-of-flight of acoustic data packets between subsea nodes, for example, between two vehicles or between a vehicle and a reference beacon of known (fixed or moving) location. Numerous studies by the authors and others report the development and extensive at-sea experimental evaluation of OWTT systems (including hardware and software) for the navigation of underwater vehicles using maximum-likelihood estimation [18,20,19], the extended Kalman filter (EKF) [68,66,67], and the extended information filter (EIF) [71,70,69].

The range-only OWTT navigation and communication system has been experimentally evaluated at sea on several different AUVs worldwide: on the *SeaBed* AUV in shallow water deployments in the Atlantic Ocean [18] and in the Mediterranean Sea [20,19]; on the *Jaguar* and *Puma* AUVs in a deep water (~5000 m) deployment at 4°S on the Southern Mid-Atlantic Ridge [66,68,67,69,70,71]; and on the *Nereus* AUV [17,11] to 10,903 m depth in the Mariana Trench [58] and to 5000 m depths in the Cayman Rise in the Caribbean [26].

Several other groups worldwide also reported results in range-only OWTT navigation. Several authors have reported least-squares methods for single-beacon range-only navigation [56,30,39,49]. Range-only target tracking has been addressed using EKFs and maximum-likelihood estimators (MLE) [54,59,1]. The use of EKFs for homing and single-beacon navigation is reported in [3,4,63,40,41,43]. Several different studies addressed the observability of single-beacon range-only navigation with linear KF estimation methods [22,23,24,25,47] and with nonlinear methods [55,32]. Several authors have addressed OWTT navigation of surface and underwater vehicles in a simultaneous localization and mapping (SLAM) framework using distributed estimators [5,6,7,21,8]. Multi-beacon, range-only navigation for terrestrial vehicles in a SLAM framework is addressed in [34,16,15,33,37,38] using radio-frequency beacons for range measurement, in [48] using audible sound, and in [45,61] using wireless sensor networks.

To the best of our knowledge, the earliest reported comprehensive study of underwater-vehicle navigation using acoustic ranging was reported by Spindel et al. in which they reported full-scale experimental evaluation of an acoustic approach to underwater-vehicle navigation in which a single underwater vehicle could detect range from a set of fixed acoustic navigation transponders whose location was known *a priori*—a method that has since been widely practiced and is now commonly known as long-baseline (LBL) acoustic navigation [31].

To the best of our knowledge, the earliest study of underwater-vehicle navigation employing acoustic detection of *both range and range-rate* was the 1978 study by Spindel et al. [60], which extends the approach reported in [31] by reporting the full-scale experimental evaluation of an approach to underwater-vehicle navigation in which a single underwater vehicle could detect both range and range rate from a set of fixed navigation transponders whose location was known *a priori*.

In [42,44], the author, apparently unaware of [31,60], reported the notion of employing acoustic range rate in addition to acoustic range for LBL navigation but did not report specific navigation algorithms for employing range rate nor any experimental evaluation. In [27], the author reports the notion of an underwater acoustic modem estimating and compensating for the Doppler shift of a received acoustic data packet transmission but does not address how a Doppler estimate might be used for navigation.

In [10], the author reports the experimental evaluation of algorithms for acoustically determining the *relative position* of two marine vehicles by employing measurement of acoustic range and acoustic range rate with specific focus on estimating relative positioning conditions, such as the closest-point-of-approach (CPA) of two vessels for the purpose of collision-avoidance.

8.3 Extended Kalman Filter

This section briefly describes the EKF and associated process model and observation models used.

8.3.1 EKF Formulation and Implementation

This study utilizes the centralized Extended Kalman Filter (CEKF), which assumes real-time access to vehicle and ship sensor data simultaneously. As such, it is suitable for post-processing simulated and experimental data. At OWTT measurement updates, the CEKF is analytically and experimentally identical to its decentralized counterpart, the decentralized extended information filter (DEIF), as reported in [69].

8.3.2 State Description

As is typical for the CEKF cooperative navigation algorithm [66,68,67,65], we define the state vector, $\mathbf{X}$, as the composite of the current vehicle state, current ship state, and n delayed states. Delayed states are required for causal processing because the range measurement occurs between the ship at time-of-launch (TOL) and the vehicle at time-of-arrival (TOA).

$$\mathbf{X} = \begin{bmatrix} x_v^T & x_s^T & x_{v-1}^T & x_{s-1}^T & \cdots & x_{v-n}^T & x_{s-n}^T \end{bmatrix}^T \tag{8.1}$$

where the current ship state is a 6-degree-of-freedom (DOF) vector containing the XY-position and heading and their respective velocities

$$x_s = \begin{bmatrix} x & y & \psi & \dot{x} & \dot{y} & \dot{\psi} \end{bmatrix}^T, \tag{8.2}$$

and the current vehicle state is a 12-DOF vector containing the local-level pose and body-frame velocities

$$x_v = \begin{bmatrix} s^T & \varphi^T & v^T & \omega^T \end{bmatrix}^T \tag{8.3}$$

$$s = \begin{bmatrix} x \\ y \\ z \end{bmatrix} \quad \varphi = \begin{bmatrix} \phi \\ \theta \\ \psi \end{bmatrix} \quad \nu = \begin{bmatrix} u \\ v \\ w \end{bmatrix} \quad \omega = \begin{bmatrix} p \\ q \\ r \end{bmatrix} \tag{8.4}$$

where s is the local-level position, φ is the local-level attitude, v is the body-frame linear velocity, and ω is the body-frame angular velocity.

8.3.3 Process Model

We assume a kinematic, nonlinear process model for both the vehicle and the ship, identical to the one reported in [67]. As is common in the literature, this process model is a purely kinematic plant model that assumes a constant-velocity second-order plant with process noise

$$\dot{x}_v = \begin{bmatrix} 0 & 0 & R(\varphi) & 0 \\ 0 & 0 & 0 & J(\varphi) \\ 0 & 0 & 0 & 0 \\ 0 & 0 & 0 & 0 \end{bmatrix} x_v + \begin{bmatrix} 0 & 0 \\ 0 & 0 \\ \mathrm{II} & 0 \\ 0 & \mathrm{II} \end{bmatrix} w_v \tag{8.5}$$

$$\dot{x}_s = \begin{bmatrix} 0 & \mathrm{II} \\ 0 & 0 \end{bmatrix} x_s + \begin{bmatrix} 0 \\ \mathrm{II} \end{bmatrix} w_s \tag{8.6}$$

where $R(\varphi)$ is the transformation between inertial and body-frame linear velocities, $J(\varphi)$ is the transformation between inertial and body-frame angular velocities, and $w_v \sim \mathcal{N}(0, Q_v)$ and $w_s \sim \mathcal{N}(0, Q_s)$ are zero-mean Gaussian process noise terms. The process model for the vehicle is linearized and discretized using standard methods [9]. The reader is referred to [67] for a full description and derivation, including the linearized discrete-time process model and the subtleties of the modified process prediction, which occurs at the top of the second when the state augmentation is performed in concert with the process-prediction step—the current state is added to the state vector and the oldest delayed state is marginalized out.

8.3.4 Observation Models

The range and range-rate observation models are nonlinear functions of the vehicle state at TOA and the ship state at TOL. Observation models of the additional sensors, including the DVL, GPS, depth sensor, and gyrocompass, are detailed in [65].

Range Observation Model As reported in [67,65], the range observation model can be written in matrix notation as

$$z_{rng} = (x^T M^T M x)^{\frac{1}{2}} + v_{rng} \tag{8.7}$$

where $v_{rng} \sim \mathcal{N}(0, R_{rng})$ and $M = \begin{bmatrix} -J_v & 0 \dots 0 & J_s & 0 \dots 0 \end{bmatrix}$, with J_v and J_s defined such that

$$J_v x_v = \begin{bmatrix} x \\ y \\ z \end{bmatrix} \quad \text{and} \quad J_s x_s = \begin{bmatrix} x_s \\ y_s \\ 0 \end{bmatrix}. \tag{8.8}$$

The measurement covariance, R_{rng}, represents the noise of the range measurement. The Jacobian of the range measurement with respect to the full state, x, is

$$H_k = \frac{\partial z_{rng}(x)}{\partial x}\bigg|_{x = \mu_{k|k-1}} = \left(\mu_{k|k-1}^T M^T M \mu_{k|k-1} \right)^{-\frac{1}{2}} \mu_{k|k-1}^T M^T M \tag{8.9}$$

Range-Rate Observation Model The range-rate observation model is the time derivative of Equation 8.7. Formally,

$$z_{rr} = \dot{z}_{rng} = \left(x^T M^T M x \right)^{-\frac{1}{2}} x^T M^T M \dot{x} + v_{rr} \tag{8.10}$$

where $v_{rr} \sim \mathcal{N}(0, R_{rr})$ represents the noise covariance of the range-rate measurement. However, $\dot{x}$ is not the state vector, but we can construct $M\dot{x}$ by defining a new constant matrix, $\hat{M}$, such that $M\dot{x} = \hat{M}x$. Explicitly, $\hat{M} = [-\hat{J}_v \quad 0 \dots 0 \quad \hat{J}_s \quad 0 \dots 0]$, with $\hat{J}_v$ and $\hat{J}_s$ defined such that

$$\hat{J}_v R(\varphi) x_v = \begin{bmatrix} \dot{x} \\ \dot{y} \\ \dot{z} \end{bmatrix} \quad \text{and} \quad \hat{J}_s x_s = \begin{bmatrix} \dot{x}_s \\ \dot{y}_s \\ 0 \end{bmatrix}$$

Thus,

$$z_{rr} = \left(x^T M^T M x \right)^{-\frac{1}{2}} x^T M^T \hat{M} x + v_{rr} \tag{8.11}$$

The Jacobian of Equation 8.11 with respect to the full state, x, is

$$H_k = \frac{\partial z_{rr}(x)}{\partial x}\bigg|_{x=\mu_{k|k-1}}$$
$$= -(\mu_{k|k-1}^T M^T M \mu_{k|k-1})^{-\frac{3}{2}} (\mu_{k|k-1}^T M^T M)(\mu_{k|k-1}^T M^T \hat{M} \mu_{k|k-1}) \tag{8.12}$$
$$+ (\mu_{k|k-1}^T M^T M \mu_{k|k-1})^{-\frac{1}{2}} \mu_{k|k-1}^T (M^T \hat{M} + \hat{M}^T M)$$

where $\mu_{k|k-1}$ is the estimated mean of the world-frame position.

8.4 Simulation Results

In this preliminary study, we utilized a numerical simulation to investigate the effect of the range-rate observation on the performance of the CEKF cooperative navigation algorithm described in Section 8.3. A full analysis of the effect the range-rate observation by varying the ship and vehicle trajectories and measurement covariances is beyond the scope of this paper; thus, preliminary results are reported herein.

We computed simulated vehicle and ship trajectories and simulated sensor data for each of the navigation sensors by generating simulated measurements with the measurement-noise characteristics outlined in Table 8.1. In the simulation presented here, the vehicle conducts a simulated survey mission of ten 1 km track lines spaced 100 m apart at a velocity

TABLE 8.1

Vehicle State Measurement Sources, Resolutions, and Accuracies

State	Source	Update Rate	Measurement Std Dev
XY Trans	modem	10 s	variable (range) 0.1 m/s (range rate)
Z Trans	Paroscientific	7 Hz	6 cm
Heading	OCTANS	3 Hz	0.10°
Pitch, Roll	OCTANS	3 Hz	0.05°
Ang Vel	OCTANS	3 Hz	0.4–0.6°/s

Source: Z.J. Harris and L.L. Whitcomb, *2016 IEEE International Conference on Robotics and Automation (ICRA)*, May 2016, pp. 2618–2624.

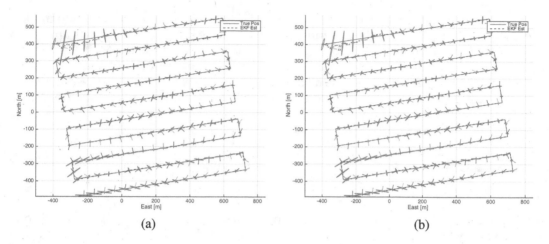

FIGURE 8.1
Vehicle XY position. (a) Range-only observation model. (b) Range and range-rate observation model. (From Z.J. Harris and L.L. Whitcomb, *2016 IEEE International Conference on Robotics and Automation (ICRA)*, May 2016, pp. 2618–2624.)

of 1 m/s and a depth of 3 m. The ship circles continuously on an 800 m radius at a velocity of 2 m/s broadcasting acoustic packets every 10 seconds. The speed of sound was assumed constant at 1500 m/s.

Figure 8.1 shows the true and estimated two-dimensional vehicle position with the filter's covariance plotted every 60 s. The arrows point from the vehicle to the ship along acoustic path with the length scaled by the angle from vertical. The standard deviations for the observations are listed in Table 8.1. Figure 8.2 shows histograms of the estimation error (that is, the difference between the estimated vehicle position and the true vehicle position) in the X- and Y-directions. These histograms indicate the X- and Y- estimation errors of the CEKF are approximately zero-mean and Gaussian.

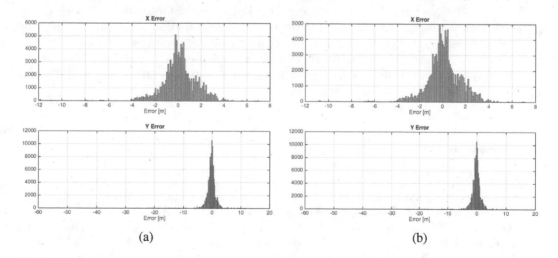

FIGURE 8.2
Error histogram in the X- and Y-directions. (a) Range-only observation model. (b) Range and range-rate observation model. (From Z.J. Harris and L.L. Whitcomb, *2016 IEEE International Conference on Robotics and Automation (ICRA)*, May 2016, pp. 2618–2624.)

For the kinematic process model described in Section 8.3.3 and the observation models described in Section 8.3.4, the addition of range-rate observations does not appear to offer a significant improvement in CEKF performance in the absence of a DVL on this simulated survey mission when utilizing high-accuracy acceleration and depth sensors with reasonable range and range-rate noise statistics. However, we note that these anecdotal results are preliminary, and more studies are needed before any broad conclusions can be made.

8.5 Conclusion

This paper reports preliminary simulation studies for cooperative navigation of one underwater vehicle client (equipped with attitude and depth sensors) with one surface vehicle server (equipped with GPS), employing both range and range-rate observations to estimate the vehicle states with a delayed-state extended Kalman filter. This preliminary study suggests that the addition of range-rate measurements may not significantly improve the performance of the CEKF cooperative navigation algorithm over the case of range-only observations with reasonable range and range-rate noise statistics. However, when range measurements are poor, the addition of acoustic range-rate observations appears to offer modest improvements in the steady-state response and a significantly smaller error in the transient response of CEKF compared to range-only navigation, for the noise statistics and geometry reported herein.

Acknowledgments

We gratefully acknowledge the support of the National Science Foundation under Award 1319667 and the National Defense Science and Engineering Graduate Fellowship. We gratefully acknowledge using and extending, with permission, a CEKF implementation originally developed by our colleague Dr. Ryan Eustice at the University of Michigan and subsequently extended by Dr. Sarah Webster who is presently with the University of Washington Applied Physics Laboratory. A version of this chapter appeared as [28] and is reproduced in part herein with permission.

References

1. J. Alleyne, "Position estimation from range only measurements." Master's thesis, Naval Postgraduate School, Monterey, CA, September 2000.
2. Anonymous, "Extremely low frequency transmitter site Clam Lake, Wisconsin." United States Navy Fact File. http://enterprise.spawar.navy.mil/UploadedFiles/fs_clam_lake_elf2003.pdf.
3. P. Baccou and B. Jouvencel, "Homing and navigation using one transponder for AUV, post-processing comparisons results with long base-line navigation." *Proceedings of the IEEE International Conference on Robotics and Automation*, Washington, DC, vol. 4, May 2002, pp. 4004–4009.

4. P. Baccou and B. Jouvencel, "Simulation results, post-processing experimentations and comparisons results for navigation, homing and multiple vehicles operations with a new positioning method using on transponder." *Proceedings of the IEEE/RSJ International Conference on Intelligent Robots and Systems*, vol. 1, October 2003, pp. 811–817.

5. A. Bahr and J. Leonard, "Cooperative localization for autonomous underwater vehicles." *Proceedings of the 10th International Symposium on Experimental Robotics (ISER)*, Rio de Janeiro, Brasil, July 2006, pp. 387–395.

6. A. Bahr, "Cooperative localization for autonomous underwater vehicles." PhD dissertation, Massachusetts Institute of Technology/Woods Hole Oceanographic Institution Joint Program, Cambridge, MA, February 2009.

7. A. Bahr, J.J. Leonard, and M.F. Fallon, "Cooperative localization for autonomous underwater vehicles," *International Journal of Robotics Research*, vol. 28, no. 6, pp. 714–728, June 2009.

8. T. Bailey, M. Bryson, H. Mu, J. Vial, L. McCalman, and H. Durrant-Whyte, "Decentralised cooperative localisation for heterogeneous teams of mobile robots." *Proceedings of IEEE International Conference on Robotics and Automation*, May 2011, pp. 2859–2865.

9. Y. Bar-Shalom, X. Rong Li, and T. Kirubarajan, *Estimation with applications to tracking and navigation*. New York: John Wiley & Sons, 2001.

10. B.S. Bourgeois, *"Using range and range rate for relative navigation."* Naval Research Laboratory, Mapping, Charting, Geodesy Branch, Marine Geosciences Division, Technical Report, September 2007.

11. A.D. Bowen, L.L. Whitcomb, and D.R. Yoegrer, "Nereus project homepage." 2012. http://www.whoi.edu/page.do?pid=10076.

12. J.A. Catipovic and L.E. Freitag, "High data rate acoustic telemetry for moving ROVs in a fading multipath shallow water environment." *Proceedings of the Symposium on Autonomous Underwater Vehicle Technology*, June 5–6 1990, pp. 296–303.

13. A. Chatfield, *Fundamentals of high accuracy inertial navigation.* AIAA (American Institute of Aeronautics & Astronautics), 1997, vol. 174.

14. M. Chitre, S. Shahabudeen, and M. Stojanovic, "Underwater acoustic communications and networking: Recent advances and future challenges," *Marine Technology Society Journal*, vol. 42, no. 1, pp. 103–116, 2008.

15. D. Djugash, S. Singh, and P. Corke, *Field and service robotics*, ser. Springer Tracts in Advanced Robotics. Berlin, Heidelberg: Springer Berlin/Heidelberg, 2006, vol. 25, ch. Further Results with Localization and Mapping Using Range from Radio, pp. 231–242.

16. J. Djugash, S. Singh, G. Kantor, and W. Zhang, "Range-only SLAM for robots operating cooperatively with sensor networks." *Proceedings of the IEEE International Conference on Robotics and Automation (ICRA)*, May 2006, pp. 2078–2084.

17. C. Dybas and S. Murphy, "The Abyss: Deepest Part of the Oceans no longer Hidden." National Science Foundation Press Release, 2009. http://www.nsf.gov/news/news_summ.jsp?cntn_id=114913.

18. R.M. Eustice, L.L. Whitcomb, H. Singh, and M. Grund, "Recent advances in synchronous-clock one-way-travel-time acoustic navigation." *Proceedings of the IEEE/MTS OCEANS Conference and Exhibition*, Boston, MA, September 2006, pp. 1–6.

19. R.M. Eustice, H. Singh, and L.L. Whitcomb, "Synchronous-clock one-way-travel-time acoustic navigation for underwater vehicles," *Journal of Field Robotics, Special Issue on State of the Art in Maritime Autonomous Surface and Underwater Vehicles*, vol. 28, no. 1, pp. 121–136, January/February 2011.

20. R.M. Eustice, L.L. Whitcomb, H. Singh, and M. Grund, "Experimental results in synchronous-clock one-way-travel-time acoustic navigation for autonomous underwater vehicles." *Proceedings of the IEEE International Conference on Robotics and Automation*, Rome, Italy, April 2007, pp. 4257–4264.

21. M.F. Fallon, G. Papadopoulos, J.J. Leonard, and N.M. Patrikalakis, "Cooperative AUV navigation using a single maneuvering surface craft," *International Journal of Robotics Research*, vol. 29, no. 12, pp. 1461–1474, October 2010.

22. A. Gadre, "Observability analysis in navigation systems with an underwater vehicle application." PhD dissertation, Virginia Polytechnic Institute and State University, Blacksburg, Virginia, January 2007.

23. A. Gadre and D. Stilwell, "Toward underwater navigation based on range measurements from a single location." *Proceedings of the IEEE International Conference on Robotics and Automation*, New Orleans, vol. 5, April 2004, pp. 4472–4477.

24. A. Gadre and D. Stilwell, "A complete solution to underwater navigation in the presence of unknown currents based on range measurements from a single location." *Proceedings of the IEEE/RSJ International Conference on Intelligent Robots and Systems*, Edmonton, Alberta, Canada, August 2005, pp. 1420–1425.

25. A. Gadre and D. Stilwell, "Underwater navigation in the presence of unknown currents based on range measurements from a single location." *Proceedings of the 2005 American Control Conference*, vol. 1, June 2005, pp. 656–661.

26. C.R. German, A. Bowen, M.L. Coleman, D.L. Honig, J.A. Huber, M.V. Jakuba, J.C. Kinsey, M.D. Kurz, S. Leroy, J.M. McDermott, B.M. de Lépinay, K. Nakamura, J.S. Seewald, J.L. Smith, S.P. Sylva, C.L.V. Dover, L.L. Whitcomb, and D.R. Yoerger, "Diverse styles of submarine venting on the ultraslow spreading Mid-Cayman Rise." *Proceedings of the National Academy of Sciences*, vol. 107, no. 32, p. 14020, 2010.

27. M. Green and K. Scussel, "Underwater data communication and instrument release management system." Patent, March, 2007, US Patent 7,187,623. [Online]. https://www.google.com/patents/US7187623

28. Z.J. Harris and L.L. Whitcomb, "Preliminary study of cooperative navigation of underwater vehicles without a DVL utilizing range and range-rate observations." *2016 IEEE International Conference on Robotics and Automation (ICRA)*, May 2016, pp. 2618–2624.

29. Z.J. Harris and L.L. Whitcomb, "Preliminary feasibility study of cooperative navigation of under-water vehicles with range and range-rate observations." *Proceedings of IEEE/MTS OCEANS 2015*, Washington, DC, October 2015.

30. J.C. Hartsfield, "Single transponder range only navigation geometry (STRONG) applied to REMUS autonomous under water vehicles." Master's thesis, Massachusetts Institute of Technology and Woods Hole Oceanographic Institution, August 2005.

31. M. Hunt, W. Marquet, D. Moller, K. Peal, W. Smith, and R. Spindel, *An acoustic navigation system.* Technical Report WHOI-74-6, December 1974.

32. J. Jouffroy and J. Reger, "An algebraic perspective to single-transponder underwater navigation." *Proceedings IEEE 2006 CCA/CACSD/ISIC*, Munich, Germany, 2006, pp. 1789–1794.

33. G. Kantor and S. Singh, "Preliminary results in range-only localization and mapping." *Proceedings of the IEEE International Conference on Robotic Automation (ICRA)*, 2002, pp. 1818–1823.

34. A. Kehagias, J. Djugash, and S. Singh, *Range-only SLAM with interpolated range data.* Robotics Institute, Pittsburgh, PA, Technical Report CMU-RI-TR-06-26, May 2006.

35. D.B. Kilfoyls and A.B. Baggeroer, "The state of the art in underwater acoustic telemetry," *IEEE Journal of Oceanic Engineering*, vol. 25, no. 1, pp. 4–27, January 2000.

36. J.C. Kinsey, R.M. Eustice, and L.L. Whitcomb, "A survey of underwater vehicle navigation: Recent advances and new challenges." *IFAC Conference of Manoeuvring and Control of Marine Craft*, Lisbon, Portugal, 2006, invited paper.

37. D. Kurth, G. Kantor, and S. Singh, "Experimental results in range-only localization with radio." *Proceedings of the IEEE/RSJ International Conference on Intelligent Robot Systems (IROS)*, October 2003, vol. 1, pp. 974–979.

38. D. Kurth, "Range-only robot localization and SLAM with radio." Master's thesis, Carnegie Mellon University, Pittsburgh, PA, May 2004.

39. C.E. LaPointe, "Virtual long baseline (VLBL) autonomous underwater vehicle navigation using a single transponder." Master's thesis, Massachusetts Institute of Technology and Woods Hole Oceanographic Institution, June 2006.

40. M. Larsen, "High performance autonomous underwater navigation: Experimental results," *Hydro International*, vol. 6, no. 1, pp. 6–9, January/February 2002.

41. M. Larsen, "Synthetic long baseline navigation of underwater vehicles." *Proceedings of the IEEE/ MTS OCEANS Conference and Exhibition*, Providence, RI, vol. 3, September 2000, pp. 2043–2050.
42. M. Larsen, "Synthetic long baseline navigation of underwater vehicles." *OCEANS 2000 MTS/ IEEE Conference and Exhibition*, vol. 3, 2000, pp. 2043–2050.
43. M. Larsen, "Autonomous navigation of underwater vehicles." PhD dissertation, Technical University of Denmark, Ørsteds Plads, Lyngby, Denmark, 2001.
44. M. Larsen, "Methods and systems for navigating under water." Patent, November 2006, US Patent 7,139,647. [Online]. https://www.google.com/patents/US7139647
45. E. Larsson, "Cramer-Rao bound analysis of distributed positioning in sensor networks," *Signal Processing Letters, IEEE*, vol. 11, no. 3, pp. 334–337, March 2004.
46. A. Lawrence, *Modern inertial technology: Navigation, guidance, and control*. Springer Verlag, 1998.
47. P.-M. Lee, B.-H. Jun, and Y.-K. Lim, "Review on underwater navigation system based on range measurements from one reference." *OCEANS 2008—MTS/IEEE Kobe Techno-Ocean*, April 2008, pp. 1–5.
48. E. Martinson and F. Dellaert, "Marco Polo localization." *Proceedings of the IEEE International Conference on Robotic Automation (ICRA)*, vol. 2, September 2003, pp. 1960–1965.
49. S. McPhail and M. Pebody, "Range-only positioning of a deep-diving autonomous underwater vehicle from a surface ship," *IEEE Journal of Oceanic Engineering*, vol. 34, no. 4, pp. 669–677, October 2009.
50. P.H. Milne, *Underwater acoustic positioning systems*. Houston: Gulf Publishing, 1983.
51. A. Palmeiro, M. Martin, I. Crowther, and M. Rhodes, "Underwater radio frequency communications." *OCEANS, 2011 IEEE*, Spain, June 2011, pp. 1–8.
52. J. Partan, J. Kurose, and B. Levine, "A survey of practical issues in underwater networks," *ACM SIGMOBILE Mobile Computing and Communications Review*, vol. 11, no. 4, pp. 23–33, 2007.
53. L. Paull, M. Seto, and J. Leonard, "Decentralized cooperative trajectory estimation for autonomous underwater vehicles." *2014 IEEE/RSJ International Conference on Intelligent Robots and Systems (IROS 2014)*, September 2014, pp. 184–191.
54. B. Ristic, S. Arulampalam, and J. McCarthy, "Target motion analysis using range-only measurements: Algorithms, performance and application to ISAR data," *Signal Processing*, vol. 82, no. 2, pp. 273–296, February. 2002.
55. A. Ross and J. Jouffroy, "Remarks on the observability of single beacon underwater navigation." *Proceedings of the International Symposium on Unmanned Untethered Submersible Technology*, Durham, New Hampshire, August 2005.
56. A. Scherbatyuk, "The AUV positioning using ranges from one transponder LBL." *Proceedings of the IEEE/MTS OCEANS Conference and Exhibition*, San Diego, CA, vol. 3, October 1995, pp. 1620–1623.
57. H. Singh, J. Catipovic, R. Eastwood, L. Freitag, H. Henriksen, F. Hover, D. Yoerger, J. Bellingham, and B. Moran, "An integrated approach to multiple AUV communications, navigation and docking." *Proceedings of the IEEE/MTS OCEANS Conference and Exhibition*, Fort Lauderdale, FL, vol. 1, September 1996, pp. 59–64.
58. S. Singh, S. E. Webster, L. Freitag, L. L. Whitcomb, K. Ball, J. Bailey, and C. Taylor, "Acoustic communication performance of the WHOI micro-modem in sea trials of the Nereus vehicle to 11,000 m depth." *Proceedings of the IEEE/MTS OCEANS Conference and Exhibition.*, Biloxi, MS, October 2009, pp. 1–6.
59. T. Song, "Observability of target tracking with range-only measurements," *IEEE Journal of Oceanic Engineering*, vol. 24, no. 24, pp. 383–387, July 1999.
60. R. Spindel, R. Porter, W. Marquet, and J. Durham, "A high-resolution pulse-doppler underwater acoustic navigation system," *IEEE Journal of Oceanic Engineering*, vol. 1, no. 1, pp. 6–13, September 1976.
61. J.R. Spletzer, "*A new approach to range-only SLAM for wireless sensor networks.*" Lehigh University, Bethlehem, PA, Technical Report, 2003.
62. D. Titterton and J. Weston, *Strapdown inertial navigation technology*, vol. 17. Peter Peregrinus Ltd, 2004.

63. J. Vaganay, P. Baccou, and B. Jouvencel, "Homing by acoustic ranging to a single beacon." *Proceedings of the IEEE/MTS OCEANS Conference and Exhibition*, Providence, RI, vol. 2, September 2000, pp. 1457–1462.

64. J.M. Walls and R.M. Eustice, "An origin state method for communication constrained cooperative localization with robustness to packet loss," *The International Journal of Robotics Research*, vol. 33, no. 9, pp. 1191–1208, 2014.

65. S.E. Webster, "Decentralized single-beacon acoustic navigation: Combined communication and navigation for underwater vehicles." PhD dissertation, Johns Hopkins University, Baltimore, MD, June 2010.

66. S.E. Webster, R.M. Eustice, C. Murphy, H. Singh, and L.L. Whitcomb, "Toward a platform-independent acoustic communications and navigation system for underwater vehicles." *Proceedings of the IEEE/MTS OCEANS Conference and Exhibition*, Biloxi, MS, October 2009, pp. 1–7.

67. S.E. Webster, R.M. Eustice, H. Singh, and L.L. Whitcomb, "Advances in single-beacon one-way-travel-time acoustic navigation for underwater vehicles," *The International Journal of Robotics Research*, vol. 31, no. 8, pp. 935–950, 2012. http://ijr.sagepub.com/content/31/8/935.

68. S.E. Webster, R.M. Eustice, H. Singh, and L.L. Whitcomb, "Preliminary deep water results in single-beacon one-way-travel-time acoustic navigation for underwater vehicles." *Proceedings of the IEEE/RSJ International Conference on Intelligent Robots and Systems*, St. Louis, MO, October 2009, pp. 2053–2060.

69. S.E. Webster, J.M. Walls, L.L. Whitcomb, and R.M. Eustice, "Decentralized extended information filter for single-beacon cooperative acoustic navigation: Theory and experiments," *IEEE Transactions on Robotics*, vol. 29, no. 4, pp. 957–974, August 2013.

70. S.E. Webster, L.L. Whitcomb, and R.M. Eustice, "Advances in decentralized single-beacon acoustic navigation for underwater vehicles: Theory and simulation." *Proceedings of the IEEE/OES Autonomous Underwater Vehicles Conference*, Monterey, CA, September 2010.

71. S.E. Webster, L.L. Whitcomb, and R.M. Eustice, "Preliminary results in decentralized estimation for single-beacon acoustic underwater navigation." *Proceedings of the Robotics: Science & Systems Conference*, Zaragoza, Spain, June 2010.

9

Cooperative Strategies for Localization and Navigation

Halgurd Sarhang Maghdid, Kayhan Zrar Ghafoor, Ali Al-Sherbaz,
Linghe Kong, and Naseer Al-Jawad

CONTENTS

9.1 Introduction

Providing the position of mobile devices from outdoors into indoors seamlessly is one of the most viable user demands in most location-based-service (LBS) applications. Mobile devices, when outdoors, use global navigation satellite systems (GNSS) information to define their location accurately. However, when indoors or in harsh environments, GNSS signals are weak or blocked and do not provide accurate location information. Therefore, an alternative technology or new solution is required to tackle such challenges [1]. Cooperative localization solutions based on a network of mobile devices/sensors are essential to provide seamless mobility and realize the full potential of the LBSs. Several strategies and hypothesis have been conducted and tested for this purpose [2].

In a simple scenario, the cooperative localization strategy (CLS) is to exchange valuable information between a team of mobile devices in which they are exploring the decentralized perception to provide position seamlessly, and to enhance the location accuracy of each mobile device. Thus, this strategy does not require a dedicated infrastructure, and capitalizes on the cooperation of mobile devices in the vicinity. This is achieved by hybridizing of the exchange information from their on-board devices such as inertial sensors, Bluetooth, WiFi transceivers, 5G and LTE signal measurements, and GNSS receivers [3,14].

This chapter begins by presenting background descriptions of the cooperative localization strategies and methodologies in Section 9.2. Section 9.3 reviews the current literature and gives an overview of the related work in the field of cooperative localization solutions. The review classifies the recent cooperative strategies for localization techniques and solutions which are adapted on mobile devices via on-board wireless and sensor technologies. The next section (Section 9.4) investigates and analyzes some of the implemented cooperative localization strategies. In the final section (Section 9.5), current issues and solutions of the implemented cooperative strategies are explained.

9.2 Background

The demand for using location information in our daily applicable services is growing. This is due to its common feature in most applications, ranging from shopping to navigation. Some examples of such applications are shopping, parking, billing, advertising, intelligent transportation, smart building, health and safety, banking, airport services, etc. The quality of such services or applications depends on the accuracy of the localization techniques utilized and the capability of the localization technologies [4]. However, there is no single localization technique or technology that offers accurate locations or high-performance solutions, and there is no a single localization technology that offers seamless positioning from outdoors to indoors. For example, GNSS provide up to 3 meters in the area of open sky with line-of-sight signals. However, this accuracy may be reduced or may not provide location information if the GNSS signals are blocked or reflected by the roofs and walls of surrounding buildings [5]. To compensate for this, there are alternative technologies to offer positioning services including cellular (4G or LTE) signals, WiFi, Bluetooth, ultra-wide-band (UWB) on-board sensors such as inertial measurement unit (IMU). However, the following are some issues of these technologies while utilizing their measurements for positioning purposes:

1. Radio communication coverage of the signals and non-existing location information of the reference points within the building such as WiFi access points or Bluetooth anchors [6].

2. There is currently no standard localization protocol [6].

3. Inaccurate measurement of the sensors' or chipset's readings [6].

4. Stand-alone localization techniques do not provide seamless positioning and do not offer high performance. For example: (a) pseudorange time-based techniques suffer from the jitters, dilution of precision (DOP) [7], and clock synchronization between the mobile device and reference station [8], (b) received-signal-strength (RSS) technique suffers from the instability of the signal power in the complex indoors infrastructures, and (c) dead-reckoning technique experiences with the drift error of the sensor readings [6].

5. Deploying and installing extra hardware or surveying the area in order to provide optimal performance incurs a large cost [9].

To overcome these issues, combining different techniques and technologies is possible by using the advantages and compensating for the limitations. To this end, there are two main approaches: an infrastructure-based approach and a cooperative-based (infrastructure-less) approach.

In the infrastructure-based approach, the exchange location messages are between the mobile devices and the base station, access point, anchors, or dedicated servers. There is no collaboration between the mobile services, that is, all the localization computation and location accuracy improvements occur at the dedicated reference device/server. However, to extend the coverage of the reference device signals, the system needs a wide-coverage reference device or installation of a large number of reference devices.

With the cooperative-based approach, the location information could be obtained only between mobile devices in the area. Thus, a connection between the mobile device and reference devices is not necessary. This means that coverage of the devices could be extended by communicating a new or already cooperated mobile device in the vicinity and there is no need to deploy a large number of reference devices. Further, the performance of such cooperative localization relies on the strategy of the solution. A good cooperative strategy provides an accurate, applicable, low-cost, on-the-go, and unconditional solution.

The cooperative strategy is based on the participating mobile devices, the environment, the localization techniques, and technology. The cooperative mobile devices could be smartphones, wearable devices, vehicles, mobile sensors (e.g., embedded on the drones), etc. These mobile devices have peer-to-peer communication to enable the cooperative strategy. The environment includes indoors and outdoors seamlessly. The well-known implemented localization techniques are rang-based, fingerprinting, angle-based, dead reckoning (relative positioning), and mapping information. Localization technologies include WiFi, Bluetooth, cellular 4G/LTE, UWB, RFID, RADAR, camera, and IMU sensors [10].

To implement and to evaluate the CLSs, the following should be considered:

1. The demand for **localization accuracy** is based on the nature of the application. For example, some applications need sub-centimeter accuracy, like human-body positioning, while shopping services need meter accuracy.

2. **Localization cost**: Some cooperative localization strategies are very costly (such as solutions that utilize extra hardware) while exchanging messages cooperatively between mobile devices, while others need only the mobile device measurements.

3. **Robustness and Seamless**: A robust and applicable localization strategy is to provide positioning from outdoors to indoors, seamlessly, on the go, even if there are some misreadings from the wireless and sensor technologies due to environmental impacts.

In the next section, the recent cooperative localization strategies are presented and are investigated in terms of their performance.

9.3 Current CLSs

A stand-alone localization technique or single localization technology have been proven not to be applicable for navigation and localization for most of the LBS applications. This is because of complex and different environments (outdoors, urban, and indoors), limitation of the signal coverage, and accumulated errors of the sensor measurements. However, a great deal of research and many strategies for combined and/or cooperative schemes have been tested to tackle these issues. In this section, a few of these strategies are briefly explained.

For example, due to the easy implementation of RSS technique and availability of a WiFi access point in the vicinity, RSS-range-based and RSS-fingerprinting techniques have been used to locate mobile devices via implementing cooperative strategy between the mobile devices. However, the accuracy of such solutions is limited due to signal propagation issues, specifically in dynamic or complex indoor environments. Recently, an alternative technology, ultra-wide-band (UWB) technology, has been used to offer accurate distance calculation based on signal-time measurements between mobile devices via the cooperative messaging method, as shown in Figure 9.1. For example, UWB distance measurements and IMU sensor measurements have been integrated cooperatively between the mobile devices via a particle filter within the infrastructure-less environment [11]. The UWB

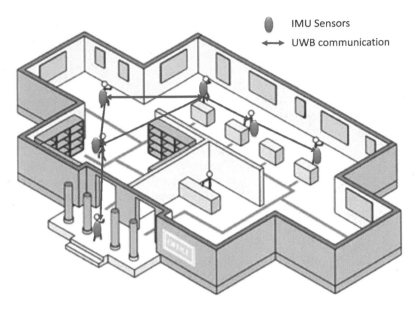

IMU Sensors
UWB communication

FIGURE 9.1
Cooperative strategy using integrated UWB signal and IMU measurements.

distance calculations are based on time-of-arrival (TOA) technique. The UWB signals are a good candidate to provide accurate positioning, since the UWB signals provide centimeter accuracy. The obtained localization accuracy via simulation and real experiments is up to 2 meters. However, such cooperative strategy suffers from the issue of signal coverage between the cooperated mobile devices. This is because the maximum coverage of a UWB signal is up to 30 meters, which is not applicable for most localization applications.

A GNSS-based solution has been proven to offer a few meters' accuracy when outdoors, but its accuracy is degraded in urban areas and indoors. To address this issue, a new cooperative strategy between a set of vehicles has been proposed [12]: implicit cooperative positioning (ICP). The ICP works by sharing information (via vehicle-to-vehicle [V2V] communication link) between the vehicles [15,16]. The ICP is to enable the cooperative-location calculation via on-board sensor devices' measurements on the vehicles. The ICP also uses the distributed Gaussian message passing (GMP) algorithm to enable the positioning technique on the vehicles. The strategy of ICP is not only relying on the V2V messaging information and GNSS measurements, it also uses the location information of non-cooperative features including people, traffic lights, trees, etc. The communication information from these non-cooperative features with the vehicles is called vehicle-to-everything (V2E), as shown in Figure 9.2. The integration GNSS location information, V2V communication, and V2E information are performed by using the Kalman filter. Simulation results of a set of experiments improved the GNSS accuracy up to 1 meter in rural areas and up to 4 meters in urban areas (harsh environments).

In another vein, a fingerprinting technique using WiFi or cellular signals as a compromised solution in urban areas and indoors has been used frequently to locate mobile devices. However, for a large-scale area, the surveying area process to construct the radio map of the reference devices and the complex/dynamic structure of the area (when compared with offline-radio map) are the two main challenges of the fingerprinting technique. To tackle such challenges, **WiFi-based non-intrusive indoors positioning system** (**WinIPS**) has been designed and implemented [13]. The aim of WinIPS is to construct the

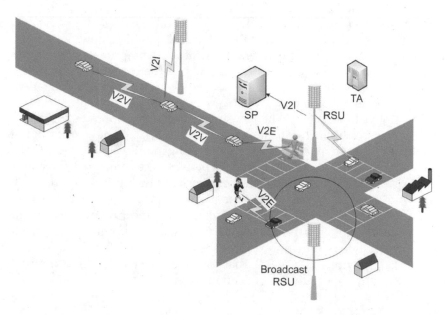

FIGURE 9.2
ICP strategy between vehicles and non-cooperative features.

online radio-map (or an online database) cooperatively in the building in a non-intrusive mode. It captures the WiFi signals in a specific area to retrieve the RSS values and MAC address of the mobile devices (which are called collaborated entities) and WiFi access points. All the capture information is transmitted and stored in a database server, as depicted in Figure 9.3. This cooperative strategy is periodically repeated to calibrate and

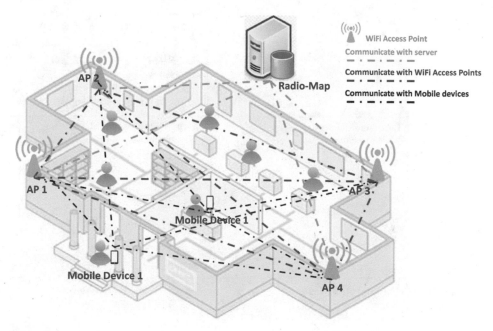

FIGURE 9.3
Cooperative strategy to construct online RSS database for indoor localizations.

update the constructed radio-map when any new mobile devices or new point of the area are recorded. Further, to solve the instability of the RSS values, a new Gaussian regression model is implemented with the cooperative strategy. Comprehensive experiments are tested in different buildings to prove the validity of the constructed radio map and to evaluate the obtained location accuracy. The results show that the localization accuracy has been improved to 2–3.5 meters.

A new cooperative localization scheme using network-smartphones has been proposed, called unconstrained indoors localization scheme (UNILS) [2]. The aim of this new scheme is to integrate multi-sensor measurement sources using multi-technologies on smartphones. The scheme uses a relative distances approach between the connected smartphones when their GNSS are enabled, especially when the majority of the SPs are outdoors. The relative distances are measured using the time-of-arrival (TOA) technique, which is very accurate when the GNSS receiver is used as the source of timestamps for the receiving and sending signals of the connected smartphones. Then the scheme combines these distances with uncertainty calculations from onboard dead-reckoning (DR) measurements using a Kalman filter (as shown in Figure 9.4), that can provide seamless and improve location accuracy significantly, especially when deep indoors.

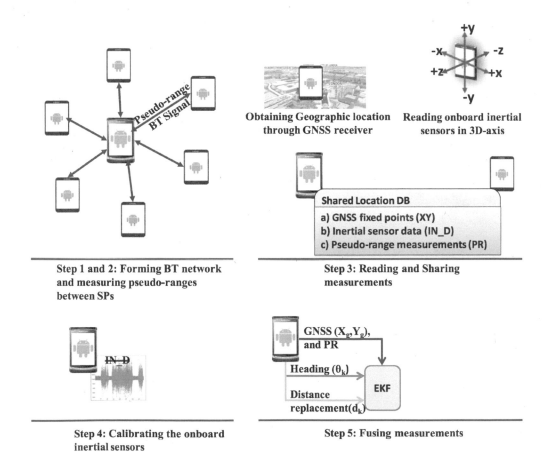

FIGURE 9.4

Steps of UNILS strategy to locate indoor-cooperated smartphones (From H.S. Maghdid et al., In *Position, Location and Navigation Symposium (PLANS), 2016 IEEE/ION*, Savannah, GA, 2016.)

This means that in deep indoors, UNILS can utilize only available devices/sensors on SPs, when communication with WAPs or BT-anchors is considered unreliable or unavailable, to offer reasonable cost and good localization performance. Results obtained from actual trials and simulations (using OPNET) of the UNILS scheme (based on sndroid-SPs network implementations for various indoors scenarios) show that around 3 meters accuracy can be achieved when locating SPs at various deep indoor situations.

9.4 Investigations

The proposed or tested cooperative strategies are doing work without the need for predefined constraints or preinstalled localization infrastructure, and do away with inaccurate stand-alone localization techniques. The uniqueness of these strategies is centered on the following points:

1. The strategies provide a new seamless outdoors-indoors localization scheme that capitalizes on the cooperation of mobile devices in the vicinity to hybridize information from their on-board devices such as inertial sensors, WiFi transceivers, cellular transceivers, Bluetooth transceivers, and GNSS receivers.
2. The strategies are a fusing of various localization techniques to provide low-cost and on-the-go solutions.
3. The strategies provide reasonable accuracy, which is enough for some of LBS applications.
4. The strategies extend the coverage of the localization, in the vicinity, and consequently offer robust localization solutions.

However, further research and experiments should be conducted to realize the proposed strategies, since most of the proposed strategies are tested under ideal circumstances (using extra hardware) or they are simulated; that is, they need further real experiments.

9.5 Summary

Cooperative localization strategies and solutions have become a huge success mainly because they provide good accuracy for LBS applications. Currently cooperative localization strategies have the advantage of reliable positioning via hybrid localization technologies, and some provide defined business models for the delivery of content to LBS users. In addition, integrating and combining the localization techniques within the schemes are included in most of the cooperative strategies. This is probably because such integration overcomes the limitations of the stand-alone techniques and takes their advantages into account.

However, current solutions still suffer from the lack continuity due to the limited coverage from outdoors into urban canyons and then to indoors. Also, current solutions use a huge number of preinstalled localization sensor networks, such as those offered by BLE beaconing and WiFi fingerprinting, and are specifically implemented to provide mobile device location while indoors in that particular vicinity. Last but not least, the current solutions with low-cost strategies suffer from accurate indoors localization, and on-the-go anywhere anytime has proven somewhat problematic to deliver thus far.

References

1. H.S. Maghdid, I.A. Lami, K.Z. Ghafoor, J. Lloret, "Seamless outdoors-indoors localization solutions on smartphones: Implementation and challenges." *ACM Computing Surveys (CSUR)*, vol. 48, no. 4, p. 53, 2016.
2. H.S. Maghdid, A. Al-Sherbaz, N. Aljawad, I.A. Lami, "UNILS: Unconstrained indoors localization scheme based on cooperative smartphones networking with onboard inertial, Bluetooth and GNSS devices." In: *Position, Location and Navigation Symposium (PLANS), 2016 IEEE/ION*, Savannah, GA, 2016.
3. I. Lami, H. Maghdid, T. Kuseler, "SILS: A smart indoors localization scheme based on on-the-go cooperative smartphones networks using onboard bluetooth, WiFi and GNSS." In: *International Technical Meeting of the Satellite Division of The Institute of Navigation, ION GNSS+ 2014*, Tampa, FL, 2014.
4. S.H. Jung, B-C. Moon, D. Han, "Performance evaluation of radio map construction methods for Wi-Fi positioning systems." *IEEE Transactions on Intelligent Transportation Systems*, vol. 18, no. 4, pp. 880–889, 2017.
5. I.A. Lami, H.S Maghdid, "Synchronising WiFi access points with GPS time obtained from smartphones to aid localisation." In: *International Conference on Computer Applications Technology (ICCAT)*, Sousse, Tunisia, 2013.
6. H.S. Maghdid. *Hybridisation of GNSS with other wireless/sensors technologies onboard smartphones to offer seamless outdoors-indoors positioning for LBS applications.* The University of Buckingham, Buckingham, UK, 2016.
7. A.T. Asaad, H.S. Maghdid, D.A. Ali, "VRPR: Virtual reference positions replacement to improve range-based techniques for smartphone localization." *International Journal of Computer Science and Information Security (IJCSIS)*, vol. 14, no. 9, pp. 345–351, 2016.
8. H.S. Maghdid, I.A. Lami, "Dynamic clock-model of Wi-Fi access-points to help indoors localisation of smartphones." In: *4th International Congress on Ultra Modern Telecommunications and Control Systems and Workshops (ICUMT)*, Saint Petersburg, Russia, 2012.
9. H.S. Maghdid, L.S. Abdulrahman, M.H. Ahmed, A.T. Sabir, "Modified WiFi-RSS fingerprint technique to locate indoors-smartphones: FENG building at Koya University as a case study." *Kurdistan Journal of Applied Research*, vol. 2, no. 3, pp. 212–217, 2017.
10. H. Naseri, V. Koivunen, "Cooperative simultaneous localization and mapping by exploiting multipath propagation." *IEEE Transactions on Signal Processing*, vol. 65, no. 1, pp. 200–211, 2017.
11. R. Liu, C. Yuen, T-N. Do, D. Jiao, X. Liu, U-X. Tan, "Cooperative relative positioning of mobile users by fusing IMU inertial and UWB ranging information." In: *Robotics and Automation (ICRA), 2017 IEEE International Conference on*, 2017.
12. G. Soatti, M. Nicoli, N. Garcia, B. Denis, R. Raulefs, H. Wymeersch, "Implicit cooperative positioning in vehicular networks." *arXiv preprint arXiv:1709.01282*, pp. 1–15, 2017.
13. H. Zou, M. Jin, H. Jiang, L. Xie, C.J. Spanos, "WinIPS: WiFi-based non-intrusive indoors positioning system with online radio map construction and adaptation." *IEEE Transactions on Wireless Communications*, vol. 16, no. 12, pp. 8118–8130, 2017.
14. S. A. Maghdid, H. S. Maghdid, S. R. Hama Salah, K. Z. Ghafoor, A. S. Sadiq, S. Khan, "Indoor human tracking mechanism using integrated onboard smartphones Wi-Fi device and inertial sensors." *Telecommunication Systems*, pp. 1–12, 2018.
15. L. Kong, G. Xue, K. Z. Ghafoor, R. Hussain, H. Sheng, "Real-time density detection in connected vehicles." *IEEE Communication Magazine.* vol. 56, no. 10, pp. 64–70, 2018.
16. K. Z. Ghafoor, Linghe Kong, Danda B. Rawat, Eghbal Hosseini, Ali Safaa Sadiq, "Quality of service aware routing protocol in software-defined Internet of vehicles." *IEEE Journal of Internet of Things.* vol. 6, no 2, pp. 2817–2828, 2018.

10

Cooperative Multi-Robot Navigation–SLAM, Visual Odometry and Semantic Segmentation

Robert G. Reid, Kai Li Lim, and Thomas Bräunl

CONTENTS

10.1 Introduction

The University of Western Australia (UWA)'s multi-robot system (MRS) [1] comprises seven Pioneer 3AT-based outdoor robots. It was designed to solve the multi-robot simultaneous localization and mapping (SLAM) problem through strongly coordinated behaviors with task allocations that are performed explicitly whereby each task is divided into subtasks that are dynamically allocated and re-allocated in response to changing conditions or failure [2]. In other words, this system is capable of structured communications while being aware of one another.

For an MRS to properly navigate an environment and perform cooperative tasks, these localization estimates need to be robust. Our system uses a contained localization system to allow for rapid deployments in unstructured/mixed indoor/outdoor environments, which is originally presented in [3]. Path planning and obstacle avoidance currently follow our implementation in [4,5]. We use a multi-robot SLAM (MR-SLAM) solution to localize robots through the construction of a shared local map. The MR-SLAM problem is complex, whereby localization is achieved through the registration of each robot in a consistent, global coordinate system, in which large amounts of sensor data fusion must occur on-line. Additionally, these robots often rely on wireless communication that is often lossy and subject to interferences with variations in latencies and bandwidth. Loop closures that are predominant in SLAM problems become more challenging, as they create larges sequences of constraint cycles, which can cause a combinatorial increase in computational complexity. This problem stems from the variations between the robots' vantage points where uncertainties in data association might arise due to the pairings between object detections and sensor measurements.

MRS projects presented in the recent literature [6–8] demonstrated a combination of low-level capabilities such as cooperative SLAM, exploration, object identification, object tracking, and object manipulation. A comprehensive review is available in [9,10]. These systems are off-line or on-line. Off-line systems collect the first sensor data that is processed at a later stage, while on-line systems perform MR-SLAM and other tasks in real time during deployment. It is worth noting that off-line systems are typically deployed based on their ease of implementation and prototyping, while often relaxing limitations imposed on computation and communication requirements. Additionally, many works are often implemented in constrained environments, such as in indoor laboratories [11,12], thereby preventing the robots from exposure to environmental irregularities, including noise, temperature, illumination, and seasonal variations. These systems can therefore be cheaper and more convenient to implement, as they usually do not require long-range mobility and sensors. Conversely, our MRS was designed for deployment in unconstrained, outdoor urban environments, requiring higher performance sensors and more rugged robots.

The incorporation of visual navigation onto the MRS stems from our motivation to solve problems relating to wheel slip in odometry and scene understanding. Works that implement visual odometry in robots [13–15] have proven that its ability to reduce the accumulating error caused by wheel slip can lead to more robust SLAM solutions. Visual SLAM algorithms such as ORB-SLAM [16] and LSD-SLAM [17] are often implemented on robots, with favorable outcomes. Likewise, scene understanding can be applied alongside obstacle detection from lidar measurements, such as classifying static and dynamic obstacles [18]. For our application, we have decided to apply semantic segmentation for scene understanding, as it offers a pixel-wise classification of a captured scene while being versatile and compatible with low-cost camera setups.

10.2 Robot Hardware Design

Each unmanned ground vehicle (UGV) in our MRS is fitted on top of a Pioneer AT3 [19] base, which provides a chassis, differential drive wheels with motor controllers and encoders, and batteries. High-level controls are performed through an Intel Core 2 Duo automotive PC that is connected to several sensors (see Figure 10.1). The sensors comprise an ibeo LUX 4 lidar [20], a SICK LMS-111 lidar [21], a Hokuyo URG-04LX lidar [22], an Xsens MTi inertial measurement unit (IMU) [23], a QStarz GPS receiver [24], wheel odometry on the Pioneer base, and a Logitech Sphere PZT camera [25]. Communications are performed between UGVs and base station through a Ubiquity Pico Station 2HP [26] over a Wi-Fi mesh, with an RF Innovations 900 MHz radio [27] as a redundant communications link. The Pico Station, LUX 4, and LMS-111 interface via 100 Mbps Ethernet, while the other sensors interface through USB 2.0.

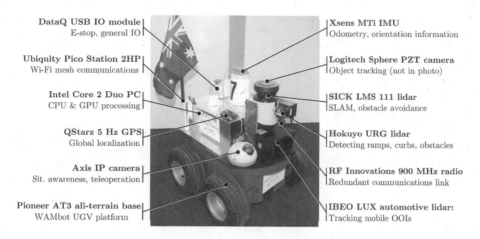

FIGURE 10.1
Photo illustrating a single MRS UGV with its hardware modules as labeled.

The sensors perform lidar-based SLAM, whereby the lidar array maps the environment horizontally and has a 20 m, 270° range at 25 Hz; the URG-04LX is mounted vertically and has a 4 m, 240° range at 10 Hz; and the LUX 4 is mounted horizontally and has a 50 m, 110° range that spans across four parallel, horizontal layers. The LMS-111 is used as the main SLAM sensor, where is it placed 0.5 m above ground to scan a single-layer horizontal plane. This results in 1080 measurements that translate to 2D "slices" of the environment around a 20 m radius. The other lidars mounted on the UGV are used for object/obstacle detection and tracking.

10.3 Cooperative Localization and Navigation

We incorporate our hybrid-decentralized and distributed MR-SLAM system onto the UGVs, which allows the decentralized UGVs to build distributed global grid-maps and navigate large urban areas [28,29]. Using this system enables the system to be deployed rapidly while allowing SLAM on the UGVs with minimal reliance on a ground control system (GCS).

10.3.1 Mapping

A typical deployment scenario of the MR-SLAM system is illustrated in Figure 10.2, showing that the back-end is executed across UGVs and GCSs. Each back-end instance stores a local copy of all submaps and constraints, which are then optimized and fused, building global maps; new loop closure constraints between submaps are also searched. Submaps are rectangular grid-maps with dynamically increasing dimensions determined by the lidar's maximum range R, the environment's shape, and a threshold heuristic described later in this section.

Local SLAM is performed independently on each UGV whereby a single-robot SLAM algorithm builds its own submap by processing its sensor data, which is then broadcasted across the mesh network. Graphical user interfaces (GUIs) are installed on GCS computers to enable operators to view and manipulate global and submaps and interact with pose graphs through a point-and-click interface.

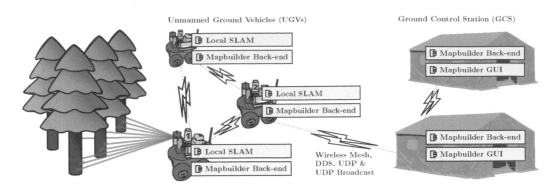

FIGURE 10.2
The MR-SLAM architecture and software development diagram showing the software components run on the GCS and UGVs.

A global grid-map is fused through several overlapping submaps that are obtained from the SLAM algorithms running on each UGV, thereby achieving a distributed map-building sequence. The fusion algorithm searches for overlaps in the grid-maps and subsequently determines if the cells in the global grid-map are occupied, free, or unknown. Each submap initializes with a local coordinate frame on the global frame with its submap pose p_a^W, which is estimated through pose graph optimization in a global Euclidean coordinate frame W. Given the UGV pose r_t^a always broadcast relative to p_a^W, and that each submap is created with the UGV at its origin, therefore $r_0^a = [0,0,0]^T$, and the UGV's time-varying pose in the global frame is thus

$$r_t^W = p_a^W \oplus r_t^a \qquad (10.1)$$

Each submap is assigned a 128-bit hexadecimal universally unique identifier (UUID) and exists either in an "open" or "closed" state, where they are always created through an "open" state to indicate that an occupied UGV is in the process of building it. Once map building is complete, the submap changes to a "closed" state to render the area immutable and non-traversable by any UGV, only allowing the back-end to update its pose, thereby fusing it onto the global grid-map. Having these states increases the robustness of the MR-SLAM system while ensuring its logic simplicity. A UGV can only occupy a single submap at any given time, and these "open" submaps are often connected to an adjoining "closed" submap. At the creation of a new submap, its map and pose uncertainty is reset. Using this approach enables the system to minimize bandwidth, storage, and redundancy across the MR-SLAM back-end.

Using a ray tracing technique based on [30], the lidar scans obtained while navigating a submap is aligned and fused into a single 2D occupancy grid-map, which maintains an accurate representation of the environment. The lidar scans are aligned with scan matching prior to ray tracing to circumvent the accumulation of minor quantization noise, which is done through a batch rounding of these lidar measurements to the nearest grid cell using the grid-map representation.

Aside from quantization noises, UGV pose uncertainties are also prevalent while it is building a submap. Although a UGV always initializes a new submap with no pose uncertainties, this uncertainty will always accumulate whenever the UGV is maneuvering, with odometric noise as its main contributor. Therefore, it is more pronounced in larger areas. To solve this problem, the algorithm initializes a new submap whenever this uncertainly surpasses a set threshold, which is determined by comparing the current angular pose uncertainty against the average distance to obstacles in the environment, estimating the amount of "blurring" in distant grid-map cells. Current lidar scans will not be fused if a new submap is triggered using this approach; using this heuristic thus minimizes distortions entering the submap grid-map, and large distortions that result in misaligned lidar scans can be prevented. The algorithm then transfers this accumulated uncertainty's covariance into the new constraint's covariance that is used to connect the old and new submaps through a maximum likelihood estimation.

By assuming an average distance between submaps D, the maximum overlap between a sequence of submaps separated by D is given by a ratio

$$Maximum\ overlap = \frac{2R}{2R+D} \tag{10.2}$$

The MRS yields a maximum submap overlap of 93%, which implies that the same UGV could create submaps that overlap the same area up to 15 times. This overlapping redundancy is required to allow the distributed back-ends to compare and align map data.

10.3.2 MR-SLAM Architecture

With reference to Figure 10.2, we have identified the functional roles of the software components as Table 10.1, which illustrates a minimalistic logical design diagram that considers a single UGV and GCS.

The class diagram in Figure 10.3 illustrates the various message types used by the system for MR-SLAM, which are all derived from the Submap message class. All messages are time-stamped with the source participant's priority and the submap UUID.

For the local SLAM front-end, we use a heuristic-driven EKF-SLAM [31] single-robot algorithm that takes all sensor data and outputs submaps, constraints, and real-time pose estimates, which are broadcasted over the mesh network. We have designed this front-end based on the following requirements:

- To estimate UGV pose and broadcast in real-time at >10 Hz locally, 1 Hz globally
- To build 10 cm submap grid-maps that are broadcasted >1 Hz locally, >0.2 Hz globally
- To robustly handle moving objects including human gaits up to 6 km/h
- To handle challenging sensing conditions such as sparse and/or featureless areas
- To detect odometric errors to minimize submap corruption

TABLE 10.1

Functional Requirements of Software Components

Component	Input	Behavior	Output
Local SLAM	Sensor data (lidar, odometry, IMU, GPS)	Performs local SLAM, creates a sequence of submaps	Broadcasts submap data, constraints, real-time UGV pose estimates
MR-SLAM back-end	Submap data from all UGVs, submap constraints, ground-truth constraints	Optimizes pose graphs, fuses submap data, searches for constraints	Global or windowed maps, submap pose estimates, submap constraints
MR-SLAM GUI	All MR-SLAM messages, GUI events; for example, keystrokes and mouse clicks	Displays global maps, interprets operator commands	Messages that alter graph structure; for example, ground-truth constraints

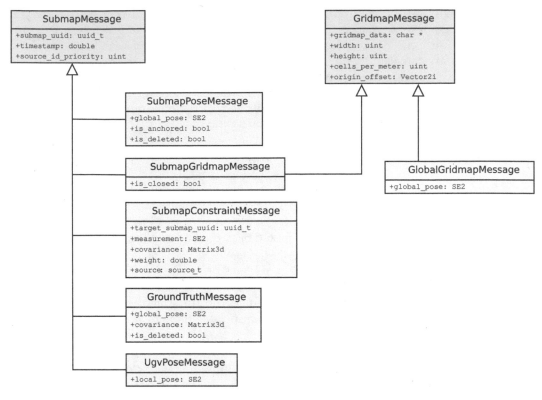

FIGURE 10.3
A class diagram showing the MR-SLAM system's message types and their variables.

- To compress submap grid-maps before broadcasting
- To use less than 25% of total computation and memory footprint

Likewise, the back-end requirements of our MRS are:

- To optimize pose graphs robustly and efficiently at <5 seconds per iteration
- To output large (5000 × 5000) grid-maps to the local partition at >1 Hz
- To match submaps to generate robust constraints at <5 seconds per match
- To broadcast **SubmapPose** messages globally to maintain decentralized pose graph
- To output **SubmapPose** updates to the local partition at >1 Hz

10.3.3 SLAM Implementation

We apply EKF-SLAM with scan matching which builds submaps by aggregating lidar scans at every 20 cm of movement or 20° of rotation, where pose estimation is achieved through an EKF. Scan matching was incorporated to reduce computation requirements by using the described threshold heuristics to decide when a current submap should be closed, which is augmented by a threshold on the percentage of lidar returns that are aligned successfully. This method enables the detection of matching failures, especially in sparse environments. EKF is used to estimate the UGV's pose by initiating each cycle to predict

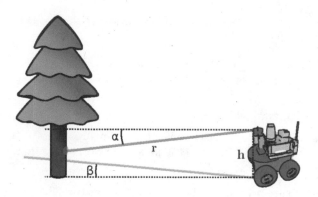

FIGURE 10.4
Lengths and angles used for calculating the local SLAM prefilter.

its current pose using the latest wheel odometry and IMU data, which aligns the lidar scan against the current submap through scan matching [32]. RANSAC [33] is also incorporated to reject outliers in the form of moving objects. The EKF and pose estimate is subsequently updated using this scan matching alignment method. Odometric noises that are present in the EKF update are adjusted according to the local ground slope gathered from the pitch and roll measurements from the IMU. An increase in slope leads the module to assume an increase in odometric noise due to wheel slip, thereby switching the filter preference for scan matching over odometry.

To filter these errors using the IMU, we assume that (1) the terrain inclination is less than β_{max}, and (2) the terrain follows the "Manhattan World" [34] assumption. With reference to Figure 10.4, each lidar measurement range r is first corrected to account for the declination α from the same measurement as $r\cos\alpha$, whereby an inequality is obtained referencing the height of the lidar mounted above the ground h

$$0 < h - r\sin\alpha - r\cos\alpha\tan\beta_{max} \tag{10.3}$$

To prevent any measurements from grazing the ground, any instance of r with declination α that dissatisfies Equation 10.3 is filtered. Lidar measurements that are corrected and filtered will then be passed to the SLAM algorithm.

10.3.4 UGV/GCS Communications

Communications between UGVs and GCSs are facilitated through a Wi-Fi mesh network over the IEEE 802.11n standard in a multi-hop configuration over a data distribution system (DDS), which provides a publisher-subscriber framework that provides robust real-time communications. Publishers are separated into partitions which are either global (all participants) or local (within a participant). Global partitions are mostly used to prevent network overloads, as the local partition is used for high-rate inter-process communications, where messages are passed over a shared memory between the front-end, back-end, and other high-level MRS software components.

Messages are broadcasted by the front-end as submaps are closing in the form of compressed grid-map data and the constraint that links the closed submap to the new one. Incomplete grid-maps for open submaps are also broadcasted to visualize real-time global maps; these maps are flagged to indicate that they are not yet immutable. The front-end broadcasts three distinct, time-stamped message types with a fixed DDS buffer size n with varying quality of service (QoS) priorities:

1. **SubmapConstraint** ($n = 1000$) defines the entire pose graph structure. A large buffer size is allocated for the series of small yet vital messages.
2. **SubmapGridmap** encodes the actual shape of the environment, constituting most of the MR-SLAM data, which are either open or closed.
   a. **Open** ($n = 0$) grid-maps are disposable as they are periodically broadcasted by the front-end, these are sent to the local partition at the lidar's scan rate.
   b. **Closed** ($n = 100$) grid-maps have higher priority as they are only transmitted once.
3. **UGVPose** ($n = 0$) are broadcasted frequently in real-time, which stales quickly and is subsequently disposable. These are also sent to the local partition at the lidar's scan rate.

A rendering algorithm ray traces the accumulated lidar scans into an empty grid map based on the methods described in [30]. The grid-map data is segregated into 32×32 cell tiles that are broadcasted over UDP on a 50:1 compression ratio.

Likewise, the back-end's publishing policies are:

1. **SubmapPose:**
   a. **Global Partition** ($n = 0$) does not require delivery guarantees since priority-based filters synchronizes this between participants.
   b. **Local Partition** ($n = 0$) does not require QoS as DDS uses shared memory to delivery real-time submap pose estimates.
2. **SubmapConstraint** ($n = 1000$) is assigned with the highest priority as they define the entire pose graph structure, hence a large buffer size is allocated.
3. **GlobalGridmap** ($n = 0$) are delivered through shared memory by DDS in real-time; QoS is therefore not required.

10.3.5 Loop Closures

A loop closure is an event that occurs when a UGV revisits a location it has previously threaded, thereby correcting its accumulated errors. The identification of loop closures occurs between overlapping submap pairs and spatially similar grid maps, which are broadcasted as new constraints that form cycles in the distributed pose graphs, bearing residual errors that require optimization. The back-end searches for local loop closures between "open" submaps in real-time, especially when multiple UGVs are operating overlapping areas to ensure proper localization and prevent the accumulation of errors. Using this method enables our system to efficiently accommodate high rates of loop closures and map changes in real-time.

The large number of loop closures generated by the system, as well as the distributed algorithm design, prompted us to utilize the graphics processing unit (GPU) to search for loop closures and merge submaps. Additionally, descriptive spatial relationships between submaps are extracted using a grid-map correlation algorithm on the GPU that calculates likelihood volumes and extracts multimodal Gaussian constraints. Using robust multimodal constraints enables the algorithm to preemptively add loop closures to the pose graph and perform outlier rejection by consensus.

An overlapping submap pair also initiates an occupancy grid-map fusion algorithm. To achieve the fusion described in Figure 10.5, we use a GPU-based approach whereby

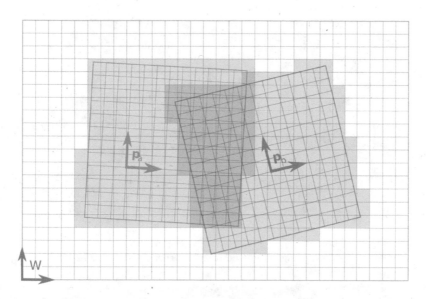

FIGURE 10.5

Occupancy grid-map fusion illustrating two submap grid-maps with their origins p_a^W and p_b^W to be fused onto the global output grid-map W shown as gray grids.

the algorithm checks for each overlapping cell in the output grid-map and performs a transformation of that cell in its 2D coordinates into the submap's coordinate frame, subsequently fusing it into the output cell based on the corresponding submap cell value.

To optimize complex multimodal Gaussian constraints, we utilized a continuous mode blending optimization technique that is based on nonlinear least-square approaches and exhibits convergence properties that are representative of the underlying multimodal constraint distributions.

10.3.6 SLAM Evaluation

For the evaluations described here, ten UGVs were deployed to explore an 80×40 m environment. The total elapsed time was 36 minutes [3]. The SLAM routine was completed with minimal user intervention whereby the UGVs autonomously explore the environment while displaying their progress in real-time onto the MR-SLAM system's GUI. This process follows our approach in [43] and is illustrated in series across Figures 10.6 through 10.8 with timestamps on the upper right corner of the images in minutes and seconds; the total odometry across all UGVs is presented in meters at the lower right corner; UGVs are shown as dots with similarly shaded lines showing their trajectories; pose graphs are shaded lines on the map, with dots, lines, and triangles representing submap poses, submap constraints, and ground truth constraints, respectively. (Refer online for colored figure and explanation.)

The UGVs start at the southeast corner of the warehouse (see Figure 10.6), where two teams split to explore the west and north sections, respectively. Both teams explore independently until a loop closure is evident, as illustrated in Figure 10.7, where the first team is about to exit via the southwestern corner. The exploration forms a total closed path length of 230 m that is measured after the separation of the UGVs in the first room. The final result is presented in Figure 10.8. This corresponds to about 70 constraints in the pose graph. Based on 88 samples collected across the surveyed area, the root-mean-square (RMS) error was calculated to be ± 0.27 m.

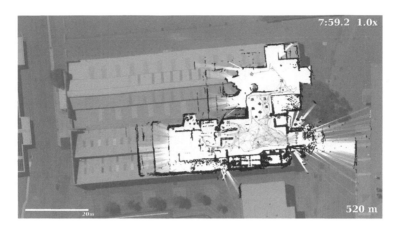

FIGURE 10.6
SLAM output showing the UGVs starting at the southeastern corner of the environment.

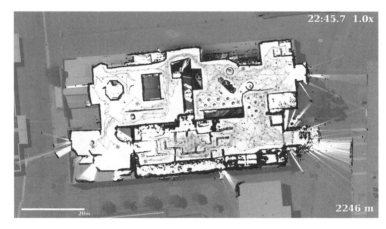

FIGURE 10.7
SLAM output showing loop closure between both teams, with the first team about to exit via the southwestern corner.

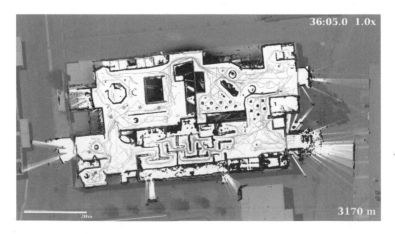

FIGURE 10.8
Completed global grid-map with ground truth overlaid as outlines.

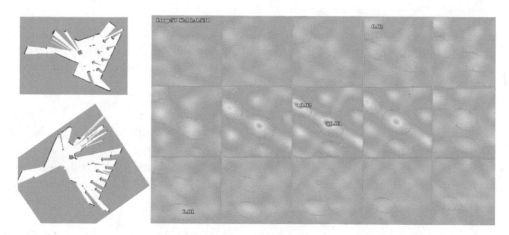

FIGURE 10.9

Example of multimodal constraint output with perceptual aliasing. The correlation results for a test area with perceptual aliasing that was caused by repetitive geometry. It shows overlapping submap pair (left) along with 15 slices through their constraint likelihood volume.

The 2D correlation between each submap in Figure 10.9 is represented by each slice, with a ± 3 m translation variation along with x- and y-axes at a fixed angular rotation. For example, the middle row represents rotations of $-6°$, $-3°$, $0°$, $3°$, and $6°$, respectively. The 2D ellipses represent three-sigma covariance modes through a Gaussian mixture model that fits the maximum likelihood volume. Occlusions have reduced the overlapping area of occupied cells (black) between the submap pair, and the matcher's output is mostly dominated by an array of columns in the environment.

10.4 Visual Odometry

While some effort was made to circumvent wheel slip accumulation in Section 10.3, other works have demonstrated that visual odometry is often an effective solution to this problem [35]. In the case of the MRS, this can be incorporated with no hardware or sensor modifications. Taking advantage of the system's software scalability, this was programmed on top of the existing software with minimal impact to the overall system.

10.4.1 Visual Odometry Method

As with most practical visual computing applications, the implementation of visual odometry in the real world comes with its own sets of challenges [36]. Environmental dynamics, including variations in seasons, light intensity, sensor occlusions, and motion blurs are capable of distorting visual odometry results that will in turn affect its accuracy. Our application will require that the algorithm is robust enough to withstand the environmental variations present in an urban outdoor environment.

Visual odometry works by tracking either features or appearances in the image frame [37,38] with appearance-based approaches usually resulting in more accurate tracking but at the cost of computation complexity. The decentralized nature of the MRS dictates visual odometry to be performed on each robot's computer as a separate subroutine to MR-SLAM, thereby requiring a feature-based method to be implemented. While visual odometry can be applied across different image features and many feature-based methods do exist [38,39], oriented FAST and rotated BRIEF (ORB) features [40] were found to be most compatible for the system due to the following.

- ORB features achieve a compromise between accuracy and system footprint. In other words, it is adequately accurate for the MRS application while having computation requirements that are low enough to be run on the individual robots.

- ORB features were proposed and tested in urban environments, whereby the structured appearance of urban environments enabled ORB features to be extracted effectively, especially on the KITTI benchmark suite [41].

Based on these rationales, an ORB-SLAM [16]-based algorithm was implemented as the visual odometry solution for this system. Once a camera has been corrected for radial distortions, the algorithm tracks ORB features across each frame to determine the displacement of each tracked pixel at every new frame, thereby localizing the robot. It functions as a separate thread and routine on the onboard computer to minimize any interference to the other routines running on the robots.

As visual odometry is effectively a visual SLAM algorithm without loop closures, the MRS' implementation of ORB-SLAM is hence used purely for odometry; loop closures and SLAM are still managed by the lidar-based front-end, independent from visual odometry.

10.4.2 Visual Odometry Evaluation

Evaluations for visual odometry were performed on individual UGVs as the implementation was fully decentralized. To optimize for performance and to reduce redundancy, a modified version of ORB-SLAM2 was proposed whereby all subroutines related to visual SLAM, such as loop closure detection and mapping, are removed. By delegating all SLAM routines to the lidar-based front-end, this modification yielded a 120% increase in performance gain in terms of output frame-rate. Additionally, maps created through the lidar-based front-end delivers more accurate point cloud measurements as compared to our monocular camera setup, and a lidar-based map requires lower computation and storage requirements than a vision-based solution.

ORB-SLAM2 achieves visual odometry by tracking ORB features, as shown in Figure 10.10. This evaluation was performed in an outdoor environment with unconstrained lighting conditions using a calibrated monocular camera. Tests were carried out on a single UGV driving along a 220 m path while generating its trajectory as shown in Figure 10.11, where it is indicated as a line trace; the black dots represent previously tracked ORB features, whereas the gray dots represent the features that are currently tracked. (Refer online for colored figure and explanation.)

10.5 Semantic Segmentation

The incorporation of semantic segmentation into the MRS enables navigation to be supplemented with scene understanding and object classification. Semantic segmentation

FIGURE 10.10
ORB features tracked by ORB-SLAM2 as shown in bounding boxes.

FIGURE 10.11
Trajectory measured using ORB-SLAM2 for dead reckoning (traced line), previously tracked ORB features (dots) and presently tracked ORB features (gray dots on right).

is a deep learning process that classifies each pixel in an image frame according to the object class it belongs to. This is especially useful in complex environments with multiple objects, with little uniformity in pose, features, and illumination.

10.5.1 Semantic Segmentation Method

For the semantic segmentation application of the MRS, SegNet [42] was selected based on its high compatibility and ease of implementation. The architecture of SegNet uses a convolution encoder and decoder setup that classifies objects from one of the following classes: sky, building, column-pole, road-marking, road, pavement, tree, sign-symbol, fence, vehicle, pedestrian and bicyclist; with a class average classification accuracy of 65.9% [42]. This MRS

FIGURE 10.12
SegNet output showing segmented pedestrian, bicycles, pathway and building. (Refer online for colored figure.)

uses SegNet whereby pedestrians, vehicles, buildings, vegetation, and pathways are classified as illustrated in Figure 10.12, and are subsequently classified into static and dynamic objects.

Static objects are stationary (with stationary positions), while dynamic objects are moving (with time varying positions). It is important for an MR-SLAM system to differentiate static and dynamic objects to devise proper navigational reactions to the environment. For example, static objects such as buildings and vegetation are permanent placements in the environment; these objects will be mapped by the MR-SLAM algorithm as part of the environment. Conversely, dynamic objects such as pedestrians and vehicles are in motion or are temporary placements in the environment; these will not be mapped by the MR-SLAM algorithm. Overall, this process of differentiating object types will ultimately result in higher mapping accuracy, especially when the ground truth does not contain dynamic objects.

The recognition of dynamic objects also enables the system to estimate the motion of a specific moving object. In other words, by segregating moving vehicles or pedestrians within an image frame, the lidar can then be utilized to estimate the motion and trajectory of the said object. This enables the robot to actively perform obstacle avoidance according to its motion, which can be implemented by comparing the robot's current speed against the lidar measurements on the classified dynamic object based on the image frame.

10.5.2 Semantic Segmentation Evaluation

Like visual odometry, semantic segmentation was also implemented in a decentralized approach onto individual UGVs. A Caffe [44] implementation of SegNet is installed onto the individual UGVs, which enables pixelwise object classification that corresponds to lidar measurements at that time instance, which can be any of 12 classifiable classes. We subsequently separate these classes into static and dynamic classes. For example, bicycles, pedestrians, and vehicles are dynamic, whereas buildings, fences, pavement, poles, road, road markings, road signs, vegetation, and sky are static. By matching the position of the classified pixel at the x-axis against that of the lidar on a fixed y-plane, dynamic objects can therefore be segregated and tracked using the lidar for motion detection. Objects in motion will be compared against the trajectory of the

FIGURE 10.13
SegNet output from a parking area showing segmented road regions, road markings, and poles. (Refer online for colored figure.)

UGV to ensure that there is no impending collision. These dynamic objects will also be ignored as part of the SLAM routine so that it does not become mapped as part of the environment.

Figure 10.12 was captured by the UGV while navigating along the path as described in Section 10.4.2. The parked bicycles on the right side and the pedestrians in the distance were properly segmented as dynamic objects, and the building and pavement as static objects. Several false detections are present due to variations in lighting and image quality, which accounts for 2.83% of the total pixels classified on the pavement region.

Figure 10.13 was also captured on the same path. While driving on roads, the road region and markings are properly classified along with the electric poles, vehicles, and pedestrians. Uniform lighting resulted in accurate classification accuracy with 0.69% of all pixels falsely classified.

10.6 Conclusion

In this chapter, we have presented a decentralized multi-robot system for SLAM in urban outdoor environments together with visual odometry and semantic segmentation techniques. The vision-based methods can be used as an alternative to lidar-based localization and object classification and may lead an overall cheaper and more powerful environmental perception system. We have presented an on-line, distributed, and decentralized MR-SLAM system that has proven to be resilient against environmental dynamics such as variations in lighting, terrain, pose, and moving objects. Evaluation results have confirmed the feasibility of using visual odometry as a viable solution to the odometry problem in an MRS, while semantic segmentation is a robust solution to object classification and scene understanding. Practical applications of this system can include search-and-rescue or reconnaissance missions in uncharted or hazardous environments, where detailed maps can be built quickly and accurately using a swarm of robots that are easily deployable, while being robust enough to cater to changes in the environment and hardware setup.

References

1. A. Boeing et al., "WAMbot: Team MAGICian's entry to the Multi Autonomous Ground-robotic International Challenge 2010," *J. Field Robot.*, vol. 29, no. 5, pp. 707–728, July 2012.
2. C.L. Ortiz, R. Vincent, and B. Morisset, "Task Inference and Distributed Task Management in the Centibots Robotic System." In: *Proceedings of the Fourth International Joint Conference on Autonomous Agents and Multiagent Systems*, New York, NY, USA, 2005, pp. 860–867.
3. R. Reid, "Large-Scale Simultaneous Localization and Mapping for Teams of Mobile Robots," PhD thesis, The University of Western Australia, Perth, Australia, 2016.
4. A. Boeing, S. Pangeni, T. Bräunl, and C.S. Lee, "Real-Time Tactical Motion Planning and Obstacle Avoidance for Multi-Robot Cooperative Reconnaissance." *2012 IEEE International Conference on Systems, Man, and Cybernetics (SMC)*, 2012, pp. 3117–3122.
5. A. Boeing, T. Bräunl, R. Reid, A. Morgan, and K. Vinsen, "Cooperative Multi-Robot Navigation and Mapping of Unknown Terrain." In: *2011 IEEE 5th International Conference on Robotics, Automation and Mechatronics (RAM)*, 2011, pp. 234–238.
6. L. Luft, T. Schubert, S.I. Roumeliotis, and W. Burgard, "Recursive Decentralized Localization for Multi-Robot Systems with Asynchronous Pairwise Communication," *Int. J. Robot. Res.*, p. 0278364918760698, vol. 37, no. 10, March 2018.
7. M. Garzón, J. Valente, J.J. Roldán, L. Cancar, A. Barrientos, and J.D. Cerro, "A Multirobot System for Distributed Area Coverage and Signal Searching in Large Outdoor Scenarios*," *J. Field Robot.*, vol. 33, no. 8, pp. 1087–1106, December 2016.
8. G. Best, O.M. Cliff, T. Patten, R.R. Mettu, and R. Fitch, "Dec-MCTS: Decentralized planning for multi-robot active perception," *Int. J. Robot. Res.*, vol. 38, no. 2–3, pp. 316–337, 2019.
9. S. Saeedi, M. Trentini, M. Seto, and H. Li, "Multiple-Robot Simultaneous Localization and Mapping: A Review," *J. Field Robot.*, vol. 33, no. 1, pp. 3–46, January 2016.
10. M.A. Abdulgalil, M.M. Nasr, M.H. Elalfy, A. Khamis, and F. Karray, "Multi-Robot SLAM: An Overview and Quantitative Evaluation of MRGS ROS Framework for MR-SLAM." *Robot Intell. Technol. Appl.*, vol. 5, pp. 165–183, 2019.
11. J. Jung et al., "Development of Kinematic 3D Laser Scanning System for Indoor Mapping and As-Built BIM Using Constrained SLAM." *Sensors*, vol. 15, no. 10, pp. 26430–26456, October 2015.
12. P. Koch et al., "Multi-Robot Localization and Mapping Based on Signed Distance Functions," *J. Intell. Robot. Syst.*, vol. 83, no. 3, pp. 409–428, September 2016.
13. Y. Kunii, G. Kovacs, and N. Hoshi, "Mobile Robot Navigation in Natural Environments Using Robust Object Tracking." In: *2017 IEEE 26th International Symposium on Industrial Electronics (ISIE)*, 2017, pp. 1747–1752.
14. C.-H. Sun, Y.-J. Chen, Y.-T. Wang, and S.-K. Huang, "Sequentially Switched Fuzzy-Model-Based Control for Wheeled Mobile Robot with Visual Odometry," *Appl. Math. Model.*, vol. 47, pp. 765–776, July 2017.
15. D.H. Kim and J.H. Kim, "Effective Background Model-Based RGB-D Dense Visual Odometry in a Dynamic Environment," *IEEE Trans. Robot.*, vol. 32, no. 6, pp. 1565–1573, December 2016.
16. R. Mur-Artal and J.D. Tardós, "ORB-SLAM2: An Open-Source SLAM System for Monocular, Stereo, and RGB-D Cameras," *IEEE Trans. Robot.*, vol. 33, no. 5, pp. 1255–1262, October 2017.
17. J. Engel, T. Schöps, and D. Cremers, "LSD-SLAM: Large-Scale Direct Monocular SLAM." In: *Computer Vision – ECCV 2014*, 2014, pp. 834–849.
18. G. Zhou, B. Bescos, M. Dymczyk, M. Pfeiffer, J. Neira, and R. Siegwat, "Dynamic Objects Segmentation for Visual Localization in Urban Environments," *arXiv:1807.02996 [cs.CV]*, July 2018.
19. Adept Technology, Inc., "Pioneer 3-AT." [Online]. Available: http://www.mobilerobots.com/Libraries/Downloads/Pioneer3AT-P3AT-RevA.sflb.ashx. [Accessed: October 5, 2018].
20. AutonomouStuff, "ibeo Standard Four Layer Multi-Echo LUX Sensor|LiDAR|Product," *AutonomouStuff, LLC.* [Online]. Available: https://autonomoustuff.com/product/ibeo-lux-standard/. [Accessed: September 10, 2017].

21. SICK AG, "LMS111-10100," Detection and Ranging Solutions. [Online]. Available: https://www.sick.com/au/en/detection-and-ranging-solutions/2d-lidar-sensors/lms1xx/lms111-10100/p/p109842. [Accessed: September 5, 2018].

22. Hokuyo Automatic Co., Ltd., "Scanning Rangefinder Distance Data Output/URG-04LX-UG01," *Product Details*. [Online]. Available: https://www.hokuyo-aut.jp/search/single.php?serial=166. [Accessed: October 5, 2018].

23. Xsens, "MTi (legacy product) – Products," *Xsens 3D motion tracking*. [Online]. Available: https://www.xsens.com/products/mti/. [Accessed: September 10, 2017].

24. QStarz International Co., Ltd., "BT-Q818XT." [Online]. Available: http://www.qstarz.com/Products/GPS%20Products/BT-Q818XT-F.htm. [Accessed: October 5, 2018].

25. Logitech, "QuickCam® Orbit AF." [Online]. Available: https://support.logitech.com/en_us/product/quickcam-sphere-af/specs. [Accessed: October 5, 2018].

26. Ubiquiti Networks, "PicoStation2HP Datasheet." [Online]. Available: https://dl.ubnt.com/pico2hp_ds.pdf. [Accessed: October 5, 2018].

27. STI Engineering Pty Ltd, "RFInnovations RFI-9256 900 MHz High Speed Data Radio," *RFInnovations*. [Online]. Available: http://www.rfinnovations.com.au/Uploads/Images/900MHz%20Data%20Radio%20Modem(2).pdf. [Accessed: October 5, 2018].

28. R. Reid, A. Cann, C. Meiklejohn, L. Poli, A. Boeing, and T. Braunl, "Cooperative Multi-Robot Navigation, Exploration, Mapping and Object Detection with ROS." *2013 IEEE Intelligent Vehicles Symposium (IV)*, 2013, pp. 1083–1088.

29. R. Reid and T. Bräunl, "Large-Scale Multi-Robot Mapping in MAGIC 2010." In: *2011 IEEE 5th International Conference on Robotics, Automation and Mechatronics (RAM)*, 2011, pp. 239–244.

30. J.E. Bresenham, "Algorithm for Computer Control of a Digital Plotter," *IBM Syst. J.*, vol. 4, no. 1, pp. 25–30, 1965.

31. M.W.M.G. Dissanayake, P. Newman, S. Clark, H F. Durrant-Whyte, and M. Csorba, "A Solution to the Simultaneous Localization and Map Building (SLAM) Problem," *IEEE Trans. Robot. Autom.*, vol. 17, no. 3, pp. 229–241, June 2001.

32. A. Censi, "An Accurate Closed-Form Estimate of ICP's Covariance." In: *Proceedings 2007 IEEE International Conference on Robotics and Automation*, 2007, pp. 3167–3172.

33. M.A. Fischler and R.C. Bolles, "Random Sample Consensus: A Paradigm for Model Fitting with Applications to Image Analysis and Automated Cartography," *Commun. ACM*, vol. 24, pp. 381–395, 1981.

34. J.M. Coughlan and A.L. Yuille, "Manhattan World: Orientation and Outlier Detection by Bayesian Inference," *Neural Comput.*, vol. 15, no. 5, pp. 1063–1088, May 2003.

35. M.O.A. Aqel, M.H. Marhaban, M.I. Saripan, and N.B. Ismail, "Review of Visual Odometry: Types, Approaches, Challenges, and Applications," *SpringerPlus*, vol. 5, no. 1, p. 1897, October 2016.

36. B. Zhao, T. Hu, and L. Shen, "Visual Odometry—A Review of Approaches." *Presented at: 2015 IEEE International Conference on the Information and Automation*, 2015, pp. 2569–2573.

37. D. Scaramuzza and F. Fraundorfer, "Visual Odometry [Tutorial]," *IEEE Robot. Autom. Mag.*, vol. 18, pp. 80–92, 2011.

38. K. Yousif, A. Bab-Hadiashar, and R. Hoseinnezhad, "An Overview to Visual Odometry and Visual SLAM: Applications to Mobile Robotics," *Intell. Ind. Syst.*, vol. 1, pp. 289–311, 2015.

39. H.J. Chien, C.C. Chuang, C.Y. Chen, and R. Klette, "When to Use What Feature? SIFT, SURF, ORB, or A-KAZE Features for Monocular Visual Odometry." *2016 International Conference on Image and Vision Computing New Zealand (IVCNZ)*, 2016, pp. 1–6.

40. E. Rublee, V. Rabaud, K. Konolige, and G. Bradski, "ORB: An Efficient Alternative to SIFT or SURF." *Presented at: The 2011 International Conference on Computer Vision*, 2011, pp. 2564–2571.

41. A. Geiger, P. Lenz, and R. Urtasun, "Are We Ready for Autonomous Driving? The KITTI Vision Benchmark Suite." In: *2012 IEEE Conference on Computer Vision and Pattern Recognition*, 2012, pp. 3354–3361.

42. V. Badrinarayanan, A. Kendall, and R. Cipolla, "SegNet: A Deep Convolutional Encoder-Decoder Architecture for Image Segmentation," *IEEE Trans. Pattern Anal. Mach. Intell.*, vol. 39, no. 12, pp. 2481–2495, December 2017.

43. S. Lopes, B. Frisch, A. Boeing, K. Vinsen, and T. Bräunl, "Autonomous Exploration of Unknown Terrain for Groups of Mobile Robots." In: *2011 IEEE Intelligent Vehicles Symposium (IV)*, 2011, pp. 157–162.

44. Y. Jia et al., "Caffe: Convolutional Architecture for Fast Feature Embedding." *Presented at: Proceedings of the 22nd ACM International Conference on Multimedia*, 2014, pp. 675–678.

11

Vehicle Localization in GNSS-Denied Environments

Ramtin Rabiee, Ian Bajaj, and Wee Peng Tay

CONTENTS

11.1 Introduction

Nowadays, if one talks about positioning, localization or navigation, the first thing that may immediately spring to mind is the popularized global positioning system (GPS), which is an example of a global navigation satellite system (GNSS). This shows how much impact GNSSs have on the field of positioning. However, depending on the application and required accuracy, GNSS-based navigation schemes may not be adequate. For example, the Dedicated Short-Range Communications (DSRC) Technical Committee of the Society of Automotive Engineers (SAE) has specified that the positioning error for at least 68% of the reported vehicle positions through basic safety messages (BSMs) must be less than 5 ft (around 1.5 m) [1]. It can be as high as lane-level precision, which usually refers to an error of 1~1.5 m. On the other hand, there is no guarantee of access to a GNSS system everywhere. Hence, vehicular navigation systems need to be designed to work continuously despite the inaccessibility of GNSS satellites. This is done by utilizing other information sources like sensors and cooperation with infrastructure or other vehicles to perform localization in GNSS-denied environments. In this chapter, we first discuss possible information sources and then methods to fuse the available information to achieve a certain level of tracking precision. It is worth mentioning that all the methods to be explained are considered from the perspective of 2D positioning, since our focus is on vehicular localization on the terrain.

The superscripts $(\cdot)^{n\cdot}$ and $(\cdot)^{b\cdot}$ denote a term in the navigation frame and the body frame, respectively. Boldfaced characters are used to denote vectors and matrices, and the superscript $(\cdot)^{T}$ denotes matrix transpose. We use $\mathbb{R}$ to denote the set of real numbers, and the symbols $\mathbf{0}_{n\times m}$ and $\mathbf{I}_n$ represent a $n \times m$ zero matrix and $n \times n$ identity matrix, respectively. Finally, the probability density function of the random variable x conditioned on y is given by $p(x|y)$.

11.1.1 Information Sources

In this section, we discuss the plausible information sources that may aid in vehicular localization. We first briefly consider on-board sources, and then discuss external sources including from the wireless infrastructure as well as from other vehicles.

11.1.1.1 *Provided by On-Board Vehicle Sources*

11.1.1.1.1 *Digital Map*

Without map information, all a positioning system might display is latitude, longitude, and altitude coordinates, which may not be very useful for an ordinary user without special navigation skills. Therefore, in any terrestrial navigation system, having a digital map is mandatory for visual understanding of one's own location. Moreover, we can use the map information in more sophisticated ways to enhance the navigation algorithm if the provided information is detailed enough [6]. Map matching (MM) is a common method to use the detailed digital information of the map as part of the tracking algorithm, where the estimated position is projected to the closest point on the map [14,15,23]. In some cities (e.g., Singapore [7]), as shown in Figure 11.1, the available map information can be as detailed as providing the position of guardrails, lane marks, and road boundaries. Therefore, one may integrate this information with other sensor measurements (e.g., radar) to achieve a higher positioning accuracy.

11.1.1.1.2 *Inertial Measurement Unit*

Thanks to the development of micro-electromechanical systems (MEMS) technology, sensors such as gyroscope and accelerometer have become affordable in common localization applications. In tracking a mobile object, movement is one of the more important parameters

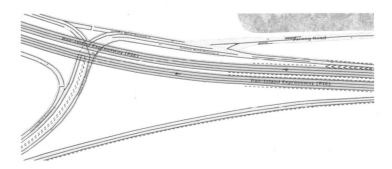

FIGURE 11.1
A map with positions of guardrails (dashed lines) and lane marks (solid lines). (From Land Transport Authority (LTA) of Singapore. Geospatial datasets. https://www.mytransport.sg/content/mytransport/home/dataMall.html.)

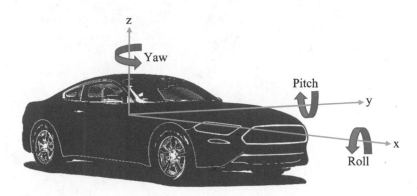

FIGURE 11.2
The 3D directions of accelerations and angular velocities provided by an IMU.

to be considered. It is more crucial for vehicles with a higher range of speeds, and when it is not necessarily driven with a constant velocity. Using inertial measurement unit (IMU), a vehicle can measure and monitor its dynamic acceleration and angular velocity in three axes as shown in Figure 11.2. Using this information, the navigation system is able to update the estimated velocity and direction of the vehicle, accordingly.

11.1.1.1.3 Radar

Radar provides distances from objects in their field of view. One of the important applications of automotive radar is in collision avoidance systems. However, radar can also be used to output the range measurements from known and unknown obstacles around the vehicle for the purpose of localization. As a simple example, information about positions of building boundaries or guardrails along a vehicle's trajectory obtained from a detailed map can be used as the reference points (i.e., anchors) together with the radar measurements to adjust and improve the estimated position of the vehicle. In practice, measurements from unknown obstacles can be challenging to filter out. However, with proper usage and fusion with other information sources, radar measurements can provide very accurate information about a target.

11.1.1.2 Provided by Cooperative Wireless Networks

Vehicle-to-everything (V2X) wireless communication has been a hot topic in the past few years. The wireless communication between vehicles (V2V), and between vehicles and related wireless networks' base stations/infrastructure (V2I) can be a source of information for the purpose of localization. By measuring the required metrics during a wireless transmission, one can extract distance information between a transmitter and receiver, and even in some cases (e.g., multiple-input multiple-output [MIMO], and millimeter wave-based communications [20]) be able to find the angle of the transmitter relative to the receiver [5]. All this information can be used for positioning through different techniques such as triangulation, trilateration, and multilateration.

11.1.1.2.1 Time of Arrival

In an ideal case, the speed of an electromagnetic wave is as high as its speed in vacuum, which is the speed of the light (i.e., $c = 3 \times 10^8$ m/s). Therefore, with this knowledge, the propagation delay, τ, is used to estimate the distance between the transmitter and the

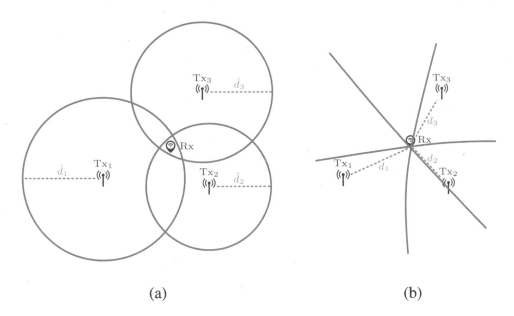

(a)　　　　　　　　　　　　　　(b)

FIGURE 11.3
Localization by using ToA and TDoA through (a) trilateration and (b) multilateration techniques, where d_i and $\hat{d_i}$ are, respectively, the true and estimated values for distance between the receiver (Rx) and the i-th transmitter (Tx_i).

receiver in a wireless transmission system (i.e., $d = c * \tau$ m). Subsequently, the unknown position of a desired point (or location of a receiver), can be found through trilateration. As shown in Figure 11.3a, the location of the desired point is estimated to be in the area of overlap or intersection of at least three circles, whose centers are the locations of known reference points (or transmitters). Note that the measured distance is erroneous, and this area of uncertainty in estimation is much higher in the case of just two time of arrivals (ToAs). The other challenges in obtaining the propagation delay from ToA are:

- Knowing the exact transmit time at the receiver side as the reference time. To extract the propagation delay from the ToA, receiver needs to know the transmit time accurately. Usually, the processing time inside the transmitter is not fixed or known, and therefore, even if the payload has a time stamp denoting the transmit time, it is not useful if the time stamp is provided by upper layers of the **Open Systems Interconnection** (OSI) model. In some wireless networks, like Wi-Fi for example, whose multiple access control (MAC) employs carrier-sense multiple access with collision avoidance (CSMA/CA) protocol, there is even more uncertainty regarding the propagation delay due to the random backoff time when the channel is busy.

- Time synchronicity between the transmitter and the receiver. Without synchronization, the propagation delay will also have a delay offset and drift (which may increase with time) in its measurement. Note that a time drift of 100 ns results in an error of 30 m in the distance measurement.

Considering the above-mentioned issues, one solution can be two-way ToA estimation, known as time of flight (ToF). In this method, device A first transmits the signal towards device B, and device B sends it back to device A upon receipt. Here, we just need to deduct the processing time inside device B from the total signal's traveling time and then divide it by two, to find the one-way propagation delay. In this method, the main challenge is to obtain the accurate processing time inside device B.

Another solution can be the multilateration technique which uses time difference of arrival (TDoA) information, as depicted in Figure 11.3b. In this method, the delay between receiving times of two simultaneous transmissions (from two different transmitters) is considered, to calculate the difference in distances to those known transmitters (considered as reference points). We need at least two TDoAs (i.e., two different pairs of simultaneous transmissions) which needs three transmitters. In this method, the location of the desired point is estimated within the intersection of hyperbolas whose foci are preknown transmitters' locations. For TDoA measurement, the receiver's clock can be unsynchronized to the transmitters', and the reference transmit time need not be known, but the transmitters themselves need to be synchronized to each other, and each pair of transmitters must send their signals simultaneously.

There are some other suggestions in the literature to tackle the challenges discussed. For example, in virtual TDoA (V-TDoA), a periodic transmission from a single transmitter has been considered to solve the issue of unknown reference time, where there is no possibility of simultaneous transmissions from two different transmitters [11]. Round-trip TDoA (RTTDoA) is another hybrid method which relieves the issue of transmitters' synchronization by combining ToF and TDoA methods [17].

11.1.1.2.2 Received Signal Strength

For wireless transmissions, the distance between the receiver and the transmitter can be estimated by measuring the signal strength at the receiver side, as an alternative to ToA-based trilateration. The received signal strength (RSS)-based localization may have higher coverage than the ToA-based method. However, it is more sensitive to the quality of channel estimation and other environmental parameters.

11.1.1.2.3 Angle of Arrival

Angle of arrival (AoA), sometimes called direction of arrival (DoA), is another source of information available for a wireless communication setup. The most common method to obtain AoA is by using an antenna array, the details of which are not covered in the scope of this chapter. But for completeness, if the orientation (or reference direction) is known, the receiver can find the absolute direction from which the transmitted signal from a known source/transmitter has been received. Therefore, as depicted in Figure 11.4a, the two absolute AoAs from two transmitters with known positions form a triangle and converge at a third vertex, which is the position of the receiver.

In many cases, the orientation is unknown. Therefore, what is generally used is one of the angle properties of a circle that says "angles in the same segment are equal," as shown in Figure 11.5. Therefore, we need to measure the difference between AoAs of two transmissions from two known transmitters. The possible positions of the receiver lie on an arc of a circular segment in which the two transmitters are the endpoints of its chord. The location of the receiver can be estimated as the intersection of at least two arcs whose angle difference of arrivals (ADoAs) need to be measured by using three transmitters, as depicted in Figure 11.4b.

11.1.2 Information Fusion

With the amount of data that can be collected through sensor measurements and observation, how we use this information is crucial. Therefore, in designing a tracking algorithm, information fusion plays a vital part in utilizing the sensor data to achieve a desired level

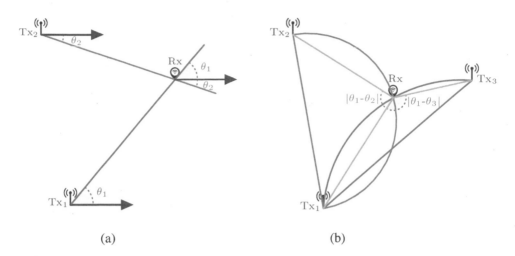

FIGURE 11.4
Localization by using AoA through triangulation with (a) known and (b) unknown reference direction, where θ_i is AoA form the i-th transmitter (Tx_i).

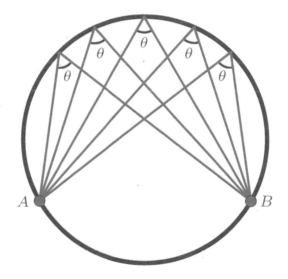

FIGURE 11.5
Angles subtended by the same arc are equal.

of accuracy in vehicle positioning. For a localization/tracking algorithm that works over time, we first need to define our motion/dynamic state-space model or the state transitions over time. Next, the set of parameters which are being estimated, the measurement models, and their relations must be expressed accordingly. The ideal scenario is to have a linear system whose noises are Gaussian distributed. However, there are many cases, like vehicle position tracking (with non-constant velocity due to the acceleration/deceleration), where the dynamic model is non-linear. Therefore, the Gaussian distributed assumption would not be valid after the initial point. Among recursive Bayesian filters, which recursively update the estimated state of the system given a sequence of observations, the extended Kalman filter (EKF) and particle filter (PF) are commonly proposed for developing tracking algorithms for non-linear models that rely on Gaussian approximation.

11.1.2.1 Extended Kalman Filter

In the EKF, the main objective is to linearize the non-linear functions of motion (state transition) and measurement (observation) models by using their Taylor series expansions around the estimate of current state.

11.1.2.2 Particle Filter

The PF models the posterior distribution of the state given a set of noisy samples (called particles) by using sequential Monte Carlo method. Although PF might be more accurate than the EKF, this precision gain would be achieved at the cost of computational complexity and processing time.

11.2 Localization via Cooperative V2I

As we have already discussed, the important part in designing a tracking algorithm is the way we fuse the accessible information to achieve the required accuracy. Since we do not want to consider any GNSS data, the developed algorithm first needs to provide a coarse estimate of the vehicle position to help figure out where it is roughly located. Later, we can use other methods and extra information to refine the estimate as much as possible.

11.2.1 Coarse Localization

To calculate a coarse estimate, we need a sufficient number of beacons from known reference points. Therefore, any of the above-explained methods of triangulation, trilateration, or multilateration can be considered as a potential solution to obtain the coarse estimation of the vehicle position where GNSS is inaccessible. However, the feasibility and performance are dependent on the wireless V2X communications platform. In our work, we consider a challenging case where the vehicular wireless communications can only provide erroneous ToAs due to unsynchronized clocks between transmitters and receivers. It should be noted that we do not consider AoA, since it may need more enhanced wireless technologies like MIMO. Although it might be available in cellular V2X (LTE-V) or future 5G enhanced V2X (eV2X) networks, it is not accessible in vehicular networks which are based on the DSRC technology with IEEE 802.11p standard for physical and MAC layers. Moreover, in our setup, we consider that only two ToAs are available at any point of time due to the resource limitations in DSRC and the infeasibility of an on-board unit (OBU) having access to more than two road side units (RSUs) in a short period of time. The estimated distance between the transmitter (e.g., RSU) and the receiver (e.g., OBU) can be obtained through any of explained methods of ToA, TDoA, or V-TDoA, and updated periodically. For example, in [15,16], V-TDoA has been applied with the periodic transmission of the 100 ms.

The assumption of a GNSS-denied scenario is situational in most cases (e.g., entering into a tunnel or turning into a highly built-up environment). Therefore, it is reasonable to assume that the position of the vehicle can be updated via GNSS until that point. If the initial position is roughly known (via either GNSS or other methods), a vehicle's kinematics provided by an IMU can be considered as another source of information together with the vehicular constraints (VCs) to track a vehicle's trajectory. In normal driving conditions, a sudden jump in the vehicle is unlikely and side slip is quite rare. Therefore, we do not expect any non-zero velocity along the y and z axes in the body frame due to VCs, which are interpreted as *nonholonomic constraints* in [21]. It is to be noted that the body frame is attached to the vehicle's body whose axes are aligned with those of the IMU, as shown in Figure 11.2.

To achieve a coarse estimate of the vehicle's location, we apply the EKF to our observations in the form of ToA, IMU measurements, and VCs. Therefore, we first need to define the state vector (x_k). In [15], an 11-dimension vector, $x_k \in \mathbb{R}^{11}$, has been considered as the state-space vector, which includes Euler angles, $\mathbf{\Psi}_k = [\phi_k, \theta_k, \psi_k]^T \in \mathbb{R}^3$, local transmit time, τ_k, ToA drift (due to the clock skew between the local oscillators [LOs] of the transmitter and the receiver), α_k, in addition to the three-axis positions, $p_k^n \in \mathbb{R}^3$, and velocities, $v_k^n \in \mathbb{R}^3$, in the navigation frame. Note that ϕ_k, θ_k, and ψ_k are, respectively, roll, pitch, and yaw, which describe the orientation of the body frame with respect to the navigation frame in the x, y, and z directions.

$$x_k \triangleq \left[\left(p_k^n\right)^T, \left(v_k^n\right)^T, (\mathbf{\Psi}_k)^T, \tau_k, \alpha_k \right]^T \in \mathbb{R}^{11}. \tag{11.1}$$

To more comprehensively monitor and track the vehicle's state, a 26-dimension vector, $x_k \in \mathbb{R}^{26}$, can be considered, as in [16,24], where three-axis acceleration in both navigation frame, $a_k^n \in \mathbb{R}^3$, and body frame, $a_k^b \in \mathbb{R}^3$, angular velocity in body frame, $w_k^b \in \mathbb{R}^3$, and their related biases, $b_{a,k} \in \mathbb{R}^3$ and $b_{w,k} \in \mathbb{R}^3$, have been taken into account in addition to those of the 11-dimension state vector as follows:

$$x_k \triangleq \left[\left(p_k^n\right)^T, \left(v_k^n\right)^T, \left(a_k^n\right)^T, (\mathbf{\Psi}_k)^T, \left(a_k^b\right)^T, \left(w_k^b\right)^T, \left(b_{a,k}\right)^T, \left(b_{w,k}\right)^T, \tau_k, \alpha_k \right]^T \in \mathbb{R}^{26}. \tag{11.2}$$

Another difference between these two methods is in the usage of the IMU data. In the first EKF-based method with 11-dimension state vector (call it EKF-11), as shown in Figure 11.6a, the IMU measurements are fed to the prediction step of the EKF as a control vector to predict the current velocity and attitude of the vehicle using measured acceleration and angular velocities. On the other hand, in the second EKF-based method with 26-dimension state vector (call it EKF-26), the IMU parameters are considered as elements of the state vector which are supposed to be estimated first before being updated with their measured values, as depicted in Figure 11.6b. The detailed expressions of the related state-space models, including state transition model and measurement model, and their process and measurement noises, can be found in [15,16].

To understand the performance of the EKF-11 and EKF-26 methods, a trajectory has been simulated along a four-lane expressway with a few lane-change maneuvers as shown in Figure 11.7. The above-mentioned EKF-based localization methods are applied to obtain coarse estimates of the vehicle's positions along the simulated trajectory, using ToA, VC, and IMU measurements. Note that there are at most two available ToAs (from two closest RSUs) at each 100 ms, which is the maximum transmission interval based on the IEEE 802.11p standard for DSRC physical layer. Figure 11.8 shows the cumulative distribution functions (CDFs) of total and lateral positioning errors and their related root-mean-square errors (RMSEs) for both the EKF-11 and EKF-26 methods. The results imply that EKF-26 can provide better coarse localization. Although the resulting lateral error meets the SAE requirement (i.e., 68% of the lateral error is less than 1.5 m), we still need to refine the obtained results to have a higher total positioning accuracy.

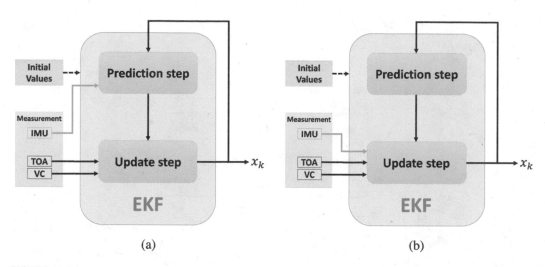

FIGURE 11.6
EKF implementation for coarse localization using ToA, VC, and IMU data where IMU measurements are used in (a) prediction step, and (b) update step.

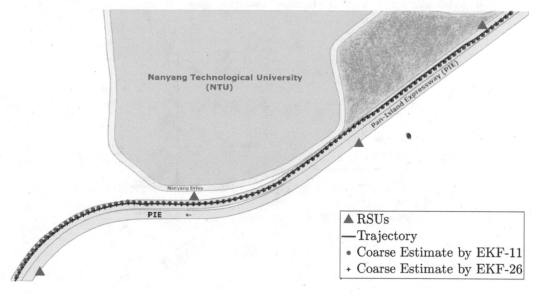

FIGURE 11.7
A trajectory along four-lane highway with lane-change maneuvers.

11.2.2 Lane-Level Localization

After finding a coarse estimate of the vehicle, we may still need to refine it to achieve a certain precision level based on the application. As mentioned before, lane-level accuracy is defined as the position error of 1.5 m and less. High-resolution digital map data and radar measurements are some of the possible sources of information that can be considered to increase the accuracy of the positioning results achieved by EKF-based methods which are discussed in Section 11.2.1.

The easiest way to implement this is to consider the map data and radar ranging which are new inputs of the update step, as shown in Figure 11.9, where the radar measurements

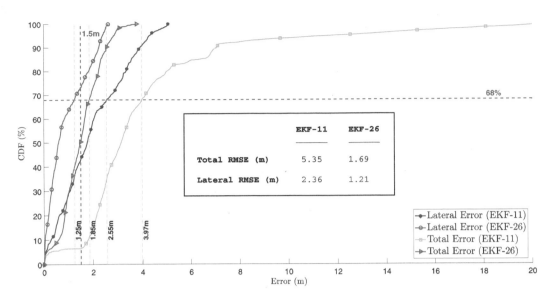

FIGURE 11.8
CDF of total and lateral errors in the coarse estimates of the positions.

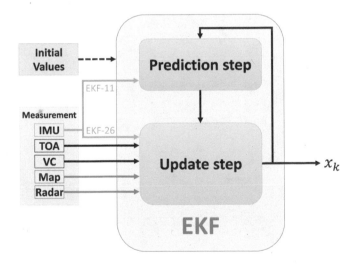

FIGURE 11.9
EKF implementation to obtain a refined estimate of the position, by using map information and radar measurement together with ToA, VC, and IMU data.

are distances from known obstacles (e.g., guardrails, building boundaries, curbs) whose positions are provided by a digital map. Figure 11.10 depicts the simulated trajectory and related estimated positions from the EKF-11 and EKF-26 schemes when map data are used together with the simulated measurements from two side-scan radars (deployed on left and right sides of the vehicle) over different scenarios (i.e., sparse and dense amounts) of unknown obstacles on the road (e.g., random vehicles driven on other lanes).

Based on the error analysis provided in Figure 11.11, there is a significant improvement in positioning accuracy as compared to that of no map and radar utilization, presented in Figure 11.8, for the EKF-11-based method. However, for the EKF-26-based scheme, this

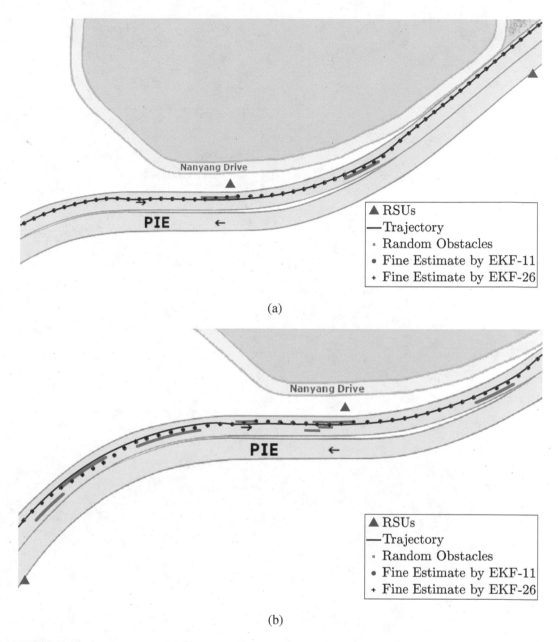

(a)

(b)

FIGURE 11.10

A trajectory along four-lane highway with lane-change maneuvers and random vehicles as unknown obstacles, where the estimated positions are refined using map information and radar measurements.

improvement is valid just for the case of sparse obstacles. Although both methods work comparably in terms of total and lateral RMSEs, EKF-26 slightly outperforms EKF-11 by considering the CDF analysis. Even so, one can see that EKF-26 is more sensitive to the density of the unknown obstacles, and its positioning errors in Figure 11.11b are worse than its coarse estimate without radar and map information as in Figure 11.8. To evaluate the recovery capability of the algorithm to an outage (i.e., when the positioning error goes beyond the desired 1.5 m), the maximum and average amount of error duration have been

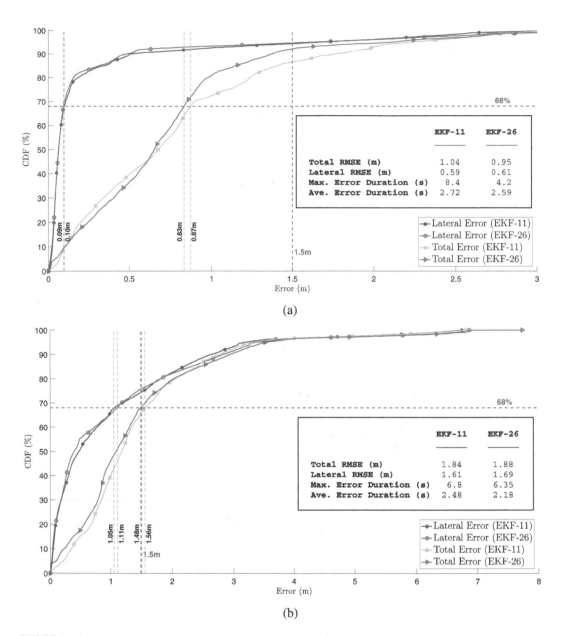

FIGURE 11.11
CDF of total and lateral errors with map information and radar measurements.

analyzed as well. Error duration implies the time delay taken to correct the positioning error of the vehicle and reduce it to below 1.5 m. Figure 11.11 reveals that EKF-26 reacts faster when positioning accuracy drops.

In general, we can see that the positioning accuracy is quite sensitive to the presence of unknown obstacles, where the range measurement is no longer from the known objects on the road (e.g., guardrails, building boundaries, curbs), and is worsened when they increase in density. Therefore, we need to address this observation to overcome the uncertainty issues in the radar measurements.

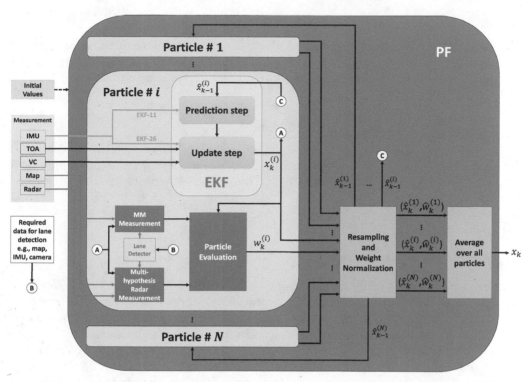

FIGURE 11.12

Particle filtering-based method to refine the EKF coarse estimate by using map and radar measurements.

On the other hand, MM is another method to refine the estimated position, as described in Section 11.1.1.1.1, by adjusting the estimated positions to the closest point in the center of the lane. However, there is no guarantee that the vehicle is being driven at the center of the lane. Therefore, this method could be useful in an application where positioning error is measured with respect to the center of the lane rather than the exact position of the vehicle along its trajectory.

To tackle the above-mentioned challenges using map and radar data, we define a multi-hypothesis model ($\mathcal{H}$) for radar measurements and utilize the map and radar data to refine the vehicle position with the help of particle filtering [16]. In this method, the EKF has been applied for each particle, and then MM is treated as a measurement, z_k^{Map}, which is used to evaluate the particles together with radar measurements, z_k^{Radar}, as depicted in Figure 11.12. The total likelihood is related to the MM and multi-hypothesis model radar measurements as follows:

$$p\left(z_k^{\text{Total}} \mid x_k\right) = p\left(z_k^{\text{Map}} \mid x_k\right) p\left(z_k^{\text{Radar}} \mid x_k\right). \tag{11.3}$$

11.2.2.1 Map Matching Measurement

Using MM, we can find the closest point on the lane center. Our expectation is to have zero displacement between estimated position, p_k^n, and its projected point onto the lane center, $\hat{p}_k^n$, in 2D (i.e., in x and y directions), $z_k^{\text{Map}} = \mathbf{0}_2$.

11.2.2.2 Multi-Hypothesis Radar Measurement

The likelihood of range measurements from side-scan radars is given by considering the fact that the measured range might be from a known guardrail or from the vehicles in possible adjacent lanes. For M radars, the likelihood for all radar measurements are given by:

$$p(z_k^{\text{Radar}} \mid x_k) = \prod_{m=1}^{M} p(z_k^{m-th\,\text{Radar}} \mid x_k), \tag{11.4}$$

where the likelihood of the range measurement from the m-th radar is obtained by summation over a set of range measurements from all possible obstacles in field of view of the m-th radar, R_m, as follows:

$$p(z_k^{m-\text{th Radar}} \mid x_k) = \sum_{r \in R_m} p(z_{r,k}^{\mathcal{H}} \mid x_k, \mathcal{H}_{r,k}) p(\mathcal{H}_{r,k}). \tag{11.5}$$

For example, the size of R_m for a side-scan radar in the highway equals the number of adjacent lanes on the same side plus the guardrail. The prior probabilities of the hypotheses can be assigned either uniformly, that is, $p(\mathcal{H}_{r,k}) = 1/|R_m|$, or non-uniformly if there exists other information such as positions of vehicles in the vicinity via cooperation.

The main challenge here is to monitor the lane index of the vehicle since both MM and multi-hypothesis radar measurements need the current lane index of the vehicle. In [16] a lane-change detector has been utilized to monitor any lane-change maneuver along the vehicle's trajectory with the initial lane index known.

In Figure 11.13, the discussed PF-based scheme has been applied to refine the coarse estimates from both EKF-11 and EKF-26, where a perfect lane detector has been considered. It reveals that the tracking result is quite stable as compared to that in Figure 11.10, and it can overcome the issue of unknown obstacles perfectly. Error analysis in Figure 11.14 implies that the PF-based refining process works well on both EKF-based methods, and their sensitivity to the density of the unknown obstacles is now tolerable. Moreover, the reaction time of the algorithm is much less than the pure EKF-based methods, and it can now support the two-second rule of the safe following distance.

Nonetheless, as discussed earlier, the performance of the PF-based method depends on the reliability of the detected lane, and hence it is sensitive to the performance of the lane detection mechanism. To have better intuition on this issue, Figures 11.15 and 11.16 depict the tracking results and performance of the PF-based method where the reliability of the lane detector is 75%. One can see that the tracking error goes higher as the quality of the lane detection drops. Note that the performance degradation in the EKF-26 is slightly less than EKF-11, especially in the case of dense obstacles, since there is a higher degree of freedom to optimize the tracking algorithm. Moreover, Figure 11.16 reveals that the PF-based method can still meet the SAE positioning requirement (i.e., 68% of the positioning errors must be less than 1.5 m), not only in lateral error, but also in total error, despite the 25% uncertainty in the lane detection.

11.3 Localization via Cooperative V2V

Though self-localization using on-board sensors, map information, GPS, or other road infrastructure can greatly reduce the likelihood of erroneous road judgments by self-neglect

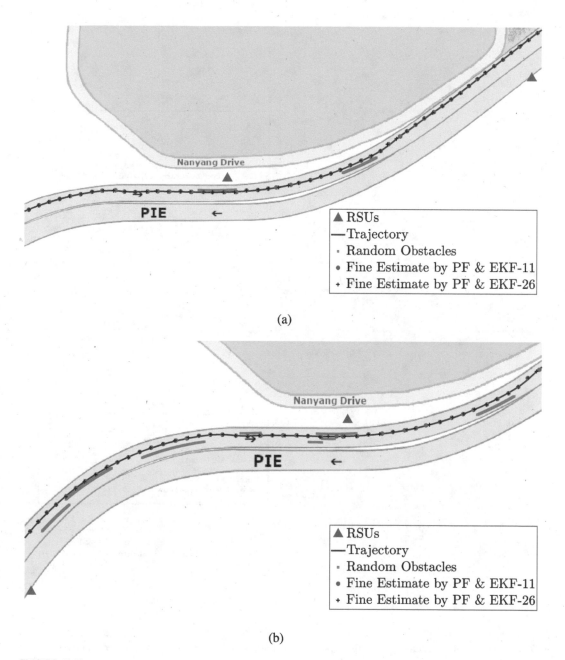

FIGURE 11.13
Refined position estimate of a trajectory along a four-lane highway with lane-change maneuvers and random unknown vehicles by using PF with a perfect lane detector.

or human error, connected vehicles addresses the issue of road safety when circumstances are out of one's hand. In today's information age, all our smart devices are reliant on a network of connected data points that enable additional services, driven by large amounts of observational data. These connected systems will always outperform an individual device that is only self-reliant on its own observable data. This extends to the domain of intelligent transport systems (ITS).

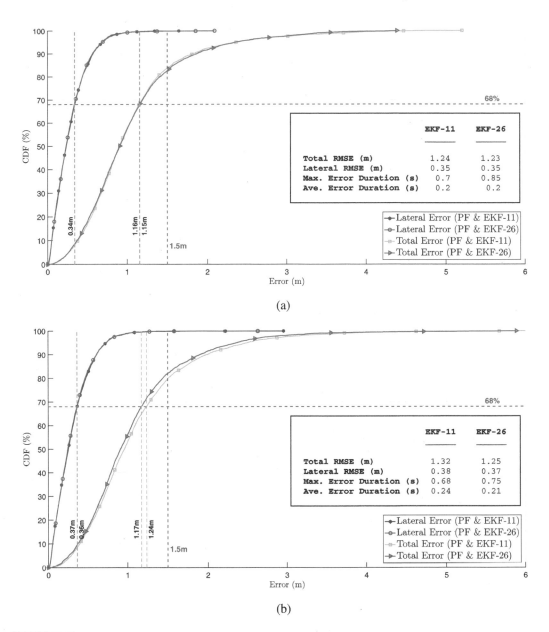

FIGURE 11.14
CDF of total and lateral errors of the refined position estimate via PF where the lane detector is perfect.

But more importantly, cooperation between vehicles is vital when we consider that 94% of all road accidents are caused by human error, of which 8% is credited to other drivers' errors [2]. In an ideal world, connected vehicles with automated vehicle assistance can safeguard against oncoming vehicle collision from rash driving or driving under the influence, blind spots and sharp turns, and other vehicle failures, by communicating with and establishing a vehicular network with their own observational and measurable data.

Apart from the obvious road safety concerns, personalized travel services can be enhanced with connected vehicles like video or data streaming, area-localized advertising

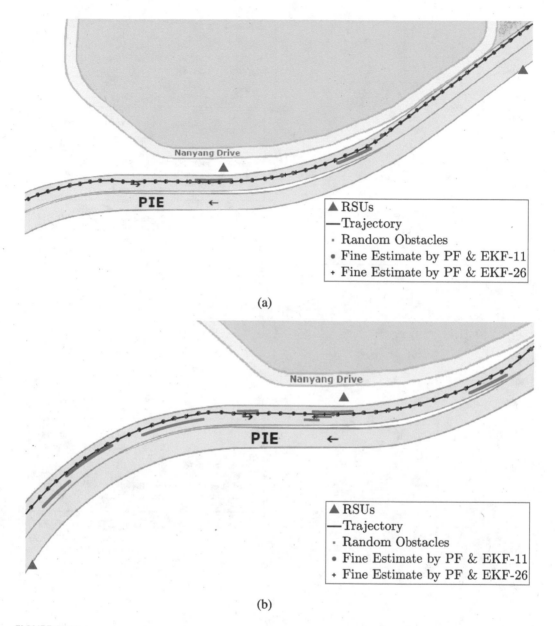

(a)

(b)

FIGURE 11.15

Refined position estimate of a trajectory along four-lane highway with lane-change maneuvers and random unknown vehicles by using PF where lane detector's reliability is 75%.

broadcasts, or social media content sharing. This focus on the end-user driving experience is what is pushing the research community and industry to actively explore the feasibility of cooperative vehicles and the means of exchanging data between them.

11.3.1 Modes of V2V Communication

So how do cooperative vehicles exchange data? This can happen through any of the previously mentioned platforms such as LTE-V, eV2X, or DSRC technologies. Note that

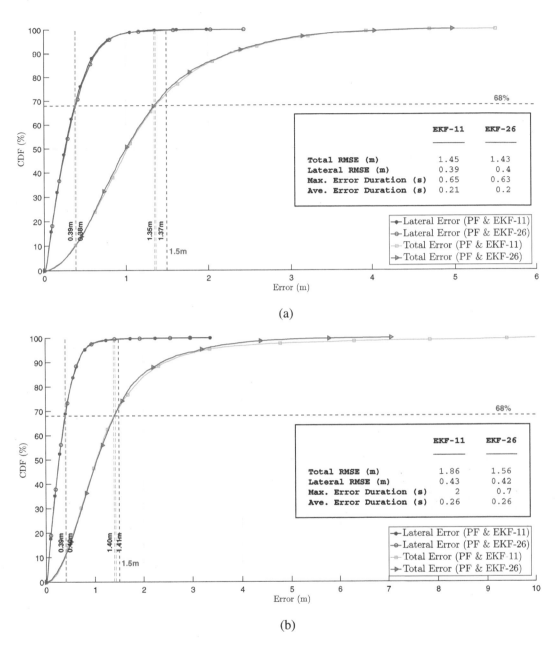

FIGURE 11.16
CDF of total and lateral errors of the refined position estimate via PF where lane detector's reliability is 75%.

eV2X is not yet finalized and DSRC has different specifications in different countries. The IEEE 802.11p standard is developed based specifically on the DSRC technologies to support ITS applications over the 5.9 GHz band, and regularized in some countries like the US and Singapore [19]. The band is divided into seven channels of 10 MHz from 5.855 to 5.925 GHz, with two dedicated safety channels, one control channel, and four service channels, as shown in Figure 11.17. Based on the standard, each vehicle is to broadcast its

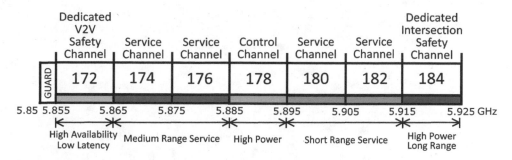

FIGURE 11.17
DSRC channel allocation.

state information in a BSM on channel 172. This message consists of its speed, acceleration, position (latitude, longitude, and elevation), and a temporary vehicle ID (VID), among other parameters. For V2V localization and other associated emergency V2V applications, each vehicle's OBU must have one radio tuned in to channel 172 as a safety protocol, while a secondary radio can switch between the control channel and service channels.

As shown in Figure 11.18, the sources of V2V information that can be communicated over DSRC include:

- Ranging measurements through ToF, radio-frequency identification (RFID) and radar. Though the distance measurements discussed in Section 11.2 were from fixed reference points, here the cooperative vehicles themselves might be moving.

- Pseudo-range measurements from sharing collected GNSS data in GNSS-denied environments. Although one's own vehicle might not be able to have full sky-view of the required 4–5 GNSS satellites for positional accuracy, connected vehicles can share pseudo-range or coarse-range measurements from the satellites that the vehicles can see, as an input for the coarse localization step in the EKF.

- The absolute but imperfect position information of other intelligent vehicles. Through the DSRC protocol, vehicles already share this information through the BSM message discussed earlier. This information can also be helpful in localizing a vehicle that has lost its own sources of self-localization (e.g., when entering a tunnel).

Though the information available from cooperative vehicles itself may be viewed as erroneous because both the transmitter and receiver are in motion, it can be argued that information from a stationary source can be worse or outdated. As in the case of ToA, TDoA, or V-TDoA information, discussed in Section 11.1.1.2.1, with a periodic transmission of the 100 ms, a vehicle moving at 20 m/s would have covered 2 m, and is receiving information regarding its previous state. This is true especially when the cooperative vehicles are traveling along the same direction with a smaller speed-differential than that of the ego vehicle relative to a stationary object (e.g., RSU). This is only to illustrate that cooperative information is not mobility-impaired. In fact, these small differences with regard to the actual position of the vehicle is relative to the position of the reference object and can be quite minuscule in comparison to other sensor measurement biases.

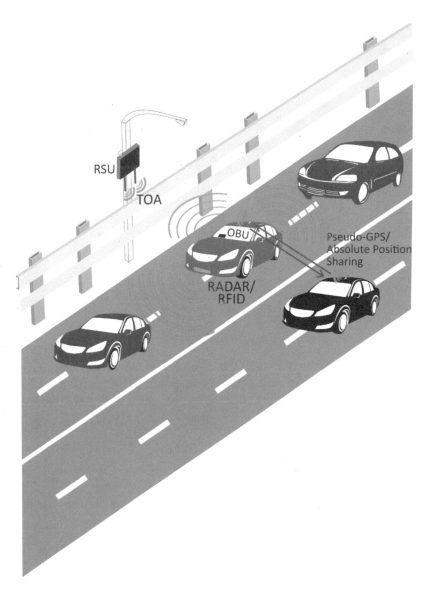

FIGURE 11.18
Cooperative information exchange.

11.3.2 Tracking Models of Cooperative Vehicles

The way the information collected through DSRC is fused depends on the cooperative model employed. The two major distinctions in literature are:

- Single-target tracking systems.
- Multi-target tracking systems.

Single-target tracking problems are generally computationally faster and involve estimating only the ego vehicle's (or one's own vehicle's) position by fusing the V2X information collected. State-of-the-art in single tracking systems have been carried out using EKF, PF, or a combination of the two, as discussed in Section 11.2.2.

Multi-target tracking problems [12], on the other hand, are usually modeled as a Bayes filtering problem to estimate the current state of all the vehicles in the network. One of the problems with multi-target tracking is that the number of targets, N_t, in the network is unknown, and would be modeled as a discrete random variable in the state-space. As the dimensionality of the problem increases with N_t, some ways to solve for the state variables are:

- Assume N_t to be constant, and other targets that are not considered as "hidden" (similar to work in clustering that considers vehicles only in the region of influence of the ego vehicle),
- Estimate N_t separately and then use the estimate to solve for the other state variables,
- Calculate the likelihood in a space of constant dimensionality, or
- Employ finite state statistics to compare state-spaces of different dimensionality, hence estimating N_t at the same time as the rest of the state variables.

Another issue with multi-target tracking is that the complexity of the problem increases with the size of the network, which makes it unfavorable for real-time implementation. Two noteworthy approximations to the model (to limit its complexity) have been studied in literature, the probability hypothesis density filter (PHD) [3,18,22], and the cubature Kalman filter (CKF) [9]. In the PHD approximation, only the first moment of the probability distribution is propagated, and its expectation calculated over a subset of the state-space reveals the number of targets in that subset. In the standard CKF, the posterior density is assumed Gaussian, and thus it is computationally easier to solve multi-dimensional Gaussian-weighted moment integrals. Though novel in their approaches, the goal in this chapter is to improve the estimate of one's position in space, and not necessarily the whole state-space of the vehicle network. We therefore focus on single-target tracking, for a computationally easier and faster approach, in our application using EKF with PF.

11.3.3 Fusion Methods for Cooperative V2V

From the possible sources of V2V information above, ranging information is the easiest to fuse. V2V-based ToA information obtained from cooperative vehicles is used in the update stage of the EKF in Figure 11.9 in the same way that V2I-based ToA information is used for self-localization. Shared position information from other vehicles can also be used as known obstacles for radar ranging in the multi-hypothesis model discussed in Section 11.2.2.2. In this way, cooperative ranging can provide both coarse and fine tuning on the ego vehicle's localization. To demonstrate the performance improvement from using cooperative V2V techniques on a single ego vehicle, we consider the case of dense obstacles with 75% lane detector reliability (as in Figure 11.16b) as the benchmark. For our simulation, we consider that the ego vehicle can communicate with a maximum of five cooperative vehicles within the communicable range for the considered vehicle trajectory. The corresponding CDF plot in Figure 11.19 demonstrates the performance from a single ego vehicle's perspective. As compared to the case without cooperative V2V (in Figure 11.16b), we notice an improvement for total and lateral RMSE, and the error durations for both the EKF-11 and EKF-26 state models.

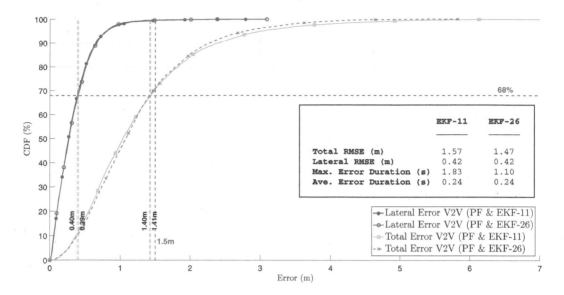

FIGURE 11.19

CDF of total and lateral errors of the refined position estimate with cooperative ranging information for dense obstacles with the lane detector's reliability of 75%.

Another source of cooperative information in the form of absolute position information of the cooperative vehicles has been studied in [4]. Here, the cooperative vehicles, known as virtual anchors, broadcast their refined position estimates, as does the ego vehicle, which is then fused with the other on-board sensors to help achieve a greater degree of positional accuracy. Finally, pseudo-range GNSS estimates can be useful in providing a differential position correction for a cooperative but distributed network of vehicles [8].

A practical challenge for single-tracking systems in cooperative localization is the issue of accuracy versus complexity. This involves having to limit the cooperative vehicles that are chosen as information sources for the ego vehicle's fusion. Cooperative clustering [10,13] is one such solution, which involves choosing a region of influence (ROI), or more aptly, the range of information fusion for the ego vehicle. A number of parameters can determine this clustering, the density of cooperative vehicles, the number of lanes, the direction of moving vehicles (for/against), the displacement or absolute distance of the cooperative vehicle from the ego vehicle, etc. This would also depend on the application for cooperative localization. For example, with collision avoidance, cooperative vehicle clusters seen only in front of the ego vehicle would be considered, and for lane-level accuracy, maybe only cooperative vehicles with their relative speed comparable to that of the ego vehicle's may be clustered to provide more precision. However, regardless of the application, cooperative localization has been shown to improve navigational accuracy, and will continue to push the boundary for end-user experience in the vehicular domain.

References

1. DSRC (Dedicated Short-Range Communication) Technical Committee. *On-Board System Requirements for V2V Safety Communications*. Standard J2945, Society of Automotive Engineers (SAE) International, Mar. 2016.

2. National Center for Statistics and Analysis. *Critical Reasons for Crashes Investigated in the National Motor Vehicle Crash Causation Survey*. Technical Report DOT HS 812506, National Highway Traffic Safety Administration (NHTSA), Mar. 2018.

3. S.A. Goli, B.H. Far, and A.O. Fapojuwo. Cooperative Multi-Sensor Multi-Vehicle Localization in Vehicular Adhoc Networks. In *2015 IEEE International Conference on Information Reuse and Integration*, pages 142–149, Aug. 2015.

4. G. Hoang, B. Denis, J. Hrri, and D.T.M. Slock. Robust Data Fusion for Cooperative Vehicular Localization in Tunnels. In *2017 IEEE Intelligent Vehicles Symposium (IV)*, pages 1372–1377, June 2017.

5. A. Kakkavas, H.M. Castaneda-Garcia, R.A. Stirling-Gallacher, and J.A. Nossek. Multi-Array 5G V2V Relative Positioning: Performance Bounds. *2018 IEEE Global Communications Converence*, Abu Dhabi, UAE, GLOBECOM, Dec. 2018.

6. S. Kuutti, S. Fallah, K. Katsaros, M. Dianati, F. Mccullough, and A. Mouzakitis. A Survey of the State-of-the-Art Localization Techniques and Their Potentials for Autonomous Vehicle Applications. *IEEE Internet of Things Journal*, 5(2):829–846, Apr. 2018.

7. Land Transport Authority (LTA) of Singapore. Geospatial Datasets. https://www.mytransport.sg/content/mytransport/home/dataMall.html.

8. K. lassoued, P. Bonnifait, and I. Fantoni. Cooperative Localization with Reliable Confidence Domains between Vehicles Sharing GNSS Pseudoranges Errors with No Base Station. *IEEE Intelligent Transportation Systems Magazine*, 9(1):22–34, Spring 2017.

9. J. Liu, B. Cai, and J. Wang. Cooperative Localization of Connected Vehicles: Integrating GNSS with DSRC Using a Robust Cubature Kalman Filter. *IEEE Transactions on Intelligent Transportation Systems*, 18(8):2111–2125, Aug. 2017.

10. W. Liu, G. Qin, Y. He, and F. Jiang. Distributed Cooperative Reinforcement Learning-Based Traffic Signal Control that Integrates V2X Networks Dynamic Clustering. *IEEE Transactions on Vehicular Technology*, 66(10):8667–8681, Oct. 2017.

11. Z. Madadi, F. Quitin, and W.P. Tay. Periodic RF transmitter geolocation using a mobile receiver. In *2015 IEEE International Conference on Acoustics, Speech and Signal Processing (ICASSP)*, pages 2584–2588, Apr. 2015.

12. R. P. S. Mahler. Multitarget Bayes Filtering via First-Order Multitarget Moments. *IEEE Transactions on Aerospace and Electronic Systems*, 39(4):1152–1178, Oct. 2003.

13. M. Minea. Cooperative V2V Clustering Algorithm for Improving Road Traffic Safety Information. In *2015 12th International Conference on Telecommunication in Modern Satellite, Cable and Broadcasting Services (TELSIKS)*, pages 369–372, Oct. 2015.

14. W.Y. Ochieng, M.A. Quddus, and R.B. Noland. Map-Matching in Complex Urban Road Networks. *Brazilian Journal of Cartography*, 55(2):1–18, Jan. 2004.

15. P. Oguz-Ekim, K. Ali, Z. Madadi, F. Quitin, and W.P. Tay. Proof of Concept Study Using DSRC, IMU and Map Fusion for Vehicle Localization in GNSS-Denied Environments. In *2016 IEEE 19th International Conference on Intelligent Transportation Systems (ITSC)*, pages 841–846, Nov. 2016.

16. R. Rabiee, X. Zhong, Y. Yan, and W.P. Tay. LaIF: A Lane-Level Self-Positioning Scheme for Vehicles in GNSS-Denied Environments. *IEEE Transactions on Intelligent Transportation Systems*, 20(8):2944–2961, Aug. 2019.

17. A. Ramirez. Time-of-flight in Wireless Networks as Information Source for Positioning. PhD thesis, Technical University of Munich (TUM), 2011.

18. H. Sidenbladh. Multi-Target Particle Filtering for the Probability Hypothesis Density. In *Proceedings of the Sixth International Conference of Information Fusion*, volume 2, pages 800–806, July 2003.

19. Telecommunications Standards Advisory Committee (TSAC). *Dedicated Short-Range Communications in Intelligent Transport Systems*. Technical Specification 1, Infocomm Media Development Authority (IMDA), Oct. 2017.

20. V. Va, T. Shimizu, G. Bansal, and R.W. Heath Jr. Millimeter Wave Vehicular Communications: A Survey. *Foundations and Trends in Networking*, 10(1):1–113, 2016.

21. G.M. Vitetta, and G. Baldini. *Theoretical framework for In-Car Navigation based on Integrated GPS/IMU Technologies*. EUR - Scientific and Technical Research Reports 27042, Joint Research Centre (JRC), Institute for the Protection and Security of the Citizen, Jan. 2015.

22. B-N. Vo, and W-K. Ma. The Gaussian Mixture Probability Hypothesis Density Filter. *IEEE Transactions on Signal Processing*, 54(11):4091–4104, Nov. 2006.

23. C.E. White, D. Bernstein, and A.L Kornhauser. Some Map Matching Algorithms for Personal Navigation Assistants. *Transportation Research Part C: Emerging Technologies*, 8(16):91–108, 2000.

24. X. Zhong, R. Rabiee, Y. Yan, and W.P. Tay. A Particle Filter for Vehicle Tracking with Lane Level Accuracy under GNSS-Denied Environments. In *2017 IEEE 20th International Conference on Intelligent Transportation Systems (ITSC)*, pages 1–6, Oct. 2017.

12

Cooperative Localization Based on Relative Measurements

Leigang Wang and Tao Zhang

CONTENTS

12.1 Cooperative Localization

12.1.1 Background

Multi-mobile robot system (MRS) can be used in collaboration for exploring, monitoring, rescuing, and so on (Nicosia 2007, Maza et al. 2010, Bernard et al. 2011). In these tasks, localization plays an important role for formation coordination, collision avoidance, multi-source information fusion, etc.

Compared with individual operation, much more accurate localization is required for MRS to complete a certain task. For example, several aperture radars on multi-satellite can be virtualized into a high-precision synthetic aperture radar, where accurate localization is the premise.

Localization is estimating the robot's pose (position and heading). The simplest localization method is dead reckoning, where using the measurement data (e.g., motion distance) from the odometers mounted on the robots, the pose of each robot is independently estimated (independent localization [IL]) as Figure 12.1a. However, due to error accumulation, the localization accuracy seriously drifts (Kelly 2002, Sharma 2011), especially in the case of long traverses.

A group of robots work collaboratively in a common situation. Meanwhile, all the robots estimate their poses by utilizing inter-robot information (e.g., connectivity,

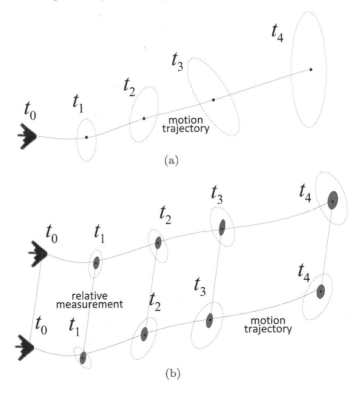

FIGURE 12.1
(a) The vehicle's position is estimated alone (IL). (b) Two vehicles' positions are estimated in cooperation manner (CL). The estimate uncertainty of IL and CL increase over time. While the increasing rate of CL is lower than that of IL. (Sharma, R. *Bearing-only Cooperative-Localization and Path-Planning of Ground and Aerial Robots*. PhD thesis, Brigham Young University, 2011, 3-5.)

relative distance). This maneuver is regarded as cooperative localization (CL) as Figure 12.1b. CL can be divided into range-free and range-based methods. In the former, the network-related information is utilized for CL. In the latter, each robot utilizes the relative-measurements among the individuals (Zhou and Roumeliotis 2008). In this chapter, the latter, that is, cooperative localization based on relative measurements, is investigated.

CL is a situation where a single robot uses other robots as dynamic landmarks. Compared with IL, because of the additional information available (e.g., relative distance/bearing), CL can achieve more reliable and precise localization than one robot alone.

12.1.2 Application Scenarios

12.1.2.1 Cost and Payload Being Limited

CL provides an approach to reduce the cost of MRS by assigning the high- and low-precision localization sensors on different robots. This strategy allows the inexpensive robots to share the effort of high-precision sensors.

Additionally, the payload of MRS can be reduced through CL. For example, in the coordinated reconnaissance, the vehicles flying at high altitude are equipped with fine localization equipment for positioning, while the vehicles at low altitude are equipped with the task load, responsible for searching or monitoring, and positioned by CL.

12.1.2.2 Specific Domain

In the urban situation or the battlefield, due to the difficult geographical environment or hostile interference, respectively, global positioning system (GPS) may be unavailable to some robots (Sharma and Taylor 2008, Qu and Zhang 2011, Lee 2012), while the other robots obtaining the absolute localization information (for example, GPS or the relative measurement to the known landmarks) can share their positions with the former in CL manner as Figure 12.2a,c,d. As a result, the positioning accuracy of all robots would be improved.

Another application domain is underwater environment as Figure 12.2b. For example, a fleet of autonomous underwater vehicles (AUVs) is implementing some special tasks, such as marine data collection, ocean meteorology investigating (Maczka et al. 2007, Song and Mohseni 2013). Usually, AUV has to surface occasionally to obtain GPS for its positioning. However, in the case of CL, only autonomous surface crafts (ASCs) are required to surface for GPS, and the rest of AUVs always implement the underwater task and their positioning accuracy is guaranteed by the cooperative strategy.

12.2 Cooperative Localization Technology

12.2.1 Hardware Requirements

Generally speaking, each individual in a MRS is equipped with a positioning system, task load (sensing the surrounding environment or target), and communication system. With regard to CL, two types of equipment are employed: measurement sensor and communication system.

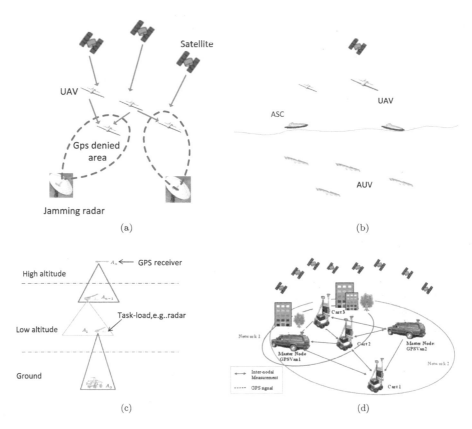

FIGURE 12.2
CL can be employed in different situations for different tasks. (a) In the task of coordinated reconnaissance, the UAVs that are interfered with by the enemy radar and cannot receive GPS signals can be cooperatively positioned by the undisturbed teammates. (b) In the task of ocean data collecting, ASCs or UAVs get GPS and act as the localization references. AUVs perform DR with the on-board IMU. Through ASC, AUVs can get more accurate localization estimates. (c) In the task of coordinated ground searching, the high-altitude flying vehicle measures its position by a GPS receiver and shares the positioning with other vehicles flying in low-altitude to cooperatively geolocate a ground target. (Adapted from Sharma, R. *Bearing-only Cooperative-Localization and Path-Planning of Ground and Aerial Robots*. PhD thesis, Brigham Young University, 2011, 48.) (d) Typical urban canyon. Each agent performs relative measurements with respect to its teammates via its inbuilt sensors, and by exchanging these data over network the whole localization accuracy is obtained. (Lee, J. K. *Journal of Navigation* 65(3), 2012: 445–457.)

1. Measurement sensor
    a. Proprioceptive sensors (e.g., the inertial equipment) that measure the motion feature of the platform. By these sensors, the motion data (e.g., linear velocity, angular velocity) are collected.
    b. Exteroceptive sensors that provide the relative measurement, for example, laser scanner and visual imaging system. By these sensors, the relative attitude (distance, azimuth, relative orientation, etc.) between platforms is obtained.
    c. Exteroceptive sensors providing the absolute measurement, for example GPS receiver and magnetometer. By these sensors, the absolute position or orientation is obtained.
2. Wireless communication devices. Each robot is equipped with wireless communication to exchange information among them.

TABLE 12.1

Symbols Used in Text

Notation	Description
N	Total number of robots
k	Time step
x_i^k	True pose of robot i, $x_i^k = [x_i^k, y_i^k, \varphi_i^k]^T$, x_i^k, y_i^k denotes the coordinate and ϕ_i^k denotes the motion heading
$\hat{x}_i^k$	Posteriori estimate vector of robot i at time step k
$\bar{x}_i^k$	Predictive estimate vector of robot i at time step k
x^k	Whole state matrix, $x_i^k = [x_1^k ..., x_N^k]$
P^k	Whole covariance of position estimate error
P_{ij}^k	Error covariance of estimate between robot i and j
z_{ij}^k	Relative measurement between robot i and j
$\nu_{i(j)}^k$	Exteroceptive measurements noise vector
w_i^k	Proprioceptive measurement noise vector
u_i^k	Motion control vector

Consider a team of N mobile robots in a 2-dimensional plane. The initial poses are known. In operation, each robot requires its pose $X_i^k = [x_i^k, y_i^k, \varphi_i^k]^T$ in a local fixed reference frame XOY. $[x_i^k, y_i^k]^T$ and ϕ_i^k denote the position coordinates and motion heading. In the following, the essence for CL, that is, the model of motion and exteroceptive measurement, is given. The symbols used throughout this chapter are defined in Table 12.1.

The proprioceptive sensors provide the data related to the robot's motion, such as velocity and acceleration, which are the input for the motion model and denoted as u_i^k. For a single robot, generally, its motion is modeled as:

$$x_i^{k+1} = f(x_i^k, u_i^k) + w_i^k \tag{12.1}$$

where f is in general a non-linear function, w_i^k is the proprioceptive measurements' noise, and $E\{w_i^k \cdot (w_i^k)^T\} = Q_i^k$ (E denotes the mathematic expectation operation).

It can be seen that if a single robot is propagating its position by itself as (12.1), the error will grow unbounded due to the noise term w_i^k.

The exteroceptive sensors measure the quantities of interest among the robots, such as the relative distance, bearing, and relative orientation etc., as Figure 12.3, where $X_b O Y_b$ denotes the body reference frame.

When robot i obtains the absolute exteroceptive measurement, the measurement can be modeled as:

$$z_i^k = h(x_i^k) + v_i^k \tag{12.2}$$

When robot i finds robot j and obtains the relative measurement between them, the relative measurement can be modeled as:

$$z_{ij}^k = h(x_i^k, x_j^k) + v_{ij}^k \tag{12.3}$$

where $v_{i(j)}^k$ is the measurement noise and $E\left\{v_{i(j)}^k \cdot (v_{i(j)}^k)^T\right\} = R_{i(j)}^k$.

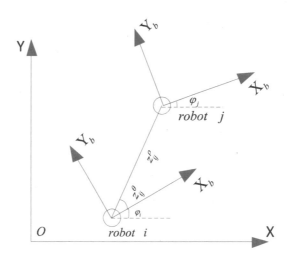

FIGURE 12.3
Representation of robot i and j with different poses.

The objective of CL is to estimate the pose by fusing the absolute/relative exteroceptive measurement data. Considering the inherent uncertainty of measurement data, the CL problem is usually formalized in two approaches: the probabilistic-based approach and the graph-based approach.

12.2.2 Probabilistic-Based Cooperative Localization

At a certain time k, let $x^k = [x_1^k, \ldots, x_N^k]$ denote the pose vector of the whole group. Let $z^k = [z_1^k, \ldots, z_N^k]$ denote all the exteroceptive measurements from all the robots. If robot i does not obtain any exteroceptive measurements, then $z_k = \phi$. All the proprioceptive and exteroceptive measurements up to time step k are denoted as $u^{0:k}$ and $z^{0:k}$, respectively.

When the probabilistic approach is employed for the CL problem, the robot's pose is modeled as a random variable in which posterior density function (PDF) is related to the measurements satisfied with a certain distribution characterization. Given all the available measurements, the PDF of all robots' poses can be represented as:

$$p[x^k \mid x^{0:k-1}, z^{0:k}, u^{0:k}] \tag{12.4}$$

The expected estimate of the whole poses and its variance are

$$\hat{x}^k = \int_{-\infty}^{\infty} x^k \, p[x^k \mid x^{0:k-1}, z^{0:k}, u^{0:k}] dx^k \tag{12.5}$$

$$P^k = \int_{-\infty}^{\infty} (x^k - \hat{x}^k)(x^k - \hat{x}^k)^{\mathrm{T}} \, p[x^k \mid x^{0:k-1}, z^{0:k}, u^{0:k}] dx^k \tag{12.6}$$

12.2.2.1 Extended Kalman Filter-Based CL

Assume that the pose estimation is regarded as Markov process, that is, each robot's future pose depends only on its current pose. Thus, CL can be regarded as the extended Kalman filter (EKF) filtering problem. The EKF-based CL is studied extensively (Roumeliotis and Bekey 2002, Ribeiro et al. 2006, Wang et al. 2008, Kia et al. 2014) which employs an excellent recursion mechanism including two phases: the predictive propagation and posteriori update.

1. Predictive propagation

   Under the EKF-based CL framework, at time step k, $\bar{x}_i^k$ denotes the predicted estimate based on the available information up to time step $k-1$. $\hat{x}_i^k$ denotes the posteriori estimate based on the information up to time step k. Accordingly, $\bar{P}_i^k$ and $\hat{P}_i^k$ represent the estimate uncertainty, respectively, and assume that $\hat{x}_i^0 \sim N(x_i^{ini}, P_i^0)$. The whole pose vector x and covariance matrix can be expressed as:

   $$x = [x_1^T, x_2^T, \ldots, x_N^T]^T, \quad P = \begin{bmatrix} P_{11} & P_{12} & \cdots & P_{1N} \\ P_{21} & \cdots & \cdots & \cdots \\ \cdots & \cdots & \cdots & \cdots \\ P_{N1} & \cdots & \cdots & P_{NN} \end{bmatrix} \tag{12.7}$$

   where the block diagonal element P_{ii} denotes the uncertainty of pose estimate, and the element P_{ij} denotes the error correlation between two pose estimates.

   Firstly, based on the posteriori estimate $\hat{x}^k$ at time step k, the predictive estimate $\bar{x}^{k+1}$ can be propagated as the following:

   $$\bar{x}^{k+1} = [f(\hat{x}_1^k, u_1^k), \ldots f(\hat{x}_N^k, u_N^k)]^T \tag{12.8}$$

   Along with the pose estimates being propagated, the whole covariance also is changed as follows:

   $$\bar{P}^{k+1} = \Phi^k \hat{P}^k (\Phi^k)^T + G^k Q^k (G^k)^T \tag{12.9}$$

   where $\Phi^k = \mathrm{diag}\{\Phi_i^k, \ldots, \Phi_N^k\}$. $G^k = \mathrm{diag}\{G_i^k, \ldots, G_N^k\}$. $Q^k = \mathrm{diag}\{Q_i^k, \ldots, Q_N^k\}$. $\mathrm{diag}\{\}$ denotes a block diagonal matrix, Φ_i^k denotes the state-error propagation matrix, and G_i is the system noise matrix.

2. Posteriori update

   For the absolute and relative measurements, generally, the measurement error models are respectively expressed as:

   $$\tilde{z}_i^k = H_i^k \cdot \tilde{x}_i^k + v_i^k \tag{12.10}$$

   $$\tilde{z}_{ij}^k = H_{ij}^k \cdot [\tilde{x}_i^k; \tilde{x}_j^k] + v_i^k \tag{12.11}$$

   where $H_{i(j)}^k$ is a linearization expression with respect to the measurement function $h(\cdot)$. Stack all the local measurement matrix $H_{i(j)}^k$ and a whole exteroceptive measurement matrix H^k can be constructed. Then the posteriori estimates and the corresponding error covariance are updated as follows:

$$\hat{\mathbf{x}}^{k+1} = \bar{\mathbf{x}}^{k+1} + \mathbf{K}(\mathbf{z}^{k+1} - \mathbf{h}(\mathbf{z}^{k+1})) \tag{12.12}$$

$$\hat{\mathbf{P}}^{k+1} = \bar{\mathbf{P}}^{k+1} - \mathbf{K}\mathbf{H}^{k+1}\bar{\mathbf{P}}^{k+1} \tag{12.13}$$

$$\mathbf{K} = \bar{\mathbf{P}}^{k+1}(\mathbf{H}^{k+1})^{\mathrm{T}}(\mathbf{H}^{k+1}\bar{\mathbf{P}}^{k}(\mathbf{H}^{k+1})^{\mathrm{T}} + R^{k})^{-1} \tag{12.14}$$

where $R^k = \mathrm{diag}\left\{R_1^k, \ldots, R_N^k\right\}$.

It should be noted that since the occurrence frequency of exteroceptive measuring is rare compared with the proprioceptive measurements, most of the time the predictive propagation is always running.

12.2.2.2 Maximum A Posteriori-Based CL

The Maximum A Posteriori (MAP)-based CL is the best pose estimate for the entire history of CL, which considers all the history data synchronously. The MAP-based CL can be formulated as (Nerurkar and Roumeliotis 2008, Nerurkar et al. 2009):

$$\hat{x}_{0:k} = \arg\max P(x_{0:k} \mid z_{0:k}, u_{0:k}) \tag{12.15}$$

Using Bayes' rule and Markov property, (12.15) can be rewritten as:

$$\hat{x}_{0:k} = \arg\max \frac{1}{P(z_{0:k})} P(z_{0:k} \mid x_{0:k}) P(x_{0:k})$$
$$= \arg\max \prod_{m=0}^{k} P(z_m \mid x_m) \prod_{m=0}^{k} P(x_{m+1} \mid x_m) P(x_0) \tag{12.16}$$

when the measurement noise satisfies the Gaussian distribution, (12.16) can be reformed as the least-squares (LS) problem as follows:

$$\hat{x}_{0:k} = \arg\min \left(\begin{array}{c} \sum_{i=1}^{N} \sum_{\substack{j=1 \\ j \neq 1}}^{N} \sum_{m=0}^{k} \| z_{ij}^m - h(x_i^m, x_j^m) \| + \sum_{i=1}^{N} \sum_{m=0}^{k} \| z_i^m - h(x_i^m) \| \\ + \sum_{i=1}^{N} \sum_{m=1}^{k} \| f(x_i^{m-1}, u_i^{m-1}) - x_i^m \| \end{array} \right) \tag{12.17}$$

Compared with the EKF-based CL where only the current poses of robots are estimated, all the historical estimated-poses $\hat{x}_{0:k}$ are re-estimated in the MAP-based CL, while a huge storage and computational consumption is required by the latter.

12.2.2.3 Maximum Likelihood Estimation-Based CL

Given the robots' pose, the joint probability of measurement is

$$P(z_{0:k}, z_{0:k}^{vir} \mid x_{0:k}) \tag{12.18}$$

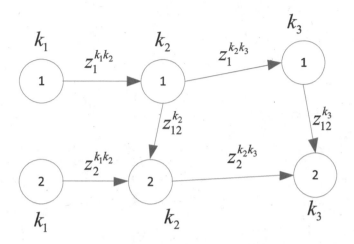

FIGURE 12.4
Demonstration of virtual relative and actual relative measurements.

where $z_{0:k}^{vir}$ denotes the virtual relative measurement from time step $k-1$ to k (Howard et al. 2002). The virtual relative measurement is derived by the proprioceptive measurements data u_i.

The relationship of measurements and robots' poses is visualized by a directed graph as Figure 12.4. The virtual relative measurement represents the same robot at two different points in time, while the true measurement represents two adjacent robots at the same point in time.

The objective of Maximum Likelihood Estimation-based CL is to find the best pose estimate that maximizes the probability of all measurement data over time, that is

$$\hat{x}_{0:k} = \arg\max P(z_{0:k}, z_{0:k}^{vir} \mid x_{0:k}) \tag{12.19}$$

In general, there is no closed-form solution for the MAP- or ML-based CL, and the numerical search methods are good choices where the solution is reached by iterating, for example, the steepest-gradient method or Newton-type methods.

12.2.3 Graph-Based Cooperative Localization

Graph-based CL is a batch processing approach where all the history poses are continuously updated along with increasing the new measurement data (Barooah and Hespanha 2007). The robot's position at different times is regarded as node, and the measurement between the nodes is regarded as the edge, which is used to represent the nonlinear constraint among the inter-robots and landmark-robot. All the nodes and the measurement edges constitute a static snapshot graph, as seen in Figure 12.5. In graph-based CL, the geometric constraint and the nodes with prior knowledge are used to estimate the unknown nodes.

The snapshot graph can be expressed as $G = (V, E)$, where three kinds of edges are included.

1. Displacement between the nodes corresponding to the same robot from the time step k to $k+1$, denoted as $E_i^{k(k+1)}$, which can be expressed as

$$x_i^{k+1} - x_i^k = f(x_i^k, u_i^k) - x_i^k = d_i^{k(k+1)} \tag{12.20}$$

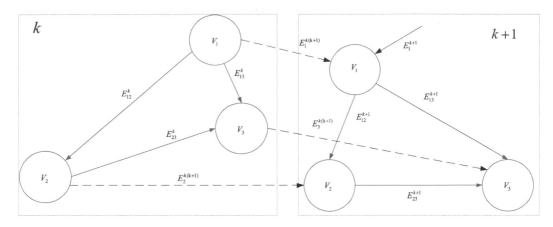

FIGURE 12.5
Snapshot graph with different types of edges.

2. For the robot obtaining the absolute positioning, the edge is denoted as E_i^k, which can be given as

$$f(x_i^k) = d_i^k \tag{12.21}$$

3. The inter-node relative measurement corresponds to different robots at the same time, denoted as E_{ij}^k, which can be given by

$$x_i^k - x_j^k = h_i(x_i^k, x_j^k) = d_{ij}^k \tag{12.22}$$

Stack all the edges to form a whole measurement vector:

$$D = [d_1^{k(k+1)}, d_1^k, d_{1*}^k, \dots d_1^{k(k+1)}, d_N^k, d_{N*}^k]^T \tag{12.23}$$

where * denotes any element belonging to the set of $\{1, \dots N\}$.

For (12.20) through (12.22), by linearizing, the following equation group is obtained as:

$$M \cdot x = D + C \tag{12.24}$$

For each row block $M_i \in \mathrm{R}^{N_i^e \times N} (i \in \{1, \dots N\})$ and N_i^e denotes the number of edges connected with the node V_i

$$M = \begin{bmatrix} 0 & \cdots & \nabla f_i - I & \cdots & 0 \cdots 0 \\ 0 & 0 & \nabla f_i' & 0 & 0 \cdots 0 \\ 0 & 0 & \nabla h_i & 0 & \nabla h_j \cdots 0 \\ & & i^{th} & & j^{th} \end{bmatrix} \tag{12.25}$$

where

$$\nabla f_i = \frac{\partial f(x_i^k, u_i^k)}{\partial x_i^k}, \quad \nabla f_i' = \frac{\partial f(x_i^k)}{\partial x_i^k}, \quad \nabla h_i = \frac{\partial h(x_i^k, x_j^k)}{\partial x_i^k}, \quad \nabla h_j = \frac{\partial h(x_i^k, x_j^k)}{\partial x_j^k}.$$

C is the constant vector generated from the linearization process. Assume that the covariance of measurement noise is $\sum$, then using the common parameter estimate technique, for example, WLS (weight least squares), the best solution can be obtained as follows:

$$\hat{x}^* = (M^T \sum^{-1} M)^{-1} M^T \sum^{-1}(D + C) \tag{12.26}$$

Compared with the EKF-based CL where the error correlation of pose estimates is quantified as the covariance term, in the MLE-based CL, MAP-based CL and graph-based CL, the correlation is no longer quantified. The various measurement relationships are recorded and preserved. When the new measurement occurs, all the historical poses are re-estimated, and the global optimal estimation is guaranteed.

12.3 Implementation of Cooperative Localization

12.3.1 Architecture of Centralized Cooperative Localization

CL can be implemented in the centralized- or distributed-architecture. In the centralized CL (CCL), there exists a processing center to execute the CL algorithm. Taking the EKF-based CL as an example, at each time step k, each robot sends its linear/rotational velocity and the exteroceptive measurement (if existing) to the center. Firstly, utilizing the linear/rotational velocity, the center propagates all the robots' positions according to Equation 12.8. Meanwhile, the corresponding covariance is also updated according to Equation 12.9. Secondly, the processing center judges whether there exists the exteroceptive measurement in the collection. If it does not exist, let

$$\hat{x}^k = \bar{x}^k \tag{12.26}$$

$$\hat{P}^k = \bar{P}^k \tag{12.27}$$

Otherwise, in the center, the exteroceptive measurement model is linearized and then used to construct the whole measurement matrix H. The predictive estimates and covariance are updated according to Equations 12.12 and 12.13. The CCL procedure is demonstrated as in Figure 12.6.

Under the centralized architecture, the algorithm design is simple. However, the communication and calculation burden usually is high for the center. The CCL is often employed for the performance analysis, for example in the literature (Huang et al. 2011). While in the application, the DCL is more popular than the CCL.

12.3.2 Architecture of Distributed Cooperative Localization

12.3.2.1 Self-Centralized Architecture

As an example of self-centralized architecture, regarding the neighboring teammates as nodes, each robot builds a local measurement sub-graph as shown in Figure 12.7a. In this case, the optimization strategy usually is adopted for the pose estimation, where each robot repeatedly sends its temporary estimate to its neighbors and receives the corresponding estimates to update itself (Costa and Patwari 2006, Barooah and Hespanha 2007). The

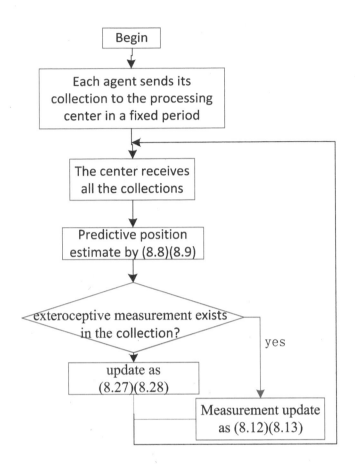

FIGURE 12.6
Procedure of CCL.

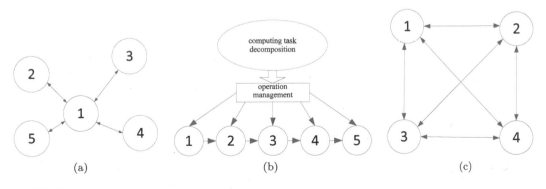

FIGURE 12.7
Architecture of DCL. (a) Self centralized-based CL. (b) Computation Task decomposed-based CL (c) Multi-centralized CL. (Wang, L.G. *Research on Distributed Cooperative Localization for Multi-mobile Platforms under Communication Constraints.* PhD thesis, Tsinghua University, Beijing, China, 2017, 9.)

algorithm has strong independence. The sub-optimization can be achieved. Additionally, the iteration and communication are required repeatedly.

12.3.2.2 Architecture Based on the Computation Task Decomposed

As shown in Figure 12.7b, the CL problem is globally analyzed via different techniques, and then the computation tasks is decomposed and assigned to each robot (Nerurkar and Roumeliotis 2008, Mu 2010). Following a certain computation order, CL would be implemented by each robot in the serial or parallel mode. Under this architecture, the exact communication sequence is required. Thus, the algorithm independence is weak.

12.3.2.3 Multi-Centralized Architecture

As shown in Figure 12.7c, each robot sends its own information to the rest, and all the robots' states are reconstructed in each robot(Trawny et al. 2009, Walls and Eustice 2014). Multi-centralized architecture is robust and can reach the centralized-equivalent accuracy. However, the communication burden is heavy.

12.3.3 EKF-Based Distributed Cooperative Localization

When robot i obtains an absolute exteroceptive measurement data z_i^k, the measurement innovation $\vartheta_i^k = z_i^k - h(\bar{\mathbf{x}}_i^k)$ and its covariance $\mathbf{S}_i^k = E\{\tilde{z}_i^k(\tilde{z}_i^k)^T\} = \mathbf{H}_i^k\bar{\mathbf{P}}_{ii}^k(\mathbf{H}_i^k)^T + \mathbf{R}_i$ can consequently be obtained by robot i independently. All the robots' poses are updated as follows.

$$\hat{\mathbf{x}}_n^k = \bar{\mathbf{x}}_n^k + \mathbf{P}_{ni}^k\Lambda(i,k) \quad (n = 1,...N);\tag{12.28}$$

$$\hat{\mathbf{P}}_{mn}^k = \bar{\mathbf{P}}_{mn}^k - \bar{\mathbf{P}}_{ni}^k\bar{\Lambda}(i,k)\bar{\mathbf{P}}_{im}^k \quad (n,m = 1,...N)\tag{12.29}$$

where $\Lambda(i,k) = (\mathbf{H}_i^k)^T(\mathbf{S}_i^k)^{-1}\upsilon_i^k$ and $\bar{\Lambda}(i,k) = (\mathbf{H}_i^k)^T(\mathbf{S}_i^k)^{-1}\mathbf{H}_i^k$, called the absolute locating message.

When robot i detects robot j and obtains a relative exteroceptive measurement z_{ij}^k, then the measurement innovation is $\upsilon_{ij}^k = z_{ij}^k - h(\bar{\mathbf{x}}_i^k, \bar{\mathbf{x}}_j^k)$ and its covariance is $\mathbf{S}_{ij}^k = E\{\tilde{z}_{ij}^k(\tilde{z}_{ij}^k)^T\} = \mathbf{H}_i^k\bar{\mathbf{P}}_{ij}^k(\mathbf{H}_j^k)^T + \mathbf{R}_{ij}$. All the robots' poses are updated as follows.

$$\hat{\mathbf{x}}_n^k = \bar{\mathbf{x}}_n^k + [\bar{\mathbf{P}}_{ni}^k, \bar{\mathbf{P}}_{nj}^k]\Lambda(i,j,k) \quad (n = 1,...N);\tag{12.30}$$

$$\hat{\mathbf{P}}_{nm}^k = \bar{\mathbf{P}}_{nm}^k - [\bar{\mathbf{P}}_{ni}^k, \bar{\mathbf{P}}_{nj}^k]\bar{\Lambda}(i,j,k)\begin{bmatrix}\bar{\mathbf{P}}_{im}^k \\ \bar{\mathbf{P}}_{jm}^k\end{bmatrix} \quad (n,m = 1,...N)\tag{12.31}$$

where $\Lambda(i,j,k) = (\mathbf{H}_{ij}^k)^T(\mathbf{S}_{ij}^k)^{-1}\upsilon_{ij}$ and $\bar{\Lambda}(i,j,k) = (\mathbf{H}_{ij}^k)^T(\mathbf{S}_{ij}^k)^{-1}\mathbf{H}_{ij}^k$, called the relative locating message The generating and utilizing of the locating message for any robot are demonstrated as Figure 12.8.

Inspecting (12.28) through (12.31), it can be found that

1. Any local measurement can cause all the pose estimates as well as the corresponding covariance to be updated.

2. The cross-covariance plays an important role for EKF-based CL, by which the locating message acts on updating the predictive estimate.

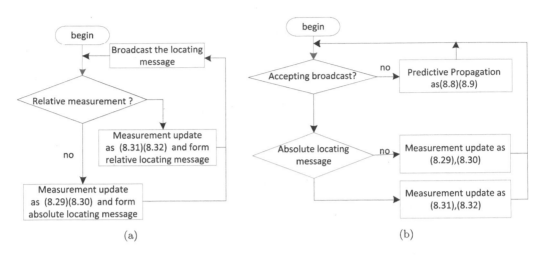

FIGURE 12.8
The generating and utilizing of the locating message are demonstrated from the perspective of the generator and the receiver, respectively. (a) Procedure for any robot in which measurement occurred. (b) Procedure for any robot which received the locating message.

Therefore, the key to EKF-based DCL is how to maintain the cross-covariance in the distributed manner.

12.3.3.1 Covariance Matrix Decomposing

The covariance matrix can be divided into two parts (called the covariance factor) by the matrix decomposition technique, for example, SVD, QR. The decomposition can be generalized as follows:

$$\overline{\mathbf{P}}_{ij}^{k} = \breve{\mathbf{P}}_{i}^{j}(k) \times \breve{\mathbf{P}}_{j}^{i}(k). \tag{12.32}$$

The covariance factors, $\breve{\mathbf{P}}_{i}^{j}(k)$ and $\breve{\mathbf{P}}_{j}^{i}(k)$ are assigned to robot i and j, respectively.

When no exteroceptive measurement occurs, the covariance factors are propagated as follows:

$$\Phi_{i}^{k+1}\breve{\mathbf{P}}_{i}^{j}(k) = \breve{\mathbf{P}}_{i}^{j}(k+1) \tag{12.33}$$

$$\breve{\mathbf{P}}_{j}^{i}(k) \times \Phi_{i}^{k+1} = \breve{\mathbf{P}}_{j}^{i}(k+1) \tag{12.34}$$

where, using only the proprioceptive measurement data, the state-error propagation matrix Φ_{i}^{k+1} and Φ_{j}^{k+1} can be obtained by each robot. Therefore (12.33) and (12.34) can be independently executed by robot i and j.

Assume that at time step $k + \Delta k$, the exteroceptive measurement occurs in a certain robot. To update the predicted positions as (12.28) and (12.30), the covariance is required again, and can be obtained as follows:

$$\breve{\mathbf{P}}_{i}^{j}(k+\Delta k) \times \breve{\mathbf{P}}_{j}^{i}(k+\Delta k) = \overline{\mathbf{P}}_{ij}^{k+\Delta k} \tag{12.35}$$

12.3.3.2 Covariance Intersection

Julier and Uhlmann (1997) proposed the covariance intersection (CI) algorithm, which is designed to fuse the estimates with the unknown covariance. Consider two correlated estimates $\hat{x}_o(1), \hat{x}_o(2)$ on the same state x_o, and the corresponding covariance matrices are $\hat{P}(1), \hat{P}(2)$. Then fusion of CI is obeyed by the convex combination of mean and covariance as follows:

$$P_o = [\omega P_o(1)^{-1} + (1 - \omega)P_o(2)^{-1}]^{-1} \tag{12.36}$$

$$\hat{x}_o = P_o \left[\omega P_o(1)^{-1}\hat{x}(1) + (1 - \omega)P_o(2)^{-1}\hat{x}(2) \right] \tag{12.37}$$

where $\omega \in [0, 1]$.

When CI is employed for the distributed covariance maintaining, the covariance does not need to be maintained all the time. When robot i detects robot j and obtains the relative exteroceptive measurement z_{ij}^k, it uses the available measurement to estimate the state of the measured robot j as follows:

$$\hat{x}_j^{k*}(i) = \hat{x}_i^k + \Gamma_{ij}^k \cdot z_{ij}^k \tag{12.38}$$

where (i) denotes that the state estimate $\hat{x}_j^{k*}(i)$ is from the standpoint of robot i. Γ_{ij}^k derives from the measurement function $h(\hat{x}_i^k, \hat{x}_j^k)$, that is

$$h(\hat{x}_i^k, \hat{x}_j^k) \approx (\Gamma_{ij}^k)^{-1}(\hat{x}_j^k - \hat{x}_i^k) \tag{12.39}$$

The uncertainty of the estimate $\hat{x}_j^{k*}$ is capsulated as follows:

$$\hat{P}_j^{k*}(i) = H_{ij}^k(\hat{P}_i^k)^{-1}(H_{ij}^k)^{\mathrm{T}} + \Gamma_{ij}^k R_{ij}(\Gamma_{ij}^k)^{\mathrm{T}}. \tag{12.40}$$

Consequently, packing $\hat{x}_j^{k*}(i)$ and $\hat{P}_j^{k*}(i)$, robot i communicates the packet to the measured robot j. Once robot j receives the packet, its predictive estimate is updated according to Equations 12.36 and 12.37, where the elements $\hat{x}_o(1), \hat{x}_o(2)$ and $\hat{P}(1), \hat{P}(2)$, are substituted by the elements $\hat{x}_j^{k*}(i), \hat{x}_j^k$ and $\hat{P}_j^{k*}(i), \hat{P}_{jj}^k$, respectively.

12.3.4 Influence Factor on Cooperative Localization

12.3.4.1 Observability Property

In CL, the observability property decides whether the pose can be estimated with the available information and reflects whether the overall positioning error is bounded. Usually, the observability property is analyzed in two strategies:

1. The linear observability for the linearized CL system.
2. The Lie derivatives theory for the nonlinear CL system.

It has been proved that

1. CL can improve the positioning accuracy. However, the error is divergent when the CL system is unobservable, unless one of the robots can obtain the absolute positioning information.

2. Compared with the relative distance and relative orientation, the relative bearing is the best measurement among all kinds of relative measurements.

The observability property can also be intuitively explained by the relative position measurement graph (RPMG) where the nodes represent the vehicle states and the edges represent the robot-to-robot measurements. The CL system is observable under the following RPMG:

1. All nodes in the RPMG have a path to at least two landmarks (position is known).
2. RPMG is connected and at least one robot obtains the absolute positioning.

12.3.4.2 CL Performance

1. The amount of exteroceptive measurements

   The relationship between the amount of exteroceptive measurements and the positioning accuracy observes a law of diminishing return. The latter cannot be improved infinitely by increasing the former (Mourikis and Roumeliotis 2006). Limited by the calculation and communication requirement, a large quantity of exteroceptive measurements are unaffordable. The balance point between the measurement amount and the accuracy requirement needs to be sought.

2. The scale of MRS

   The relationship between the positioning accuracy and the scale of MRS is investigated by Roumeliotis and his team, and valuable conclusions are obtained. (i) For a group of robots equipped with homogeneous sensors, the growth of localization uncertainty is inversely proportional to the number of robots. (ii) The growth rate of localization uncertainty depends only on the number of robots and the accuracy of proprioceptive sensor, and not on the exteroceptive sensor.

12.4 Challenge for Cooperative Localization

The progress of CL includes recognition, measuring, and communicating. Accordingly, the effectiveness of CL is affected by the measurement and communication environment. In practice, the ideal conditions cannot be guaranteed in consideration of the following challenges.

12.4.1 Challenge from Complex Measurement Environment

The acquisition progress of relative measurement, especially in the dynamic environment, is not easy when speed and accuracy are demanded.

12.4.1.1 Dynamic Measurement Environment

The process of relative measuring is completed as follows: by the inbuilt sensor (e.g., radar, camera), robot A finds and identifies its teammate B by some features. Then robot A obtains the relative measurement by some measured parameters (e.g., arrival time difference). In thisprocess, firstly, when the robot is moving and its pose is changing, then it is difficult to find and identify the teammates due to the time-variant spatial relationship, especially in the case of high maneuver. Secondly, considering other tasks, it is impossible to change the formation or pose deliberately for CL. The above factors cause the relative measurements to be scarce.

12.4.1.2 Non-Gaussian Measurement Model

In current research or application, the measurement noise is usually assumed to be in normal distribution. However, this assumption cannot be met. For instance, in the marine environment, considering the factors, for example, temperature variations or salinity discontinuity, the noise characterization is a complex multi-modal distribution and is constantly changing. Therefore, it is difficult to build an accurate measurement model. In addition, it is usually assumed that the measurement is not dependent on the measured target. However, the assumption may be invalid in some cases. For instance, in the radar measurement system, the slope error may be dependent on the distance between target and radar.

12.4.2 Challenge from Complex Communication Environment

The essence of CL is using communication to reduce the localization error. Communication plays an important role for CL. Free connection communication is a common assumption for most CL research. However, it cannot be guaranteed in realistic scenarios, that is, communication constraint problem.

12.4.2.1 Limitation from External Environment

In the context of some special applications, especially in the battlefield, communication may be destroyed. In addition, for security considerations, it is also necessary to reduce the frequency of communication among the group as much as possible so as not to be exposed. Even in the city, due to high buildings or tunnels, the signal is often obscured and the communication link is broken.

12.4.2.2 Limitation from Intrinsic Ability

The communication capacity is proportional to the equipment power. When heavy communication equipment is employed, it is in conflict with the finite payload capacity. The conflict is more prominent in long voyages without energy supply. On the other hand, when robots are dispersed in a wide space, the space may exceed the communication range.

12.4.2.3 Limitation from Application Pattern

CL may be a secondary level task compared with the other tasks (e.g., cooperative detecting). When the resource requirements of different tasks are conflicting, the requirement of CL should yield to the high-level tasks. This means that: (i) the communication topology is determined by the high-level tasks, and CL must passively adapt, and (ii) the high-level tasks preferentially occupy more bandwidth, and the resources allocated to CL are limited.

References

Barooah, P., Hespanha, J. P. Estimation on graphs from relative measurements. *Control Systems* 27(4), 2007: 57–74.

Bernard, M., Kondak, K., Maza, I., and Ollero, A. Autonomous transportation and deployment with aerial robots for search and rescue missions. *Journal of Field Robotics* 28(6), 2011: 914–931.

Costa, J. A., Patwari, N. Distributed weighted-multidimensional scaling for node localization in sensor networks. *ACM Transactions on Sensor Networks* 1(2), 2006: 39–64.

Howard, A., Matark, M. J., and Sukhatme, G. S. Localization for mobile robot teams using maximum likelihood estimation. *International Conference on Intelligent Robots and Systems*, Lausanne, Switzerland, 2002, pp. 434–439.

Huang, G. P., Trawny, N., Mourikis, A. I., and Roumeliotis, S. I. Observability-based consistent EKF estimators for multi-robot cooperative localization. *Autonomous Robots* 30(1), 2011: 99–122.

Julier, S. J., Uhlmann, J. K. A non-divergent estimation algorithm in the presence of unknown correlations. *American Control Conference*, 1997, pp. 2369–2373.

Kelly, A. General solution for linearized stochastic error propagation in vehicle odometry. *International Conference on Intelligent Robots and Systems*, Maui, HI, 2002, pp. 25–30.

Kia, S. S., Rounds, S. F., and Martinez, S. A centralized-equivalent decentralized implementation of Extended Kalman Filters for cooperative localization. *International Conference on Intelligent Robots and Systems*, Chicago, IL, 2014, pp. 3761–3766.

Lee, J. K. Network-based collaborative navigation in GPS-denied environment. *Journal of Navigation* 65(3), 2012: 445–457.

Maczka, D. K., Gadre, A. S., and Stilwell, D. J. Implementation of a cooperative navigation algorithm on a platoon of autonomous underwater vehicles. *Oceans*, Vancouver, BC, 2007, pp. 1–6.

Maza, I., Caballero, F., Capitan, J., Martinez-De-Dios, J. R., and Ollero, A. Firemen monitoring with multiple UAVs for search and rescue missions. *IEEE International Workshop on Safety Security and Rescue Robotics*, Bremen, Germany, 2010, pp. 1–6.

Mourikis, A. I., Roumeliotis, S. I. Performance analysis of multirobot cooperative localization. *IEEE Transactions on Robotics* 22(4), 2006: 666–681.

Mu, H. *Decentralized algorithms of cooperative navigation for mobile platforms*. PhD thesis, National University of Defense Technology, Changsha, China, 2010.

Nerurkar, E. D., Roumeliotis, S. I. *Distributed MAP Estimation Algorithm for Cooperative Localization*. MARS Lab, Department of Computer Science & Engineering, University of Minnesota, Minneapolis, MN, 2008.

Nerurkar, E. D., Roumeliotis, S. I., and Martinelli, A. Distributed maximum a posteriori estimation for multi-robot cooperative localization. *IEEE International Conference on Robotics and Automation*, Kobe, Japan, 2009, pp. 1375–1382.

Nicosia, J. Decentralized cooperative navigation for spacecraft. *Aerospace Conference*, Big Sky, MT, 2007, pp. 1–6.

Qu, Y., Zhang, Y. Cooperative localization against GPS signal loss in multiple UAVs flight. *Journal of Systems Engineering and Electronics* 22(1), 2011: 103–112.

Ribeiro, A., Giannakis, G. B., and Roumeliotis, S. I. SOI-KF: Distributed Kalman filtering with low-cost communications using the sign of innovations. *IEEE Transactions on Signal Processing* 54(12), 2006: 4782–4795.

Roumeliotis, S. I., Bekey, G. A. Distributed multirobot localization. *IEEE Transactions on Robotics and Automation* 18(5), 2002: 781–795.

Sharma, R. *Bearing-only Cooperative-Localization and Path-Planning of Ground and Aerial Robots*. PhD thesis, Brigham Young University, 2011.

Sharma, R., Taylor, C. Cooperative navigation of MAVs in GPS denied areas. *IEEE International Conference on Multisensor Fusion and Integration for Intelligent Systems*, Seoul, South Korea, 2008, pp. 481–486.

Song, Z., Mohseni, K. Cooperative underwater localization in ocean currents. *AAA Guidance, Navigation, and Control*, 2013, pp. 1090–1096.

Trawny, N., Roumeliotis, S. I., and Giannakis, G. B. Cooperative multi-robot localization under communication constraints. *IEEE International Conference on Robotics and Automation*, Kobe, Japan, 2009, pp. 4394–4400.

Walls, J. M., Eustice, R. M. An origin state method for communication constrained cooperative localization with robustness to packet loss. *The International Journal of Robotics Research* 33(9), 2014: 1191–1208.

Wang, L.G. *Research on Distributed Cooperative Localization for Multi-mobile Platforms under Communication Constraints. PhD thesis*, Tsinghua University, Beijing, China, 2017.

Wang, L., Wan, J. W., Liu, Y. H., and Shao, J. X. Cooperative localization method for multi-robot based on PF-EKF. *Science in China* 51(8), 2008: 1125–1137.

Zhou, X. S., Roumeliotis, S. I. Robot-to-robot relative pose estimation from range measurements. *IEEE Transactions on Robotics* 24(6), 2008: 1379–1393.

13

Multi-Vehicle Cooperative Range-Based Navigation

Daniela De Palma, Giovanni Indiveri, and Gianfranco Parlangeli

CONTENTS

13.1 Introduction

Localization and target tracking are the primary abilities of a mobile robot in many land [22,34], aerial [18], and marine [4,21] applications. The capability of an agent to localize itself with the necessary accuracy is a key feature for a wide number of applications of multi-agent robots.

In most land and air applications, localization and navigation algorithms can exploit advanced measurement solutions as, for example, the global positioning system (GPS), radar-based tracking systems, or computer vision techniques in conjunction with attitude heading reference systems (AHRS), gyros, accelerometers, and compass devices. However, several restrictions of the above techniques are often experienced in practice. For example, GPS solutions cannot be adopted in indoor environments. Moreover, in other important emerging scenarios, as underwater, trilateration-based systems exploiting electromagnetic-based communication can be prevented, and agents can rely only on acoustics. Underwater acoustic-based trilateration solutions as long baseline (LBL) systems require complex deployment operations.

For teams of cooperating vehicles, single-range localization algorithms have recently been studied for different reasons [14,29,32]. Autonomous vehicles moving in a common unstructured area can opportunistically measure their relative distances through time-of-flight (ToF) techniques associated with the acoustic/electromagnetic signals exchanged for communicating.

From a theoretic point of view, single-range measurements are described by a nonlinear algebraic map of the vehicles' positions, hence preventing the observability analysis from being performed with the well-known tools for linear systems. Observability analysis approaches available in the literature can be divided into two categories. The first one builds on a nonlinear dynamical systems approach. This leads to the analysis of local weak observability of the nonlinear system with the tools of differential algebraic geometry. A major contribution to observability for nonlinear systems is [15] where the fundamental ideas and results about local and weakly local observability are described: single-range localization studies building on differential geometric tools need to tackle the difficulties related to local and weakly local observability as opposed to the global observability concept known for linear systems. Such issues are clearly addressed, by example, in references [12,17,25]. A second approach consists in re-elaborating the nonlinear system equations to recast the observability problem as for a linear time varying system. This latter approach allows study of the global observability of the original system [6,9].

Cooperative navigation solutions exploiting single-range information have recently been described within underwater robotics scenarios in [11,32,33] as well as in more general settings in [7,29]. Indeed, since the milestone paper [28], the area of multirobot navigation and localization has received considerable attention (see, e.g., [2,3,5,14,26,27], among others) and is still a challenging topic, particularly in unstructured and harsh environments such as underwater. As illustrated in [23], relative position measurements can be exploited to improve dead reckoning localization of a team of unicycle robots.

Inspired by the above considerations, the goal of the present work is to extend the results in [16] to a multi-vehicle framework. More precisely, we consider a scenario where vehicles use a communication network to exchange range-only measurements to improve localization. As opposed to the solution presented in [23], we exploit the geometrical constraint associated with the relative positions of the vehicles. Indeed, such geometrical constraint provides additional information that can be incorporated in the state estimation framework, reducing uncertainty. Contrary to the approach in [23], in the present work vehicles performing the localization are assumed to measure only intra-vehicle ranges rather than relative position. Possible applications to different scenarios involving vehicles moving on land, in water, or in air are also discussed. The main results and methods described in this chapter are based on [8]. This chapter is structured as follows: the description of the system model and the problem formulation are addressed in Section 13.2. Observer design is illustrated in Section 13.3, while numerical results are reported in Section 13.4. Section 13.5 addresses possible applications in different scenarios. Finally, concluding remarks are reported in Section 13.6.

13.2 Modeling

Prior to formalizing the problem under investigation, a few details regarding the adopted notation and terminology are introduced. The symbol $\otimes$ is used for the Kronecker product [20] between two matrices, namely given $A \in \mathbb{R}^{n \times m}$ and $B \in \mathbb{R}^{p \times q}$, the matrix $A \otimes B \in \mathbb{R}^{np \times mq}$ is defined as:

$$A \otimes B = \begin{bmatrix} a_{11}B & \cdots & a_{1m}B \\ \vdots & \ddots & \vdots \\ a_{n1}B & \cdots & a_{nm}B \end{bmatrix}. \tag{13.1}$$

A block diagonal matrix is represented as diag($A_1, \ldots, A_n$). A graph is defined as $\mathcal{G} = (\mathcal{V}, \mathcal{E})$, where $\mathcal{V} = \{1, \ldots, n\}$ is the set of nodes and $\mathcal{E} \subseteq \mathcal{V} \times \mathcal{V}$ is the set of edges. A path $\mathcal{P} = (\mathcal{V}_\mathcal{P}, \mathcal{E}_\mathcal{P})$ is a subgraph of $\mathcal{G}$ where $\mathcal{V}_\mathcal{P} = \{i_1, \ldots, i_k\}$ and $\mathcal{E}_\mathcal{P} = \{(i_\ell, i_{\ell+1}) : \ell = 1, \ldots, k-1, (i_\ell, i_{\ell+1}) \in \mathcal{E}\}$. A graph $\mathcal{G} = (\mathcal{V}, \mathcal{E})$ is connected if there exists a path connecting each $h, g \in \mathcal{V}$. A cycle is defined as a path where $i_k = i_1$ and $|\mathcal{V}_\mathcal{P}| \geq 3$. A cycle $\bar{C}$ is linearly independent from a given a set of cycles if there exists one edge in $\bar{C}$ that does not belong to any cycle of the set. A maximal set of linearly independent cycles is called a cycle basis. Given a connected graph $\mathcal{G} = (\mathcal{V}, \mathcal{E})$, the number of independent cycles is equal to $|\mathcal{E}| - |\mathcal{V}| + 1$. For further details about graph theory, refer to [13].

13.2.1 Problem Formulation

Consider n agents with position $\mathbf{x}_i \in \mathbb{R}^3$ for $i = 1, 2, \ldots, n$ and velocity $\mathbf{v}_i \in \mathbb{R}^3$ expressed in a common frame $\mathcal{I}$. The agents are assumed to know (eventually through a standard on-board navigation system) their own velocity and their attitude with reference to the common frame $\mathcal{I}$; namely, the rotation matrix ${}^{\mathcal{I}}R_i \in SO(3)$ from the body frame i to $\mathcal{I}$. The knowledge of ${}^{\mathcal{I}}R_i$ allows each agent to map its measured velocity as acquired in body frame to the common frame $\mathcal{I}$. Hence in the following, unless otherwise specified, all vectors are assumed to be expressed in the common frame $\mathcal{I}$. The evolution of the agents' positions is modeled as follows:

$$\dot{\mathbf{x}}_i = \mathbf{v}_i \quad : \quad i = 1, 2, \ldots, n \tag{13.2}$$

$$\mathbf{z}_{ij} := \mathbf{x}_i - \mathbf{x}_j \tag{13.3}$$

$$\mathbf{v}_{ij} := \mathbf{v}_i - \mathbf{v}_j \tag{13.4}$$

such that

$$\dot{\mathbf{z}}_{ij} = \mathbf{v}_{ij} \tag{13.5}$$

$$y_{ij} = \| \mathbf{z}_{ij} \|^2. \tag{13.6}$$

Variables $\mathbf{z}_{ij}$ in Equation 13.3 represent relative positions among agents that are able to communicate bidirectionally with each other: such agents share their own velocity vector information through their communication channel, and they can acquire measurements of their relative Euclidean distance y_{ij} in Equation 13.6. The problem addressed in this chapter is the estimation of $\mathbf{z}_{ij}$ on behalf of the agent i (or j) based on relative range measurements y_{ij}. For a single pair of agents $(i\,j)$, the problem of estimating $\mathbf{z}_{ij}$ in the model (13.5–13.6) can be solved, by example, through the method described in [16]. Yet in the scenario under investigation, each agent may estimate its relative position with respect to several other agents, namely all those sharing a common communication link. Indeed, the different relative positions possibly estimated are not independent. In fact, the relative positions are subject to geometric constraints, associated with the connection topology. The connection topology of the agents can be represented through a relative position measurement graph (RPMG) introduced in [23]. In the following, a RPMG is intended as a simple graph $\mathcal{G}$, where the nodes set $\mathcal{V} = \{1, \ldots, n\}$ represents the n agents and the edges set $\mathcal{E}$ represents the communication links. The number of such communication links, denoted with m,

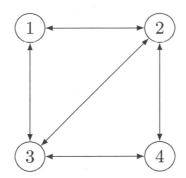

FIGURE 13.1
RPMG with $n=4$ agents and $m=5$ communication links.

corresponds to the number of relative positions $\mathbf{z}_{ij}$ to be estimated. Therefore, each cycle in the graph corresponds to a geometric constraint on the relative positions, and the cycle basis will map to a set of $m - n + 1$ independent geometric constraints. Consider, for example, a team of $n=4$ agents with $m=5$ communication links as illustrated in the graph in Figure 13.1. The relative positions involved in a such configuration are $\mathbf{z}_{12}, \mathbf{z}_{24}, \mathbf{z}_{43}, \mathbf{z}_{31}, \mathbf{z}_{23}$. As noticed, they should satisfy $m - n + 1 = 2$ independent geometric constraints, namely

$$\mathbf{z}_{12} + \mathbf{z}_{23} + \mathbf{z}_{31} = \mathbf{0}_{3\times1} \tag{13.7}$$

$$-\mathbf{z}_{23} + \mathbf{z}_{24} + \mathbf{z}_{43} = \mathbf{0}_{3\times1} \tag{13.8}$$

that can be rewritten as $D\mathbf{z}^* = \mathbf{0}_{3\times1}$ being

$$D = \begin{bmatrix} I_{3\times3} & 0_{3\times3} & 0_{3\times3} & I_{3\times3} & 0_{3\times3} \\ 0_{3\times3} & I_{3\times3} & I_{3\times3} & 0_{3\times3} & -I_{3\times3} \end{bmatrix} \in \mathbb{R}^{6\times15} \tag{13.9}$$

$$\mathbf{z}^* = (\mathbf{z}_{12}^\top \ \mathbf{z}_{24}^\top \ \mathbf{z}_{43}^\top \ \mathbf{z}_{31}^\top \ \mathbf{z}_{23}^\top)^\top \in \mathbb{R}^{15} \tag{13.10}$$

For the sake of generality, given a group of n agents with its connection topology, the associated independent constraint equations can be expressed as follows:

$$D\mathbf{z}^* = \mathbf{0}_{3(m-n+1)\times1} \tag{13.11}$$

being

$$\mathbf{z}^* = (\cdots \mathbf{z}_{ij}^\top \cdots)^\top \in \mathbb{R}^{3m} \quad \text{with } (i,j) \in \mathcal{E}, \tag{13.12}$$

$$D = A \otimes \mathbf{I}_{3\times3}, \tag{13.13}$$

$$A = (a_{lk}) \quad (l=1,\ldots,m-n+1; k=1,\ldots,m) \tag{13.14}$$

where $D \in \mathbb{R}^{3(m-n+1)\times3m}$, $A \in \mathbb{R}^{(m-n+1)\times m}$ and the generic element $a_{lk} \in \{-1, 0, +1\}$ depending on the constraint. It is worth highlighting that, given the independency of the geometric constraints, the matrix D can be assumed to be full rank; moreover, Equation 13.11 exists only if $n \geq 3$ and $m > n - 1$; that is, if there is at least one cycle in the RPMG. The agents connected through common RPMG cycles are assumed to share measurements and estimated values; hence, as illustrated in the following, all agents belonging to an RPMG cycle are able to perform the estimate of the corresponding $\mathbf{z}^*$ in Equation 13.12. It is worth noting that the solution of the problem stated above can be implemented resorting to a centralized architecture or, eventually, seeking a decentralized one. Details are provided in the following.

13.2.2 Projection Approach

If the estimates of the m vectors $\mathbf{z}_{ij}$ in Equation 13.12 are known, the constraint (13.11) can be incorporated in the estimation scheme through a projection approach as in [31]. In particular, given a symmetric positive definite weight matrix $W \in \mathbb{R}^{3m \times 3m}$, this would lead to computing an estimate $\hat{\mathbf{z}}_p^*$ satisfying the constraint (13.11) as

$$\hat{\mathbf{z}}_p^* = U\hat{\mathbf{z}}^* \tag{13.15}$$

where U is the projection operator

$$U := I_{3m \times 3m} - [W^{-1}D^\top (DW^{-1}D^\top)^{-1}]D \tag{13.16}$$

such that $U^2 = U$, $DU = 0_{3(m-n+1) \times 3m}$. As shown in [30], assuming $\hat{\mathbf{z}}^*$ to be a Kalman state estimate with posterior covariance $\mathrm{cov}(\mathbf{z}^* - \hat{\mathbf{z}}^*)$, if in Equations 13.15–13.16 the weight matrix W is chosen as $W = (\mathrm{cov}(\mathbf{z}^* - \hat{\mathbf{z}}^*))^{-1}$, then the estimate $\hat{\mathbf{z}}_p^*$ in Equation 13.15 is minimum variance, namely

$$\mathrm{cov}(\mathbf{z}^* - \hat{\mathbf{z}}_p^*) \leq \mathrm{cov}(\mathbf{z}^* - \hat{\mathbf{z}}^*); \tag{13.17}$$

whereas, if the weight matrix W is chosen as $W = I$, then the constrained estimate $\hat{\mathbf{z}}_p^*$ in Equation 13.15 is always closer to the true state than the unconstrained estimate, namely

$$\left\| \mathbf{z}^* - \hat{\mathbf{z}}_p^* \right\| \leq \left\| \mathbf{z}^* - \hat{\mathbf{z}}^* \right\|.$$

13.3 Observer Design

With reference to the problem of estimating the relative position $\mathbf{z}_{ij}(t)$ in the model (13.5–13.6), suppose the methods described in [16] are used to design an observer. Consider the integral of Equation 13.5

$$\mathbf{z}_{ij}(t) = \mathbf{z}_{ij}(t_0) + \int_0^t \mathbf{v}_{ij}(\tau)d\tau = \mathbf{z}_{ij}(t_0) + \mathbf{d}_{ij}(t) \tag{13.18}$$

having defined $\mathbf{d}_{ij}(t)$ as

$$\mathbf{d}_{ij}(t) := \int_{t_0}^t \mathbf{v}_{ij}(\tau)d\tau \in \mathbb{R}^{3 \times 1}. \tag{13.19}$$

Equation 13.18 allows to compute

$$\mathbf{z}_{ij}^\top(t_0)\,\mathbf{z}_{ij}(t_0) = y_{ij}(t_0) =$$

$$= y_{ij}(t) + \|\,\mathbf{d}_{ij}(t)\,\|^2 - 2\mathbf{d}_{ij}^\top(t)\mathbf{z}_{ij}(t) \tag{13.20}$$

that can be rewritten as follows

$$\overline{y}_{ij}(t) := \frac{1}{2}[y_{ij}(t) - y_{ij}(0) + \|\,\mathbf{d}_{ij}(t)\,\|^2] \tag{13.21}$$

$$\overline{y}_{ij}(t) = \mathbf{d}_{ij}^\top(t)\mathbf{z}_{ij}(t) \tag{13.22}$$

where the output signal $\overline{y}_{ij}(t)$ on the left-hand side of Equation 13.22 is a known quantity as well as the time varying vector $\mathbf{d}_{ij}(t)$ on the right side. Having introduced the time varying output Equation 13.22, the original non-linear system in Equations (13.5–13.6) can be equivalently represented by the linear time varying (LTV) system

$$\dot{\mathbf{z}}_{ij}(t) = \mathbf{v}_{ij}(t) \tag{13.23}$$

$$\overline{y}_{ij}(t) = \mathbf{d}_{ij}^\top(t)\mathbf{z}_{ij}(t). \tag{13.24}$$

The problem of estimating $\mathbf{z}_{ij}(t)$ in Equations (13.23–13.24) can thus be solved through a standard linear estimator (as a Kalman filter) for LTV systems. Notice that the output matrix $C(t) = \mathbf{d}_{ij}^\top(t)$ is a function of the input, hence observability depends on the velocity of the input $\mathbf{v}_{ij}$. For a detailed discussion on the observability properties of this system refer to [16]. Assuming to explicitly account for noise, and discretizing the system with a sampling time T_s, the resulting discrete time version of the LTV system is

$$\mathbf{z}_{ij}(k+1) = \mathbf{z}_{ij}(k) + \mathbf{v}_{ij}(k)T_s + \boldsymbol{\omega}(k) \tag{13.25}$$

$$\overline{y}_{ij}(k) = \mathbf{d}_{ij}^\top(k-1)\mathbf{z}_{ij}(k) + \epsilon(k) \tag{13.26}$$

where

$$\overline{y}_{ij}(k) = \frac{1}{2}[y_{ij}(k) - y_{ij}(0) + \|\,\mathbf{d}_{ij}(k-1)\,\|^2], \tag{13.27}$$

$$\mathbf{d}_{ij}(k-1) = \sum_{l=0}^{k-1}\mathbf{v}_{ij}(l)T_s, \tag{13.28}$$

$\boldsymbol{\omega}(k)$ and $\epsilon(k)$ are assumed to be zero mean white noises mutually independent with covariances $Q(k)$ and $R(k)$, respectively.

The Kalman filter estimate $\hat{\mathbf{z}}_{ij}$ for the model in Equations 13.25–13.26 results in

$$\hat{\mathbf{z}}_{ij}(k+1|k) = \hat{\mathbf{z}}_{ij}(k|k) + \mathbf{v}_{ij}(k)T_s \tag{13.29}$$

$$P_{ij}(k+1|k) = P_{ij}(k|k) + Q(k) \tag{13.30}$$

$$K = (P_{ij}^{-1}(k+1|k) + \mathbf{d}_{ij}(k)R(k+1)^{-1}\mathbf{d}_{ij}^{\mathsf{T}}(k))^{-1}\mathbf{d}_{ij}(k)R(k+1)^{-1} \tag{13.31}$$

$$\hat{\mathbf{z}}_{ij}(k+1|k+1) = \hat{\mathbf{z}}_{ij}(k+1|k) + K(\bar{y}(k+1) - \mathbf{d}_{ij}^{\mathsf{T}}(k)\mathbf{z}_{ij}(k+1|k)) \tag{13.32}$$

$$P_{ij}(k+1|k+1) = (P_{ij}^{-1}(k+1|k) + \mathbf{d}_{ij}(k)R(k+1)^{-1}\mathbf{d}_{ij}^{\mathsf{T}}(k))^{-1}. \tag{13.33}$$

In the scenario under investigation, owing to the information shared by the agents on the network, every agent has potentially access to the estimation of the m relative positions $\mathbf{z}_{ij}$ together with the respective covariances of the estimate error. Therefore, every agent can exploit the additional constraint (13.11) in order to obtain a more accurate estimation resorting to the projection approach described in the previous section. Defining the vector of the m Kalman estimates $\hat{\mathbf{z}}_{ij}$ as

$$\hat{\mathbf{z}}^*(k) = (\cdots \hat{\mathbf{z}}_{ij}^{\mathsf{T}}(k) \cdots)^{\mathsf{T}} \in \mathbb{R}^{3m} : (i,j) \in \mathcal{E}, \tag{13.34}$$

and the posterior covariance of the Kalman estimate $\hat{\mathbf{z}}^*$ as

$$P^*(k) = \mathrm{diag}(\cdots P_{ij}(k) \cdots) \in \mathbb{R}^{3m \times 3m} : (i,j) \in \mathcal{E} \tag{13.35}$$

and choosing a weight matrix $W = P^{*-1}(k)$, the projection approach defined in Equation 13.15 leads to the estimate $\hat{\mathbf{z}}_p^*(k)$ satisfying the constraint (13.11)

$$\hat{\mathbf{z}}_p^*(k) = \hat{\mathbf{z}}^*(k) - [P^*(k)D^{\mathsf{T}}(DP^*(k)D^{\mathsf{T}})^{-1}]D\hat{\mathbf{z}}^*(k). \tag{13.36}$$

The resulting error covariance is given by

$$P_p^*(k) = P^*(k) - P^*(k)D^{\mathsf{T}}(DP^*(k)D^{\mathsf{T}})^{-1}DP^*(k). \tag{13.37}$$

Interestingly, with the proposed formulation of the problem, even if an agent does not know all the m Kalman estimates $\hat{\mathbf{z}}_{ij}$, but only a subset, namely $\bar{m}$ Kalman estimates $\hat{\mathbf{z}}_{ij}$ (with $\bar{m} < m$), the $\bar{m}$ available Kalman estimates can still be projected considering only those constraints (cycles on the RPMG) that involve such relative positions. In any case, the resulting estimates $\hat{\mathbf{z}}_p^*$ are characterized by a minimum variance as highlighted in Equation 13.17.

13.3.1 Remark

It is worth noting that the new output $\bar{y}_{ij}(t)$ in Equation 13.21 depends on the very first measurement $y_{ij}(t_0)$. This dependency can impact on the robustness of the solution as a single bad measurement (as an outlier) at $t = t_0$ would affect significantly the output. A remedy to this issue can be found by periodically resetting the initial measurement $y(t_0)$ with $y(t)$. This would also prevent possible uncertainties in the knowledge of $\mathbf{v}_{ij}(t)$ from causing an unbounded bias in the displacement $\mathbf{d}_{ij}(t)$ in Equation 13.19 used to compute $\bar{y}_{ij}(t)$. In the discrete time case this would correspond to periodically mapping $y_{ij}(0) \rightarrow y_{ij}(k^*)$ as if the measurement had started at step k^* while the state estimate $\hat{z}_{ij}(k+1|k+1)$ continues its update dynamics.

The results presented in the following section refer to the discrete time case with periodic mapping of the initial measurement $y_{ij}(0)$ with $y_{ij}(k-1)$ (i.e., $k^* = k-1$). Consequently, the displacement in Equation 13.28 becomes $\mathbf{d}_{ij}(k-1) = \Sigma_{l=k^*}^{k-1} \mathbf{v}_{ij}(l)T_s = \mathbf{v}_{ij}(k-1)T_s$.

13.4 Results

The presented multi-vehicle relative localization technique has been tested through several numerical experiments. Different RPMG topologies have been considered showing the effectiveness of the proposed method. In particular, a simulation relative to a group of $n=4$ agents is reported here. Two different connection topologies have been considered with $m=4$ and $m=6$ communication links, respectively, as illustrated in Figure 13.2. The geometric constraints corresponding to the RPMG in Figure 13.2a are

$$Dz^* = \begin{bmatrix} I_{3\times3} & 0_{3\times3} & 0_{3\times3} & -I_{3\times3} & I_{3\times3} & 0_{3\times3} \\ 0_{3\times3} & I_{3\times3} & I_{3\times3} & 0_{3\times3} & -I_{3\times3} & 0_{3\times3} \\ I_{3\times3} & I_{3\times3} & 0_{3\times3} & 0_{3\times3} & 0_{3\times3} & I_{3\times3} \end{bmatrix} \begin{bmatrix} \mathbf{z}_{12} \\ \mathbf{z}_{24} \\ \mathbf{z}_{43} \\ \mathbf{z}_{13} \\ \mathbf{z}_{23} \\ \mathbf{z}_{41} \end{bmatrix} = \mathbf{0}_{9\times1} \qquad (13.38)$$

whereas for the RPMG in Figure 13.2b

$$Dz^* = [I_{3\times3} \quad I_{3\times3} \quad I_{3\times3} \quad -I_{3\times3}] \begin{bmatrix} \mathbf{z}_{12} \\ \mathbf{z}_{24} \\ \mathbf{z}_{43} \\ \mathbf{z}_{13} \end{bmatrix} = \mathbf{0}_{3\times1}. \qquad (13.39)$$

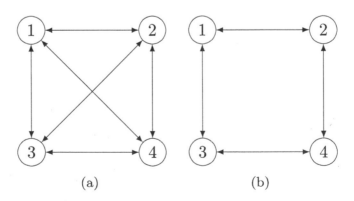

(a) (b)

FIGURE 13.2
(a) RPMG with $n=4$ agents and $m=6$ communication links. (b) RPMG with $n=4$ agents and $m=4$ communication links.

The following velocity inputs have been assigned to the 4 agents, considering a sampling time $Ts = 0.02$ [s]:

$\mathbf{v}_1(k) = (0.5 \cos(0.3\,k\,T_s),\, -0.5 \sin(0.3\,k\,T_s),\, -0.5)^\top$[m/s];
$\mathbf{v}_2(k) = (0,\, 1,\, 0)^\top$ [m/s];
$\mathbf{v}_3(k) = (-3 \sin(0.08\,k\,T_s) + 0.1,\, -3 \sin(0.07\,k\,T_s) + 0.1,\, \cos(0.5\,k\,T_s))^\top$ [m/s];
$\mathbf{v}_4(k) = (-5 \sin(0.1\,k\,T_s),\, 2 \cos(0.2\,k\,T_s),\, 2 \sin(0.3\,k\,T_s))^\top$ [m/s].

The initial positions of the agents are $\mathbf{x}_1(0) = (2,\, 2,\, 15)^\top$ [m]; $\mathbf{x}_2(0) = (12,\, 10,\, 12)^\top$ [m]; $\mathbf{x}_3(0) = (-5,\, 30,\, -5)^\top$ [m]; $\mathbf{x}_4(0) = (5,\, -20,\, 20)^\top$ [m]. The resulting trajectories are shown in Figure 13.3. Notice that the above velocity profiles $\mathbf{v}_i$ satisfy the observability conditions illustrated in [16] for the system (13.23–13.24). Therefore, the states $\mathbf{z}_{ij}$ can be estimated using the Kalman observer in Equations (13.29–13.33). The covariances of the state $\mathbf{z}_{ij}$ and the output $\bar{y}$ employed in the Kalman filter are $Q = 10^{-4} \cdot \mathrm{diag}(1, 1, 1)$ [m²] and $R = 0.25$ [m²], respectively. The initial Kalman filter state estimate is given by

$$\hat{\mathbf{z}}_{ij}(0) \sim \mathcal{N}(\mathbf{z}_{ij}(0), P_{ij}(0)), P_{ij}(0) = 0.9 \cdot \mathrm{diag}(1,1,1)[\mathrm{m}^2], \tag{13.40}$$

namely, $\mathbf{z}_{ij}(0)$ is the initial true state, and the initial condition $\hat{\mathbf{z}}_{ij}(0)$ of the filter is assigned randomly with covariance $P_{ij}(0)$.

With reference to the RPMG in Figure 13.2a, the estimate $\hat{\mathbf{z}}^* = (\hat{\mathbf{z}}_{12}^\top \ \hat{\mathbf{z}}_{24}^\top \ \hat{\mathbf{z}}_{43}^\top \ \hat{\mathbf{z}}_{13}^\top \ \hat{\mathbf{z}}_{23}^\top \ \hat{\mathbf{z}}_{41}^\top)^\top \in \mathbb{R}^{18}$ has been defined using the Kalman filter estimates. This estimate $\hat{\mathbf{z}}^*$ does not satisfy exactly the constraint (13.38). An estimate satisfying such a constraint is computed projecting $\hat{\mathbf{z}}^*$ in the direction P^{*-1} relative to the constraint surface through Equations 13.36–13.37. Figure 13.4 shows the evolution of the norm of the geometric constraints applied on both the Kalman estimates and the projected estimates, namely $\| D * \hat{\mathbf{z}}^* \| / \| \mathbf{z}^* \|$. As expected, the unconstrained Kalman estimate (dotted line) violates the equality constraint, whereas the projection onto the constraint surface provides estimates (continuous line) satisfying the equality constraint; that is $\| D * \hat{\mathbf{z}}^* \| / \| \mathbf{z}^* \| = 0$. Furthermore, the projection approach has the effect of reducing the covariance of the estimation. Indeed, $P_p^* - P^* < 0$. These results are clearly illustrated in Figure 13.5, where the maximum eigenvalue of the matrix $P_p^* - P^*$ is shown. Such eigenvalue

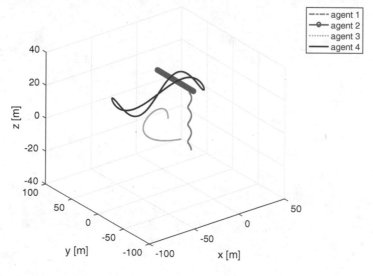

FIGURE 13.3
Trajectories of the agents.

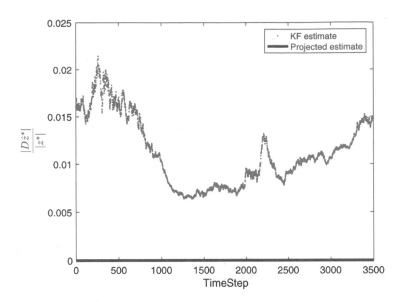

FIGURE 13.4
Equally constraint $\| D^* \hat{\mathbf{z}}^* \| / \| \mathbf{z}^* \| = 0$ for the RPMG in Figure 13.2a.

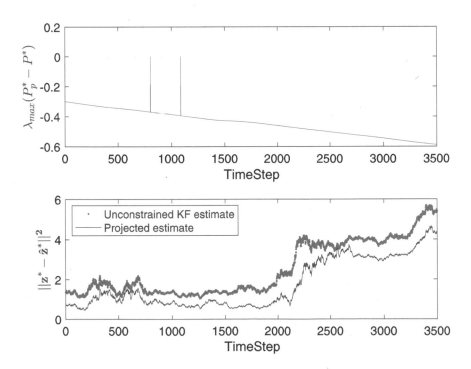

FIGURE 13.5
Maximum eigenvalue of the matrix $P_p^*(k) - P^*(k)$ (top plot) and norm of the estimation errors (bottom plot) for the RPMG in Figure 13.2a.

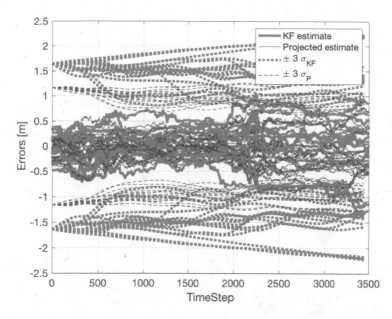

FIGURE 13.6
Errors and $\pm 3\sigma$ values of the covariance in the position estimates for the relative positions $\mathbf{z}^*$ for the RPMG in Figure 13.2a.

is always negative (not positive), confirming the reduction in the uncertainty achieved by exploiting the geometric constraints. Note that, in terms of norm of the estimation error, the projected estimates provide results comparable with the ones corresponding to the unconstrained Kalman estimates, if not even better in most cases, as visible in Figure 13.5. In order to provide an overview of the effects of including the geometric constraints in the relative localization problem, Figure 13.6 illustrates the estimation errors of each component of the state $\mathbf{z}^* \in \mathbb{R}^{18}$ for both estimates. The unconstrained Kalman estimates are plotted with a thick continuous line and the projected ones with a continuous line. On the same figure, the $\pm 3\sigma$ values of the errors' covariance are depicted with dotted lines for the unconstrained Kalman estimates and with dashed lines for the projected ones. It is worth highlighting that the estimation errors are bounded by the respective $\pm 3\sigma$ values and that, in accordance with the theoretical analysis, the uncertainty region described by the $\pm 3\sigma$ bounding lines of the projected estimates is smaller than the one associated to the unconstrained Kalman estimate. For the sake of clarity, the estimation errors and the $\pm 3\sigma$ bounding lines for the first component of the state $\mathbf{z}^* \in \mathbb{R}^{18}$ are reported in Figure 13.7. The same numerical experiment has been performed for the RPMG graph in Figure 13.2b characterized by a single geometric constraint. The results are similar to the ones obtained for the graph in Figure 13.2a. Figure 13.8 illustrates the evolution of the maximum eigenvalue of the matrix $P_p^* - P^*$ and the norm of the estimation errors. It is worth emphasizing that even the inclusion of a single geometric constraint in the estimation problem provides a reduction of the uncertainty region for the relative localization. Although the network topology, and consequently the D matrix of the geometric constraint (13.11), has been considered constant throughout the chapter, should the RPMG change during a mission, the proposed filter estimation architecture could still be adopted by dynamically redefining the overall state vector $\mathbf{z}^*$ and updating the D constraint matrix in compliance with the new RPMG. Details are not included for the sake of brevity.

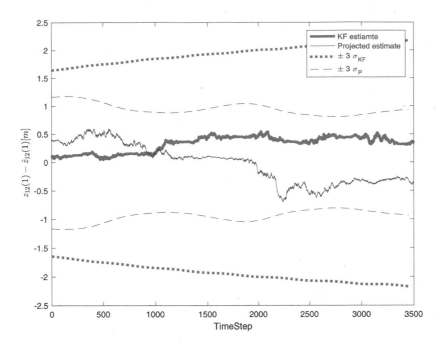

FIGURE 13.7
Errors and $\pm 3\sigma$ values of the covariance in the position estimates for the first component of the state $\mathbf{z}^*$ for the RPMG in Figure 13.2a.

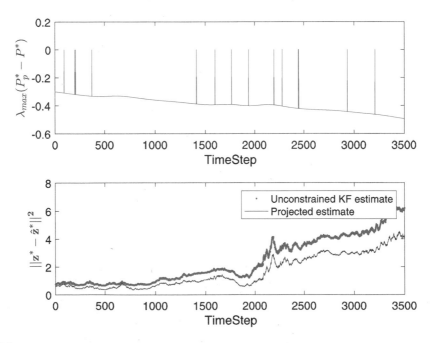

FIGURE 13.8
Maximum eigenvalue of the matrix $P_p^*(k) - P^*(k)$ (top plot) and norm of the estimation errors (bottom plot) for the RPMG in Figure 13.2b.

13.5 Application Scenarios: Land, Aerial, and Underwater Environment

The proposed navigation solution may be applied to different scenarios involving vehicles possibly moving on land, in water, or in air. In all cases, the implementation of described cooperative algorithms requires equipping the vehicles with range and velocity sensors and sharing the information through a communication network among the agents.

The technology available for communicating and measuring relative ranges depends on the specific scenario. Indeed, for land and aerial agents, radio frequency-based technologies can be exploited, whereas in an underwater environment no electromagnetic signals can be used and only lower bandwidth acoustic-based solutions can be used.

Moreover, in terms of how to share the necessary information among the agents to implement the proposed cooperative navigation solution, different approaches can be adopted. In essence, they can be classified as centralized or decentralized communication architectures. In the centralized approach, a *leader* agent collects all the information required for the estimation schema and shares the resulting relative positions among the agents. In the decentralized approach, each agent communicates only with its neighborhood and estimates only the relative positions associated to the RPMG cycle where it is involved. In both cases, all agents involved should be equipped with synchronized clocks so as to use one-way travel time (OWTT) range measurement schemas. Then, a time-division multiple access (TDMA) scheme can be employed to access the communication channel. It should be noted that the duration of the time slots depends on the available bit rate and the specific communication protocol. In the following analysis, the connection topology of the agents is assumed to be unknown and time-invariant (or slowly time-varying).

In the centralized approach, the needed range information to run the filter can be distributed in the team of n members as follows: agents sequentially (one in each time slot) broadcast a data packet containing their identifying label and a time stamp. When all agents but one (i.e., after $n-1$ time slots) will be done, the knowledge of all m relative Euclidean distances (13.6) is distributed among the agents. At this point, each agent pings to the leader agent a data packet containing its own velocity expressed in the common frame $\mathcal{I}$ and the set of relative Euclidean distances collected during the previous $n-1$ communication slots. After $n-1$ time slots, the knowledge of all m relative Euclidean distances (13.6) is distributed among the agents. At this point, each agent pings to the leader agent a data packet containing its own velocity expressed in the common I frame and the set of relative Euclidean distances collected during the previous $n-1$ communication slots. Hence, after n time slots, the leader agent knows all the information—that is, relative velocity (13.4) and relative distances (13.6)—required to solve the estimation problem, taking into account the additional geometric constraints (13.11) associated to the connection topology. Once the estimates $\hat{\mathbf{z}}_p^*$ of the m relative positions are computed using the collected information, the leader agent broadcasts to all agents a data packet containing such estimates. Overall, $2n$ time slots are required to complete one estimation step. This kind of scaling appears to be most likely acceptable for most applications involving a limited number of vehicles.

In the decentralized approach, considering that the communication graph is unknown to the agents, a preliminary step is implemented with the aim of retrieving the graph topology, that is, the set of all its edges. Each agent during its time slot broadcasts a data packet containing its label and the set of links already identified, if any. All agents receiving the ping decode the data packet to obtain the list of identified links, the label of the transmitter agent, and identify the link between itself and the transmitter agent. This is repeated until all agents collect knowledge about the whole connection topology. In the worst case, $2n-1$ communication slots are required to ensure that all agents have identified the connection topology.

Then each agent can individually identify a cycle basis. Among the possible cycle bases, the basis with cycles involving the minor number of agents are preferred so as to reduce the amount of communications and the complexity. Once each agent has identified the cycles where it is involved, it can start a communication round with the neighborhoods belonging to such cycles. In such a communication round, each agent during its time slot broadcasts a data packet containing its label, its velocity, and the set of relative distances, if any. All agents receiving the ping decode the data packet to obtain the label of the transmitter agent, the velocity of the transmitter agent, the time of the ping, and the set of relative distances, and compute its relative Euclidean distance to the transmitter agent.

Once an agent has collected all the information related to its own cycles (in terms of agents' velocity and relative distances), the projection-based estimation can be performed. For example, for a cycle having 3 agents, 2 communication slots are required to collect all the information and proceed with the estimation, whereas after 3 communication slots all agents of the cycle can proceed with the estimation. Similarly, it can be proven that for a cycle having 4 agents, 4 communication slots are required to collect all the information and proceed with the estimation, whereas after 6 communication slots all agents of the cycle can proceed with the estimation.

As for typical measurement and communication rates, these are strongly correlated to the specific technology adopted. A brief overview of current technologies for the different environments is addressed in the following subsection.

13.5.1 Land/Aerial Environment

In an aerial scenario, typical velocities for small unmanned aerial vehicles are about 3–4 [m/s], these velocities can increase considerably for ground vehicles. The communications are based on electromagnetic signals. In ground environment a common wireless communication standard is ITS-G5 (5.9 GHz) characterized by a nominal bit rate of 6 Mbps [10]. Other sensors commonly used in aerial environment are the ultra-wideband (UWB) radio modules (3.1–10.6 GHz). They have been used, for example, for quadrocopter state estimation in [24].

13.5.2 Underwater Environment

In an underwater scenario, the velocities for small autonomous underwater vehicles are typically low and do not exceed 1 or 2 [m/s]. Communications are based on acoustic signals, and the associated channel bandwidth depends on the specific acoustic modem. Moreover, the delay related to the travel time of acoustic signals may be significant for larger distances and should be accounted for. Indeed, sound speed underwater is approximately 1500 [m/s], namely about six orders of magnitude lower than the speed of electromagnetic signals in air.

Examples of acoustic modems commonly used in underwater environment are the middle frequency (MF) modems (18–34 kHz) [19]. They have been recently used for underwater positioning during geotechnical survey experiments performed within a European project [1]. Such modems are characterized by a nominal bit rate in the range 3.10–3.85 kbps.

13.6 Conclusions

Building on a single range localization solution described in [16] for a single vehicle, a cooperative localization filter has been proposed for a networked group of vehicles. The distinctive features of the proposed solution are (i) the use of intra-vehicle range (rather than relative position) measurements, and (ii) the explicit inclusion in the estimation algorithm of geometric constraints associated to the unknown relative vehicle positions. The geometric constraints provide additional information with respect to the available measurements and to the state equations, hence they are able to improve the performance of the estimation by reducing the estimate covariance. The results of numerical simulations are described showing the effectiveness of the proposed solution. Possible applications to different scenarios, that is, land, aerial, and underwater, are also discussed.

References

1. P. Abreu, G. Antonelli, F. Arrichiello, A. Caffaz, A. Caiti, G. Casalino, N.C. Volpi et al. Widely scalable mobile underwater sonar technology: An overview of the H2020 WiMUST project. *Marine Technology Society Journal (Research Initiatives in Europe: Cooperation for Blue Growth)*, 50(4):42–53, July/August 2016. Guest Editors: Andrea Caiti, Giuseppe Casalino and Andrea Trucco.
2. B. Allotta, R. Costanzi, E. Meli, L. Pugi, A. Ridolfi, and G. Vettori. Cooperative localization of a team of AUVs by a tetrahedral configuration. *Robotics and Autonomous Systems*, 62(8):1228–1237, August 2014.
3. T. Arai, E. Pagello, and L.E. Parker. Editorial: Advances in multi-robot systems. *IEEE Transactions on Robotics and Automation*, 18(5):655–661, 2002.
4. F. Arrichiello, G. Antonelli, A.P. Aguiar, and A. Pascoal. Observability metric for the relative localization of AUVs based on range and depth measurements: Theory and experiments. In *Intelligent Robots and Systems (IROS), 2011 IEEE/RSJ International Conference on*, pages 3166–3171, September 2011.
5. A. Bahr, J.J. Leonard, and M.F. Fallon. Cooperative localization for autonomous underwater vehicles. *International Journal of Robotics Research*, 28(6):714–728, 2009.
6. P. Batista, C. Silvestre, and P. Oliveira. Single range aided navigation and source localization: Observability and filter design. *Systems & Control Letters*, 60:665–673, 2011.
7. M. Cao, C. Yu, and B.D.O. Anderson. Formation control using range-only measurements. *Automatica*, 47(4):776–781, 2011.
8. D. De Palma, G. Indiveri, and G. Parlangeli. Multi-vehicle relative localization based on single range measurements. In *3rd IFAC Workshop on MultiVehicle System - MVS 2015*, volume 48, pages 17–22, Genova, Italy, May 18, 2015.
9. D. De Palma, F. Arrichiello, G. Parlangeli, and G. Indiveri. Underwater localization using single beacon measurements: Observability analysis for a double integrator system. *Ocean Engineering*, 142:650–665, September 2017.
10. F. de Ponte Müller, E.M. Diaz, and I. Rashdan. Cooperative positioning and radar sensor fusion for relative localization of vehicles. In *IEEE Intelligent Vehicles Symposium (IV 2016)*, pages 1060–1065, June 2016.
11. M.F. Fallon, G. Papadopoulos, J.J. Leonard, and N.M. Patrikalakis. Cooperative AUV navigation using a single maneuvering surface craft. *International Journal of Robotics Research*, 29(12):1461–1474, 2010.

12. A.S. Gadre and D.J. Stilwell. Toward underwater navigation based on range measurements from a single location. In *Proceedings of IEEE International Conference on Robotics and Automation, 2004 (ICRA 2004)*, volume 5, pages 4472–4477, New Orleans, LA, USA, April 26 to May 1, 2004.

13. C.D. Godsil and G. Royle. *Algebraic graph theory*, volume 207. Springer New York, 2001.

14. T. Halsted and M. Schwager. Distributed multi-robot localization from acoustic pulses using euclidean distance geometry. In *2017 International Symposium on Multi-Robot and Multi-Agent Systems (MRS)*, pages 104–111. IEEE, December 2017.

15. R. Hermann and A.J. Krener. Nonlinear controllability and observability. *IEEE Transactions on Automatic Control*, 22(5):728–740, October 1977.

16. G. Indiveri, D. De Palma, and G. Parlangeli. Single range localization in 3-D: Observability and robustness issues. *Control Systems Technology, IEEE Transactions on*, 24(5):1853–1860, 2016.

17. J. Jouffroy and J. Reger. An algebraic perspective to single-transponder underwater navigation. In *Computer Aided Control System Design, 2006 IEEE International Conference on Control Applications, 2006 IEEE International Symposium on Intelligent Control*, pages 1789–1794. IEEE, 2006.

18. Z.M. Kassas and T.E. Humphreys. Observability analysis of opportunistic navigation with pseudorange measurements. In *AIAA Guidance, Navigation, and Control Conference, AIAA GNC*, 2012.

19. K.G. Kebkal, O.G. Kebkal, V.K. Kebkal, A.M. Pascoal, J. Ribeiro, G. Indiveri, and S. Jesus. Performance assessment of underwater acoustic modems operating simultaneously at different frequencies in the presence of background impulsive noise emitted by a sparker. In *4th Underwater Acoustics Conference and Exhibition (UACE2017)*, pages 325–334, 2017.

20. A.J. Laub. *Matrix Analysis for Scientists and Engineers*. SIAM, 2005.

21. T. Maki, T. Matsuda, T. Sakamaki, T. Ura, and J. Kojima. Navigation method for underwater vehicles based on mutual acoustical positioning with a single seafloor station. *IEEE Journal of Oceanic Engineering*, 38(1):167–177, 2013.

22. A. Martinelli and R. Siegwart. Observability analysis for mobile robot localization. In *2005 IEEE/RSJ International Conference on Intelligent Robots and Systems*, pages 1471–1476, Aug. 2005.

23. A.I. Mourikis and S.I. Roumeliotis. Performance analysis of multirobot cooperative localization. *Robotics, IEEE Transactions on*, 22(4):666–681, 2006.

24. M.W. Mueller, M. Hamer, and R. D'Andrea. Fusing ultra-wideband range measurements with accelerometers and rate gyroscopes for quadrocopter state estimation. In *IEEE International Conference on Robotics and Automation (ICRA 2015)*, pages 1730–1736, Seattle, Washington, May 2015.

25. A. Ross and J. Jouffroy. Remarks on the observability of single beacon underwater navigation. In *International Symposium on Unmanned Untethered Submersible Technology (UUST 05)*, Durham, NH, August 2005.

26. S.I. Roumeliotis and G.A. Bekey. Collective localization: A distributed Kalman filter approach to localization of groups of mobile robots. In *Robotics and Automation, 2000. Proceedings. ICRA'00. IEEE International Conference on*, volume 3, pages 2958–2965. IEEE, 2000.

27. S.I. Roumeliotis and G.A. Bekey. Distributed multirobot localization. *Robotics and Automation, IEEE Transactions on*, 18(5):781–795, 2002.

28. A.C. Sanderson. A distributed algorithm for cooperative navigation among multiple mobile robots. *Advanced Robotics*, 12(4):335–349, 1997.

29. I. Sarras, J. Marzat, S. Bertrand, and H. Piet-Lahanier. Collaborative multiple micro air vehicles' localization and target tracking in gps-denied environment from range-velocity measurements. *International Journal of Micro Air Vehicles*, 10(2):225–239, 2018.

30. D. Simon and T.L. Chia. Kalman filtering with state equality constraints. *Aerospace and Electronic Systems, IEEE Transactions on*, 38(1):128–136, January 2002.

31. D. Simon. *Optimal State Estimation: Kalman, H Infinity, and Nonlinear Approaches*. John Wiley & Sons, 2006.

32. J.M. Soares, A. Pedro Aguiar, A.M. Pascoal, and A. Martinoli. Joint ASV/AUV range-based formation control: Theory and experimental results. In *2013 IEEE International Conference on Robotics and Automation (ICRA 2013)*, pages 5579–5585, Karlsruhe, Germany, 6-10 May, 2013.

33. S.E. Webster, J.M. Walls, L.L. Whitcomb, and R.M. Eustice. Decentralized extended information filter for single-beacon cooperative acoustic navigation: Theory and experiments. *IEEE Transactions on Robotics*, 29(4):957–974, 2013.

34. X.S. Zhou and S.I. Roumeliotis. Robot-to-robot relative pose estimation from range measurements. *IEEE Transactions on Robotics*, 24(6):1379–1393, 2008.

14

Precise Pedestrian Positioning by Using Vehicles as Mobile Anchors

Suhua Tang and Sadao Obana

CONTENTS

14.1 Introduction and Background

On the roads dominated by high speed vehicles, pedestrians are susceptible to severe injury or even death in crashes with vehicles, and are known to be "weak in traffic." An annual report of traffic accident statistics [1] shows that more than one-third of people killed in fatal vehicle crashes in Japan are pedestrians.

A great deal of effort has been devoted to reducing pedestrian accidents, such as (i) a vehicle itself is equipped with many on-board sensors such as millimeter-wave-radar or stereo cameras, by which it can independently detect a nearby pedestrian, and (ii) a pedestrian estimates his own position and sends it to nearby vehicles by pedestrian-to-vehicle communication (PVC) [2]. This helps a driver learn the presence of a pedestrian behind a building and take actions to avoid a potentially fatal crash. Here, PVC, not requiring roadside infrastructure and working well even without line of sight (LOS), is a good supplement to radar, etc.

PVC is a necessity even in the era of self-driving. The first fatal accident involving a pedestrian caused by a self-driving car, occurred on March 18, 2018, in Tempe, Arizona [3]. Just before the crash, a female pedestrian was walking a bicycle across the road. The car approaching her was in the self-driving mode. There was a driver aboard for the purpose of handling emergent situations (in other words, this is a level-3 driving automation ["eyes off"] according to the Society of Automotive Engineer's [SAE's] definitions). Although the on-board radar and LiDAR systems detected the pedestrian 6 seconds before the crash, the accident still happened due to both design flaws and technical defects. As for the latter, the pedestrian was first detected as an unknown object, then a vehicle, and later a bicycle, but not as a pedestrian at all. In such cases, an explicit notification of pedestrian presence to vehicles via PVC may help a self-driving vehicle make a better decision.

PVC consists of two main components: one is positioning, and the other is transmission control. In this system, each pedestrian periodically transmits his position to notify nearby vehicles of his presence. At large intersections crowded with pedestrians, if all pedestrians transmit frequently with the same interval, this may lead to congestion in the shared communication channel. As a result, pedestrians truly facing potential risks may fail to deliver their messages to nearby vehicles. One of the solutions to this problem is context-aware transmission control [2]. The degree of risk of a pedestrian is computed based on his context (e.g., the venue [position] where he is, and the time-to-collision [position, speed] with respect to vehicles), and on this basis transmission control is performed, causing pedestrians at high risk to transmit with high priority. In this way, transmission control for a pedestrian also heavily depends on his positioning function.

The position of a pedestrian is usually computed by his mobile device, based on GPS signals, which relies on measuring the direct distances between satellites and the GPS receiver. However, in urban canyons, direct paths between satellites and a GPS receiver might be obstructed by tall buildings, and a GPS receiver may not find four satellites for computing its position, which leads to an outage of positioning. Although different satellite systems (GPS, GLONASS, Beidou, Galileo, etc.) can be combined to increase the number of satellites, the directly visible satellites usually are overhead, with a poor distribution. This limits the effect of improving positioning precision. A high-sensitivity GPS receiver helps to partially mitigate the outage of positioning but tends to leverage reflected signals, which, however, have large errors (known as multipath error) in the estimated distance and greatly degrade positioning precision. Some methods choose not to use reflected signals at all. This, however, again leads to the outage of positioning. Pedestrian dead-reckoning (PDR) [4] is a practical method that relies on other means, such as pedestrian speed. To remove the accumulated error, however, the position must be corrected by using satellite signals.

In this chapter, we show how to compute pedestrian position more precisely by using wireless signals from vehicles. Vehicles periodically exchange position information to learn the presence of nearby vehicles so as to avoid collisions [5], and their position precision will be very high in the era of self-driving. It is assumed that a pedestrian device can overhear position messages from vehicles. By measuring pedestrian-vehicle distances, a pedestrian device can use vehicles as anchors (pseudo-satellites) for computing the position, which helps to solve the problem of satellite shortage and improve the distribution of anchors for positioning. Figure 14.1 shows the system framework. Signals from vehicles (*A*, *B*, and *C*) with LOS (line-of-sight) paths are used to compute the position of a pedestrian with high precision. Then, the position information of this pedestrian is sent to all vehicles including those (*D*) without LOS paths, by the reflection and/or diffraction of wireless signals. It should be noted that, in addition to vehicles, roadside units (RSUs) sharing the same communication protocols can also be used as anchors. Here, the focus is (i) how to accurately estimate the pedestrian-vehicle distance and (ii) how to exploit the sequential measurements of distances in position computing.

The rest of this chapter discusses the following topics: (i) why it is necessary and feasible to use vehicles as anchors in computing pedestrian position, (ii) how to compute

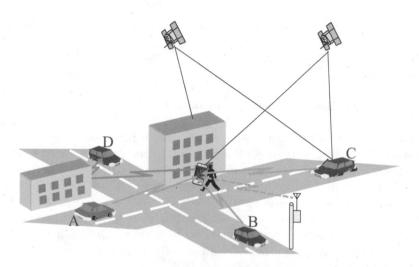

FIGURE 14.1

System framework of pedestrian positioning in urban canyons: using vehicles (and roadside units) as anchors. Signals from vehicles (*A*, *B*, and *C*) with LOS paths are used to compute the position of a pedestrian with high precision. Then, the position information of this pedestrian is sent to all vehicles including those (*D*) without LOS paths.

pedestrian-vehicle distance, (iii) how to detect whether there is a LOS component in the received signal, and (iv) how to compute pedestrian position, in the standalone mode or using a Kalman filter.

14.2 Why Use Vehicles as Anchors

As for using vehicles as anchors, the necessity is explained in Section 14.1—to solve the outage of positioning. Actually, it is also feasible to use vehicles as anchors. Compared with pedestrians with only simple portable devices, vehicles have other auxiliary means to improve their position precision with different sensors, as follows.

1. *Speedometer*. Different from pedestrians, vehicles move along roads (lanes), and in the moving direction, vehicle speed can be accurately measured by a speedometer (from rotational speed of wheels) and used for dead-reckoning (current position is estimated based on an initial position and moving speed/direction).

2. *Map matching*. In the vertical (lateral) direction, vehicle positions are not arbitrary but limited to roads, and often matched to the nearest road when a digital map is available.

3. *Camera/LiDAR*. Vehicle position can be further constrained to lanes by using cameras or LiDAR for lane detection.

In [6], a method is suggested to integrate a 3D map (for detecting LOS path of satellite signals and removing satellites that are not directly visible), inertial measurement units (for measuring moving direction of vehicles), speedometer (for vehicle speed), and camera-based lane detection via a particle filter, for the purpose of computing accurate vehicle position. This method achieves an average positioning error of 0.75 m in the urban area of Tokyo, although its instantaneous error may be as large as 3 meters. The development of self-driving surely will further improve the positioning precision of vehicles.

Compared with satellites, using vehicles as anchors has other merits, as follows.

1. *Better anchor placement.* High precision positioning requires a good (balanced) placement of anchors, which decides dilution of precision (DOP). DOP is a metric that reflects how errors in the distance measurement will affect the final position estimation. From the viewpoint of DOP, satellites with low elevation angles are desired, but they are more susceptible to obstructions in urban canyons. Satellites with high elevation angles are less susceptible to obstructions but limiting the distribution of satellites overhead will lead to a high DOP, and potentially large errors in the computed position. In comparison, vehicles on the road have nearly zero elevation angles, with a much smaller DOP.

2. *High availability.* Usually there are more vehicles available than satellites in urban canyons, where most satellite signals may be reflected ones and cause large multipath error in their estimated distances. Here, using vehicles as anchors make it safer to remove invisible satellites from position computation.

3. *More dynamics.* Because of a very long distance between a satellite and a pedestrian, the relative direction of a satellite with respect to a pedestrian does not change very much within a short time. In other words, the obstruction of a satellite by roadside buildings cannot be removed quickly. In contrast, the relative position between vehicles and pedestrians may change greatly within a short time (Figure 14.2), and the obstruction state of vehicles can be removed quickly. Meanwhile, vehicles transmit position information periodically (default period is 100 ms). Then, the frequent measurements of pedestrian-vehicle distances at different locations can be used to smoothen the estimated pedestrian position, by applying a Kalman filter.

FIGURE 14.2
Comparison between vehicles and satellites in the role as anchors.

14.3 Distance Estimation by Received Signal Strength Indicator

The estimation of pedestrian-vehicle distance is a key factor when using vehicles as anchors. A conventional method for distance estimation is to exploit the attenuation property of signal strength, received signal strength indicator (RSSI). According to [7], the path loss $PL(d)$ at a particular distance d is defined by the log-normal distribution, as follows:

$$PL(d)[dB] = \overline{PL(d_0)} + 10n \cdot \log_{10}\left(\frac{d}{d_0}\right) + X_\sigma, \tag{14.1}$$

where $\overline{PL(d_0)}$ is the average path loss at reference point d_0, and n is the path loss exponent. X_σ is a zero-mean Gaussian distributed variable reflecting the random attenuation caused by shadowing. The signal strength $P_r(d)$ is defined as the difference between the transmission power P_t (including antenna gain) and the path loss.

$$P_r(d)[dBm] = P_t[dBm] - PL(d)[dB]. \tag{14.2}$$

The above model reflects the variation of RSSI over distance due to path loss and shadowing. Because RSSI is the overall signal strength, which includes potential LOS component, reflected and/or diffracted components with different attenuations and random phases, in multipath-rich environments, RSSI fluctuates on the order of signal wavelength and affects the precision of estimated distance.

Due to log distance, RSSI attenuates quickly at short distances and more slowly at long distances. Substituting (14.1) into (14.2) and taking the differentiation operation on both sides leads to the following equation.

$$\Delta d[m] = -\frac{\ln 10}{10n} \cdot d[m] \cdot \Delta P_r[dBm] \tag{14.3}$$

with a given variation (measurement error) in signal strength (ΔP_r), the error in distance estimation (Δd) linearly increases with the distance. This should be taken into account in computing positions.

14.4 Distance Estimation by Channel State Information

The large error in distance estimated from RSSI is caused by the multipath components inside RSSI. Then, a natural idea to reduce distance error is to separate the direct (LOS) component from other reflected ones and use it for distance estimation. This requires obtaining the channel state information (CSI).

14.4.1 What Is CSI

The signal from a vehicle arrives at a pedestrian device as different components via different paths, each with its own propagation delay and signal strength, as shown in Figure 14.3. Here, CSI describes how a signal propagates and represents the combined effect

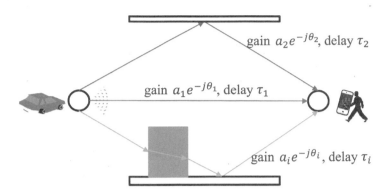

FIGURE 14.3
Multipath propagation of wireless signals.

of scattering, fading, and power decay with distance [8]. In the time domain it is defined as channel impulse response (CIR)

$$h(t) = \sum_{i=1}^{N} a_i e^{-j\theta_i} \delta(t - \tau_i), \tag{14.4}$$

where a_i, θ_i, and τ_i are the amplitude, phase, and time delay of the ith path, respectively. N is the number of multipath components, and $\delta(t)$ is the Dirac delta function. Each impulse represents a multipath component at delay τ_i, with a complex gain $a_i e^{-j\theta_i}$.

The received signal $r(t)$ is the convolution of transmitted signal $s(t)$ and the CIR $h(t)$:

$$r(t) = s(t) \otimes h(t) = \sum_{i=1}^{N} a_i e^{-j\theta_i} s(t - \tau_i). \tag{14.5}$$

In the frequency domain, the received signal spectrum $R(f)$ is the multiplication of the transmitted signal spectrum $S(f)$ and the channel frequency response (CFR) $H(f)$.

$$R(f) = S(f) \times H(f). \tag{14.6}$$

Theoretically, CIR can be derived directly from the deconvolution of received and transmitted signals. However, the calculation of CIR via deconvolution is costly, while CFR as the ratio of the received and the transmitted spectrums can be computed easily. Therefore, a common trick to derive CIR is to first compute CFR and then perform the inverse Fourier transform.

14.4.2 CSI from Off-the-Shelf WLAN Module

Vehicles usually use the OFDM modulation to transmit their messages, where a wide band is divided into orthogonal subcarriers. For the purpose of channel equalization, there are special training symbols that carry predefined data in all subcarriers. This is used to estimate the channel gain per subcarrier, which is simply the CFR.

Although CFR is exploited in the hardware for the low-level signal processing, there is no interface for accessing this information in the era of IEEE 802.11a/g. The function of beamforming has been introduced since IEEE 802.11n, where the WLAN driver may require

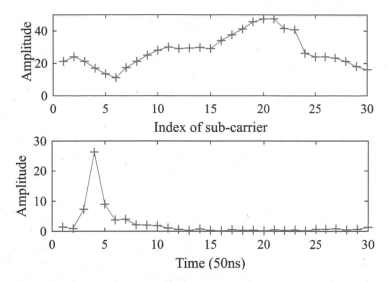

FIGURE 14.4
CFR (top) obtained by off-the-shelf WLAN module and the corresponding CIR (bottom).

collection of CFR information. A CSI tool, built on the Intel Wi-Fi wireless link (IWL) 5300 802.11n MIMO radios and using a custom modified firmware and open source Linux wireless drivers, has been made open for public use [9]. The IWL5300 device reports the CFR information for 30 subcarrier groups, which is about one group for every two subcarriers when the bandwidth is 20 MHz (IEEE 802.11a/g with 64 subcarriers). Each CFR entry is a complex number, with signed 8-bit resolution each for the real and imaginary parts. It specifies the gain and phase of the signal path between a single transmit-receive antenna pair. Figure 14.4 shows an example of the CFR amplitude (top), and the CIR amplitude (bottom).

CSI information from off-the-shelf WLAN modules at first was used for indoor positioning [10], where stationary Wi-Fi access points are used as anchors. In such cases, the position precision can be further improved by using fingerprints with CSI information [11] or using multiple antennas to get the angle of arrival information [12]. In comparison, this chapter will investigate how to use mobile vehicles as anchors for the outdoor positioning of pedestrians (with single antenna) in urban canyons.

Estimating CSI is similar to the direct multipath estimation [13] for GPS positioning, but it becomes simpler with the help of the OFDM modulation in Wi-Fi systems.

14.4.3 The Impact of Time Resolution

It is expected that in the signal received from a vehicle, the LOS component can be separated from the subsequent reflected/diffracted ones so that the strength of the LOS component can be used to compute the pedestrian-vehicle distance, following the same methods as in Section 14.3. However, the time resolution of off-the-shelf WLAN modules is limited; 50 ns when 20 MHz bandwidth is used.

Assume the time resolution of a receiver is T_s. All components that arrive with a time difference less than T_s will be regarded as one component, and the equivalent CSI can be computed as

$$h(t) = \sum_k b_k \delta(t - kT_s), \quad b_k = \sum_{kT_s \leq \tau_i < (k+1)T_s} a_i e^{-j\theta_i}. \tag{14.7}$$

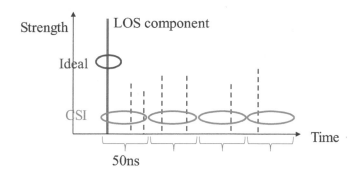

FIGURE 14.5
CSI under finite time resolution.

For example, in Figure 14.5, the LOS component and two reflected ones are regarded as one equivalent component whose signal strength is the overall strength of the three components. Signal strength of the equivalent component, although dominated by the LOS component, may be affected by the subsequent reflected components.

Several methods can be used to increase the time resolution of CSI information. One is to increase the bandwidth, e.g., 40 MHz in times of IEEE 802.11n, 80 MHz in times of IEEE 802.11ac. An alternative solution is to let a mobile device and its associated AP hop over all Wi-Fi channels to collect CFR over a very wide band [14]. This, however, affects the communication performance.

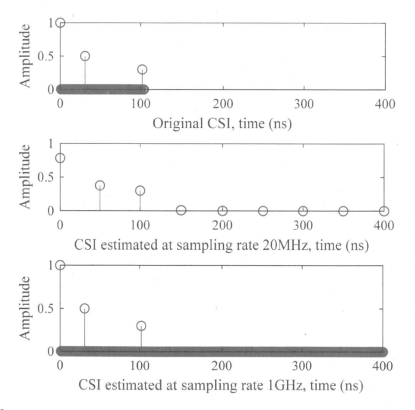

FIGURE 14.6
Improving time resolution by increasing the sampling rate at the receiver.

With the bandwidth fixed, directly increasing the sampling rate at a receiver also improves the time resolution. Figure 14.6 shows an example, where the original CSI cannot be recovered at the sampling rate 20 MHz but is recovered at the sampling rate 1 GHz. In such cases, besides the training symbols, the pulse shaping function should also be taken into account.

14.4.4 Distance Estimation by Support Vector Regression (SVR)

Distance estimation depends on the attenuation property of wireless signals, and usually the LOS component should be used. When the LOS component is overlapped by subsequent non-LOS component in the same bin, the overall strength fluctuates and affects the distance estimation. Although the LOS component and subsequent non-LOS components propagate via different paths, generally the longer the distance is, the smaller all CSI bins tend to be. Therefore, a model is trained to estimate the distance from all bins of a CSI vector, instead of merely using the bin (including the LOS component) with the greatest strength. Initial experiments show that distance estimation is insensitive to the phase information of the CSI. Therefore, only the amplitude of CSI is used in distance estimation. To better fit the propagation model, the CSI amplitude is converted to the log scale (dBm), and the pair of <log(distance), log (CSI amplitude)> is used to train a non-linear SVM regression model [15] for predicting the distance from a CSI vector.

The non-linear SVM regression model [15] relies on the kernel function, and a Gaussian kernel is defined as follows, where x_i and x_j are CSI vectors.

$$K(x_i, x_j) = \exp\left(-\|x_i - x_j\|^2\right) \tag{14.8}$$

with N pairs of CSI measurement x_i and distance y_i, the distance prediction y, given a new CSI measurement x, is computed as follows:

$$y = \sum_{i=1}^{N} \left(a_i - a_i^*\right) \cdot K(x_i, x) + b, \tag{14.9}$$

where a_i and a_i^* are parameters for each CSI measurement x_i, and can be found by minimizing the Lagrangian function

$$L(a) = \frac{1}{2} \sum_{i=1}^{N} \sum_{j=1}^{N} \left(a_i - a_i^*\right)\left(a_j - a_j^*\right) K(x_i, x_j) + \varepsilon \sum_{i=1}^{N} \left(a_i + a_i^*\right) - \sum_{i=1}^{N} y_i \left(a_i - a_i^*\right), \tag{14.10}$$

subject to the constraints

$$\sum_{i=1}^{N} \left(a_i - a_i^*\right) = 0, \quad \forall i, \ 0 \le a_i \le C, \ 0 \le a_i^* \le C. \tag{14.11}$$

Figure 14.7 shows the average error in distance estimation, by using RSSI (with linear regression) or CSI (with SVR). The RSSI/CSI information was collected on the road using PCs with IWL5300 modules. Generally, the error in distance increases with distance, which

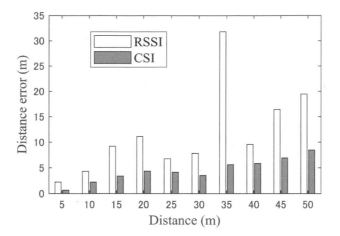

FIGURE 14.7
Average error in distance estimation by using RSSI (linear regression) or CSI (support vector regression), by using IWL 5300 (time resolution = 50 ns).

is consistent with (14.3). Using CSI can effectively reduce distance errors compared with using RSSI. Here, the distance error when CSI is used is still relatively large, which is due to the limit of time-resolution in the experiment device IWL 5300.

14.5 Recognition of LOS

The LOS component, if present in a signal, is the first one that arrives at a pedestrian device, with a relatively greater strength than other components which have extra attenuation per reflection, diffraction, etc. Figure 14.8a shows an example of CSI where the solid line represents the LOS path. On the other hand, in the absence of a LOS path (Figure 14.8b), the

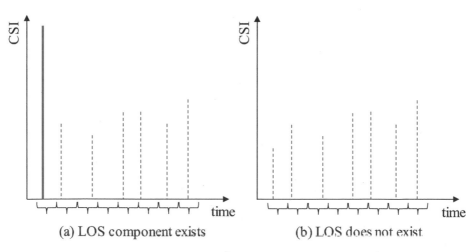

FIGURE 14.8
Distinguishing CSI with/without LOS component.

TABLE 14.1

Success Rate of Distinguishing the Existence
of LOS Component

	With LOS	Without LOS
With LOS	0.912	0.088
Without LOS	0.080	0.920

Note: Table 2 in [16].

first component is also a reflected or diffracted one, and its signal strength is comparable to the subsequent ones. Correctly measuring vehicle-pedestrian distance requires using the strength of the LOS component. To this end, it is necessary to first detect whether a LOS component exists in the signal received from a vehicle. Then, a pedestrian device uses only signals with LOS components, and uses the strength of the LOS component to estimate pedestrian-vehicle distance.

Both CSI with LOS and without LOS are collected and annotated, and a SVM-based recognition model is trained by supervised learning. Table 14.1 shows the success rate of distinguishing the existence of LOS component. The probability of correctly recognizing LOS is relatively high, over 90%.

14.6 Collecting Messages from Vehicles

In satellite-based positioning, all satellites transmit positioning signals at the same time, using different spreading (orthogonal) codes to enable a receiver to simultaneously receive signals from all satellites. In vehicular communications, vehicles transmit messages with different modulations, and simultaneous transmissions lead to message collisions and receiving failure. To avoid this problem, carrier sense multiple access with collision avoidance (CSMA/CA) is used to share the same channel among vehicles. As a result, there will be a random delay before each transmission. Although positions are computed at the timing synchronized to GPS, position messages are transmitted at different timings, with a later transmission leading to a longer delay. As a result, the position information included in vehicle messages is not the position at the timing of transmission. But in the positioning computation, the position of a vehicle at the timing of transmission is needed. Fortunately, a pedestrian and nearby vehicles are roughly synchronous with the help of GPS, and each vehicle includes the positioning timing and the speed information in its message. Based on the difference between the transmission timing and positioning timing, a pedestrian can estimate the position of a vehicle at the transmission timing, meanwhile measuring the pedestrian-vehicle distance. In this way, a pedestrian takes a snapshot of each vehicle at their respective transmission timing. Although this timing is different for vehicles, it does not affect the position computation.

A pedestrian himself may also move. As a result, a pedestrian receives messages from vehicles at different positions. But the pedestrian speed is low. If the positioning period is set to 100 ms, within this period, the distance a pedestrian moves is no more than 20 cm, much less than the positioning error. Therefore, the pedestrian is assumed to have the same position, and all messages received from vehicles within the predefined period are used to compute his position.

14.7 Single-Point Positioning

After pedestrian-vehicle distance is estimated per vehicle, the pedestrian position can be computed as in the conventional GPS positioning, using the vehicles (detected within a predefined period) as extra anchors [16].

Assume $r^{(s)}$ and r are the 3D coordinates of satellite s and the pedestrian, respectively, and their true distance is

$$d^{(s)} = | r^{(s)} - r |. \tag{14.12}$$

Considering the distance error ε due to clock drift of pedestrian device and other errors ($\xi^{(s)}$) caused by extra propagation delay (such as ionosphere, troposphere, etc.), the measured distance $p^{(s)}$ is a sum of the true distance $d^{(s)}$ and the error factors.

$$p^{(s)} = d^{(s)} + \varepsilon + \xi^{(s)}. \tag{14.13}$$

The estimated distance ($p^{(v)}$) to a vehicle (v) with a LOS path is the sum of the true distance ($d^{(v)} = | r^{(v)} - r |$) and the measurement error ($\xi^{(v)}$)

$$p^{(v)} = d^{(v)} + \xi^{(v)}. \tag{14.14}$$

These equations in (14.13) and (14.14) can be rewritten in a vector form

$$p = d(r) + \varepsilon + \xi. \tag{14.15}$$

This is approximately linearized at an initial position (r_0) as

$$p = d(r_0) + H(r_0) \cdot \begin{bmatrix} r - r_0 \\ \varepsilon - \varepsilon_0 \end{bmatrix} + \varepsilon_0 + \xi, \tag{14.16}$$

where H is composed of directional vectors.

$$H(r) = \begin{bmatrix} -\dfrac{r^{(1)} - r}{| r^{(1)} - r |} & 1 \\ -\dfrac{r^{(2)} - r}{| r^{(2)} - r |} & 1 \\ \vdots & 0 \\ -\dfrac{r^{(n+m)} - r}{| r^{(n+m)} - r |} & 0 \end{bmatrix}. \tag{14.17}$$

Then, the pedestrian position can be solved iteratively, as follows

$$\begin{bmatrix} r_{i+1} - r_i \\ \varepsilon_{i+1} - \varepsilon_i \end{bmatrix} = (H_i^T H_i)^{-1} H_i^T \cdot (p - d(r_i) - \varepsilon_i). \tag{14.18}$$

As discussed earlier, the measurement errors in pedestrian-vehicle distances and pedestrian-satellite distances are different. Let the covariance matrix of ξ be R. Then, R^{-1} can be assigned to distances as weights, with a small weight for distances that have large errors.

Typically R is a diagonal matrix and the diagonal elements of R^{-1} are weights (w). But these diagonal elements are unknown and change over time. As for satellites, signals with a LOS path typically have high SNR and small ranging errors, but reflected signals with low SNR have large ranging errors. Therefore, a heuristic method is to associate SNR (γ) with weights. First, a SNR threshold (γ_{th}) is set to remove all signals whose SNR is below this threshold (potentially without a LOS path). Then, for all usable satellites, a weight is set to be proportional to the SNR difference ($\gamma - \gamma_{th}$), a larger value for satellites with larger SNR. As for vehicles, errors in distance estimation increase with distances. Therefore, a large weight is used for a short estimated distance.

Then, the position can be computed by

$$\begin{bmatrix} r_{i+1} - r_i \\ \varepsilon_{i+1} - \varepsilon_i \end{bmatrix} = (H_i^T R^{-1} H_i)^{-1} \cdot H_i^T R^{-1} \cdot (p - d(r_i) - \varepsilon_i). \tag{14.19}$$

The imperfect LOS recognition may lead to the use of non-LOS signal in position computing. The introduction of weights helps to alleviate this problem. This is because the signal strength without the LOS component is weak and leads to a large estimated distance. This corresponds to a small weight, which decreases the impact of the error in distance estimation.

14.7.1 Impact on Horizontal DOP

H in (14.17) can be represented by elevation angle (α) and orientation angle (β) of anchors.

$$H = \begin{bmatrix} -\cos\beta_1\cos\alpha_1 & -\sin\beta_1\cos\alpha_1 & -\sin\alpha_1 & 1 \\ -\cos\beta_2\cos\alpha_2 & -\sin\beta_2\cos\alpha_2 & -\sin\alpha_2 & 1 \\ \vdots & \vdots & \vdots & \vdots \\ -\cos\beta_n\cos\alpha_n & -\sin\beta_n\cos\alpha_n & -\sin\alpha_n & 0 \end{bmatrix}. \tag{14.20}$$

Assume all the anchors have the same elevation angles (α_i) but with equal space in orientation angle (β_i), the horizontal DOP (HDOP) under different numbers of anchors is computed and shown in Figure 14.9. It is clear that using anchors (vehicles) with low elevation angles leads to low HDOP, which helps to reduce the positioning error.

14.7.2 Simulation and Experiment Results

With GPS trace data (raw measurements including pseudo-ranges and satellite positions, collected near Tokyo station where there are many high buildings) collected using a u-blox receiver (EVK-6T) and CSI trace data collected on the road using PCs with IWL5300 modules, the positioning precision of a pedestrian is evaluated by trace-based simulation. Figure 14.10 shows the simulation scenario. The average inter-vehicle distance is variable, adjusted to simulate different vehicle densities. A large inter-vehicle distance means fewer vehicles around a pedestrian. Satellite positions, pedestrian-satellite distances, vehicle

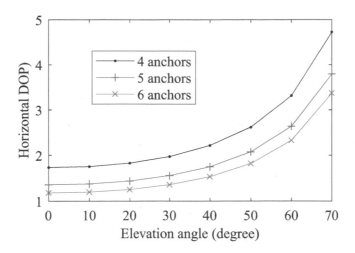

FIGURE 14.9
HDOP under different elevation angles.

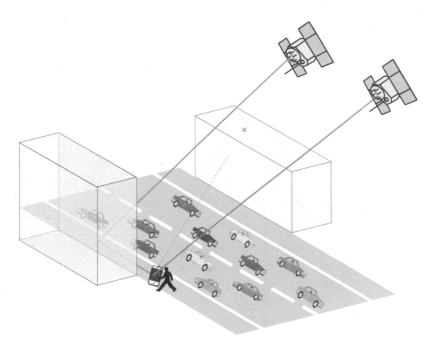

FIGURE 14.10
Scenario for trace-based simulation evaluation. Satellite positions, pedestrian-satellite distances, vehicle positions, and pedestrian-vehicle distances are randomly selected from the trace data.

positions, and pedestrian-vehicle distances are randomly selected from the trace data and used to compute pedestrian position.

Figure 14.11 shows the result of fixing rate under different inter-vehicle distances. The fixing rate is defined as the probability that the iteration in (14.19) converges. Due to distance errors, this iteration does not always converge. The fixing rate, when RSSI is used, is the lowest. Using CSI greatly increases the fixing rate by improving the distance accuracy.

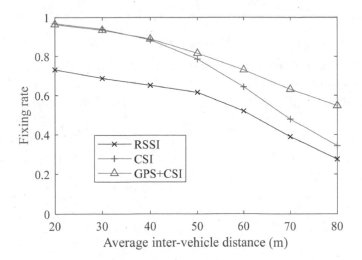

FIGURE 14.11
Fixing rate at different inter-vehicle distances (Fig. 6(b) in [16]).

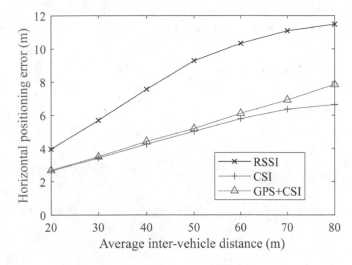

FIGURE 14.12
Average horizontal positioning error under different inter-vehicle distances (Fig. 6(a) in [16]).

As average inter-vehicle distance gets larger, the fixing rate decreases with vehicle density. When average inter-vehicle distance is 80 m, the fixing rate of using CSI is 0.343. The fixing rate of GPS alone (not shown in the figure) is about 0.3. The combination of GPS and CSI improves the fixing rate to 0.547, which confirms the effect of using vehicles as anchors in alleviating the outage of positioning.

Figure 14.12 shows the result of average horizontal positioning error under different inter-vehicle distances. Positioning error increases with average inter-vehicle distance (lower vehicle density). GPS alone (not shown in the figure) has an average positioning error of 46 m. Even directly using RSSI greatly reduces positioning error compared with GPS. CSI further reduces positioning error, by improving the distance precision. Due to the constraint of time resolution of WLAN receivers, the effect of using CSI is limited.

14.8 Benefiting from Vehicle Movement

With frequent measurement of pedestrian-vehicle distances, the precision of pedestrian position can be improved by using a Kalman filter [16].

Let the state of a Kalman filter be $X_t = [r_t, v_t, \dot{v}_t, \varepsilon_t]^T$, which consists of pedestrian position r_t, speed v_t, acceleration $\dot{v}_t$, and clock error ε_t at time t. Then, the evolution of state X_t can be represented by a matrix Φ, with a random vector $w_{t-\Delta t}$ indicating potential variations, as follows

$$X_t^- = \Phi X_{t-\Delta t}^+ + w_{t-\Delta t} \tag{14.21}$$

$$\Phi = \begin{bmatrix} I & \Delta t \cdot I & & \\ & I & \Delta t \cdot I & \\ & & I & \\ & & & 1 \end{bmatrix} \tag{14.22}$$

The variance of w_t, denoted as Q, is nearly stationary.

The measurement $Y_t = [p_t, v_t]^T$ contains pedestrian-satellite pseudo-range and pedestrian-vehicle distance (p_t) and pedestrian speed (v_t), and is indirectly associated with the state X_t by the following equation

$$Y_t = \begin{bmatrix} d_t + \varepsilon_t \\ v_t \end{bmatrix} + \xi_t. \tag{14.23}$$

Y_t can be linearized near a state X_t, where a small change ∂X_t leads to a change $H_t \cdot \partial X_t$ in Y_t. R, the variance of ξ_t, changes with time and is set empirically.

The process of updating state X_t and its variance P_t for a pedestrian is shown in Figure 14.13 and is briefly described as follows:

- KF0: Measure pedestrian-satellite pseudo-range and pedestrian-vehicle distance.

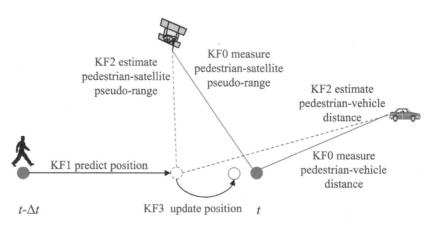

FIGURE 14.13
Flowchart of estimating pedestrian position by using a Kalman filter.

- KF1: Predict pedestrian state. New state X_t^- (and position r_t^-) of a pedestrian is predicted from the past position and moving speed.

$$X_t^- = \Phi X_{t-\Delta t}^+ \tag{14.24}$$

$$P_t^- = \Phi P_{t-\Delta t}^+ \Phi^T + Q \tag{14.25}$$

- KF2: Estimate pedestrian-satellite pseudo-range and pedestrian-vehicle distance. From the predicted pedestrian position, pseudo-ranges to satellites and distances to vehicles are estimated. $p_t^{(s)-}$ is computed from $d_t^{(s)} = \left| r_t^- - r_t^{(s)} \right|$ by adding clock error for satellites.
- KF3: Update pedestrian state. Kalman gain K_t is computed based on the variance (P_t^-) of predicted information and the variance R of measured information. Then, the predicted state X_t^- is updated to its new value X_t^+ by adding the prediction error $Y_t - (p_t^-, v_t)^T$ weighted by the Kalman gain, and pedestrian position r_t^+ is obtained from X_t^+.

$$K_t = P_t^- \cdot H_t^T \cdot \left[H_t \cdot P_t^- \cdot H_t^T + R \right], \tag{14.26}$$

$$X_t^+ = X_t^- + K_t \cdot \left[Y_t - \begin{pmatrix} p_t^- \\ v_t^- \end{pmatrix} \right], \tag{14.27}$$

$$P_t^+ = (I - K_t H_t) P_t^-. \tag{14.28}$$

Figure 14.14 shows the cumulative distribution of horizontal position error under different transmission intervals of vehicles. It is clear that position error decreases with

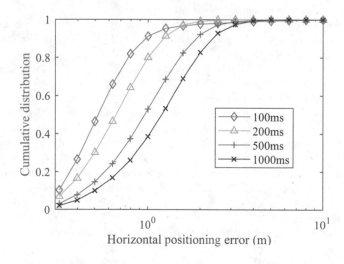

FIGURE 14.14
Cumulative distribution of average positioning error (average inter-vehicle distance is 40 m, vehicle transmission interval is 100 ms, 200 ms, 500 ms, or 1 s) (Fig. 9(a) in [16]).

the transmission interval. This is because reducing the transmission interval increases the number of measurements of pedestrian-vehicle distances. Together with a Kalman filter, this helps to smoothen the computed position to have a smaller error (less than 1 m on average).

14.9 Conclusions

This chapter has discussed how to use vehicles, besides GPS satellites, as anchors in computing pedestrian position. This helps to improve the fixing rate and solve the outage of positioning, especially in urban canyons that have high buildings on roadsides shielding positioning signals from satellites. In addition, by detecting whether a received signal contains LOS component and using only the strength of LOS component in computing pedestrian-vehicle distance, the positioning error is greatly reduced compared with the GPS alone method. Smoothening the positioning results by a Kalman filter helps to further reduce the average positioning error to be less than 1 m. All these results confirm the potential and feasibility of using signals from vehicles in improving positioning precision of pedestrians. The positioning precision is limited by the distance precision, which is further constrained by the time-resolution of pedestrian receivers. More efforts are needed to overcome this to further improve the precision of pedestrian position.

References

1. Institute for Traffic Accident Research and Data Analysis, Annual report of traffic accident statistics (2017 edition, in Japanese), Technical Report, 2018. https://www8.cao.go.jp/koutu/taisaku/h30kou_haku/pdf/zenbun/1-1-1.pdf

2. S. Tang, K. Saito, and S. Obana, Transmission control for reliable pedestrian-to-vehicle communication by using context of pedestrians, *Proc. IEEE Int. Conf. Vehicular Electronics and Safety (ICVES)*, 2015, pp. 41–47.

3. NTSB, Preliminary report highway HWY18MH010, https://www.documentcloud.org/documents/4483190-NTSBuber.html, 2018.

4. A.R. Pratama and R. Widyawan Hidayat, Smartphone-based pedestrian dead reckoning as an indoor positioning system, *Proc. Int. Conf. System Engineering and Technology (ICSET)*, 2012, pp. 1–6.

5. Association of Radio Industries and Businesses. 700 MHz band intelligent transport systems, ARIB Std. STD-T109, Tokyo, Japan, 2012.

6. Y. Gu, L.-T. Hsu, S. Kamijo, Passive sensor integration for vehicle self localization in urban traffic environment, *Sensors*, 2015, 15, (12), pp. 30199–30220.

7. T.S. Rappaport, *Wireless communications: Principles and practice*, Prentice Hall, Upper Saddle River, NJ, 2002, 2nd edn.

8. Z. Yang, Z. Zhou, Y. Liu, From RSSI to CSI: Indoor localization via channel response, *ACM Comput. Surv.*, 2013, 46, (2), pp. 25:1–25:32.

9. D. Halperin, W. Hu, A. Sheth, and D. Wetherall, Tool release: Gathering 802.11n traces with channel state information, *SIGCOMM Comput. Commun. Rev.*, 2011, 41, (1), pp. 53–53. https://dhalperi.github.io/linux-80211n-csitool/

10. S. Sen, J. Lee, K.-H. Kim, and P. Congdon, Avoiding multipath to revive inbuilding WiFi localization, *Proc. MobiSys'13*, 2013, pp. 249–262.
11. X. Wang, L. Gao, S. Mao, and S. Pandey, CSI-based fingerprinting for indoor localization: A deep learning approach, *IEEE Trans. Veh. Technol.*, 2017, 66, (1), pp. 763–776.
12. M. Kotaru, K. Joshi, D. Bharadia, and S. Katti, Spotfi: Decimeter level localization using WiFi, *Proc. SIGCOMM'15*, 2015, pp. 269–282.
13. R.D.J. van Nee, J. Siereveld, P.C. Fenton, and B.R. Townsend, The multipath estimating delay lock loop: Approaching theoretical accuracy limits, *Proc. IEEE Position Location and Navigation Symp.*, 1994, pp. 246–251.
14. Y. Xie, Z. Li, and M. Li, Precise power delay profiling with commodity WiFi, *Proc. MobiCom'15*, 2015, pp. 53–64.
15. V. Cherkassky and Y. Ma, Practical selection of SVM parameters and noise estimation for SVM regression, *Neural Netw.*, 17, (1), pp. 113–126, 2004.
16. S. Tang and S. Obana, Improving performance of pedestrian positioning by using vehicular communication signals, *IET Intell. Transp. Syst.*, 2018, 12, (5), pp. 366–374.

15

Cross-Provider Cooperation for Network-Based Localization

Shih-Hau Fang

CONTENTS

15.1 Introduction

The location information of users has become a key issue for many location-based services in both indoor and outdoor environments [1–3]. Positioning methods generally include device-based and network-based architectures, depending on whether they adopt an integrated or stand-alone positioning infrastructure [4–6]. For device-based positioning, the mobile device determines its location using signals it receives while protecting privacy of users. For network-based positioning, a network of base stations locates the device by measuring different characteristics such as power [7,8], angle [9], and time [10,11]. Although this approach requires a network connection, it consumes little power and does not require any extra hardware modification of the device [12]. The Enhanced 911 system is the initial driver of such network-based positioning and requires network operators to provide user location within specified accuracies [13,14].

Among various networking technologies, global system for mobile (GSM) communications is an attractive option for positioning because it is a popular standard for cellular phones and has widespread infrastructures around the world [15–18]. A typical GSM phone is served by only a single network provider according to the inserted subscriber identification module (SIM) card. In recent years, the technology for the development of mobile devices has progressed tremendously. The dual-SIM mobile phones have been developed so that users can incorporate two SIM cards into one handset. These phones can connect to multiple networks and allow users to separate private and professional calls, or to select the most cost-efficient network. This cross-provider architecture should enable a future computing environment. From the viewpoint of network-based localization, connecting a mobile device to multiple operators makes it possible to combine multiple location estimations from different network providers. When location estimations from multiple providers are available, combining their individual advantages can achieve better performance than any individual provider, and can avoid selecting a poorly-performing network provider. However, the cross-provider information is still unexplored for localization. Although previous studies present some hybrid positioning schemes, most of them focus on the signal or model combination [19–24].

To efficiently exploit the complementary information from network providers, this study proposes two cross-provider localization algorithms via a cooperative approach. The first algorithm, called linear regression (LR), chooses the provider weights that minimize the mean-square positioning errors. The second algorithm, called locally cooperative embedding (LCE), further considers the local performance of providers. It first builds a *cooperation map* to store the confidence at different reference positions from different network providers. Then, it dynamically embeds the cooperation according to each provider's local accuracy surrounding a user and iteratively refines the positioning process through the cooperation map.

The experiments reported in this study applied the proposed positioning algorithm to a real GSM environment at the Yuan-Ze University (YZU) campus. A dual-SIM mobile phone was fitted with two SIM cards to connect to two GSM network providers in Taiwan, including Chung-Hwa Telecommunication (CHT) and Far EasTone Telecommunication (FET) companies. The experiments implemented the single-provider location system via a fingerprinting-based method [25,26] and compared three cross-provider approaches, the uniform weighting (UW), LR, and LCE.

The on-site experiments considered two different scenarios. When the performance of each provider is similar, results show the effectiveness of all cooperative methods, where LCE achieves the best. The UW, LR, and LCE approaches reduced the median error by 1.96%, 8.82%, and 31.37%, respectively, compared to the best single provider. When the performance of each provider is diverse, results show that the LCE approach still offers consistent accuracy improvement, whereas UW and LR approaches are not necessarily better than the single provider. This is because the LCE method uses additional information regarding the provider's local accuracy such that a network provider with poor performance on average can still contribute to the location estimation. The results show that the mean positioning error of LCE approaches the Cramer-Rao lower bound of single-provider. The experiments in this study also investigate the effects of the density of the reference locations to the proposed cross-provider cooperation mechanism.

15.2 Related Works

To compensate for the drawbacks of GPS, GSM always comes to mind because it is a popular standard for cellular phones. Industry standards in GSM systems, such as the uplink time difference of arrival (U-TDOA), enhanced observed time difference (E-OTD), and assisted global navigation satellite systems (A-GNSS) incorporate geolocation techniques [27,28]. The physical characteristics that these network-based methods use can be classified as received signal strength (RSS) [25], time of arrival (TOA) [11], and angle of arrival (AOA) [6] characteristics. Each characteristic has its own limitations and advantages. For instance, using AOA requires line-of-sight propagation, obtaining accurate TOA requires extra synchronization, and converting RSS to distance requires that many parameters must be solved. Note that RSS is commonly employed in indoor positioning systems due to the popularity of Wi-Fi networks [29,30].

Many researchers have studied hybrid positioning schemes from the signal level, including AOA/TDOA [9], TOA/TDOA [24], and TOA/RSS [20,22]. A few studies try to combine multiple wireless technologies, such as GPS-based methods [32] and GSM/UMTS [33]. Unlike these combination schemes, the proposed approach attempts to combine the location estimations from each network provider. Connecting multiple providers is possible based on current hardware. In an emergency situation, traditional GSM devices can connect to an alternative operator other than the one for the inserted SIM card. Recently, the dual-SIM mobile phones have been developed so that users can incorporate two SIM cards into one handset. These phones have become more and more popular, and even triple-SIM phones are available from the Chinese market. This inspires us to develop cross-provider cooperation approaches to improve mobile positioning.

Although there have been numerous works on cooperative localization, they have mainly focused on the signal levels in sensor networks [34–37]. When some sensors cannot localize themselves due to physical limitations (the limited energy and communication range, or connection less than three base stations), they can connect with each other and measure the signals between themselves to achieve localization [38–41]. In the considerable literature, such systems have alternatively been described as "multi-hop," "ad-hoc," "distributed," or "network calibration" [42–44]. We clarify that, although using the term "cooperative," traditional works which emphasize signal levels in sensor networks differ from our work which focuses on the decision level between network providers. To the best of the author's knowledge, such cooperation is not present in the literature.

15.3 Cross-Provider Cooperation

15.3.1 Overview

Different network providers offer complementary information for mobile phone positioning tasks. If a phone can be served by multiple providers, a more accurate location estimation could be obtained by combining multiple providers in some way, instead of just using a single provider. Assume the estimated location coordinate $\tilde{p} \in \Re^D$ is a D-dimensional vector, where D equals two for a plane or three for a solid. Let us use the notation $\tilde{p}^{(d)}$ to refer to the the d-th component of $\tilde{p}$. Then, our goal is to devise a cross-provider cooperation that can extract the complementary information from providers to generate an improved position estimation, based on the estimates from the individual provider:

$$\tilde{p}^{(d)} = \sum_{i=1}^{K} w_i^{(d)} \cdot \hat{p}_i^{(d)} \tag{15.1}$$

where K is the number of network providers, $\tilde{p}^{(d)}$ is the d-th combined result, $w_i^{(d)}$ and $\hat{p}_i^{(d)}$ are, respectively, the weight and the estimation of the i-th provider for the d-th coordinate. The weight in Equation 15.1 represents the confidence given by a specified network provider. Determining the weights in an intelligent manner is an important issue to achieve a higher positioning accuracy. A simple average rule, called uniform weighting (UW), is usually used to determine the weights as $w_i^{(d)} = 1/K$ to reduce some random errors. However, this method treats each provider in an equal way. The contribution of providers on the location estimation has not been differentiated.

To efficiently exploit the complementary information from network providers, this study proposes two cross-provider localization algorithms via a cooperative approach. The first algorithm, LR, chooses the weights that minimize the mean-square positioning errors. The second algorithm, LCE, takes a further step to consider the regional performance, and recursively determines the cooperation according to each provider's local accuracy surrounding a user.

15.3.2 Linear Regression

The first method uses multiple LR to model the D relationships between estimated positions and their true coordinates. First, Equation 15.1 can be rewritten in a matrix form as

$$\tilde{p}^{(d)} = [\mathbf{w}^{(d)}]^T \cdot \tilde{\mathbf{p}}^{(d)} \tag{15.2}$$

where $\hat{p}^{(d)} \in \Re^K$ is a column vector containing the d-th coordinate of K providers' outputs as $[\hat{p}_1^{(d)}, \ldots, \hat{p}_K^{(d)}]^T$, and $\mathbf{w}^{(d)} \in \Re^K$ is a column vector containing K provider's weights contributing to the d-th coordinate as $\left[w_1^{(d)}, \ldots, w_K^{(d)}\right]^T$.

LR is based on least squares: the weights are determined such that the sum-of-square of positioning errors is minimized. Thus, this approach requires training data, $\{p^{(d)}(j), \hat{\mathbf{p}}^{(d)}(j)\}_{j=1}^N$, where N is the number of training samples, $p^{(d)}(j)$ is the d-th dimension of the j-th sample's true position, and $\hat{\mathbf{p}}^{(d)}(j) \in \Re^K$ is the j-th sample's estimated positions from K providers. Using these training data, the LR approach solves the following problem:

$$\mathbf{w}^{(d)} = \underset{\mathbf{w}^{(d)} \in \Re^K}{\arg\min} \frac{1}{N} \sum_{j=1}^N \left\| p^{(d)}(j) - [\mathbf{w}^{(d)}]^T \hat{\mathbf{p}}^{(d)}(j) \right\|^2 \tag{15.3}$$

with

$$\sum_{i=1}^K w_i^{(d)} = 1 \tag{15.4}$$

Taking the derivation of Equation 15.2 with respect to $\mathbf{w}^{(d)}$, we may find the optimum weight vector as:

$$[\mathbf{w}^{(d)}]^T = \{E[\hat{\mathbf{p}}^{(d)} (\hat{\mathbf{p}}^{(d)})^T]\}^{-1} E[p^{(d)} (\hat{\mathbf{p}}^{(d)})^T] \tag{15.5}$$

By evaluating Equation 15.5, LR estimates the provider weights that can be stored for cross-provider cooperation. When a mobile device requests services, it combines each provider's estimation by Equation 15.2 for ultimate positioning. Compared to UW, LR allows a more reliable provider to play a more significant role during the localization process. This gives us reason to believe that LR may produce a more accurate location estimation.

If we carefully examine the positioning errors at different locations, it will be discovered that each network provider has expertise in different local regions. The following section presents another novel algorithm which takes the provider's local accuracy into account such that the cooperation can be more intelligently constructed.

15.3.3 Locally Cooperative Embedding

The second method, LCE, takes advantage of the fact that each provider has expertise in different regions of the network covered space. By recording the local errors at different reference positions, this approach embeds a larger confidence to the provider with more accurate estimations in the area where a user is located. Unlike the static weighting in LR, LCE is different in two steps: locally cooperative embedding and iterative cooperation refinement. The basic idea of the first step is that the cooperative weightings should be also location dependent. This step incorporates the local performance of providers into the cooperative process as:

$$\tilde{p}^{(d)} = [\mathbf{w}(\tilde{p}^{(d)})^{(d)}]^T \cdot \hat{\mathbf{p}}^{(d)} \tag{15.6}$$

where the weights $\mathbf{w}^{(d)}$ in Equation 15.6 is a function of $\tilde{p}^{(d)}$ rather than a fixed value in Equation 15.2. This mechanism allows cooperative positioning systems to dynamically design the weights from the estimated location. For convenience, we use $\mathbf{w}^{(d)}$ representing $\mathbf{w}(\tilde{p}^{(d)})^{(d)}$ in the following derivation.

Next, LCE proposes how to find the function that most accurately depicts the relationship between the weights and the estimated locations. To achieve this goal, we partition the training data into different reference locations and consider the provider diversity in these local regions. Assume N training data are collected from R reference locations as

$$\{p^{(d)}(j)\}_{j=1}^{N} \in \{l_1^{(d)}, l_2^{(d)}, \ldots, l_R^{(d)}\} \tag{15.7}$$

where $l_r^{(d)}$ is the d-th dimension of the r-th reference location's true position. Considering the provider diversity in these R regions, LCE builds the *cooperation map* (CM) or each dimension d to store the optimum weights $w_{ri}^{(d)}$ for the i-th provider and at the r-th reference position as:

$$\mathbf{CM}^{(d)} = \begin{bmatrix} w_{11}^{(d)} & \cdots & w_{1K}^{(d)} \\ \vdots & w_{ri}^{(d)} & \vdots \\ w_{R1}^{(d)} & \cdots & w_{RK}^{(d)} \end{bmatrix} \tag{15.8}$$

with

$$\sum_{i=1}^{K} w_{ri}^{(d)} = 1 \tag{15.9}$$

where the weights $w_{ri}^{(d)}$ can be obtained by performing linear regression with each reference location's training data and are normalized to add up to one for K providers. For each dimension d, the rows of CM indicate the confidence at a particular reference position from K network providers, while the columns of CM represent the confidence given by a specified provider to R reference positions.

After constructing the cooperation map, LCE adopts the nearest-neighbor rule to estimate appropriate weights. Given an estimated position $\tilde{p}$, we can find the nearest reference position $l_{\hat{r}}$ by comparing the distance between $\tilde{p}$ and l_r in the physical spatial space. The index of the nearest reference location $\hat{r}$ is computed as

$$\hat{r} = \underset{r \in \{1, \ldots R\}}{\arg\min} \|\mathbf{1}_r, \tilde{\mathbf{p}}\| \tag{15.10}$$

where $\|\cdots\|$ is a distance metric function that gives a generalized scalar distance between $\tilde{\mathbf{p}}$ and $\mathbf{1}_r$. This study uses the Euclidean metrics for the D-dimensional vectors as

$$\|\mathbf{1}_r, \tilde{\mathbf{p}}\| = \left(\sum_{d=1}^{D} |\mathbf{1}_r^{(d)} - \tilde{\mathbf{p}}^{(d)}|^2 \right)^{1/2} \tag{15.11}$$

After determining the nearest reference position $\hat{r}$ by Equation 15.10, LCE determines $\mathbf{w}^{(d)}$ as

$$\mathbf{w}^{(d)} = \mathrm{CM}_{\hat{r},*}^{(d)} = [w_{\hat{r}1}^{(d)}, \ldots, w_{\hat{r}k}^{(d)}]^T \tag{15.12}$$

where $\mathrm{CM}_{\hat{r},*}^{(d)}$ is the $\hat{r}$-th row of $\mathrm{CM}(d)$. In Equation 15.12, LCE adopts the weights of the $\hat{r}$-th reference location for the cross-provider cooperation because $\mathbf{1}_r$ is located nearest to the estimated result. That is, LCE dynamically adjusts the cooperative mechanism based on the provider's local confidence and the corresponding cooperation map. However, the errors contained in the estimated location $\tilde{\mathbf{p}}$ may incur an inaccurate $\hat{r}$. Thus, LCE adopts the second step to avoid this problem. This step iteratively refines the cooperation, as shown in the following iterative equations.

$$\tilde{p}^{(d)}[m] = [\mathbf{w}^{(d)}[m]]^T \cdot \hat{\mathbf{p}}^{(d)} \tag{15.13}$$

$$\mathbf{w}^{(d)}[m+1] = f(\tilde{p}^{(d)}[m], \mathbf{w}^{(d)}[m]) \tag{15.14}$$

where the iteration index m is introduced in this mechanism. Equation 15.13 uses the weight at iteration m, $\mathbf{w}^{(d)}[m]$, to compute the final location estimation $\tilde{p}^{(d)}[m]$ while Equation 15.14 calculates the weights in the next iteration $\mathbf{w}^{(d)}[m+1]$ using the previously estimated result $\tilde{p}^{(d)}[m]$ and weights $\mathbf{w}^{(d)}[m]$. LCE updates the weights as

$$\mathbf{w}^{(d)}[m+1] = \eta \cdot \mathbf{w}^{(d)}[m] + (1 - \eta) \cdot \mathrm{CM}_{\hat{r},*}^{(d)} \tag{15.15}$$

Equation 15.15 shows how the stored local weights $\mathrm{CM}_{\hat{r},*}^{(d)}$ combine with the previous weightings $\mathbf{w}^{(d)}[m]$ to obtain the new weights $\mathbf{w}^{(d)}[m+1]$. Equation 15.15 also shows that $\mathbf{w}^{(d)}[m+1]$ is a linear combination of $\mathbf{w}^{(d)}[m]$ and $\mathrm{CM}_{\hat{r},*}^{(d)}$, with coefficients η and $1 - \eta$. These coefficients balance the confidence between the previous weight and the relative local weights. If $\eta = 0$, we have a degenerate case in which $\mathbf{w}^{(d)}[m+1]$ is determined by only the stored local weights from CM. At the other extreme of $\eta = 1$, only the previous weight is used. Note that the proposed method normalized the updated weights to K providers at each iteration. The distance between the reference location and the estimated result in Equation 15.11 determines the parameter η to be

$$\eta = \alpha \|\mathbf{1}_r, \tilde{\mathbf{p}}\| \tag{15.16}$$

TABLE 15.1

Main Steps of the Proposed Positioning Algorithm

Summary of the locally cooperative embedding approach

---
initialize:
1. Calculate initial weights $\mathbf{w}^{(d)}[1]$ using LR (Equation 15.5)
2. Estimate initial location estimation (Equation 15.2)
3. Construct the Cooperation Map (Equation 15.8)

iterative cooperation refinement:
4. **for** $m = 1{:}100$
5.   Find the nearest neighbors $\hat{r}$ from CM (Equation 15.10)
6.   Estimate the local accuracy of $\hat{r}$ (Equation 15.12)
7.   Compute the confidence (Equations 15.11 and 15.16)
8.   Update the cooperative weights $\mathbf{w}^{(d)}[m+1]$ (Equation 15.15)
9.   Localization using updated weights (Equation 15.13)
10.   **if** $\|\,\tilde{\mathbf{p}}[m+1] - \tilde{\mathbf{p}}[m]\,\| < \theta$
11.   **then** (goto 14.)
12.   **endif**
13. **end**
14. **return** $\tilde{\mathbf{p}}[m]$

---

where α is a bias gain adjusted to make η smaller than 1. Equation 15.16 shows that the decreasing $\|\,\mathbf{1}_r, \tilde{\mathbf{p}}\,\|$ results in a larger $1 - \eta$. This gives more confidence to the local weights when the estimated position is near the reference location. Substituting Equation 15.16 into Equation 15.15 obtains the update weights $\mathbf{w}^{(d)}[m+1]$ at the next iteration. Feeding the updated weights to Equation 15.13 allows it to recalculate the final result iteratively. The iteration repeats until the resulting location difference $\|\,\tilde{\mathbf{p}}[m+1] - \tilde{\mathbf{p}}[m]\,\|$ falls below the specified threshold θ.

The combination of these two steps, locally cooperative embedding (Equation 15.6) and iterative cooperation refinement (Equations 15.13 and 15.14), forms LCE, which embeds the local provider's cooperation and refines the location estimation iteratively. This is unlike the static combination methods such as UW and LR, which fix the weights after training. Compared to the traditional methods, LCE devises a more intelligent cross-provider cooperation using the complementary information from each network provider. Table 15.1 summarizes the main steps of LCE for cross-provider positioning.

Since LCE iteratively refines the cooperation, the variation of the d-th coordinate resulted from the corresponding changing weights as $\Delta \tilde{p}^{(d)} = \Delta w^{(d)} \cdot \hat{p}^{(d)}$. Substituting $w^{(d)}[m]$ into Equation 15.15 obtains the change in weight as $\Delta w^{(d)}[m+1] = (\eta - 1)w^{(d)}[m] + (1 - \eta)\mathrm{CM}_{\hat{r},*}^{(d)}$. Similarly, we can find the new weight change at the next iteration as $\Delta w^{(d)}[m+2] = (\eta - 1)w^{(d)}[m+1] + (1 - \eta)\mathrm{CM}_{\hat{r},*}^{(d)}$. Substituting $w^{(d)}[m+1]$ and $\Delta w^{(d)}[m+1]$ into the above equation obtains that $\Delta w^{(d)}[m+2] = \eta \cdot \Delta w^{(d)}[m+1]$. This shows that the change in weight at the next iteration is directly proportional to the multiplication of η and current changing weights. This achieves convergence since η is smaller than one to balance the confidence between the previous weight and the relative local accuracy. The above analysis shows the sufficient condition for the convergence of the LCE iterations.

Figure 15.1 shows the block diagram of the proposed cross-provider positioning algorithm. To localize a mobile device, each network provider localizes the device independently of the others. Then, the estimated outputs are cooperated by LCE to derive a final location estimation.

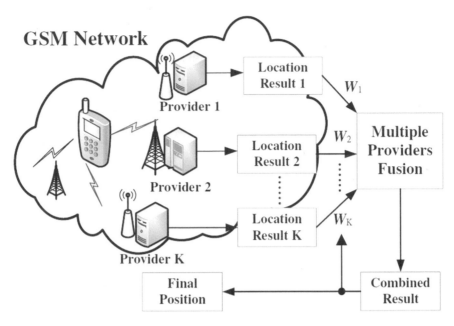

FIGURE 15.1
The architecture of the cross-provider cooperation for network-based positioning.

15.4 Experimental Result and Analysis

15.4.1 Experimental Setup

This section describes the training data collection methodology adopted in this study. We used a commercially available dual- subscriber identification module (SIM) mobile phone for GSM network access. This phone was equipped with two SIM cards from two GSM network providers in Taiwan, including CHT and FET companies ($K=2$). The device can connect to CHT and FET simultaneously and report the location estimations from each provider. The CHT and FET estimated their locations independently by a fingerprinting approach, which relies on matching the received power from base stations in a database and ignores the height information ($D=2$). Currently, the dual-SIM phone market is growing exponentially and there are even triple-SIM phones available. Moreover, several dual-SIM mobile phones have been launched by leading brands in the market, such as Nokia, Motorola, and HTC since these phones have become more popular. Although connecting multiple providers is not so popular, these phones can still obtain the measurements through automatic switch such that the cooperation can be performed by the service providers.

Figure 15.2 shows the YZU campus at which the experiments were performed, where the balloons and tacks, respectively, indicate the locations of CHT's and FET's base stations (BSs). We first define the location of BSs in a two-dimensional Cartesian space. In this stage, the GPS system is used as the ground truth to record the latitude and longitude for each BS. Then, the Cartesian coordinate is obtained by WGS84 transformations where the distance between locations is calculated by the great-circle distance formula [14]. The cross and plus sign represent where we collect realistic testing and training data, respectively. We define the coordinate of these locations based on the Euclidean distances between locations and BSs.

FIGURE 15.2
The YZU campus at which the experiments were performed. The balloons and tacks, respectively, indicate the locations of CHT's and FET's base stations.

YZU is located in Chung-Li city of Tao-Yuen country in Taiwan. This campus consists of seven academic buildings, and the size of the test-bed is 176,400 m². The training data was measured at 20 different reference locations ($R = 20$) and the distance between two neighboring training locations ranged from 80 to 150 m. For every location, we collected 36 samples ($N = 36*20$) and recorded the estimated results from each provider. To obtain the LR model, the whole data-collection process requires several hours, including the moving time between reference locations and the operating time. Although increasing measurements may obtain a better LR model, this certainly increases the time-consuming effort for data-collection. By performing linear regression to each reference location's training data, we can obtain the provider's local weights and construct the cooperation map during the offline stage. In the test mode, a dual-SIM mobile phone was fitted two SIM cards to connect to two GSM network providers. We collected testing data from alternative 16 locations in this area. The device is static at each location in the testing stage. We assume that the mobile phone connects multiple providers during a short time period to obtain the individual location estimations at the same location. The test details were stored into the log file in the phone and then the cooperation and analysis were performed in a PC. The proposed algorithm sets the threshold θ to 10 m and the bias gain α is 1/200.

15.4.2 Performance Evaluation

This section compares the performance of single provider with that of three cross-provider mechanisms in terms of the positioning accuracy. This study defines positioning accuracy as the cumulative percentage of estimations within specified errors. The positioning error is defined as the Euclidean distance between the estimated result and the true coordinate.

Figure 15.3 compares the accuracy of three cross-provider cooperation methods, UW, LR and LCE, and two single-provider results, CHT and FET. This figure shows that FET performed better than CHT. Although the same positioning algorithm is adopted in each provider, the estimated results may be influenced by the distributions of base stations. Figure 15.4 depicts the average power from FET and CHT networks in the targeted area,

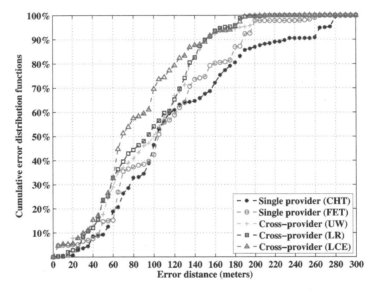

FIGURE 15.3
Cumulative percentage of error for different algorithms, where single-provider results are CHT and FET, and cross-provider results are UW, LR, and LCE.

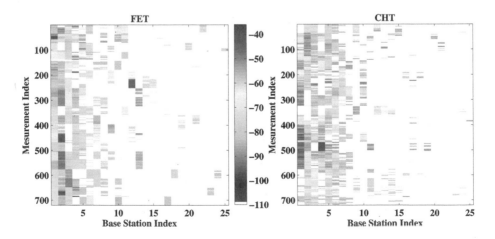

FIGURE 15.4
Power measurements from 25 base stations to 20 reference locations in FET and CHT GSM networks. Thirty-six measurements were taken at each location.

showing that the base stations of FET had a stronger power while those of CHT had a uniformly low power. This observation agrees well with Figure 15.2, where the FET's base stations distributed in the central part of YZU. This may explain why FET produced a slightly better accuracy than CHT. Note that although the mobile device can sense 25 BSs in the test-bed, Figure 15.2 shows only the five serving BSs in the YZU campus, while the other BSs may locate outside the plotted region.

The results in Figure 15.3 also show that the cross-provider methods, UW and LR, perform better than any individual network provider. This shows that the cooperation of multiple network providers can improve accuracy, even via a simple UW method. This is because some random errors in multiple location estimations may cancel each other out, increasing estimation accuracy. More importantly, Figure 15.3 shows that the proposed LCE algorithm outperforms UW and LR. This figure shows that the accuracy improvement of LCE is more significant towards the middle of the distribution.

The results also show that the positioning errors are relatively larger in the northwest region. One possible reason may be that the distribution of BSs is extremely asymmetrical in the realistic test. We can observe from the map that BSs are sparse in the northwest region, thus resulting large-error estimations from individual network providers. When both providers have poor accuracy, the large-error inputs may limit the improvement of the cooperation mechanism.

Note that similar results of UW and LR are expected. This is because when providers have similar performance, the weights of LR are approaching uniformly distributed. In this case, the LR approach does not effectively exploit the cross-provider location information.

Figure 15.5 further reports five numerical error measures, including mean error, standard deviation of error, median error, 67%, and 90% circular error probability (CEP). CEP is the

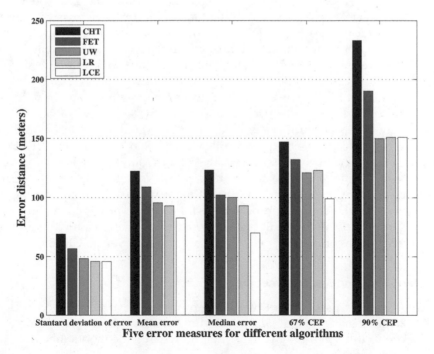

FIGURE 15.5
A comparison of five numerical error measures for different algorithms, where single-provider results are CHT and FET, and cross-provider results are UW, LR, and LCE.

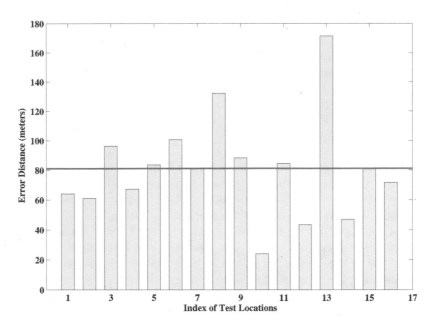

FIGURE 15.6
The positioning Cramer-Rao lower bound of single-provider at different test locations.

radius of a circle whose center is at the true location and contains the location estimates within a probability. From Figure 15.5, LCE reduces the median error and 67% CEP by 24.73%–30% and 18.18%–19.51%, respectively, compared to UW and LR. Note that only the 67% CEP of LCE decreases to 99 m, satisfying the FCC accuracy requirement for network-based positioning techniques. Figure 15.5 shows that the proposed LCE method provides the best positioning performance among all the compared approaches. This can be explained by the LCE mechanism's ability to dynamically embed the provider weights according to the estimated location and iteratively refine the cooperative process. Compared to LR, the dominant provider in LCE is the one with greater local accuracy surrounding the user, rather than the one with the best performance on average.

Next, we follow the procedure of [45,46] to derive the standard deviation of the distance estimation errors and calculate the positioning Cramer-Rao lower bound (CRLB) of single-provider. Figure 15.6 shows the CRLB values of FET single-provider at different test locations, where the line represents the average value. Note that the average error of CHT single-provider is 94 m. We observe an interesting result that the mean positioning error of cross-provider in Figure 15.5 approaches CRLB of single-provider in Figure 15.6. This figure shows that, by using the additional provider's information, only the proposed LCE algorithm can provide comparable positioning accuracy to average CRLB.

15.4.3 Alternative Cooperative Scheme

This section considers a scenario in which the providers have diverse accuracies. Unlike Section 15.4.2, this scenario changes the positioning scheme of CHT from the fingerprinting to an enhanced Cell-ID method [31]. This method regards the mobile phone's position as the interpolation of the base station locations.

Figure 15.7 illustrates the cumulative error distribution in this scenario. In this case, FET performs much better than CHT because the enhanced Cell-ID that CHT adopted performs

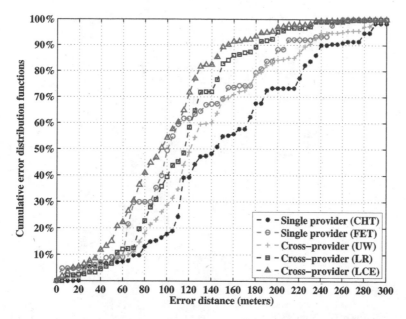

FIGURE 15.7
The performance comparison with an alternative positioning algorithm for CHT with diverse performance on average.

worse than a fingerprinting approach. This figure shows the interesting result that UW and LR are not necessarily better than FET. This is because these combined methods only consider the performance of each single provider independently. Since FET performs much better than CHT, the performance of UW is even worse than that of a single provider due to the blind average. FET also dominates the weighted result of LR. In this case, the LR approach does not effectively exploit the location information provided by CHT to compensate for FET due to CHT's much smaller weights. This explains why UW and LR cannot offer consistent improvement when providers have fluctuating performance. Nevertheless, this experiment still shows the performance improvement of LCE. In Figure 15.7, the median error of LCE falls from 100 to 88 m, whereas those of LR and UW increase to 102 and 110 m, respectively. This is because LCE can efficiently exploit the complementary information from each provider.

Figure 15.8 shows the spatial distribution of CHT's cooperative weights stored in CM. The weight ranges from zero to one, and a larger value represents a higher confidence of CHT's estimation. This figure is geometrically consistent with Figure 15.2, where the circle indicates the origin of the coordinates. The optimum cooperation weights clearly show a spatial correlation. Figure 15.8 shows that the CHT's weights are relatively larger in the middle campus, indicating these regions prefer CHT to FET. This means that a provider with poor performance on average may still achieve good performance in certain regions and contribute to location estimation. These results strongly support that the proposed approach can improve positioning accuracy because it considers the spatial diversity and complementary advantages of each network provider This finding is of great practical importance in the design of a cross-provider location system because different providers in real environments inevitably produce unequal performance. We can conclude that if the performance of each provider is similar, all cross-provider cooperative methods can improve the accuracy and LCE achieves the best performance. If the performance of each provider is diverse, LCE again provides the accuracy enhancement, whereas traditional

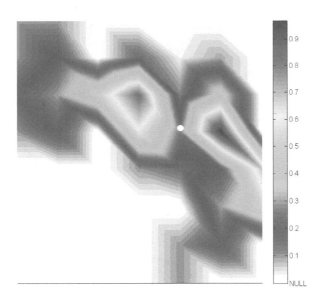

FIGURE 15.8
The cooperative weights of CHT that LCE uses at different reference locations.

methods do not work well. Thus, the proposed LCE algorithm is a general cross-provider approach that can be applied to different network-based location systems. These results, with an alternative positioning algorithm for CHT, confirm the advantages of LCE.

15.4.4 Effects of the Density of Reference Locations

This section studies the effects of the density of reference locations on positioning accuracy. A critical point in the proposed LCE approach is how much we trust the local weights that stored in CM. One of the most important factors is the density of reference locations. In high-density situations, it is reasonable to place more confidence in the local weights because more precise estimation is available in a local region. Thus, varying the number of data collection points can simulate different densities of reference locations at YZU. Figure 15.9 shows the positioning accuracy versus three kinds of densities: low, medium, and high. The original experiment setup is a medium density ($R = 20$). We dropped 50% of the reference locations ($R = 10$) to simulate a low density and doubled the numbers to simulate a high density ($R = 40$). LR can be viewed as a medium density since it follows the original setup.

Figure 15.9 demonstrates that as the density of reference location increases, the positioning accuracy of LCE clearly increases. The median error decreases to 62 m while doubling the density. Conversely, the median error increases, approaching that of LR while using only 50% of the reference locations. This is because the nearest neighbor reporting local weights is selected from the reference locations. In a high-density environment, the distance to the nearest neighbor is usually small, producing a smaller η. In this case, we can update the provider's weights with more confidence to the local weights stored in CM, as Equation 15.16 indicates. Although the local information provided by the extra reference locations improves positioning accuracy, increasing the density of the reference locations requires site surveys to collect provider performance data in the initialization phase. This could be a tradeoff between system cost and accuracy. It is also possible to allocate resources to calibrate a multiple-provider location system to support accuracy requirements of the intended services.

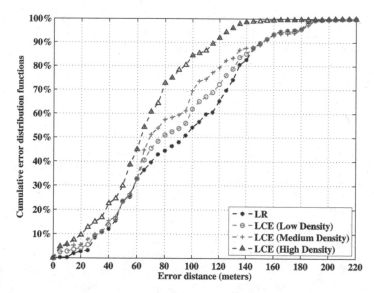

FIGURE 15.9
Positioning accuracy of LCE versus three kinds of densities of reference locations.

15.5 Conclusion

This study proposes two cross-provider cooperation algorithms, LR and LCE, capable of obtaining a combined location estimation that is more accurate than the estimations made by any one of the individual providers. LR chooses the provider weights that minimize the mean-square positioning errors. LCE dynamically embeds the cooperation and iteratively refines the positioning process according to each provider's local accuracy surrounding the user. The proposed cross-provider algorithms were evaluated on the campus of YZU using a dual-SIM mobile phone. Experimental results show that the cooperation approaches improve the positioning accuracy of each network provider, where LCE achieves the best performance.

Acknowledgment

The authors would like to thank the financial support provided by Ministry of Science and Technology, Taiwan (MOST 108-2634-F-155-001) and Qualcomm Taiwan University Research Program.

References

1. N. Ghaboosi and A. Jamalipour, "The geometry of overhearing and its application for location estimation in cellular networks," *IEEE Transactions on Vehicular Technology*, vol. 60, no. 7, pp. 3324–3331, 2011.
2. W.-Y. Chiu and B.-S. Chen, "Mobile location estimation in urban areas using mixed Manhattan/ Euclidean norm and convex optimization," *IEEE Transactions on Wireless Communications*, vol. 8, no. 1, pp. 414–423, 2009.

3. N. Alam, A. Tabatabaei Balaei, and A. Dempster, "A DSRC doppler-based cooperative positioning enhancement for vehicular networks with GPS availability," *IEEE Transactions on Vehicular Technology*, vol. 60, no. 9, pp. 4462–4470, 2011.

4. F. Gustafsson and F. Gunnarsson, "Mobile positioning using wireless networks: Possibilities and fundamental limitations based on available wireless network measurements," *IEEE Signal Processing Magazine*, vol. 22, no. 4, pp. 41–53, 2005.

5. G. Sun, J. Chen, W. Guo, and K. Liu, "Signal processing techniques in network-aided positioning: A survey of state-of-the-art positioning designs," *IEEE Signal Processing Magazine*, vol. 22, no. 4, pp. 12–23, 2005.

6. C. Botteron, A. Host-Madsen, and M. Fattouche, "Effects of system and environment parameters on the performance of network-based mobile station position estimators," *IEEE Transactions on Vehicular Technology*, vol. 53, no. 1, pp. 163–180, 2004.

7. R. Haeb-Umbach and S. Peschke, "A novel similarity measure for positioning cellular phones by a comparison with a database of signal power levels," *IEEE Transactions on Vehicular Technology*, vol. 56, no. 1, pp. 368–372, 2007.

8. B.-C. Liu and K.-H. Lin, "Accuracy improvement of SSSD circular positioning in cellular networks," *IEEE Transactions on Vehicular Technology*, vol. 60, no. 4, pp. 1766–1774, 2011.

9. L. Cong and W. Zhuang, "Hybrid TDOA/AOA mobile user location for wideband CDMA cellular systems," *IEEE Transactions on Wireless Communications*, vol. 1, no. 3, pp. 439–447, 2002.

10. M. McGuire, K. Plataniotis, and A. Venetsanopoulos, "Location of mobile terminals using time measurements and survey points," *IEEE Transactions on Vehicular Technology*, vol. 52, no. 4, pp. 999–1011, 2003.

11. Y. Qi, H. Kobayashi, and H. Suda, "On time-of-arrival positioning in a multipath environment," *IEEE Transactions on Vehicular Technology*, vol. 55, no. 5, pp. 1516–1526, 2006.

12. D.-B. Lin, R.-T. Juang, H.-P. Lin, and C.-Y. Ke, "Mobile location estimation based on differences of signal attenuations for GSM systems," *IEEE Transactions on Vehicular Technology*, vol. 54, no. 4, pp. 1447–1454, 2005.

13. J. Bull, "Wireless geolocation," *IEEE Vehicular Technology Magazine*, vol. 4, no. 4, pp. 45–53, 2009.

14. S.-H. Fang, J.-C. Chen, H.-R. Huang, and T.-N. Lin, "Is FM a RF-based positioning solution in a metropolitan-scale environment? A probabilistic approach with radio measurements analysis," *IEEE Transactions on Broadcasting*, vol. 55, no. 3, pp. 577–588, 2009.

15. M. Chen, T. Sohn, D. Chmelev, D. Haehne, J. Hightower, J. Hughes, A. LaMarca, F. Potter, I. Smith, and A. Varshavsky, "Practical metropolitan-scale positioning for GSM phones." *Ubicomp*, pp. 225–242, 2006.

16. M. Ibrahim and M. Youssef, "CellSense: An accurate energy-efficient GSM positioning system," *IEEE Transactions on Vehicular Technology*, vol. 61, no. 1, pp. 286–296, 2012.

17. S.-H. Fang, B.-C. Lu, and Y.-T. Hsu, "Learning location from sequential signal strength based on GSM experimental data," *IEEE Transactions on Vehicular Technology*, vol. 61, no. 2, pp. 726–736, 2012.

18. L. Anthony, C. Yatin, C. Sunny, H. Jeffrey, S. Ian, S. James, S. Timothy, H. James, H. Jeff, P. Fred, T. Jason, P. Pauline, B. Gaetano, and S. Bill, "Place Lab: Device positioning using radio beacons in the wild," *Lecture Notes in Computer Science*, vol. 3468, pp. 116–133, 2005.

19. M. Caceres, F. Penna, H. Wymeersch, and R. Garello, "Hybrid cooperative positioning based on distributed belief propagation," *IEEE Journal on Selected Areas in Communications*, vol. 29, no. 10, pp. 1948–1958, 2011.

20. C.-Y. Yang, B.-S. Chen, and F.-K. Liao, "Mobile location estimation using fuzzy-based IMM and data fusion," *IEEE Transactions on Mobile Computing*, vol. 9, no. 10, pp. 1424–1436, 2010.

21. B. Sieskul, F. Zheng, and T. Kaiser, "A hybrid SS-ToA wireless NLoS geolocation based on path attenuation: ToA estimation and CRB for mobile position estimation," *IEEE Transactions on Vehicular Technology*, vol. 58, no. 9, pp. 4930–4942, 2009.

22. A. Catovic and Z. Sahinoglu, "The Cramer-Rao bounds of hybrid TOA/RSS and TDOA/RSS location estimation schemes," *IEEE Communications Letters*, vol. 8, no. 10, pp. 626–628, 2004.

23. S.-H. Fang and T.-N. Lin, "Cooperative multi-radio localization in heterogeneous wireless networks," *IEEE Transactions on Wireless Communications*, vol. 9, no. 5, pp. 1547–1551, 2010.

24. T. Kleine-Ostmann and A. Bell, "A data fusion architecture for enhanced position estimation in wireless networks," *IEEE Communications Letters*, vol. 5, no. 8, pp. 343–345, 2001.
25. Z. li Wu, C. hung Li, J.-Y. Ng, and K.R. Leung, "Location estimation via support vector regression," *IEEE Transactions on Mobile Computing*, vol. 6, no. 3, pp. 311–321, 2007.
26. Y. Jin, W.-S. Soh, and W.-C. Wong, "Indoor localization with channel impulse response based fingerprint and nonparametric regression," *IEEE Transactions on Wireless Communications*, vol. 9, no. 3, pp. 1120–1127, 2010.
27. K.W. Kolodziej and J. Hjelm, *Local Positioning Systems: LBS Applications and Services*. CRC Taylor & Francis, 2006.
28. K. Axel, *Location-Based Services: Fundamentals and Operation*. John Wiley & Sons, 2005.
29. C. Figuera, L. Rojo-Alvarez, Jose, I. Mora-Jimenez, A. Guerrero-Curieses, M. Wilby, and J. Ramos-Lopez, "Time-space sampling and mobile device calibration for WiFi indoor location systems," *IEEE Transactions on Mobile Computing*, vol. 10, no. 7, pp. 913–926, 2011.
30. S.-P. Kuo and Y.-C. Tseng, "Discriminant minimization search for large-scale RF-based localization systems," *IEEE Transactions on Mobile Computing*, vol. 10, no. 2, pp. 291–304, 2011.
31. T. Wigren, "Adaptive enhanced cell-ID fingerprinting localization by clustering of precise position measurements," *IEEE Transactions on Vehicular Technology*, vol. 56, no. 5, pp. 3199–3209, 2007.
32. C.-L. Chen and K.-T. Feng, "Hybrid location estimation and tracking system for mobile devices." *Vehicular Technology Conference*, vol. 4, pp. 2648–2652, 2005.
33. P. Kemppi and S. Nousiainen, "Database correlation method for multi-system positioning." *Vehicular Technology Conference*, vol. 2, pp. 866–870, 2006.
34. H. Wymeersch, J. Lien, and M. Win, "Cooperative localization in wireless networks," *Proceedings of the IEEE*, vol. 97, no. 2, pp. 427–450, 2009.
35. Z. Wang and S. Zekavat, "Omni-directional mobile NLOS identification and localization via multiple cooperative nodes," *IEEE Transactions on Mobile Computing*, vol. 99, no. 12, p. 1, 2011.
36. J. Rantakokko, J. Rydell, P. Stromback, P. Handel, J. Callmer, D. Tornqvist, F. Gustafsson, M. Jobs, and M. Gruden, "Accurate and reliable soldier and first responder indoor positioning: Multisensor systems and cooperative localization," *IEEE Wireless Communications*, vol. 18, no. 2, pp. 10–18, 2011.
37. O. Abumansoor and A. Boukerche, "A secure cooperative approach for nonline-of-sight location verification in VANET," *IEEE Transactions on Vehicular Technology*, vol. 61, no. 1, pp. 275–285, 2012.
38. N. Patwari, J. Ash, S. Kyperountas, I. Hero, A.O., R. Moses, and N. Correal, "Locating the nodes: Cooperative localization in wireless sensor networks," *IEEE Signal Processing Magazine*, vol. 22, no. 4, pp. 54–69, 2005.
39. A. Conti, M. Guerra, D. Dardari, N. Decarli, and M. Win, "Network experimentation for cooperative localization," *IEEE Journal on Selected Areas in Communications*, vol. 30, no. 2, pp. 467–475, 2012.
40. R. Ouyang, A.-S. Wong, and C.-T. Lea, "Received signal strength-based wireless localization via semidefinite programming: Noncooperative and cooperative schemes," *IEEE Transactions on Vehicular Technology*, vol. 59, no. 3, pp. 1307–1318, 2010.
41. S. Mazuelas, Y. Shen, and M. Win, "Information coupling in cooperative localization," *IEEE Communications Letters*, vol. 15, pp. 737–739, 2011.
42. S. Zhu and Z. Ding, "Distributed cooperative localization of wireless sensor networks with convex hull constraint," *IEEE Transactions on Wireless Communications*, vol. 10, no. 7, pp. 2150–2161, 2011.
43. M. Win, A. Conti, S. Mazuelas, Y. Shen, W. Gifford, D. Dardari, and M. Chiani, "Network localization and navigation via cooperation," *IEEE Communications Magazine*, vol. 49, no. 5, pp. 56–62, 2011.
44. H. Chen, Q. Shi, R. Tan, H. Poor, and K. Sezaki, "Mobile element assisted cooperative localization for wireless sensor networks with obstacles," *IEEE Transactions on Wireless Communications*, vol. 9, no. 3, pp. 956–963, 2010.
45. J. Zhou, J. Shi, and X. Qu, "Landmark placement for wireless localization in rectangular-shaped industrial facilities," *IEEE Transactions on Vehicular Technology*, vol. 59, pp. 3081–3090, 2010.
46. Y. Qi and H. Kobayashi, "On relation among time delay and signal strength based geolocation methods." *IEEE Global Telecommunications Conference*, vol. 7, pp. 4079–4083, December 2003.

Section III

Estimation Methodologies for Cooperative Localization and Navigation

16

Cooperative Ranging-Based Detection and Localization: Centralized and Distributed Optimization Methods

Reza Shahbazian, Francesca Guerriero, and Seyed Ali Ghorashi

CONTENTS

16.1 Introduction to Localization

The availability of location information is of great importance in many commercial and governmental applications. According to the Cambridge English dictionary, the word "localize" means to *make local, fix in or assign or restrict to a particular place*. The localization task is to determine the spatial coordination of objects, in two-dimensional (2D), that is, $[x, y]$ or three-dimensional (3D), that is, $[x, y, z]$ environments. The first efforts on modern localization systems go back to the previous century where preliminary radar systems performed continuous detection, localization, and tracking. The tracking task is to localize the objects continuously in time based on the required time resolution. Normally the target coordinates have correlation during tracking and therefore the algorithms used are different from those implemented in localization. In this chapter, we focus on detection and mainly localization techniques.

16.1.1 The Necessity of Localization

Why with improvements in global positioning systems such as GPS or GLONASS, is localization still an open issue of research? The answer is that equipping all electronic devices and sensors, such as network nodes with GPS, is costly and not applicable in practice because of the high battery usage (Gazestani et al. 2017). The GPS also does not operate well in harsh environments, such as indoors, since the signals cannot propagate through obstacles (Ghari et al. 2017). On the other hand, in some new applications like intelligence transportation systems (ITS), the accuracy needed is far beyond what GPS can offer (Papadimitratos et al. 2009). In such environments or applications, localization errors can be unacceptably large. This is why localization algorithms are needed. To elaborate, consider the sample diagram of an ITS depicted in Figure 16.1, where the smart vehicles

FIGURE 16.1
A sample diagram of a vehicular network in the intelligent transportation system (ITS) where the vehicles are equipped with many different sensors including GPS but need to have a precise awareness of location (centimeter level) for autonomous driving and collison avoidance.

should be aware of their own location with high precision (centimeter level) to avoid accidents in autonomous driving systems.

Localization is a difficult problem to solve because several issues need to be considered, including accuracy, computational cost, power consumption, security, coverage range, scalability, and complexity of hardware or software (Shahbazian and Ghorashi 2017a,b). Depending on the application, the network may be on the 2D plane, 3D volume (e.g., for underwater and atmospheric monitoring) (Teymorian et al. 2009), or 3D surface (e.g., monitoring in mountains) (Zhao et al. 2012). The problem in 3D surfaces is more challenging compared to 2D, and it is not always straightforward to apply 2D approaches to localization on 3D surfaces. A general 3D surface is not always localizable, given only the surface distance constraints. A 3D surface is localizable if it has a unique embedding under given distance and height constraints. Only a special case of 3D surfaces is proven to be localizable based on distance and height information, in which the projection of 3D points to the 2D plane would not cause any overlap (Hornung et al. 2013).

16.1.2 Classification of Localization Methods

From a processing perspective, localization methods may be performed in a centralized or distributed manner. In centralized methods, all the information is transferred and processed in a center, while in distributed methods the localization of targets is performed in a decentralized manner with no center involved. The performance of processing in centralized algorithms highly depends on the center, and any wreck in the center may fail the whole task. The security issues in transferring data and robustness of localization are some of the great challenges in centralized algorithms. Transferring a huge amount of data to the center for decision making and waiting for the estimated coordinates takes time, which cannot be tolerated in some applications. All these reasons have encouraged researchers to contribute more on distributed algorithms and methods to solve the localization problem (Shahbazian and Ghorashi 2017a,b).

Localization can be performed in a non-cooperative or cooperative manner. In cooperative methods, the informative elements collaborate to improve the estimation precision (Wymeersch et al. 2009). The reasons that localization accuracy increases in cooperative methods are explained in Section 16.2.3.

Localization can be performed by range-free or ranging-based methods. We consider range-based techniques in which the information is normally gained by the measurement of received signal strength (RSS) (Shirahama and Ohtsuki 2008), the time-based range extraction including time of arrival (TOA) or time difference of arrival (TDOA) (Wang et al. 2011), and signal phase methods including angle of arrival (AOA), also called direction of arrival (DOA) (Fardad et al. 2014). Range-free localization algorithms use connectivity information among the nodes to determine the positions of unknown nodes (He et al 2003). Since the range-based methods require a hardware setup that might be complex and costly, the range-free methods could be a possible solution to hardware-limited problems of localization accuracy (Dil et al. 2006).

When only one object is located, the problem goes by target or node localization (Shahbazian and Ghorashi 2017a,b). When the problem aims to localize all the nodes of a network—for instance, the location of all nodes in the wireless sensor network (WSN)—the problem is called network localization (Ghari et al. 2017). The categorization mentioned for localization is presented in Figure 16.2.

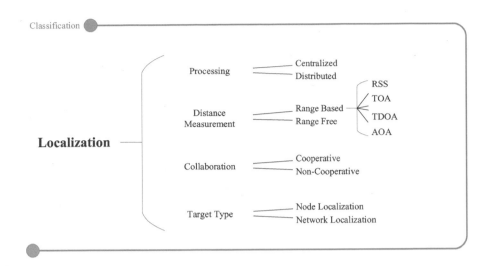

FIGURE 16.2
The classification of localization methods; a centralized or distributed processing could be used in a cooperative and non-cooperative manner using the range-based measurements or range-free techniques for network localization or to localize a target.

16.1.3 The Difference between Network and Target Localization

In network localization problems, generally two kinds of nodes, namely anchors and agents, exist (Figure 16.3). Anchors have known positions, while the location of agents is unknown (Ghari et al. 2017). Each agent is localizable alone (could be performed in agent or center) based on range measurements from at least three distinct anchors in 2D and four distinct anchors in 3D environments. One example is the GPS, in which an agent can determine its location based on the signals received from GPS satellites. In GPS systems, the target localizes itself based on the signals received from the satellite.

In network localization problem one of the following cases occurs:

- The pairwise distance measurements between agents and anchors (agent to agent and agent to anchor) are sent to the center for further processing and applying the proper localization algorithm, usually optimization-based methods.

- The agents localize themselves, applying a proper algorithm. In a non-cooperative manner, the agent needs to have measurements between itself and at least three anchors with a known location. In a cooperative manner, this number of direct connections is not a necessity, and the pairwise measurements between agents compensate for that.

The constellation of sample WSN with 100 agents (sensors) and 5 anchors is depicted in Figure 16.3.

In a target localization problem, it is generally assumed that the agents know their own location (for instance by using GPS) and try to estimate the location of one or several target nodes (Cheng et al. 2012). The location awareness of agents is noisy and affects the performance of localization techniques. For instance, single-frequency commercial GPS systems provide the location information with 10 m^2 of error (Drawil et al. 2013). A 3D

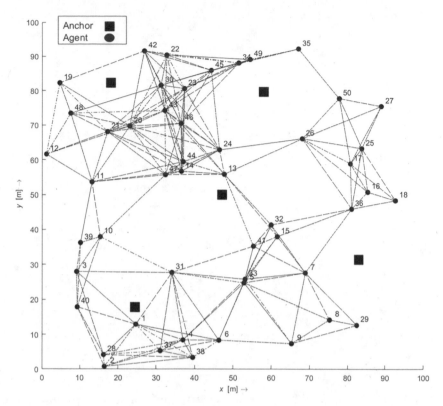

FIGURE 16.3
Diagram of a sample WSN used for network localization in a 2D environment with 100 sensors and 5 anchors. The anchors are shown with squares and agents with small circles.

sensor network is illustrated in Figure 16.4, in which agents are aware of their location (with a noise variance) and try to estimate the location of the target (shown with square).

16.1.4 Distance Measurements Based on Ranging Techniques

The localization process typically consists of two operations, in which agents make intra- and inter-node measurements using different sensors, and a location update, in which agents infer their own positions using an algorithm that incorporates both prior pieces of knowledge of their positions and new measurements.

In this section, we introduce ranging-based measurement techniques, which are normally affected by impairments such as topology, multipath propagation, environmental conditions, interference, noise, and clock drift. The inter-node measurements include waveform, ranging, and direction, while intra-node measurements include data provided by acceleration, angular velocity, and Doppler effects (Wilhjelm and Pedersen 1993).

The distance measurements can be inferred with from received waveforms. Commonly used techniques include TOA, TDOA, AOA (also called DOA) and RSS.

By using TOA, the possible positions of an agent are given on a circle with the anchor at the center. It can be obtained by either the one-way time of flight of a signal in a synchronized network or the round-trip time of flight in a non-synchronized network (Wang et al. 2011).

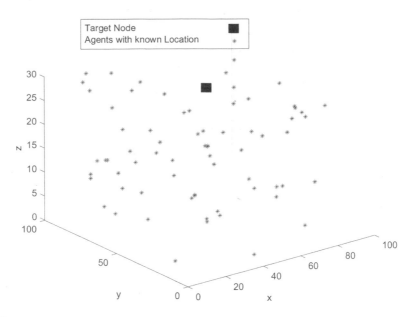

FIGURE 16.4
Sample diagram of target localization in a 3D environment. All agents know their own location with known error variance.

The TDOA provides possible positions of an agent on the hyperbola determined by the difference in the TOAs from two anchors located at the foci. TDOA requires synchronization among anchors but not necessarily with the agents (Kaune et al. 2011). The AOA can be obtained using an array of antennas, based on the signals received time at different antennas. AOA usually requires more expensive hardware equipment compared with other measurement techniques, while it achieves high accuracy (Fardad et al. 2014).

RSS is also used to estimate the propagation distance between nodes. This technique can be implemented using low-complexity circuits. RSS is the most common method of range; however, it has limited accuracy due to the difficulty in precisely modeling the relationship between the RSS and the propagation distance. The receiving agent converts the measured RSS into a distance estimate using path loss models. The choice of model strongly affects the ranging accuracy. A widely used model of pass loss is given as follows (Shirahama and Ohtsuki 2008):

$$P_r(d) = P_0 - 10\gamma \log_{10} d + S \tag{16.1}$$

where $P_r(d)$ (dBm) is the received power, P_0 is the received power (dBm) at 1 m, d (meters) is the receiver's distance from the transmitter, γ is the path loss exponent, and S (dB) is the large-scale fading (shadowing) that usually is modeled as a Gaussian random variable with zero mean and standard deviation σ_S. RSS-based ranging suffers from a mismatch between distance and signal attenuation leading to inaccurate distance estimates, especially in cluttered environments. The comparison between different ranging techniques is summarized in Table 16.1.

TABLE 16.1

Comparison of Different Ranging Techniques Used for Localization

Distance Measurement Technique	Advantage	Disadvantage
TOA	Accurate	Needs synchronization between anchors and agents Expensive
TDOA	Synchronization only between anchors	Could be Expensive
RSS	Simple hardware circuit	Affected by environment Low accuracy
AOA	Accurate	Very expensive equipment

Note: RSS is the cheapest in implementation, highly affected by the signal's propagation environment. Time-based methods need some kind of synchronization. The angle-based methods are the most precise with expensive implementation.

16.2 Cooperative Detection and Localization

16.2.1 Cooperative Target Detection

In many applications, such as surveillance, it is crucial to detect the presence of the target—for instance, an intruder in a short time—and then find its location for further actions. The first action is called detection, that is a necessity for localization while often omitted in localization research problems. In target detection, it is very important to satisfy the quality of service (QoS) in terms of high detection and very low false alarm probability while the algorithm performs in almost real time (Guerriero et al. 2010).

In a general detection, the problem model could be expressed in Equation 16.2. It is assumed that the target transmits a signal, say $s_i(t)$, that is received in agents, say $y_i(t)$. The received signal is not equal to transmitted $s_i(t)$, as a cause of receiver noise, $n_i(t)$ and communication channel effect, h_i. We assume that the nodes can exchange information with other nodes in their communication range called neighbors. In the mathematical model, normally two hypotheses are considered, H_0 and H_1. The former occurs when the target is absent, and H_1 occurs when the target is present. The noise could be modeled as many available types, normally as an additive white Gaussian noise (AWGN) where $n_i(t) \sim N(0, \sigma_k^2)$. To reduce the problem complexity, it is usually assumed that $s_i(t)$ and $\{n_i(t)\}$ are independent and therefore (Shahbazian and Ghorashi 2017a,b),

$$y_i(t) = \begin{cases} n_i(t), & H_0 \\ h_i s_i(t) + n_i(t), & H_1 \end{cases} \tag{16.2}$$

The detection process can be performed both in a centralized and distributed manner. In a centralized manner, the measurements or decision of each agent regarding the energy of the target is transferred to the center that makes the final decision on the presence of the target. In a distributed manner, each agent decides based on its own and neighbors' measurements according to a distributed algorithm. At the start, the decision of agents might be different. However, considering an efficient algorithm and cooperation, their decision could converge. Meaning that without any center, the network could decide and reach an agreement based on that decision. Assume that each node measures m samples in a predefined time and then calculates the energy of the received signal. A number of samples m are chosen according to the required resolution. By comparing the Y_i with a predefined or threshold (could be chosen dynamically) denoted by λ each node decides on the presence of the target.

$$Y_i = \sum_{t=0}^{m-1} |y_i(t)|^2 \tag{16.3}$$

Considering AWGN channels, the test statistic of the ith node is given by Y_i. Since Y_i is the sum of the squares of m Gaussian random variables, it can be shown that Y_i^2/σ_i^2 follows a central chi-square, χ^2 distribution with m degrees of freedom if H_1 is true; otherwise, it would follow a non-central χ^2 distribution with m degrees of freedom and parameter h_i. Defining local signal-to-noise (SNR) ratio with η_i and E_s as transmitted signal energy, and according to the central limit theorem, if the number of samples m is large enough (e.g., 10), the test statistics Y_i are asymptotically normally distributed with the statistics given as follows:

$$\eta_i = \frac{E_s |h_i|^2}{\sigma_i^2},$$

$$E_s = \sum_{t=0}^{m-1} |s(t)|^2 \tag{16.4}$$

$$E(Y_i) = \begin{cases} m\sigma_i^2, & H_0 \\ (m+\eta_i)\sigma_i^2, & H_1 \end{cases}$$

$$\text{Var}(Y_i) = \begin{cases} 2m\sigma_i^4, & H_0 \\ 2(m+2\eta_i)\sigma_i^2, & H_1 \end{cases} \tag{16.5}$$

After the measurements, each node enters the collaborative data exchange and decision-making process. In centralized and distributed algorithms, it is possible to weight the measurement or decision of nodes. Let $\omega_i \geq 0$, be the weight for measurements of ith the node, then the final decision on target detection in a centralized manner is achieved as presented in Equation 16.6, where Y_F is the combined received energy in nodes, $\vec{\omega} = [\omega_1, \omega_2, \ldots, \omega_N]^T$ and $\mathbf{Y} = [Y_1, Y_2, \ldots, Y_N]^T$.

$$Y_F = \sum_{i=1}^{N} \omega_i Y_i = \vec{\omega}^T \mathbf{Y}. \tag{16.6}$$

In distributed algorithms, the simple weight vector in centralized form turns into a $N \times N$ weighting matrix, $\mathbf{W} = [\omega_{il}]$ where ω_{il}, is the weight given to the measurement of a node l by node i. Defining k as iteration index and $\mathbf{N_i}$ as the set of neighbor nodes for the node i we have:

$$Y_i(k) = \sum_{l \in \mathbf{N_i}} \omega_{il} Y_l(k-1) \tag{16.7}$$

In distributed algorithms, normally, two objectives of cooperation and decentralization are required. At first, nodes should combine the received measurements of their own and neighbor nodes and then update their decisions based on an algorithm. It is obvious that all the weights of measurements should be non-negative, $\omega_i \geq 0$ and $\sum_{l \in N_i} \omega_{il} = 1$. This means that the weighting matrix is left stochastic (Tu and Sayed 2012).

16.2.2 Cooperative Localization Algorithms

In centralized localization, every node sends ranging measurements to a central processing unit that runs the localization algorithm. If the coordinates are required only at the center, the operation is complete; in cases where nodes need to know their position, the center has to communicate to every node, which presents several problems. The main challenge in central processing is traffic, since each node must send data to the center. This traffic grows while the network expands. Therefore, in a large-scale network, solving the localization problem is very complex. Other challenges are resilience to failure, security, and privacy issues that are typical when a centralized localization is considered.

The distributed localization is well suited because the problem is naturally decentralized, and nodes are distributed in the space. In distributed localization, no central exists and all nodes perform the required computations. These distributed and centralized methods are both applicable to node and network localization problems.

16.2.3 The Effect of Cooperation on Localization Accuracy

Localization accuracy is bounded due to noise, fading, shadowing, and multipath propagation. The Cramér–Rao bound (CRB) sets a lower bound on the variance of estimates for deterministic parameters (Win et al. 2011). Put simply, the minimum achievable localization error is bounded based on their characteristics. There is an incorrect belief that high-accuracy localization can only be achieved using high-power anchors or a high-density anchor deployment, which are cost-prohibitive and impractical in realistic scenarios. A practical and promising solution to address the need for high-accuracy algorithms is cooperative localization, where agents cooperate and share their information regarding the target.

The Fisher Information Matrix (FIM) is defined as the covariance of score function. It is a curvature matrix and has an interpretation as the negative expected Hessian of log-likelihood function. Employing the cooperation in the network, the FIM becomes richer. Simply stated, cooperation adds information, and this information decreases the localization error. It is known that the accuracy of localization is affected by two factors: the quality of pair-wise measurements and the network topology (Patwari et al. 2003).

Considering the WSN, the localization information from anchors can be expressed in a canonical form as one-dimensional information from individual anchors. The cooperation always improves the localization accuracy since it adds information in comparison with non-cooperative localization. To illustrate the effect of cooperation on FIM, a sample network with three anchors is considered that corresponds to three individual information presented in FIM, say k_1, k_2, k_3 as illustrated in Figure 16.5. In this matrix, every $C_{i,j}$ corresponds to cooperation between the agent i and j. In non-cooperative networks, the off-diagonal elements $C_{i,j}$ become 0. Therefore, the $C_{i,j}$ is the extra information that cooperation adds to FIM. It is also proven that the lower bound of localization error is related to the inverse of the FIM. Therefore, richer the FIM gets, the more accuracy increases.

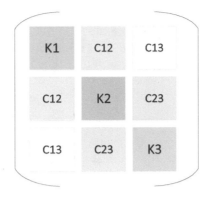

FIGURE 16.5
The Fisher Information Matrix (FIM) for a sample cooperative WSN where diagonal elements correspond to anchors and off-diagonal elements correspond to cooperation in the network.

16.3 Optimization Based Centralized Network Localization

In the past, localization problems were solved algebraically and computed by least-squares solution (Guvenc et al. 2007) to hyperbolic equations called multi-alteration (Zhou et al. 2012). Nowadays, the optimum solution for sensor network localization is provided using optimization. One of the first proposed centralized convex optimization models for wireless sensor network localization is second-order cone programming (SOCP). However, this relaxation method needs a large number of anchors on the area boundary to achieve acceptable performance. Semi-definite programming (SDP) relaxation method was later proposed for WSN localization (Biswas et al. 2006). Other SDP-based methods proposed to improve the performance of localization. The maximum likelihood (ML)-based SDP methods are one of these methods that provide higher accuracy in the price of a longer solution time compared with original SDP (Simonetto and Leus 2014). To reduce the solution time, a method called smaller SDP (SSDP) was proposed, which further relaxes the original SDP, in which a single semi-definite matrix cone is relaxed into a set of small-size semi-definite matrix cones (Wang et al. 2006). The research on improving the performance of SDP is still an open issue. In recent developments, researchers improved the performance of WSN localization in terms of robustness and accuracy, especially in non-line of sight (NLOS) and harsh environments (Ghari et al. 2017).

However, in all SDP solutions, when the size of the SDP problem increases, the dimension of the matrix cone increases simultaneously, and the amount of unknown variables increases non-linearly. It is known that the arithmetic operation complexity of the SDP is at least $O(n^2)$ to obtain an approximate solution. This complexity prevents solving large size problems. On the other hand, the impact of noise on distance measurement and estimation error is important, and this effect varies inversely with problem size (Biswas et al. 2006).

16.3.1 Cooperative Centralized Network Localization Using SDP Relaxation

In this section, we present the system model for cooperative centralized network localization (Biswas et al. 2006). We consider a WSN with m anchors (known positions) and n nodes (unknown positions) in a two-dimensional (2D) environment. Some notations are as follows. **I**, **e**, and **0** denote the identity matrix, the vector of all ones and the vector of all zeros, respectively. The 2-norm of a vector **x** is denoted by $\|\mathbf{x}\|$. A positive semi-definite matrix **X** is represented by $\mathbf{X} \succ 0$. The position of anchor nodes is presented by the vector $\mathbf{V}_a = \{a_1, a_2, \ldots, a_m\}$ and the Euclidean distance between x_j and x_i is denoted as d_{ij} and between a_k and x_j is denoted by d_{jk} as $d_{ij} = \|x_i - x_j\|$ and $d_{jk} = \|x_j - x_k\|$.

$$\text{Find} \quad \mathbf{X} \in R^{2 \times n}$$
$$\text{S.t.} \quad Y_{ii} - 2Y_{ij} + Y_{jj} = \bar{d}_{ij}^2, \forall (i,j) \in N_s$$
$$Y_{jj} - 2X_j^T a_k + a_k^2 = \bar{d}_{jk}^2, \forall (j,k) \in N_a \tag{16.8}$$
$$N_s = \left\{(i,j)\big| x_i - x_j < r\right\}, \quad N_a = \left\{(j,k)\big| x_j - x_k < r\right\}$$
$$\mathbf{Y} = \mathbf{X}^T \mathbf{X}$$

where r is the communication range, and $\mathbf{X} = [x_1, \ldots, x_n]$, $\bar{d}_{ij}$ and $\bar{d}_{jk}$ are noisy range measurements. The problem presented in Equation 16.8 is non-convex and may be relaxed by using SDP relaxation, presented as follows (Biswas et al. 2006):

$$\mathbf{Y} \succeq \mathbf{X}^T \mathbf{X} \rightarrow \mathbf{Z} \succeq 0; \mathbf{Z} = \begin{pmatrix} \mathbf{I}_2 & \mathbf{X}^T \\ \mathbf{X} & \mathbf{Y} \end{pmatrix} \tag{16.9}$$

The solution to a problem introduced in (16.9) could be gained by numerical values applied to CVX. CVX is a MATLAB®-based modeling system for convex optimization. CVX allows constraints and objectives to be specified using standard MATLAB expression syntax (Grant et al. 2008). An example of network localization using SDP is presented in Section 16.2.2.

16.3.2 SDP-Based Network Localization Evaluation

In this section, we present the simulation results of network localization by using original SDP relaxation. The parameters used for the simulation are presented in Table 16.2.

To perform this simulation, it is assumed that the pairwise distance measurements are available but in noisy format. We have deployed 100 sensor agents and 7 anchors in a normalized environment, randomly. The exact location of agents and anchors, including the estimated location by implemented SDP relaxation and the localization error, is depicted in Figure 16.6. The noise factor is also a parameter to add random noise (AWGN) to the

TABLE 16.2

The Simulation Parameters Used for Network Localization Using SDP Relaxation

Simulation Parameter	Value
Simulation area	Normalized 1×1
Number of sensors	100
Number of anchors	7
Communication range	0.3
Noise factor	0.15

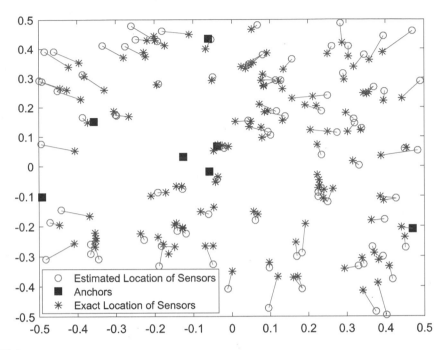

FIGURE 16.6

A sample network localization using SDP relaxation in a normalized simulation area. The anchors are shown with squares, the sensor's exact location is shown with start, and the estimated location of agent sensors is shown with circles. The line between exact and estimated locations corresponds to localization error.

normalized distance measurements. It means that to model the distance measurements as practice, we first deploy the sensors in the network, knowing their exact location. The pairwise measurements are extracted from these coordinates, corrupted by noise. The normalized noise distances are produced as follows:

$$\bar{d}_{ij} = d_{ij} \cdot (1 + randn(1) \times \text{noise factor})$$
$$\bar{d}_{jk} = d_{jk} \cdot (1 + randn(1) \times \text{noise factor})$$

(16.10)

As can be seen in Figure 16.6, the localization error of agents is different based on their communications or distance to the anchors. To evaluate and compare the algorithms, the average error of network localization is used defined in Equation 16.10.

$$\text{Average Error} = \frac{1}{n} \cdot \sum_{j=1}^{n} \| \bar{x}_j - a_j \|$$

(16.11)

16.4 Cooperative Distributed Target Localization

Similar to centralized methods, it is assumed that the target node transmits a radio frequency or acoustic signal that is received in the agents. As mentioned earlier, in the target localization problem, it is assumed that the agents know their own location and try to estimate the target's location based on their own computational power. In range-based methods, distance measurements toward the target are achieved by using RSS, TOA, or TDOA techniques. The rest is updating the estimated location based on their own measurements in non-cooperative methods and both their own and neighbor agents' measurements in cooperative ones. In practice, the location awareness in agent nodes is noisy because of non-exact location estimators such as GPS. Many distributed algorithms are studied in the literature. The most important parameters in distributed algorithms are their convergence guarantee and rate and estimation accuracy. The available research works differ by considering different practical parameters such as distance measurement techniques, optimality or sub-optimality of the solution, communication link errors in cooperative methods, and many others.

In the mathematical model, the target localization problem is considered to be a network as an undirected graph with N vertices as agents. The neighborhood of the ith agent is N_i which denotes the placed nodes in communication range of the ith node. The node degree is defined as the number of neighbors for each node, say $\deg(i) = |N_i|$. The target transmits a signal, say $S_i(t)$, and the received signal in agents, say $Y_i(t)$, is not equal to transmitted one, because of noise, say $n_i(t)$ and communication channel effect, say h_i. In cooperative methods, the nodes can exchange information with other nodes in their communication range called neighbors. Besides the receiver noise, the imperfect communications—for instance, link failure probability—may influence the problem. A simple diagram of the modeled network is shown in Figure 16.7.

The circles in Figure 16.7 show the communication range of agents. For simplicity, the circle is drawn only for a few numbers of nodes. The localization can be defined as an

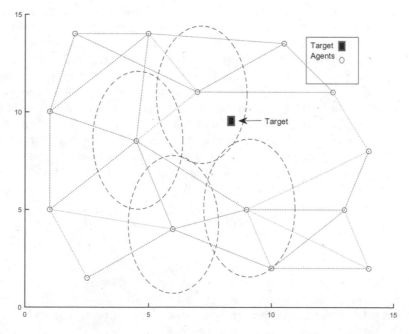

FIGURE 16.7

Sample diagram of the sensor network in which agents with known location cooperate in communication range to localize the target. The target is shown with square and agents are shown with a small circle. The dashed lines of large circles stand for communication range.

optimization problem, in which the goal is to minimize the estimation error, called cost function. The error can be modeled as a cost function, $Cost(\mathbf{x})$ where $\mathbf{x}$ is the estimated. The optimization problem presented in Equation 16.12 needs to be solved where $\mathbf{x} = [x, y]$.

$$\mathbf{x} \triangleq \arg\min_x Cost(\mathbf{x}) \qquad (16.12)$$

We may categorize the distributed localization algorithms into stochastic and classic methods. The Gauss-Newton based, consensus-based and diffusion adaptation-based methods are considered as stochastic optimization solutions. On the other hand, the solutions based on ADMM lie in the classic optimization solutions. The alternating direction method of multipliers (ADMM) is a simple but powerful algorithm that solves optimization problems, decomposing them into smaller local sub-problems, which are easier to handle. The solutions to these local sub-problems are coordinated to find the solution to a global problem. This algorithm is well suited for distributed optimization. Today this approach is on the center of attention, owing to the presence of large-scale distributed computing systems and the need to solve massive optimization problems. ADMM can be applied to both convex and non-convex problems (Boyd et al. 2011).

In all distributed methods, a common assumption is that the problem's cost function can be separated among agents. This means that if $Cost(\mathbf{x})$ is the cost function of target localization in the centralized approach, it is equal to $cost(\mathbf{x}) = \sum_j cost_j(\mathbf{x})$. Therefore, the linear summation of cost functions used to solve the optimization problem in each agent separately is equal to the main cost function of the network. It is obvious that this equality is not necessarily correct, and therefore the upper band of localization accuracy in distributed methods is what can be achieved in a centralized manner with the same information. In other words, the solution provided in distributed solutions is sub-optimal compared to the solution provided by the central solution to the same problem. However, some recent developments in distributed algorithms provide almost the same accuracy with high reliability and scalability and less sensitivity to network changing parameters.

The distributed algorithms are normally solved in an iteration manner to reach an agreement on the predefined objective. For instance, considering Equation 16.13, the objective is to estimate X in an iterative cooperative manner. Based on a simple consensus-based algorithm (Tu and Sayed 2012), the updating variable of each agent, say i in iteration k, depends on the decision of the same agent and its neighbors in previous iterations. It is obvious that there is no general guarantee on the convergence of network. In the case of convergence, the agreement speed (number of iterations before convergence) and its precision are also important. All these parameters provide the criteria to evaluate the performance of a distributed localization algorithm.

$$X_i(k) = \sum_{l \in N_i} \omega_{il} X_l(k-1) - \mu_i \nabla_{X} \cdot J_i(X_i(k-1)) \qquad (16.13)$$

In Equation 16.13, k denotes the iteration index, ∇ is the gradient vector, $J(X)$ is the Jacobian, μ_i is the step size satisfying, $X_i(k)$ is the updated decision variable for a node i in the iteration k, and ω is the weight given by an agent to the decision of neighbor nodes. Choosing this setting as an initial point to start the algorithm for each node is a necessity. The Jacobian matrix is normally defined as follows:

$$\nabla_X J_i(X) \triangleq \begin{bmatrix} \dfrac{\partial J(X)}{\partial X_1} & \dfrac{\partial J(X)}{\partial X_2} & \cdots & \dfrac{\partial J(X)}{\partial X_N} \end{bmatrix} \qquad (16.14)$$

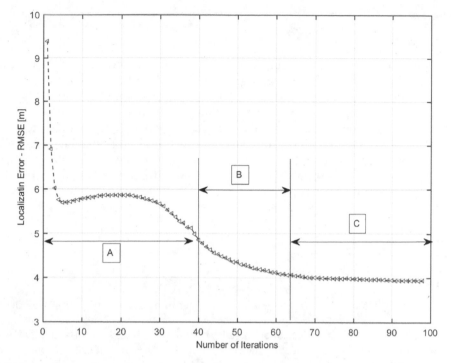

FIGURE 16.8
The localization error in terms of root mean square error (RMSE) versus a number of iterations for a stochastic distributed localization algorithm. The area is divided into three parts, A, B, and C. In area labeled A, the network is far from agreement on the location of the target and this is why the localization error is high. In area B, the cooperation has affected the network and the localization error has decreased significantly. Area C shows the convergence of the distributed algorithm.

After a predefined number of iterations, the whole network reaches an agreement on the objective function. A general localization error versus a number of iterations for a stochastic distributed algorithm is shown in Figure 16.8. The localization error is shown in root mean square error (RMSE) defined in Equation 16.15 for 2D environments.

$$\mathrm{RMSE} = \sqrt{E\left[(x - \hat{x})^2 + (y - \hat{y})^2\right]} \qquad (16.15)$$

As can be seen in Figure 16.7, we have divided the area to three separate parts, called A, B, and C. Consider the first area, A. At first iterations (iteration = 1) the localization error is high. This was predictable because at first every node decides almost based on its own measurements, and the cooperation has not started. After passing a few iterations—for this example 40—we reach the second area, B, in which the localization error has significantly decreased to an acceptable range. We can clearly see the effect of cooperation between agents on localization error; however, the convergence is not provided for the problem. In area C, the convergence of the network is achieved. For this case, the minimum achievable error by this network topology the implemented algorithm and measurements is almost 4 meters. The minimum error, convergence speed, and computational complexity are the varying parameters that put a value on different distributed localization algorithms.

16.5 Summary

Location awareness is of great importance in many surveillance, commercial, and governmental applications. The normal global positioning systems are not adequate for all environments and applications. This is why finding efficient localization algorithms is still an open research issue. In the literature, many centralized or distributed algorithms are proposed, each proper for a case of localization problems. In this chapter, different classifications of localization algorithms, in terms of processing, distance measurements, cooperation, and the goal of localization task, are presented. The localization problem has been studied more thoroughly by introducing optimization-based centralized and distributed localization algorithms.

References

Biswas, P., Lian, T.C., Wang, T.C. and Ye, Y., 2006. Semidefinite programming based algorithms for sensor network localization. *ACM Transactions on Sensor Networks (TOSN)*, 2(2), pp. 188–220.

Boyd, S., Parikh, N., Chu, E., Peleato, B. and Eckstein, J., 2011. Distributed optimization and statistical learning via the alternating direction method of multipliers. *Foundations and Trends® in Machine Learning*, 3(1), pp. 1–122.

Cheng, L., Wu, C., Zhang, Y., Wu, H., Li, M. and Maple, C., 2012. A survey of localization in wireless sensor network. *International Journal of Distributed Sensor Networks*, 8(12), p.9625215.

Dil, B., Dulman, S. and Havinga, P., 2006, February. Range-based localization in mobile sensor networks. In *European Workshop on Wireless Sensor Networks* (pp. 164–179). Springer, Berlin, Heidelberg.

Drawil, N.M., Amar, H.M., Basir, O.A. et al., 2013. GPS localization accuracy classification: A context-based approach. *IEEE Transactions on Intelligent Transportation Systems*, 14(1), pp. 262–2715.

Fardad, M., Ghorashi, S.A. and Shahbazian, R., 2014. May. A novel maximum likelihood based estimator for bearing-only target localization. In *Electrical Engineering (ICEE), 2014 22nd Iranian Conference on* (pp. 1522–1527). IEEE.

Gazestani, A.H., Shahbazian, R. and Ghorashi, S.A., 2017. Decentralized consensus based target localization in wireless sensor networks. *Wireless Personal Communications*, 97(3), pp. 3587–3599.

Ghari, P.M., Shahbazian, R. and Ghorashi, S.A., 2017. Wireless sensor network localization in harsh environments using SDP relaxation. *IEEE Communications Letters*, 20(1), pp. 137–140.

Grant, M., Boyd, S. and Ye, Y., 2008. CVX: Matlab software for disciplined convex programming. [Online] Available at: https://web.stanford.edu/~boyd/software.html

Guerriero, M., Svensson, L. and Willett, P., 2010. Bayesian data fusion for distributed target detection in sensor networks. *IEEE Transactions on Signal Processing*, 58(6), pp. 3417–3421.

Guvenc, I., Chong, C.C. and Watanabe, F., 2007, April. Analysis of a linear least-squares localization technique in LOS and NLOS environments. In *Vehicular Technology Conference, 2007. VTC2007-Spring. IEEE 65th* (pp. 1886–1890). IEEE.

He, T., Huang, C., Blum, B.M., Stankovic, J.A. and Abdelzaher, T., 2003, September. Range-free localization schemes for large scale sensor networks. In *Proceedings of the 9th Annual International Conference on Mobile Computing and Networking* (pp. 81–95). ACM.

Hornung, A., Wurm, K.M., Bennewitz, M., Stachniss, C. and Burgard, W., 2013. OctoMap: An efficient probabilistic 3D mapping framework based on octrees. *Autonomous Robots*, 34(3), pp. 189–206.

Kaune, R., Hörst, J. and Koch, W., 2011, July. Accuracy analysis for TDOA localization in sensor networks. In *Information Fusion (FUSION), 2011 Proceedings of the 14th International Conference on IEEE* (pp. 1–8). IEEE.

Papadimitratos, P., De La Fortelle, A., Evenssen, K., Brignolo, R. and Cosenza, S., 2009. Vehicular communication systems: Enabling technologies, applications, and future outlook on intelligent transportation. *IEEE Communications Magazine*, 47(11), pp. 84–95.

Patwari, N., Hero, A.O., Perkins, M., Correal, N.S. and O'dea, R.J., 2003. Relative location estimation in wireless sensor networks. *IEEE Transactions on Signal Processing*, 51(8), pp. 2137–2148.

Shahbazian, R. and Ghorashi, S.A., 2017a. Distributed cooperative target detection and localization in decentralized wireless sensor networks. *The Journal of Supercomputing*, 73(4), pp. 1715–1732.

Shahbazian, R. and Ghorashi, S.A., 2017b. Localization of distributed wireless sensor networks using two stage SDP optimization. *International Journal of Electrical and Computer Engineering (IJECE)*, 7(3), pp. 1255–1261.

Shirahama, J. and Ohtsuki, T., 2008, May. RSS-based localization in environments with different path loss exponent for each link. In *Vehicular Technology Conference, 2008. VTC Spring 2008. IEEE* (pp. 1509–1513). IEEE.

Simonetto, A. and Leus, G., 2014. Distributed maximum likelihood sensor network localization. *IEEE Transactions on Signal Processing*, 62(6), pp. 1424–1437.

Teymorian, A.Y., Cheng, W., Ma, L., Cheng, X., Lu, X. and Lu, Z., 2009. 3D underwater sensor network localization. *IEEE Transactions on Mobile Computing*, 8(12), pp. 1610–1621. DOI: 10.1109/TMC.2009.80.

Tu, S.Y. and Sayed, A.H., 2012. Diffusion strategies outperform consensus strategies for distributed estimation over adaptive networks. *IEEE Transactions on Signal Processing*, 60(12), pp. 6217–6234.

Wang, Y., Ma, X. and Leus, G., 2011. Robust time-based localization for asynchronous networks. *IEEE Transactions on Signal Processing*, 59(9), pp. 4397–4410.

Wang, Z., Zheng, S., Boyd, S. and Ye, Y., 2006. Further relaxations of the semidefinite programming approach to sensor network localization. *SIAM Journal Optimization*, 19(2), pp. 655–673.

Wilhjelm, J.E. and Pedersen, P.C., 1993. Target velocity estimation with FM and PW echo ranging Doppler systems I. Signal analysis. *IEEE Transactions on Ultrasonics, Ferroelectrics, and Frequency Control*, 40(4), pp. 366–372.

Win, M.Z., Conti, A., Mazuelas, S., Shen, Y., Gifford, W.M., Dardari, D. and Chiani, M., 2011. Network localization and navigation via cooperation. *IEEE Communications Magazine*, 49(5), pp. 56–62, 2011.

Wymeersch, H., Lien, J. and Win, M.Z., 2009. Cooperative localization in wireless networks. *Proceedings of the IEEE*, 97(2), pp. 427–450.

Zhao, Y., Wu, H., Jin, M. and Xia, S., 2012, March. Localization in 3D surface sensor networks: Challenges and solutions. In *INFOCOM* (pp. 55–63).

Zhou, Y., Li, J. and Lamont, L., 2012, December. Multilateration localization in the presence of anchor location uncertainties. In *Global Communications Conference (GLOBECOM), 2012 IEEE* (pp. 309–314). IEEE.

17

Theory and Algorithms in Distributed Localization for Multi-Vehicle Networks Using Graph Laplacian Techniques

Zhiyun Lin

CONTENTS

Distributed localization aims to solve the coordinate of each vehicle in a network in a distributed manner, given the coordinates of a small set of vehicles as the anchors and local inter-vehicle measurements. This is extremely important for distributed systems such as swarm robotics in GPS-denied or partially GPS-denied environment. This chapter presents several recent results on distributed localization for multi-vehicle networks by using graph Laplacian technique, which is a linear approach and thus ensures global convergence. According to different types of local measurements, distributed localization theory, and algorithms based on range, bearing, or both, information is discussed in a unified linear framework, which holds eminent promises for many applications.

17.1 Introduction

Location information is a piece of very important information in position-based operations and services of autonomous multi-vehicle systems such as unmanned vehicles and satellite formations. However, location is often not known *a priori*; therefore, localization is the task of determining the position (e.g., Euclidean coordinates) of a vehicle in the world coordinate system or the spatial relationships among objects. GPS is a global positioning system and gives the coordinates of receivers. However, when a network of vehicles is in a partially GPS-denied environment such as in urban areas or underwater, not every vehicle in the network is able to localize itself by using GPS. Thus the cooperative localization problem occurs, for which the objective is to determine the Euclidean coordinates of all the vehicles

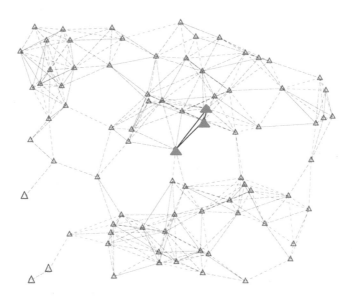

FIGURE 17.1
Cooperative localization: to solve the coordinates of the nodes (hollow triangles) given the coordinates of a small set of nodes (solid triangles) and a collection of inter-vehicle measurements indicated by edges in the graph.

given a collection of inter-vehicle measurements and the Euclidean coordinates of a small number of vehicles in the network (Figure 17.1).

Centralized localization algorithms collect all inter-vehicle measurements and the known Euclidean coordinates of a small number of vehicles in the network at a computing center and then compute every vehicle's location. Some existing techniques include maximum likelihood estimators [7,21], multidimensional scaling (MDS) methods [14,23], semi-definite programming [3,24], optimization-based methods [4,10], etc. Distributed localization algorithms, however, solve every vehicle's location by itself using its own local measured information and the information received from its neighbors. This methodology is much more attractive, especially in distributed systems. Distributed localization includes nonlinear approaches that may not guarantee global convergence, and linear approaches that do ensure global convergence. This chapter mainly presents recent progress on distributed localization with linear approaches, for which the schemes can be run in a concurrent way. The sequential method, which begins with a set of nodes with known coordinates and computes the locations of other nodes one by one or group by group according to certain sequential orders [1,8,12], is not discussed here. Cooperative localization using motion information such as [2,5,16,22,25] is also not covered in this chapter.

The setup for distributed localization is given in the following. For a multi-vehicle network, consider a sensing graph $\mathcal{G} = (\mathcal{V}, \mathcal{E})$ (either directed or undirected) with each node $i \in \mathcal{V}$ representing a vehicle in the network and each edge $(i, j) \in \mathcal{E}$ indicating that node i and j are able to measure some local information such as range, bearing, or both, between i and j. Let $\mathcal{V} = \mathcal{A} \cup \mathcal{S}$ with $\mathcal{A} = \{1, \dots, m\}$ denoting the set of anchor nodes, whose Euclidean coordinates are known in a global coordinate system Σ_g, and $\mathcal{S} = \{m + 1, \dots, n\}$ representing the set of free nodes, whose Euclidean coordinates in Σ_g need to be computed. For each node i, we denote by $p_i \in \mathbb{R}^d$ ($d = 2$ or 3) the coordinate of node i in Σ_g. Localization is usually related to solve p_i, for $i \in \mathcal{S}$, from some linear or nonlinear equations, which come from the constraints in terms of the Euclidean coordinates of all the vehicles and all

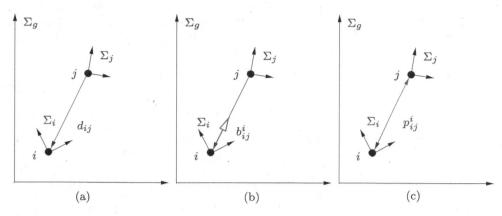

FIGURE 17.2
Three different types of measurements. (a) Range; (b) bearing; (c) relative position.

the locally available inter-vehicle measurements. We then summarize the three types of measurements as follows.

1. Range information: Node i measures the distance d_{ij} about its neighbor $j \in \mathcal{N}_i$, where $d_{ij} = \|p_j - p_i\|$ (Figure 17.2a).

2. Bearing information: Node i measures the bearing b_{ij}^i about its neighbor $j \in \mathcal{N}_i$ in the local coordinate system Σ_i (Figure 17.2b).

3. Range and bearing information: Node i measures the relative position p_{ij}^i about its neighbor $j \in \mathcal{N}_i$ in the local coordinate system Σ_i (Figure 17.2c).

Some notion and notations are given below.

A *configuration* in $\mathbb{R}^d$ of a set of n points is defined by their coordinates in the Euclidean space $\mathbb{R}^d$, denoted as $p = \left[p_1^T, \ldots, p_n^T\right]^T \in \mathbb{R}^{nd}$, where each $p_i \in \mathbb{R}^d$ for $1 \leq i \leq n$. A *framework* in $\mathbb{R}^d$ is a graph $\mathcal{G}$ equipped with a configuration p, denoted as $\mathcal{F} = (\mathcal{G}, p)$. We say that two frameworks $(\mathcal{G}, p)$ and $(\mathcal{G}, q)$ are *congruent*, and we write $(\mathcal{G}, p) \equiv (\mathcal{G}, q)$, if $\|p_i - p_j\| = \|q_i - q_j\|$, $\forall i, j \in \mathcal{V}$. We say that two frameworks $(\mathcal{G}, p)$ and $(\mathcal{G}, q)$ are *similar*, and we write $(\mathcal{G}, p) \sim (\mathcal{G}, q)$, if for some constant $\gamma > 0$, $\|p_i - p_j\| = \gamma \|q_i - q_j\|$, $\forall i, j \in \mathcal{V}$.

$\mathbb{R}$ denotes the set of real numbers and $\mathbb{C}$ stands for the set of complex numbers. ι is the imaginary unit $\sqrt{-1}$. $\mathbf{1}_n$ represents the n-dimensional vector of ones and I_n represents the identity matrix of order n. The symbol $\otimes$ denotes the Kronecker product.

17.2 Distributed Localization Using Range Information

For a multi-vehicle network, if the range information is measured by the vehicles according to the sensing graph, then the distributed localization problem becomes the one using range information. It is also equivalent to the so-called *graph realization problem*: Given an undirected graph $\mathcal{G} = (\mathcal{V}, \mathcal{E})$ and sets of non-negative weights, say $\{d_{ij} : (i, j) \in \mathcal{E}\}$ on edges, the goal is to compute a *realization* of $\mathcal{G}$ in the Euclidean space $\mathbb{R}^d$ for a given dimension d. That is, to place the nodes of $\mathcal{G}$ in $\mathbb{R}^d$ such that the Euclidean distance between a pair of adjacent nodes (i, j) equals to the prescribed weight d_{ij}.

We write down all the range measurements together as a constraint for the multi-vehicle network, that is,

$$\vdots$$
$$\| p_i - p_j \| = d_{ij}, \quad \forall (i,j) \in \mathcal{E}$$
$$\vdots$$

Thus, the localization problem is to solve the above nonlinear equations. In another way, it can also be formulated as an optimization problem

$$\min_{p_{m+1},\dots,p_n} J(p_{m+1},\dots,p_n)$$

where

$$J(p_{m+1},\dots,p_n) = \sum_{i=1}^{n} \sum_{j \in \mathcal{N}_i} \left(\| p_i - p_j \|^2 - d_{ij}^2 \right)^2.$$

However, the cost function is not strictly convex, and indeed it has a large number of saddle points, particularly when the number of vehicles in the network is large, which makes the optimization problem hard to solve or at least hard to devise a globally convergent algorithm.

To overcome this challenge, a new technique based on the barycentric coordinates and graph Laplacian is proposed. The main idea is to derive linear equation constraints by using the barycentric coordinates and then solve the linear equations in a distributed manner.

As the first step, we expect to obtain the following barycentric coordinate representation for each vehicle in the network:

$$p_i = \sum_{j \in \mathcal{N}_i} a_{ij} p_j \tag{17.1}$$

with a_{ij} for $j \in \mathcal{N}_i$ satisfying $\Sigma_{j \in \mathcal{N}_i} a_{ij} = 1$, called the *barycentric coordinate* of vehicle i with respect to its neighbors. The computation of the barycentric coordinate for each node can be locally done by using the range measurements available to vehicle i and its neighbors. We show in the sequel how we compute the barycentric coordinates in 2D and 3D, respectively.

In 2D, as shown in Eren et al. [11], all the vehicles in a network are localizable if and only if the sensing graph is globally rigid and there are at least three anchor nodes. This implies every free node has at least three neighbors. Now, let us consider that a vehicle has exactly three neighbors in the plane, as some examples in Figure 17.3. In this case, the Euclidean coordinate p_i satisfies the following equation [9]

$$\begin{cases} p_i = a_{ij} p_j + a_{ik} p_k + a_{il} p_l, \\ a_{ij} + a_{ik} + a_{il} = 1. \end{cases} \tag{17.2}$$

The barycentric coordinate $\{a_{ij}, a_{ik}, a_{il}\}$ can be obtained by noticing that it is invariant for congruent frameworks [6]. In other words, we can construct a congruent framework

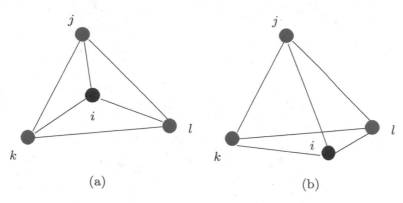

FIGURE 17.3
Two configurations in 2D: vehicle i has three neighbors, namely, j, k, and l.

with the Euclidean coordinates q_i, q_j, q_k, q_l in a specific coordinate system according to the following optimization problem [6]:

$$\min_{q_i,q_j,q_k,q_l} \sum_{r,s\in\{i,j,k,l\}} \left(\| q_r - q_s \|^2 - d_{rs}^2\right)^2$$

$$\text{subject to: } q_i = \begin{bmatrix} 0 \\ 0 \end{bmatrix}; \; q_j = \begin{bmatrix} d_{ij} \\ 0 \end{bmatrix}; \; q_k = \begin{bmatrix} q_{kx} \\ q_{ky} \end{bmatrix} \text{ with } q_{ky} > 0.$$

Next, we compute the barycentric coordinate $\{a_{ij}, a_{ik}, a_{il}\}$ by using the Euclidean coordinates q_i, q_j, q_k, q_l as follows.

$$a_{ij} = \frac{S_{\triangle ikl}}{S_{\triangle jkl}}, \; a_{ik} = \frac{S_{\triangle ijl}}{S_{\triangle jkl}}, \; a_{il} = \frac{S_{\triangle ijk}}{S_{\triangle jkl}}, \tag{17.3}$$

where

$$S_{\triangle ikl} = \frac{1}{2}\begin{vmatrix} q_i & q_k & q_l \\ 1 & 1 & 1 \end{vmatrix}, \; S_{\triangle ijl} = \frac{1}{2}\begin{vmatrix} q_i & q_j & q_l \\ 1 & 1 & 1 \end{vmatrix}, \; S_{\triangle ijk} = \frac{1}{2}\begin{vmatrix} q_i & q_j & q_k \\ 1 & 1 & 1 \end{vmatrix} \text{ and}$$

$$S_{\triangle jkl} = \frac{1}{2}\begin{vmatrix} q_j & q_k & q_l \\ 1 & 1 & 1 \end{vmatrix}$$

are the signed areas of the corresponding triangles $\triangle ikl$, $\triangle ijl$, $\triangle ijk$, and $\triangle jkl$.

In 3D, likewise, every free node has at least four neighbors for localizability. Let us consider that a vehicle has exactly four neighbors in 3D. An example is given in Figure 17.4. Applying the same idea as in 2D, we firstly construct a congruent framework with the Euclidean coordinates q_i, q_j, q_k, q_l, q_h. A multidimensional scaling (MDS) method is recalled to construct a congruent framework. Define $H = [d_{rs}^2] \in \mathbb{R}^{5\times5}$ that contains d_{rs}^2 for $r, s \in \{i, j, k, l, h\}$ as its entry and denote $J = I - (1/5)\mathbf{1}_5\mathbf{1}_5^T$. A general algorithm for computing a congruent framework with five nodes in 3D is given as follows [13]. The columns of Q are the Euclidean coordinates q_i, q_j, q_k, q_l, and q_h, respectively. The algorithm is also applicable in 2D.

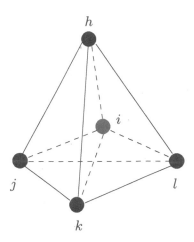

FIGURE 17.4
A configurations in 3D: vehicle i has four neighbors, namely, j, k, l, and h.

Then we are able to get the barycentric coordinate $\{a_{ij}, a_{ik}, a_{il}, a_{ih}\}$ in 3D by the following equations:

$$a_{ij} = \frac{V_{iklh}}{V_{jklh}}, \quad a_{ik} = \frac{V_{jilh}}{V_{jklh}},$$

$$a_{il} = \frac{V_{jkih}}{V_{jklh}}, \quad a_{ih} = \frac{V_{jkli}}{V_{jklh}},$$

Input: H

Output: Q
1. Compute $X = -(1/2)JHJ$;
2. Do singular value decomposition on X as

$$X = V\Lambda V^{\mathrm{T}}$$

where $V = [v_1, v_2, v_3, v_4, v_5]$ is a 5×5 unitary matrix and Λ is a diagonal matrix whose diagonal elements $\lambda_1 \geq \lambda_2 \geq \lambda_3 \geq \lambda_4 \geq \lambda_5 \geq 0$ are singular values;
3. Make $\Lambda_* = \mathrm{diag}(\lambda_1, \lambda_2, \lambda_3) \in \mathrm{R}^{3\times3}$ and $V_* = [v_1, v_2, v_3] \in \mathrm{R}^{5\times3}$;
4. Obtain $Q = \Lambda_*^{1/2} V_*^{\mathrm{T}}$.

---

where

$$V_{iklh} = \frac{1}{6}\begin{vmatrix} q_i & q_k & q_l & q_h \\ 1 & 1 & 1 & 1 \end{vmatrix}, V_{jilh} = \frac{1}{6}\begin{vmatrix} q_j & q_i & q_l & q_h \\ 1 & 1 & 1 & 1 \end{vmatrix},$$

$$V_{jkih} = \frac{1}{6}\begin{vmatrix} q_j & q_k & q_i & q_h \\ 1 & 1 & 1 & 1 \end{vmatrix}, V_{jkli} = \frac{1}{6}\begin{vmatrix} q_j & q_k & q_l & q_i \\ 1 & 1 & 1 & 1 \end{vmatrix},$$

$$V_{jklh} = \frac{1}{6}\begin{vmatrix} q_j & q_k & q_l & q_h \\ 1 & 1 & 1 & 1 \end{vmatrix}.$$

That is, the barycentric coordinate $\{a_{ij}, a_{ik}, a_{il}, a_{ih}\}$ satisfies

$$\begin{cases} p_i = a_{ij}p_j + a_{ik}p_k + a_{il}p_l + a_{ih}p_h, \\ a_{ij} + a_{ik} + a_{il} + a_{ih} = 1. \end{cases}$$

In general, a vehicle may have more than three neighbors in 2D or have more than four neighbors in 3D. In such a case, taking 3D as an example, vehicle i can choose any subset consisting of exactly four neighbors and find its barycentric coordinate with respect to its four neighbors, that is,

$$p_i^{(r)} = a_{ij}^{(r)}p_j + a_{ik}^{(r)}p_k + a_{il}^{(r)}p_l + a_{ih}^{(r)}p_h,$$

where r enumerates the possible combinations of any four neighbors. Finally, the barycentric coordinate of vehicle i with respect to all its neighbors can be obtained via

$$p_i = \sum_r \gamma_r p_i^{(r)} = \sum_{j \in \mathcal{N}_i} a_{ij}p_j,$$

where $0 < \gamma_r < 1$ is a random value satisfying $\Sigma_r \gamma_r = 1$.

In the presence of measurement noises, the available range measurement becomes $\hat{d}_{ij} = d_{ij} + \Delta d_{ij}$, where Δd_{ij} is the noise. For this situation, what we can do is to construct a slightly deformed congruent framework with the Euclidean coordinates q_i, q_j, q_k, q_l, q_h. Thus, we construct $\hat{H} = \left[\hat{d}_{rs}^2\right] \in \mathbb{R}^{5\times5}$ that contains $\hat{d}_{rs}^2$ for r, s $\in \{i, j, k, l, h\}$ as its entry. However, it should be noted that for the noisy case, $X = -(1/2)JHJ$ may not be positive semi-definite and thus the algorithm of constructing $Q = [q_i\, q_j\, q_k\, q_l\, q_h]$ is modified as follows [13].

Input: $\hat{H}$

Output: Q
1. Compute $X = -(1/2)J\hat{H}J$;
2. Do singular value decomposition on X as

$$X = U\Lambda V^T$$

where $U = [u_1, u_2, u_3, u_4, u_5]$ and $V = [v_1, v_2, v_3, v_4, v_5]$ are 5×5 unitary matrices, and Λ is a diagonal matrix whose diagonal elements $\lambda_1 \geq \lambda_2 \geq \lambda_3 \geq \lambda_4 \geq \lambda_5 \geq 0$ are singular values;

3. Make $\Lambda_* = \text{diag}(\lambda_1, \lambda_2, \lambda_3) \in \mathbb{R}^{3\times3}$ and $U_* = [u_1, u_2, u_3] \in \mathbb{R}^{5\times3}$;
4. Obtain $Q = \Lambda_*^{1/2}U_*^T$.

For the overall network, we put all barycentric coordinate representation (17.1) together and obtain a linear constraint in the matrix form:

$$p = (A \otimes I_d)p \tag{17.4}$$

where $p = \left[p_a^T, p_s^T\right]^T = \left[p_1^T, \ldots, p_{m+n}^T\right]^T$ with p_a representing the vector of the Euclidean coordinates of all anchor nodes, p_s denoting the vector of the Euclidean coordinates of all free nodes, and I_d is the identity matrix with $d = 2$ or 3 corresponding to the localization problem 2D or 3D. Here A is the matrix with the (i, j)th entry being a_{ij}, which has the following form

$$A = \left[\begin{array}{c|c} I_{m\times m} & 0 \\ \hline * & * \end{array}\right]$$

Notice that all row sums of A equal to one. So it is clear that $L := I - A$ is a Laplacian matrix satisfying $L1 = 0$. As the weights corresponding to the Laplacian matrix can either be positive or negative, it is called the *signed Laplacian*. The Equation 17.4 can be re-written as

$$(L \otimes I_d)p = 0, \tag{17.5}$$

where L has the form

$$L = \left[\begin{array}{c|c} 0 & 0 \\ \hline L_a & L_s \end{array}\right].$$

As the second step, we expect to solve for p_s from the linear Equation 17.5 in a distributed manner. When the linear Equation 17.5 has a unique solution, all the vehicles in the network are localizable. It is certain that p_s can be solved uniquely if and only if L_s is nonsingular. We provide distributed iteration schemes both in continuous time and discrete time to solve the linear Equation 17.5.

In the continuous time, the following iteration scheme is proposed.

$$\dot{\tilde{p}}(t) = -(KL \otimes I_d)\tilde{p}(t), \tag{17.6}$$

where K is a diagonal matrix making the eigenvalues of KL in the right-half complex plane except for m zero eigenvalues. Such a diagonal matrix K exists as shown in [20]. For the continuous-time iteration scheme 17.6, $\tilde{p}(t)$ tends to the true Euclidean coordinates p of the vehicles starting from any initial condition $\tilde{p}(0)$ when t approaches to ∞. Write the continuous-time iteration scheme 17.6 in the individual form:

$$\dot{\tilde{p}}_i(t) = -k_i \sum_{j \in \mathcal{N}_i} (\tilde{p}_i(t) - \tilde{p}_j(t)), \tag{17.7}$$

for which $\tilde{p}_k(t) = p_k$ for the anchor nodes (namely, $i = 1, \ldots, m$). The formula 17.7 requires the vehicles that are neighbors in the sensing graph to exchange their estimates about themselves locations. That is, the communication graph should be the same as the sensing graph.

In the discrete time, an iteration scheme is proposed by Diao et al. [9]. A few steps of derivation are given below. From Equation 17.5, it can be obtained that

$$(L_s \otimes I_d)p_s = -(L_a \otimes I_d)p_a, \tag{17.8}$$

where p_s and p_a are the aggregated vectors of the Euclidean coordinates for the free nodes and the anchor nodes, respectively. Thus, p_s can be solved from Equation 17.8 by the Richardson iteration:

$$\tilde{p}_s(k+1) = \left(I - \epsilon\left(L_s^T L_s\right) \otimes I_d\right)\tilde{p}_s(k) - \epsilon\left(\left(L_s^T L_a\right) \otimes I_d\right)p_a, \tag{17.9}$$

where ϵ should satisfy $0 < \epsilon < \left(2/\left(\lambda_{\max}\left(L_s^T L_s\right)\right)\right)$. The iteration achieves the fastest convergence rate for $\epsilon = \left(2/\left(\lambda_{\max}\left(L_s^T L_s\right) + \lambda_{\min}\left(L_s^T L_s\right)\right)\right)$. By introducing an auxiliary variable $\zeta_i \in \mathbb{R}^d$ for each node and considering one step delay of exchanging the auxiliary variable ζ_i, Equation 17.9 is modified to become [17]

$$\begin{cases} \zeta(k+1) = (L_s \otimes I_d)\tilde{p}_s(k) + (L_a \otimes I_d)p_a, \\ \tilde{p}_s(k+1) = \tilde{p}_s(k) - \epsilon(L_s^T \otimes I_d)\zeta(k), \end{cases} \tag{17.10}$$

where ζ is the aggregated vector of all ζ_i's and ϵ should satisfy $0 < \epsilon < \left(1/\left(\lambda_{\max}\left(L_s^T L_s\right)\right)\right)$. Writing in the individual form, Equation 17.10 turns out to become

$$\begin{cases} \zeta_i(k+1) = \sum_{j \in \mathcal{N}_i} a_{ij}(\tilde{p}_i(k) - \tilde{p}_j(k)), \\ \tilde{p}_i(k+1) = \hat{p}_i(k) - \epsilon\zeta_i(k) + \epsilon\sum_{j \in \mathcal{N}_i} a_{ji}\zeta_j(k). \end{cases} \tag{17.11}$$

The formula 17.11 requires the vehicles that are neighbors in the sensing graph to exchange their estimates about themselves locations as well as the auxiliary variable ζ_i's. Again, the communication graph should be the same as the sensing graph.

The approach described above is called the *real barycentric coordinate approach*, which has been developed in [15,9,6,13], etc. Reference [15] discusses the approach with a convexity assumption. That is, each free node lies inside the convex hull spanned by its neighbors and all free nodes lie inside the convex hull spanned by the anchor nodes. With this assumption, the barycentric coordinate of each free node i with respect to its neighbors is positive. Reference [9] removes the convexity assumption [15] by providing a general formula for the distributed localization problem. However, it requires accurate range measurements. References [6] and [13] then propose a robust way of computing the barycentric coordinate for each free node in either 2D or 3D.

17.3 Distributed Localization Using Bearing Information

For a multi-vehicle network, if the bearing information is measured by the vehicles according to the sensing graph, then the distributed localization problem becomes the one using bearing information. That is, the constraints between the neighboring vehicles are presented in terms of bearing angles.

Bearing measurements are usually represented in two ways, namely, an angle or a unit vector. For a vehicle i in 2D, let $p_i \in \mathbb{R}^2$ denote its Euclidean coordinate in Σ_g and let p_i^j denote its coordinate in Σ_j, where the superscript j of p_i^j for $j = 1, \ldots, n$ is used throughout the paper to represent the value in the local coordinate system Σ_j. Note that $p_j^j = 0$. Let δ_{ij} be the angle between p_i^j and the x-axis of Σ_i, and δ_{ji} be the angle between p_j^i and the x-axis of Σ_j. Denote θ_i and θ_j the angle between the local coordinate system Σ_i, Σ_j and the global coordinate

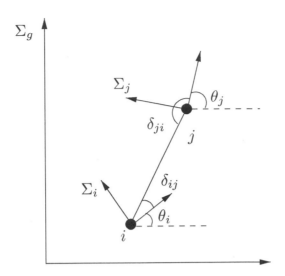

FIGURE 17.5
Global coordinate system vs. local coordinate system and bearing measurements.

system Σ_g, respectively (Figure 17.5). Take node i for example; the bearing measurement is represented by

$$\delta_{ij} \quad \text{or} \quad b_{ij}^i := \begin{bmatrix} \cos(\delta_{ij}) \\ \sin(\delta_{ij}) \end{bmatrix} = \frac{p_j^i}{\| p_j^i \|}.$$

For the localization purpose, it is expected that the bearing measurements can provide some constraints on the global Euclidean coordinates, particularly linear constraints. However, the formula for the bearing angle in general is not like what we expect as it also depends on the orientation of the local frames. Towards this objective, Lin et al. [18] propose a novel complex Laplacian based method, for which a complex barycentric coordinate is introduced. The basic idea is to apply the bearing information to construct a linear constraint like Equation 17.1 as in the range-based localization problem. In 2D, a two-dimensional vector is equivalent to a complex number in the complex plane. Thus, in the following, p_i and p_i^j are all written as complex numbers in $\mathbb{C}$, denoting the coordinate in the global coordinate system Σ_g and the local coordinate system Σ_j. Likewise, the bearing measurement is represented by complex numbers as well, that is,

$$b_{ij}^i = \frac{p_j^i}{\left| p_j^i \right|}.$$

The key idea in [18] is to find complex barycentric coordinate for each vehicle with respect to its neighbors, that is,

$$p_i = \sum_{j \in \mathcal{N}_i} c_{ij} p_j,$$

where c_{ij}'s are complex numbers satisfying $\sum_{j \in \mathcal{N}_i} c_{ij} = 1$. In 2D, when there are at least two neighbors for a vehicle, it is able to derive the complex barycentric coordinate. Equivalently, we look for complex weights $a_{ij} \in \mathbb{C}, j \in \mathcal{N}_i$, by using only bearing measurements in Σ_i, such that

$$\sum_{j \in \mathcal{N}_i} a_{ij}(p_j - p_i) = 0. \tag{17.12}$$

Once a_{ij} in Equation 17.12 can be obtained, it is certain that $\left\{ c_{ij} = \left(a_{ij} / \left(\sum_{j \in \mathcal{N}_i} a_{ij} \right) \right) \right\}$ is a complex barycentric coordinate for vehicle i with respect to its neighbors.

It can be verified that the complex barycentric coordinates are invariant for similar frameworks. This observation plays a key role in computing the linear constraints 17.12 of complex variables by using bearing information. That is to say, we could use the bearing measurements of vehicle i and its neighbors' bearing measurements to construct a similar framework with the Euclidean coordinates in some specific coordinate system and then derive the complex barycentric coordinates using these Euclidean coordinates. First, we consider that vehicle i has exactly two neighbors, say j and k. In this case, with respect to the local coordinate system Σ_i, we construct a similar framework for which the Euclidean coordinate of node i, say q_i, is at the origin (i.e., $q_i = 0$), and moreover the Euclidean coordinate of node j, say q_j, is on the line b_{ij}^i of unit length, that is, $q_j = \left(b_{ij}^i / |b_{ij}^i| \right)$. Finally, the intersection of the line b_{ik}^i passing through q_i and the line b_{jk}^i passing through q_j uniquely defines the Euclidean coordinate q_k in Σ_i. See Figure 17.6 for an illustration. With q_i, q_j, q_k obtained by using bearing measurements, one pair of complex weights a_{ij} and a_{ik} can be easily solved from the following equation

$$a_{ij}(q_j - q_i) + a_{ik}(q_k - q_i) = 0.$$

Thus, we obtain that

$$a_{ij}(p_j - p_i) + a_{ik}(p_k - p_i) = 0$$

also holds.

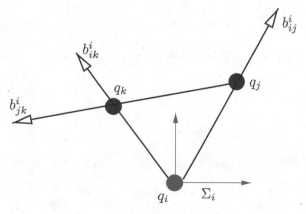

FIGURE 17.6
Constructing a similar framework.

When there are more than two neighbors for some vehicle, the same idea of enumerating all possible combinations of two neighbors and using the convex combination of all derived complex barycentric coordinates as discussed in the range-based localization section is applied to compute the general linear constraints 17.12.

Thus, by putting all these linear constraints together for the whole network, it is obtained that

$$L_p = 0, \tag{17.13}$$

where p is the aggregated vector of all complex numbers corresponding to the Euclidean coordinates of both anchor nodes and free nodes, and L is a Laplacian matrix satisfying $L\mathbf{1} = 0$ and having the form

$$L = \left[\begin{array}{c|c} 0 & 0 \\ \hline L_a & L_s \end{array}\right].$$

As the weights corresponding to the Laplacian matrix are complex numbers, it is called the *complex Laplacian*. By taking away the trivial constraints for the anchor nodes, 17.13 becomes

$$L_s p_s = -L_a p_a \tag{17.14}$$

where p_a and p_s are the aggregated vectors of all anchor nodes and all free nodes, respectively.

Checking that Equations 17.13 and 17.14 are the same as Equations 17.5 and 17.8 except that I_d now becomes 1 and all constants and variables become complex numbers and variables. Therefore, the distributed continuous-time iteration scheme 17.7 and distributed discrete-time iteration scheme 17.11 can also be used to solve for the solution starting from any initial condition.

The approach presented above is called the *complex barycentric coordinate approach*, which does not require to have a common sense of NORTH. In contrast, Zhu et al. [28] assume that all the nodes have access to the global orientation, under which a matrix-valued Laplacian based constraint is constructed using locally available bearing measurements. Moreover, [26] extends the idea to 3D under the same premise. Zhong et al. attempt to address different orientations of local coordinate systems with a matrix-valued Laplacian based constraint in [27]. The main idea in [27] is to adopt a simultaneous orientation consensus strategy such that all the orientations of local coordinate systems become eventually aligned.

17.4 Distributed Localization Using Range and Bearing Information

For a multi-vehicle network, if both range and bearing information are measured by the vehicles according to the sensing graph, then the distributed localization problem becomes the one using range and bearing information [17]. For this case, again the complex variables are used to represent equivalent 2D vectors.

Denote by $p_i \in \mathbb{C}$ the coordinate of node i (either anchor or free node) in a global coordinate system Σ_g. For every free node i, suppose that it measures the range and bearing (relative position) information of its neighbors in its own frame Σ_i, which can be represented as

$$p_j^i = e^{-\imath\theta i}(p_j - p_i), j \in \mathcal{N}_i.$$

With both range and bearing information available, it is much easier to obtain the complex barycentric coordinate for each vehicle than the case with only the bearing measurements. In other words, with a sufficient number of neighbors, vehicle i can simply solve a set of complex coefficients a_{ij} to satisfy

$$\sum_{j \in \mathcal{N}_i} a_{ij} p_j^i = 0. \tag{17.15}$$

Notice that for the same set of complex coefficients a_{ij}, linear equation constraint 17.15 implies

$$\sum_{j \in \mathcal{N}_i} a_{ij}(p_j - p_i) = 0. \tag{17.16}$$

Writing down all these equations for the whole network, it is obtained that

$$Lp = 0, \tag{17.17}$$

where $p = \left[p_a^T, p_s^T\right]^T$ with p_a and p_s being the aggregated coordinate vectors of all anchor nodes and all free nodes respectively, and L is an $(m + n) \times (m + n)$ complex Laplacian, which associates to the graph with complex weights a_{ij} solved from Equation 17.15. Note that the complex Laplacian must be of the following form

$$L = \left[\begin{array}{c|c} 0 & 0 \\ \hline L_a & L_s \end{array}\right].$$

It is certain that p_s can be solved uniquely if and only if L_s is nonsingular. A graphical condition is obtained in [17], that is, in the associated graph $\mathcal{G}(L)$ every free node is 2-reachable from the anchor nodes. The distributed continuous-time iteration scheme 17.7 and distributed discrete-time iteration scheme 17.11 can also be applied to solve for the solution starting from any initial condition.

17.5 Concluding Remarks and Open Problems

Recent progress on distributed localization with linear approaches has been discussed in this chapter. The main idea is to establish a linear constraint on each individual vehicle by using either real barycentric coordinate or complex barycentric coordinate. Thus, a linear system equation in the form of graph Laplacian is obtained to contain the Euclidean coordinates of free nodes as unknowns. In this way, the distributed localization problem turns out to be the problem of how to solve a linear equation with certain structure properties in a distributed manner. It is certain that the solvability of the linear equation (namely, localizability using this linear approach) relies on some graph connectivity properties. As it is a linear system, the developed distributed algorithms ensure global convergence, which is a significant step towards practical applications. Moreover, since for either range measurements, bearing measurements, or both, the same linear framework is used, it is then also possible to solve the distributed localization problem with hybrid measurements in the network [19], for which different nodes may have different types of measurements.

For distributed localization problems of multi-vehicle networks, there remain a number of open issues that have not been completely solved. For instance, it is still unclear how the complex barycentric coordinate in 2D can be extended to deal with the localization problem in 3D with bearing measurements or relative position measurements when the individuals in the network do not share a common sense of direction. Moreover, if both sensing and communication are unidirectional rather than bidirectional between the members in the network, new information flow structure and new techniques need to be explored to address the distributed localization problems in such settings. As in many applications, distributed localization may need to be carried out when the whole network of vehicles is in fast motion in their working space. In this case, distributed localization should take into consideration individual motion information and develop some new dynamic estimators. Other open research issues include localization performance analysis in presence of noise, fast iteration/non-iteration localization algorithms, distributed localization with measurements relative to indistinguishable neighbors, etc.

References

1. B. D. O. Anderson, P. N. Belhumeur, T. Eren, D. K. Goldenberg, A. S. Morse, W. Whiteley, and Y. R. Yang. Graphical properties of easily localizable sensor networks. *Wireless Networks*, 15(2):177–191, 2009.

2. A. Bahr, J. J. Leonard, and M. F. Fallon. Cooperative localization for autonomous underwater vehicles. *The International Journal of Robotics Research*, 28(6):714–728, 2009.

3. P. Biswas, T. Lian, T. Wang, and Y. Ye. Semidefinite programming based algorithms for sensor network localization. *ACM Transactions on Sensor Networks*, 2(2):188–220, 2006.

4. M. Cao, B. D. O. Anderson, and A. S. Morse. Sensor network localization with imprecise distances. *Systems & Control Letters*, 55(11):887–893, 2006.

5. G. Chai, C. Lin, Z. Lin, and M. Fu. Single landmark based collaborative multi-agent localization with time-varying range measurements and information sharing. *Systems & Control Letters*, 87:56–63, 2016.

6. P. Cheng, T. Han, X. Zhang, R. Zheng, and Z. Lin. A single-mobile-anchor based distributed localization scheme for sensor networks. In *Proceedings of the 35th Chinese Control Conference*, IEEE, 2016, pages 8028–8031.

7. G. Destino and G. Abreu. On the maximum likelihood approach for source and network localization. *IEEE Transactions on Signal Processing*, 59(10):4954–4970, 2011.

8. Y. Diao, M. Fu, Z. Lin, and H. Zhang. A sequential cluster-based approach to node localizability of sensor networks. *IEEE Transactions on Control of Network Systems*, 2(4):358–369, 2015.

9. Y. Diao, Z. Lin, and M. Fu. A barycentric coordinate based distributed localization algorithm for sensor networks. *IEEE Transactions on Signal Processing*, 62(18):4760–4771, 2014.

10. Y. Ding, N. Krislock, J. Qian, and H. Wolkowicz. Sensor network localization, Euclidean distance matrix completions, and graph realization. *Optimization and Engineering*, 11(1):45–66, 2010.

11. T. Eren, D. K. Goldenberg, W. Whiteley, Y. R. Yang, A. S. Morse, B. D. O. Anderson, and P. N. Belhumeur. Rigidity, computation, and randomization in network localization. In *Proceedings of the 23rd Annual Joint Conference of the IEEE Computer and Communications Societies*, 2004, pages 2673–2684.

12. J. Fang, M. Cao, A. S. Morse, and B. D. O. Anderson. Sequential localization of sensor networks. *SIAM Journal on Control and Optimization*, 48(1):321–350, 2009.

13. T. Han, Z. Lin, R. Zheng, Z. Han, and H. Zhang. A barycentric coordinate based approach to three-dimensional distributed localization for wireless sensor network. In *Proceedings of the 13th IEEE International Conference on Control & Automation*, IEEE, 2017, pages 600–605.

14. X. Ji and H. Zha. Sensor positioning in wireless ad-hoc sensor networks using multidimensional scaling. In *Proceedings of the 23rd Annual Joint Conference of the IEEE Computer and Communications Societies*, volume 4, IEEE, 2004, pages 2652–2661.

15. U. Khan, S. Kar, and J. M. F. Moura. Distributed sensor localization in random environments using minimal number of anchor nodes. *IEEE Transactions on Signal Processing*, 57(5):2000–2016, 2009.

16. C. Lin, Z. Lin, R. Zheng, G. Yan, and G. Mao. Distributed source localization of multi-agent systems with bearing angle measurements. *IEEE Transactions on Automatic Control*, 61(4):1105–1110, 2016.

17. Z. Lin, M. Fu, and Y. Diao. Distributed self localization for relative position sensing networks in 2d space. *IEEE Transactions on Signal Processing*, 63(14):3751–3761, 2015.

18. Z. Lin, T. Han, R. Zheng, and M. Fu. Distributed localization for 2-D sensor networks with bearing-only measurements under switching topologies. *IEEE Transactions on Signal Processing*, 64(23):6345–6359, 2016.

19. Z. Lin, T. Han, R. Zheng, and C. Yu. Distributed localization with mixed measurements under switching topologies. *Automatica*, 76:251–257, 2017.

20. Z. Lin, L. Wang, Z. Chen, M. Fu, and Z. Han. Necessary and sufficient graphical conditions for affine formation control. *IEEE Transactions on Automatic Control*, 61(10):2877–2891, 2016.

21. R. L. Moses, D. Krishnamurthy, and R. M. Patterson. A self-localization method for wireless sensor networks. *EURASIP Journal on Advances in Signal Processing*, 2003(4):1–11, 2003.

22. P. N. Pathirana, N. Bulusu, V. V. Savkin, and S. Jha. Node localization using mobile robots in delay-tolerant sensor networks. *IEEE Transactions on Mobile Computing*, 4(3):285–296, 2005.

23. Y. Shang and W. Rum. Improved MDS-based localization. In *Proceedings of the 23rd Annual Joint Conference of the IEEE Computer and Communications Societies*, volume 4, IEEE, 2004, pages 2640–2651.

24. A. M. So and Y. Ye. Theory of semidefinite programming for sensor network localization. *Mathematical Programming*, 109(2–3):367–384, 2007.

25. M. Ye, B. D. O. Anderson, and C. Yu. Multiagent self-localization using bearing only measurements. In *Proceedings of the 52nd IEEE Annual Conference on Decision and Control*, IEEE, 2013, pages 2157–2162.

26. S. Zhao and D. Zelazo. Bearing-only network localization: Localizability, sensitivity, and distributed protocols. *arXiv Preprint ArXiv:1502.00154*, 2015.

27. J. Zhong, Z. Lin, Z. Chen, and W. Xu. Cooperative localization using angle-of-arrival information. In *Proceedings of the 11th IEEE International Conference on Control and Automation*, IEEE, 2014, pages 19–24.

28. G. Zhu and J. Hu. A distributed continuous-time algorithm for network localization using angle-of-arrival information. *Automatica*, 50(1):53–63, 2014.

18

Range-Based Navigation Algorithms for Marine Applications

David Moreno-Salinas, Naveen Crasta, António M. Pascoal, and Joaquín Aranda

CONTENTS

18.1 Introduction

The oceans cover a large part of the Earth's surface, are home to a wide diversity of living beings, and have a major influence on the weather and the climate. In addition, the ocean floor is known to contain a large accumulation of mineral resources which, according to current estimates, exceed those that can be found on land. A compelling example are rare earth elements (REEs), indispensable to high-tech technology equipment and emerging clean energy devices, which have been predicted to exist in vast amounts on the seafloor. Due to the fast depletion of land resources, it has been proposed to exploit this untapped potential in a cautious manner, with a view to reducing the negative impact of exploitation activities. Because of the harsh conditions imposed by the ocean environment, ocean exploration and exploitation for scientific and commercial purposes meet with formidable technical challenges. For these reasons, marine robots are steadily becoming the tool par excellence to collect ocean data at unprecedented temporal and spatial scales and to execute tasks that require interaction with underwater infrastructures.

From a broad perspective, marine robotics research aims to design, develop, and provide support to the operation of advanced underwater systems for scientific and academic research purposes, as well as commercial applications. Remotely operated vehicles (ROVs), autonomous surface vehicles (ASVs), gliders, autonomous underwater vehicles (AUVs), and vertical profilers are some examples of marine robots. From an economic perspective, autonomous marine vehicles (AMVs) provide a cost-effective alternative to other available technologies, such as manned submersibles and ROVs. AMVs have made significant inroads into a variety of complex but challenging missions of increasing complexity levels such as geophysical surveying, monitoring of cold-water coral reefs, and marine habitat mapping, to name a few. In recent years, there has been a paradigm shift from single-agent to multiple-agent operations, in view of the fact that a given task may be too complex to be accomplished by a single autonomous agent. While a group of autonomous agents may provide increased robustness when compared to a single agent, this raises tremendous challenges due to the need to coordinate the different agents involved. Clearly, present and future developments in marine systems, including autonomous robots, will impact strongly on the maritime economy and the shipping industry and on the techniques required to monitor and map the underwater environment, maintain and repair offshore wind, energy, and fish farm installations, and patrol critical infrastructures. Among other enabling factors, reduced human supervision and the ability of marine robots to perform accurate navigation in the presence of stringent environmental conditions are two key factors central to the development of innovative underwater systems for ocean exploration and exploitation.

18.1.1 An Overview of Marine Robotics Technology

Recent reports on the "future of the sea" emphasize the tremendous opportunities available to capitalize on existing technologies and the diverse ocean industries. In view of this, in recent times marine robotic systems have emerged as standard tools to explore the ocean and to carry out a wide range of missions in spite of many challenges, especially in what concerns navigation algorithms, communications, control, and sensing. In addition, the increased level of autonomy of the vehicles that are becoming available offers increased scope for areas such as sensing and improved data transfer between autonomous vehicles and satellites that are of paramount importance to the marine economy. Amidst the promising future of marine robotics, there are a number of challenges that need to be addressed at both theoretical and practical levels.

The availability of advanced systems for for motion planning and control, underwater positioning and localization, and communications is a key factor for the successful execution of missions at sea. Motion planning comprises path generation and trajectory generation in the presence of environmental disturbances such as ocean currents, winds, etc., by taking into account explicitly the capabilities of each vehicle. Methods designed for land and aerial autonomous vehicles meet with difficulties for AMV-related applications, especially when they are required to operate over large geographical areas or in the vicinity of complex 3D structures underwater. Point stabilization, trajectory tracking, and path following are some of the main problems in motion control. A variety of methods available for motion control are rooted in firm nonlinear control theory and include Lyapunov-based techniques, passivity-based approaches, back-stepping methods [1], and model-predictive control [2].

Common methods used for navigation include non-acoustic (inertial and geomagnetic [3–5]) and acoustic-based navigation [6]. Classical examples of acoustic-based underwater navigation systems rely on the measurements of the ranges between an AUV and a number of transponders in a baseline configuration (long baseline systems [LBLs]), or on the computation of the relative delays in the times of arrival of an acoustic wave emitted by a beacon and received at the different elements of an array of hydrophones (short baseline systems [SBLs]) [7, Chapters 3 through 5], [6, Sections 4.2.6 through 4.2.7]. Other systems that are variants of SBL systems rely on the computation of the differences in phase of the incoming waves (ultra-short baseline systems [USBLs]). For decades, LBL systems have been used for positioning underwater equipment and navigation of AUVs. From a practical standpoint, LBL systems provide higher position accuracy, subjected to proper calibration. However, LBL systems require deployment and calibration, which are time consuming, and for missions that must be carried out at different locations, the cost of these operations becomes a key factor. USBL systems are a viable alternative, but their cost remains high.

Finally, underwater wireless communication systems used by marine vehicles rely mostly on acoustic networks. However, the communication channels exhibit limited bandwidth and often cause severe multipath dispersion and time-variability. A vast amount of research exists on the design of acoustic communication systems in the underwater environment to address issues such as time delays, ray bending, Doppler shifts, communication losses, and low bandwidth. These topics take central stage in the development of advanced systems for networked multiple vehicle operations at sea.

The diversity of areas that must be brought together for successful development and operation of autonomous robots at sea is at the root of the tremendous interest in marine robotic systems worldwide. In this chapter, we focus on the specific sub-problem of underwater vehicle navigation which, stated in simple terms, aims to develop the systems required to compute the position of an object underwater. Among the plethora of methods available for this purpose we will consider methods that rely on the computation of ranges between the objects (targets) to be localized and one or more stationary or moving units, called trackers, equipped with range-measuring devices, as explained next.

18.1.2 Range-Based Localization

Estimating the position of single or multiple fixed or moving objects using complementary motion-related sensor information is of the utmost importance in many engineering and scientific applications. Among the methodologies available for this purpose, position estimation techniques that exploit the information on the ranges between the object(s) and one or more reference units have been receiving widespread attention due to the simplicity of implementation and the low cost of range-measuring devices. There are many other types of problems where range measurements play a central role, such as those that can be tackled via the analysis of so-called Euclidean distance matrices (EDMs). In this context, given a set of points $\{x_1, \ldots, x_m\} \subset \mathbb{R}^n$, the EDM, denoted by $D_{m,n}$, is an $m \times m$ symmetric matrix whose ijth element is the squared distance between x_i and x_j. For instance, referring to Figure 18.1, the EDM matrix $D_{3,2} \in \mathbb{R}^{3 \times 3}$ for three points $x_1, x_2, x_3 \in \mathbb{R}^2$ is given by

$$D_{3,2} = \begin{bmatrix} 0 & \|x_1 - x_2\|^2 & \|x_1 - x_3\|^2 \\ \|x_1 - x_2\|^2 & 0 & \|x_2 - x_3\|^2 \\ \|x_1 - x_3\|^2 & \|x_2 - x_3\|^2 & 0 \end{bmatrix}.$$

Using this framework, a classical problem that arises is that of computing, based on the information contained in a EDM, the geometric configuration of points, $\{x_1, \ldots, x_m\} \subset \mathbb{R}^n$ apart from a rotation and a translation; see for example [8]. EDMs have found application in molecular conformation problems occurring in bioinformatics [9], dimensionality reduction in machine learning and statistics [10,11], and wireless sensor network (WSN) localization [12], to name but a few. In contrast with the above, in the present chapter we focus on techniques that allow for the computation of the absolute positions of a number of objects by measuring their ranges (distances) to a set of points with known inertial coordinates. This occurs, for example, in situations where GPS is ineffective, as in indoor and underwater robotic applications. Stated in simple terms, a range measurement is obtained by computing the so-called time of arrival (TOA), often called time of flight (ToF),

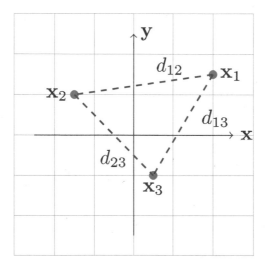

FIGURE 18.1
Illustration of EDM with three points in $\mathbb{R}^2$.

defined as the travel time of a signal (acoustic or electromagnetic) from a transmitter to a receiver using the knowledge of the speed of propagation of the signal in the respective medium.

Classically, a unit in charge of computing the distance to another unit interrogates the latter (called a transponder) which in turn replies back to the interrogating unit (receiver). The receiver can then compute the round-trip travel time from receiver to transmitter and back to receiver, from which the TOA can be determined. If the clocks of the two units are synchronized, then there is no need to interrogate the transponder if the times of emission are agreed upon, and the receiver can measure directly the TOA. The classical technique of computing the position of an object based on measurements of its distances to a number of units with known positions is called *trilateration* [12, Chapter 10], which traditionally refers to the situation where the reference units with known positions are stationary but the object to be positioned may be moving.

Another closely related problem arises in situations when the times of emission are unknown. This leads to the problem of object localization using range differences or time differences of arrival (TDOA) between the object to be localized and a number of receivers. The TDOA localization problem is also known as *hyperbolic localization*, due to the fact that in its 2D form, each range difference equation defines a hyperbola where the position of the source might lie.

Both TOA- and TDOA-based object localization methods using a number of receivers at known positions are by now classical; see for example [12, Chapter 10], [7], and the references therein. Over the past few years, new techniques of object localization have come to the fore that rely on the use of a single receiver (range-measuring device) and are often referred as *single-beacon navigation* problem [13–15] in the context of marine robotics. In this case, it becomes necessary to impart sufficiently exciting motions to the receiver, thus affording it the spatial diversity required to acquire different ranges to the object being localized (target), from which the position of the latter, which may be moving, can be estimated using range-based localization techniques. In this context, it becomes important to clearly identify what kinds of trajectories must be imparted to the receiver to ensure observability of the target's motion, i.e. derive conditions under which it is possible to reconstruct the target's state. See for instance [16–21] for different methods of observability analysis. Even more important, it is crucial to study the types of motions that increase the amount of range-related information available for target localization. This problem is even more challenging when simultaneous localization of multiple targets must be achieved by resorting to a reduced number of mobile range-measuring units (receivers). The literature on these topics is vast and it is not our objective in this chapter to provide even a brief overview of it. Instead, we refer the reader to [22,23] and the references therein for a fast-paced introduction to the subject. At this point, it is important to remark that because the ranges being measured are nonlinear functions of the relative position between emitters and receivers, the observability analysis mentioned above is non-trivial. Identical comments apply to the problem of optimal sensor (receiver) motion planning to maximize the range-related information available for target localization and the design of range-based efficient estimators using linear and nonlinear techniques.

18.1.3 Outline of the Chapter

The objective of this chapter is to systematically introduce the theoretical tools leading to the building blocks required for the development of range-based algorithms to solve two classes of problems. Namely, *vehicle positioning* and *target localization*, which are deeply rooted in many marine applications. So far, we have essentially been using the terms positioning and localization interchangeably. There is no general agreement on the exact definition of each one, but we found it appropriate to adopt definitions (followed by a number of authors) that will allow us to clearly identify two types of important and distinct problems in the area of marine robotics. Accordingly, in what follows, and in the context of range-based navigation, underwater (self) vehicle positioning refers to the problem of estimating the coordinates of the vehicle in a given reference frame, based on measurements of the distances of the vehicle with respect to one or more landmarks with known locations. Underwater vehicle localization, in contrast, is concerned with the problem of estimating, using one or more external devices, the coordinates of the vehicle. In this situation the underwater vehicle is viewed as a target and the external devices, henceforth referred to as trackers, measure the ranges to the target and compute its position. In this work, the trackers may be fixed sensors or autonomous surface vehicles equipped with range-measuring units. We remark that in this chapter we will focus our discussion on the target localization problem only, though the same methods can be extended to some problems arising in vehicle positioning.

In our context, depending on the nature of the targets, i.e. fixed or moving, the problem thus defined can be viewed as *static target localization* or *moving target localization*. The latter is more challenging compared to the former and hence requires some prior knowledge about the target motions. See for example [24–26] for an introduction to this topic and various types of target models used in practice. Further, each of the above problems can be addressed using either a network of fixed sensors or one or more moving sensors installed on-board a group of surface vehicles. Based on this, two main problems can be formulated: (i) fixed target localization using a static sensor network and (ii) fixed/moving target localization using a single or multiple moving vehicles.

A simple but naive approach to the problem would involve placing fixed sensors in some ad-hoc geometric configuration or imparting simple motions to one or more tracker vehicles, collecting ranges to the target(s), and estimating the position of the latter. This simple approach meets with considerable difficulties. Namely,

a. By resorting to ad-hoc procedures, it may not be possible to estimate the target or targets' positions. Establishing the conditions under which this is possible to do lead to interesting *observability* problems. Since ranges are nonlinear functions of the targets' positions, the study of observability is a challenging issue.

b. Often, the accuracy of the target's position estimate is critical to most of the applications and significantly depends on the geometrical sensor configuration or the motion imparted to the trackers. Thus, a critical question arises as to what is the optimal geometric configuration that will maximize the range-related information for target localization. This problem is known as *optimal sensor placement*.

c. In the context of moving targets, it is not possible to consider that their motions are completely unknown, and some appropriate assumptions must be made regarding the type of motion that they undergo, based on prior information. For instance, in the case of cooperative targets, it may be known that their trajectories are lines, arcs, or a combination thereof. On the other hand, one must also adopt a model for the moving trackers as well.

d. Finally, even when good observability conditions can be guaranteed, the question arises as to what kind of range-based estimators should be used. The literature on this topic is extensive, with most estimation methods adopting a stochastic setting. In this respect, it is interesting to call attention to minimum energy estimators

(MEEs) [27,28] which, by adopting a deterministic setup, produce estimates of the state of a given system that are most compatible with its dynamics (in the sense that they require the least amount of weighted process and measurement noise energies to explain the measured outputs). See [23,29,30] for an application of MEE estimation theory to range-based navigation problems.

With this brief background, in what follows we provide a summary of selected results from our past research work in an attempt to answer the aforementioned questions in the context of fixed and moving target localization, using fixed and moving sensors. We complement the theoretical results on single moving target localization using a single tracker with experimental results using the MEDUSA class of vehicles developed at the Institute for Systems and Robotics (ISR), Instituto Superior Técnico (IST), University of Lisbon, Portugal.

18.1.4 Organization of the Chapter

The organization of the chapter is as follows. In Section 18.2 we introduce the required notation, the key concepts of observability of nonlinear systems, and the Fisher information matrix (FIM) that plays a fundamental role in estimation theory. Section 18.3 addresses the problems of optimal sensor placement in 2D and 3D for single and multiple static target localization using static sensors. Finally, in Section 18.4 we consider the problem of stationary or moving, single or multiple target localization using one or more moving trackers. Section 18.5 contains the main conclusions and a brief discussion of challenging topics that warrant future research.

18.2 Notation and Preliminaries

18.2.1 Notation

For $m, n \in \mathbb{N}$ with $m < n$, we let $\mathbb{I}_{[m,n]} := \{s \in \mathbb{N} : m \leq s \leq n\}$. We denote vectors in $\mathbb{R}^n$ by lower-case boldface letters such as $\mathbf{x}$, matrices in $\mathbb{R}^{m \times n}$ by upper-case letters such as A, and frames in $\mathbb{R}^n$ by upper-case letters such as $\mathcal{F}$. We denote by I_n the identity matrix of size n and by $0_{m \times n}$ the zero matrix of size $m \times n$. A positive definite matrix $A \in \mathbb{R}^{n \times n}$ is simply written $A \succ 0$. We further let $\det(\cdot)$ and $\mathrm{trace}(\cdot)$ denote the determinant and the trace of a square matrix, respectively. Given $A \in \mathbb{R}^{m \times n}$, $\mathrm{vec}(A) \in \mathbb{R}^{mn}$ denotes the column vector obtained by stacking the columns of the matrix A on top of one another. Given $\mathbf{w} \in \mathbb{R}^n$, $\mathrm{diag}(\mathbf{w}) \in \mathbb{R}^{n \times n}$ denotes the diagonal matrix whose diagonal elements are the components of vector $\mathbf{w}$. In a similar manner, given $s, n \in \mathbb{N}$ and $A_j \in \mathbb{R}^{n \times n}$, $j \in \mathbb{I}_{[1,s]}$, we can define $\mathrm{diag}(A_1, \ldots, A_s)$, and the direct sum of $A_1, \ldots, A_s$ as

$$\bigoplus_{j \in \mathbb{I}_{[1,s]}} A_j := \mathrm{diag}(A_1, \ldots, A_s).$$

We denote the Euclidean norm in $\mathbb{R}^n$ by $\|\cdot\|$, the unit sphere in $\mathbb{R}^{n+1}$ by S^n, that is, $S^n := \{\mathbf{x} \in \mathbb{R}^{n+1} : \|\mathbf{x}\| = 1\}$, and the special group of orthogonal matrices of dimension n by $\mathrm{SO}(n) := \{R \in \mathbb{R}^{n \times n} : RR^\mathsf{T} = I_n \text{ and } \det(R) = +1\}$. For simplicity we denote $C_\phi = \cos \phi$ and

$S_\phi = \sin\phi$, and we define two orthogonal unit vectors $\mathbf{g}: [0, 2\pi) \to S^1$ and $\mathbf{g}^\perp: [0, 2\pi) \to S^1$ by $\mathbf{g}(\phi) := [C_\phi, S_\phi]^T$ and $\mathbf{g}^\perp(\phi) := [-S_\phi, C_\phi]^T$, $\phi \in [0, 2\pi)$, respectively. Note that for every $\phi \in [0, 2\pi)$, $\mathbf{g}(\phi)$ and $\mathbf{g}^\perp(\phi)$ are the columns of the planar rotation matrix, that is, $\mathcal{R}(\phi) = [\mathbf{g}(\phi) \quad \mathbf{g}^\perp(\phi)] \in SO(2)$. Given a smooth function $g: \mathbb{R}^n \to \mathbb{R}$, the gradient of g is denoted by ∇g. Unless indicated otherwise, gradient vectors in this chapter will be column vectors. Furthermore, we denote by $f_X(\mathbf{x}; \boldsymbol{\theta})$ the probability density function (pdf) of a random variable X parametrized by a parameter $\boldsymbol{\theta}$. For convenience, we denote it by $X \sim f_X(\mathbf{x}; \boldsymbol{\theta})$.

We use the following conventions in the rest of the chapter. We let $\{\mathcal{A}\} := \{\mathbf{x}_\mathcal{A}, \mathbf{y}_\mathcal{A}, \mathbf{z}_\mathcal{A}\}$ denote a reference frame with origin $O_\mathcal{A}$ and orthonormal unit vectors $\mathbf{x}_\mathcal{A}$, $\mathbf{y}_\mathcal{A}$, and $\mathbf{z}_\mathcal{A}$ that form a right-handed coordinate system in the usual sense. Given two reference frames $\{\mathcal{A}\}$ and $\{\mathcal{B}\}$, $Q \in SO(3)$ is the rotation matrix from $\{\mathcal{B}\}$ to $\{\mathcal{A}\}$, i.e. for every coordinate free vector $\mathbf{u}$, the components $\mathbf{u}_\mathcal{B} \in \mathbb{R}^3$ and $\mathbf{u}_\mathcal{A} \in \mathbb{R}^3$ of $\mathbf{u}$ in frames $\{\mathcal{B}\}$ and $\{\mathcal{A}\}$, respectively are related by $\mathbf{u}_\mathcal{A} = Q\mathbf{u}_\mathcal{B}$.

In this chapter, we let $d \in \{2, 3\}$ denote the dimension of a Cartesian coordinate system. We use $\mathbf{p} \in \mathbb{R}^d$ and $\mathbf{q} \in \mathbb{R}^d$ to denote the inertial sensor and target positions, respectively. Further, we reserve $i \in \mathbb{I}_{[1,p]}$ and $\alpha \in \mathbb{I}_{[1,q]}$ to denote the ith sensor and αth target, respectively.

18.2.2 Preliminaries

We now recall the notions of observability and Fisher information and provide a connection between the two notions for the class of linear time invariant (LTI) systems.

18.2.2.1 Observability of Nonlinear Systems

Consider the general nonlinear system

$$\left.\begin{array}{l} \dot{\mathbf{x}} = \mathbf{f}(\mathbf{x}, \mathbf{u}) \\ \mathbf{y} = \mathbf{h}(\mathbf{x}) \end{array}\right\} \tag{18.1}$$

where $\mathbf{x} \in \mathbb{R}^n$ is the state, $\mathbf{u}$ is the input vector taking values in a compact subset Ω of $\mathbb{R}^p$ containing zero in its interior, $\mathbf{f}$ is a complete and smooth vector field on $\mathbb{R}^n$, and the output function $\mathbf{h}: \mathbb{R}^n \to \mathbb{R}^q$ has smooth components. We recall the following definitions from [31]. To capture the physical constraints of the underlying system, we assume that $\mathbf{u}$ belongs to a (possibly large) set $\mathcal{U}_{ad}$ of admissible inputs.

Definition 18.1

(*Indistinguishability*): Two initial states $\mathbf{z}, \mathbf{z}' \in \mathbb{R}^n$ of (18.1) are indistinguishable in $[t_0, t_f)$ if, for every input $\mathbf{u}$ in the set of admissible inputs $\mathcal{U}_{ad}$ the solutions of (18.1) satisfying the initial conditions $\mathbf{x}(t_0) = \mathbf{z}$ and $\mathbf{x}(t_0) = \mathbf{z}'$ produce identical output-time histories in $[t_0, t_f)$.

For every $\mathbf{z} \in \mathbb{R}^n$, we let $\mathcal{I}(\mathbf{z}) \subseteq \mathbb{R}^n$ denote the set of all states that are indistinguishable from $\mathbf{z}$. Note that indistinguishability is an equivalence relation.

Definition 18.2

(*Observability*): The system (18.1) is observable at $\mathbf{z} \in \mathbb{R}^n$ if $\mathcal{I}(\mathbf{z}) = \{\mathbf{z}\}$ and is observable if $\mathcal{I}(\mathbf{z}) = \{\mathbf{z}\}$ for every $\mathbf{z} \in \mathbb{R}^n$.

Definition 18.3

(*Weak observability*): The system (18.1) is weakly observable at $\mathbf{z} \in \mathbb{R}^n$ if $\mathbf{z}$ is an isolated point of $\mathcal{I}(\mathbf{z})$ and is weakly observable if it is weakly observable at every $\mathbf{z} \in \mathbb{R}^n$.

It is important to remark that the above definitions, though elegant, may prove to be quite restrictive in a number of applications. To show this, notice that if a system is weak observable at a point $\mathbf{z} \in \mathbb{R}^n$, then there is an open neighborhood N_z of $\mathbf{z}$ such that every initial condition $\mathbf{z}' \in N_z$ different from $\mathbf{z}$ is distinguishable from $\mathbf{z}$ itself. However, the computation of a particular input that will distinguish $\mathbf{z}$ and $\mathbf{z}'$ may, for a fixed $\mathbf{z}$, depend on the initial condition $\mathbf{z}'$. It is therefore natural to ask whether, for a given system, there is a specific class of admissible inputs that are simple to characterize and yet can be used to distinguish the state $\mathbf{z}$ from any other state $\mathbf{z}' \in N_z$ by forcing the system with a particular, *fixed input* in that class. Here, we are strongly motivated by the concept of uniform universal inputs introduced by Sontag in [32].

As we show later, the answer to the above question may be affirmative in the context of systems that describe the motions of a large class of autonomous vehicles if the outputs (measurements) are chosen appropriately. In the latter case, there is a reduced class $\mathcal{U}_c \subseteq \mathcal{U}_{\mathrm{ad}}$ of admissible inputs, with elements denoted $\mathbf{u}^*$, that can be parameterized in terms of a small number of parameters but are sufficiently general to generate maneuvers of interest in a wide range of applications. One such example consists of classical AUV trimming trajectories that are obtained by holding the physical inputs to the vehicle constant. As we show later, such trajectories are fully parametrized by total speed, yaw rate, and flight path angle and correspond to helices in 3D space that may degenerate into circumferences and straight lines [33]. In the context of this chapter, such parameters play the role of inputs to the model adopted for AUV trajectory generation. Interestingly enough, each element in this reduced class of inputs (that generate helicoidal trajectories) is sufficiently rich to yield, under well-defined conditions, useful observability properties for the models whose outputs consist of range or range and depth measurements. This result affords system designers an effective way of selecting simple and yet effective maneuvers from an observability standpoint. With this motivational background, we recall a weaker notion of observability originally proposed in [16] that, as we shall see, will allow for the derivation of observability conditions for the localization system studied in this chapter that are easy to interpret physically.

Definition 18.4

($\mathbf{u}^*$-*Indistinguishability*): Let $\mathbf{u}^*$ be an admissible input in a given set $\mathcal{U}_c$. We say that two initial states $\mathbf{z}, \mathbf{z}' \in \mathbb{R}^n$ of 18.1 are $\mathbf{u}^*$-indistinguishable in $[t_0, t_f)$, if the solutions of (18.1) satisfying the initial conditions $\mathbf{x}(t_0) = \mathbf{z}$ and $\mathbf{x}(t_0) = \mathbf{z}'$ produce identical output-time histories in $[t_0, t_f)$ for $\mathbf{u}^*$.

For every $\mathbf{z} \in \mathbb{R}^n$, let $\mathcal{I}^{\mathbf{u}^*}(\mathbf{z}) \subseteq \mathbb{R}^n$ denote the set of all states that are $\mathbf{u}^*$-indistinguishable from $\mathbf{z}$.

Definition 18.5

($\mathbf{u}^*$-*Observability*): The system (18.1) is $\mathbf{u}^*$-observable at $\mathbf{z} \in \mathbb{R}^n$ if $\mathcal{I}^{\mathbf{u}^*}(\mathbf{z}) = \{\mathbf{z}\}$ and is $\mathbf{u}^*$-observable if $\mathcal{I}^{\mathbf{u}^*}(\mathbf{z}) = \{\mathbf{z}\}$ for every $\mathbf{z} \in \mathbb{R}^n$.

Definition 18.6

($\mathbf{u}^*$-*Weak observability*): The system (18.1) is $\mathbf{u}^*$-weakly observable at $\mathbf{z} \in \mathbb{R}^n$ if $\mathbf{z}$ is an isolated point of $\mathcal{I}^{\mathbf{u}^*}(\mathbf{z})$ and is $\mathbf{u}^*$-weakly observable if it is $\mathbf{u}^*$-weakly observable at every $\mathbf{z} \in \mathbb{R}^n$.

Remark 18.1

Note that observability (*O*) implies weak observability (*WO*), while $\mathbf{u}^*$-observability ($\mathbf{u}^*$-*O*) implies $\mathbf{u}^*$-weak observability ($\mathbf{u}^*$-*WO*).

18.2.2.2 Fisher Information Matrix

We next recall the concept of Fisher information in a general setup. Consider the problem of estimating an unknown but fixed parameter $\boldsymbol{\theta} \in \mathbb{R}^n$ using a sequence of measurements $\mathbf{z}_k \in \mathbb{R}^m$ corrupted by additive noise $\boldsymbol{\eta}_k \in \mathbb{R}^m$, $k \in \mathbb{I}_{[1,N]}$, according to the measurement model

$$\mathbf{z}_k = \mathbf{h}(\boldsymbol{\theta}) + \boldsymbol{\eta}_k, \, k \in \mathbb{I}_{[1,N]},$$

where $\mathbf{h}: \mathbb{R}^n \to \mathbb{R}^m$ is a known static map of $\boldsymbol{\theta}$. Define the data vector $\mathbf{z} = \left[\mathbf{z}_1^{\mathrm{T}}, \ldots, \mathbf{z}_N^{\mathrm{T}} \right]^{\mathrm{T}} \in \mathbb{R}^{mN}$ and let $f_{Z|\theta}(\mathbf{z}|\boldsymbol{\theta})$ denote the conditional pdf of $\mathbf{z}$ given $\boldsymbol{\theta}$. The *likelihood function* $\mathcal{L}_\theta(\mathbf{z})$ for the measurement vector with respect to the unknown parameter is given by

$$\mathcal{L}_\theta(\mathbf{z}) := f_{Z|\theta}(\mathbf{z}|\boldsymbol{\theta}).$$

For later purposes, we find it convenient to work with the log-likelihood function $\Lambda_\theta := \ln \mathcal{L}_\theta(\mathbf{z})$ rather than the likelihood function.

Suppose $\mathbf{z} \mapsto \hat{\boldsymbol{\theta}}(\mathbf{z})$ be any unbiased estimator of $\boldsymbol{\theta}$. Then,

$$\mathrm{Cov}\left(\hat{\boldsymbol{\theta}}(\mathbf{z}) - \boldsymbol{\theta}\right) \succeq (FIM(\boldsymbol{\theta}))^{-1}, \tag{18.2}$$

where

$$FIM(\boldsymbol{\theta}) := \mathbb{E}\left[(\nabla_\theta \Lambda_\theta)(\nabla_\theta \Lambda_\theta)^{\mathrm{T}} \right] \in \mathbb{R}^{n \times n}$$

is symmetric and positive semi-definite and $\mathbb{E}$ is the expectation operator. The result given in (18.2) is the celebrated Cramér-Rao lower bound (CRLB) [35]. The inverse of the FIM is instrumental in computing a lower bound on the covariance of the parameter estimation error that can possibly be achieved with any unbiased estimator. In practice, the accuracy of the estimates can be ascertained by computing the determinant of the FIM or, preferably, by estimating how close the FIM is to being singular.

18.2.2.3 Connection between Observability and FIM Analysis for Linear Systems

In what follows, for the sake of completeness, we state a general result that connects the so-called observability Gramian of a linear system with measurements corrupted by white Gaussian noise and the corresponding FIM. The result is well known, but its proof is not pervasive in the literature.

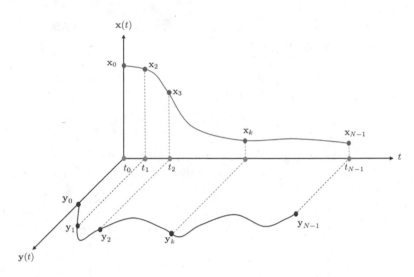

FIGURE 18.2
State and output trajectories.

Consider an unactuated[*] LTI system

$$\left.\begin{aligned}\dot{\mathbf{x}}(t) &= A\mathbf{x}(t)\\ \mathbf{y}(t_k) &= C\mathbf{x}(t_k) + \mathbf{w}(t_k)\end{aligned}\right\} \tag{18.3}$$

where $t \in [0, t_f]$, $\mathbf{x} \in \mathbb{R}^n$, $\mathbf{y} \in \mathbb{R}^q$, $A \in \mathbb{R}^{n \times n}$, $C \in \mathbb{R}^{q \times n}$, $\mathbf{h}(\mathbf{x}) = C\mathbf{x}$, $\mathbf{w} \in \mathbb{R}^q$ is $\mathcal{N}(0, \Sigma)$ with $\Sigma = \Sigma^{\mathsf{T}} \succ 0$.

Suppose one wishes to estimate the unknown but deterministic initial condition $\mathbf{x}_0 \in \mathbb{R}^n$ using a sampled sequence of independently and identically distributed (i.i.d.) output random signals $\mathbf{y}(t_i) \sim \mathcal{N}(\mu_i(\mathbf{x}_0), \Sigma)$, $i \in \mathbb{I}_{[0,N-1]}$, as depicted in Figure 18.2, where $0 = t_0 < t_1 < \cdots < t_{N-1} = t_f$ and $\mu_i(\mathbf{x}_0) := Ce^{At_i}\mathbf{x}_0$, $i \in \mathbb{I}_{[0,N-1]}$. For (18.3), we define the *(stochastic) observability Gramian* for the discrete time system, given by

$$W_{\mathrm{d}}(0, t_f) := \sum_{i=0}^{N-1} e^{A^{\mathsf{T}} t_i} C^{\mathsf{T}} \Sigma^{-1} C e^{At_i}.$$

Note that in $W_{\mathrm{d}}(\cdot)$, the inverse of the covariance matrix Σ^{-1}, appears explicitly. Naturally, this is not the case for the deterministic observability Gramian. At this point, we recall that the LTI system is observable if and only if the observability Gramian is full rank [36]. The following result holds true.

Theorem 18.1

Consider the system (18.3) and let $\mathbf{x}_0 \in \mathbb{R}^n$ be an unknown but deterministic initial condition that must be estimated from the independently sampled output random sequence $\mathbf{y}(t_i); i \in \mathbb{I}_{[0,N-1]}$, corrupted by white noise. Then,

$$W_d(0, t_f) = FIM(\mathbf{x}_0).$$

---

[*] This is done for the sake of simplicity, but without any loss of generality.

Proof The solution to (18.3) at time $t \geq 0$ starting from an initial condition $\mathbf{x}_0 \in \mathbb{R}^n$ is given by $\mathbf{x}(t) = e^{At}\mathbf{x}_0$. Consequently, $\mathbf{h}(\mathbf{x}(t)) = Ce^{At}\mathbf{x}_0$. Define

$$
Y := \begin{bmatrix} \mathbf{y}(t_0) \\ \vdots \\ \mathbf{y}(t_{N-1}) \end{bmatrix} \quad \text{and} \quad \boldsymbol{\mu} := \begin{bmatrix} \boldsymbol{\mu}_0(\mathbf{x}_0) \\ \vdots \\ \boldsymbol{\mu}_{N-1}(\mathbf{x}_0) \end{bmatrix},
$$

where $\mathbf{y}(t_0), \dots, \mathbf{y}(t_{N-1})$ are Gaussian and independent, i.e. $\mathbf{y}(t_i) \sim \mathcal{N}(\boldsymbol{\mu}_i(\mathbf{x}_0), \Sigma)$ and $\mathbf{y}(t_i)$ and $\mathbf{y}(t_j)$ are independent for every distinct i and j. Then, the log-likelihood function is given by

$$
\log f_{Y|X_0}(Y; \mathbf{x}_0) = \log K - \frac{1}{2}(Y - \boldsymbol{\mu})^{\mathsf{T}} \Sigma^{-1} (Y - \boldsymbol{\mu}),
$$

where $K = 1/\sqrt{(2\pi)^N \det(\Sigma)}$. Notice now that

$$
FIM(\mathbf{x}_0) = \mathbb{E}_Y[(\nabla \ln f_{Y|X_0}(Y; \mathbf{x}_0))(\nabla \ln f_{Y|X_0}(Y; \mathbf{x}_0))^{\mathsf{T}}]
$$

or, equivalently,

$$
FIM(\mathbf{x}_0) = \frac{1}{4} \mathbb{E}_Y \left[\{\nabla([Y - \boldsymbol{\mu}]^{\mathsf{T}} \Sigma^{-1} [Y - \boldsymbol{\mu}])\} \{\nabla([Y - \boldsymbol{\mu}]^{\mathsf{T}} \Sigma^{-1} [Y - \boldsymbol{\mu}])\}^{\mathsf{T}} \right].
$$

A simple computation yields

$$
\nabla([Y - \boldsymbol{\mu}]^{\mathsf{T}} \Sigma^{-1} [Y - \boldsymbol{\mu}]) = [\nabla(Y - \boldsymbol{\mu})]^{\mathsf{T}} \tilde{\Sigma} [Y - \boldsymbol{\mu}],
$$

where $\tilde{\Sigma} = \Sigma^{-1} + (\Sigma^{-1})^{\mathsf{T}}$. Since Σ is symmetric, we have that

$$
\nabla([Y - \boldsymbol{\mu}]^{\mathsf{T}} \Sigma^{-1} [Y - \boldsymbol{\mu}]) = -2(\nabla \boldsymbol{\mu})^{\mathsf{T}} \Sigma^{-1} (Y - \boldsymbol{\mu}).
$$

Therefore,

$$
\begin{aligned}
FIM(\mathbf{x}_0) &= \mathbb{E}_Y[(\nabla \boldsymbol{\mu})^{\mathsf{T}} \Sigma^{-1} (Y - \boldsymbol{\mu})(Y - \boldsymbol{\mu})^{\mathsf{T}} \Sigma^{-1} (\nabla \boldsymbol{\mu})] \\
&= (\nabla \boldsymbol{\mu})^{\mathsf{T}} \Sigma^{-1} \{\mathbb{E}_Y[(Y - \boldsymbol{\mu})(Y - \boldsymbol{\mu})^{\mathsf{T}}]\} \Sigma^{-1} (\nabla \boldsymbol{\mu}) \\
&= (\nabla \boldsymbol{\mu})^{\mathsf{T}} \Sigma^{-1} (\nabla \boldsymbol{\mu}) \\
&= \sum_{i=0}^{N-1} (Ce^{At_i})^{\mathsf{T}} \Sigma^{-1} Ce^{At_i} \\
&= W_{\mathrm{d}}(0, t_f).
\end{aligned}
$$

This completes the proof.

Remark 18.2

In the context of LTI systems, Theorem 18.1 connects the two important notions, observability Gramian on one hand and FIM on the other hand. FIM is a very important concept due to the celebrated result embodied in the CRLB. From this perspective, the connection between the observability Gramian and the Fisher information under the conditions stated is very significant.

The computation of the (stochastic) observability Gramian introduced earlier requires explicit computation of the solution of the underlying LTI system, which is straightforward to do. In the case of nonlinear systems, a possible generalization of this methodology involves the computation of the empirical observability Gramian (EOG), see [54]. See also [34] for the connection between the EOG and the FIM for nonlinear systems. In the context of nonlinear regression problem a partial result is known, that is, nonsingularity of FIM implies local observability [37].

18.3 Optimal Sensor Placement: Fixed Sensors and Targets

For an adequate introduction to the problem of multiple underwater target positioning and also for clarity of comprehension, we first address the problem of optimal sensor placement for underwater target positioning based on measurements of the ranges between the targets and a set of sensors (acoustic ranging devices) for the case when both sensors and targets are static. The geometry of the optimal sensor configuration depends strongly on the constraints imposed by the given target positioning problem itself (e.g., maximum number of sensors available or possible constraints on their placement) and the environment (e.g., characteristics of the acoustic transmission channel). In fact, an inadequate sensor configuration may yield large positioning errors no matter what positioning system is used. For this reason, there exists considerable interest in the analysis and development of underwater positioning systems based on sensor networks. See for example [38] for a survey of wireless sensor networks for underwater target positioning.

The optimal sensor placement problem seeks inspiration from the GPS Intelligent Buoy (GIB) system, see [39], but allows for its extension to the case where the surface buoys can be replaced by ASVs carrying acoustic sensors that measure the ranges to a target. With this setup, the system can compute the position of the target and adopt the most appropriate sensor configuration in accordance with the estimated position of the target and the noise characteristics of the range sensors. Moreover, the formation can reconfigure itself in response to on-line detected changes in the mission conditions so as to yield the best positioning accuracy possible. Therefore, the optimality conditions determined in the following will be at the core of the algorithms to compute the best geometric formation to be adopted as the mission unfolds.

Let $\mathbf{p}^{[i]} \in \mathbb{R}^d$, $i \in \mathbb{I}_{[1,p]}$, be the inertial position of the ith sensor and let $\mathbf{q}^{[\alpha]} \in \mathbb{R}^d$, $\alpha \in \mathbb{I}_{[1,q]}$, be the inertial position of the αth target. Each sensor output function is given by

$$y_{i,\alpha} = h(\mathbf{p}^{[i]}, \mathbf{q}^{[\alpha]}) + \eta_{i,\alpha}, \, \alpha \in \mathbb{I}_{[1,q]}, \tag{18.4}$$

where $h(\mathbf{p}^{[i]}, \mathbf{q}^{[\alpha]}) = \|\mathbf{p}^{[i]} - \mathbf{q}^{[\alpha]}\|$ is the true range, $\eta_{i,\alpha} \sim \mathcal{N}(0, \sigma_{i,\alpha}^2)$, and $\mathbf{y}^{[i]} = [y_{i,1} \ldots y_{i,q}]^{\mathrm{T}} \in \mathbb{R}^q$. For a given target localization problem, there are two basic questions:

i. Given range measurements and the sensor coordinates, compute the unknown target position.

ii. Find the best strategy to place the sensors around the target so as to maximize the range-related information for target localization.

The first question is a classical estimation problem, which can be solved by least squares (LS), maximum likelihood (ML), or maximum a priori (MAP) estimation methods. For instance, the least squares estimator for the αth target is given by

$$\hat{\mathbf{q}}_{LS}^{[\alpha]} = \arg \min_{x \in \mathbb{R}^{pd}} L(\mathbf{q}^{[\alpha]}) = \frac{1}{2} \sum_{i=1}^{p} \left(y_{i,\alpha} - h\left(\mathbf{p}^{[i]}, \mathbf{q}^{[\alpha]}\right) \right)^2.$$

The second question goes beyond the mechanization of an estimation procedure, as it requires finding the best sensor configuration by optimizing a cost function, such as one related to the FIM. This problem, as mentioned above, is referred as the *optimal sensor placement* problem. To simplify the notation, we let $\boldsymbol{\theta} = \mathbf{q}, \mathbf{x} = [(\mathbf{p}^{[1]})^{\mathrm{T}}, \ldots, (\mathbf{p}^{[p]})^{\mathrm{T}}]^{\mathrm{T}} \in \mathbb{R}^{pd}$, and $\sigma_{i,\alpha} = \sigma$ for all $(i,\alpha) \in \mathbb{I}_{[1,p]} \times \mathbb{I}_{[1,q]}$. With this setup, the problem at hand can be formulated as that of computing

$$\mathbf{x}^* = \arg \min_{\mathbf{x} \in \mathbb{R}^{pd}} J(\mathbf{x}) := -\ln \det(FIM(\boldsymbol{\theta}; \mathbf{x}))$$

where

$$FIM(\boldsymbol{\theta}; \mathbf{x}) = \frac{1}{\sigma^2} \sum_{i=1}^{p} (\bar{\mathbf{u}}^{[i]})(\bar{\mathbf{u}}^{[i]})^{\mathrm{T}} \tag{18.5}$$

with $\bar{\mathbf{u}}^{[i]} = \nabla_{\boldsymbol{\theta}}(\|\mathbf{q} - \mathbf{p}^{[i]}\|), i \in \mathbb{I}_{[1,p]}$.

Using this mathematical framework, we first address the problem of finding the optimal geometric configuration of a sensor formation for single underwater target positioning, based on target-to-sensor range measurements in 2D and 3D. Afterwards, the analysis is extended to the multiple target localization problem, also in 2D and 3D. The results presented in this section are a summary of the algorithms and solutions developed by the authors in previous works. See [40–43] for complete details.

It is important to stress that even though the optimal sensor placement problem presented is strongly motivated by an underwater application scenario, the methodology developed is not restricted to this area. In fact, the methodology developed is sufficiently general to be applied to any positioning problem in 2D or 3D.

18.3.1 Single-Target Localization in $\mathbb{R}^2$

In this scenario we may consider that the target depth is known, and only the $\{\mathbf{x}_{\mathcal{I}}, \mathbf{y}_{\mathcal{I}}\}$ target coordinates must be estimated. For the single target localization problem, let $\mathbf{q} = [q_x, q_y]^{\mathrm{T}}$ be the position of an arbitrary target, $\mathbf{p}^{[i]} = [p_x^{[i]}, p_y^{[i]}]^{\mathrm{T}}, i \in \mathbb{I}_{[1,p]}$, the position of the ith ranging sensor, and $\eta_i \sim \mathcal{N}(0, \sigma_i^2)$, the corresponding measurement noise. Further, let d_i be the actual distance between target $\mathbf{q}$ and the ith sensor. For the sake of simplicity and without loss of generality, the target is considered to be placed at the origin of the inertial coordinate frame and $\sigma_i = \sigma$. Therefore, the FIM becomes

$$FIM(\boldsymbol{\theta}; \mathbf{x}) = \frac{1}{\sigma^2} \sum_{i=1}^{p} \begin{bmatrix} \dfrac{(p_x^{[i]})^2}{d_i^2} & \dfrac{p_x^{[i]} p_y^{[i]}}{d_i^2} \\ \dfrac{p_x^{[i]} p_y^{[i]}}{d_i^2} & \dfrac{(p_y^{[i]})^2}{d_i^2} \end{bmatrix} = \frac{1}{\sigma^2} \sum_{i=1}^{p} \begin{bmatrix} C_{\phi^{[i]}}^2 & C_{\phi^{[i]}} S_{\phi^{[i]}} \\ C_{\phi^{[i]}} S_{\phi^{[i]}} & S_{\phi^{[i]}}^2 \end{bmatrix} \tag{18.6}$$

where $\phi^{[i]}$ is the angle that the ith range vector forms with the $\{\mathbf{x}_I\}$ axis of the inertial coordinate frame.

Following the mathematical procedure detailed in [41], the FIM that provides the maximum possible (logarithm of the) FIM determinant yields

$$FIM^*(\boldsymbol{\theta}; \mathbf{x}) = \left(\frac{p}{2\sigma^2} \right) I_2 \tag{18.7}$$

with $\det(FIM^*(\boldsymbol{\theta}; \mathbf{x})) = (\sigma^{-2} p/2)^2$.

Comparing the optimal FIM in (18.7) with the generic one in (18.6) gives an implicit characterization of the conditions that the sensor network must satisfy in order for it to be optimal:

$$\sum_{i=1}^{p} \left(\frac{p_x^{[i]}}{d_i} \right)^2 = \frac{p}{2}, \quad \sum_{i=1}^{p} \left(\frac{p_y^{[i]}}{d_i} \right)^2 = \frac{p}{2}, \quad \text{and} \quad \sum_{i=1}^{p} \left(\frac{p_x^{[i]}}{d_i} \right) \left(\frac{p_y^{[i]}}{d_i} \right) = 0. \tag{18.8}$$

Notice, from (18.8), how the optimality conditions that characterize the sensor formation geometry that maximizes the FIM determinant depend only on the angles that the range vectors form between them. Thus, the optimal formations can be obtained analytically from the system (18.8), which can be rewritten as:

$$\sum_{i=1}^{p} \left(\frac{p_x^{[i]}}{d_i} \right)^2 = \sum_{i=1}^{p} C_{\phi^{[i]}}^2 = \frac{p}{2}, \quad \sum_{i=1}^{p} \left(\frac{p_y^{[i]}}{d_i} \right)^2 = \sum_{i=1}^{p} S_{\phi^{[i]}}^2 = \frac{p}{2}, \quad \text{and}$$

$$\sum_{i=1}^{p} \frac{p_x^{[i]} p_y^{[i]}}{d_i^2} = \sum_{i=1}^{p} S_{\phi^{[i]}} C_{\phi^{[i]}} = 0 \tag{18.9}$$

Using by now classical terminology, from (18.9), the sensor formation must be first- and second-moment balanced. One simple and intuitive configuration arises by noticing the following orthogonality relations for sines and cosines from Fourier analysis [44]:

$$\sum_{i=0}^{p-1} C_{2i\pi/p}^2 = \sum_{i=0}^{p-1} S_{2i\pi/p}^2 = \frac{p}{2},$$

$$\sum_{i=0}^{p-1} C_{2i\pi/p} S_{2i\pi/p} = \sum_{i=0}^{p-1} C_{2i\pi/p} = \sum_{i=0}^{p-1} S_{2i\pi/p} = 0. \tag{18.10}$$

It follows from the above that the maximum FIM determinant is obtained when the sensors are regularly distributed around the target position. Moreover, an infinite number of solutions may be obtained by rotating the sensors rigidly around the target position, that is, by allowing the above angles to become $2\pi i/p + \phi_c$, $i \in \mathbb{I}_{[0,p-1]}$, where $\phi_c \in [0, 2\pi]$ is a fixed but arbitrary angle, and multiplying the distance of any of the sensors to the target by an arbitrary positive number. This fact shows clearly that the optimal solutions depend directly on the angles formed between the range vectors. Notice also one important feature about the optimal solutions that can be computed based on the analysis explained above: if two disjoint sets of $\mathbf{p}_1^*$ and $\mathbf{p}_2^*$ sensors each are optimally placed, the resulting formation of combining $\mathbf{p}_1^*$ and $\mathbf{p}_2^*$ sensors is also optimal. Therefore, new higher-order optimal solutions can be obtained by combining reduced-order optimal configurations. This fact can be seen in the example below.

In order to better understand the efficacy of the solutions obtained in the different examples studied, we now introduce the concept of $\det(FIM(\boldsymbol{\theta}; \mathbf{x})_{\mathcal{D}})$. To clarify, this notation, consider a finite spatial region $\mathcal{D}$ around the target, where the determinant of the FIM is computed for a number of hypothetical target points by allowing these points to be on a grid in $\mathcal{D}$, given an optimal sensor configuration computed for the nominal situation where the target is at the origin. A plot of $\det(FIM(\boldsymbol{\theta}; \mathbf{x})_{\mathcal{D}})$ over a region $\mathcal{D}$ will allow for a quick assessment of how good the given sensor formation is in terms of yielding accurate positioning of the real target, in comparison with the positioning accuracy that is possible for any hypothetical target (different from the real target) positioned anywhere in a finite spatial region $\mathcal{D}$.

Example 18.1

In this example, two possible optimal sensor formations of seven sensors are computed. In the first one, shown in Figure 18.3a, the sensors are regularly distributed around the target position. In the second one, shown in Figure 18.3b, it can be seen how the combination of a four sensor optimal formation with a three sensor optimal formation

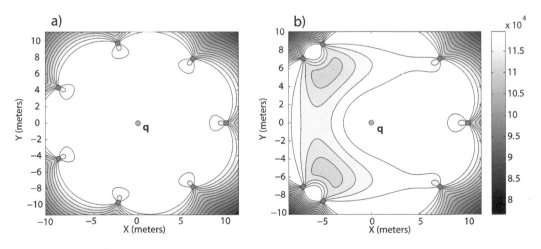

FIGURE 18.3
Optimal sensor configuration for seven sensors formed by the regular distribution of sensors around the target position $\mathbf{q}$ at the origin, and $\det(FIM(\boldsymbol{\theta}; \mathbf{x})_{\mathcal{D}})$, with $\boldsymbol{\theta}$ over the region of interest (a). Optimal sensor configuration for seven sensors formed by the combination of a four-sensor optimal formation and a three-sensor optimal formation, and $\det(FIM(\boldsymbol{\theta}; \mathbf{x})_{\mathcal{D}})$ over the region of interest (b). Lighter regions correspond to areas where $\det(FIM(\boldsymbol{\theta}; \mathbf{x})_{\mathcal{D}})$ is larger.

also yields an optimal formation of seven sensors for the situation where the target is at the origin. For both optimal formations, the maximum theoretical FIM determinant is obtained at the target position, that is, det($FIM(\theta; x)$) = 12.25 × 10⁴ m⁻⁴, and thus, they are equivalent. However, they are quite different in what concerns the possibility of yielding good target position accuracy when, for a frozen sensor formation, the target is allowed to be anywhere in $\mathcal{D}$. Clearly. solution (a) leads to a better "coverage" of a small area centered at the origin, in comparison with solution (b).

18.3.2 Single-Target Localization in $\mathbb{R}^3$

The problem of optimal sensor placement in 3D is considerably more complex than that in 2D. In the 3D case, the characterization of all possible solutions is not a simple generalization of that offered in the 2D case and, as we will show, there are a large number of optimal sensor configurations. The following analysis summarizes the presentation in [42], which was inspired by the work in [41].

Similarly to the analysis in the previous section, for the sake of simplicity and without loss of generality, the target is considered to be placed at the origin of an inertial coordinate frame. It can be shown that the optimal FIM is given by

$$FIM^*(\theta; x) = \left(\frac{p}{3\sigma^2}\right) I_3. \tag{18.11}$$

The proof of this result can be found in [42, Appendix A].

Comparing the optimal FIM in (18.11) with the generic one in (18.5) gives an implicit characterization of the conditions that the sensor network must satisfy to be optimal, as follows:

$$\sum_{i=1}^{p} \frac{(p_x^{[i]})^2}{d_i^2} = \sum_{i=1}^{p} \frac{(p_y^{[i]})^2}{d_i^2} = \sum_{i=1}^{p} \frac{(p_z^{[i]})^2}{d_i^2} = \frac{p}{3}, \tag{18.12}$$

$$\sum_{i=1}^{p} \frac{p_x^{[i]} p_y^{[i]}}{d_i^2} = \sum_{i=1}^{p} \frac{p_x^{[i]} p_z^{[i]}}{d_i^2} = \sum_{i=1}^{p} \frac{p_z^{[i]} p_y^{[i]}}{d_i^2} = 0. \tag{18.13}$$

Notice that the FIM (18.11) is diagonal and its eigenvalues are equal. Therefore, the optimality conditions derived not only maximize the determinant of the FIM (D-optimality criterion) but minimize also its maximum eigenvalue (E-optimality criterion) and the trace of the CRLB (A-optimality criterion); see [42, Appendices B and C]. Similarly to the 2D problem, all optimal sensor configurations are characterized in terms of the angles that the range vectors form with the unit axes of the inertial frame adopted, i.e., there is no explicit dependence on the ranges themselves and, by rewriting $\cos(\varrho_x^{[i]}) = p_x^{[i]}$, $\cos(\varrho_y^{[i]}) = p_y^{[i]}$, and $\cos(\varrho_z^{[i]}) = p_z^{[i]}$, it can be shown that optimal sensor formations correspond to having the sensors placed on the surface of a unit sphere centered at the origin. This is because in the formulation adopted it was tacitly assumed that the covariance of the range measurements is distance-invariant. See [42] for a discussion of this assumption.

Once an optimal solution is found in terms of the direction cosines referred above, an infinite number of optimal solutions can be generated by: (i) multiplying the range of each sensor to the target by an arbitrary positive number (this scaling of the ranges may depend on practical physical constraints) and (ii) rotating the sensor formation rigidly around an arbitrary axis passing through the target position. The first statement is trivial to prove. The proof of the second statement can be outlined as follows: let an initial sensor formation on the unit sphere be described by vectors $\mathbf{p}^{[i]}$, $i \in \mathbb{I}_{[1,p]}$, and let $\tilde{\mathbf{p}}^{[i]} = R\mathbf{p}^{[i]}$, $i \in \mathbb{I}_{[1,p]}$, be the formation that is obtained by applying the same rotation matrix R to all vectors. Equations (18.12) and (18.13), with $d_i = 1$, can be written in compact form as

$$\sum_{i=1}^{p} (\mathbf{p}^{[i]})(\mathbf{p}^{[i]})^{\mathrm{T}} = \left(\frac{p}{3}\right) I_3. \tag{18.14}$$

It then follows that

$$\sum_{i=1}^{p} (\tilde{\mathbf{p}}^{[i]})(\tilde{\mathbf{p}}^{[i]})^{\mathrm{T}} = \sum_{i=1}^{p} R(\mathbf{p}^{[i]})(\mathbf{p}^{[i]})^{\mathrm{T}} R^{\mathrm{T}} = \left(\frac{p}{3}\right) I_3 \tag{18.15}$$

because $R \in SO(3)$; thus, the new sensor positions verify (18.14) and therefore, (18.12) and (18.13).

A procedure to generate a sufficiently rich set of optimal configurations is presented next. Let $\boldsymbol{\xi}_p = \mathbf{p}^{[1]} + \cdots + \mathbf{p}^{[p]}$, $p > 2$, denote the geometric center of an optimal sensor formation and let $\mathbf{d}_q$ denote the vector directed from the origin (target position) to $\boldsymbol{\xi}_p$. Clearly, given (18.15) the formation obtained by moving the sensor rigidly with $\mathbf{d}_q$ and aligning it with the $\{\mathbf{z}_I\}$ axis of the inertial coordinate frame is also optimal. The center of the resulting formation will be denoted by $\mathbf{z}_q$. In this situation, again without loss of generality, the sensor positions satisfy

$$\sum_{i=1}^{p} p_x^{[i]} = 0, \quad \sum_{i=1}^{p} p_y^{[i]} = 0, \quad \text{and} \quad \sum_{i=1}^{p} p_z^{[i]} = p z_q, \tag{18.16}$$

where $\mathbf{p}^{[i]} = [p_x^{[i]}, p_y^{[i]}, p_z^{[i]}]^{\mathrm{T}}$ is the ith sensor position, and $\mathbf{z}_q = [0, 0, z_q]^{\mathrm{T}}$.

To better characterize some classes of optimal solutions, we now restrict the types of solutions to not lie only on the unit sphere but also on a general quadratic surface that intersects the unit sphere. Consider a quadratic surface described by

$$\bar{\mathbf{x}}^{\mathrm{T}} A \bar{\mathbf{x}} = 0, \tag{18.17}$$

where $\bar{\mathbf{x}} = [x, y, z, 1]^{\mathrm{T}} \in \mathbb{R}^4$ and $A \in \mathbb{R}^{4 \times 4}$.

To define the possible quadratic surfaces of interest, the equation of the unit sphere and also (18.17) must be satisfied by the sensor positions, together with (18.12), (18.13), and (18.16) with $d_i = 1$. Adding p equations, one for each sensor, the following constraint defines the possible quadratic surfaces of interest:

$$\operatorname{tr}(A) + 2a_{44} + 6a_{34} z_q = 0. \tag{18.18}$$

In the above equation, a_{44} and a_{34} are (4,4) and (3,4) entries of A.

There are of course an infinite number of degrees of freedom in the choice of a particular solution. See Figure 18.4 where an optimal sensor formation of four sensors is defined over the intersection of a hyperbolic cylinder, described by $4x^2 - y^2 = 1$ and the unit sphere.

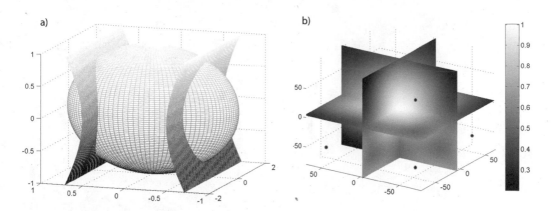

FIGURE 18.4
Determination of the loci of optimal sensor networks obtained by intersecting the unit sphere and a quadratic surface: intersection between the unit sphere and a hyperbolic cylinder (a). Plot of det($FIM(\theta; \mathbf{x})_D$) in the three main planes of region $\mathcal{D}$, using four sensors marked as circles. Lighter regions correspond to larger values of det($FIM(\theta; \mathbf{x})$) (b).

Moreover, as in the 2D analysis, for two distinct optimal formations $\mathbf{p}^*_1$ and $\mathbf{p}^*_2$, the resulting formation of combining $\mathbf{p}^*_1$ and $\mathbf{p}^*_2$ is also optimal. Therefore, new higher-order optimal solutions can be computed by combining reduced-order optimal configurations.

By adopting this method one can compute specific optimal configurations in 3D space. Furthermore, it becomes possible to incorporate physical constraints on the sensor placement problem by using appropriate quadratic surfaces. For example, many applications require that the sensors be positioned at the ocean surface, on the ocean bottom, or a combination thereof. This is done by further restricting the types of quadratic surfaces used, as explained next.

The above problem may be cast in the following equivalent form: given a target at the origin of an inertial reference frame $\{\mathcal{I}\}$, a unit sphere centered at the target, and two horizontal parallel planes (one of the possible quadratic surfaces), compute the distance z_q from the target to one of the planes and determine an optimal formation that lies entirely on the intersection of the unit sphere and that plane. Once a solution is found, scale the ranges based on the physical constraints on the sensor placement while preserving the direction cosines of the range vectors. Clearly, the scaling factor in the practical application at hand, is d_t/z_q, where d_t is the target depth. These results yield straightforward solutions to the problem of underwater target positioning, as follows.

The computation of the optimal formation on the desired plane unfolds in two steps: first, the z_q coordinate of the plane is computed; the geometric configuration of the sensors on the plane is then derived. The first step is straightforward: it follows from (18.12) that

$$p_z^{[1]} = \cdots = p_z^{[p]} = z_q = \pm 1/\sqrt{3}. \tag{18.19}$$

Solutions of (18.19) correspond to two horizontal planes that intersect the unit sphere along two circumferences of radii r', as depicted in Figure 18.5, and thus the sensors may be distributed on the circumference of the plane specified by $z_q = 1/\sqrt{3}$ or on that of the plane given by $z_q = -1/\sqrt{3}$. The formation also lies on the unit sphere, so that $(r')^2 + z_q^2 = 1$

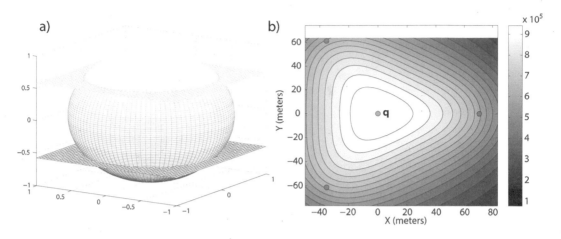

FIGURE 18.5
Intersections between two horizontal parallel planes and the unit sphere to define two possible loci where the sensors can be located for optimal sensor network placement on a plane (a). Optimal sensor formation for underwater target positioning in 3D with three sensors placed on a horizontal plane and $\det(FIM(\theta; \mathbf{x})_\mathcal{D})$, where $\mathcal{D}$ belongs to the target plane (b).

and thus the optimal radii become $r' = \sqrt{1-z_q^2} = \sqrt{2/3}$. Let now d_t be the target depth and assume that the p sensors are constrained to lie at the sea surface. From the discussion above, an optimal solution corresponds to distributing the sensors on a circumference of radius $r_s = \sqrt{2}d_t$. Similarly, if the target is at a distance (altitude) h_t above the seabed, and the p sensors are constrained to lie on the sea bottom, an optimal solution corresponds to distributing the sensors on a circumference of radius $r_s = \sqrt{2}h_t$. In the above two situations, the center of the circumferences is positioned either directly above or under the target. Finally, the geometric configuration of the sensors on one of the above circumferences should be determined. To this end, the sensor positions are rewritten in polar coordinates as $\mathbf{p}^{[i]} = r'\mathbf{g}(\phi^{[i]})$ and $p_z^{[i]} = z_q$, where $\phi^{[i]}$ is the angle that the projection of the ith range vector on the $\{\mathbf{x}_\mathcal{I}, \mathbf{y}_\mathcal{I}\}$ plane forms with the $\{\mathbf{x}_\mathcal{I}\}$ axis, $r' = \sqrt{2/3}$, and $z_q = 1/\sqrt{3}$. Therefore, (18.12) and (18.13) yield

$$\sum_{i=1}^{p} r'^2 C_{\phi^{[i]}}^2 = \frac{p}{3} \rightarrow \frac{2}{3}\sum_{i=1}^{p} C_{\phi^{[i]}}^2 = \frac{p}{3}, \quad \sum_{i=1}^{p} r'^2 S_{\phi^{[i]}}^2 = \frac{p}{3} \rightarrow \frac{2}{3}\sum_{i=1}^{p} S_{\phi^{[i]}}^2 = \frac{p}{3},$$

$$\sum_{i=1}^{p} z_q^2 = \frac{p}{3} \rightarrow \sum_{i=1}^{p} \left(\frac{1}{\sqrt{3}}\right)^2 = \frac{p}{3}, \quad \sum_{i=1}^{p} r'^2 C_{\phi^{[i]}} S_{\phi^{[i]}} = 0 \rightarrow \sum_{i=1}^{p} C_{\phi^{[i]}} S_{\phi^{[i]}} = 0,$$

$$\quad\quad\quad\quad\quad\quad\quad\quad\quad\quad\quad\quad\quad\quad\quad\quad\quad\quad\quad\quad\quad\quad\quad\quad\quad\quad\quad (18.20)$$

$$z_q \sum_{i=1}^{p} r' C_{\phi^{[i]}} = 0 \rightarrow \sum_{i=1}^{p} C_{\phi^{[i]}} = 0, \quad z_q \sum_{i=1}^{p} r' S_{\phi^{[i]}} = 0 \rightarrow \sum_{i=1}^{p} S_{\phi^{[i]}} = 0.$$

A possible optimal solution on one of these circumferences may be defined through the orthogonality relations for sines and cosines (18.10).

Thus, from (18.20), the optimal solutions are obtained by distributing the p sensors uniformly along the circumferences, the vectors from the sensors to the center of the circumferences making angles $2\pi i/p$, $i \in \mathbb{I}_{[0,p-1]}$, with the $\{\mathbf{x}_\mathcal{I}\}$-axis. Again, an infinite number of solutions may be defined by rotating the sensors rigidly along the circumferences a fixed but arbitrary angle ϕ_c in $[0, 2\pi]$, allowing the above angles to become $2\pi i/p + \phi_c$, $i \in \mathbb{I}_{[0,p-1]}$. See Figure 18.5 for a graphical representation of the determination of the loci where the sensors must be located when two parallel planes are considered for the intersection with the unit sphere (a), and a possible solution over one of the circumferences obtained from the above intersection for a three sensor formation (b).

The above results imply that once a solution is obtained, an infinite number of solutions can be generated in three steps: (i) compute the vectors from the target to the sensor positions, (ii) rotate them rigidly about a same axis, and (iii) find (if they exist) the intersections of the extensions of the rotated vectors with the horizontal plane. These intersections will become the new optimal sensor positions.

18.3.3 Multiple Target Localization in $\mathbb{R}^2$: Analytical Solutions

We now tackle the problem of optimal sensor placement for multiple underwater target localization. We start with the analysis of the optimal sensor placement problem for multiple target localization in 2D, since analytical solutions that provide the maximum FIM determinant for each of the targets simultaneously can actually be computed in this case for some situations of interest. In the next section, the general problem of sensor placement for any number of targets and sensors in 2D and 3D is studied, since in a general setup the maximum FIM determinant cannot be achieved for every target, and tradeoff solutions must be adopted. Optimal solutions for which the maximum FIM determinant is not achievable for all targets simultaneously are searched using numerical optimization tools.

As described, there are situations in 2D scenarios for which it is possible to define sensor configurations that provide the maximum FIM determinant for all of the targets simultaneously, as explained next.

Recall that for the target α, the FIM yields

$$FIM^{[\alpha]}(\boldsymbol{\theta}_\alpha;\mathbf{x}) = \frac{1}{\sigma^2}\sum_{i=1}^{p}\begin{bmatrix} C_{\phi^{[i,\alpha]}}^2 & C_{\phi^{[i,\alpha]}}S_{\phi^{[i,\alpha]}} \\ C_{\phi^{[i,\alpha]}}S_{\phi^{[i,\alpha]}} & S_{\phi^{[i,\alpha]}}^2 \end{bmatrix} \tag{18.21}$$

for $i \in \mathbb{I}_{[1,p]}$ and $\alpha \in \mathbb{I}_{[1,q]}$, and with $\phi^{[i,\alpha]}$ being the angle that the range vector from target α to sensor i forms with the $\{\mathbf{x}_\mathcal{I}\}$-axis. The determinant of (18.21) may be written as

$$\det(FIM^{[\alpha]}(\boldsymbol{\theta}_\alpha;\mathbf{x})) = \frac{1}{\sigma^4}\begin{bmatrix} \sum_{i=1}^{p}C_{\phi^{[i,\alpha]}}^2 & -\sum_{i=1}^{p}C_{\phi^{[i,\alpha]}}S_{\phi^{[i,\alpha]}} \end{bmatrix}\begin{bmatrix} \sum_{i=1}^{p}S_{\phi^{[i,\alpha]}}^2 \\ \sum_{i=1}^{p}C_{\phi^{[i,\alpha]}}S_{\phi^{[i,\alpha]}} \end{bmatrix}. \tag{18.22}$$

To shed light on how analytical solutions may be defined, the simplest case of two targets and an arbitrary number of sensors is studied first. Afterwards, the methodology is extended to the general problem of an arbitrary number of targets. In the multiple target scenario we have $\boldsymbol{\theta} = [\boldsymbol{\theta}_1, \ldots, \boldsymbol{\theta}_q] = [\mathbf{q}^{[1]}, \ldots, \mathbf{q}^{[q]}]$, and for the two targets case, the cost function becomes

$$\mathbf{p}^* = \arg\max_{\mathbf{p}} \left(\log\det(FIM^{[1]}(\boldsymbol{\theta}_1; \mathbf{x})) + \log\det(FIM^{[2]}(\boldsymbol{\theta}_2; \mathbf{x})) \right). \qquad (18.23)$$

Equation (18.23) can be rewritten as:

$$\mathbf{p}^* = \arg\max_{\mathbf{p}} \log(\det(FIM_T(\boldsymbol{\theta}; \mathbf{x}))) \qquad (18.24)$$

where $\det(FIM_T(\boldsymbol{\theta}; \mathbf{x})) := \det(FIM^{[1]}(\boldsymbol{\theta}_1; \mathbf{x})) \det(FIM^{[2]}(\boldsymbol{\theta}_2; \mathbf{x}))$. From (18.22) and (18.24), the cost function yields

$$\mathbf{p}^* = \arg\max_{\mathbf{p}} \log\det(FIM_T(\boldsymbol{\theta}; \mathbf{x})) = \arg\max_{\mathbf{p}} \log\left(\frac{1}{\sigma^{4q}} \prod_{\alpha=1}^{q} \mathbf{a}_\alpha^T \mathbf{b}_\alpha \right) \qquad (18.25)$$

with

$$\mathbf{a}_\alpha = \begin{bmatrix} \sum\limits_{i=1}^{p} C_{\phi^{[i,\alpha]}}^2 \\ -\sum\limits_{i=1}^{p} C_{\phi^{[i,\alpha]}} S_{\phi^{[i,\alpha]}} \end{bmatrix} \quad \text{and} \quad \mathbf{b}_\alpha = \begin{bmatrix} \sum\limits_{i=1}^{p} S_{\phi^{[i,\alpha]}}^2 \\ \sum\limits_{i=1}^{p} C_{\phi^{[i,\alpha]}} S_{\phi^{[i,\alpha]}} \end{bmatrix}.$$

Following the analysis described in [41], the optimality condition that yields the maximum value of the FIM determinant for each of the targets is given by

$$\sum_{i=1}^{p} \mathbf{g}(2\phi^{[i,\alpha]}) = \mathbf{0}. \qquad (18.26)$$

Condition (18.26) is valid for any number of sensors and allows for the computation of optimal sensor formations that provide the theoretical maximum accuracy for both targets simultaneously.

Once the analytical solution for the two-target positioning problem has been defined, the analysis can be extended for an arbitrary number of targets, and the cost function becomes:

$$\mathbf{p}^* = \arg\max_{\mathbf{p}} \sum_{\alpha=1}^{q} \log\det(FIM^{[\alpha]}(\boldsymbol{\theta}_\alpha; \mathbf{x})) = \arg\max_{\mathbf{p}} \log\det(FIM_T(\boldsymbol{\theta}; \mathbf{x})). \qquad (18.27)$$

From (18.27), following a similar procedure to that used for two targets, detailed in [41], it is possible to obtain the gradient equations (with respect to the sensor positions) that define the optimality conditions that optimal sensor configurations must satisfy, defined as

$$\nabla_{\mathbf{p}^{[i]}} \log\det(FIM_T(\boldsymbol{\theta}; \mathbf{x})) = -\sum_{\alpha=1}^{q} \frac{1}{\det(FIM^{[\alpha]}(\boldsymbol{\theta}_\alpha; \mathbf{x}))} \left(\frac{f^{[i,\alpha]}}{d_{i,\alpha}} \right) \mathbf{g}^\perp(\phi^{[i,\alpha]}) \qquad (18.28)$$

with

$$f^{[i,\alpha]} = S_{2\phi^{[i,\alpha]}} \sum_{s=1}^{p} C_{2\phi^{[s,\alpha]}} - C_{2\phi^{[i,\alpha]}} \sum_{s=1}^{p} S_{2\phi^{[s,\alpha]}}.$$

Equation (18.28) must hold for all sensors in the network. Notice that (18.28) may have multiple solutions, with the solution in (18.26) being a valid one; this sensor configuration will yield the maximum FIM determinant $n^2/(4\sigma^4)$ for all targets simultaneously. However, (18.26) may or may not be a solution for a given multiple-target localization problem, depending on the configuration of the targets and on the number of sensors involved in the task. If it is not possible to define a sensor configuration for which (18.26) will be satisfied for all the targets, then a tradeoff solution must be adopted, where tradeoffs may be mission-dependent. These solutions are obtained by resorting to an optimization algorithm that is described in the next section.

Example 18.2

The present example deals with two different cases: (a) six sensors and two targets, and (b) six sensors and three targets. It is possible to compute, for both cases, optimal configurations for which the maximum FIM determinant is obtained for all of the targets simultaneously, i.e., (18.26) holds for each of the targets. For case (a) the target positions and optimal sensor positions are described in Table 18.1 and the optimal sensor configuration is shown in Figure 18.6a. For case (b), the target and optimal sensor

TABLE 18.1

Target Positions and Optimal Sensor Position Coordinates

	$q^{[1]}$	$q^{[2]}$	$p^{[1]}$	$p^{[2]}$	$p^{[3]}$	$p^{[4]}$	$p^{[5]}$	$p^{[6]}$
$x(m)$	−10	0	21.79	0	−21.79	−21.79	0	21.79
$y(m)$	10	0	12.34	24.95	12.34	−12.34	−24.95	−12.34

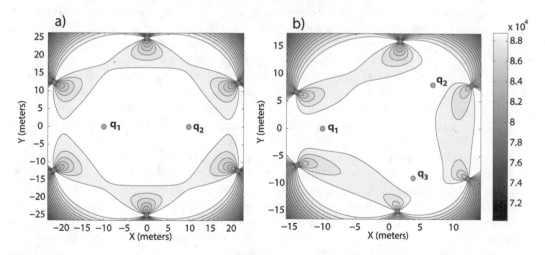

FIGURE 18.6
(a) Optimal sensor formation for six sensors and two targets, and det($FIM(\theta; x)_D$) for the region of interest. (b) Optimal sensor formation for six sensors and three targets, and det($FIM(\theta; x)_D$). Targets are indicated with labels q_1, q_2, and q_3.

TABLE 18.2

Target Positions and Optimal Sensor Position Coordinates

	$q^{[1]}$	$q^{[2]}$	$q^{[3]}$	$p^{[1]}$	$p^{[2]}$	$p^{[3]}$	$p^{[4]}$	$p^{[5]}$	$p^{[6]}$
$x(m)$	−10	7	4	12.86	1.81	−14.0	−13.85	1.43	12.34
$y(m)$	0	8	−9	7.94	15.97	6.62	−6.73	−15.07	−9.10

positions are described in Table 18.2 and the optimal sensor configuration is shown in Figure 18.6b.

In Figure 18.6a and b, it can be seen how the theoretical maximum FIM determinant is obtained at the target positions, with $\det(FIM^{[\alpha]}(\boldsymbol{\theta}_\alpha;\mathbf{x})) = 5^2/(4\sigma^4) = 9 \times 10^4$ m^{-4}, and the optimality condition (18.26) holds for each and every target.

18.3.4 Multiple Target Localization: A Global Numerical Approach for $\mathbb{R}^2$ and $\mathbb{R}^3$

Based on the previous results, in this section we address the generic problem of computing the optimal geometric configuration of a surface sensor network that will maximize the range-related information available for *multiple target localization* in 2D and 3D space. The following analysis is based on a summary of the methods described in [41,43]. For a detailed analysis and proofs of the results described next, the reader is referred to the above references.

There will be tradeoffs involved in the accuracy with which each of the targets can be localized, since, in a general setup, it is not possible to define a sensor configuration for which the maximum FIM determinant can be obtained for all of the targets simultaneously. To analyze these tradeoffs, we resort to techniques that borrow from estimation theory and Pareto optimization [45–47]. Stated briefly, using classical concepts on Pareto-optimality, combinations of the logarithms of the determinants of the FIMs for each of the targets are maximized in order to compute the Pareto-optimal surface or front that gives a clear image of the tradeoffs involved in the multi-objective optimization problem. This will allow determination of the sensor configuration that yields, if possible, a proper tradeoff for the accuracy with which the position of the different targets can be computed.

It is interesting to point out that in the literature on economics a Pareto optimal outcome is one such that *no person can be made better off without having someone else worse off*. When Pareto optimal solutions exist, in general they are not unique. However, the determination of the Pareto optimal set of solutions for a given multi-objective problem allows for a thorough study of the tradeoffs involved in the problem at hand. For example, when several targets are involved in some practical application, different levels of accuracy may be required for each of them. Moreover, in most of the practical problems, improving the accuracy in the estimate of one target may at times be done only in detriment of the accuracy of the other estimates, and thus the tradeoffs involved must be studied carefully. The set of Pareto optimal solutions is often called the Pareto front or Pareto boundary. The computation of the Pareto front showing these tradeoffs allows for the election of the adequate "operation" points.

Notice that the multi-objective optimization problem at hand involves the logarithms of the determinants of the FIMs of the targets. The FIM belongs to the set of positive definite matrices. It is well known that $\log \det(F)$, where F is a symmetric positive definite matrix, is a "well behaved" function, that is concave in the set of symmetric positive definite matrices; see for example [48, Chapter 3]. However, it is not necessarily true that $\log \det(F)$ is concave in the parameter search space (in the present case the sensor position coordinates that define the geometric configuration of the sensor network). Thus, the solutions should be computed with non-convex optimization tools, and each case must be examined carefully. Therefore, strictly speaking it is not correct to name the representation of the tradeoffs as a Pareto front, as defined classically in the context of multiple optimization theory. For this reason, in what follows we will refer to such surfaces (fronts) as pseudo-Pareto fronts. However, the weights, i.e. the coefficients in the convex combination of the individual cost functions, will still be named Pareto weights for simplicity of explanation.

In the present problem, the scalar functions used to construct the pseudo-Pareto front are related to the logarithms of the determinants of the FIMs corresponding to each of the targets. The objective is to maximize these functions jointly in a well-defined sense. Formally, this can be done by computing

$$\mathbf{p}^*(\lambda) = \arg\max_{\mathbf{p}} \sum_{\alpha=1}^{q} \lambda^{[\alpha]} \log \det(FIM^{[\alpha]}(\boldsymbol{\theta}_\alpha; \mathbf{x})) \tag{18.29}$$

where $\lambda \in \Lambda := \{\lambda \in \mathbb{R}_+^q : \lambda^{[1]} + \cdots + \lambda^{[q]} = 1\}$ are Pareto weights and studying the corresponding pseudo-Pareto-optimal front. Notice that for each value of λ the above optimization process yields the corresponding values for the objective functions, with $\det(FIM^{[\alpha]}(\boldsymbol{\theta}_\alpha; \mathbf{x})) = \det(FIM^{[\alpha]}(\boldsymbol{\theta}_\alpha; \mathbf{x})(\lambda))$. In the case of two targets, for example, the corresponding pseudo-Pareto-optimal front is a $\det(FIM^{[2]}(\boldsymbol{\theta}_2; \mathbf{x})(\lambda))$ versus $\det(FIM^{[1]}(\boldsymbol{\theta}_1; \mathbf{x})(\Lambda))$ curve that captures the tradeoffs between the two criteria and allows, for example, to compute that maximum possible value of $\det(FIM^{[2]}(\boldsymbol{\theta}_2; \mathbf{x}))$ given a desired, achievable lower bound for $\det(FIM^{[1]}(\boldsymbol{\theta}_1; \mathbf{x}))$. These considerations can be easily generalized to the case of an arbitrary number of targets.

Let $\det(FIM^{[\alpha]*}(\boldsymbol{\theta}_\alpha; \mathbf{x}))$ denote the optimal (maximum) value of $\det(FIM^{[\alpha]}(\boldsymbol{\theta}_\alpha; \mathbf{x}))$, corresponding to the αth target, achievable with optimal sensor positions specified by $\mathbf{p}^{[\alpha]*}$ In general, it is not possible to define optimal sensor positions $\mathbf{p}^*$ (the same for all targets) that will make simultaneously all $\det(FIM^{[\alpha]}(\boldsymbol{\theta}_\alpha; \mathbf{x})); \alpha \in \mathbb{I}_{[1,q]}$, equal or close to the individually optimal ones due to a number of causes that include the geometry of the multiple-target positioning problem and possible hard constraints on sensor placement. As a consequence, tradeoff solutions must be examined. For example, one may seek to maximize, by proper choice of sensor positions $\mathbf{p}^*$, the accuracy with which one target can be localized, while keeping the accuracies of the other targets above desired, achievable lower bounds.

From a practical standpoint, the Pareto-related optimization problem posed in (18.29) yields two possible optimal sensor placement strategies that, to some extent, correspond to two different interpretations of the weight vector λ. The first strategy consists of solving the optimization problem for all values of λ and studying the tradeoffs involved in optimizing each of the cost functions given by $\log \det(FIM^{[\alpha]}(\boldsymbol{\theta}_\alpha; \mathbf{x}))$. This is in principle possible to do if one grids the search parameter space described by λ and plots an approximation to the corresponding pseudo-Pareto-optimal front. The latter is instrumental in capturing the tradeoffs involved, thus helping a designer choose a specific point on the Pareto front for the particular problem being solved. Clearly, this task is difficult for more than three objective functions, for in this case it may not be simple to visualize the Pareto-optimal front of interest. The second strategy consists of viewing λ as a collection of weights $\lambda^{[\alpha]}; \alpha \in \mathbb{I}_{[1,q]}$, where the appearance of a weight, say $\lambda^{[\alpha]}$, that is larger than the other weights captures the fact that the priority is to obtain a very good estimate of the position of the αth target, possibly in detriment of the accuracy with which the other targets can be positioned. Again, the final choice for a reasonable solution involves repeating the procedure for a number of choices of λ and examining the corresponding solutions and how they meet the performance objectives. Therefore, in order for the information about the optimal configurations to be useful, one must check if the determinants of the individual FIMs for each target meet desired specifications. In addition, with the knowledge that the maximum $\det(FIM^*(\boldsymbol{\theta}; \mathbf{x}))$ is unique, and that this maximum is well defined for a single target, we obtain tools to determine if the solution provided by the optimization process achieves desired conditions and if it is close to the optimal value that would be obtained for a single target working in isolation.

For the computation of the sensor positions, since we deal with a non-convex optimization problem, we resorted to a simulated annealing optimization method. The advantage of this meta-heuristics algorithm resides in the fact that it can avoid possible local minima and reach the global optimal, combined with small computation times. The simulated annealing algorithm is run using the Global Optimization Toolbox of MATLAB®. Once a solution is provided, a new optimization process is carried out with a constrained gradient optimization algorithm with the Armijo rule [49, Chapter 8] to refine the optimal solution.

In what follows, an example for two-target positioning in 3D is studied to show graphically the tradeoffs involved in the multiple-target localization task. For more than two targets the procedure would be exactly the same, but the representation of the pseudo-Pareto front may be not possible due to their high dimensionality.

Example 18.3

In this example, two targets are localized by a six-sensor network. In Example 18.2 it was shown that for two targets in 2D it is possible to obtain for each of the targets the same optimal FIM determinant that would be obtained for a single target. In 3D it is not possible to define a sensor configuration that provides the maximum FIM determinant for both targets simultaneously, so a tradeoff solution must be adopted based on the importance (Pareto weight) assigned to each of the targets. In the example at hand, the region where the sensors can be placed is bounded, being a square area of 600 m × 600 m with its center aligned vertically with the origin of the inertial coordinated frame $\{\mathcal{I}\}$, and the targets are over the $\{\mathbf{x}_{\mathcal{I}}\}$-axis, with 300 meters between them and centered at the origin. In this particular case, a single Pareto weight $\lambda^{[1]}$ can be considered, since $\lambda^{[2]} = 1 - \lambda^{[1]}$.

The optimal formation is computed for a couple of values of the Pareto weight: $\lambda^{[1]} = 0.2$, and $\lambda^{[1]} = 0.5$. The corresponding optimal formations are listed in Table 18.3, and the values of the FIM determinants obtained for each of the targets and for each Pareto weight are shown in Table 18.4.

In Figure 18.7, the level curves of the FIM determinant over the region $\mathcal{D}$ of interest for each optimal formation are shown, together with the sensor positions. In Figure 18.8, the pseudo-Pareto front for the problem at hand is shown. Notice how the accuracies of the targets vary depending on the Pareto weight chosen.

TABLE 18.3

Optimal Sensor Position Coordinates for a 2-Target Positioning Problem with 6 Sensors for Different Pareto Weights

Sensor Positions	Pareto Weights	
(x, y) [m]	$\lambda^{[1]} = 0.2$	$\lambda^{[1]} = 0.5$
$\mathbf{p}^{[1]}$	(159.05, 23.69)	(159.57, 18.18)
$\mathbf{p}^{[2]}$	(−61.01, 300)	(0, 300)
$\mathbf{p}^{[3]}$	(−159.70, 12.41)	(−159.57, 18.18)
$\mathbf{p}^{[4]}$	(−159.70, −12.41)	(−159.57, −18.18)
$\mathbf{p}^{[5]}$	(−61.01, −300)	(0, −300)
$\mathbf{p}^{[6]}$	(159.05, −23.69)	(159.57, −18.18)

TABLE 18.4

FIM Determinants Obtained in a 2-Target Positioning
Problem Using 6 Sensors, for Different Pareto Weights

Pareto Weights	$\det(FIM^{[1]}(\theta_1; x))$	$\det(FIM^{[2]}(\theta_2; x))$
$\lambda^{[1]} = 0.2$	7.3835×10^6	7.9561×10^6
$\lambda^{[1]} = 0.5$	7.7551×10^6	7.7551×10^6

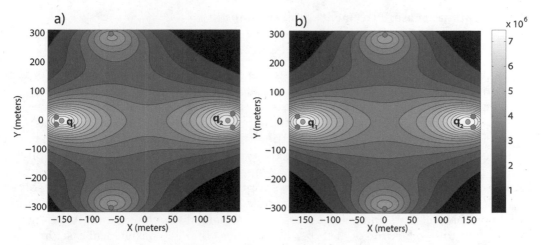

FIGURE 18.7
Optimal formations and level curves of $\det(FIM(\theta; \mathbf{x})_{\mathcal{D}})$ for two-target positioning with six sensors, for different Pareto weights, (a) $\lambda^{[1]} = 0.2$; and (b) $\lambda^{[1]} = 0.5$. Targets are indicated with labels $\mathbf{q}_1$ and $\mathbf{q}_2$.

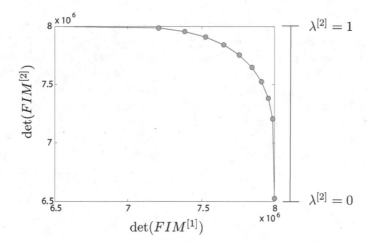

FIGURE 18.8
Pseudo-Pareto front for six sensors and two targets.

It is important to remark that in the extreme cases for the computation of the pseudo-Pareto front, $\lambda^{[1]} = 0$ and $\lambda^{[1]} = 1$, the optimal formation was searched as follows: by keeping the maximum theoretical accuracy possible in the corresponding target, $\det(FIM^{[\alpha]}(\theta_\alpha; \mathbf{x}))^* = 8 \times 10^6$, maximize the accuracy of the remaining one. We adopted this procedure because multiple optimal sensor configurations may be defined for the single target localization problem. Therefore, in a practical case, we must choose among these optimal sensor configurations the one that yields the maximum FIM determinant for the remaining target.

In summary, with the methodology proposed we can compute optimal sensor configurations in 2D and 3D for any number of sensors and targets, where the tradeoffs involved in the localization of the multiple targets are established using their corresponding Pareto weights.

18.3.5 Optimal Sensor Placement with Probabilistic Uncertainty in the Target Location

It is important to remark that the above analysis of optimal sensor configurations was developed with the assumption that the target positions were known in advance, as is usual in the literature; see for example [50] and the references therein. It may be argued that this assumption defeats the purpose of devising a method to compute the target position. The rationale for the study of this seemingly unrealistic problem stems from the need to first fully understand the simpler situation where the position of the target is known and to characterize, in a rigorous manner, the types of solutions obtained for the optimal sensor placement problem. In a practical scenario, the prior target position is always known with uncertainty and the problem of optimal sensor placement cannot be solved analytically; as a consequence, it is not possible to obtain a geometrical characterization of all possible solutions and one must resort to numerical search methods. At this stage, an in-depth understanding of the types of solutions obtained for the ideal case provides possible initial guesses for the optimal sensor placement algorithm adopted, based on an initial target position estimation. These issues are discussed in detail in [41,42,51], effectively establishing the core theoretical tools to address and solve the case when there is uncertainty in the position of an underwater target. The following analysis for a single target here is a brief exposition of results available in the above references.

The objective is similar to the one described previously, but we now consider a region instead of a single point as a possible location for the target, i.e., the objective is to maximize, by proper sensor placement, the average value of the FIM determinant over the region that describes the uncertainty in the target position. Therefore, for the sake of completeness, in this section we formulate the optimal sensor placement problem considering that the prior target position is defined by a probability density function (pdf).

Let $\mathbf{p}^{[i]} \in \mathbb{R}^d$ denote the position of the ith sensor and $\mathbf{p} = [(\mathbf{p}^{[1]})^{\mathsf{T}}, \ldots, (\mathbf{p}^{[p]})^{\mathsf{T}}]^{\mathsf{T}}$. Further let $\varphi(\mathbf{q}); \mathbf{q} \in \mathbb{R}^d$ be a pdf that describes the uncertainty of the target position in region $\mathcal{D} \in \mathbb{R}^d$. With this notation, the optimal sensor placement problem can be cast in the form of finding a vector $\mathbf{p}^*$ such that

$$\mathbf{p}^* = \arg\max_{\mathbf{p}} \int_{\mathcal{D}} \det(FIM(\mathbf{p}, \varphi(\mathbf{q}))) d\mathbf{q} \tag{18.30}$$

where $\det(FIM(\mathbf{p}, \varphi(\mathbf{q})))$ represents the dependence of the FIM on the assumed sensor position $\mathbf{p}$ and the possible target location $\mathbf{q}$, given by prior information on its pdf. The latter is obviously dependent on the type of mission executed by the target.

Conceptually, the procedure to determine an optimal sensor configuration is similar to that explained in the previous sections. In the setup adopted, this can be done by combining an analytical solution for a known target position with an optimization algorithm, for example, a gradient optimization procedure that relies on a Monte Carlo method. In the initial step, the optimal sensor positions **p** that maximize (18.30) for a sample target position in the region of interest are computed analytically, similarly to the procedure described in the previous sections. This solution is then used in a second step as an initial guess for the Monte Carlo gradient algorithm in which we include explicitly the pdf that describes the uncertainty in the target location. In this case, the optimal formation depends largely on the pdf that defines the target position.

The above procedure may be used in the cases of static and/or mobile target and sensors. The most challenging problem is definitely the one where the underwater target is mobile and the sensor network, also mobile, estimates its position on-line and computes the next optimal sensor positions for the sensors to move to, this cycle repeating itself as the mission unfolds. Often, however, the sensor network must be stationary. In this case, a possible course of action is to use prior information about the area/volume where the target is expected to operate and compute the sensor positions that will yield good positioning accuracy of the target, no matter where it is inside the region of uncertainty. Another interesting situation is the one in which there is prior information about a specific path that an underwater vehicle will follow. In this case, the region of uncertainty is reduced, and the sensor network will be placed so as to yield good target positioning accuracy for all points along the path. This circle of ideas is not restricted to underwater positioning problems and can of course be applied to a large number of positioning problems in constrained 3D space. An illustrative example is presented next.

Example 18.4

In this example, for the sake of clarity and completeness, the optimal formation obtained for single-target localization when the target position is known with uncertainty is compared with the one that would be obtained for a known target position. We consider a formation of four sensors placed at the sea surface, and a target at a constant depth of $d_t = 50$ meters that operates inside a circular region of 100 meters of radius. In this particular example, the probabilistic distribution considered for the target position is a step-like function, taking the value 1 inside and on the circumference and the value 0 outside; therefore the target may be placed at any point inside the circumference of

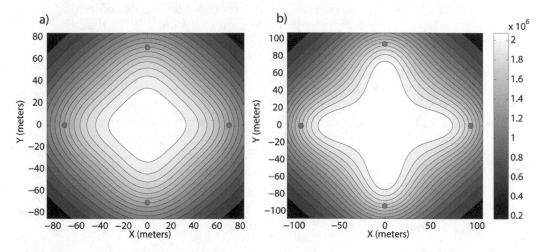

FIGURE 18.9
Optimal formations and level curves of $\det(FIM(\theta;x)_{\mathcal{D}})$ for single-target positioning with four sensors (a) for known target position, and (b) with uncertainty in the target position.

interest. In Figure 18.9a the optimal formation for a known target position is shown, where the target position is the center of the circumference. Notice how for this case, the sensors are placed over the optimal circumference of radius $r' = \sqrt{2}d_t = 70.71$ meters, as described in Section 18.3.2. In Figure 18.9b the optimal sensor formation when the target position is known with uncertainty is shown. Notice in this case how the formation size is larger, with a radius of $r' = 97.33$ meters, although the solutions are close to each other, thus making the solution for known target position an adequate initial guess for the numerical algorithm. The larger size of the formation shown in Figure 18.9b has the effect of increasing the average FIM determinant inside the work area. As a consequence, the shape of the plot $\det(FIM(\boldsymbol{\theta}; \mathbf{x})_{\mathcal{D}})$ in Figure 18.9b is flatter over a larger area than that of Figure 18.9a. Therefore, the sensors are placed in such a way as to tradeoff optimal performance at a point against slightly reduced performance, albeit uniformly over a large area around that point in the presence of uncertainty.

18.4 Optimal Sensor Placement: Moving Sensors and Targets

In this section, we take the more challenging path of using a group of vehicles as a mobile sensor network to collectively localize one or more targets underwater using range-related information. In the setup adopted, the vehicles are ASVs, henceforth referred to as *trackers*, whose task is to collectively estimate the positions of moving targets by using measurements of the ranges between the targets and the trackers, the latter undergoing trajectories that are known in real time. In some cases, to reduce operational costs, we exploit the use of a single tracker. We refer to this problem as underwater *target localization* using moving sensors. We consider at the outset the more challenging situation where the target or targets are moving. From a practical standpoint, the problem thus defined is especially attractive when the number of ASVs (trackers) is small, or even when only one ASV is used. In the latter case, and for a single target, the problem bears close connection with that of single-beacon navigation, which has been the subject of intensive research. Here, however, we will consider the multiple target case and exploit the use of more than one mobile sensor, while keeping the number of sensors at a minimum possible. The use of mobile trackers is suited for non-stationary targets and provides more flexibility than that afforded by a fixed sensor network but requires addressing far more challenging problems from a theoretical and practical standpoint. In fact, a possible solution will necessarily require that the following algorithms be implemented and run simultaneously using a receding-horizon type of strategy: (i) tracker motion planning, based on prior information about the motion of the target(s), with a view to obtaining a sequence of range measurements that will yield sufficiently rich information for target localization in a parameter estimation setting, (ii) motion control, to guarantee that the tracker or trackers undergo the motions planned by the tracker motion planner, and (iii) on-line target estimation based on the range measurements acquired. These processes are intertwined and "feed" on each other.

We start by addressing the topic of process modeling.

18.4.1 Process Modeling

In what follows, we develop a continuous-time kinematic model for the trackers and a discrete-time measurement model for the measurements of the ranges between the trackers and the targets.

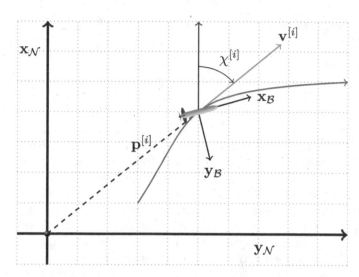

FIGURE 18.10
Kinematics of a single tracker.

18.4.1.1 Tracker Model

Let $p \in \mathbb{N}$ be the number of trackers. Consider Figure 18.10, where $\{\mathbf{x}_\mathcal{N}, \mathbf{y}_\mathcal{N}\}$ and $\{\mathbf{x}_\mathcal{B}, \mathbf{y}_\mathcal{B}\}$ are an inertial frame and a body-frame attached to the vehicle, respectively. For simplicity of exposition, we assume that water current is zero. In this case, the kinematic model for a generic ith tracker; $i \in \mathbb{I}_{[1,p]}$, is given by

$$\left.\begin{array}{l} \dot{\mathbf{p}}^{[i]}(t) = \mathrm{v}^{[i]}(t)\mathbf{g}(\chi^{[i]}(t)) \\ \dot{\chi}^{[i]}(t) = r^{[i]}(t) \end{array}\right\} \tag{18.31}$$

where $t \in [0, t_f]; t_f > 0, \mathbf{p}^{[i]} \in \mathbb{R}^2$ is the inertial position vector of the ith tracker, $\chi^{[i]} : [0, t_f] \to [0, 2\pi)$ is the course angle of the ith tracker that gives the orientation of its body-frame with respect to the inertial frame, $v^{[i]} : [0, t_f] \to \mathbb{R}$ is the speed of the ith tracker defined as $v^{[i]} \equiv \|\mathbf{v}^{[i]}\|$, where $\mathbf{v}^{[i]} \in \mathbb{R}^2$ is the inertial velocity vector and $r^{[i]} : [0, t_f] \to \mathbb{R}$ is its course-rate. Using a state-space formulation $\mathbf{x}^{[i]} = (\mathbf{p}^{[i]}, \chi^{[i]}) \in \mathcal{M} := \mathbb{R}^2 \times [0, 2\pi)$ is the state vector and $\mathbf{u}^{[i]} = (v^{[i]}, r^{[i]}) \in \mathcal{U} := \mathbb{R} \times \mathbb{R}$ is the input vector. The solution to (18.31) at time $t \in [0, t_f]$, given the initial condition $\mathbf{x}_0^{[i]} = (\mathbf{p}_0^{[i]}, \chi_0^{[i]}) \in \mathcal{M}$ and the input function $\mathbf{u}^{[i]}(t) = (v^{[i]}(t), r^{[i]}(t)), t \in [0, t_f]$, is given by

$$\mathbf{x}^{[i]}(t) = \mathbf{x}_0^{[i]} + \left(\int_0^t \mathrm{v}^{[i]}(\tau)\mathbf{g}(\chi^{[i]}(\tau))\mathrm{d}\tau, \int_0^t r^{[i]}(\tau)\mathrm{d}\tau \right). \tag{18.32}$$

In order to avoid collisions, the trackers must satisfy the condition

$$\|\mathbf{p}^{[i]}(t) - \mathbf{p}^{[j]}(t)\| \ge R, \quad (i, j) \in \mathbb{I}_{[1,p]} \times \mathbb{I}_{[1,p]}, \quad i < j, \tag{18.33}$$

for all $t \in [0, t_f]$, where $R > 0$ is a safety radius.

18.4.1.2 *Target Motion Description*

For optimal sensor placement (i.e., optimal tracker motion planning) with a view to single- or multiple-range-based target localization, some assumptions must be made regarding the motion of the targets themselves. See for example [52] for a discussion of this topic. At this point, a simple and yet important fact deserves clarification: for pure motion planning purposes, and in accordance with the three-step procedure proposed in the receding-horizon approach for target localization described before, only prior information on the types of trajectories executed by the targets over a finite time horizon is required in the first step. In other words, no kinematic models for the targets are required at this step. This stems from the fact that for motion planning purposes, all that matters are the computed (not measured) distances (ranges) between the trackers and the motions of the targets, as embodied in the prior information available about the latter. However, kinematic target models play a key role in the design of the range-based estimators required in the third step of the procedure. This topic is not at the core of the present chapter and is only briefly mentioned later in the section devoted to field experiments.

In line with the above reasoning, and with a view to simplifying the presentation, one may consider that under certain circumstances that have to do with the types of missions being executed, the target motions correspond to straight line segments and arcs of circumferences with known radii, or concatenations of these two types of spatial paths, as illustrated in Figure 18.11. We may also assume, for clarity of exposition, that the linear speed of each target along its path is constant. In this case, the admissible trajectories (spatial paths plus linear speed along the paths) for each target are completely parameterized by their initial positions (starting points) and velocity vectors. Given the latter, the initial linear speeds, course angles, and angular speeds can easily be computed. Clearly, the initial value of the angular speed is equal to zero for straight lines. Stated equivalently, prior information about the targets includes descriptions whereby their trajectories are fully specified, apart from possible translations and rotations. The latter interpretation allows for the consideration of very general types of trajectories. In this setup, some or all of the trajectory parameters (initial conditions) may not be known in advance and need to be estimated, as explained later. For example, in the case of straight lines in 2D one may consider the cases

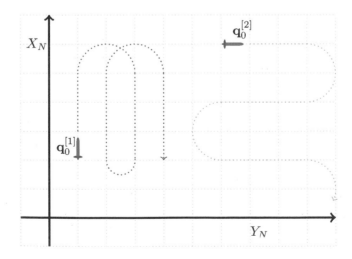

FIGURE 18.11
Illustration of the paths of two targets.

where: (i) the velocity of each target is known but its initial position is not, and (ii) the speed of each target is known, but its course angle and initial position are not.

Let $q \in \mathbb{N}$ be the number of targets and, for each $\alpha \in \mathbb{I}_{[1,q]}$, let $\mathbf{q}^{[\alpha]} \in \mathbb{R}^2$ denote the inertial position of the αth target. For each $(i,\alpha) \in \mathbb{I}_{[1,p]} \times \mathbb{I}_{[1,q]}$, we let $\mathbf{d}^{[i,\alpha]} \in \mathbb{R}^2$ denote the relative position vector of the ith tracker with respect to the αth target, i.e. $\mathbf{d}^{[i,\alpha]} := \mathbf{p}^{[i]} - \mathbf{q}^{[\alpha]}, (i,\alpha) \in \mathbb{I}_{[1,p]} \times \mathbb{I}_{[1,q]}$, and we denote the corresponding distance by $d_{i,\alpha} := ||\mathbf{d}^{[i,\alpha]}||, (i,\alpha) \in \mathbb{I}_{[1,p]} \times \mathbb{I}_{[1,q]}$.

18.4.1.3 Measurement Model

We assume each tracker is equipped with a sensor that measures distances to all the targets at the same discrete instants of time, as illustrated in Figure 18.12 for the case of two trackers and two targets. Thus, the measurements are naturally modeled in a discrete-time setting.

Let $m \in \mathbb{N}$ and consider a finite, strictly monotonically increasing time sequence $\{t_k\}_{k=0}^{m-1} \subseteq [0, t_f]$ of length m defined over a time-interval $[0, t_f]$, i.e. $0 = t_0 < t_1 < \cdots < t_{m-2} < t_{m-1} = t_f$, where $t_k; k \in \mathbb{I}_{[1,m-1]}$, are sampling instants at which range measurements (samples) are obtained. For each $k \in \mathbb{I}_{[0,m-1]}$, let $D(t_k) \in \mathbb{R}^{p \times q}$ denote the matrix of true distances at time t_k, i.e. $[D(t_k)]_{i,j} := d_{i,j}(t_k)$, where $d_{i,j}(t_k); i \times j \in \mathbb{I}_{[1,p]} \times \mathbb{I}_{[1,q]}$, denotes the distance between tracker i and target j at the sampling time t_k. Note that in the matrix $D(\cdot)$, columns represent the vector of ranges from a given target to all the trackers, while rows represent the vector of ranges from a given tracker to all the targets. The discrete-time measurements of distances collected at time $t_k; k \in \mathbb{I}_{[0,m-1]}$, denoted $Y(t_k)$, are corrupted by independent and identically distributed additive Gaussian noise sequences according to the model

$$Y(t_k) = D(t_k) + \eta(t_k), \tag{18.34}$$

where $\eta(t_k) \in \mathbb{R}^{p \times q}$ is given by $[\eta(t_k)]_{i,\alpha} := \eta_{i,\alpha}(t_k)$ with $\eta_{i\alpha}(t_k) \sim \mathcal{N}(0, \sigma_{i,\alpha}^2), (i,\alpha) \in \mathbb{I}_{[1,p]} \times \mathbb{I}_{[1,q]}$.

In what follows we use $i, j \in \mathbb{I}_{[1,p]}, \alpha, \beta \in \mathbb{I}_{[1,q]}$, and $k, s \in \mathbb{I}_{[0,m-1]}$ to denote the ith or jth tracker, the αth or βth target, and the kth or sth sample, respectively.

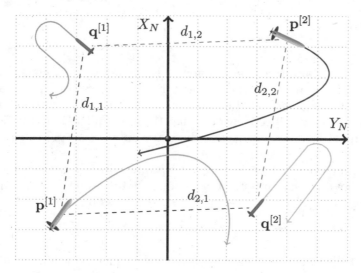

FIGURE 18.12
Illustration of two trackers $\mathbf{p}^{[1]}$ and $\mathbf{p}^{[2]}$ and two targets $\mathbf{q}^{[1]}$ and $\mathbf{q}^{[2]}$.

18.4.2 Optimal Sensor Motion Planning: Problem Formulation

To simplify the notation, for each $i \in \mathbb{I}_{[1,p]}, \alpha \in \mathbb{I}_{[1,q]}$, and $k \in \mathbb{I}_{[0,m-1]}$, we let $\mathbf{p}_k^{[i]} := \mathbf{p}^{[i]}(t_k), \mathbf{g}_k^{[i]} := \mathbf{g}(\chi^{[i]}(t_k)), \chi_k^{[i]} := \chi^{[i]}(t_k), \mathbf{d}_k^{[i,\alpha]} := \mathbf{d}^{[i,\alpha]}(t_k), d_{i,\alpha,k} := d_{i,\alpha}(t_k), D_k := D(t_k), \eta_k := \eta(t_k)$, and $Y_k := Y(t_k)$.

We make the following assumption for the trackers.

Assumption 18.1

We assume that the linear speed and course rate of each tracker are piecewise constant functions of time bounded above and below, that is, for each $i \in \mathbb{I}_{[1,p]}$ and $k \in \mathbb{I}_{[0,m-2]}$,

$$v^{[i]}(t) = \bar{v}_k^{[i]} \in [\bar{v}_{\min}^{[i]}, \bar{v}_{\max}^{[i]}], \quad t \in [t_k, t_{k+1}),$$

$$r^{[i]}(t) = \bar{r}_k^{[i]} \in [-\bar{r}_{\max}^{[i]}, \bar{r}_{\max}^{[i]}], \quad t \in [t_k, t_{k+1}),$$

where $\bar{v}_{\min}^{[i]}, \bar{v}_{\max}^{[i]} > 0$ and $\bar{r}_{\max}^{[i]} > 0$ are the minimum and maximum tracker speed, and maximum course rate, respectively, for the ith tracker.

We denote the set of admissible inputs by $\mathcal{U}_{\mathrm{adm}}^{[i]} := [\bar{v}_{\min}^{[i]}, \bar{v}_{\max}^{[i]}] \times [-\bar{r}_{\max}^{[i]}, \bar{r}_{\max}^{[i]}]$ which, in case the trackers have the same bounds for speed and course rates, we simply denote by $\mathcal{U}_{\mathrm{adm}}$. In light of Assumption 18.1, for each $k \in \mathbb{I}_{[0,m-2]}$, we have

$$\chi^{[i]}(t) = \chi_k^{[i]} + (t - t_k)\bar{r}_k^{[i]}, \quad t \in (t_k, t_{k+1}). \tag{18.35}$$

Consequently, for $k \in \mathbb{I}_{[0,m-2]}$,

$$\mathbf{p}_{k+1}^{[i]} = \mathbf{p}_0^{[i]} + \sum_{s \in \mathbb{I}_1} \left(\frac{\bar{v}_s^{[i]}}{\bar{r}_s^{[i]}} \right) \left\{ \left(\mathbf{g}_s^{[i]} \right)^\perp - \left(\mathbf{g}_{s+1}^{[i]} \right)^\perp \right\} + \sum_{s \in \mathbb{I}_2} \bar{v}_s^{[i]}(t_{s+1} - t_s)\mathbf{g}_s^{[i]},$$

where $\mathbb{I}_1 \subseteq \mathbb{I}_{[0,k]}$ and $\mathbb{I}_2 \subseteq \mathbb{I}_{[0,k]}$ denote the indices of nonzero course rate and zero course rate, respectively. The key problem that we address in this chapter is motivated by the following general question:

Problem 18.1

Given a set of trackers and prior information on some of the parameters that characterize the motion of a finite number of targets, what is the time-history of the trackers over a fixed time interval, i.e., the sequence of actions for each of the trackers (in terms of trackers' speed and course rates) that will collectively maximize the range-related information available to estimate the targets' initial positions and thus their subsequent positions while avoiding tracker collisions?

Due to space limitations and for simplicity of exposition, in view of the previous discussion we make the following assumption on the motions of the targets. The assumption can be lifted to include more general cases. See [52] for further details.

Assumption 18.2

We assume that the linear and angular speeds as well as the initial headings of all the targets are known as prior information to all the trackers, while their initial positions $\mathbf{q}_0^{[\alpha]} \in \mathbb{R}^2, \alpha \in \mathbb{I}_{[1,q]}$, are assumed to be unknown.

From a conceptual standpoint, knowledge of the initial targets positions is in these circumstances sufficient to estimate their positions at all times. Motivated by these considerations, we formulate the following problem.

Problem 18.2

Consider a set of targets and a set of trackers. Assume that the motions of the targets are characterized by a number of parameters that are known, except for the initial positions of the targets. Given prior information on the latter, compute the admissible trajectories of the trackers (as specified in Assumption 18.1) that will maximize the range-related information available to estimate the targets' initial positions while avoiding tracker collisions.

18.4.3 Fisher Information Matrix

In this section we formulate, using a mathematical setting that borrows from classical estimation theory, the problem of estimating the initial positions of the targets. To this end, we derive the corresponding FIM that builds on the general model for the trackers, the prior information available for the motions for the targets, and the range-based measurement model. Generalizing the strategy rationale adopted for the case of stationary sensors and targets, the final objective is to plan the motions of the trackers so as to maximize log $\det(FIM(\theta))$, with θ identified as the vector of initial targets' positions, by proper choice of the trackers' motion parameters over a finite time horizon. For the sake of simplicity, in the sequel we use the following compact notation:

$$\hat{\mathcal{P}}_\alpha^{[i]} := \left[(d_{i,\alpha,0})^{-1}\mathbf{d}_0^{[i,\alpha]} \quad \cdots \quad (d_{i,\alpha,m-1})^{-1}\mathbf{d}_{m-1}^{[i,\alpha]} \right] \in \mathbb{R}^{2\times m},$$

$$\boldsymbol{v}^{[i]} := [\overline{v}_0^{[i]},\dots,\overline{v}_{m-2}^{[i]}] \in [0,\overline{v}_{\mathrm{ub}}]^{m-1},$$

$$\boldsymbol{r}^{[i]} := [\overline{r}_0^{[i]},\dots,\overline{r}_{m-2}^{[i]}] \in [-\overline{r}_{\mathrm{ub}},\overline{r}_{\mathrm{ub}}]^{m-1},$$

$$\mathbf{U}_i := \left[\mathbf{u}_0^{[i]} \quad \cdots \quad \mathbf{u}_{m-1}^{[i]} \right] = \left[\mathbf{v}^{[i]},\boldsymbol{r}^{[i]}\right] \in \mathbb{R}^{m\times 2}.$$

$$\mathbf{U} := \left[\mathbf{U}_1 \quad \cdots \quad \mathbf{U}_p \right] \in \mathbb{R}^{m\times 2p}.$$

A simple algebraic manipulation shows that

$$\mathrm{trace}(\hat{\mathcal{P}}_\alpha^{[i]}\,(\hat{\mathcal{P}}_\alpha^{[i]})^\mathsf{T}) = m. \tag{18.36}$$

18.4.3.1 FIM for a Single Target

We first derive the FIM for a single target, i.e. $q = 1$, with p trackers. Consider the ith tracker motion described by (18.31) with the output equation given by

$$\mathbf{y}_k^{[i]} = \mathbf{d}_k^{[i]} + \boldsymbol{\eta}_k^{[i]},$$

where $\mathbf{d}_k^{[i]}$ and $\boldsymbol{\eta}_k^{[i]}$ are the ith rows of D_k and η_k, respectively. In this particular case $\theta = \mathbf{q}_0$ and $FIM_{\mathbf{U}}^{[\alpha]}(\theta) \in \mathbb{R}^{2\times 2}$ denotes the corresponding FIM. Following a by now standard procedure, we have

$$FIM_{\mathbf{U}}^{[\alpha]}(\theta) = \sum_{i\in\mathbb{I}_{[1,p]}} \sum_{k\in\mathbb{I}_{[0,m-1]}} \sigma_{i,\alpha}^{-2}(\nabla_\theta d_{i,\alpha,k})(\nabla_\theta d_{i,\alpha,k})^\mathsf{T}, \tag{18.37}$$

where

$$\nabla_\theta d_{i,\alpha,k} = (d_{i,\alpha,k})^{-1} \mathbf{d}_k^{[i,\alpha]} \in \mathbb{R}^2. \tag{18.38}$$

Substituting (18.38) into (18.37) and simplifying further yields

$$FIM_{\mathsf{U}}^{[\alpha]}(\boldsymbol{\theta}) = \sum_{i\in\mathbb{I}_{[1,p]}} \sigma_{i,\alpha}^{-2} \hat{\mathcal{P}}_\alpha^{[i]} (\hat{\mathcal{P}}_\alpha^{[i]})^\mathsf{T}.$$

Consequently, (18.36) implies that

$$\mathrm{trace}(FIM_{\mathsf{U}}^{[\alpha]}(\boldsymbol{\theta})) = m \sum_{i\in\mathbb{I}_{[1,p]}} \sigma_{i,\alpha}^{-2}.$$

18.4.3.2 FIM for Multiple Targets

We next derive the FIM for more than one vehicle. In this case $\mathbf{U} := (\mathbf{U}_1, \dots, \mathbf{U}_p)$, $\boldsymbol{\theta} := (\boldsymbol{\theta}_1, \dots, \boldsymbol{\theta}_q)$, and $FIM_{\mathsf{U}}(\boldsymbol{\theta}) \in \mathbb{R}^{2q\times 2q}$ denotes the corresponding FIM. The FIM for the complete system is given by

$$FIM_{\mathsf{U}}(\boldsymbol{\theta}) = \sum_{i\in\mathbb{I}_{[1,p]}} \sum_{\alpha\in\mathbb{I}_{[1,q]}} \sum_{k\in\mathbb{I}_{[0,m-1]}} \sigma_{i,\alpha}^{-2} (\nabla_\theta d_{i,\alpha,k})(\nabla_\theta d_{i,\alpha,k})^\mathsf{T},$$

where

$$\nabla_\theta d_{i,\alpha,k} = \frac{1}{d_{i,\alpha,k}} \begin{bmatrix} \mathbf{0}_{2(i-1)\times 1} \\ \nabla_{\theta_\alpha} d_{i,\alpha,k} \\ \mathbf{0}_{2(p-i)\times 1} \end{bmatrix} \in \mathbb{R}^{2q}.$$

Simplifying further yields

$$FIM_{\mathsf{U}}(\boldsymbol{\theta}) = \bigoplus_{\alpha\in\mathbb{I}_{[1,q]}} FIM_{\mathsf{U}}^{[\alpha]}(\boldsymbol{\theta}_\alpha).$$

Note that the overall FIM depends on $v^{[1]}, \dots, v^{[p]}$ and $r^{[1]}, \dots, r^{[p]}$. We now compute the maximum possible value of the FIM determinant.

18.4.4 Optimal FIM Determinant

In this section we examine the optimum value of the cost functional described by

$$J(\mathbf{U}) = \ln\det(FIM_{\mathsf{U}}(\boldsymbol{\theta})) = \sum_{\alpha\in\mathbb{I}_{[1,q]}} \ln\det(FIM_{\mathsf{U}}^{[\alpha]}(\boldsymbol{\theta}_\alpha)).$$

In the above equation, the second equality follows by noting that

$$\det\left(\bigoplus_{\alpha\in\mathbb{I}_{[1,q]}} FIM_{\mathsf{U}}^{[\alpha]}(\boldsymbol{\theta}_\alpha)\right) = \prod_{\alpha\in\mathbb{I}_{[1,q]}} \det(FIM_{\mathsf{U}}^{[\alpha]}(\boldsymbol{\theta}_\alpha)).$$

Thus, it suffices to maximize the FIM associated with each of the targets in order to maximize the overall FIM. The following result is obtained.

Proposition 18.1

Consider $\alpha \in \mathbb{I}_{[1,q]}$ and assume that $\sigma_{i,\alpha} := \sigma$ for all $i \in \mathbb{I}_{[1,p]}$. Let

$$\sum_{i \in \mathbb{I}_{[1,p]}} \hat{\mathcal{P}}_\alpha^{[i]} (\hat{\mathcal{P}}_\alpha^{[i]})^\mathrm{T} = \left(\frac{pm}{2}\right) I_2.$$

Then, $\det(FIM_U^{[\alpha]}(\theta_\alpha))$ is maximum and given by $(2^{-1}\sigma^{-2}pm)^2$; consequently, the optimal value of $\det(FIM_U(\theta))$ is given by $(2^{-1}\sigma^{-2}pm)^{2q}$.

Proof We refer to [53] for the proof.

Proposition 18.1 gives the maximum possible value for the cost functional adopted, as a function of the assumed target motions, in accordance with prior information available. At this stage, however, it does not shed light into the optimal trajectories of the trackers. This is because in the formulation adopted only the positions of the trackers at discrete instants of time appear explicitly in the FIM, without taking into consideration explicitly whether the trackers will actually be able to go through the optimal sequences of assigned points. Still, the optimal FIM determinant derived in Proposition 18.1 can be used as an absolute yardstick against which to evaluate the performance of a particular set of admissible trajectories that will meet the tracker model constraints. Furthermore, in a number of representative cases (number of trackers and targets less than or equal to two and three, respectively), it is possible to fully characterize the solutions to the problem thus formulated, in terms of the desired positions for the trackers as functions of time. This is done by exploiting an obvious analogy with the fixed-sensor placement problem, as if the successive positions (waypoints) of the moving trackers were viewed as a collection of fixed sensors, see [40]. As described later, this adds considerable geometric insight into the desired relative motion between the trackers and the targets.

An alternative route would be to formulate at the outset an optimization problem with the objective of maximizing the FIM-related cost criterion by taking explicitly into account the models adopted for the trackers, which include naturally linear and rotational speed constraints. However, this procedure has the disadvantage of not providing geometric insight into the types of trajectories executed by the trackers. There are obviously pros and cons involved in both approaches. For the representative cases mentioned above, the following strategy to generate optimal motions for the trackers is simple and may be adequate:

i. First, a sequence of possible optimal waypoints is computed by solving the unconstrained problem referred before. Multiple solutions are possible, as explained in [40].

ii. Next, check if the trackers can be steered through a computed sequence of waypoints (to collect the range measurements at each of the waypoints) by taking into account the rather simple kinematic tracker constraints. Dynamical constraints can also be included at this stage. Should the computed waypoints be feasible to reach, then the corresponding FIM is optimal.

It is important to stress that in the case of an arbitrary number of trackers and targets or when the trackers' kinematic constraints are violated in step (ii) above, then we must necessarily resort to numerical optimization methods to take into account collision and vehicle maneuvering constraints. This subject is not addressed here.

18.4.5 Single Tracker/Single Target

18.4.5.1 Observability Analysis

A central topic underlying the design of a single or multiple target localization system using range-based measurements is that of (nonlinear) system observability, in the spirit of the concepts pioneered in [31] and mentioned briefly in Section 18.2.2.1. The study of this topic would take us quite far from the main purpose of this chapter. However, for the sake of completeness we summarize some key results obtained by the authors. Some of the results apply to problem of range-based underwater vehicle positioning with respect to a fixed beacon at a known location, which can be viewed as the dual problem of underwater fixed target localization using a moving tracker.

i. Reference [14] considers the problem of vehicle positioning using a fixed beacon. The setup consists of an underwater vehicle (described as a planar kinematics in 2D) performing single-beacon navigation for two specific classes of maneuvers, whereby the vehicle measures its distance to a fixed transponder located at a known position using an acoustic ranging device. It is shown that, in the presence of known ocean currents, the system is globally observable for constant relative course and constant (nonzero) relative course rate inputs in the sense of [31].

ii. References [16,56] analyze the observability properties of the kinematic model of an AUV moving in 3D, under the influence of ocean currents, using range and depth measurements under the assumption that the AUV undergoes maneuvers commonly known as *trimming trajectories* [33]. Trimming trajectories are obtained when the inputs (thruster RPMs and control surface deflections) are held constant and the family of trimming trajectories are completely characterized by three variables: (a) linear body speed $\|\mathbf{v}\|$, (b) flight-path angle γ, and (c) yaw rate $\dot{\psi}$. In the above references, it is assumed that $\|\mathbf{v}\| > 0$, γ, and $\dot{\psi}$ are constant but otherwise arbitrary (within the constraints of the vehicle capabilities) and the observability of the resulting system with the two above-mentioned sensor suites is examined. Consider the single-beacon case. For range measurements only, it is shown that in the absence of ocean currents the 3D kinematic model of an AUV undergoing trimming trajectories with nonzero flight-path angle and yaw rate is observable. On the other hand, with both range and depth measurements, under the assumption that the yaw rate is different from zero, observability is obtained even when the flight-path angle is zero (vehicle moving in a horizontal plane) and there are nonzero unknown currents. These obvious advantages are lost if yaw rate is equal to zero, for in this case the model is only weakly observable. Finally, it is shown that the extended model obtained by considering multiple (at least two) transponders is observable in all situations if the yaw rate is different from zero.

iii. Finally, reference [52] derives some sufficient conditions on the tracker's course rate so that the design model used to study the problem of range-based target localization problem becomes observable. In this context, the tracker's motion is governed by a simple kinematic model. Furthermore, the target's motion is assumed to be fully parametrized by its initial position, constant linear speed, initial course angle, and course rate. For this setup, it is shown that when the target is moving along a straight line, for most of the unknown target parameter combinations, observability can be achieved by a nonzero constant tracker's course rate. In the case where the target moves along a circular path, with the knowledge of the angular speed, some nontrivial sufficient conditions are derived on the tracker input to achieve observability.

18.4.5.2 Field Experiments

We now discuss the results of field tests performed using two MEDUSA-class AMVs (see Figure 18.13).[*] Each vehicle has two side thrusters, which can be independently controlled to impart longitudinal and rotational motions about the $\{z_T\}$-axis and two vertical thrusters for depth control. In addition, the vehicles are equipped with attitude and heading reference systems (AHRS) that provide measurement of body orientation and body fixed-angular velocity for control purposes. Each vehicle carries an acoustic Blueprint Seatrac data modem and ranging unit[†] that is used for communications and range measurements. During the tests, we operated two MEDUSA vehicles: one of them was used as a target operating at a constant depth underwater, while the other was used as a tracker operating at the surface (equipped with GPS), while interrogating the target. Starting from an unknown initial position, the target executed a lawnmowing motion with a constant body-speed and performed dead reckoning navigation using a DVL and the AHRS. In the tests, for the sake of simplicity, the tracker had access not only to the range to the target but also to the velocity vector of the latter (communicated via the acoustic communications channel) every 1.5 [s]. However, we remark that we can relax this requirement. The target parameters are summarized in Table 18.5.

The integrated three-step *motion planning*, *control*, and *estimation* strategy adopted for range-based target localization is depicted in the block diagram of Figure 18.14. By running

FIGURE 18.13
The MEDUSA-class of autonomous marine vehicles (AMVs).

---

* The MEDUSA AMVs are robotic platforms capable of operating as AUVs or ASVs, designed, built, and operated by the Institute for Systems and Robotics (ISR) of IST, Univ. de Lisboa, Portugal [57].
† http://blueprintsubsea.com/seatrac/index.php

TABLE 18.5

Target Parameters

Depth	Linear (Inertial) Speed	Range Measurements Sampling Period
1 m	0:2 m/s	1:5 s

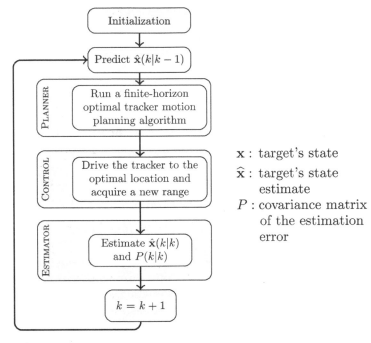

FIGURE 18.14

Integrated motion planning, control, and estimation for range-based target localization: flow chart of the algorithm.

the underlying algorithms over a sliding time window, the tracker effectively recruits the following three systems in an intertwined manner:

- *Step 1*: The *planner*, using prior information about the target's motion, computes optimal positions for the tracker for the next six samples of time (time window adopted). For details about the algorithm, we refer the reader to [53].

- *Step 2*: The closed-loop *control* law steers the tracker to reach the next waypoint, at which a new range measurement is acquired.

- *Step 3*: A range-based target localization filter (EKF *estimator*) is run in parallel on-board the tracker to update the information about the target motion. By assuming that over the next time horizon the trajectory of the target is a straight line, the current values of the estimator are used to obtain prior information about the target's motion, as required for Step 1. The process is then repeated.

Figure 18.15 shows the optimal trajectory of the tracker and the estimated trajectory of the target. For comparison purposes, the target trajectory was estimated using two independent sources of information: (i) the output of an EKF target localization filter and (ii) the relative

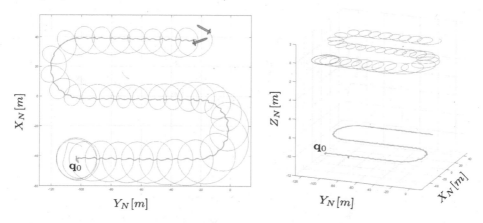

FIGURE 18.15
Optimal circular tracker motion and estimated target trajectories (experimental data).

position of the target with respect to the tracker as measured by a USBL unit installed on board the latter. Figure 18.16 shows the time-history of the optimal and measured tracker speed (top) and course angle (center). Note that at the beginning of the mission, the measured speed was saturated at the nominal speed 1 (m/s) of the MEDUSA. This was due to the fact that motion planning was done by assuming a maximum possible speed of 1.5 (m/s), but the particular tracker used in the field tests could not reach speeds higher than 1 (m/s).

Finally, recall that the optimal FIM for the problem at hand is given by $(m\sigma^{-2}/2)I_2$, that is, the off-diagonal elements of the FIM are zero, while the diagonal elements are equal and given by $m\sigma^{-2}/2$. Consequently, the optimal determinant of FIM is $m^2\sigma^{-4}/4$. Figure 18.16 (bottom) shows the plot of the evolution of the normalized FIM versus the number of samples, which is consistent with the theoretical findings. For the computation of the normalized FIM, the FIM at each sampling time k is divided by the theoretical maximum FIM determinant for these given k samples.

Remark 18.3

Due to space limitations, we have omitted several details. We remark that in [53,58] we have provided a complete analysis and the construction of a family of optimal solutions that maximize the FIM. In contrast to the strategy outlined in Section 18.4.4, for this particular case we have adopted a direct approach of constrained optimization. Nevertheless, one can follow the strategy of Section 18.4.4 to arrive at a similar solution. In a nutshell, the tracker tries to "encircle" the target as shown in Figure 18.15. However, the single-tracker approach has two limitations:

   i. The tracker motion may be quite demanding in terms of maneuverability required.

   ii. For multiple target localization, this approach is inadequate.

In view of these two limitations, in the following subsection we consider multiple trackers and explore optimal tracker trajectories for more than two targets.

18.4.6 Multiple Trackers/Multiple Targets

Motivated by the experimental results described above, in which a single tracker must execute demanding maneuvers to localize the target even when the latter describes a

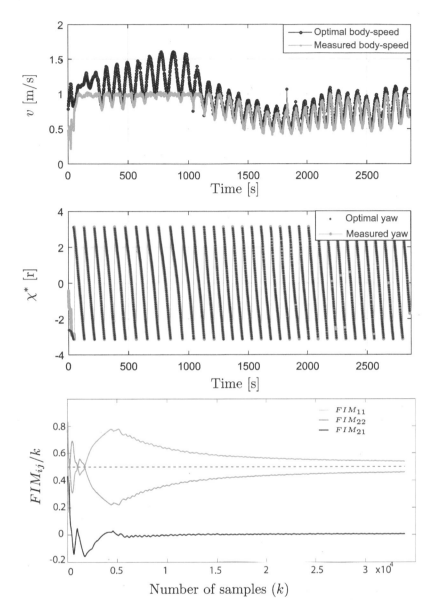

FIGURE 18.16
Tracker's speed (top) and optimal course angle (center). Components of normalized FIM (bottom).

straight-line path, we now address the problem of target localization when more than one tracker is used.

It is important to remark that although in a general setting we may consider an arbitrary number of tracking vehicles, the use of a large number of trackers would increase exponentially the complexity of the cooperative operation of vehicles, and would defeat the purpose of reducing the number of sensors used for target localization. Therefore, in the following analysis, which is rooted in the results presented in [59], we restrict the localization system to two trackers to limit the complexity of the solutions and keep the cost low.

As explained in [41], a formation of p sensors must be regularly distributed around a target's position in order for it to be optimal. Under this condition, the optimal FIM for the target α yields

$$FIM^{[\alpha]}(\mathbf{p}, \mathbf{q}^{[\alpha]}) = \left(\frac{p}{2\sigma^2}\right) I_2,$$

and the optimality conditions for the sensor-target configuration are given by (18.8). For the specific case of two sensors, the optimality conditions stated in (18.8) are satisfied when the relative position vectors of the two sensors with respect to the target position are orthogonal. Furthermore, for a multi-target problem, if this orthogonal configuration can be defined with respect to each of the targets simultaneously, then all targets may be optimally localized.

In the following, we provide insight into the problem of defining optimal sensor configurations to localize at most three targets simultaneously using two trackers, since optimal solutions for the desired locations of the trackers' waypoints can be explicitly computed. For more than three targets, the approach described is only valid under several well-defined conditions, and in general, one must resort to numerical optimization methods similar to those explained in Section 18.3.4.

18.4.6.1 Analytical Solutions Using Two Trackers (At Most Three Targets)

For a moving target α, optimal sensor placement is obtained by keeping the range vectors from the sensors to the target orthogonal at each of the m sampling times, i.e. by defining an optimal sensor formation with respect to the target position for every sampling time, so that the optimal FIM yields

$$FIM_U^{[\alpha]}(\mathbf{q}_0^{[\alpha]}, \mathbf{p}) = \left(\frac{mp}{2\sigma^2}\right) I_2. \tag{18.39}$$

Following the approach described in Section 18.4.4, once the assumed motions of the targets are given over a fixed time-horizon, the above orthogonality condition may be used to generate a set of optimal waypoints for the different trackers. We next study the optimal motions of two trackers for one, two, and three targets without any constraints on them.

18.4.6.1.1 Case 1: One target

For the particular case of single-target localization with two trackers, the trajectories followed by the trackers, and thus their maneuvers, are quite less demanding than the case with a single tracker, making it a very attractive solution from a practical standpoint. The analysis for single-target localization with two trackers is omitted due to space limitations, but in [53,55] we provide optimal analytical solutions to this particular problem that show the practical interest and advantage of using two trackers.

Example 18.5

As stated above, the optimal condition implies that at each instant of time t_k the relative position vectors of the trackers with respect to the assumed target position must be

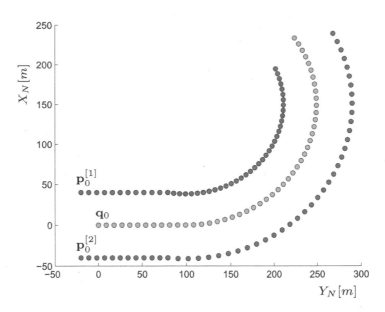

FIGURE 18.17
Optimal tracker trajectories for single target ($\mathbf{q}$) localization using two trackers $\mathbf{p}^{[1]}$ and $\mathbf{p}^{[2]}$.

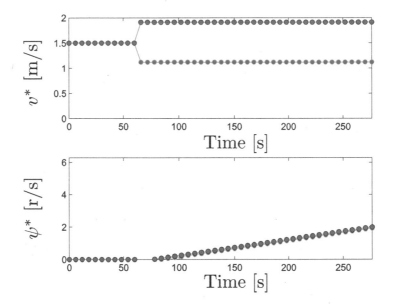

FIGURE 18.18
Evolution of optimal trackers' speeds (top) and course angles (below).

orthogonal. In this example, the target moves along a trajectory consisting of a straight line followed by an arc of circumference, with a constant (linear) speed of 1.5 (m/s) along all the path, and a target (angular) speed of 0.01 (rad/s) on the circular path. Figure 18.17 is a plot of the trajectories of all vehicles (target and trackers). Figure 18.18 shows the corresponding speeds and course angles of the trackers. The initial simulation parameters are $\mathbf{p}_0^{[1]} = [-20, 40]^{\mathrm{T}} \, \mathrm{m}$, $\mathbf{p}_0^{[2]} = [-20, -40]^{\mathrm{T}} \, \mathrm{m}$, and $\mathbf{q}_0 = \mathbf{0} \, \mathrm{m}$.

From this example, a clear and extremely important advantage of using two trackers over a single tracker emerges. In the case of a single tracker, the vehicle is required to undergo circular maneuvers around the moving target, as shown in the experimental test before. As a consequence, the tracker may be requested to maneuver at large linear and rotational speeds that in some scenarios cannot be reached. With two trackers, the maneuvers are definitely less demanding, thus making this a very attractive solution from a practical standpoint.

18.4.6.1.2 *Case 2: Two targets*

In this scenario we have two sensors and two targets, i.e. $p = q = 2$ and $i, \alpha \in \{1, 2\}$. Let $\mathbf{b}_c \in \mathbb{R}^2$ be the midpoint of the line joining the two targets, i.e. $\mathbf{b}_c := 0.5(\mathbf{q}^{[1]} + \mathbf{q}^{[2]})$, and $\mathbf{b}_c^{\perp}$ be the orthogonal vector to the line joining the two targets, which is given by $\mathbf{b}_c^{\perp} := \begin{bmatrix} \mathbf{q}_y^{[1]} - \mathbf{q}_y^{[2]} & \mathbf{q}_x^{[2]} - \mathbf{q}_x^{[1]} \end{bmatrix}^{\mathrm{T}}$. The two sensors must keep an orthogonal configuration with respect to each of the targets simultaneously. Recall that if A, B, and C are any three points on the circumference, where A and B are diametrically opposite, then the line segments AC and BC are orthogonal. Moreover, for two arbitrary points (the target positions) it is possible to define an infinite number of circumferences passing over these positions, and any pair of points defining a diameter in any of these circumferences provides an optimal sensor configuration with respect to each of the targets. Note that the centers of all these circumferences lie along a line that is orthogonal to the line joining the two targets and passes through the midpoint of the latter line. The line where the centers lie is given by $\mathbf{p}_c(\lambda) = \mathbf{b}_c + \lambda \|\mathbf{b}_c^{\perp}\|^{-1} \mathbf{b}_c^{\perp}, \lambda \in \mathbb{R}$, where $\mathbf{p}_c(\cdot)$ is the possible center of any of the circumferences. Then, choosing one of the possible centers and the circumference associated, we can define the optimal sensor configuration of interest (see Figure 18.19a).

Therefore, for the case of two targets, we have an infinite number of circumferences to define optimal configurations of sensors, and there is considerable freedom to design and select the optimal sensor formations depending on the mission and tracker constraints. We show next a simple example of two sensors tracking two targets while keeping an optimal configuration along the mission.

Example 18.6

For the sake of simplicity, we do not consider any constraints on the placement of the sensors (trackers), which allows us to define the radius/diameter of a circumference of interest in advance. Notice that in order to compute a valid solution, the diameter must be larger than the distance between the targets, so for the example at hand we consider a diameter of 1.5 times the distance between targets. The initial target positions are $\mathbf{q}_0^{[1]} = [0, 10]^{\mathrm{T}}$ m and $\mathbf{q}_0^{[2]} = [0, -20]^{\mathrm{T}}$ m. In this scenario, shown in Figure 18.19b, the first target moves with velocity $\mathbf{v}^{[1]} = [0.2, 0.2]^{\mathrm{T}}$ m/s and the second one with $\mathbf{v}^{[2]} = [0.2, 0]^{\mathrm{T}}$ m/s. The time interval between any two successive range measurements is 6 s. We consider that the diameter chosen to place the sensors is the one parallel to the $\{\mathbf{y}_\mathcal{I}\}$-axis and that the initial sensor positions are already optimal, so the optimal configuration must be kept along the mission. Figure 18.19b shows the paths followed by the sensors to keep an optimal formation with respect to the target positions, providing the maximum FIM determinant for each of the targets.

18.4.6.1.3 *Case 3: Three targets*

We now consider the case of two sensors and three targets, i.e. $p = 2$ and $q = 3$ with $i \in \{1, 2\}$ and $\alpha \in \{1, 2, 3\}$. As mentioned above, the two sensors must keep an orthogonal configuration with respect to each of the targets to define an optimal formation. However, we can define one circumference that passes over the three target positions, i.e. there exists

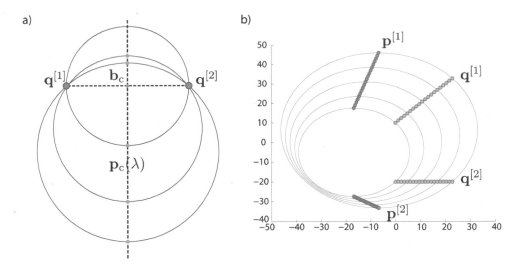

FIGURE 18.19
(a) Circumferences used to define optimal sensor configurations for the localization of two targets. (b) Optimal localization of two moving targets ($\mathbf{q}^{[1]}$ and $\mathbf{q}^{[2]}$) with two sensors ($\mathbf{p}^{[1]}$ and $\mathbf{p}^{[2]}$).

a circumference that circumscribes the triangle defined by the targets. If the two sensors are placed over any of the diameters of this circumference, then they will define an optimal configuration with respect to each of the targets simultaneously. The first step is to compute the circumference that passes through the three target positions, which is given by solving the system of nonlinear equations

$$\left\| \mathbf{q}^{[\alpha]} - \mathbf{p}_c \right\|^2 = R_c^2, \quad \alpha \in \{1, 2, 3\}. \tag{18.40}$$

Subtracting the equation with $\alpha = 1$ from $\alpha = 2$ and $\alpha = 3$, respectively, yields

$$\| \mathbf{q}^{[\alpha]} - \mathbf{p}_c \|^2 - \| \mathbf{q}^{[1]} - \mathbf{p}_c \|^2 = 0, \quad \alpha \in \{2, 3\}.$$

On further simplification, the above two equations yield the system $A\mathbf{p}_c = \mathbf{a}$, where

$$A := \begin{bmatrix} (\mathbf{q}^{[1]} - \mathbf{q}^{[2]})^{\mathrm{T}} \\ (\mathbf{q}^{[1]} - \mathbf{q}^{[3]})^{\mathrm{T}} \end{bmatrix} \quad \text{and} \quad \mathbf{a} := \frac{1}{2} \begin{bmatrix} \| \mathbf{q}^{[1]} \|^2 - \| \mathbf{q}^{[2]} \|^2 \\ \| \mathbf{q}^{[1]} \|^2 - \| \mathbf{q}^{[3]} \|^2 \end{bmatrix}.$$

Since the target positions are distinct, it follows that the matrix A is nonsingular and hence $\bar{\mathbf{p}}_c = A^{-1}\mathbf{a}$ is the unique solution. Finally, R_c can be obtained by substituting for $\bar{\mathbf{p}}_c$ in one of the equations in (18.40); see Figure 18.20a for an example of how to draw this circumference.

Since the sensors can only be placed over a single circumference, the possible optimal sensor positions and thus the optimal waypoint sequences for the trackers are limited.

Example 18.7

Similarly to the previous example, for each sampling time the circumference passing over the target positions is computed, and one diameter is selected for sensor placement. The

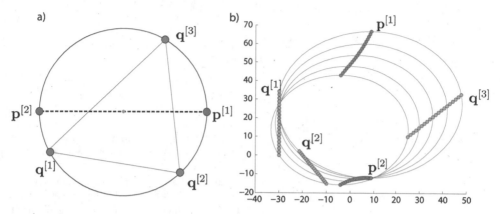

FIGURE 18.20

(a) Circumference defined by three targets (green) positions. The sensors (red) are placed at the extremes of an arbitrary diameter making an orthogonal configuration with respect to each target. (b) Optimal localization of three moving targets ($\mathbf{q}^{[1]}$, $\mathbf{q}^{[2]}$, and $\mathbf{q}^{[3]}$) with two sensors ($\mathbf{p}^{[1]}$ and $\mathbf{p}^{[2]}$).

initial targets' positions are $\mathbf{q}_0^{[1]} = [-30,0]^T$ m, $\mathbf{q}_0^{[2]} = [-10,-15]^T$ m, and $\mathbf{q}_0^{[3]} = [25,10]^T$ m, with velocities $\mathbf{v}^{[1]} = [0, 0.3]^T$ m/s, $\mathbf{v}^{[2]} = [-0.1, 0.15]^T$ m/s, and $\mathbf{v}^{[3]} = [0.2, 0.2]^T$ m/s. The time interval between any two successive range measurements is 6 s. For the sake of simplicity, the diameter chosen to place the sensors is the one parallel to the $\{\mathbf{y}_I\}$-axis, and the initial sensor positions are already over the diameter of interest of the circumference defined by the initial target positions. In Figure 18.20b, the path followed by the sensors to keep the optimal sensor configuration, as a means to yield obtain the maximum FIM determinant for each of the targets, is shown.

Clearly, for two and three targets it is possible to compute optimal sequences of waypoints for the trackers so as to keep an optimal sensor formation along the mission being performed. As mentioned earlier, for more than three targets the methodology proposed is only valid in very specific scenarios, and in general, numerical optimization methods must be used. To tackle this problem, one may resort to the algorithms presented in Section 18.3.4 simply by extending them to take into account the successive, assumed target positions with a view to defining sets of waypoints for the different trackers to go through.

Based on the results obtained in the chapter, Table 18.6 offers a general overview of some interesting properties related to the problem of target localization using single and multiple trackers.

18.5 Conclusions and Future Directions of Research

In this chapter we provided a summary of some theoretical results leading to a number of building blocks that are essential to the development of range-based algorithms for underwater single- and multiple-target localization using groups of static or mobile sensors. The types of problems addressed are shown pictorially in Figure 18.21. A representative example includes the situation where a number of underwater targets must be localized using surface units (trackers) equipped with range-measuring devices. In this and other examples, the need arises to optimally place the surface sensors (static or mobile) as a means to increase the range-based information available for underwater target localization.

TABLE 18.6

Target Localization Using Single and Multiple Trackers: General Results

No. of Targets	Target Feature	No. of Trackers	Remark
1	Fixed or moving	1	Demanding tracker motions (for a moving target)
		2	Less demanding tracker motions
		≥3	Less demanding tracker motions but expensive to implement
2	Fixed or moving	1	Cannot be localized simultaneously with the same accuracy as that achievable for each target independently
		2	Localization with the best possible accuracy achievable for each target independently
		≥3	Less demanding tracker motions but expensive to implement
3	Fixed or moving	1	Cannot be localized simultaneously with the same accuracy as that achievable for each target independently
		2	Localization with the best possible accuracy achievable for each target independently
		≥3	Less demanding tracker motions but expensive to implement

Building on classical estimation theory, the methodology adopted for sensor placement amounts to maximizing the determinant of an appropriately defined FIM. In the case of static sensors, the FIM is a function of the sensor positions, while in the case of mobile sensors the FIM is a function of the particular trajectories adopted for the trackers. In the case of mobile targets, the FIM is also a function of the assumed motions of the targets, given as prior information available.

For static sensor and target scenarios, solutions for optimal sensor placement were described for single and multiple target localization in 2D and 3D. In some cases, the optimal sensor distributions obtained lend themselves to clear and useful geometrical interpretations. Optimal sensor placement for multiple-target localization was tackled in the context of Pareto optimization theory as a means to address the tradeoffs that arise naturally when multiple, possibly conflicting objectives must be taken into consideration. A methodology was also described to address the problem of optimal sensor placement with probabilistic uncertainty in the target location.

The challenging problem of using a group of vehicles as a mobile sensor network to collectively localize one or more moving targets underwater using range-related information was also addressed. A methodology was introduced to compute the appropriate FIMs, which embody in their structure prior information available regarding the assumed motions of the targets. Algorithms were proposed to generate reference waypoints for the mobile trackers to go through as a means to optimize a FIM-related cost criterion. Similarly to the procedure adopted in the case of static sensors and targets, the objective is once again to increase the range-based information available for underwater target localization. Simple and useful geometric interpretations of the desired set of waypoint vectors for the trackers (with respect to the assumed motion of the targets) were given for the case when the number of trackers and targets is less than or equal to two and three, respectively. The general case of an arbitrary number of trackers and targets and the inclusion of possibly tight constraints on the motions of the

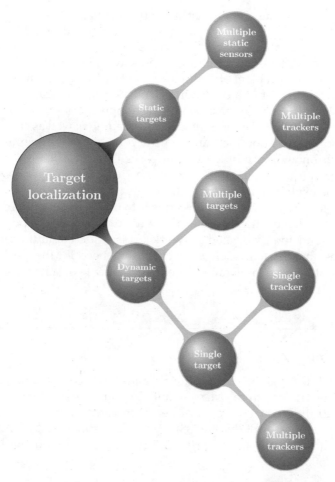

FIGURE 18.21
Target localization: a graphical illustration of representative problems.

trackers, which was beyond the scope of the presentation, can be solved using numerical optimization tools.

Experimental results with two MEDUSA-class AMVs playing the roles tracker and target were described. In the experiments, three algorithms run simultaneously, implementing a receding-horizon type of strategy: (i) tracker motion planning, based on prior information about the motion of the target, with a view to obtaining a sequence of range measurements capable of yielding sufficiently rich information for target localization in a parameter estimation setting, as per the methodology introduced in the chapter, (ii) motion control, to guarantee that the tracker or trackers undergo the motions planned by the tracker motion planner, and (iii) on-line target estimation, based on the range measurements acquired by resorting to a classical filtering structure. From the results, it is clear that the tracker may be required to execute maneuvers that are quite demanding from a dynamic standpoint. In order to alleviate this problem, an analysis was done, and simulations were run to help assess the performance expected when more than one tracker is used for single-target localization. Interestingly enough, the use of two trackers yields far smoother tracker trajectories. The analysis was extended to include the problem of localizing three targets simultaneously.

The results described are but a small contribution to the vibrant field of underwater target localization and positioning. In spite of the progress done collectively by many authors, the road ahead is quite challenging and will certainly require the proper combination of theoretical and practical field work to address and solve efficiently a number of problems. Among these, the general problem of multiple-target localization with the help of multiple trackers, where both are part of a distributed acoustic communication network, appears to be formidable. Here, the challenge lies in the development of proper theoretical tools to address explicitly temporal and energy-related constraints, as well as stringent constraints imposed by the communication medium and the possibly time-varying topology of the underlying asynchronous communication networks that are often plagued with latency and temporary communication failures. The connections with cooperative motion control must necessarily be exploited in this context.

From a practical standpoint, the main objective is to transition from the drawing board to the field and actually demonstrate the efficacy of the methods developed using a sizeable number of AMVs and the associated network of acoustic communication and range-measuring devices. In particular, there is a tremendous need for the development of new, distributed cooperative motion planning algorithms capable of running in real-time.

Acknowledgments

We thank the DSOR team at ISR/IST for their support and collaboration on the planning and execution of the trials with the Medusa autonomous marine vehicles. This work was funded by the H2020 EU Marine Robotics Research Infrastructure Network (Project ID 731103), "Projecto Mobilizador" (No. 24508), OceanTech sponsored by ANI, Lisboa 2020, Cresc Algarve 2020, Compete 2020, Portugal 2020 and Feder, and the Portuguese FCT Project UID/EEA/5009/2013.

References

1. H.K. Khalil, *Nonlinear Control*. Prentice Hall, 2014.
2. J.B. Rawlings, D.Q. Mayne, and M. Diehl, *Model Predictive Control: Theory, Computation, and Design*. Nob Hill Publishing, 2017.
3. J. Leonard, A. Bennett, C. Smith, and H. Feder, "*Autonomous underwater vehicle navigation*," *Technical Report*, MIT Marine Robotics Laboratory Technical Memorandum, 1998.
4. F. Teixeira, *Terrain-Aided Navigation and Geophysical Navigation of Autonomous Underwater Vehicles*. PhD thesis, Instituto Superior Técnico, Lisbon, Portugal, 2007.
5. S. Carreno, P. Wilson, P. Ridao, and Y. Petillot, "A survey on terrain based navigation for AUVs," in *IEEE OCEANS*, Seattle, Washington, pp. 1–7, 2010.
6. R. Christ and S.R. Wernli, *The ROV Manual: A User Guide for Remotely Operated Vehicles*. Butterworth-Heinemann, 2 ed., 2013.
7. P.H. Milne, *Underwater Acoustic Positioning Systems*. Houston: Gulf Publishing, 1983.
8. J. Dattorro, *Convex Optimization & Euclidean Distance Geometry*. Meboo, 2010.
9. T.F. Havel, *Distance Geometry: Theory, Algorithms and Chemical Applications*. John Wiley & Sons, 1998.
10. M. Turk and A. Pentland, "Eigenfaces for recognition," *Journal of Cognitive Neuroscience*, vol. 3, pp. 71–86, 1991.

11. A.K. Jain and D. Zongker, "Representation and recognition of handwritten digits using deformable templates," *IEEE Transactions on Pattern Analysis and Machine Intelligence*, vol. 19(12), pp. 1386–1391, 1997.

12. W. Dargie and C. Poellabauer, *Fundamentals of Wireless Sensor Networks: Theory and Practice*. John Wiley & Sons, 2010.

13. A. Scherbatyuk, "The AUV positioning using ranges from one transponder LBL," in *Proceedings of the IEEE/MTS OCEANS Conference Exhibition*, vol. 4, pp. 1620–1623, 1995.

14. N. Crasta, B. Bayat, A.P. Aguiar, and A.M. Pascoal, "Observability analysis of 2D single-beacon navigation in the presence of constant currents for two classes of maneuvers," in *Proceedings of the 9th IFAC Conference on Control Applications in Marine Systems*, vol. 46(33), pp. 227–232, 2013.

15. P. Batista, C. Silvestre, and P. Oliveira, "Single range aided navigation and source localization: Observability and filter design," in *Systems & Control Letters*, vol. 60(8), pp. 665–673, 2011.

16. N. Crasta, M. Bayat, A. Aguiar, and A. Pascoal, "Observability analysis of 3D AUV trimming trajectories in the presence of ocean currents using single beacon navigation," in *Proceedings of the 19th IFAC World Congress*, Cape Town, South Africa, vol. 47(3), pp. 4222–4227, 2014.

17. P. Batista, C. Silvestre, and P. Oliveira, "Single beacon navigation: Observability analysis and filter design," in *American Control Conference, Marriott Waterfront*, Baltimore, MD, USA, pp. 6191–6196, June 2010.

18. A.S. Gadre, *Observability Analysis in Navigation Systems with an Underwater Vehicle Application*. PhD thesis, Virginia Polytechnic Institute and State University, 2007.

19. J. Jouffroy and A. Ross, "Remarks on the observability of single beacon underwater navigation," in *14th International Symposium on Unmanned Untethered Submersible Technology (UUST'05)*, 2005.

20. G. Parlangeli and G. Indiveri, "Single range observability for cooperative underactuated underwater vehicles," in *Proceedings of the 19th IFAC World Congress*, vol. 47(3), pp. 5127–5138, 2014.

21. D.D. Palma, F. Arrichiello, G. Parlangeli, and G. Indiveri, "Underwater localization using single beacon measurements: Observability analysis for a double integrator system," *Ocean Engineering*, vol. 142, pp. 650–665, 2017.

22. S. Webster, J. Walls, L. Whitcomb, and R. Eustice, "Decentralized extended information filter for single-beacon cooperative acoustic navigation: Theory and experiments," *IEEE Transactions on Robotics*, vol. 29(4), pp. 957–974, 2013.

23. M. Bayat, *Nonlinear Robust Adaptive State Estimation*. PhD thesis, Instituto Superior Técnico, Universidade de Lisboa, Portugal, April 2015.

24. Y. Bar-Shalom, R.X. Li, and T. Kirubarajan, *Estimation with Applications to Tracking and Navigation: Theory Algorithms and Software*. New York: John Wiley & Sons, 2004.

25. X.R. Li and P.V. Jilkov, "Survey of maneuvering target tracking—Part I. dynamic models," *IEEE Transactions on Aerospace and Electronic Systems*, vol. 39, no. 4, pp. 1333–1364, 2003.

26. X.R. Li and P.V. Jilkov, "A survey of maneuvering target tracking-Part II: Ballistic target models," in *SPIE Proceedings Series*, pp. 559–581, 2001.

27. A.P. Aguiar and J.P. Hespanha, "Minimum-energy state estimation for systems with perspective outputs," *IEEE Transactions on Automatic Control*, vol. 51, no. 2, pp. 226–241, 2006.

28. A.J. Krener, "The convergence of the minimum energy estimator," in *New Trends in Nonlinear Dynamics and Control and their Applications*, pp. 187–208, Springer, 2003.

29. B. Bayat, N. Crasta, A.P. Aguiar, and A.M. Pascoal, "Range-based underwater vehicle localization in the presence of unknown ocean currents: Theory and experiments," *IEEE Transactions on Control Systems Technology*, vol. 24(1), pp. 122–139, 2016.

30. M. Pedro, D. Moreno-Salinas, N. Crasta, and A.M. Pascoal, "Underwater single-beacon localization: Optimal trajectory planning and minimum energy estimation," in *Proceedings of the IFAC Workshop on Navigation, Guidance, and Control of Underwater Vehicles (NGCUV)*, vol. 48(2), pp. 155–160, 2015.

31. R. Hermann and A. J. Krener, "Nonlinear controllability and observability," *IEEE Transactions on Automatic Control*, vol. 22, pp. 728–740, 1977.

32. E. Sontag and Y. Wang, "Uniformly universal inputs," *Analysis and Design of Nonlinear Control Systems*, pp. 9–24, 2008. Springer.
33. M.R. Elgersma, *Control of Nonlinear Systems Using Partial Dynamic Inversion*. PhD thesis, University of Minnesota, Minneapolis, MN, 1988.
34. N.D. Powel and K.A. Morgansen, "Empirical observability Gramian rank condition for weak observability of nonlinear systems with control," *In 54th IEEE Conference on Decision and Control (CDC)*, Osaka, Japan, pp. 6342–6348, 2015.
35. H.L. Van Trees, *Detection, Estimation, and Modulation Theory*. Wiley-Interscience, 2001.
36. J.S. Bay, *Fundamentals of Linear State Space Systems*. WCB/McGraw-Hill, 1999.
37. C. Jauffret, "Observability and Fisher information matrix in nonlinear regression," *IEEE Transactions on Aerospace and Electronic Systems*, vol. 43, no. 2, pp. 756–759, 2007.
38. S. Wang and H. Hu, "Wireless sensor networks for underwater localization: A survey," *Technical Report* 521, University of Essex, 2012.
39. A. Alcocer, *Positioning and Navigation Systems for Robotic Underwater Vehicles*. PhD thesis, Instituto Superior Técnico, Lisbon, Portugal, 2009.
40. D. Moreno-Salinas, *Adaptive Sensor Networks for Mobile Target Localization and Tracking*. PhD thesis, Universidad Nacional de Educación a Distancia, Spain, 2013.
41. D. Moreno-Salinas, A.M. Pascoal, and J. Aranda, "Optimal sensor placement for multiple target positioning with range-only measurements in two-dimensional scenarios," *Sensors*, vol. 13, pp. 10674–10710, 2013.
42. D. Moreno-Salinas, A.M. Pascoal, and J. Aranda, "Optimal sensor placement for acoustic underwater target positioning with range-only measurements," *IEEE Journal of Oceanic Engineering*, vol. 41, no 3, pp. 620–643, 2016.
43. D. Moreno-Salinas, A.M. Pascoal, and J. Aranda, "Multiple underwater target positioning with optimally placed acoustic surface sensor networks," *International Journal of Distributed Sensor Networks*, vol. 14, no 5, 2018. doi:10.1177/1550147718773234.
44. K.B. Howell, *Principles of Fourier Analysis*. CRC Press, 2001.
45. P. Khargonekar and M. Rotea, "Multiple objective optimal control of linear systems: The quadratic norm case," *IEEE Transactions on Automatic Control*, vol. 36, pp. 14–24, 1991.
46. N.O. Cunha and E. Polak, "Constrained minimization under vector-valued criteria in topological spaces," in *Mathematical Theory of Control, Proceedings of the USC Conference, A.V. Balakrishnan and L.W. Neustad*, editors, pp. 96–108, 1967.
47. T.L. Vincent and W.J. Grantham, *Optimality in Parametric Systems*. New York: Wiley, 1981.
48. S. Boyd and L. Vandenberghe, *Convex Optimization*. Cambridge University Press, 2004.
49. D.H. Luenberger and Y. Ye, *Linear and Nonlinear Programming*. Springer, 2016.
50. S. Martinez and F. Bullo, "Optimal sensor placement and motion coordination for target tracking," *Automatica*, vol. 42, no. 4, pp. 661–668, 2006.
51. D. Moreno-Salinas, A.M. Pascoal, and J. Aranda, "Sensor networks for optimal target localization with bearings-only measurements in constrained three-dimensional scenarios," *Sensors*, vol. 13, no. 8, pp. 10386–10417, 2013.
52. N. Crasta, D. Moreno-Salinas, B. Bayat, A.M. Pascoal, and J. Aranda, "Range-based underwater target localization using an autonomous surface vehicle: Observability analysis," in *IEEE/ION Position, Location and Navigation Symposium (PLANS)*, Monterey, California, pp. 487–496, 2018.
53. N. Crasta, D. Moreno-Salinas, A.M. Pascoal, and J. Aranda, "Multiple autonomous surface vehicle motion planning for cooperative range-based underwater target localization," *Annual Reviews in Control*, vol. 46, pp. 326–342, 2018.
54. S. Lall, J.E. Marsden, and S. Glavaški, "Empirical model reduction of controlled nonlinear systems," in *Proceedings of the IFAC World Congress*, New York, pp. 473–478, 1999.
55. N. Crasta, D. Moreno-Salinas, A. Pascoal, and J. Aranda, "Range-based cooperative underwater target localization," *20th IFAC World Congress*, vol. 50, no. 1, pp. 12366–12373, 2017.
56. N. Crasta, M. Bayat, A.P. Aguiar, and A.M. Pascoal, "Observability analysis of 3D AUV trimming trajectories in the presence of ocean currents using range and depth measurements," *Annual Reviews in Control*, vol. 40, pp. 142–156, 2015.

57. P.C. Abreu, J. Botelho, P. Góis, A. Pascoal, J. Ribeiro, M. Ribeiro, M. Rufino, L. Sebastião, and H. Silva, "The MEDUSA class of autonomous marine vehicles and their role in EU projects," IEEE OCEANS, Shanghai, China, pp. 1–10, April 2016.

58. D. Moreno-Salinas, N. Crasta, M. Ribeiro, B. Bayat, A.M. Pascoal, and J. Aranda, "Integrated motion planning, control, and estimation for range-based marine vehicle positioning and target localization," in *10th IFAC Conference on Control Applications in Marine Systems*, vol. 49, no. 23, pp. 34–40, 2016.

59. D. Moreno-Salinas, N. Crasta, A. Pascoal, and J. Aranda, "Optimal multiple underwater target localization and tracking using two surface acoustic ranging sensors," in *11th IFAC Conference on Control Applications in Marine Systems, Robotics, and Vehicles, Opatija, Croatia*, vol. 51(29), pp. 177–182, 2018.

19

Energy-Efficient Distributed Localization for Wireless Sensor Networks

Nuha A.S. Alwan and Zahir M. Hussain

CONTENTS

19.1 Introduction

A wireless sensor network (WSN) is an ad hoc communication network that consists of a large number of small, low-cost, low-energy sensors with limited processing capability. The sensors or nodes are intended to observe and communicate spatiotemporal information of the environment in which they are deployed. WSNs serve a wide range of applications such as monitoring (temperature, humidity, etc.), military surveillance, process control, routing, target tracking, and so forth. The sensor deployment of a WSN does not necessitate the existence of an infrastructure or pre-planning. In fact, in many instances, the sensors are scattered by aircraft onto harsh fields or disaster areas, resulting in uncontrolled landing positions. In all applications, localization of the sensor nodes is crucial for the information collected by the sensors to be meaningful.

Recently, WSN localization has emerged as a very important and a most intensively studied issue and sparked a considerable amount of innovative research. The powerful and widely accessible global positioning system (GPS) cannot be employed for WSN node localization as it is impractical to install in each sensor due to being costly and energy consuming. Only a few reference nodes, called anchors or beacons, are equipped with GPS modules. Position information of the other nodes can be obtained using the known anchor locations through anchor-based localization methods. The material of the present chapter hinges around this particular class of localization methods. According to the dependence on anchor nodes, localization algorithms are classified into anchor-based and anchor-free. Anchor-free techniques do not assume knowledge of any node positions. For example, one mobile anchor node whose position is known may be used instead of a number of static anchors (Mesmoudi et al. 2013). This mobile anchor broadcasts its known location coordinates as it moves within the sensor network such that a number of virtual anchor nodes are generated and subsequently used to estimate the location of the rest of the network nodes.

Localization methods in WSNs generally utilize various inter-sensor distance measurement techniques which, in turn, can be classified into time of arrival (TOA), time difference of arrival (TDOA), angle of arrival (AOA), and received signal strength (RSS) measurement techniques (Patwari et al. 2003, 2005). Radio frequency (RF) or ultra-wide band (UWB) signals may be used for sensing. Localization methods can also be classified into two main categories: centralized and distributed. In the former, all measurements are sent to a fusion center where locations are calculated. In the latter, sensor nodes calculate their own locations by interacting with other nodes. While both face the high cost of communication, centralized localization results in more positioning accuracy, whereas distributed localization is more resilient to node and link failures. Both categories are further classified into range-based and range-free techniques. In range-based techniques, the above-mentioned measurements are used to estimate the distance between nodes and to calculate position coordinates. They are characterized by high accuracy but also high energy consumption and extra hardware requirements. In contrast, range-free techniques are simple and energy-efficient, depending solely on the contents of the received packet to achieve localization using geometric interpretations, constraint minimization, and resident area formation (Paul and Sato 2017, Singh and Khilar 2017).

The focus of this chapter is on distributed anchor-based range-based localization, employing the TOA method of distance measurement from knowledge of time and speed of light. The TOA is the time measured between transmission from the target node to be localized and reception at the anchor node. In the case of lack of synchronism between the target node and the anchors, which often occurs, two-way TOA measurements can be obtained by which the anchor sends a signal to the target, which immediately replies. In this case, the anchor determines the TOA as the delay between transmission and reception divided by two (Patwari et al. 2005). The TOA measurement has the merit of being less sensitive to inter-device distances than the RSS measurement (Patwari et al. 2005), provided that no restriction of indoor localization is assumed. The reason is that the significant multipath effect in indoor environments affects the TOA measurements so that the RSS method of distance measurement is rather employed in such cases. The TOA distance measurements considered in this chapter are line-of-sight (LOS) arrivals contaminated by additive measurement noise. These measurement errors can be modeled by a zero-mean additive Gaussian noise component.

19.1.1 Multilateration Localization versus Iterative Optimization

Trilateration and multilateration are analytical localization techniques implemented in two-dimensional (2D) and three-dimensional (3D) WSNs, respectively (Zhang et al. 2011). In trilateration, the node position is estimated analytically using distance measurements from three anchors by computing the intersection point of three circles, as shown in Figure 19.1. Localization in three dimensions is more practical and yields more accurate results, despite the fact that it is less studied in the literature. Multilateration is the 3D counterpart of trilateration, where at least four anchors are needed (Bachrach and Taylor 2005), which is the case considered in this chapter. These methods are characterized by decreased accuracy and computational complexity, especially in the 3D case. Their suboptimal performance becomes most evident with noisy and fluctuating measurements, and localization becomes difficult and uncertain as the intersection in Figure 19.1 becomes an overlapped region.

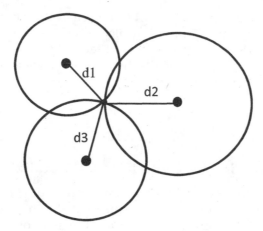

FIGURE 19.1
Trilateration. The centers of the three circles are the anchor positions. The intersection of the three circles is the target location.

Iterative optimization techniques constitute an attractive alternative solution for the localization problem due to their accuracy, low computational complexity, and suitability to moving-target tracking (Alwan and Hussain 2017). The Kalman filter can be used as an iterative state estimator for localization in the presence of noisy measurements. However, its computational complexity is usually incompatible with the limited resources of the WSN nodes. Therefore, the most common iterative algorithm is the computationally efficient gradient descent (GD) algorithm, which has been extensively used in literature on 2D localization (Garg et al. 2010, Qiao and Pang 2011).

19.1.2 Energy Considerations and Reliability

All nodes in a WSN should practically consume as little energy as possible to prolong battery life, thereby achieving cost-effectiveness. WSN energy efficiency has been achieved in the context of centralized moving target tracking by using compressive sensing (CS) (Alwan and Hussain 2018). CS, aka random sampling, has been recently introduced to perform data sensing and compression simultaneously by random undersampling at a sub-Nyquist rate; the resultant sampled data is completely recoverable if it is sufficiently sparse in the frequency domain (Candes and Wakin 2008). In Alwan and Hussain (2018), randomly sampled TOA data are sent consecutively by the anchor nodes to the fusion center in which GD iterative localization calculations are carried out, as the target moves along its path, after linearly interpolating the TOA data. Linear interpolation at the fusion center enables iterative localization at a higher rate to recover the path of the target node. In addition to the advantage of energy efficiency, CS and linear interpolation have been shown to achieve a remarkably improved performance over other energy consumption-minimizing techniques such as reduced rate uniform sampling or sample rate conversion. In the present chapter we extend the study to achieve a distributed version of the work in Alwan and Hussain (2018) by utilizing the distributed WSN localization technique introduced in Alwan and Mahmood (2015). The key point in the latter technique is the implementation of gossip algorithms to perform distributed averaging or summing, as GD localization involves the computation of sums.

The remainder of the chapter is organized as follows. Section 19.2 deals with compressive sensing principles in view of the localization/tracking application under consideration assuming centralized localization. Section 19.3 describes distributed GD localization in WSNs employing distributed averaging gossip algorithms, and at the same time introduces CS to this scenario for achieving energy-efficient target tracking. The simulation results and discussion are also presented. Finally, Section 19.4 recapitulates the chapter with conclusions and suggestions for future work.

19.2 Compressive Sensing for Centralized GD Localization in WSNs

GD localization in 3D WSNs is an iterative optimization method for which the associated objective error function to be minimized is defined as the sum of the squared distance errors from all anchors. This error function at the estimated point p, designated by $f(p)$, is expressed as:

$$f(p) = \sum_{i=1}^{N} \left\{ [(x - x_i)^2 + (y - y_i)^2 + (z - z_i)^2]^{1/2} - d_i \right\}^2 \tag{19.1}$$

with

$$d_i = c\,\varepsilon_i = c\,(t_i - t_o) \tag{19.2}$$

where $p = [x, y, z]^T$ is the vector of estimated position coordinates, N is the number of anchor nodes, (x_i, y_i, z_i) are the position coordinates of the i-th anchor node, ε_i is the TOA computed at the i-th anchor as the difference between the receive time t_i of the sensing signal by the i-th anchor and the transmit time t_o of the target node, and c is the speed of light. The anchors are assumed to be within the radio range of the moving target.

The error function is minimized iteratively by GD to produce the position estimate of the target node to be localized as follows:

$$p_{k+1} = p_k - \alpha\,g_k \tag{19.3}$$

where k is the iteration number, α is the step size or convergence rate of the GD algorithm, and g_k is the gradient vector of the objective error function at the k-th iteration. In the case of tracking a moving target, k would also represent the time index. The gradient vector can be expressed as:

$$g_k = \left[\frac{\partial f}{\partial x} \quad \frac{\partial f}{\partial y} \quad \frac{\partial f}{\partial z} \right]^T \tag{19.4}$$

If uniformly sampled TOA data are sent consecutively to the fusion center, the gradient computations can be carried out with each time index or iteration number using Equations 19.1–19.4. To achieve energy efficiency, however, CS may be employed by randomly undersampling the TOA data. Note that this is only virtual undersampling; the intermediate TOA values are not generated in the first place, enabling a reduction in the energy consumption requirement of the sensing system between target and anchors in addition to that of the communication system between anchors and fusion center. At this point, it is convenient to demonstrate how, and under what conditions, CS or random undersampling guarantees perfect reconstruction of the received TOA data at the fusion center. CS, in this case, has been proven in Alwan and Hussain (2018), in the context of mobile node tracking, to outperform uniform undersampling for the same value of average sampling time interval.

CS theory can be summarized by considering a signal vector x the elements of which are to be sub-Nyquist-sampled to produce another shorter signal vector $\mathbf{y}$. We will see that $\mathbf{x}$ is perfectly recoverable from $\mathbf{y}$ under certain conditions. Let the discrete-time signal vector $\mathbf{x} \in R^M$ be K-sparse in the M-dimensional space spanned by the set of M basis vectors $\{\psi_i\}_{i=1}^M$,

$$\mathbf{x} = \sum_{i=1}^{M} \psi_i s_i = \Psi \mathbf{s} \tag{19.5}$$

where $\Psi = [\psi_1 \ldots \psi_M]^T$ is the matrix whose rows are the M basis vectors.

$\mathbf{s} = [s_1 \ldots s_M]^T$ is the transform vector which contains exactly K non-zero elements, $K \ll M$. In the case under consideration, Ψ is the DFT. Therefore, we are assuming that we have sparsity in the frequency domain.

CS comprises the acquisition or sampling of $\mathbf{x}$ by a measurement matrix Φ to yield the vector $\mathbf{y} \in R^V$, $V < M$, given by:

$$\mathbf{y} = \Phi \mathbf{x} = \Phi \Psi \mathbf{s} = \Theta \mathbf{s} \tag{19.6}$$

where

$$\Theta = \Phi \Psi. \tag{19.7}$$

Equation 19.6 clearly represents an underdetermined system of equations since $V < M$. Provided that the sparsity information of $\mathbf{x}$ is preserved in $\mathbf{y}$, it is possible to reconstruct $\mathbf{x}$ exactly from $\mathbf{y}$ if the sparsity can somehow be restored. It has been proven in Candes and Romberg (2007) that it can be arranged for the sparsity information to be preserved following CS acquisition if the matrix Θ can be made to satisfy the restricted isometry property (RIP) (Candes et al. 2006, Candes and Romberg 2007, Candes and Wakin 2008). As such, the sparsity information would be preserved, though hidden, in $\mathbf{y}$. For Θ to satisfy the RIP conditions, the measurement matrix Φ must be incoherent with the basis vectors matrix Ψ. This is guaranteed when Φ is chosen as a random matrix. Only then would $\mathbf{x}$ be fully recoverable from $\mathbf{y}$ providing that V satisfies: $CK \log (M/K) < V < M$, where C is a constant, and using l_1-minimization for the solution of the convex optimization problem associated with the underdetermined system of equations (Candes et al. 2006, Candes and Wakin 2008).

The above analysis justifies the application of random sampling or CS to our localization problem. In general, since TOA measurements have low-pass narrow-band frequency content—that is, sparsity in the frequency domain—then the computationally simple and almost real-time linear interpolation would be sufficient to recover the original sparse TOA signal with acceptable error instead of resorting to the computationally expensive l_1-minimization. The rationale behind this approach, which was followed in Alwan and Hussain (2018), is that linear interpolation itself is a low-pass filtering operation. Thus, linear interpolation is well-suited for the range-based localization problem at hand but is certainly not always a good substitute for l_1-minimization (Candes and Wakin 2008). It was demonstrated in Alwan and Hussain (2018) that random undersampling of TOA data, obeying a uniform probability distribution, performed better in terms of localization error function than uniform undersampling for the same average sampling interval. Besides, the performance improvement was shown to be more accentuated when more randomness was introduced in terms of greater maximum sampling interval. The greater the maximum sampling interval for random undersampling, the more the incoherence of the measurement matrix Φ, and the better the sparsity is preserved, all of which are factors contributing to enhanced reconstruction performance and subsequent localization accuracy.

In this work, as in Alwan and Hussain (2018), random sampling is achieved by selecting samples separated by random periods that are represented by a uniformly distributed random variable over a time interval $[T_{\min}, T_{\max}]$. Thus, the random sampling time instants, indexed by the integer $n \geq 0$, can be expressed as:

$$t_n = t_o + T_s \sum_{k=1}^{n} T_k; \quad T_k = U(T_{\min}, T_{\max}) \tag{19.8}$$

where t_o is the initial sampling instant, T_s is the original sampling time interval, and $U(K,L)$, with $L > K$, is the discrete uniform distribution over the integer interval $[K,L]$.

19.3 Distributed GD Localization in WSNs with Compressive Sensing

The GD localization problem is very well tailored to distributed computing such as the distributed localization technique proposed in Alwan and Mahmood (2015) with the aim of achieving resilience against node and link failures. The key idea is to employ the well-established gossip-based distributed averaging/summing techniques to compute the sums inherent in GD localization in a distributed fashion. In this work, we attempt to gain further considerable advantages by enhancing the algorithm in Alwan and Mahmood (2015) by using compressive sampling of the sensing or TOA data. Let us first review the above-mentioned algorithm briefly.

Partially differentiating Equation 19.1 with respect to each of the three dimensions, the gradients in Equation 19.4 can be expressed as:

$$\left.\frac{\partial f}{\partial x}\right|_k = \sum_{i=1}^{N} 2 \frac{\left\{[(x_k - x_i)^2 + (y_k - y_i)^2 + (z_k - z_i)^2]^{1/2} - d_i\right\}}{[(x_k - x_i)^2 + (y_k - y_i)^2 + (z_k - z_i)^2]^{1/2}} \cdot (x_k - x_i) \tag{19.9}$$

$$\left.\frac{\partial f}{\partial y}\right|_k = \sum_{i=1}^{N} 2 \frac{\left\{[(x_k - x_i)^2 + (y_k - y_i)^2 + (z_k - z_i)^2]^{1/2} - d_i\right\}}{[(x_k - x_i)^2 + (y_k - y_i)^2 + (z_k - z_i)^2]^{1/2}} \cdot (y_k - y_i) \tag{19.10}$$

$$\left.\frac{\partial f}{\partial z}\right|_k = \sum_{i=1}^{N} 2 \frac{\left\{[(x_k - x_i)^2 + (y_k - y_i)^2 + (z_k - z_i)^2]^{1/2} - d_i\right\}}{[(x_k - x_i)^2 + (y_k - y_i)^2 + (z_k - z_i)^2]^{1/2}} \cdot (z_k - z_i) \tag{19.11}$$

We also rewrite Equation 19.1 at the k-th iteration as:

$$f(p)\big|_k = \sum_{i=1}^{N} \left\{ [(x_k - x_i)^2 + (y_k - y_i)^2 + (z_k - z_i)^2]^{1/2} - d_i \right\}^2 \tag{19.12}$$

The above Equations 19.9–19.12 indicate that there are four N-term sums to be computed in each iteration of GD localization. Each sum involves variables such as the distance measurements, the anchor positions, and the estimated (k-th) position. For N equal to 4, the variables in each of the four terms constituting a sum are resident in one of the four anchors. This implies that by sharing information (gossiping) among the anchor nodes, each of the four sums can be computed in a distributed manner with the final results residing in all four anchors, with the possibility of slight differences. When the distributed summing tasks are achieved within one localization iteration, each anchor will have estimates of all four sums residing within. Thereafter, Equation 19.3 can be computed in each anchor to yield the estimated position of the target node at the k-th iteration.

At the heart of many signal processing applications is the averaging or summing problem. This problem can be conveniently solved by using gossip-based algorithms. These randomized distributed algorithms compute averages through a sequence of pairwise averages (Dumard and Riegler 2009). In our present problem, the gossiping nodes are the anchors, and we assume they are within the radio range of one another. The simple gossip-based synchronous averaging algorithm known as the push-sum (PS) algorithm (Dumard and Riegler 2009, Strakova and Gansterer 2013) will be used in our distributed GD localization application. The following subsection introduces the PS algorithm.

19.3.1 The Push-Sum Distributed Summing Algorithm

The PS distributed summing algorithm can be employed to iteratively compute a summation whose terms each reside in one of a number of nodes in the WSN (the anchor nodes in our case) through information exchange, or "gossip", between the nodes. Upon convergence, every node will have an estimate of the sum that differs slightly from those of the other nodes since the PS iterative algorithm is not exact. It is assumed that the nodes are within the radio range of one another and that they operate synchronously. Henceforth, to avoid confusion, we will preserve the term "iteration" for the GD algorithm and the term "round" for the time step of the PS algorithm, t. During each external GD iteration, a number of internal rounds are executed, until convergence of the PS algorithm, to compute each of the four sums of that iteration.

For the computation of each of the four sums, the PS algorithm works as follows. For every round t, each node i, $i = 1, 2, \ldots, N$, is assigned a weight $\omega(i)$ that is initialized in all nodes to the value $1/N$, where N is the number of nodes (anchors). Also, an initial sum $s(i)$ is made to reside in each node that is initially set to $s(i) = x(i)$ where $x(i)$ is one of the terms to be summed that is resident in that node. At round $t = 0$, each node i sends the pair $[s(i), \omega(i)]$ to itself, and for rounds $t = 1, 2, \ldots, T$, where T is the total number of rounds until convergence, each node i executes the following algorithmic steps.

Algorithm 19.1   The PS method of distributed summing
(based on Dumard and Riegler 2009)

Input: N and T
Initially: $t = 0$, $s(i) = x(i)$ and $\omega(i) = 1/N$ for $i = 1, \ldots, N$.

repeat

Let $\{\hat{s}(r), \hat{\omega}(r)\}$ be the set of all pairs sent to node i at round $t - 1$.

Set $s(i) \equiv \sum_r \hat{s}(r)$ and $\omega(i) \equiv \sum_r \hat{\omega}(r)$.

Choose a target node uniformly at random.

Send the pair $[1/2\, s(i), 1/2\, \omega(i)]$ to target node and yourself (node i).

$[s(i)/\omega(i)]$ is the estimate of the sum at round t and node i.

$t = t + 1$

until $t = T$.

Output:

At node i and round t:

$[s(i)/\omega(i)]$ is the sum.

It is also worth noting that at all rounds t, where $t = 0, 1, 2, \ldots, T$, the following applies:

$$\sum_{i=1}^{N} \omega(i) = 1 \ \text{ and } \ \sum_{i=1}^{N} s(i) = \text{the sum}.$$

It is proved in (Kempe et al. 2003) that the number of rounds T needed such that the relative error of the PS algorithm described above is less than ρ with probability of at least $(1 - \delta)$ is of order:

$$T(\delta, N, \rho) = O\left(\log_2 N + \log_2 \frac{1}{\rho} + \log_2 \frac{1}{\delta}\right) \tag{19.13}$$

Taking the exact value instead of the order in Equation 19.13, we find that the number of rounds T needed for convergence of the PS algorithm is:

$$T = \log_2\left(\frac{N}{\rho\delta}\right) \tag{19.14}$$

The final accuracy of localization that we can achieve in the present application is dependent on the accuracy level ρ that we set in the PS algorithm. If we set $\rho = \delta = 2^{-6}$, the number of rounds would be $T = 14$ for $N = 4$. Therefore, we expect a relative error of $\rho \leq 2^{-6} = 0.0157$ with probability higher than 0.9843 (Dumard and Riegler 2009, Alwan and Mahmood 2015). In the absence of link failures, this would almost attain the accuracy of non-distributed (exact) summing.

At this point, it is helpful to demonstrate the operation of the PS algorithm and its resilience against link failures by a simple MATLAB® experiment. If we take the number of nodes $N = 4$, and the terms to be summed, and that are each resident in one of the four nodes as [0.2 0.4 0.15 0.25], then by simulating the PS summing steps of Algorithm 1, we obtain the convergence behavior shown in Figure 19.2. This figure is a plot of the progressing sum in one anchor versus the number of rounds up to $T = 14$. Link failures are also taken into consideration for comparison. These occur owing to several factors such as channel congestion, message collisions, or dynamic topology (Sluciak et al. 2012) and can be modeled by the omission of a bidirectional connection between two nodes. For four nodes, there are six possible links. Link failures of up to three (50%) are experimented with. This is achieved at each round by removing a percentage of the links between the anchor nodes at random. The randomly chosen missing links differ with every round, but their number is kept fixed for a certain tested percentage of link failures. Since there is

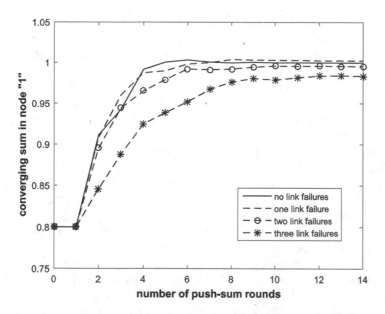

FIGURE 19.2

Convergence of the push-sum summing algorithm, the four initial elements resident in the anchors at $t = 0$ are [0.2 0.4 0.15 0.25]. The exact sum is unity.

randomness in choosing faulty links in each iteration, as well as in the execution of the PS algorithm, the results are averaged over 100 runs. It is clear from Figure 19.2 that the case with no link failures converges to the exact sum of unity, while the cases with link failures converge to values around unity.

It can be deduced by inspecting Algorithm 1 that the PS method requires considerable communication overhead between the nodes, thereby contributing to more energy consumption. In distributed GD localization, however, this disadvantage is outweighed by the achieved resilience against link failures (Alwan and Mahmood 2015). Using CS for the acquisition of the TOA data for the case of moving target tracking, as proposed in this chapter, would increase the energy efficiency of the tracking process since it would considerably reduce the energy consumption associated with the sensing mechanism.

19.3.2 Simulation Results

The distributed energy-efficient moving target tracking under consideration is implemented in MATLAB. Four anchor nodes are positioned randomly in a 3D space of $100 \times 100 \times 100$ m³, all of which are assumed to be within the radio range of one another and of the moving target. As in Alwan and Hussain (2018), the moving node follows a 3D helical path described by:

$$
\begin{aligned}
x &= r\cos\theta \\
y &= r\sin\theta \\
z &= \kappa\theta
\end{aligned}
\tag{19.15}
$$

where the constants r and κ take on the values of 40 and 20 respectively, and the angle θ ranges between 0 and 2π. In a similar scenario, the resulting TOA signal was found in

Alwan and Hussain (2018) to have low-frequency content in the time domain and sufficient sparsity in the frequency domain. For the purpose of comparison, we first illustrate the performance of the centralized energy-efficient GD tracking. For centralized GD tracking, one would normally send uniformly-sampled TOA data consecutively to the fusion center where the gradient descent computations are carried out with each new data sample (time index or iteration number), using Equations 19.1–19.4. However, due to the sparsity of the TOA data in the frequency domain, CS can be used by randomly undersampling the TOA signal to achieve both energy efficiency of the sensing system as well as better reconstruction of the TOA data at the fusion center, and consequently better tracking when compared to uniform undersampling of TOA data (Alwan and Hussain 2018). TOA data reconstruction at the fusion center can in this case be adequately performed by linear interpolation. The initial node position is taken to be (0, 0, 0). The random sampling intervals vary according to a uniform distribution with values between $T_{min} = 1$ and $T_{max} = 7$. The average sampling interval is 4. The GD step size α is optimized as in Alwan and Hussain (2017) and the references therein and fixed to the value of 0.25. Linear interpolation and GD localization of a 56-point helical path are achieved as in Figure 19.3, with one faulty link (out of four) chosen at random in each iteration. The four links are those between each of the four anchors and the fusion center. So this means that there is a percentage of faulty links of 25% which causes one anchor to be isolated from the fusion center. Thus, we are left with only three anchors, though randomly chosen at each iteration. These and all subsequent results are averaged over 100 runs.

Distributed CS-based localization performance of the same 56-point helical path is demonstrated in Figure 19.4 for a percentage of faulty links of 50% which is equivalent to three faulty links among the six possible inter-anchor links. Linear interpolation is also employed as the TOA data reconstruction method. The localized path of Figure 19.4 is that computed by one of the anchor nodes that we designate as node "1". A convergent PS algorithm is implemented within each GD iteration to compute the four GD algorithmic sums. The GD step size is 0.25, and the PS number of rounds is $T = 14$.

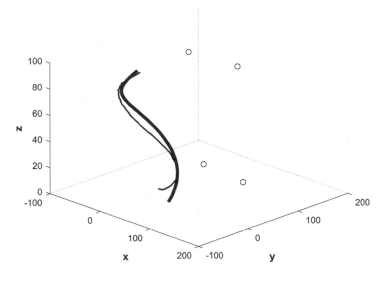

FIGURE 19.3

Centralized GD localization of 56-point helical path with random sampling of TOA data and linear interpolation reconstruction. The true path is shown in bold, and the small circles represent the four anchors. Link failure percentage is 25%.

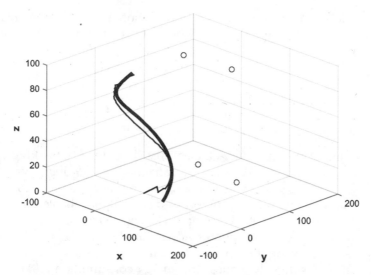

FIGURE 19.4
Distributed CS-based GD localization (as computed in anchor node "1") of 56-point helical path with random sampling of TOA data and linear interpolation reconstruction. The true path is shown in bold, and the small circles represent the four anchors. Link failure percentage is 50%.

The localization error function at each iteration, as given by Equation 19.1, is shown in Figure 19.5 for the specific centralized and distributed experiments of Figures 19.3 and 19.4, respectively. It is clear that the localization error for the centralized case is greater than that of the distributed localization all along the 56-point path, especially in the tracking region of the plot, despite the fact that the percentage of link failures is even higher in the distributed case. This clearly demonstrates the resilience of distributed localization to link

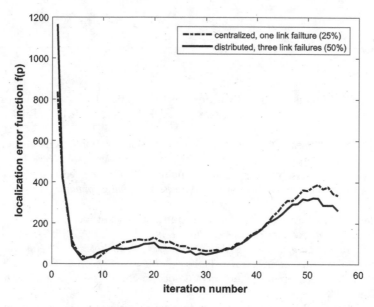

FIGURE 19.5
Localization/tracking error function for CS-based centralized and distributed localization with 25% and 50% link failures respectively, corresponding to the two experiments of Figures 19.3 and 19.4.

failures, a fact which, when combined with the additional advantage of energy efficiency due to using CS, highlights the importance of the localization/tracking technique proposed in the present work.

19.4 Recapitulation and Future Work

In this work, we have achieved energy-efficient distributed target tracking using iterative gradient descent in the context of WSN node localization. The double advantage of energy efficiency and reliability is gained, as opposed to previous work focusing either solely on energy-efficient localization using compressed sensing or reliable localization against node and link failures using distributed computing. The distributed localization algorithm employed requires more communication overhead and more computational complexity since the localization is computed in all gossiping anchor nodes. However, the use of compressed sensing considerably reduces the energy consumption of the sensing system that comprises the anchors and the moving target node.

As a future prospect, we contemplate replacing the random sampling in the present work by deterministically chaotic sequences, an approach already tested successfully in Alwan and Hussain (2019) for the centralized counterpart of the system described in this chapter. Deterministic chaos approximates randomness in just a few algorithmic steps, enabling the use of this pseudo-random sampling, and has first appeared in Nguyen et al. (2008). The advantage gained is ease of implementation on hardware design. If the chaos parameters are known to the anchor nodes, the chaotic sequence can be regenerated in the anchors with minimum hardware and computational requirements. In contrast, with random sampling, the sequence of random sampling instants must be stored in the anchors that operate synchronously.

References

Alwan, N.A.S. and Z.M. Hussain. Gradient descent localization in wireless sensor networks. In *Wireless Sensor Networks – Insights and Innovations*, P. Sallis (Ed.). InTech, Rijeka, Croatia, 2017: 39–59.

Alwan, N.A.S. and Z.M. Hussain. Compressive sensing for localization in wireless sensor networks: An approach for energy and error control. *IET Wireless Sensor Systems* 8(3), June 2018: 116–120.

Alwan, N.A.S. and Z.M. Hussain. Compressive sensing with chaotic sequences: An application to localization in wireless sensor networks. *Wireless Personal Communications* 105(3), April 2019: 941–950.

Alwan, N.A.S. and A.S. Mahmood. Distributed gradient descent localization in wireless sensor networks. *Arabian Journal for Science and Engineering* 40(3), March 2015: 893–899.

Bachrach, J. and C. Taylor. *Localization in Sensor Networks: Handbook of Sensor Networks: Algorithms and Architectures.* John Wiley and Sons Inc., Hoboken, New Jersey, USA, 2005, vol. 2005.

Candes, E.J. and J. Romberg. Sparsity and incoherence in compressive sampling. *Inverse Problems* 23(3), 2007: 969–985.

Candes, E.J., J. Romberg and T. Tao. Robust uncertainty principles: Exact signal reconstruction from highly incomplete frequency information. *IEEE Transactions on Information Theory* 52(2), 2006: 489–509.

Candes, E.J. and M. Wakin. An introduction to compressive sampling. *IEEE Signal Processing Magazine* 25(2), 2008: 21–30.

Dumard, C. and E. Riegler. Distributed sphere decoding. *International Conference on Telecommunications (ICT'09), IEEE*, Marrakech, 25–27 May 2009, pp. 172–177.

Garg, R., A.L. Varna and M. Wu. Gradient descent approach for secure localization in resource constrained wireless sensor networks. *International Conference on Acoustics, Speech and Signal Processing (ICASSP), IEEE*, Dallas, Texas, USA, 14 March 2010, pp. 1854–1857.

Kempe, D., A. Dobra and J. Gehrki. Gossip-based computation of aggregate information. *The 44th Annual IEEE Symposium on Foundations of Computer Science*, 2003, pp. 482–491.

Mesmoudi, A., M. Feham and N. Labraoui. Wireless sensor networks localization algorithms: A comprehensive survey. *International Journal of Computer Networks and Communications (IJCNC)* 5(6), November 2013: 45–64.

Nguyen, L.T., D.V. Phong, Z.M. Hussain, H.T. Huynh, V.L. Morgan and J.C. Gore. Compressed sensing using chaos filters. *Australasian Telecommunication Networks and Applications Conference (ATNAC 2008)*, 2008.

Patwari, N., J.N. Ash, S. Kyperoutas et al. Locating the nodes: Cooperative localization in wireless sensor networks. *IEEE Signal Processing Magazine* 22(4), 2005: 54–69.

Patwari, N., A.O. Hero III, M. Perkins et al. Relative localization estimation in wireless sensor networks. *IEEE Transactions on Signal Processing* 51(8), 2003: 2137–2148.

Paul, A.K. and T. Sato. Localization in wireless sensor networks: A survey on algorithms, measurement techniques, applications and challenges. *Journal of Sensor and Actuator Networks* 6(4), 2017: 24 pages.

Qiao, D. and G.K.H. Pang. Localization in wireless sensor networks with gradient descent. *Proceedings of the IEEE Pacific Rim Conference on Communications, Computers and Signal Processing (Pac-Rim). IEEE*, Victoria, BC, Canada, 2011, pp. 91–96.

Singh, M. and P.M. Khilar. Mobile beacon based range free localization method for wireless sensor networks. *Wireless Networks* 23, 2017: 1285–1300.

Sluciak, O., H. Strakova, M. Rupp and W.N. Gansterer. Distributed Gram-Schmidt orthogonalization based on dynamic consensus. *The 46th Asilomar Conference on Signals, Systems and Computers. IEEE*, Pacific Grove, CA, 4–7 November 2012, pp. 1207–1211.

Strakova, H. and W.N. Gansterer. A distributed Eigensolver for loosely coupled networks. *The 21st Euromicro International Conference on Parallel, Distributed and Network-Based Processing(PDP). IEEE*, Belfast, 27 February–1 March, 2013, pp. 51–57.

Zhang, L., C. Tao and G. Yang. Wireless positioning: Fundamentals, systems and state of the art signal processing techniques. In *Cellular Networks-Positioning, Performance Analysis, Reliablility*, A. Melikov (Ed.). InTech, Rijeka, Croatia, 2011: 3–50.

20

Diffusion Kalman Filtering Based on Covariance Intersection

Jinwen Hu, Yang Lyu, Zhao Xu, Chunhui Zhao, and Quan Pan

CONTENTS

This chapter addresses the distributed target estimation problem for linear time-varying systems in Mobile Sensor Network (MSN). Following the work of [3], which showed that the diffusion of the estimates of local Kalman filters can improve the estimation performance of the whole network, a (covariance and intersection) CI-based diffusion Kalman filtering algorithm (CI-DKF) is proposed which allows each agent to obtain a stable estimate by sharing information only with its neighbors, where the entire system and network topology can be time-varying, and the system may be unobservable via each agent together with its neighbors. Different from the diffusion Kalman filtering (DKF) algorithm proposed in [3], which fuses the estimates of local Kalman filters by a convex combination regardless of the error covariance information, our estimates are fused by the CI algorithm, which incorporates the error covariance information as an important factor for stability assurance.

First, the system model is given and the DKF algorithm is recalled in Section 20.1. In Section 20.2, choice rules of adaptive weights of the CI-DKF algorithm are designed based on the CI method. Then, in Section 20.3, a CI-DKF algorithm is proposed for the case where local observability may be lost. The effectiveness of the CI-DKF algorithm is testified by simulation in Section 20.4. Section 20.5 gives the conclusions.

20.1 Background

20.1.1 System Description

Consider a set of N agents with limited communication ranges which are spatially distributed over a surveillance region. Agent i takes measurement $y_{i,k} \in \mathbb{R}^q$ of a common environment state $x_k \in \mathbb{R}^p$ independently at time k. The state-space model associated with the environment and the measurement of agent i are respectively of the form:

$$x_{k+1} = F_k x_k + w_k,$$
$$y_{i,k} = H_{i,k} x_k + v_{i,k}, \tag{20.1}$$

where w_k is the process noise and $v_{i,k}$ the measurement noise of agent i at time k. The matrices F_k and $H_{i,k}$ are allowed to be time-varying but bounded (A matrix is bounded if each element of the matrix has a lower bound and an upper bound). w_k and $v_{i,k}$ are assumed to be zero-mean, uncorrelated, and white with

$$E \begin{bmatrix} w_k \\ v_{i,k} \end{bmatrix} \begin{bmatrix} w_l \\ v_{j,l} \end{bmatrix}^T = \begin{bmatrix} Q_k \delta_{kl} & 0 \\ 0 & R_{i,k} \delta_{kl} \delta_{ij} \end{bmatrix}, \tag{20.2}$$

where Q_k and $R_{i,k}$ are assumed to be positive definite and bounded. Further, w_k and $v_{i,k}$ are uncorrelated with the initial state x_0. If the measurement information is processed in a centralized manner, the augmented form is applied:

$$y_k = \begin{bmatrix} y_{1,k} \\ \vdots \\ y_{N,k} \end{bmatrix}, \quad H_k = \begin{bmatrix} H_{1,k} \\ \vdots \\ H_{N,k} \end{bmatrix}, \quad v_k = \begin{bmatrix} v_{1,k} \\ \vdots \\ v_{N,k} \end{bmatrix}. \tag{20.3}$$

The estimate of x_k obtained by agent i based on local observations up to time l is denoted as $\hat{x}_{i,k|l}$. The estimation error is denoted as $\tilde{x}_{i,k|l} \triangleq x_k - \hat{x}_{i,k|l}$. $P_{i,k|l}$ is an estimate of the error covariance matrix $E\left[\tilde{x}_{i,k|l} \tilde{x}_{i,k|l}^T\right]$ kept by agent i. Specifically, the estimate of agent i is said to be uniformly stable if there exists a bounded positive definite matrix $\bar{P}_i$ such that $E\left[\tilde{x}_{i,k|k} \tilde{x}_{i,k|k}^T\right] \leqslant \bar{P}_i$ (i.e., $\bar{P}_i - E\left[\tilde{x}_{i,k|k} \tilde{x}_{i,k|k}^T\right]$ is positive semidefinite for all k. The graph of the network and other elementary notations are defined in Appendix. Let the set $\{i_{m_k}\}(m_k = 1, \dots, d_{i,k})$ denote the indices of the neighbors of agent i at time k. Then, the local observation matrix for each i can be defined as:

$$H_{i,k}^{\text{loc}} \triangleq \left[H_{i_1,k}^T, H_{i_2,k}^T, \dots, H_{i_{d_i,k},k}^T \right]^T. \tag{20.4}$$

20.1.2 Diffusion Kalman Filtering

Recently, a diffusion Kalman filtering (DKF) algorithm was proposed in [3] as shown in Figure 20.1a. Since our algorithm adopts the same information processing procedure as the DKF algorithm, which includes an incremental update (standard Kalman filtering update)

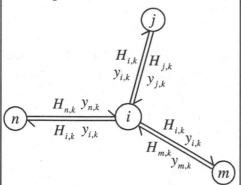
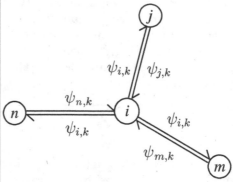

Diffusion update

- Exchange local data

- Iterate the KF with estimate $\hat{x}_{i,k|k-1}$ and the measurement information to get estimate $\psi_{i,k}$.

Incremental update

- Exchange local data

- Calculate weighted average:
$$\hat{x}_{i,k|k} = \sum_{j \in \mathcal{N}_{i,k}} C_{i,j,k} \psi_{j,k}$$

(a) The DKF algorithm.

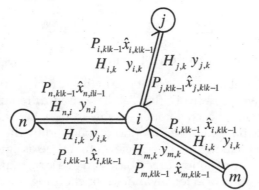

Diffusion update

- Exchange local data

- Calculate weighted average:
$$\psi_{i,k} = \sum_{j \in \mathcal{N}_{i,k}} C_{i,j,k} \hat{x}_{j,k|k-1}$$

where $C_{i,j,k}$ is calculated by the CI algorithm.

Incremental update

- Iterate the KF with estimate $\psi_{i,k}$ and the measurement information to get estimate $\hat{x}_{i,k|k}$.

(b) The CI-DKF algorithm.

FIGURE 20.1

Comparison between the DKF and CI-DKF algorithms.

step and a diffusion update step, it will be illustrated following the introduction of the DKF algorithm by showing the differences between the two algorithms.

The objective of the DKF is for every agent i in the network to compute a stable estimate of the unknown state x_k, while sharing data only with its neighbors. In the DKF algorithm, at every time instant k, agent i sends the quantities $H_{i,k}^T R_{i,k}^{-1} H_{i,k}$ and $H_{i,k}^T R_{i,k}^{-1} y_{i,k}$ to its neighbors for incremental update and the intermediate estimate $\psi_{i,k}$ for diffusion update. Having received the intermediate estimates from its neighbors, agent i combines the intermediate estimate $\psi_{j,k}, j \in \mathcal{N}_{i,k}$ in the diffusion update step by a $p \times p$ diffusion matrix $C_{i,j,k}, k, j \in \nu$ as a weight which is subject to

$$\sum_{j \in \mathcal{N}_{i,k}} C_{i,j,k} = I, \quad C_{i,j,k} = 0 \quad \text{for} \quad l \notin \mathcal{N}_{i,k}. \tag{20.5}$$

On the basis of Kalman filtering, which incorporates the real-time measurement information, the DKF algorithm improves the estimate of each agent by fusing the local estimates of its neighbors. A simplified convex combination is used for fusion of estimates with unknown correlations, which makes the algorithm applicable in a general distributed multi-agent network. Though an adaptive method of choosing combination weights is developed in [2], no conclusions have been made on how to guarantee the estimation stability for a general time-varying system. Furthermore, the assumptions in [2,3] are restrictive since they require that F_k and $H_{i,k}^{loc}$ be time-invariant and $\{F, H_{i,k}^{loc}\}$ be detectable for each agent in the stability proof. This motivates us to develop a DKF algorithm with an effective diffusion strategy suitable for more general applications.

20.2 CI-Based Adaptive Diffusion Matrix

The diffusion matrices $C_{i,j,k}$ play an important role in the diffusion update which influences the performance of the whole network through assigning different weights to different local Kalman filter estimates at each iteration. In this section, the choice rules of diffusion matrices for fusion of estimates will be introduced, followed by the design of the filtering algorithm. Heuristically speaking, a larger weight should be assigned to a local Kalman filter estimate with better estimation accuracy. Based on this idea, the CI algorithm is applied to incorporate the information of error covariance for choosing the diffusion matrices.

Suppose agent i is going to fuse the estimates $\hat{x}_{j,k|k-1}, j \in \mathcal{N}_{i,k}$ from its neighbors with the estimates of their error covariance matrices $P_{j,k|k-1} > 0$. The fusion based on the CI algorithm proposed in [5] is given by

$$\psi_{i,k} = \sum_{j \in \mathcal{N}_{i,k}} C_{i,j,k} \hat{x}_{j,k|k-1}, \tag{20.6}$$

where

$$C_{i,j,k} = \beta_{i,j,k} \Lambda_{i,k} P_{j,k|k-1}^{-1},$$

$$\Lambda_{i,k} = \left(\sum_{j \in \mathcal{N}_{i,k}} \beta_{i,j,k} P_{j,k|k-1}^{-1} \right)^{-1}, \tag{20.7}$$

and $\beta_{i,j,k}$ are subject to $0 \leqslant \beta_{i,j,k} \leqslant 1$ and $\sum\limits_{j \in \mathcal{N}_{i,k}} \beta_{i,j,k} = 1$ which are chosen such that the trace or determinant of $\Lambda_{i,k}$ is minimized. However, such optimization is nonlinear, and a high computation load is required for computing the optimal $\beta_{i,j,k}$. Several fast CI algorithms that produce suboptimal solutions have been proposed in terms of trace or determinant minimization [4,6,12]. For the sake of computational simplicity, the simplified algorithm proposed in [6] is used to calculate $\beta_{i,j,k}$ in Equation 20.7. First, there always exists an agent $g_{i,k} \in \mathcal{N}_{i,k}$ at time k such that $tr(P_{g_{i,k},k|k-1}) = \min\limits_{j \in \mathcal{N}_{i,k}} tr(P_{j,k|k-1})$.

Then, $\beta_{i,j,k}$ is calculated as follows:

$$
\beta_{i,j,k} = \begin{cases} \dfrac{1/tr(P_{j,k|k-1})}{\sum\limits_{m \in \mathcal{N}_{i,k}} 1/tr(P_{m,k|k-1})}, & \text{if } tr(\Lambda_{i,k}) < tr(P_{g_{i,k},k|k-1}); \\ 1, & \text{if } tr(\Lambda_{i,k}) \geq tr(P_{g_{i,k},k|k-1}), j = g_{i,k}; \\ 0, & \text{otherwise.} \end{cases} \tag{20.8}
$$

If $tr(\Lambda_{i,k})$ is ignored in the above equation for simplification of computation and let $\beta_{i,g_{i,k},k} = 1$, $\beta_{i,j,k} = 0 \ \forall j \neq g_{i,k}$, the following 0–1 weighting rule can be obtained:

$$
C_{i,j,k} = \begin{cases} I, & \text{if } j = g_{i,k}; \\ 0, & \text{otherwise,} \end{cases} \tag{20.9}
$$

$$
\Lambda_{i,k} = P_{g_{i,k},k|k-1}.
$$

The weight choice rule (20.9) means that agent i takes the estimate of the neighbor which has the minimum trace of estimated prediction error covariance matrix as the best estimate. It is actually a special case of rule (20.7) and also provides a suboptimal solution. However, it greatly simplifies the computation. In Section 20.4, performances of Equations 20.7 and 20.9 will be compared by simulation.

It should be noted that $P_{j,k|k-1}$ does not have to be the true error covariance matrix of $\hat{x}_{j,k|k-1}$, and its influence on the estimation stability will be discussed later. By implementing the CI algorithm with rule (20.7) or (20.9), the following conclusion can be obtained.

Lemma 20.1 ([5]).

If $P_{j,k|k-1} > 0$ and $P_{j,k|k-1} \geq E\left[\tilde{x}_{j,k|k-1} \tilde{x}_{j,k|k-1}^{\mathsf{T}}\right]$ for $j \in \mathcal{N}_{i,k}$, it holds that $\Lambda_{i,k} \geq E\left[\tilde{x}_{i,k|k-1} \tilde{x}_{i,k|k-1}^{\mathsf{T}}\right]$.

20.3 CI-Based Diffusion Kalman Filtering Algorithm

In [3], the stability of the DKF algorithm is proven under the assumption that the local Kalman filter for each agent is stable based on its local measurement information, that is, the measurements from its neighbors which include the measurement of itself according to the definition of neighbors. This assumption is restrictive in general. Our aim is to seek methods such that each agent can obtain a stable estimate even when the estimation based only on local measurements is unstable.

Before designing the algorithm, first recall the definition of uniform observability of linear time-varying systems [1]. Consider the system with time-varying matrix F_k and global measurement matrix H_k defined in Equation 20.3 and let the observability Gramian be given by

$$\bar{W}_{k+\delta,k} = \sum_{t=k}^{k+\delta} \Phi_{t,k}^T H_t^T H_t \Phi_{t,k} \tag{20.10}$$

for some integer $\delta > 0$, where $\Phi_{k,k} = I$ and

$$\Phi_{t,k} = F_{t-1} \cdots F_k$$

for $t > i$. The matrices F_k and H_k are said to satisfy the uniform observability condition and the entire system is said to be uniformly observable, if there are real numbers $\underline{\eta}, \bar{\eta} > 0$ and an integer $\delta > 0$, such that

$$\underline{\eta} I \leqslant \bar{W}_{k+\delta,k} \leqslant \bar{\eta} I. \tag{20.11}$$

In the same way, the local uniform observability can be defined for each agent. Consider agent i with time-varying matrix F_k and local measurement matrix $H_{i,k}^{\text{loc}}$ defined in Equation 20.4, and let the local observability Gramian be given by

$$W_{k+\delta_i,k}^k = \sum_{t=k}^{k+\delta_i} \Phi_{t,k}^T \left(H_{i,t}^{\text{loc}} \right)^T H_{i,t}^{\text{loc}} \Phi_{t,k} \tag{20.12}$$

for same integer $\delta_i > 0$ The matrices F_k and $H_{i,k}^{\text{loc}}$ are said to satisfy the uniform observability condition and the local subsystem of agent i is said to be uniformly observable, if there are real numbers $\underline{\eta_i}, \bar{\eta}_i > 0$ and an integer $\delta_i > 0$, such that

$$\underline{\eta_i} I \leq W_{k+\delta_i,k}^k \leq \bar{\eta}_i I. \tag{20.13}$$

In a network, (20.13) may not hold for all agents. First, denote by Ω the set of agents for which (20.13) holds, that is, $i \in \Omega$ if F_k and $H_{i,k}^{\text{loc}}$ satisfy the local uniform observability condition, and $i \notin \Omega$. otherwise In the sequel, the problem will be discussed under two scenarios, partial local uniform observability ($\Omega \neq \varnothing$ and at least one agent i can determine $i \in \Omega$) and no local uniform observability ($\Omega \neq \varnothing$ or each agent i cannot determine $i \in \Omega$).

Remark 20.1

In the case of partial local uniform observability, there exists at least one agent in the network that can obtain stable estimates merely based on the measurement information from its neighbors without the need of acquiring information from other agents. However, in the case of no local uniform observability, that is, none of the agents satisfies the local uniform observability condition, there may not exist an agent that can give a stable estimate.

Note that the uniform observability condition may be weakened. For example, for linear time-invariant systems, the observability condition can be replaced by detectability condition.

A system is detectable if and only if all of its unobservable modes are stable. For a detectable system, it can be decomposed into two subsystems, one with all observable modes, which may not be a stable system and the other with all unobservable modes, which is a stable system. Then, it is needed only to consider the estimation stability of the subsystem with all observable modes since the estimation error covariance matrix for the stable subsystem is always bounded no matter what initial estimate is made.

20.3.1 Partial Local Uniform Observability

In this case, each agent $i \in \Omega$ can give a stable estimate only based on the measurement information from its neighbors. Our main task is to let all the other agents $j \notin \Omega$ obtain stable estimates by diffusion of local estimates.

To this end, the CI-DKF algorithm is proposed for the case of partial local uniform observability as shown in Figure 20.1b and Algorithm 20.1. The CI-DKF algorithm requires that at every instant k, each agent communicate to its neighbors the quantities $H_{i,k}^T R_{i,k}^{-1} H_{i,k}$, $H_{i,k}^T R_{i,k}^{-1} y_{i,k}$, $\hat{x}_{i,k|k-1}$ and $P_{i,k|k-1}$ by one message. After receiving the messages from neighbors, each agent first fuses the estimates $\hat{x}_{j,k|k-1}$ for $j \in \mathcal{N}_{i,k}$ in the diffusion update step based on the CI algorithm introduced in Section 20.2. Then, the measurement information is incorporated by Kalman filtering in the incremental update step. Different from the DKF algorithm, the CI-DKF algorithm (Algorithm 20.1) does not require each agent to communicate an intermediate estimate, which reduces the communication load, as each message contains significant overhead information. However, the estimation error covariance matrix is communicated in the CI-DKF algorithm in order to get a stable estimate for each agent. The following theorem shows that the estimation stability can be obtained by implementing the CI-DKF algorithm in the case of partial local uniform observability.

Algorithm 20.1 CI-based diffusion Kalman filtering for partial local uniform observability

---

Start with $\hat{x}_{i,0|-1} = 0$, $P_{i,0|-1} = \Pi_0 > 0$ and $i = 0$ for all i:

Step 1: Diffusion Update:

    Take measurement and communicate $H_{i,k}^T R_{i,k}^{-1} H_{i,k}$, $H_{i,k}^T R_{i,k}^{-1} y_{i,k}$, $\hat{x}_{i,k|k-1}$ and $P_{i,k|k-1}$ to neighbors;

    Calculate $\Lambda_{i,k}$ and the diffusion matrix $C_{i,j,k}$ by Equations 20.7 or 20.9;

    If $i \in \Omega$ & $P_{i,k|k-1} \geq \Lambda_{i,k}$,

$$\hat{x}_{i,k|k-1} \leftarrow \sum_{j \in \mathcal{N}_{i,k}} C_{i,j,k} \hat{x}_{j,k|k-1}$$

$$P_{i,k|k-1} \leftarrow \Lambda_{i,k}$$

    else if $k \notin \Omega$,

$$\hat{x}_{i,k|k-1} \leftarrow \sum_{j \in \mathcal{N}_{i,k}} C_{i,j,k} \hat{x}_{j,k|k-1}$$

$$P_{i,k|k-1} \leftarrow \Lambda_{i,k}$$

    end if

Step 2: Incremental Update:

$$S_{i,k} = \sum_{j \in \mathcal{N}_{i,k}} H_{j,k}^T R_{j,k}^{-1} H_{j,k}$$

$$q_{i,k} = \sum_{j \in \mathcal{N}_{i,k}} H_{j,k}^T R_{j,k}^{-1} y_{j,k}$$

$$P_{i,k|k}^{-1} = P_{i,k|k-1}^{-1} + S_{i,k}$$

$$\hat{x}_{i,k|k} = \hat{x}_{i,k|k-1} + P_{i,k|k}[q_{i,k} - S_{i,k}\hat{x}_{i,k|k-1}]$$

$$\hat{x}_{i,k+1|k} = F_k \hat{x}_{i,k|k}$$

$$P_{i,k+1|k} = F_k P_{i,k|k} F_k^T + Q_k$$

$$k \leftarrow k+1$$

Theorem 20.1

With $P_{j,0|-1} > 0$ and $P_{j,0|-1} \geq E[\tilde{x}_{j,0|-1}\tilde{x}_{j,0|-1}^T]$ for $j \in \mathcal{V}$, if there exists at least one agent $i \in \Omega$ and the network is connected all the time, then the estimates of all agents are uniformly stable under Algorithm 20.1.

Proof. First, consider agent $i \in \Omega$ which satisfies the local uniform observability condition. $P_{i,k|k-1}$ is bounded if only the incremental update is executed for agent i based on the measurement information from its neighbors with bounded matrices F_k and Q_k, that is, there exists a matrix such that $P_{i,k|k-1} \leq \bar{P}_i$ for all k. Since the diffusion update for $i \in \Omega$ replaces $P_{i,k|k-1}$ with the matrix $\Lambda_{i,k} \leq P_{i,k|k-1}$, $P_{i,k|k-1}$ is bounded during each iteration of Algorithm 20.1 with the diffusion update step. Thus, for all k we have

$$P_{i,k|k} = (P_{i,k|k-1}^{-1} + S_{i,k})^{-1} \leq P_{i,k|k-1} \leq \bar{P}_i.$$

Now, consider agent $j_1 \in \mathcal{N}_{i,k}$. According to Lemma 20.1, after the replacement in diffusion update, that is, $P_{j,k|k-1} = \Lambda_{j,k}$ for $j \in \mathcal{V}$, we have

$$tr(P_{j_1,k|k-1}) = tr(\Lambda_{j_1,k}) \leq tr(F_{i-1}P_{i,k-1|k-1}F_{i-1}^T + Q_{i-1})$$
$$\leq tr(F_{i-1}F_{i-1}^T)tr(P_{i,k-1|k-1}) + tr(Q_{i-1})$$
$$\leq tr(F_{i-1}F_{i-1}^T)tr(\bar{P}_i) + tr(Q_{i-1}).$$

Since F_k and Q_k are bounded, there exist finite real numbers π_F and π_Q such that $tr\left(F_k^T F_k\right) \leq \pi_F$ and $tr(Q_k) \leq \pi_Q$ for all k. Thus, we can get

$$P_{j_1,k|k-1} \leq [tr(F_{i-1}^T F_{i-1})tr(\bar{P}_i) + tr(Q_{i-1})]I \leq [\pi_F tr(\bar{P}_i) + \pi_Q]I$$

and

$$P_{j_1,k|k} = (P_{j_1,k|k-1}^{-1} + S_{i,k})^{-1} \leq P_{j_1,k|k-1} \leq [\pi_F tr(\bar{P}_i) + \pi_Q]I \triangleq \bar{P}_{j_1}.$$

Similarly, for agent $j_2 \in \mathcal{N}_{j_1,k}$, we have

$$tr(P_{j_2,k|k-1}) = tr(\Lambda_{j_2,k}) \leq tr(F_{i-1}^T F_{i-1})tr(\bar{P}_{j_1}) + tr(Q_{i-1})$$

which implies

$$P_{j_2,k|k-1} \leqslant [\pi_F tr(\bar{P}_i) + \pi_Q] I$$

and

$$P_{j_2,k|k} \leq P_{j_2,k|k-1} \leq [\pi_F tr(\bar{P}_i) + \pi_Q]I \triangleq \bar{P}_{j_2}.$$

In the same way, an upper bound $\bar{P}_j$ of the estimated error covariance matrix $P_{j,k|k}$ can be found for any other agent $j \in \mathcal{V}$.

On the other hand, if $P_{j,k|k-1} \geq E\left[\tilde{x}_{j,k|k-1}\tilde{x}_{j,k|k-1}^T\right]$ holds for $j \in \mathcal{N}_{i,k}$ before the diffusion update during the k-th iteration, the CI algorithm guarantees that $\Lambda_{i,k} \geq E\left[\tilde{x}_{j,k|k-1}\tilde{x}_{j,k|k-1}^T\right]$ according to Lemma 20.1, which implies that $P_{j,k|k-1} \geq E\left[\tilde{x}_{j,k|k-1}\tilde{x}_{j,k|k-1}^T\right]$ still holds after the diffusion update within the same iteration. Following that, the incremental update in Algorithm 20.1 gives

$$P_{i,k|k}^{-1} = P_{i,k|k-1}^{-1} + S_{i,k}$$
$$\hat{x}_{i,k|k} = \hat{x}_{i,k|k-1} + P_{i,k|k}\left[q_{i,k} - S_{i,k}\hat{x}_{i,k|k-1}\right]. \tag{20.14}$$
$$P_{i,k+1|k} = F_k P_{i,k|k} F_k^T + Q_k$$

Then, we have

$$\tilde{x}_{i,k|k} = (I - P_{i,k|k}S_{i,k})\,\tilde{x}_{i,k|k-1} - P_{i,k|k}(q_{i,k} - S_{i,k}x_k)$$
$$= P_{i,k|k}P_{i,k|k-1}^{-1}\tilde{x}_{i,k|k-1} - P_{i,k|k}\sum_{j \in \mathcal{N}_{i,k}} H_{j,k}^T R_{j,k}^{-1} v_{j,k}.$$

$\tilde{x}_{i,k|k-1}$ and $v_{j,k}$ are uncorrelated, which implies

$$E\left[\tilde{x}_{i,k|k}\tilde{x}_{i,k|k}^T\right] = P_{i,k|k}P_{i,k|k-1}^{-1}E\left[\tilde{x}_{i,k|k-1}\tilde{x}_{i,k|k-1}^T\right]P_{i,k|k-1}^{-1}P_{i,k|k}$$
$$+ P_{i,k|k}\sum_{j \in \mathcal{N}_{i,k}} H_{j,k}^T R_{j,k}^{-1} H_{j,k} P_{i,k|k} \tag{20.15}$$
$$\leq P_{i,k|k}P_{i,k|k-1}^{-1}P_{i,k|k} + P_{i,k|k}S_{i,k}P_{i,k|k} = P_{i,k|k}.$$

We can further get

$$E\left[\tilde{x}_{i,k+1|k}\tilde{x}_{i,k+1|k}^T\right] = F_k E\left[\tilde{x}_{i,k|k}\tilde{x}_{i,k|k}^T\right]F_k^T + Q_k \leq F_k P_{i,k|k}F_k^T + Q_k = P_{i,k+1|k}.$$

with the initial condition $E\left[\tilde{x}_{i,0|-1}\tilde{x}_{i,0|-1}^T\right] \leq P_{i,0|-1}$ for $j \in \mathcal{V}$, $P_{j,k|k} \geq E\left[\tilde{x}_{j,k|k}\tilde{x}_{j,k|k}^T\right]$ holds through all iterations. Consequently, $E\left[\tilde{x}_{j,k|k}\tilde{x}_{j,k|k}^T\right] \leq \bar{P}_j$ holds for all k, which implies Theorem 20.1 holds. ∎

20.3.2 No Local Uniform Observability

In the case of no local uniform observability, one choice to make the DI-DKF algorithm applicable is to build the local uniform observability for some agents in the network. The key to building the observability is to get enough measurement information of the state.

In Algorithm 20.1, the local measurement information of agent i is contained in $S_{i,k}$ and $q_{i,k}$, and the global information matrix and information vector which include the measurement information of all agents are respectively:

$$\bar{S}_k = \sum_{i=1}^{N} H_{i,k}^T R_{i,k}^{-1} H_{i,k} = H_k^T R_k^{-1} H_k$$

$$\bar{q}_k = \sum_{i=1}^{N} H_{i,k}^T R_{i,k}^{-1} y_{i,k} = H_k^T R_k^{-1} y_k$$

(20.16)

To obtain a stable estimate, each agent should obtain sufficient measurement information through one-hop and/or multi-hop communications. However, communicating and storing raw measurement information from each agent would bring heavy communication and storage burdens for agents.

In recent years, many consensus approaches have been proposed for distributed information diffusion [7,8,9,10,13]. By these approaches, the information matrix $S_{i,k}$ and the information vector $q_{i,k}$ based on the information collected by agent i are updated iteratively with the initial values respectively as $H_{i,k}^T R_{i,k}^{-1} H_{i,k}$ and $H_{i,k}^T R_{i,k}^{-1} y_{i,k}$ during each sampling interval, and converge to $\bar{S}_k/N$ and $\bar{q}_k/N$ respectively in an asymptotic manner as the number of communication cycles within each sampling interval goes to infinity. During each communication cycle, only the updated $S_{i,k}$ and $q_{i,k}$ are sent and stored in the memory instead of the raw measurements information from other agents, which helps the agents lower the communication burden and save much storage space. The finite-time consensus for continuous-time systems is shown to be achievable in [11]. However, there has been no effective method to achieve the finite-time consensus for discrete-time systems. A consensus protocol is used for distributed Kalman filtering in [7], where each agent implements a consensus protocol to gather measurement information between two successive Kalman filter updates as shown in Figure 20.2. However, by this approach, $S_{i,k}$ and $q_{i,k}$ may not have achieved consensus, that is, converge to $\bar{S}_k/N$ and $\bar{q}_{i,k}/N$ before the next Kalman filter update, and the error caused by treating $S_{i,k}$ and $q_{i,k}$ as $\bar{S}_k$ and $\bar{q}_{i,k}$ respectively may destroy the estimation stability.

In this chapter, the distributed consensus protocol proposed in [13] is adopted for information diffusion between two successive Kalman filter updates (Figure 20.2) but scale the updated information by multiplying an appropriate coefficient at the end of the consensus implementation. In [13], each agent aims to let $S_{i,k}$ and $q_{i,k}$ reach a consensus by the following protocol:

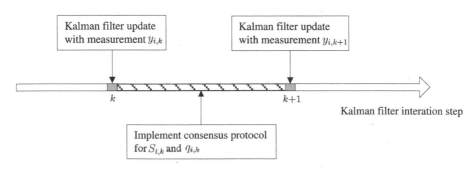

FIGURE 20.2
Implementation of consensus protocol embedded in the Kalman filter for agent i.

$$S_{i,k}(l+1) = \left(1 - \frac{d_{i,k}(l+1)-1}{N}\right)S_{i,k}(l) + \frac{1}{N}\sum_{\substack{j\in\mathcal{N}_{i,k}(l+1)\\ j\neq i}} S_{j,k}(l)$$

$$q_{i,k}(l+1) = \left(1 - \frac{d_{i,k}(l+1)-1}{N}\right)q_{i,k}(l) + \frac{1}{N}\sum_{\substack{j\in\mathcal{N}_{i,k}(l+1)\\ j\neq i}} q_{j,k}(l)$$

(20.17)

where k refers to the k-th sampling interval and l the l-th communication cycle, $d_{i,k}(l+1)$ is the number of neighbors of agent i including itself at the time instant of the $(l+1)$-th communication cycle, and $S_{i,k}(0) = H_{i,k}^T R_{i,k}^{-1} H_{i,k}$, $q_{i,k}(0) = H_{i,k}^T R_{i,k}^{-1} y_{i,k}$. Then, we can get

$$S_{i,k}(l) = \sum_{j=1}^{N} \rho_{i,j,k}(l)S_{i,k}(0)$$

$$q_{i,k}(l) = \sum_{j=1}^{N} \rho_{i,j,k}(l)q_{i,k}(0)$$

(20.18)

where $\rho_{i,i,k}(0)=1$ and $\rho_{i,j,k}(0)=0$ for $j \neq i$. From Equation 20.17, it is easy to get

$$\rho_{i,j,k}(l+1) = \left(1 - \frac{d_{i,k}(l+1)-1}{N}\right)\rho_{i,j,k}(l) + \sum_{\substack{m\in\mathcal{N}_{i,k}(l+1)\\ m\neq k}} \frac{\rho_{m,l,k}(l)}{N}$$

(20.19)

and $\rho_{i,j,k}$ are the weights satisfying

$$0 \leq \rho_{i,j,k}(l) \leq 1, \quad \rho_{i,j,k}(l) = \rho_{j,k,k}(l), \quad \sum_{j=1}^{N} \rho_{i,j,k}(l) = 1.$$

Since $\rho_{i,j,k}(l)$ for $j \in \mathcal{N}_{i,k}(l)$ can be obtained by agent i at time $l+1$ through communications with its neighbors, $\rho_{i,j,k}(l)(i,j \in \mathcal{V})$ can be computed iteratively according to 20.19. It has been proven [13] that $\lim_{j\to\infty} \rho_{i,j,k}(l) = 1/N$ for $i,j \in \mathcal{V}$, thus $S_{i,k}(l) \to \bar{S}_k/N$ and $q_{i,k}(l) \to \bar{q}_k/N$ as $l \to \infty$.

Lemma 20.2

If the network is connected all the time, then there exists a time $0 < t_o \leq N-1$ such that for all k and $j \geq t_o$,

$$\rho_{i,j,k}(l) \geq \left(\frac{1}{N}\right)^{t_o} \geq \left(\frac{1}{N}\right)^{N-1}$$

where $i,j \in \mathcal{V}$.

Proof. Let $t_{i,j,k}$ denote the first time at which $\rho_{i,j,k}(j)$ becomes non-zero. Since the information is transmitted in a diffusion way and the network is connected all the time, $t_{i,j,k}$ must exist and is less than N, that is, $\max\limits_{i,j} t_{i,j,k} \leq N-1$ for all k. Moreover, by the definition of $t_{i,j,k}$, we have $\rho_{i,j,k}(t_{i,j,k}) = (1/N)^{t_{i,j,k}}$ for $i \neq j$. Then, for $t \geq t_{i,j,k}$ and $i \neq j$, we have

$$\rho_{i,j,k}(t) \geq \left(\frac{1}{N}\right)^{t_{i,j,k}} \prod_{l=t_{i,j,k}+1}^{t} \left(1 - \frac{d_{k,i}(l)-1}{N}\right) \geq \left(\frac{1}{N}\right)^{t}.$$

For $i = j$, we have

$$\rho_{i,i,k}(t) \geq \prod_{l=1}^{t} \left(1 - \frac{d_{k,i}(l)-1}{N}\right) \geq \left(\frac{1}{N}\right)^{t}.$$

Hence, by letting $t_o = \max\limits_{i,j \in \mathcal{V}} t_{i,j,k}$, we have $\rho_{i,j,k}(t_o) \geq (1/N)^{t_o} \geq (1/N)^{N-1}$ for all i and j, which implies $\rho_{i,j,k}(l) \geq (1/N)^{N-1}$ for $l \geqslant t_o$ according to 20.17. ∎

Remark 20.2

Lemma 20.2 shows that one can always find a finite time length t_o for running the consensus protocol during each communication cycle such that the scaled measurement information from all agents can be collected by each agent, though the collected information from each agent is scaled by weight $\rho_{i,j,k}(t_o)$ rather than the original. In addition, it shows that the choice of $t_o = N-1$ guarantees that $\rho_{i,j,k}(t_o) > 0$ for all i and j.

Summarizing the results above, an information diffusion scheme (Algorithm 20.2) is designed, where t_o is a given time length for running the consensus protocol during each communication cycle. Then, we can define a set Υ, where $i \in \Upsilon$ if $\min\limits_{j \in \mathcal{V}} \rho_{i,j,k}(t_o) > 0$ for all k, and $i \notin \Upsilon$ otherwise. With Algorithm 20.2 embedded, the CI-DKF algorithm for the case of no local observability is shown in Algorithm 20.3 which includes an information diffusion step, a diffusion update step, and an incremental update step. At every time instant k, each agent i first takes measurement and collects measurement information from other agents to get $S_{i,k}$ and $q_{i,k}$ by implementing the information diffusion scheme in the information diffusion step. In the meantime, each agent i obtains the prediction estimates $\hat{x}_{i,k|k-1}$ and $P_{i,k|k-1}$ from its neighbors, which is followed by the same steps as in Algorithm 20.1 for diffusion and incremental updates.

Algorithm 20.2 Information diffusion scheme

Start with $S_{i,k}(0) = H_{i,k}^T R_{i,k}^{-1} H_{i,k}$, $q_{i,k}(0) = H_{i,k}^T R_{i,k}^{-1} y_{i,k}$, $\rho_{i,i,k}(0) = 1$,
$\rho_{i,j,k}(0) = 0 (j \in \mathcal{V}, j \neq i)$ and $j = 1$ for all i:
    Repeat the following steps until $j > t_o$:
        Calculate $S_{i,k}(l)$ and $q_{i,k}(l)$ by Equation 20.17;
        Calculate $\rho_{i,j,k}(l)(j \in \mathcal{V})$ by Equation 20.19;
        $l \leftarrow l+1$
    end
$$S_{i,k} \leftarrow \frac{1}{\max\limits_{j \in \mathcal{V}} \rho_{i,j,k}(t_o)} S_{i,k}(t_o)$$

$$q_{i,k} \leftarrow \frac{1}{\max\limits_{j \in \mathcal{V}} \rho_{i,j,k}(t_o)} q_{i,k}(t_o).$$

Theorem 20.2

With $P_{j,0|-1} > 0$ and $P_{j,0|-1} \geqslant E\left[\tilde{x}_{j,0|-1}\tilde{x}_{j,0|-1}^T\right]$ for $j \in \mathcal{V}$, if there exists at least one agent $i \in \Upsilon$ and the network is connected all the time, then the estimates of all agents are uniformly stable under Algorithm 20.3 if F_k and H_k satisfy the uniform observability condition.

Proof. With $P_{i,k|k-1} \geq E\left[\tilde{x}_{i,k|k-1}\tilde{x}_{i,k|k-1}^T\right]$ before the diffusion update, it still holds after the diffusion update according to Lemma 20.1. Following that, the incremental update in Algorithm 20.3 gives

$$P_{i,k|k}^{-1} = P_{i,k|k-1}^{-1} + S_{i,k}$$
$$\hat{x_{i,k|k}} = x_{i,k|\hat{k-1}} + P_{i,k|k}\left[q_{i,k} - S_{i,k}x_{i,k|k-1}^{\hat{}}\right]. \tag{20.20}$$
$$P_{i,k+1|k} = F_k P_{i,k|k} F_k^T + Q_k$$

Algorithm 20.3 CI-based diffusion Kalman filtering for no uniform local observability

---

Start with $x_{i,0|-1} = 0$, $P_{i,0|-1} = \Pi_0 > 0$ and $i = 0$ for all i:

Step 1: Information Diffusion:

Take measurement, communicate $\hat{x}_{i,k|k-1}$ and $P_{i,k|k-1}$ to neighbors and implement Algorithm 20.2
to get $S_{i,k}, q_{i,k}$ and $\rho_{i,j,k}(j \in \mathcal{V})$;

Step 2: Diffusion Update:

Calculate $\Lambda_{i,k}$ and the diffusion matrix $C_{i,j,k}$ by Equations 20.7 or 20.9;

If $i \in \Upsilon$ & $P_{i,k|k-1} \geqslant \Lambda_{i,k}$

$$\hat{x}_{i,k|k-1} \leftarrow \sum_{j \in \mathcal{N}_{i,k}} C_{i,j,k}\hat{x}_{j,k|k-1}$$

$$P_{i,k|k-1} \leftarrow \Lambda_{i,k}$$

else if $i \notin \Upsilon$,

$$\hat{x}_{i,k|k-1} \leftarrow \sum_{j \in \mathcal{N}_{i,k}} C_{i,j,k}\hat{x}_{j,k|k-1}$$

$$P_{i,k|k-1} \leftarrow \Lambda_{i,k}$$

end if

Step 3: Incremental Update:

$$P_{i,k|k}^{-1} = P_{i,k|k-1}^{-1} + S_{i,k}$$
$$\hat{x}_{i,k|k} = \hat{x}_{i,k|k-1} + P_{i,k|k}[q_{i,k} - S_{i,k}\hat{x}_{i,k|k-1}]$$
$$\hat{x}_{i,k+1|k} = F_k\hat{x}_{i,k|k}$$
$$P_{i,k+1|k} = F_k P_{i,k|k} F_k^T + Q_k$$
$$k \leftarrow k + 1.$$

Then, we have

$$\tilde{x}_{i,k|k} = (I - P_{i,k|k}S_{i,k})\tilde{x}_{i,k|k-1} - P_{i,k|k}(q_{i,k} - S_{i,k}x_k)$$

$$= P_{i,k|k}P_{i,k|k-1}^{-1}\tilde{x}_{i,k|k-1} - P_{i,k|k}\sum_{j=1}^{N}\frac{\rho_{i,j,k}(t_o)}{\max_{j \in \mathcal{V}}\rho_{i,j,k}(t_o)}H_{j,k}^T R_{j,k}^{-1}v_{j,k}.$$

Since $\tilde{x}_{i,k|k-1}$ and $v_{j,k}$ are uncorrelated, it is implied that

$$
\begin{aligned}
E\left[\tilde{x}_{i,k|k}\tilde{x}_{i,k|k}^{\mathrm{T}}\right] &= P_{i,k|k}P_{i,k|k-1}^{-1}E\left[\tilde{x}_{i,k|k-1}\tilde{x}_{i,k|k-1}^{\mathrm{T}}\right]P_{i,k|k-1}^{-1}P_{i,k|k} \\
&\quad + P_{i,k|k}\sum_{j=1}^{N}\left[\frac{\rho_{i,j,k}(t_o)}{\max\limits_{j\in\mathcal{V}}\rho_{i,j,k}(t_o)}\right]^2 H_{j,k}^T R_{j,k}^{-1}H_{j,k}P_{i,k|k} \\
&\leqslant P_{i,k|k}P_{i,k|k-1}^{-1}P_{i,k|k} + P_{i,k|k}\sum_{j=1}^{N}\frac{\rho_{i,j,k}(t_o)}{\max\limits_{j\in\mathcal{V}}\rho_{i,j,k}(t_o)}H_{j,k}^T R_{j,k}^{-1}H_{j,k}P_{i,k|k} \\
&= P_{i,k|k}P_{i,k|k-1}^{-1}P_{i,k|k} + P_{i,k|k}S_{i,k}P_{i,k|k} = P_{i,k|k}.
\end{aligned}
\tag{20.21}
$$

We can further get

$$
E\left[\tilde{x}_{i,k+1|k}\tilde{x}_{i,k+1|k}^{T}\right] = F_k E\left[\tilde{x}_{i,k|k}\tilde{x}_{i,k|k}^{T}\right]F_k^T + Q_k \leqslant F_k P_{i,k|k}F_k^T + Q_k = P_{i,k+1|k}.
$$

Therefore, with the initial condition $E\left[\tilde{x}_{k,0|-1}\tilde{x}_{k,0|-1}^{T}\right] \leqslant P_{k,0|-1}$, $E\left[\tilde{x}_{i,k|k-1}\tilde{x}_{i,k|k-1}^{T}\right] \leqslant P_{i,k|k-1}$ holds for all k.

On the other hand, for $i \in \Upsilon$, the diffusion update replaces $P_{i,k|k-1}$ with the matrix $\Lambda_{i,k} \leqslant P_{i,k|k-1}$, which means $P_{i,k|k-1}$ must be bounded by the one without diffusion update. Hence, we ignore the diffusion update and write the algebraic Riccati iteration of $P_{i,k|k-1}$ by only considering the Kalman filter update as

$$
\begin{aligned}
P_{i,k+1|k} &= F_k\left[P_{i,k|k-1}^{-1} + S_{i,k}\right]^{-1}F_k^T + Q_k \\
&= F_k\left[P_{i,k|k-1}^{-1} + \sum_{j=1}^{N}\frac{\rho_{i,j,k}(t_o)}{\max\limits_{j\in\mathcal{V}}\rho_{i,j,k}(t_o)}H_{j,k}^T R_{j,k}^{-1}H_{j,k}\right]^{-1}F_k^T + Q_k \\
&= F_k\left[P_{i,k|k-1}^{-1} + H_k^T \hat{R}_k^{-1}H_k\right]^{-1}F_k^T + Q_k,
\end{aligned}
\tag{20.22}
$$

where

$$
\hat{R}_k \triangleq \max_{j\in\mathcal{V}}\rho_{i,j,k}(t_o)\mathrm{diag}\left\{\frac{1}{\rho_{i,1,k}(t_o)}R_{1,i},\ldots,\frac{1}{\rho_{i,N,k}(t_o)}R_{1,i}\right\}
$$

and $\hat{R}_k$ satisfies

$$
R_k \leqslant \hat{R}_k \leqslant \frac{\max\limits_{j\in\mathcal{V}}\rho_{i,j,k}(t_o)}{\min\limits_{j\in\mathcal{V}}\rho_{i,j,k}(t_o)}R_k \leqslant \frac{1}{\min\limits_{j\in\mathcal{V}}\rho_{i,j,k}(t_0)}R_k.
$$

According to Lemma 20.2, we have $\min_{j\in\mathcal{V}} \rho_{i,j,k}(t_0) \geqslant (1/N)^{N-1}$, which implies $\hat{R}_k \leqslant N^{N-1} R_k$. Hence, 20.22 can be seen as a centralized algebraic Riccati iteration for the entire system with bounded noise covariance matrices $\hat{R}_k$ and Q_k. According to the conclusions in [1], there exists a positive definite matrix $\bar{P}_i$ such that $P_{i,k|k-1} \leqslant \bar{P}_i$ for all k if the bounded matrices F_k and H_k satisfy the uniform observability condition. Then, it follows that

$$E\left[\tilde{x}_{i,k|k}\tilde{x}_{i,k|k}^T\right] \leqslant P_{i,k|k} = \left(P_{i,k|k-1}^{-1} + S_{i,k}\right)^{-1} \leqslant P_{i,k|k-1} \leqslant \bar{P}_i.$$

which implies that Theorem 20.2 holds. ∎

Remark 20.3

Based on the definition of Υ, each agent $i \in \Upsilon$ can obtain the measurement information from all agents within each sampling interval. For a general case with time-varying topologies, it is hard to require an agent to determine its identity if not all agents are the members of Υ. In this case, the only method is to set a large enough to such that each agent can obtain the measurement information from all agents as illustrated in Remark 20.2, which implies that all agents are members of Υ. However, for time-invariant topologies, it is still possible to have a smaller t_o, under which not all agents have to be members of Υ. In this case, each agent i can determine its identity only by checking if $\min_{j\in\mathcal{V}} \rho_{i,j,k}(t_o) > 0$ in the first communication cycle because $\min_{j\in\mathcal{V}} \rho_{i,j,k}(t_o)$ is a fixed value for all k under a fixed topology. For example, in a connected network with a time-invariant topology, there always exists a tree that connects all agents. Then, the height (or an upper bound of it) of the tree can be set as t_o, which guaranties that there exists at least one agent (e.g., the root agent of the tree) $i \in \Upsilon$ with $\min_{j\in\mathcal{V}} \rho_{i,j,k}(t_o) > 0$ instead of all agents. Up to now, our selection of t_o is merely for the sake of estimation stability. It should be noted that a sufficiently large t_o can not only help more agents become uniformly observable, but can also lead to an estimate that is close to the optimal one given by the centralized method, since $S_{i,k}(t_o) \to \bar{S}_k$ and $q_{i,k}(t_o) \to \bar{q}_k$ as $t_o \to \infty$ under Algorithm 20.2. Hence, users should make a tradeoff between the communication energy consumption and the estimation performance for a suitable t_o. The influence of t_o on the estimation performance will be shown by simulation in Section 20.4.

20.4 Simulation

20.4.1 Simulation Environment

A time-invariant unstable system model is considered for the ease of simulation, though the proposed algorithm is not restricted to it. A stationary sensor network is estimating the dynamic energy intensity of two stationary sources, the positions of which are known. Each agent has a sensing range $R_s = 20$ m. Two exponential functions are used to denote the energy intensity of two sources spreading over a surveillance region and the system model is given by:

$$F = \begin{bmatrix} 1 & 0.005 \\ 0 & 1 \end{bmatrix}, \quad G = I, \quad Q = 5I, \quad R_i = 20,$$

$$H_i = \left[e^{-\lambda(s_i-\mu_1)^2}\mathbf{1}_{\{\|s_i-\mu_1\|\leqslant R_s\}} \quad e^{-\lambda(s_i-\mu_2)^2}\mathbf{1}_{\{\|s_i-\mu_2\|\leqslant R_s\}}\right],$$

where s_i is the position of agent i, μ_1 and μ_2 are the positions of the two sources, $\lambda = 0.02$ is the attenuation factor, and $\mathbf{1}_{\{\|s_i - \mu_1\| \leqslant R_s\}}$ is the indication function defined as

$$\mathbf{1}_{\{\|s_i - \mu_1\| \leqslant R_s\}} = \begin{cases} 1, & \text{if } \|s_i - \mu_1\| \leqslant R_s; \\ 0, & \text{otherwise.} \end{cases}$$

In the following simulations, different values will be set for the source positions and communication range R_c to get different local observability for each agent.

The same performance index adopted in [3], that is, the mean-square deviation (MSD), is used to evaluate the algorithm performance. The MSD for agent i at time k is defined as

$$\text{MSD}_{i,k} \triangleq E\left[\tilde{x}_{i,k|k}^T \tilde{x}_{i,k|k}\right].$$

Then, the MSD of the whole network is calculated as

$$\text{MSD}_k \triangleq \frac{1}{N} \sum_{i=1}^{N} \text{MSD}_{i,k}.$$

The results are averaged over 100 independent experiments. Besides, to show that $E\left[\tilde{x}_{i,k|k}\tilde{x}_{i,k|k}^T\right]$ is bounded by $P_{i,k|k}$, the following averaged trace is defined:

$$Tr_k \triangleq \frac{1}{N} \sum_{i=1}^{N} tr(P_{i,k|k}).$$

We first implement simulations in two scenarios to compare the performance of six different algorithms: DKF algorithm with adaptive weights, no-diffusion algorithm (i.e., each agent only implements the incremental update), CI algorithm, CI-DKF algorithm for partial local observability (Algorithm 20.1) with weight choice rule 20.7 and 20.9, and the centralized algorithm (i.e., each agent can obtain the original measurement information of all agents), which provides the optimal estimate. In Scenario I (Figure 20.3), $N = 25$ agents are uniformly deployed over a 50×50 m^2 square region and the positions of the two sources are fixed at $\mu_1 = [20\ 30]^T$ and $\mu_2 = [30\ 20]^T$, respectively. The communication range is set as $R_c = 20$ m such that each agent can obtain enough measurement information to become uniformly observable. In Scenario II (Figure 20.4), the communication range is changed to $R_c = 15$ m such that not all agents can collect enough measurement information from their neighbors and thus become unobservable. In addition, ten more agents are added randomly within $[-10, 0] \times [-10, 0]$ which are not uniformly observable since they are too far away from the sources and cannot get enough measurement information from their neighbors as well.

Then, in Scenario III, the performance of the CI-DKF algorithm is tested for no local observability (Algorithm 20.3) with weight choice rule 20.7. In this case, each agent needs to collect the measurement information using Algorithm 20.2. After obtaining enough measurement information to become uniformly observable for some agents, the condition will be the same with that in Scenario I or II. Hence, the above algorithms are not compared in Scenario III. Instead, we examine the influence of t_o on the estimation performance of Algorithm 20.3. In this simulation, again $N = 25$ agents are deployed uniformly over a 50×50 m^2 square region, but place the sources at $\mu_1 = [0\ 50]^T$ and

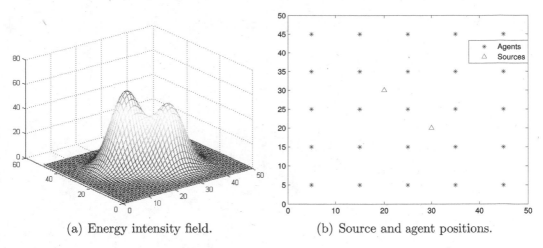

(a) Energy intensity field.   (b) Source and agent positions.

FIGURE 20.3
Simulation setup of Scenario I: full local uniform observability.

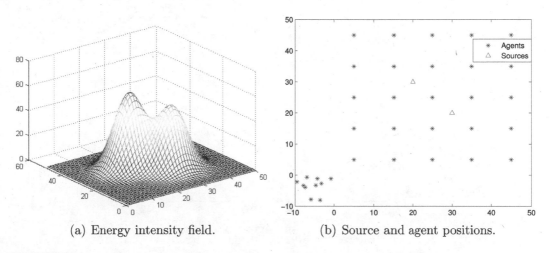

(a) Energy intensity field.   (b) Source and agent positions.

FIGURE 20.4
Simulation setup of Scenario II: partial local uniform observability.

$\mu_2 = [50\ 0]^T$ respectively, and set $R_c = 15$ m such that no agent is uniformly observable as shown in Figure 20.5. In this case, the smallest time length that makes only one agent (i.e., the agent at the center of the region) be the member of Υ is $t_o = 2$. To show the influence of t_o on the algorithm performance, several different values are selected for testing, that is, $t_o = 2$, $t_o = 15$ and $t_o = 30$. The results are all compared with the centralized algorithm.

20.4.2 Simulation Results

In Scenario I, all six algorithms can provide stable estimation, as shown in Figure 20.6. However, the results of our algorithm are closest to the optimal estimate given by the centralized algorithm, which illustrates that the diffusion and incorporation of error covariance information can both improve the estimation performance. In Scenario II (Figure 20.7), the results of the DKF algorithm with adaptive weights and the no-diffusion

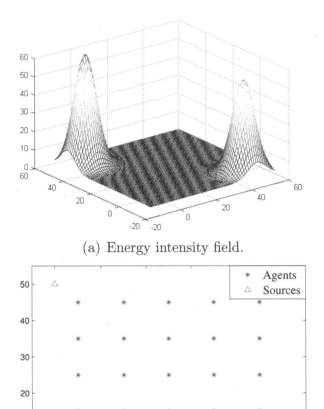

(a) Energy intensity field.

(b) Source and agent positions.

FIGURE 20.5
Simulation setup of Scenario III: no local uniform observability.

algorithm both diverge. Although the CI algorithm still provides a stable estimate, its result is much worse than that of the centralized algorithm, while the results of our algorithm are close to the optimal estimate. The comparison between the algorithms with diffusion and the algorithm without diffusion in the two scenarios illustrates that diffusion can improve the network average performance. Another point to note is that Algorithm 20.1 with rules 20.7 and 20.9 has nearly the same performance. This suggests us to use rule 20.9 in real applications due to its simplified computation.

In Scenario III, Figure 20.8 shows that Algorithm 20.3 can produce a stable estimate for each agent with only two consensus steps, and a sufficiently large t_o can lead to a near-optimal result. It is also shown that getting the optimal estimate requires many more communications than getting a stable estimate. In all, our algorithm provides users with an adjustable parameter through which they can obtain a better tradeoff between their needs of performance and the communication cost.

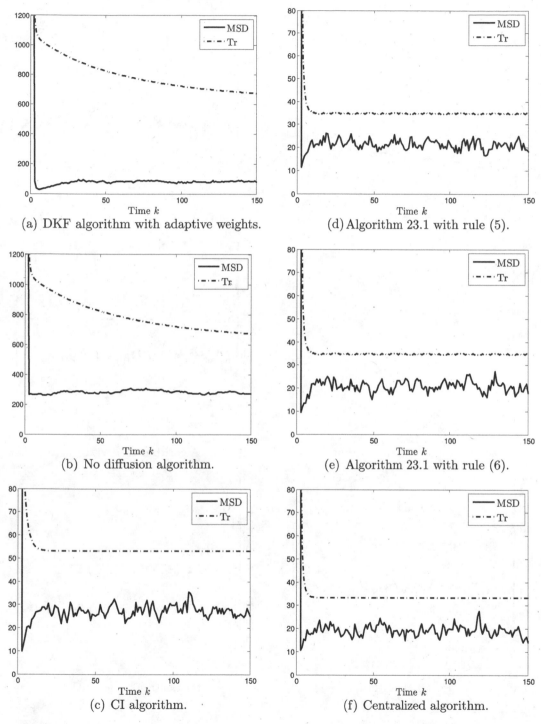

FIGURE 20.6
Results of Scenario I: full local uniform observability.

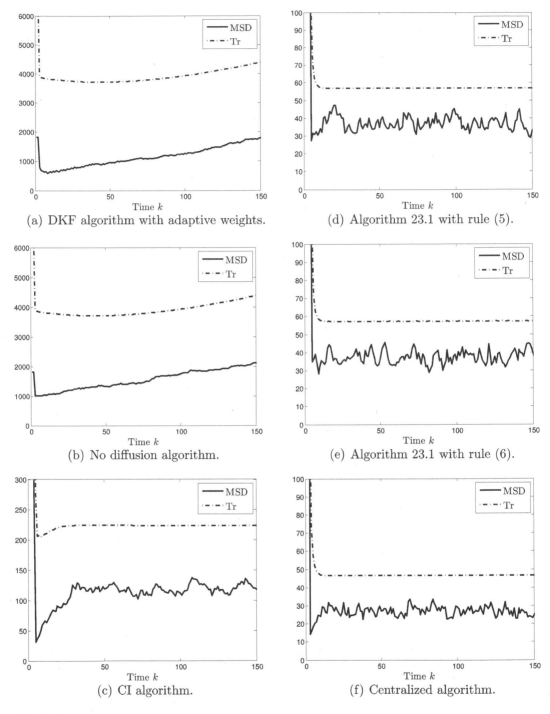

(a) DKF algorithm with adaptive weights.

(b) No diffusion algorithm.

(c) CI algorithm.

(d) Algorithm 23.1 with rule (5).

(e) Algorithm 23.1 with rule (6).

(f) Centralized algorithm.

FIGURE 20.7
Results of Scenario II: partial local uniform observability.

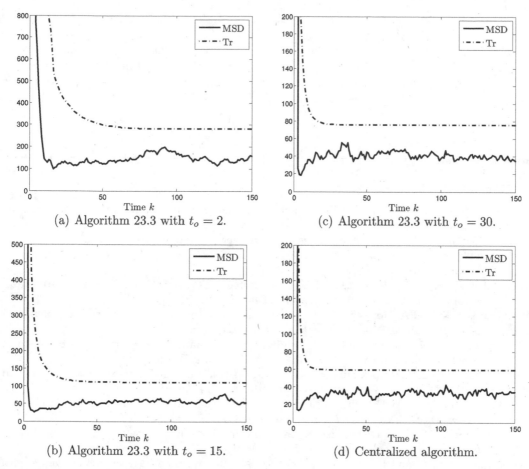

(a) Algorithm 23.3 with $t_o = 2$.

(c) Algorithm 23.3 with $t_o = 30$.

(b) Algorithm 23.3 with $t_o = 15$.

(d) Centralized algorithm.

FIGURE 20.8
Results of Scenario III: no local uniform observability.

20.5 Conclusions

In this chapter, a CI-based diffusion Kalman filtering (CI-DKF) algorithm is proposed by incorporating the covariance information. The CI-DKF algorithm can be applied in the case of lacking local observability. A consensus-based information diffusion scheme is embedded when no single agent can observe the state. Simulation shows that the CI-DKF algorithm has better performance than that by the original DKF and those by local Kalman filters.

Appendix

The network topology of N agents with limited communication range Rc is modeled by a time-varying undirected graph G_k which consists of a constant vertex set $V = \{1, 2, \ldots, N\}$

and a time-varying edge set $E_k = \{\{i, j\} : i, j \in V; //\mu_{i,k} - \mu_{j,k}// \leq Rc\}$ at each time step k, i.e., $G_k = (E_k, V)$. For $i, j \in V$ and $i \neq j$, $\{i, j\} \in E_k$ is an unordered pair of vertices. G_k is connected at time k if for any two vertices i and j there exists a sequence of edges (a path) $\{i, \nu_1\}, \{\nu_1, \nu_2\}, \ldots, \{\nu_{n-1}, \nu_n\}, \{\nu_n, j\}$ in E_k. Let $N_{i,k} = \{j \in V \mid \{i, j\} \in E_k\} \cup \{i\}$ denote the set of neighbors of vertex i at each time k, where a vertex is assumed to be a neighbor of itself. The degree (number of neighbors) of vertex i at time k is denoted as $d_{i,k} = |N_{i,k}|$.

References

1. Anderson, B. and J. Moore 1981. Detectability and stabilizability of time-varying discrete-time linear systems. *SIAM Journal on Control and Optimization* 19 (1), 20–32.
2. Cattivelli, F. and A. Sayed 2009. Diffusion distributed Kalman filtering with adaptive weights. In *2009 Conference Record of the 43rd Asilomar Conference on Signals, Systems and Computers*, Pacific Grove, CA, USA.
3. Cattivelli, F. and A. Sayed 2010. Diffusion strategies for distributed Kalman filtering and smoothing. *IEEE Transactions on Automatic Control* 55 (9), 2069–2084.
4. Franken, D. and A. Hupper 2005. Improved fast covariance intersection for distributed data fusion. In *International Conference on Information Fusion*, pp. 7.
5. Julier, S. and J. Uhlmann 1997. A non-divergent estimation algorithm in the presence of unknown correlations. In *The American Control Conference*, Volume 4.
6. Niehsen, W. 2002. Information fusion based on fast covariance inter-section filtering. In *International Conference on Information Fusion*, pp. 901–904 Volume 2.
7. Olfati-Saber, R. 2005. Distributed Kalman filter with embedded consensus filters. In *The 44th IEEE Conference on Decision and Control*, pp. 8179–8184.
8. Olfati-Saber, R. 2007. Distributed Kalman filtering for sensor networks. In *The 46th IEEE Conference on Decision and Control*, pp. 5492–5498.
9. Olfati-Saber, R. and R. M. Murray 2004. Consensus problems in networks of agents with switching topology and time-delays. *IEEE Transactions on Automatic Control* 49 (9), 1520–1533.
10. Ren, W., R. Beard, and E. Atkins 2007. Information consensus in multivehicle cooperative control. *IEEE Control Systems Magazine* 27 (2), 71–82.
11. Wang, X., V. Yadav, and S. Balakrishnan 2007. Cooperative UAV formation flying with obstacle/collision avoidance. *IEEE Transactions on Control Systems Technology* 15 (4), 672–679.
12. Wang, Y. and X. R. Li 2009. A fast and fault-tolerant convex combination fusion algorithm under unknown cross-correlation. In *International Conference on Information Fusion*, pp. 571–578.
13. Xiao, L., S. Boyd, and S. Lall 2005. A scheme for robust distributed sensor fusion based on average consensus. In *International Symposium on Information Processing in Sensor Networks*, pp. 9.

21

Cooperative Integrity Monitoring for Vehicle Localization: A GNSS/DSRC Integrated Approach

Jiang Liu and Baigen Cai

CONTENTS

21.1 Introduction

The connected vehicles solution, which enables various wireless connectivities for the vehicles, is expected to be the next frontier for an automotive revolution and a key to the evolution to the next-generation intelligent transportation systems (ITSs) [1]. Like other important instantiations of the Internet of Things (IoT), the Internet of Vehicles (IoV) enables a distributed transport fabric capable of making its own decisions about driving customers to destinations [2]. Under the cooperative operation environment, any task of the vehicle is based on its moving state and the driving situation that describes its relationship with neighborhood vehicles and other transport system elements. Therefore, accurate and trustworthy positioning has been a challenge facing the implementation of intelligent mobility [3]. In recent years, GNSS (global navigation satellite system) has been widely used in IoV applications and advanced systems to provide the positioning, navigation, and timing (PNT) services. The increasing cost efficiency of GNSS receivers makes it possible to achieve a high precision level with mass market devices. In addition, many technologies have been introduced in IoV to compensate the drawbacks of GNSS, that is, signal-in-space (SIS) unavailability, multi-path and signal interference. Different from conventional integrated positioning solutions, V2X communication has been improved to provide an additional capability of positioning using dedicated short-range communication (DSRC), including the received signal strength (RSS) ranging and carrier frequency offset (CFO) range-rating solutions [4,5].

For the GNSS-based PNT solutions, the measure of trust that can be given to the derived positioning results, which refers to the integrity, has to be highly concerned, especially for safety-related applications such as self-driving and collision prevention. Different from the ground/satellite-based augmentation solutions that require external integrity data, the receiver autonomous integrity monitoring (RAIM) technique provides a simple and cost-efficient approach for integrity monitoring and quantifies the concept of using the self-consistency check of redundant measurements [6]. Great efforts have been made in developing advanced RAIM algorithms and improving the performance of the snapshot residual-based (RB) and filter innovation-based (IB) solutions [7]. However, the availability of conventional RAIM scheme for fault detection and exclusion (FDE) is constrained by the requirement of satellite visibility. Some sensor-aided RAIM solutions, like the inertial sensor-based approach [8], have been proposed to break the limitations. These approaches may fail to achieve an enhanced RAIM capability in a cost-efficient way due to the requirement of additional sensors.

In this chapter, we use the DSRC range-rating information to realize cooperative integrity monitoring. Extended GNSS measurements based on DSRC and map data are utilized under a solution separation RAIM scheme. Without relying on additional sensors, the availability of FDE is improved even under challenged GNSS observing conditions. Simulations are involved to illustrate the potentials in future advanced transportation system applications.

21.2 Architecture of Cooperative Vehicle Positioning

The general architecture of a cooperative vehicle positioning system is shown in Figure 21.1. Different from conventional on-board subsystem solutions in cooperative applications, the DSRC transceiver is expected to be an effective sensor that provides position-related measurement beyond the role of the wireless communication node. Thus, the system input consists of sensor data including raw satellite pseudo-ranges, DSRC range rates, and the data from the road map database, which can be seen as a "static sensor," provides useful prior information for cooperative positioning. As the output of the system, we determine the state of the vehicle in terms of the coordinate location, the occupied road/lane segment, and the integrity indicator representing the satisfactory of integrity requirements of the users. The core of this system is the position-processing unit with an integrity monitoring

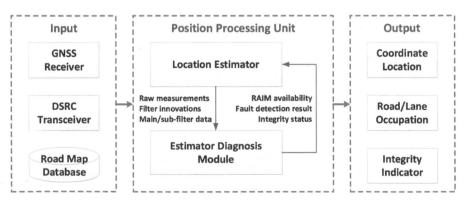

FIGURE 21.1
Block diagram of the cooperative positioning architecture.

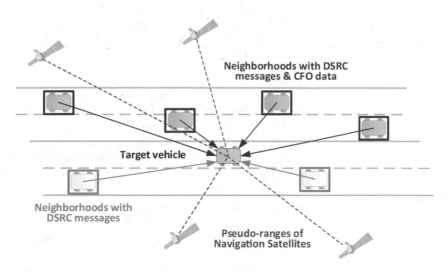

FIGURE 21.2
Principle of DSRC-enabled cooperative positioning and integrity monitoring.

capability. This unit mainly involves two parts, including the location estimator and the integrity module.

As can be seen in the figure, the location estimator is in charge of position computing, using specific state estimation algorithms with multi-source data inputs. Different from ordinary solutions that only focus on the estimator, the presented solution in this research involves a diagnosis module monitoring the integrity status of the positioning system in real time, which adjusts the output using the integrity indicator according to the fault detection results. When faults or failures are detected by this module, a fault exclusion operation will be activated. If fault exclusion is not adopted, the estimation results will be marked as invalid and switched off in the output messages.

With this architecture, vehicle positioning is achieved without relying too much on additional sensors and components of the on-board equipment. Each vehicle within the vehicular ad-hoc network (VANET) environment can benefit from the cooperation with its neighborhood vehicles, as in Figure 21.2, and the capability of DSRC is deeply exploited in ensuring a higher integrity level over conventional solutions.

21.3 GNSS/DSRC-Based Cooperative RAIM Solution

21.3.1 DSRC-Assisted GNSS Measurement Model

This is a well-known 3D GNSS measurement model using the pseudo-range set from N visible satellites, where the total number of satellites fulfills $N \geq 4$. With the ith satellite's 3D coordinate location $\boldsymbol{p}^{(i)}(t-\tau)$, the model can be written as

$$\rho^{(i)}(t) = \left\| \boldsymbol{p}_u(t) - \boldsymbol{p}^{(i)}(t-\tau) \right\|_2 + c \cdot (\delta l_u(t) - \delta l^{(i)}(t-\tau)) + \xi^{(i)}(t)$$
$$i = 1, \ldots, N(t)$$

$$(21.1)$$

where $\rho^{(i)}(t)$ denotes the pseudo-range from the ith satellite, $\boldsymbol{p}_u(t)$ is 3D location of the GNSS antenna in the target vehicle, c indicates the light speed, the components $\delta l_u(t)$ and $\delta l^{(i)}(t-\tau)$ represent the user clock bias and the satellite clock offset respectively, and $\xi^{(i)}(t)$ denotes the range measurement noise.

In order to determine the 3D location and the clock error, at least four satellites are required to be visible at instant t. The implementation of multi-constellation scheme makes great contributions to an enhanced availability of GNSS positioning services. However, the problem of signal-in-space (SIS) unavailability still exits due to constraints from the urban environment, which leads to limited availability of the RAIM technique under specific vehicular application scenarios. In order to release the limitations in RAIM operations, assistant information from DSRC is considered in our design, where DSRC CFO-enabled range-rate data from multiple neighborhood vehicles are integrated to expand the GNSS measurement set.

To build pseudo-range-like measurements for an estimator using the current estimation architecture, DSRC measurements and the road map data are utilized to rebuild the pseudo-range of an invisible satellite. With the estimation result $\hat{\boldsymbol{p}}_u(t-1)$ at instant $t-1$, the current location within the map-defined coordinate is predicted as

$$\tilde{\boldsymbol{p}}_u(t) = \arg map(\hat{s}_{u,\mathrm{MM}}(t-1) \pm \Delta \tilde{s}_u(t,t-1)) \tag{21.2}$$

where $map(*)$ is a function projecting the 3D position frame into 1D road/lane coordinate, $\hat{s}_{u,\mathrm{MM}}(t-1)$ indicates the map matched result of the last estimate $\hat{\boldsymbol{p}}_u(t-1)$ using a specific principle like the perpendicular projection, $\Delta \tilde{s}_u(t,t-1)$ is the 1D location increment during the time interval from $\hat{s}_{u,\mathrm{MM}}(t-1)$, which is calculated using a constant acceleration kinematical model.

Taking the CFO-based range-rates corresponding to the cooperative neighborhoods into account, the predicted result $\tilde{\boldsymbol{p}}_u(t)$ can be calibrated by DSRC measurements to obtain a reliable estimation $\tilde{\boldsymbol{p}}_u^{\mathrm{D}}(t)$, where the DSRC range-rate $\dot{r}^{(k)}(t)$ is modeled as

$$\dot{r}^{(k)}(t) = \frac{dr^{(k)}(t)}{dt} = \frac{d\left(\left\|\tilde{\boldsymbol{p}}_u^{\mathrm{D}}(t) - \boldsymbol{p}_{\mathrm{nv}}^{(k)}(t)\right\|_2 + \varepsilon^{(k)}(t)\right)}{dt} \tag{21.3}$$
$$k = 1, \ldots, Q(t)$$

where $\boldsymbol{p}_{\mathrm{nv}}^{(k)}(t)$ is the 3D location of the kth neighborhood vehicle, $\varepsilon^{(k)}(t)$ represents the range-rating error, and $Q(t)$ indicates the total number of the involved neighborhood vehicles.

A specific nonlinear estimator is utilized to precalculate the location $\tilde{\boldsymbol{p}}_u^{\mathrm{D}}(t)$ with the prediction $\tilde{\boldsymbol{p}}_u(t)$ and range-rate observations $\{\dot{r}^{(k)}(t)\}$. Thus, by using the GNSS ephemeris data, the pseudo-range of a specific blocked satellite j can be simulated with a similar model as (21.1), which means

$$\rho^{(j)}(t) = \left\|\tilde{\boldsymbol{p}}_u^{\mathrm{D}}(t) - \boldsymbol{p}^{(j)}(t)\right\|_2 + \xi^{(j)}(t) \tag{21.4}$$
$$j = 1, \ldots, M(t)$$

where pseudo-range measurements of totally $M(t)$ invisible satellites are simulated to realize an extended measurement vector $\boldsymbol{y}_t = (\ldots, \rho^{(i)}(t), \ldots \rho^{(j)}(t), \ldots)^{\mathrm{T}}$ for integrity monitoring.

An optimization method is necessary to determine which satellite's pseudo-range should be rebuilt. The HDOP of the extended constellation with $N(t)+M(t)$ satellites is adopted as the objective index to improve the satellite geometry and the corresponding protection levels in integrity monitoring.

21.3.2 Solution Separation-Based RAIM Method

To derive the position-domain solution separation RAIM approach, we first concern the location estimator involved in the position processing unit. Based on the extended measurement model, a nonlinear cubature Kalman filter (CKF) is adopted to realize the state estimation, where the system model is built as

$$\begin{cases} x_t = f(x_{t-1}) + \omega_t \\ y_t = h(x_t) + \xi_t \end{cases} \tag{21.5}$$

where x_t denotes the state vector as $x_t = (x_t, y_t, z_t, \dot{x}_t, \dot{y}_t, \dot{z}_t, \Delta l_t)^{\mathrm{T}}$, Δl_t is clock error, $f(*)$ and $h(*)$ represent nonlinear system and measurement functions determined by the models, ω_t and ξ_t are independent system and measurement noise vectors.

The CKF allows us to achieve a systematic solution for high-dimensional nonlinear filtering problems using a third-degree spherical-radial cubature rule [7]. The estimation is derived combining the state prediction and the weighted innovation as

$$\hat{x}_{t|t} = \hat{x}_{t|t-1}^{-} + K_t r_t = \hat{x}_{t|t-1}^{-} + K_t(y_t - h(\hat{x}_{t|t-1}^{-})) \tag{21.6}$$

Conventional innovation-based RAIM method is designed based on the assumption that the normalized innovations $\tilde{r}_t(i) = r_t(i)/\sqrt{W(i)}$ follow the Gaussian distribution [8]. With the normalized innovation square (NIS), the test statistic is usually built as $T_{\mathrm{IB},t} = r_t^{\mathrm{T}} W_t^{-1} r_t$ under a Chi-square distribution.

Different from conventional innovation-based RAIM considering the measurement set from all-in-view satellites, the solution separation method is drawn from the difference of the components of interest in state estimation [9]. Based on the CKF-based architecture, a solution separation vector is built using a main filter and $N(t)+M(t)$ sub-filters. The nth sub-filter refers to the nth measurement-excluded case, and that means a reduced measurement vector $y_{i,t}$ with only $N(t)+M(t)-1$ pseudo-ranges. Thus, the solution separation vector $\delta x_{n,t}$ corresponding to the sub-filter n is

$$\delta x_{n,t} = \hat{x}_{0,t} - \hat{x}_{n,t}, \quad n = 1, \ldots, N(t) + M(t) \tag{21.7}$$

where $\hat{x}_{0,t}$ and $\hat{x}_{n,t}$ represent the estimates using a full measurement set and the nth measurement-excluded set, respectively.

We consider the horizontal position of the vehicle. Hence the horizontal test statistic is defined using two components in $\delta x_{n,t}$ as

$$T_{\mathrm{SS},n,t} = \sqrt{(\delta x_{n,t}(1))^2 + (\delta x_{n,t}(2))^2} \tag{21.8}$$

where $\delta x_{n,t}(i)$ indicates the ith component of the solution separation vector $\delta x_{n,t}$.

According to the definition of solution separation vector, distribution of $T_{SS,n,t}\,|\,H_0$ is normal under the fault-free hypothesis H_0. Hence, the threshold $T_{D,n}$ for fault detection corresponding to the nth sub-filter can be derived using a false alarm probability p_{FA}, where the probability allocation principle is adopted to all the sub-filters.

$$T_{D,n} = \sigma_{\max,n} \cdot T_d \tag{21.9}$$

where $\sigma_{\max,n}$ denotes the maximum standard deviation of horizontal estimation errors of the sub-filter, T_d is an original detection threshold, and they are calculated as

$$\sigma_{\max,n} = \max_{ii}\left\{\sqrt{P_{n,t,ii}}\right\}, \quad ii = 1,2 \tag{21.10}$$

$$\frac{p_{FA}}{N(t)+M(t)} = P(\lambda > T_d) = \int_{T_d}^{\infty} \lambda e^{-(\lambda^2/2)} d\lambda \tag{21.11}$$

where $P_{n,t,ii}$ denotes the ith diagonal element of the error covariance $\boldsymbol{P}_{n,t}$, p_{FA} is the given false alarm probability, and T_d is derived by the Rayleigh distribution according to the definition of test statistic $T_{SS,n,t}$ as (21.8).

With all the derived test statistics $\{T_{SS,n,t}\}$, a fault may be detected when any $T_{SS,n,t}$ exceeds its threshold $T_{D,n}$. In other words, it is determined that there is no fault only when all the test statistics satisfy the condition $T_{SS,n,t} \leq T_{D,n}$. Before we perform fault detection, RAIM availability has to be identified first. The horizontal protection level (HPL) is calculated and compared with the given horizontal alert limit (HAL). HPL consists of a maximum threshold of sub-filter test statistics and the protection level under the single-fault case, which means

$$\mathrm{HPL}_t = \max_{n}(T_{D,n}) + \mathrm{HPL}_{0,t}$$
$$n = 1,\dots,N(t)+M(t) \tag{21.12}$$

where $\mathrm{HPL}_{0,t}$ is determined by p_{HMI}, the probability of the hazardously misleading information (PHMI), and the standard deviation of the horizontal position error from the main filter. It is updated as the following expression.

$$\mathrm{HPL}_{0,t} = \sqrt{\sigma_x^2 + \sigma_y^2} \cdot Q^{-1}\left(1 - \frac{p_{HMI}}{2p_{H_1}}\right) \tag{21.13}$$

where $Q^{-1}(*)$ represents the inverse of cumulative distribution function of a Gaussian distribution $N(0,1)$, and p_{H_1} is the prior probability of the single fault hypothesis.

21.3.3 Summary and Discussions

According to the aforementioned methods and steps, the flow chart of the presented cooperative integrity monitoring solution can be summarized as Figure 21.3. It can be seen that the multiple sub-filters are involved in the original CKF-based position estimation, and the fault detection takes place between the measurement update of the main filter and

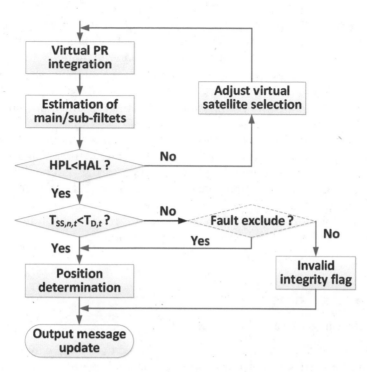

FIGURE 21.3
Flow chart for DSRC-enabled cooperation integrity monitoring.

the position determination step. Only when the fault detection does not raise an alarm with the comparison between $T_{SS,n,t}$ and $T_{D,n}$, the extended measurements, from both the practically visible satellites and DSRC-enabled virtual satellites, are utilized in achieving a final position estimation result.

As can be seen in Figure 21.3, the DSRC-enabled measurement extension and the introduction of the CKF estimators make it possible of performing RAIM calculation effectively under different satellite observing conditions with the solution separation method. The availability of fault detection is significantly enhanced over conventional RAIM solutions. At the same time, performance of integrity monitoring is guaranteed with a relatively conservative fault detection and measurement isolation principle. The fault exclusion step is given in the flow chart, but it is not discussed in this paper since it requires further level sub-filters to identify the fault measurement if we can afford an increased computation cost. To summary the presented solution, the distinct features are given as follows.

1. Compared to traditional RAIM solutions using GNSS measurement only or the integrated measurements from additional sensors such as the inertial unit, odometer, and radar, this solution takes advantage of the V2V cooperation through DSRC channels. No additional sensor is required in the fundamental cooperative on-board system with GNSS and DSRC devices.

2. This solution breaks the inherent limitation of ordinary RAIM methods, where at least five visible satellites are required for fault detection. DSRC enables virtual satellite observations and greatly improves the availability of RAIM operations.

3. The fault exclusion is an optional step in this solution. In order to ensure safe and trustworthy positioning, this solution opts to strictly isolate risky measurements if "fault detected" is returned to the position-processing logic.

4. This solution establishes an open architecture for involving additional sensors. When redundant sensor information is available, this solution can be easily adopted simply by adjusting the measurement model for the main/sub-filters.

It should be noted that the continuity of the positioning function under this RAIM architecture would be affected due to a conservative measurement isolation principle. Furthermore, performance of cooperative integrity monitoring will be determined by the quality of the DSRC CFO data and the penetration rate of the vehicular cooperative mode. Accurate DSRC range-rate results and a high V2V penetration rate, which indicates a good geometry distribution of neighborhood vehicles, will greatly contribute to the achieved availability and performance level of the RAIM calculation.

21.4 Simulation and Analysis

In order to demonstrate the performance of the presented cooperative integrity monitoring solution, simulations are carried out based on the real road geometry in the city of Langfang in China. Vehicle trajectory from the field experiments is taken to build the original scenario. Using a specific traffic simulator, the neighborhood vehicle sets of a target vehicle are obtained, and the DSRC range-rate measurements are simulated with a fitted error model from field tests. For the selected trajectory, only four GPS satellites are observable during the whole operation, which lasts 120 s, which means conventional RAIM solutions fail to calculate the test statistic and realize FDE. The involvement of DSRC CFO data and the road map database makes it possible to carry out RAIM operations by establishing extended satellite pseudo-range measurements for four invisible satellites. Therefore, totally eight satellites are "visible" with the presented solution under DSRC-based the cooperative operational environment. A step fault with a bias level of 100 m is injected to the original pseudo-range of a satellite. For comparison purposes, the fault-free scenario is also involved in the simulation.

Figure 21.4 depicts the fault detection results of all the sub-filters under the fault-free scenario, and Figure 21.5 shows the results under the fault-injected case, where sub-filter 3 corresponds to a measurement subset that excludes a satellite (PRN [pseudo-random noise] number = 25) with a step fault of 100 m from the instant $t = 60$. The sub-filters 5–8 denote that the subsets exclude a virtual satellite enabled by the DSRC and road map data.

According to the equations for evaluating the HPL, Figures 21.6 and 21.7 show the comparison results of HPE and HPL under both the fault-free and fault-injected cases. Different components of HPL as (21.12) are given in these figures to show the capability of the presented approach in bounding the horizontal errors.

To further evaluate the performance of the presented solution, a wide range of bias levels is considered to check the performance of fault detection. By injecting different bias errors ranging from 10 to 100 m from the instant $t = 60$, the latency of detection, which denotes a time delay from the beginning of the injection to the instant when the test statistic firstly exceeds its detection threshold, and the missed-detection rate have been summarized as Figures 21.8 and 21.9.

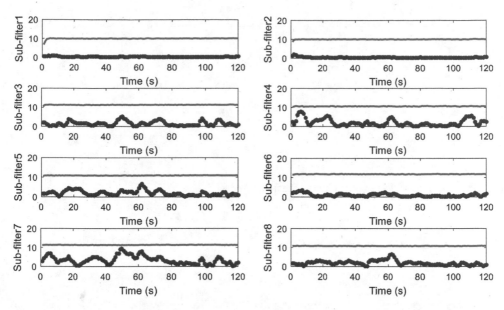

FIGURE 21.4
Test statistics of all the sub-filters under the fault-free case.

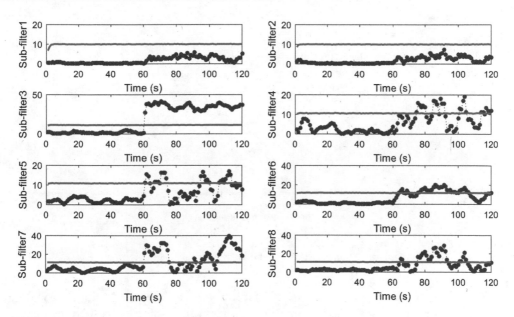

FIGURE 21.5
Test statistics of all the sub-filters under the fault-injected case.

Based on the results in these figures, some discussions are given as follows.

1. The presented solution is capable of achieving RAIM calculation under challenged GNSS observing conditions, where the operation of the conventional RAIM methods is constrained. The extended pseudo-range measurement set using DSRC and the road map data provide a great deal of help to utilizing the solution separation RAIM algorithm.

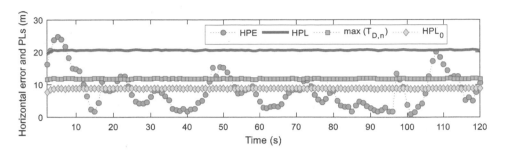

FIGURE 21.6
Horizontal error and protection levels under the fault-free case.

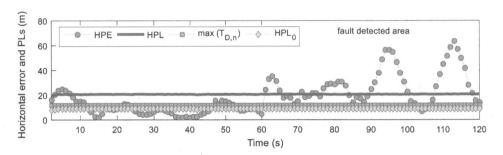

FIGURE 21.7
Horizontal error and protection levels under the fault-injected case.

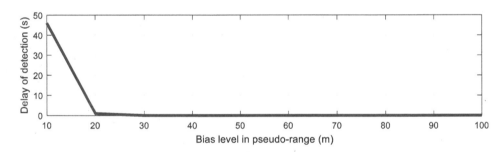

FIGURE 21.8
Comparison of delay of fault detection under different bias fault levels.

2. False alarm does not occur under the fault-free case using this solution. When a fault is injected, test statistics from the corresponding sub-filter, which means the sub-filter 3 as Figure 21.5, exceeds its threshold obviously. It validates the design of the test statistic using the error between estimations with all-set and sub-set measurements.

3. By integrating the two different components as (21.12), both the derived HPLs under the fault-free and fault-injected cases provide effective over-bounds to horizontal errors. It should be noted that HPL in the "fault detected area" in Figure 21.7 does not actually exist according to the fault detection and response strategy, where the estimation results from the instant $t = 61$ s have been marked as invalid and switched off. It is obvious that the continuity of the positioning system is sacrificed to preventing from

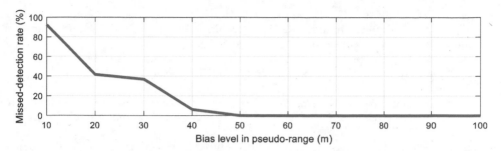

FIGURE 21.9
Comparison of missed-detection rate under different bias fault levels.

falling into the hazardous misleading zone. Hence, safe and trustworthy positioning is ensured to satisfy the specific requirements in practical applications.

4. From the evaluations under different bias levels, good detectability of the step fault is illustrated. There is no delay and missed detection when the bias exceeds 50 m.

21.5 Conclusion and Future Plans

A cooperative GNSS integrity monitoring method using DSRC/map-enabled extended pseudo-ranges has been proposed. We designed the solution of establishing virtual satellite measurements under the cooperative environment. Moreover, the solution separation RAIM scheme is established to utilize this solution. Simulations under the GNSS-challenged condition illustrate the capability in fault detection under a single fault case. In future works, the fault exclusion logic under this scheme will be studied. Multi-fault hypothesis also will be examined and tested with experiments.

Acknowledgment

This research was supported by Beijing Natural Science Foundation (4182053) and National Natural Science Foundation of China (61873023, 61403021).

References

1. Lu N, Cheng N, Zhang N, Shen X, Mark JW. Connected vehicles: Solutions and challenges. *IEEE Internet of Things Journal* 1(4), 2014: 289–299.
2. Gerla M, Lee E-K, Pau G, Lee U. Internet of vehicles: From intelligent grid to autonomous cars and vehicular clouds. In: *2014 IEEE World Forum on Internet of Things (WF-IoT)*, Seoul, South Korea, 2014: 241–246.

3. Meng X, Roberts S, Cui Y, Gao Y, Chen Q, Xu C, He Q, Sharples S, Bhatia P. Required navigation performance for connected and autonomous vehicles: Where are we now and where are we going? *Transportation Planning and Technology* 41(1), 2018: 104–118.

4. Alam N, Balaei T, Dempster A G. A DSRC doppler-based cooperative positioning enhancement for vehicular networks with GPS availability. *IEEE Transactions on Vehicular Technology* 60(9), 2011: 4462–4470.

5. Ansari K, Wang C, Wang L, Feng Y. Vehicle-to-vehicle real-time relative positioning using 5.9 GHz DSRC media. In: *2013 IEEE 78th Vehicular Technology Conference (VTC Fall)*, Las Vegas, NV, 2013: 1–7.

6. Rakipi A, Kamo B, Cakaj S, Kolici V, Lala A, Shinko I. Integrity monitoring in navigation systems: Fault detection and exclusion RAIM algorithm implementation. *Journal of Computer and Communications* 3, 2015: 25–33.

7. Grosch A, Crespillo O G, Martini I, Gunther C. Snapshot residual and Kalman filter based fault detection and exclusion schemes for robust railway navigation. In: *2017 European Navigation Conference*, Lausanne, Switzerland, 2017: 36–47.

8. Zhao L, Huang P, Li L, Liu J, Yang Y. RAIM analysis of GPS/DR tightly coupled system for vehicle navigation. In: *2013 China Satellite Navigation Conference (CSNC 2013)*, Wuhan, China, 2013: 1–6.

9. Blanch J, Ene A, Walter T, Enge P. An optimized multiple hypothesis RAIM algorithm for vertical guidance. In: *ION GNSS 2007*, Fort Worth, TX, 2007: 2924–2933.

22

Flexible and Steady Control for Cooperative Target Observation

Miguel Aranda and Youcef Mezouar

CONTENTS

This chapter considers the problem of coordinating multiple mobile robots to sense and track a moving target. For this task, it is equally fundamental to take into account the quality of the acquired target localization information and the suitability of the motions of the robotic team. A multirobot control methodology that simultaneously addresses these two concerns is described. Accurate and complete cooperative observation of the target is obtained by driving the robots to achieve a set of prescribed relative viewing angles, encapsulated by a default desired enclosing pattern, with respect to the target. In particular, relative robot position regulation and efficient target tracking are integrated via a formation-based controller that relies on global information and incorporates an optimal pattern rotation. The control framework allows each robot to select freely its desired distance to the target. By doing so, the robots can optimize their individual perception quality and avoid collisions during navigation without affecting the desired cooperative target observation diversity. It is shown that even with these distributed distance adjustments, the team movements remain steady with the proposed controller. This noteworthy property results in efficient agent robot motions and facilitates navigation and stable perception of the target by the team. An additional advantage of the described controller is that each robot can implement it using its independent local reference frame. Simulation tests in different cooperative target enclosing and navigation scenarios are presented to illustrate the performance of the methodology.

22.1 Introduction

It is well known that a multirobot system can allow completion of many real-world tasks efficiently and reliably. It is, by definition, more powerful and capable than a single robot. Considerable efforts have therefore been dedicated to exploring what specific coordination mechanisms can make the most of these systems' potential. Particular areas of interest have been cooperative sensing or perception, target tracking, or navigation for transportation tasks. This chapter concerns specifically the problem of tracking a mobile element (i.e., a *target*) with a team of robots. We describe recent research advances in control design that provide the robots with the ability to collectively perceive the target with a suitable quality, while preserving the team's ability to move with flexibility and steadiness. Such integration of desirable traits on both counts (perception and motion) is the most relevant feature of the presented contents.

In order to observe a phenomenon, it is advantageous to have different viewpoints provided by multiple sensors placed at diverse positions; this can prevent occlusions and allows improvement of the quality of perception via data integration (e.g., position triangulation). With these multi-view perception capabilities one can generate a suitable model of an object that needs to be, for example, manipulated, or transported by a robotic system. Surveillance and escorting problems can also greatly benefit from these capabilities. Motion capture systems [11] are an important multi-sensor application along these lines. However, the sensors in motion capture setups are normally static, covering an operational space that is fixed.

Considering other more general potential application scenarios, it interesting to extend the concept of motion capture to highly dynamic targets moving across non-fixed environments; in this case, it is required to use sensors mounted on robots, which have to move to concurrently observe the target and track its motion [27]. There are multiple constraints that are essential to consider in order to solve the resulting tracking/observation problem. One type of constraint concerns the need to identify and prescribe team geometries and motion policies that optimize the quality of the collective perception of the target. There has been substantial work in this area. The approaches that have been introduced typically focus on uncertainty minimization using metrics such as mutual information or Fisher information [6,13,22,23,28,29,31].

However, aside from such target perception constraints, there are other equally essential requirements: these concern the suitability of the robots' motions, at both the individual and team levels. The robots' relative positions must always be such that they can perceive and/ or communicate with each other and avoid colliding with other robots or the environment. Formation control is a vast of field of research devoted to addressing these issues. It deals with controlling robot teams towards suitable arrangements called formations, defined in general by distances and/or angles [7,9,12,17,19,24,26,30]. Still, standard formation controllers focus on steering the robots' relative states but do not address collective perception and target tracking objectives, as in this chapter. For flexible motion and optimal target perception, exploring alternatives beyond the scope of reaching a fixed-shape formation can be clearly advantageous. This is the approach taken in the methodology described here.

Existing works, for example, [3,20,25], have already explored the use of formation-based formalisms for target observation. These motion control frameworks have lower dimensionality and/or adaptability of the robotic team's geometry compared with the method described in this chapter. One can also use strategies based on persistent motion, such as target encirclement, to solve the tracking problems we consider. Approaches along these lines include [8,16,21]. In these methods, however, the robots remain gyrating and thus the team does not move steadily. Steady motions—which the method described here generates—can provide increased efficiency and stability of perception data. The control policy presented in the chapter integrates the two goals mentioned above: it facilitates appropriate target perception diversity and allows efficient control of the robots' relative states. It does so by prescribing a default target-enclosing formation, which the robots are driven towards. In particular, the robots move to reach an optimal—in terms of a global shape alignment metric—rotated version of that prescribed formation, while they track the target's motion.

Crucially, the control approach examined in this chapter allows flexibility in the formation shape—to, for example, avoid obstacles, adapt to the size of the environment, or improve perception quality. This is achieved by enabling each individual robot to freely select its desired distance to the target. Even with these individual adjustments, the pursued perceptual diversity is ensured: in particular, this is the case because the robots' relative positions evolve towards a pattern in which they maintain the same relative target-viewing angles as in the default formation. Via the study of the pattern rotation dynamics, we show that the behaviors of the formation and target tracking remain stable and steady, regardless of the distance selection procedure used by the robots. An additional important feature is that the method presented here applies to general 3D motions and can be implemented with local measurements, not requiring common reference frames. This independence of global coordinate references, which provides great advantages in flexibility and simplicity, is known to make team coordination harder to study and guarantee [24].

The contents of the chapter build on work presented in [4]. The chapter is characterized by an emphasis on applying the control approach that is examined to collective navigation tasks. We believe that the approach is particularly useful for these tasks [1,2,5,18], which have been and remain of great importance for practical applications. The key reason is that the *target* entity considered in the proposed controller can be regarded as a *leader* for the purposes of navigation. In this context, several properties of the method reveal themselves as especially interesting, as will be discussed throughout the chapter. We use simulation tests to exemplify the usefulness of the presented control approach. In particular, results from simulation tests of a target enclosing behavior in a 3D setting and a navigation task in a 2D environment are described.

22.2 Problem Formulation

We consider a team of $N > 2$ robots in $\mathbb{R}^3$ modeled as point masses. Each robot is assigned an index $i \in \{1, \ldots, N\}$. We assume the robots move according to a single integrator model, that is, each of them satisfies:

$$\dot{\mathbf{q}}_i = \mathbf{u}_i,$$

(22.1)

where $\mathbf{q}_i \in \mathbb{R}^3$ is robot i's position, and $\mathbf{u}_i \in \mathbb{R}^3$ is its control input. Further, we denote as $\mathbf{q}_t \in \mathbb{R}^3$ the position of the *target*. The robots are tasked with collectively observing this target, which is assumed to displace with finite-norm arbitrary velocity $\mathbf{v}_t$:

$$\dot{\mathbf{q}}_t = \mathbf{v}_t. \tag{22.2}$$

The positions of the robots and the target are expressed in an arbitrary global reference frame. We will refer to relative position vectors using the following notation: $\mathbf{q}_{ij} = \mathbf{q}_i - \mathbf{q}_j = -\mathbf{q}_{ji}$.

The problem addressed consists in ensuring that the target is observed from a suitable diversity of viewpoints all through its movement. Next, we describe the framework we consider. By doing so, we will be able to define the problem in more precise terms. We define a *default desired configuration (or formation)*, which is a reference layout of the N robots in their ambient space. This configuration is encapsulated via relative position vectors, in such a way that $\mathbf{c}_{ji} \in \mathbb{R}^3$, $\forall i, j \in \{1, ..., N\}$ denotes the default desired vector from robot i to robot j. In addition, we denote as $\mathbf{c}_{ti}$ the desired vector from robot i to the target. An objective of the task will be for the robots to *enclose* the target. In agreement with this objective, we assume that the target's desired position is right at the centroid of the desired pattern. Therefore, $\Sigma_{i=1}^{N}\mathbf{c}_{ti} = \mathbf{0}$. For the purposes of controller analysis, we make the assumption that the desired geometry is generic [12] and it has no exact symmetries. In reality, this is not a restrictive constraint; one can modify infinitesimally any starting desired geometry to make it satisfy the constraint.

Given the presented geometric description, we can directly extract from it an angle-based configuration, encoded by the relative angles at which pairs of robots observe the target. Specifically, for any pair of robots $i, j \in \{1, ..., N\}$, their desired relative viewing angle with respect to the target is defined as $\alpha_{ij} = \angle(\mathbf{c}_{it}, \mathbf{c}_{jt})$. Thus, a given desired target-observation diversity can be prescribed by defining a suitable default enclosing formation, as the one we have specified; such a formation directly encapsulates the desired relative target-viewing angles. We illustrate the default formation definition in Figure 22.1.

We define the control goal as follows. The objective is for the system to reach a state where there exist a point $\mathbf{p}_a \in \mathbb{R}^3$, a rotation matrix $\mathbf{R}_a \in SO(3)$, and a set of N scalars $s_{ai} > 0$ such that:

$$\mathbf{q}_i = \mathbf{p}_a + \mathbf{R}_a\mathbf{c}_{it}^{s_{ai}}, \quad \forall i \in \{1, ..., N\}, \tag{22.3}$$

with $\mathbf{c}_{it}^{s_{ai}} = s_{ai}\mathbf{c}_{it}$, and where simultaneously $\mathbf{p}_a$ remains suitably close to $\mathbf{q}_t$.

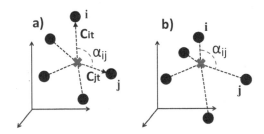

FIGURE 22.1
(a) Default formation of robots (circles) which encloses a target (cross) in 3D space. In (b), the formation has the same relative viewing angles α_{ij} with respect to the target as in (a), with different robot-target distances. Observe that if we apply a given rotation to every vector between the target and a robot, the relative angles α_{ij} for all pairs of robots do not change.

The condition (22.3) implies the desired relative target-viewing angles are achieved with respect to $\mathbf{p}_a$, as shown next. For this fact to hold, the angle—denoted as β_{ij}—between a pair of vectors $\mathbf{q}_i - \mathbf{p}_a$ and $\mathbf{q}_j - \mathbf{p}_a$ must be equal to α_{ij} $\forall i, j$. We rotate all such vectors by the common matrix $\mathbf{R}_a^{-1}$. This clearly does not change the angles between them. We obtain a set of vectors as follows: $\mathbf{v}_i = \mathbf{R}_a^{-1}(\mathbf{q}_i - \mathbf{p}_a) = \mathbf{c}_{it}^{s_{ai}}$ $\forall i$. Then one can see that for each i, the angle between $\mathbf{v}_i$ and $\mathbf{c}_{it}$, which we denote as β_{vci}, is zero, because $\cos(\beta_{vci}) = <\mathbf{v}_i, \mathbf{c}_{it}> /(\| \mathbf{v}_i \| \cdot \|\mathbf{c}_{it}\|) = s_{ai} < \mathbf{c}_{it}, \mathbf{c}_{it} > /(s_{ai}(\| \mathbf{c}_{it}\| \cdot \| \mathbf{c}_{it}\|)) = 1$. Thus, $\beta_{ij} = \alpha_{ij}$ $\forall i$, $j \in \{1, ..., N\}$.

The default desired configuration captures the relative angular constraints that define the problem we consider. Given the fact that the problem's constraints are angular, one may wonder why it can be useful or appropriate to use, as we do, a desired configuration which encodes both angles and distances. The idea is that this configuration represents a pattern that, in the absence of other task constraints, is a preferred configuration for the team to maintain. The reasons can be that the distances in that configuration are such that they facilitate the robots' interactions, and that the team's geometrical shape and size in that configuration are particularly favorable for the specific environment of operation. Nevertheless, the robots can deform the shape of the default configuration while still satisfying the relative angular constraints, as explained in subsequent sections. This capability to deform the team is very relevant as it provides great flexibility in the performance of the observation and navigation tasks addressed.

22.3 Control Strategy

We describe in this section the proposed multirobot control approach. The strategy will allow one to carry out the task of target enclosing for observation according to the definition put forward in the preceding section. A way to solve the addressed problem is to steer the robots towards a geometric pattern that satisfies the condition in Equation 22.3. The default desired configuration is such a pattern. Given that a common rotation of all desired target-robot vectors leaves the angles α_{ij} unchanged, one can try to move the robots towards an *optimally* rotated version of the default formation. By doing so, the efficiency of their motions will be increased. To this end, we define the rotation by solving the problem of optimally aligning two *shapes*: the set of current positions and the set of desired points. A cost function that expresses the error of this alignment can be given by the following sum of squared distances:

$$\gamma = \sum_i \sum_j \left\| \mathbf{q}_{ij} - \mathbf{R}_c \mathbf{c}_{ij} \right\|^2 ,$$
(22.4)

where i, j both go from 1 to N, and $\mathbf{R}_c \in SO(3)$ is a rotation matrix that acts on the desired vectors. This cost function is equivalent to the one considered in orthogonal Procrustes shape alignment problems [10].

22.3.1 Rotation Matrix Computation

The rotation matrix is a key element of the presented control method. It enables the goals of optimality—because it minimizes γ—and independence of global coordinate systems, as will be detailed later. We can stack the position vectors between robots and obtain the following $N^2 \times 3$ matrices:

$$
\begin{aligned}
\mathbf{Q} &= [\mathbf{q}_{11} \cdots \mathbf{q}_{1N} \; \mathbf{q}_{21} \cdots \mathbf{q}_{2N} \cdots \mathbf{q}_{N1} \cdots \mathbf{q}_{NN}]^T \\
\mathbf{C} &= [\mathbf{c}_{11} \cdots \mathbf{c}_{1N} \; \mathbf{c}_{21} \cdots \mathbf{c}_{2N} \cdots \mathbf{c}_{N1} \cdots \mathbf{c}_{NN}]^T .
\end{aligned}
\tag{22.5}
$$

Let us define the matrix $\mathbf{A} = \mathbf{C}^T \mathbf{Q}$ from these two sets. For optimality, we wish to find $\mathbf{R}$ such that when one chooses $\mathbf{R}_c = \mathbf{R}$ in Equation (22.4), γ takes it minimum possible value for two given matrices $\mathbf{Q}$ and $\mathbf{C}$. It is known that this rotation can be computed using the Kabsch algorithm [14], which employs the singular value decomposition (SVD) of $\mathbf{A}$, $\mathbf{A} = \mathbf{U}\Lambda\mathbf{V}^T$, as follows:

$$
\mathbf{R} = \mathbf{V}\mathbf{D}\mathbf{U}^T = \mathbf{V}
\begin{pmatrix}
1 & 0 & 0 \\
0 & 1 & 0 \\
0 & 0 & d
\end{pmatrix}
\mathbf{U}^T ,
\tag{22.6}
$$

where $d = sign(det(\mathbf{V}\mathbf{U}^T))$. The solution thus obtained for $\mathbf{R}$ is unique unless $rank(\mathbf{A}) < 2$ or $\mathbf{A}$ has a degenerate smallest singular value [15]. This uniqueness will be a precondition of our stability analysis presented in a subsequent section. Each robot follows the described operations to obtain at each time instant this rotation matrix, which is equal to all of them, and uses it to compute its control law.

22.3.2 Selection of Desired Robot-Target Distances

A salient aspect of the method proposed is described next. Specifically, we propose to enable each robot to select independently its desired distance to the target. This is an attractive capability: it can allow, for example, optimization of target perception quality and enhancement of navigation performance. In particular, it can be used to avoid collisions with environmental obstacles and to increase the target's comfort and safety during navigation. Thus, a robot i can choose to define a distance from the target equal to $\| \mathbf{c}_{ti}^s \| = s_i \| \mathbf{c}_{ti} \|$, at which the robot wants to position itself. $s_i > 0$ is an upper-bounded and constant factor, which we term the *control scale* of i. This scalar is chosen freely by i and is unknown to all other robots.

22.3.3 Control Law

Every robot $i \in \{1, \ldots, N\}$ moves according to the following closed-loop control law:

$$
\mathbf{u}_i = \dot{\mathbf{q}}_i = K_c \left(\mathbf{q}_{ti} - \mathbf{R}\mathbf{c}_{ti}^s \right),
\tag{22.7}
$$

where $K_c > 0$ is a control gain and $\mathbf{c}_{ti}^s = s_i \mathbf{c}_{ti}$ represents i's desired position relative to the target, weighted by its control scale. The interpretation of this law is that the robot i moves towards a position that is situated at its desired distance to the target, and meanwhile the robot is also taking into account the team coordination goal via $\mathbf{R}$ (22.6), the optimal rotation that robot i computes at each time instant.

22.3.4 Information Requirements and Implementation Details

In order to compute (22.7), a robot i needs to know the positions of the other robots relative to itself, that is, $\mathbf{q}_{ji}\ \forall j \neq i$. Hence, given that $\mathbf{q}_{jk} = \mathbf{q}_{ji} - \mathbf{q}_{ki}$, it is clear that i can compute the matrix $\mathbf{Q}$ (22.5) and then $\mathbf{R}$ (22.6). This relative position information required by the robot can be obtained via sensing, or by integration of the data received from other robots—through communications—with its own measurements. Note that the robots need to have global information of the team; we believe this to be a reasonable condition, given that the number of robots for a cooperative target perception task will typically be small in practical scenarios. This means that the data processing requirements will be low. It would be reasonable to consider the use of a central unit of computation, as this would increase the efficiency of processing. The downside of this choice is that it would create a central point of failure and make the use of communications mandatory. Robot i must also be able to perceive the target and its relative position, $\mathbf{q}_{ti}$. We do not elaborate on the issue of how the sensing of the target is performed, as this problem lies outside the scope of the chapter. Special dedicated sensors of different types could be employed for this purpose. We define the steps a robot i can follow to implement the proposed approach in Algorithm 22.1, where, for generality, we assume no central unit is present in the system.

The proposed control can be computed by each robot using a local and arbitrarily oriented coordinate frame. This is a prominent trait of the presented methodology, which we illustrate next for a robot k. Notice first that $\mathbf{q}_{tk}$ and $\mathbf{q}_{jk}$, $\forall j \in \{1, \ldots, N\}$, are all *relative* measurements; this means that there is no need for the robots to share a common coordinate origin. Moreover, we show next that the specific orientation of the reference frame of each robot is irrelevant. Let us denote as $\mathbf{P}_k \in SO(3)$ the rotation matrix between the global frame and the local frame which k uses, that is, $\mathbf{q}_k^P = \mathbf{P}_k \mathbf{q}_k$, being $\mathbf{q}_k^P$ the position of k in a frame aligned with the local frame and with its center in the global origin. Consider (22.4) and assume the robots have fixed positions $\mathbf{q}_i$. We denote with a superscript Lk the variables that are expressed in k's local frame. Let us write down the cost function (note that $\mathbf{q}_{ij}^{Lk} = \mathbf{P}_k \mathbf{q}_{ij}$):

$$
\begin{aligned}
\gamma^{Lk}(\mathbf{R}_c^{Lk}) &= \sum_i \sum_j \| \mathbf{q}_{ij}^{Lk} - \mathbf{R}_{cij}^{Lk} \|^2 \\
&= \sum_i \sum_j \| \mathbf{q}_{ij} - \mathbf{P}_k^{-1} \mathbf{R}_c^{Lk} \mathbf{c}_{ij} \|^2 = \gamma\left(\mathbf{P}_k^{-1} \mathbf{R}_c^{Lk} \right),
\end{aligned}
\tag{22.8}
$$

for any $\mathbf{R}_c^{Lk} \in SO(3)$. It is thus clear that the unique optimal rotations in the two frames must be such that, $\mathbf{R} = \mathbf{P}_k^{-1} \mathbf{R}^{Lk}$, that is, $\mathbf{R}^{Lk} = \mathbf{P}_k \mathbf{R}$. Then, since:

$$
\mathbf{u}_k^{Lk} = K_c \left(\mathbf{P}_k \mathbf{q}_{tk} - \mathbf{P}_k \mathbf{R} \mathbf{c}_{tk}^s \right) = \mathbf{P}_k \mathbf{u}_k,
\tag{22.9}
$$

the same motion is obtained when the control law is computed in each of the two frames.

Algorithm 22.1 Robot i's motion control loop

---

1. Initial Data: Template configuration ($\mathbf{c}_{ji}\ \forall j \neq i$).
2. **While** control executes **do:**
   a. *Data collection:* Using onboard sensing/communications with other robots:
      i. Acquire relative position data of other robots and target, $\mathbf{q}_{ji}\ \forall j \neq i$, $\mathbf{q}_{ti}$

    ii.   Acquire environment constraints data

    iii. Acquire target observation data

  b. *Control computation*: From a), compute rotation matrix $\mathbf{R}$ (22.6), determine s_i using a desired distance selection algorithm, and compute control law $\mathbf{u}_i$ (22.7)

  c. *Motion execution*: Execute physical motor commands corresponding to control $\mathbf{u}_i$

22.4 Stability Analysis

We inspect the stability properties of the target enclosing strategy. Our analysis employs the following assumption:

    A1: The rotation matrix (22.6) is at all times unique and differentiable as a function of the robots' positions.

Let us explain the assumption. When $\mathbf{R}$ has multiple solutions (see Section 22.3.1), this rotation is not differentiable with respect to time. Such differentiability is a prerequisite of the stability analysis we present below. The scenarios where these undesirable situations can arise, expressed by conditions on the rank and the SVD of matrix $\mathbf{A}$, are linked with perfect symmetries or singular geometries of the current and desired relative positions of the robots. These configurations have measure zero, and an infinitesimal perturbation of the robot positions brings $\mathbf{A}$ out of them. The conditions under which non-uniqueness or non-differentiability can appear correspond with specific degenerate arrangements of the robots' positions that are not attracting configurations under the proposed controller, and we disregard them via assumption A1. Further explanations on these matters are provided in Remark 22.1.

22.4.1 Rotation Matrix Dynamics

We study next the time evolution of the rotation matrix $\mathbf{R}$, which is a key element of the proposed approach.

Theorem 22.1

Under the controller (22.7) and if assumption A1 is satisfied, the rotation matrix $\mathbf{R}$ remains constant.

Proof. Let us first write down the inter-robot dynamics, from Equation 22.7, as follows:

$$\dot{\mathbf{q}}_{ij} = \dot{\mathbf{q}}_i - \dot{\mathbf{q}}_j = -K_c\left(\mathbf{q}_{ij} - \mathbf{R}\left(\mathbf{c}_{it}^s - \mathbf{c}_{jt}^s\right)\right). \tag{22.10}$$

We define a constant $N^2 \times 3$ matrix $\mathbf{C_s} = [\mathbf{c}_{1t}^s - \mathbf{c}_{1t}^s \ \mathbf{c}_{1t}^s - \mathbf{c}_{2t}^s \ ... \ \mathbf{c}_{Nt}^s - \mathbf{c}_{Nt}^s]^T$. We can then give this expression for the dynamics of $\mathbf{Q}$ (22.5):

$$\dot{\mathbf{Q}}(t) = -K_c[\mathbf{Q} - \mathbf{C_s}\mathbf{R}^T]. \tag{22.11}$$

The dynamics of $\mathbf{A}$ is, hence, as follows:

$$\dot{\mathbf{A}} = \mathbf{C}^T\dot{\mathbf{Q}} = -K_c(\mathbf{A} - \mathbf{C}^T\mathbf{C_s}\mathbf{R}^T). \tag{22.12}$$

We will examine the evolution of $\mathbf{R}$ by studying how $\mathbf{A}$ evolves. Observe that it is clear from the definitions of $\mathbf{A}$ and $\mathbf{R}$ (Section 22.3.1) that matrix $\mathbf{AR}$ is symmetric, and therefore:

$$\mathbf{R}^T\mathbf{A}^T - \mathbf{AR} = 0, \tag{22.13}$$

and as a consequence:

$$\dot{\mathbf{R}}^T\mathbf{A}^T + \mathbf{R}^T\dot{\mathbf{A}}^T - \dot{\mathbf{A}}\mathbf{R} - \mathbf{A}\dot{\mathbf{R}} = 0. \tag{22.14}$$

One can now substitute (22.12) in this last equation and obtain:

$$\dot{\mathbf{R}}^T\mathbf{A}^T - K_c\mathbf{R}^T\mathbf{A}^T + K_c\mathbf{R}^T\mathbf{R}\mathbf{C_s}^T\mathbf{C}$$
$$+K_c\mathbf{AR} - K_c\mathbf{C}^T\mathbf{C_s}\mathbf{R}^T\mathbf{R} - \mathbf{A}\dot{\mathbf{R}} = 0. \tag{22.15}$$

As $\mathbf{AR}$ is symmetric, it is possible to write:

$$\dot{\mathbf{R}}^T\mathbf{A}^T - \mathbf{A}\dot{\mathbf{R}} + K_c\left(\mathbf{C_s}^T\mathbf{C} - \mathbf{C}^T\mathbf{C_s}\right) = 0. \tag{22.16}$$

Let us now define $\mathbf{P_c} = \mathbf{C_s}^T\mathbf{C}$. We express $\mathbf{C_s} = \mathbf{C_{s1}} - \mathbf{C_{s2}}$ and, from Equation 22.5, $\mathbf{C} = \mathbf{C_1} - \mathbf{C_2}$, where $\mathbf{C_{s1}} = [\mathbf{c}_{1t}^s \dots \mathbf{c}_{1t}^s\ \mathbf{c}_{2t}^s \dots \mathbf{c}_{2t}^s \dots \mathbf{c}_{Nt}^s]^T$ and $\mathbf{C_{s2}} = [\mathbf{c}_{1t}^s \dots \mathbf{c}_{Nt}^s\ \mathbf{c}_{1t}^s \dots \mathbf{c}_{Nt}^s \dots \mathbf{c}_{Nt}^s]^T$. Note that analogous expressions, without the s superscripts, apply to $\mathbf{C_1}$ and $\mathbf{C_2}$. Then, one can see that an individual element of $\mathbf{P_c}$ can be expressed as:

$$\mathbf{P_c}[i, j] = \sum_{k=1}^{N^2}\mathbf{C_{s1}}[k, i]\mathbf{C_1}[k, j] + \mathbf{C_{s2}}[k, i]\mathbf{C_2}[k, j]$$
$$-\mathbf{C_{s1}}[k, i]\mathbf{C_2}[k, j] - \mathbf{C_{s2}}[k, i]\mathbf{C_1}[k, j], \tag{22.17}$$

for $i, j = 1, 2, 3$. Thanks to the structure of the matrices, we can write: $\Sigma_{k=1}^{N^2}\mathbf{C_{s1}}[k, i]\mathbf{C_2}[k, j] = \Sigma_{k=1}^{N}\mathbf{c}_{kt}^s[i]\Sigma_{l=1}^{N}\mathbf{c}_{lt}[j] = 0$, because every sum along a given coordinate (x, y or z) of the N vectors from the target ($\mathbf{c}_{lt}$) is equal to zero. We reiterate that this is the case because the target lies at the centroid of the default desired configuration. In a similar fashion, one can also see that the last sum of terms in Equation 22.17 is zero. Thus, only the two sums in the first line of Equation 22.17 are different from zero. One then has:

$$\mathbf{P_c}[i, j] = \sum_{k=1}^{N^2}\mathbf{C_{s1}}[k, i]\mathbf{C_1}[k, j] + \mathbf{C_{s2}}[k, i]\mathbf{C_2}[k, j]$$
$$= \sum_{k=1}^{N^2}s_{p1}(k)\mathbf{C_1}[k, i]\mathbf{C_1}[k, j] + s_{p2}(k)\mathbf{C_2}[k, i]\mathbf{C_2}[k, j], \tag{22.18}$$

where $s_{p1}(k)$ and $s_{p2}(k)$ are scale values dependent only on the index k. From Equation 22.18, one clearly sees that matrix $\mathbf{P_c}$ is symmetric. Thus, (22.16) becomes:

$$\dot{\mathbf{R}}^T\mathbf{A}^T - \mathbf{A}\dot{\mathbf{R}} = 0. \tag{22.19}$$

We now refer to the analysis of an identical equation and its possible solutions given in Proposition 1 of [3]. One can use the properties of the time-derivative of the rotation matrix and the SVD of $\mathbf{A}$, and consider assumption A1, to conclude that the correct solution to (22.19) is $\dot{\mathbf{R}} = \mathbf{0}$. Hence, the rotation matrix remains constant over time under the action of the proposed controller.   □

Let us highlight why this result is relevant. The control scales s_i determine in a very direct manner the directions of motion—see (22.7). These scales are chosen freely by each robot, without coordination among them. Even though they are not used in the computation of the rotation $\mathbf{R}$, one could therefore reasonably expect that the time-evolution of $\mathbf{R}$ would be dependent on the values of these scales. However, the presented analysis has shown that $\mathbf{R}$ stays constant. A consequence of this fact is that the motion of a given robot is not influenced by the other robots' desired distances to the target. Thus, the team's motions turn out to be steady and predictable. In the context of the tracking and observation task that is addressed, these are desirable and advantageous properties.

Remark 22.1

Observe from Equation (22.12) that, as $\mathbf{R}$ is constant when assumption A1 holds, one can define a constant matrix $\mathbf{A}_f = \mathbf{C}^T \mathbf{C}_s \mathbf{R}^T$ such that:

$$\dot{\mathbf{A}} = -K_c(\mathbf{A} - \mathbf{A}_f). \tag{22.20}$$

Hence, $\mathbf{A}$ converges exponentially to $\mathbf{A}_f$. One can assume $\mathbf{A}$ satisfies A1 at the start of the execution. Also, given that there are no exact symmetries and alignments of the robots' positions in the desired configuration (as discussed in Section 22.2), $\mathbf{C}^T \mathbf{C}_s$ satisfies A1, and the same holds for $\mathbf{A}_f$. Thus, $\mathbf{A}$ is attracted exponentially by a configuration in which A1 holds true. This provides support to the validity of the assumption that the degenerate cases can be disregarded when analyzing the controller. An alternative option would be to formulate an almost-global stability result where all possible cases would be contemplated. Regarding the geometric spaces where the robots may lie, note that the proposed method can be equally applied if the robots and target operate in 2D space. This case is explored in detail in Section 22.5.1.   □

22.4.2 Formation and Tracking Behaviors

Let us address the analysis of the inter-robot dynamics, and characterize the target tracking capabilities of the proposed control approach. Let us define the *central point of observation* associated with the current positions of the robots as the following *weighted centroid*:

$$\mathbf{p}_{wq} = \frac{\displaystyle\sum_{i=1}^{N} s_i^{-1} \mathbf{q}_i}{\displaystyle\sum_{i=1}^{N} s_i^{-1}}. \tag{22.21}$$

It can be seen that when the positions of the robots satisfy the condition (22.3), with scales s_i, and relative to a certain point in space, then it holds that such point is equal to $\mathbf{p}_{wq}$. The dynamic behavior of this point is used in the following to study the formation and tracking dynamics.

Proposition 22.1

Under the action of controller (22.7) and if A1 holds, the robots converge exponentially to a configuration where the desired relative viewing angles, and individually selected desired distances, are attained with respect to the team's weighted centroid.

Proof. From Equation (22.7), we express the dynamics of the relative position between two robots i and j as follows:

$$
\begin{aligned}
\dot{\mathbf{q}}_{ij} = \dot{\mathbf{q}}_i - \dot{\mathbf{q}}_j &= K_c(\mathbf{q}_{ti} - \mathbf{Rc}_{ti}^s) - K_c(\mathbf{q}_{tj} - \mathbf{Rc}_{tj}^s) \\
&= -K_c(\mathbf{q}_{ij} - \mathbf{R}(s_j\mathbf{c}_{tj} - s_i\mathbf{c}_{ti})).
\end{aligned}
\tag{22.22}
$$

Owing to Theorem 22.1, $\mathbf{R}$ remains constant over time, and given that the control scales are constant too, the final term of Equation (22.22) is a constant. This allows one to directly deduce the exponential convergence of each position vector $\mathbf{q}_{ij}$ to a vector $\mathbf{q}_{ij}^f = \mathbf{R}(s_j\mathbf{c}_{tj} - s_i\mathbf{c}_{ti})$. One can now define $\mathbf{p}_{wq^f} = \mathbf{q}_i^f + \mathbf{R}(s_i\mathbf{c}_{ti}) = \mathbf{q}_j^f + \mathbf{R}(s_j\mathbf{c}_{tj})$. If we isolate in these expressions $\mathbf{q}_i^f$ and substitute, for all i, in Equation 22.21, we can find that $\mathbf{p}_{wq^f}$ is the weighted centroid of the final robot positions. Then, as $\mathbf{q}_i^f = \mathbf{p}_{wq^f} + \mathbf{R}s_i\mathbf{c}_{it}$ and all other robots satisfy an analogous expression, if one considers (22.3) and the contents in Section 22.3.2 it is possible to directly conclude the stated result. □

Proposition 22.2

Under the action of the controller (22.7) and if A1 holds, the weighted centroid tracks the position of the target at all times, and the velocity of every robot becomes exponentially equal to the velocity of the weighted centroid.

Proof. The weighted centroid's dynamics are as follows:

$$
\begin{aligned}
\dot{\mathbf{p}}_{wq} &= \frac{\sum_{i=1}^{N} s_i^{-1}\dot{\mathbf{q}}_i}{\sum_{i=1}^{N} s_i^{-1}} = \frac{\sum_{i=1}^{N} s_i^{-1}K_c(\mathbf{q}_{ti} - s_i\mathbf{Rc}_{ti})}{\sum_{i=1}^{N} s_i^{-1}} \\
&= K_c \frac{(\sum_{i=1}^{N} s_i^{-1})\mathbf{q}_t - \sum_{i=1}^{N} s_i^{-1}\mathbf{q}_i - \mathbf{R}\sum_{i=1}^{N} \mathbf{c}_{ti}}{\sum_{i=1}^{N} s_i^{-1}}.
\end{aligned}
\tag{22.23}
$$

Recall that $\sum_{i=1}^{N} \mathbf{c}_{ti} = 0$ because the target is in the centroid of the default desired pattern. Thus, we can directly reach:

$$
\dot{\mathbf{p}}_{wq} = K_c(\mathbf{q}_t - \mathbf{p}_{wq}),
\tag{22.24}
$$

that is, the weighted centroid tracks the target. Moreover, given that each robot's position converges exponentially to a constant position relative to the time-varying weighted centroid (Proposition 22.1), it follows that every one of the robots' velocities converges to that of the weighted centroid.   □

This result is interesting because it implies that the central point of observation is always tracking the target's motion, even when the inter-robot formation is still far from being reached. This central point is the point that will be eventually viewed with the desired set of relative observation angles by the multirobot team. This suggests that an appropriate tracking performance will be achieved and is a desirable type of behavior for the proposed target enclosing system.

22.5 Usage in Navigation

The flexibility allowed by the method in defining each robot's motion is very handy for navigation tasks. A natural way to explore and analyze the proposed controller's application to navigation scenarios is to regard the *target* entity as a *leader* for the multirobot team. Let us elaborate on this issue next.

This leader will have a sense of global localization and environment awareness, and it will essentially direct the team's navigation towards the desired destination in the environment. In this scenario, the robots do not need to have a similar global awareness to successfully play their assigned role of tracking and perceiving the target. They can simply implement the proposed controller, which requires only local sensing capabilities, and navigate alongside the target. The target, in this case, can be considered a member of the multirobot team, rather than an external element. Therefore, the target can take into account the aim of helping the robots' navigation when making its own motion decisions. For instance, it can avoid traversing too-narrow passages. The proposed control method is particularly well suited to the described navigation scenario: its flexibility in terms of the team shape allows it to adapt to a changing environment, and the steady motions it creates can increase the safety and comfort of navigation from the perspective of the target. These aspects are illustrated in a practical fashion via a simulation example (Section 22.6.2).

22.5.1 2D Formulation of the Controller

A case that is very relevant in practical navigation scenarios is one where the robots and the target all move in 2D (i.e., on a planar ground). It turns out that for this situation, the rotation matrix (22.6) admits a closed-form expression. In particular, the single angle α_o of the optimal rotation matrix $\mathbf{R}(\alpha_o) \in SO(2)$ can be computed as follows:

$$\alpha_o = atan2(T^\perp, T),\tag{22.25}$$

where the definitions of the terms used are:

$$T = \sum_i \sum_j \mathbf{q}_{ij}^\mathsf{T}\mathbf{c}_{ij}, \quad T^\perp = \sum_i \sum_j \mathbf{q}_{ij}^\mathsf{T}\mathbf{c}_{ij}^\perp,\tag{22.26}$$

with $\mathbf{c}_{ij}^\perp = [(0,1)^T,(-1,0)^T]\mathbf{c}_{ij}$ and the sum indices going from 1 to N.

This simpler closed-form formulation makes it easier to identify the dynamic properties of the controller. It can be shown that the fact that the rotation matrix remains constant (Theorem 22.1) also holds true for this 2D case. Also, from (22.25) it is clear that in this case the possible degeneracies of the controller are reduced to situations where $T = T^\perp = 0$, which are zero-measure configurations.

The 2D case has peculiar features. First, it can be more appropriate to use a kinematic modeling—such as (22.1)—than in the 3D case, due to the inferior agility of the ground platforms. Also, avoidance of obstacles becomes a more pressing issue and therefore it needs to be given a greater importance when selecting the desired distances to the target (Section 22.3.2). The simulation tests provided in the section that follows will illustrate these points.

22.6 Simulation Study

We describe in this section simulation tests, performed using MATLAB®, that illustrate the presented control methodology. We introduce a useful performance metric:

$$e_a = \sum_i \sum_j |\beta_{ij} - \alpha_{ij}| , \qquad (22.27)$$

in which the sums are taken over all N robots. This function measures, at each time instant, the total error in the relative angles at which the robots view the weighted centroid, with respect to the desired ones.

22.6.1 Tracking of a Target in 3D Space

The first example includes a team of four robots, with a default configuration having the shape of a regular tetrahedron. The target's dynamics were modeled as a sum of sinusoids. Gaussian noise was added to the relative position measurements the robots used, in order to test the method's robustness. Each robot selected a different individual control scale. The values of these scales were [1.21 0.75 0.82 1.43]. The results are illustrated in Figure 22.2. One can see that despite the visible effects of the perturbations, the behavior is as desired, with the metric e_a being suitably regulated. The rotation matrix also remains fairly steady—up to the presence of noise. The fact that the weighted centroid tracks the target closely can also be visualized.

22.6.2 Navigation in a 2D Environment

In this example, a mobile target navigates in a 2D environment and is being enclosed and tracked by a team of six robots. The default desired configuration had the shape of a regular hexagon. As discussed in Section 22.5, in such a scenario one can exploit the adaptability of the team's shape around the target and the local character of the provided controller. The target, acting in this case as a *navigation leader*, follows a trajectory in the environment. The target has navigation capabilities that allow it to reach its destination while suitably handling the constraints posed by the environment. The robots, meanwhile, are able to operate without such global knowledge, and without a common coordinate frame of reference. In the example we describe—whose results

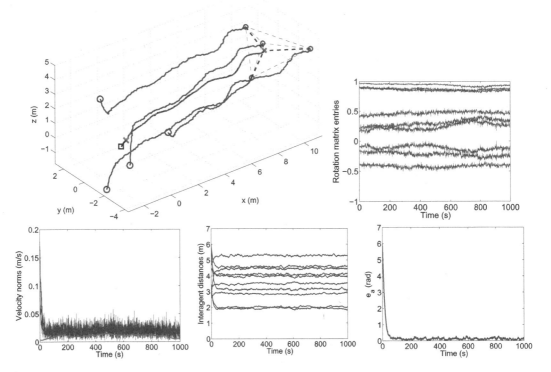

FIGURE 22.2

Results for the 3D simulation example. *Top left*: Paths followed by robots and target. Robots are marked as circles, target as a cross. Larger markers are used for the initial positions. Final positions are also marked. The initial weighted centroid is marked with a square and its path is shown, along with dashed lines joining it with each robot at the final configuration. *Top right*: Evolution of rotation matrix for one of the robots. *Bottom, left to right*: Time evolution of velocity norms for robots (thinner lines) and target (thicker line), inter-robot distances, and error metric e_a.

are illustrated in Figure 22.3—the target encounters a narrow passage it has to traverse. The robots adapt their distances to the target, as required in order to avoid colliding with the environment. What is interesting to note is that the observation quality is preserved even when these changes occur, and that a suitable performance in terms of target tracking is also retained.

22.7 Conclusion

In this chapter, we have described a novel method that a team of robots can use to enclose and observe a moving target. This motion coordination policy relies on an optimal rotation matrix and allows suitable control of the relative directions from which the different robots observe the target. Several features that make the proposed strategy interesting have been highlighted. In particular, the controller can be implemented on local coordinate frames, and it produces steady motions of the team of robots while allowing each of them to control individually its distance to the target. In this way, the approach naturally lends itself to collective navigation tasks where the robots surround the target, which acts as a

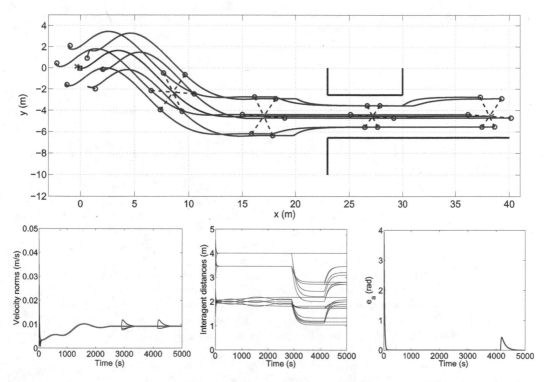

FIGURE 22.3
Results for the 2D navigation test. *Top*: Paths followed by robots and target. Direction of motion is left to right. Robots are marked as circles, target as a cross. The initial, final, and three intermediate configurations are marked. The initial weighted centroid is marked with a square and its path is shown, along with dashed lines joining it with each robot at each of the marked configurations. Environmental obstacles are marked with straight solid lines. *Bottom, left to right*: Time evolution of velocity norms for robots (thinner lines) and target (thicker line), inter-robot distances, and error metric e_a.

team leader, while it moves across an environment. These appealing properties have been demonstrated with simulation examples.

Many improvements and additions are possible, building on the foundations laid by the presented methodology. For instance, an interesting issue to explore would be how to adapt the control framework to tasks requiring interaction with physical objects or human agents. Certain aspects, such as the robustness to disturbances and model uncertainties, can deserve a deeper formal study. It would be important to incorporate in the system model more complex and realistic robot dynamics. Finally, it would be necessary to exploit the system's degrees of freedom (i.e., the desired distances to the target) via algorithms that provide adaptivity, safety, and robustness.

Acknowledgment

This work was supported by the French Government via programs Fonds Unique Interministériel (project Aerostrip) and Investissements d'Avenir (IDEX-ISITE initiative 16-IDEX-0001 (CAP 20-25, project MaRoC)).

References

1. L. Adouane. *Autonomous Vehicle Navigation: From Behavioral to Hybrid Multi-Controller Architectures*. CRC Press, 2016.
2. J. Alonso-Mora, S. Baker, and D. Rus. Multi-robot navigation in formation via sequential convex programming. In *IEEE/RSJ International Conference on Intelligent Robots and Systems*, pages 4634–4641, 2015.
3. M. Aranda, G. López-Nicolás , C. Sagüés, and M. M. Zavlanos. Three-dimensional multirobot formation control for target enclosing. In *IEEE/RSJ International Conference on Intelligent Robots and Systems*, pages 357–362, 2014.
4. M. Aranda and Y. Mezouar. Multirobot target enclosing with freely selected observation distances. In *European Control Conference*, pages 1405–1410, 2018.
5. N. Ayanian and V. Kumar. Decentralized feedback controllers for multiagent teams in environments with obstacles. *IEEE Transactions on Robotics*, 26(5):878–887, 2010.
6. N. Bishop, B. Fidan, B. D. O. Anderson, K. Doganay, and P. N. Pathirana. Optimality analysis of sensor-target localization geometries. *Automatica*, 46(3):479–492, 2010.
7. D. V. Dimarogonas and K. J. Kyriakopoulos. A connection between formation infeasibility and velocity alignment in kinematic multi-agent systems. *Automatica*, 44(10):2648–2654, 2008.
8. A. Franchi, P. Stegagno, and G. Oriolo. Decentralized multi-robot encirclement of a 3D target with guaranteed collision avoidance. *Autonomous Robots*, 40(2):245–265, 2016.
9. H. García de Marina, M. Cao, and B. Jayawardhana. Controlling rigid formations of mobile agents under inconsistent measurements. *IEEE Transactions on Robotics*, 31(1):31–39, 2015.
10. J. C. Gower and G. B. Dijksterhuis. *Procrustes Problems*. Oxford University Press, 2004.
11. G. Guerra-Filho. Optical motion capture: Theory and implementation. *Journal of Theoretical and Applied Informatics*, 12(2):61–89, 2005.
12. Z. Han, L. Wang, Z. Lin, and R. Zheng. Formation control with size scaling via a complex Laplacian-based approach. *IEEE Transactions on Cybernetics*, 46(10):2348–2359, 2016.
13. K. Hausman, J. Müller, A. Hariharan, N. Ayanian, and G. S. Sukhatme. Cooperative multi-robot control for target tracking with onboard sensing. *The International Journal of Robotics Research*, 34(13):1660–1677, 2015.
14. W. Kabsch. A solution for the best rotation to relate two sets of vectors. *Acta Crystallographica*, 32:922–923, 1976.
15. K. Kanatani. Analysis of 3-D rotation fitting. *IEEE Transactions on Pattern Analysis and Machine Intelligence*, 16(5):543–549, 1994.
16. T.-H. Kim and T. Sugie. Cooperative control for target-capturing task based on a cyclic pursuit strategy. *Automatica*, 43(8):1426–1431, 2007.
17. L. Krick, M. E. Broucke, and B. A. Francis. Stabilisation of infinitesimally rigid formations of multi-robot networks. *International Journal of Control*, 82(3):423–439, 2009.
18. X. Li, D. Sun, and J. Yang. A bounded controller for multirobot navigation while maintaining network connectivity in the presence of obstacles. *Automatica*, 49(1):285–292, 2013.
19. Z. Lin, L. Wang, Z. Han, and M. Fu. A graph Laplacian approach to coordinate-free formation stabilization for directed networks. *IEEE Transactions on Automatic Control*, 61(5):1269–1280, 2016.
20. G. López-Nicolás, M. Aranda, and Y. Mezouar. Formation of differential-drive vehicles with field-of-view constraints for enclosing a moving target. In *IEEE International Conference on Robotics and Automation*, pages 261–266, 2017.
21. A. J. Marasco, S. N. Givigi, and C.-A. Rabbath. Model predictive control for the dynamic encirclement of a target. In *American Control Conference*, pages 2004–2009, 2012.
22. S. Martínez and F. Bullo. Optimal sensor placement and motion coordination for target tracking. *Automatica*, 42(4):661–668, 2006.
23. D. Moreno-Salinas, A. M. Pascoal, and J. Aranda. Optimal sensor placement for multiple target positioning with range-only measurements in two-dimensional scenarios. *Sensors*, 13(8):10674–10710, 2013.

24. K.-K. Oh, M.-C. Park, and H.-S. Ahn. A survey of multi-agent formation control. *Automatica*, 53:424–440, 2015.
25. F. Poiesi and A. Cavallaro. Distributed vision-based flying cameras to film a moving target. In *IEEE/RSJ International Conference on Intelligent Robots and Systems*, pages 2453–2459, 2015.
26. S. Ramazani, R. Selmic, and M. de Queiroz. Rigidity-based multiagent layered formation control. *IEEE Transactions on Cybernetics*, 47(8):1902–1913, 2017.
27. C. Robin and S. Lacroix. Multi-robot target detection and tracking: Taxonomy and survey. *Autonomous Robots*, 40(4):729–760, 2016.
28. Z. Yang, X. Shi, and J. Chen. Optimal coordination of mobile sensors for target tracking under additive and multiplicative noises. *IEEE Transactions on Industrial Electronics*, 61(7):3459–3468, 2014.
29. S. Zhao, B. M. Chen, and T. H. Lee. Optimal sensor placement for target localisation and tracking in 2D and 3D. *International Journal of Control*, 86(10):1687–1704, 2013.
30. S. Zhao and D. Zelazo. Bearing rigidity and almost global bearing-only formation stabilization. *IEEE Transactions on Automatic Control*, 61(5):1255–1268, 2016.
31. K. Zhou and S. I. Roumeliotis. Multirobot active target tracking with combinations of relative observations. *IEEE Transactions on Robotics*, 27(4):678–695, 2011.

23

Motion Control of Multiple AUVs for Simultaneous Mapping and Navigation

Marcelo Borges Nogueira and Fernando Lobo Pereira

CONTENTS

23.1 Introduction

Navigation is a central problem in robotics, and localization plays a central role in navigation. There are two general kinds of sensors used for navigation: proprioceptive (gyros, accelerometers, etc.) and external or exteroceptive (global positioning system [GPS], vision, etc.). Independently of the quality of the sensors used, the covariance of the dead reckoning-based position estimate always grows with time. Therefore, they must be periodically reset using information from external sensors.

In the underwater environment, a submerged autonomous underwater vehicle (AUV) usually cannot rely on optical instruments (cameras and laser range finders), mainly for navigation in open spaces, and GPS data can be used to update its position only after surfacing. However, since, in general, surfacing is often too costly, navigation data from exteroceptive sensors is needed. Range sensors, based on time of flight (TOF) measurements from communications with beacons or other devices, have been used for this purpose.

The underwater environment is typically feature-poor, and this leads us to consider the use of artificial features. Some of these might be static, and some others might be mobile. An appropriate number of static beacons located in known positions forming a long baseline (LBL) system is a well-known way to obtain range measurements for localization. However, this imposes constraints that might be undesirable for many applications, especially in deep water, such as constraints in the operation area, and substantial deployment effort before operations (Bahr and Leonard 2006).

To overcome the above difficulties, static beacons can be replaced by a group of AUVs, thus creating a moving LBL system (Vaganay et al. 2004). An important drawback of this approach is that now the landmarks have some uncertainty associated with their position, which may change over time. This scenario configures a simultaneous localization and mapping (SLAM) problem with moving features, and thus the concept of mapping an environment involving mobile features has to be defined. In cartography, according to Wikipedia, the term "mapping" means constructing a representation of an area, that is, a symbolic depiction highlighting relationships between elements of that space such as objects, regions, and themes. The notion of a map usually requires some sort of persistence over time (for example, stationary), as it should preserve the ability of motion reference.

However, in a SLAM with moving features, a spatial component x no longer suffices to specify a map M, but a time component t is also required, that is, $M = M(x, t)$. It is well known that in the conventional SLAM with static features, the map converges to the true map with some offset (Bailey and Durrant-Whyte 2006; Durrant-Whyte and Bailey 2006). This means that an accurate relative map of the environment is generated, differing from the real one only by a fixed homogeneous transformation. Now, such a guarantee does not exist in the case of SLAM with dynamic features.

In this chapter, we present a centralized SLAM-like scheme for cooperative localization using the Extended Kalman Filter (EKF) to improve the AUV pose estimates. The proposed algorithm can support the low bandwidth and communication faults and still build a consistent cross-correlation matrix of the AUV's pose estimation.

We show that, under certain assumptions, this scheme generates a relative map that will converge to an accurate one defined relatively to an environment reference frame determined by the observed features, and that as time elapses, this relative map will diverge from the real one. In order to prevent the divergence of the generated relative map relative to the real map, a reset with absolute positioning measurements is required from time to time (Nogueira et al. 2010; Roumeliotis and Bekey 2002). This can be done via GPS fixes after surfacing, or via any other underwater absolute reference at a predefined location (say, a fixed LBL-like structure). Observe that there are applications of interest for which relative positioning suffices. For example, determining the source of a chemical plume. In this scenario, it is possible that under appropriate assumptions, a group of vehicles generates a successful strategy to search the source of a chemical plume by using concentration readings at their location estimated on a relative frame.

We also investigate what should be the optimal spatial distribution of the vehicles so as to maximize the information given by the range-only sensors. We demonstrate that the optimal spatial distribution of the vehicles depends independently on the bearing between each vehicle and a fixed point. Then, we propose a control law that drives the AUVs to their target positions while saving energy and introducing less noise into the system. Simulations show that if the vehicles are commanded to their optimal positions, the results of the cooperative localization are improved.

In the next section, 23.2, we provide a short overview of pertinent research work developed earlier and show how our results go beyond the current state of art. In Section 23.3, the problem is stated in detail, and some classical approaches to similar problems are examined. Based on the EKF, a moving SLAM-centralized algorithm for a group of AUVs equipped with the same kind of dead reckoning and external sensors is presented in Section 23.4. The problem of the information-driven motion coordination based only on range measurements, and the demonstration of some of its properties are presented in Section 23.5. Here, we also derive a control strategy that minimizes the error introduced in the system. Some simulations and experimental results are presented in Section 23.6, followed by some conclusions along with future work in Section 23.7.

23.2 Related Work

23.2.1 Underwater Cooperative Localization

When a group of vehicles is very heterogeneous, possibly carrying different proprioceptive and exteroceptive sensors and executing different trajectories in the environment, the quality of the localization estimates may vary significantly across the individual members at any given time. If any vehicle can sense and communicate with the other members of the group at all times, then each one will have an uncertainty about its position lower than that of the vehicle with the best localization results (Roumeliotis and Bekey 2002). In Roumeliotis and Rekleitis (2004), it is shown that in a group of N robots performing cooperative localization, the maximum expected rate of positioning uncertainty increase at steady state is inversely proportional to the number of robots. In Mourikis and Roumeliotis (2006), the authors provide analytical expressions for the upper bound on the uncertainty of the vehicles estimated position, and show that the rate of uncertainty increase depends only on the accuracy of the proprioceptive and orientation sensors on the robots. In Nettleton and Durrant-Whyte (2001) and Grime and Durrant-Whyte (1994), the authors use an information filter to fuse the information in a decentralized sensor network in a non-conservative and exact way.

Kurazume et al. (1996) proposed a method that organizes ground operating robots into two groups, A and B. Group A remains stationary and acts as a landmark, while group B moves. Then, group B stops and acts as a landmark for group A. Unfortunately, for most classes of AUVs it is very hard to stay stationary unless the vehicle rests on the seafloor, as proposed in Matsuda et al. (2014). Besides, it is still necessary to incorporate into the coordination control scheme algorithm the fact that underwater communication is slow, limited, and unreliable. In Paull et al. (2015) the authors perform a cooperative SLAM reducing the communication between vehicles, however also considering the vehicles are near the seafloor and observe it for features.

These challenges have been attracting the interest of many researchers for the problem of cooperative localization with underwater vehicles (Dajun et al. 2017; Lichuan et al. 2009; Yao, Xu and Yan 2009; Yao, Xu, Yan and Gao 2009). In Bahr and Leonard (2006), Vaganay et al. (2004), Chen et al. (2017), and Li and Zhang (2016), the authors proposed a solution for an underwater cooperative localization problem in which a subgroup of vehicles, called Communication and Navigation Aid-AUVs, or CNAs, could maintain an accurate estimate of their position and would help the other ones with poor navigation systems to update their position. In Huang et al. (2018), the authors also assume the existence of a CNA and develop an algorithm suitable for inaccurate noise covariance matrices. In Webster et al. (2010), the authors propose an information-based, vehicle-based algorithm to perform localization between a vehicle and a single moving beacon. All these methods also take into account that underwater communication is slow.

The main challenge in a decentralized cooperative localization algorithm concerns communications constraints which make it hard to create and update covariance information between vehicles as they communicate with one another. This deficit of covariance information between the vehicles may lead to a repetitive use of the same evidence and to an overconfidence of the robots' pose, as discussed in Fox et al. (2001). In Roumeliotis and Bekey (2002), the authors present what they call collective localization, where each robot collects data regarding its own motion and shares information during update cycles. This exchange of information is only necessary when two vehicles measure their relative position, but it is required that all the vehicles must successfully communicate with one another whenever there is an observation. They show that by exchanging only individual estimates of pose and covariance, inter-vehicle correlation can still be maintained. In Bahr et al. (2009), the authors propose a decentralized solution by running in each vehicle a bank of filters and keeping a table of measurements that prevents using any of the measurements more than once. The vehicles share their bank of filters and then the multiple estimates are combined in a consistent manner, yielding conservative covariance estimates.

The methods presented in the literature review rely on either fast and/or faultless communication, or do not keep a consistent cross-correlation matrix, leading to overconfidence or conservative covariance estimates. The goal of this work is to develop a consistent algorithm for cooperative localization, considering the slow and faulty communications. We developed a centralized cooperative AUV localization algorithm based on the EKF that takes into account the low bandwidth. Unreliability, delays, and very low bandwidth make acoustic communications an important challenge in the design of a moving SLAM system for a group of AUVs. We derived a way for the vehicles to share the least possible information necessary to execute the SLAM algorithm, creating a consistent cross-correlation matrix, which is very important for the SLAM convergence properties.

Another issue that arises in the cooperative localization setup considered in this chapter concerns finding the spatial distribution of the vehicles that yields the best possible information from the observations and optimizes the vehicles' uncertainty. In Whaite and Ferrie (1995), it is shown that a locally optimal strategy for observations is to maximize the determinant of the prediction variance in the EKF equations. This strategy is used by Sim (2005) to determine the trajectory of a vehicle equipped with bearings-only sensors observing known features. In Bahr and Leonard (2007), the authors analyze the best position for a vehicle to be relative to two features with some associated errors, taking range measurements to each feature. The range information is used to update the position of the vehicle only. In Zhang et al. (2016), an optimal formation is proposed for cooperative localization of AUVs in a leader-follower scheme using geometry triangulation. We will provide an analysis of the problem where the are several vehicles involved, and the range information is used to update the position of all the vehicles. Based on each vehicle's uncertainty and its correlation with other vehicles, we determine the best relative vehicle formation that maximizes the information extracted from observations.

23.2.2 Extended Kalman Filter

Estimation of the localization is needed whenever a GPS is not available. This may be done by making use of some variant of the celebrated classical Kalman Filter (KF). Developed by R.E. Kalman, the KF recursively combines the dynamics system noisy state or output measurements with the currently available information to obtain the best state estimate

(Maybeck 1979). It yields optimal values for linear systems contaminated with Gaussian white noise.

For nonlinear systems, the EKF can be used. A nonlinear discrete time system may be modeled as shown in Equation 23.1, where $X(k)$ (notice the upper case so that the whole state X does not get confused with the x-axis part of the state, x, later on), $u(k)$, $y(k)$, $q(k)$, and $r(k)$ are, at time instant k, respectively, the state vector (robot pose—position and orientation—in our case), the control signal, the output measurement or observation, and the process and the output measurement noises (both Gaussian and with zero mean).

$$X(k+1) = f(X(k), u(k), k) + q(k)$$
$$y(k) = h(X(k), u(k), k) + r(k)$$
(23.1)

The EKF equations, 23.2, involve the linearizations of the functions f and h ($D_X f$ and $D_X h$ denote, respectively, the Jacobians of f w.r.t. X and the Jacobian of h w.r.t. X). The filter equations are given by:

$$e(k) = y(k) - h(\hat{X}(k), u(k), k)$$
$$L(k) = D_X h P_p(k) D_X h^T + R(k)$$

Prediction:

$$\hat{X}_p(k+1) = f(\hat{X}(k), u(k), k)$$
$$P_p(k+1) = D_X f P(k) D_X f^T + Q(k)$$
(23.2)

Update:

$$K(k) = P_p(k) D_X h^T L(k)^{-1}$$
$$\hat{X}(k) = \hat{X}_p(k) + K(k) e(k)$$
$$P(t) = P_p(t) - K(k) L K(k)^T,$$

where $e(k)$ is called the innovation, $K(k)$ is the Kalman gain, $\hat{X}_p(k)$ and $\hat{X}(k)$ are the predicted and the filtered (estimated) state estimates, $y(k)$ and $y_p(k)$ are the observation and predicted observation, and $P(k)$ and $P_p(k)$ are the filtered and the predicted error covariance matrix. The matrices $Q(k)$, $R(k)$ are given by:

$$E\left\{\begin{bmatrix} q(k) \\ r(k) \end{bmatrix} [q^T(k) \quad r^T(k)]\right\} = \begin{bmatrix} Q(k) & 0 \\ 0 & R(k) \end{bmatrix}.$$

The algorithm has to be initialized with values $x_p(0)$ and $P_p(0)$. The update is possible only if an observation is available, otherwise we have $\hat{X}(k) = X_p(k)$ and $P(k) = P_p(k)$.

23.2.3 Simultaneous Localization and Mapping with EKF

As discussed previously, a robot working in a certain environment may use an existing map of this environment and a set of observations of some of its features to improve the estimate of its position. However, if such a map is not available, then the SLAM technique may be used: a robot can localize itself using a map of landmarks and an onboard sensor which senses these landmarks, and, conversely, from accurate localization of the robot, it is possible to build a map of the landmarks (determine their spatial coordinates). The SLAM problem consists in performing both tasks simultaneously (Kehagias et al. 2006). This problem has been of great interest in the robotics community since the seminal papers of Smith et al. (1990) and Durrant-Whyte (1987). A SLAM algorithm builds a consistent estimate of both environment map and vehicle trajectory using noisy proprioceptive and some exteroceptive information (Bailey and Durrant-Whyte 2006); Durrant-Whyte and Bailey 2006).

It is possible to use again an EKF to solve a SLAM problem with nonlinear system models (Dissanayake et al. 2001; Newman (2004). In this case, the state vector $X(k)$ is composed of the vehicle pose, $X_v(k)$, and map feature parameters, that is

$$X(k) = [X_v(k) X_{f1}(k) X_{f2}(k)...X_{fn}(k)]^T,$$

where $X_{fi}(k)$ are the parameters of the ith feature at time instant k. The system will be described by Equation 23.3, where f_v describes the motion of the vehicle, and, since the features are static, their position remains unchanged over time in the system model (but their position estimate may be changed by the KF when there is an observation).

$$\begin{aligned} X_v(k+1) &= f_v(X_v(k), u(k), k) + q_v(k) \\ X_{fi}(k+1) &= X_{fi}(k), \end{aligned} \tag{23.3}$$

Notice that the features are not contaminated by noise. The process noise matrix Q and the Jacobian of the function f w.r.t. X are given by Equation 23.4, where Q_v is the process noise of the vehicle motion function f_v and $D_X f_v$ is the Jacobian of f_v.

$$Q = \begin{bmatrix} Q_v & 0 & \cdots & 0 \\ 0 & 0 & & \vdots \\ \vdots & & \ddots & 0 \\ 0 & \cdots & 0 & 0 \end{bmatrix} \quad D_X f = \begin{bmatrix} D_X f_v & 0 & \cdots & 0 \\ 0 & I & & \vdots \\ \vdots & & \ddots & 0 \\ 0 & \cdots & 0 & I \end{bmatrix} \tag{23.4}$$

The covariance matrix P, shown in Equation 23.5, will incorporate all the terms in the vector state x. Hence, besides the covariance matrix of the vehicle P_{vv} and each feature P_{fifi}, it will also include components representing the cross-covariance between the vehicle and the features $P_{vfi} = P_{fiv}$ and between the features $P_{fifj} = P_{fjfi}$, $i \neq j$.

$$P = \begin{bmatrix} P_{vv} & P_{vf1} & P_{vf2} & \cdots & P_{vfn} \\ P_{f1v} & P_{f1f1} & P_{f1f2} & \cdots & P_{f1fn} \\ P_{f2v} & P_{f2f1} & P_{f2f2} & \cdots & P_{f2fn} \\ \vdots & \vdots & \vdots & \cdots & \vdots \\ P_{fnv} & P_{fnf1} & P_{fnf2} & \ddots & P_{fnfn} \end{bmatrix} \tag{23.5}$$

As shown by Dissanayake et al. (2001), the structure of the SLAM problem is critically dependent on maintaining complete knowledge of the cross-correlation between landmark estimates. As the vehicle travels around the environment and observes features, the correlation of the errors in their estimates increases until the errors of the estimate of all features are fully correlated. At this point, the map of relative locations of the features is known with absolute precision. This means that given the real position of any feature, the whole map can be computed with zero error. It is also shown that the absolute error of the relative map reaches a lower bound determined only by the error that existed when the first feature was observed.

23.2.4 Range-Only SLAM

A special case of SLAM is when only range measurements to landmarks are available, like TOF measurements underwater. If there are n features in the environment, and the vehicle observes feature i at time t, the observation equation is given by

$$h(X) = \sqrt{(x_v - x_{fi})^2 + (y_v - y_{fi})^2},$$

where (x_v, y_v) and (x_{fi}, y_{fi}) are the position of the vehicle and of the ith landmark in the global reference frame, respectively. In this case, a single measurement does not contain enough information to determine the location of a landmark. If there is no *a priori* information about the position of the landmark, then this partial observability implies the need to fuse data of several observations from different vehicle positions (at least three non-collinear measurements are needed in a 2D scenario) in order to determine the landmark location (Newman and Leonard 2003; Olson et al. 2006). But, once the positions of the landmarks have been initialized, or *a priori* positions are known, a SLAM scheme may use the set of observations to improve the position estimates of both the vehicle and landmark.

23.3 The Problem

Static artificial features, like beacons, limit the operation area and require a substantial deployment effort. But what if other AUVs are used as features? In this scenario, we could have several vehicles cooperating with each other in that, while executing its own mission, each AUV uses navigation data from the others to improve the estimate of its own position, as shown in Figure 23.1. Given the strict constraints on onboard resources, this concept of having several AUVs cooperating in order to fulfill a given collective mission is very interesting and promising. For example, the search of a particular area can be performed more efficiently by using multiple AUVs. Another good context is that of dangerous missions in very hostile environments in which it is unlikely that one AUV will be able to survive long enough. In this case, it may be better to use several inexpensive but prone to failure AUVs instead of a single and much more expensive one.

If we consider, for simplicity, a 2D unicycle model for all AUVs in the system (by using pressure sensors it is possible to reduce the cooperative localization from a 3D to a 2D problem) (Bahr 2009), the state vector X_i for vehicle i will be composed by the position (x_i, y_i) on the plane and the orientation θ_i relative to the global reference frame, as show in Equation 23.6.

$$X_i = [x_i \ y_i \ \theta_i]^T \tag{23.6}$$

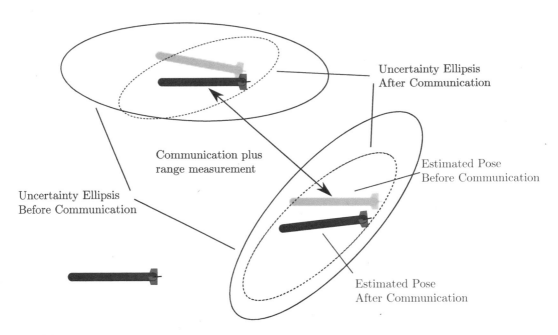

FIGURE 23.1
Cooperative localization: Vehicles communicate when possible and will obtain a new pose estimate with a new, smaller uncertainty after they share information fused with the range measurement.

The control input would be:

$$u_i^{Vi}(k) = \left[v_x^{Vi}(k)\, v_y^{Vi}(k)\, v_\Theta^{Vi}(k) \right]^T,$$

where v_x^{Vi} and v_y^{Vi} are the *velocity* in the x and y axes of the vehicle's local reference frame $\{Vi\}$ and v_Θ^{Vi} is the angular velocity of the vehicle. The discrete motion model of the vehicle is given by Equation 23.7, where $q_i(k) = [q_x(k)\, q_y(k)\, q_\theta(k)]$ represents the vehicle process noise.

$$X_i(k+1) = f_i\left(X_i(k), u_i^V(k) \right) + q_i(k) \tag{23.7}$$

$$\begin{bmatrix} x_i(k+1) \\ y_i(k+1) \\ \theta_i(k+1) \end{bmatrix} = \begin{bmatrix} x_i(k) + v_x^{Vi}(k)c\theta_i(k) - v_y^{Vi}(k)s\theta_i(k) + q_x(k) \\ y_i(k) + v_x^{Vi}(k)s\theta_i(k) + v_y^{Vi}(k)c\theta_i(k) + q_y(k) \\ \theta_i(k) + v_\theta^{Vi}(k) + q_\theta(k) \end{bmatrix}$$

If we put together the vector state of all the n vehicles into a single augmented state vector, we will get

$$X = [X_1 X_2 \ldots X_n],$$

which is close to the to the SLAM structure presented in Section 23.2.3, but now instead of one vehicle and several static features we will have a group of moving vehicles. So this scenario constitutes a problem similar to the SLAM problem, but with moving features.

Range calculations based on underwater communication (TOF) usually are contaminated by several noise sources, some of them not Gaussian, like the fluctuating speed of sound in water or multipath. Both these sources are related to environmental conditions, which do not change rapidly and can be non-Gaussian (Olson et al. 2006). Since KFs give optimal solution for Gaussian noise only, the outliers' measurements must be removed before the filtering stage. In this work we do not consider those sources of non-Gaussian noise on the simulations, but in a real application a technique to remove the outliers, such as the one presented in Olson et al. (2006), must be used. Anyway, according to Bahr et al. (2009), many experiments have shown that the error in the range measurement can be modeled as a Gaussian with mean r and a fixed variance σ^2.

For a group of vehicles to perform cooperative localization and improve their estimated state, the vehicles must share navigation data among themselves. However, underwater communication channels are known to be problematic. Underwater acoustic communications have a limited small bandwidth, the channel is subject to time variations and multi-path propagation, and there is a strong signal attenuation. These difficulties impose a cooperative algorithm with small data transmission between the AUVs, besides a capacity to tolerate faulty communications.

23.4 Algorithm

In this section, we present an algorithm based on the EKF designed for the low communication bandwidth in the underwater environment.

23.4.1 Algorithm Development

In our approach, we present a centralized algorithm that successfully generates and updates the cross-covariance information between the vehicles while minimizing the communication traffic. The developed algorithm, designated by cooperative localization using EKF based on SLAM (CEKOLM), first presented in Nogueira et al. (2010), considers one AUV as the central processor unit (from now on called CAUV), and the other AUVs as features on the environment (called FAUVs), which communicate with the CAUV. The communication between the CAUV and the FAUVs consists of the CAUV first "inviting" an FAUV to communicate. This FAUV then transmits its pose, pose uncertainty (error covariance matrix $P(t)$) and some additional data (to be detailed later). Then, the CAUV processes the received data and responds, sending a new pose estimate and covariance matrix to the FAUV. A disadvantage of this communication scheme is that the frequency a particular FAUV communicates increases with the number of vehicles on the system. However, as noted in Roumeliotis and Rekleitis (2004), the uncertainty growth is inversely proportional to the number of vehicles, so there is not much advantage in localization by increasing their number. If the application requires a large number of vehicles to be employed, we could create several subgroups of robots, each group using the CEKOLM algorithm.

One major drawback of the centralized system is that if the CAUV fails or gets out of communication range the whole system is compromised. In order to prevent getting out of range, the CAUV, knowing the estimated position of all the other vehicles, can position itself or even command the other vehicles to avoid such situations. However, this alone is no guarantee that the CAUV will always be available for communication. One way to solve this problem would be to assign to the vehicles involved in the mission a hierarchy (which can be decided prior to the mission) to choose which one of them will became the next CAUV in case the current one "disappears." The absence of the current CAUV can be determined by the next vehicle in line, say $FAUV_i$, if it does not receive any signal from the current CAUV for a certain amount of time t_{max}. $FAUV_i$ can then broadcast a message asking if the other AUVs have heard from the CAUV lately ($t < t_{max}$). If it receives any positive answer, it comes back to its normal operation. If it does not receive any acknowledgment, it will become the new CAUV and broadcast a message with this information.

When synchronized clocks are available, the AUVs can estimate the range to each other by calculating the communications flight time. Otherwise, the communications flight time can be estimated by the round-trip ping time. Therefore, the observation represents range information only. Since each FAUV transmits its current position's own estimate, there is always *a priori* knowledge of the "landmark" location. This eliminates the partial observability of the range-only sensor, as discussed in Section 23.3. While no observation is available, each AUV uses an EKF and proprioceptive information obtained, for example, through a compass or an inertial navigation system (INS), to estimate its current position and update its error covariance matrix.

We assume that once in a while it is possible for one of the vehicles to obtain a low uncertainty estimate of its pose. This could be materialized by, for example, surfacing in order to get a GPS fix, being the underlying decision part of the overall resources management activity: tradeoff between the priority of assigned tasks, "cost" of obtaining the estimate, and impact on the propagation of AUV's position estimates is a possible criterion to select the AUV that will obtain a GPS estimate of its pose. In our simulations, we assumed that the CAUV has to be submerged and is executing a high-priority task, and thus it cannot surface to update its pose.

The proposed SLAM-based algorithm runs on the CAUV and uses an EKF to update the state vector, which is the Cartesian product of the pose of all AUVs in the system, that is,

$$X_{SEKF}(k) = [X_C(k)X_{F1}(k)X_{F2}(k)...X_{Fn}(k)]^T, \tag{23.8}$$

where X_C and X_{Fi} are, respectively, the state vector of the CAUV and of the ith FAUV. We call this SLAM EKF running in the CAUV by SEKF. The SEKF takes a step every time an observation is available. The covariance matrix $P_{SEKF}(k)$ of the SEKF filter, shown in Equation 23.9, contains the covariance matrix of all the vehicles in the system, and also their cross-correlation terms. The discrete time index k is asynchronous and will increment every time the CAUV successfully communicates with a FAUV.

$$P_{SEKF}(k) = \begin{bmatrix} P_{CC} & P_{CF1} & \cdots \\ P_{CF1} & P_{F11} & \cdots \\ \vdots & \vdots & \ddots \end{bmatrix} \tag{23.9}$$

In between observations, the CAUV also runs a regular EKF algorithm (designated by REKF) in which the state vector is the pose of the CAUV, $X(j)_C = [x(j)\,y(j)\,\theta(j)]$. The goal of the REKF, also present in all FAUVs, is to predict the AUV pose and update its error covariance matrix $P(j)$ by using proprioceptive information. Notice that the discrete time index used for the REKF is j. This index is a synchronous time, incremented every Δt seconds (given by the update frequency of the proprioceptive sensors in each vehicle). Since there is only one SEKF present in the system (running on the CAUV), its state vector and covariance matrix will be represented by X_{SEKF} and P_{SEKF}. By X_i and P_i we are designating the vector state and covariance matrix of the REKF running in vehicle i.

In CEKOLM, when an observation is available, the CAUV uses the last computed pose given by the SEKF, $\hat{X}_{SEKF_C}(k)$, and the current predicted pose by the REKF, $\hat{X}_C(j)$, to calculate the control signal $u_C(k)$ that took the CAUV from $\hat{X}_{SEKF_C}(k)$ to $\hat{X}_C(j)$. Similarly, the CAUV uses the last pose received by the current communicating FAUV (designated by CCFAUV), $\hat{X}_{CCF}(k_{CCF})$, which was transmitted at time k_{CCF} (in the SEKF), and the current received one, $\hat{X}_{CCF}(j)$, to compute a control signal, $u_{CCF}(k)$ that took the CCFAUV from $\hat{X}_{CCF}(k_{CCF})$ to $\hat{X}_{CCF}(j)$. These two control signals represent the accumulated motion described by the vehicles since their last communication. In order to compute the control signal of the CCFAUV, the CAUV contains a table with the last communicated pose $\hat{X}_i(k_i), i = 1 \ldots n$ for each FAUV.

It is possible to consider that all other FAUVs, other than the CCFAUV, are communicating their last communicated pose with a random range measurement, but with infinite measurement variance. Hence, their control signal will be zero, their uncertainty during the predict step of the SEKF filter will not increase (the uncertainty of the whole motion they did in between communications will be considered when each one of them becomes the CCFAUV), and their range measurement will not be used by the SEKF update equations. In this way, the control signal of the SEKF is given by $u_{SEKF}(k) = [u_C(k)\,0\,0\ldots u_{CCF}(k)\ldots 0]^T$.

The SEKF process noise covariance matrix Q_{SEKF} is given by the error covariance matrix calculated by the REKF running on the CAUV and on the CCFAUV, as shown in Equation 23.10. For all the other FAUVs the process noise is zero.

$$Q_{SEKF}(k) = \begin{bmatrix} P_C(j) & 0 & 0 & \ldots & 0 \\ \vdots & \ddots & \vdots & \ldots & 0 \\ 0 & \ldots & P_{CCF}(j) & \ldots & 0 \\ \vdots & \vdots & \ddots & \ldots & 0 \\ 0 & 0 & 0 & \ldots & 0 \end{bmatrix} \qquad (23.10)$$

In order to ensure that $P_i(j)$ corresponds, for each AUV, only to the noise occurring from the time of last observation until the current time, this error covariance matrix is set to zero both in the CAUV and in the CCFAUV, at every time an observation is taken.

According to the prediction equation of the EKF for the REKF running in each AUV we have

$$P_p(j+1) = D_X f P(j) D_X f^T + D_u f Q(j) D_u f^T.$$

If there is no observation at time step j, then we simply set $P(j) = P_p(j)$. Now, by reiterating, we have

$$P(j+2) = P_p(j+2) = D_X f P(j+1) D_X f^T + D_u f Q(j+1) D_u f^T,$$
$$P(j+2) = D_X f (D_X f P(j) D_X f^T + D_u f Q(j) D_u f^T) D_X f^T + D_u f Q(j+1) D_u f^T,$$
$$P(j+2) = D_X f^2 P(j)(D_X f^T)^2 + D_X f D_u f Q(j) D_u f^T D_X f^T + D_u f Q(j+1) D_u f^T.$$

Thus, we conclude that, if there are no observations between times steps j and $j+m$

$$P(j+m) = (D_X f)^m P(j)\left(D_X f^T\right)^m$$
$$+ (D_X f)^{m-1} D_u f Q(j) D_u f^T (D_X f)^{m-1} + \cdots + D_u f Q(j) D_u f^T \qquad (23.11)$$

where, to simplify the notation, $(D_X f)^m$ means

$$(D_X f)^m = D_X f(j+m) D_X f(j+m-1) \ldots D_X f(j). \qquad (23.12)$$

Since every time vehicle i is involved in an observation we set $P_i(j) = 0$, the covariance equation for the REKF running in each vehicle can be written as

$$P(j+m) = (D_X f)^{j+m-1} D_u f Q(j) D_u f^T (D_X f)^{j+m-1} + \cdots + D_u f Q(j) D_u f^T. \qquad (23.13)$$

When the CCFAUV communicates with the CAUV, at time $j+m$, the SEKF prediction equation, considering only a single vehicle, for some matrix $M(m)$, is given by

$$P_{pSEKF}(k+1) = M(m) P_{SEKF}(k) M(m)^T + Q_{SEKF}(k). \qquad (23.14)$$

Since we defined that the process noise covariance matrix, Q_{SEKF}, is composed by the accumulated uncertainty of the vehicles in between communications, as shown in Equation 23.10, we have

$$P_{pSEKF}(k+1) = M(m) P_{SEKF}(k) M(m)^T + P(j+m). \qquad (23.15)$$

Notice that the SEKF predict equations are supposed to give the same result as the REKF equations, if we were not setting the term $P(j) = 0$ when there is an observation, as shown in Equation 23.11. By comparing Equations 23.15, 23.11, and 23.13, observing that $P_{SEKF}(k)$ plays the role of $P(j)$, we conclude that in order for Equation 23.15) to represent the total uncertainty for the CCFAUV at the current time (uncertainty at the last observation, $P_{SEKF}(k)$, added with the uncertainty accumulated since this last observation, $P(j)$, until now, $P(j+m)$), we must have

$$M(m) = D_x f(m) D_x f(m-1) \ldots D_x f(0) = D_x f(m) M(m-1), \qquad (23.16)$$

where m represents the number of time steps executed by the REKF since the last communication and the present time. Recall that the matrix $D_x f$ actually depends on $u(j)$ and $X(j)$, which vary over time. This means that each vehicle must compute the matrix M recursively during each step of the execution of the REKF and send it to the CAUV, together with $P(j)$, at the observation time. Notice that now, right after an observation, besides setting $P(j) = 0$, we must also set $M(j) = I$ (identity matrix) so that m, in Equation 23.16, will represent the desired number of time steps.

In this way, and by considering for simplicity that we have only two vehicles in the system (the CAUV and one FAUV), the SEKF Prediction equation is:

$$P_{pSEKF}(k+1) = M_{12} P_{SEKF}(k) M_{12}^T + Q_{SEKF}(k)$$

where $M_{12} = diag(M_1, M_2)$ and M_i is the M matrix computed by the ith vehicle. In the case of more than two vehicles, M_i is the identity matrix if the ith vehicle is not the CCFAUV. The matrix M_i is a $s \times s$ matrix, where s is the state vector dimension of each AUV. However, if, for example, the vehicles are modeled by a 2D unicycle model, this matrix can be fully described by only two parameters.

The reason why we cannot simply use the current covariance matrix transmitted by CCFAUV and substitute it for the corresponding terms of P_{SEKF} is because as the vehicles move, the change in the cross-correlation terms (due to initial nonzero correlation) would then be ignored. Since the entire structure of the SLAM problem critically depends on maintaining complete knowledge of the cross-correlation between landmark estimates, this miss-update of the cross-correlation terms could result in inconsistent and divergent solutions to the map construction.

After receiving this information from the CCFAUV, the SEKF can compute the estimated position for all AUVs in the system using the update equations of the EKF. After this computation, the CAUV sends to the CCFAUV its new estimated pose and pose uncertainty. The modifications of the pose of the FAUVs other than the CCFAUV will be taken into account at the time at which each one of them becomes the CCFAUV. The CEKOLM algorithm for the CAUV is shown in Algorithm 23.1, and a block diagram is shown in Figure 23.2.

Algorithm 23.1 CEKOLM running on the CAUV

---

1. Initialize State Vector and Error Covariance Matrix
2. While true
   (a) If an observation is available, then
      i. Compute the control signal $u_C(k)$ that took the CAUV from $\hat{X}_{SEKF_C}(k)$ to $\hat{X}_C(j)$
      ii. Compute the control signal $u_{CCF}(k)$ that took the CCFAUV from $\hat{X}_{CCF}(k_{CCF})$ to $\hat{X}_{CCF}(j)$ and compose the augmented control signal $u_{SEKF}(k) = [u_C(k)\ 0\ 0 \ldots u_{CCF}(k) \ldots 0]^T$
      iii. Construct the augmented process noise matrix $Q_{SEKF}(k)$ with $P_C(j)$ and $P_{CCFV}(j)$
      iv. Update the state vector $\hat{X}_{SEKF}(k)$ and uncertainty $P_{SEKF}(k)$ using SEKF prediction and update equations
      v. Extract the computed pose by the SEKF and set it in the REKF of the CAUV, $\hat{X}_{REKF}(j) = \hat{X}_{SEKF_C}(k)$
      vi. Extract and transmit to the CCFAUV its new estimated pose $\hat{X}_{CCF}(k)$ and uncertainty $P_{CCF}(k)$
      vii. Reset the covariance matrix $P_C(j)$ and set $M_C(j) = I$ (the CCFAUV should do the same in its own REKF)
      viii. $k = k + 1$
   (b) Predict the CAUV's next pose $\hat{X}_{pREKF}(j+1)$ and uncertainty $P_{pREKF}(j+1)$ using REKF prediction equations.
   (c) Make $\hat{X}_{REKF}(j+1) = \hat{X}_{pREKF}(j+1)$ and $P_{REKF}(j+1) = P_{pREKF}(j+1)$.
   (d) Compute the matrix $M(j) = D_x f(j)M(j-1)$
   (e) $j = j + 1$
3. Endwhile

The proposed algorithm can be initialized whenever the CAUV (chosen prior to the mission start) communicates with any other vehicle for the first time (the SEKF state and covariance matrix will be increased to add the data of the new vehicle observed). The only thing needed to be known by the vehicles are their current estimated pose and pose uncertainty, given by their REKF that runs in each one of them.

23.4.2 Handling Communication Failures in CEKOLM

Since underwater communication is unreliable, it is important that the CEKOLM is able to handle communications failures. The communication between CAUV and FAUVs can be organized in three steps:

1. CAUV invites the communication of FAUV*i*.

2. FAUV*i* sends information to the CAUV.

3. After updating both position estimates, the CAUV sends back information to the FAUV*i*.

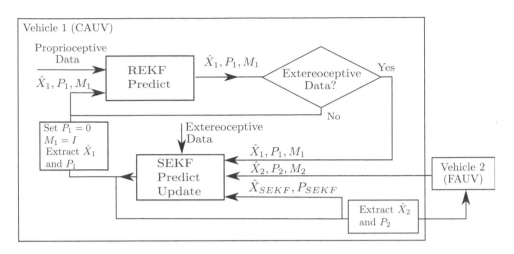

FIGURE 23.2
Block diagram of the CEKOLM algorithm ruining in the CAUV.

A communication failure can happen in any one of these three steps. Now, let us analyze how to solve a failure in each of them.

If there is a failure in step 1, then the FAUV*i* will not receive its invitation to communicate. Therefore, the CAUV will not hear any response from it, and, after waiting a certain fixed time interval, it simply quits waiting for the FAUV*i*. Then, the CAUV chooses another FAUV to communicate with, and repeats this process until a successful reception of a response.

A failure in the second step means that the FAUV received the invitation, and then sent its data to the CAUV, but the CAUV did not received it. The CAUV will consider this as a failure in the first step and will quit communication with FAUV*i*. Then, it will choose another FAUV to communicate with. The FAUV*i* will not know whether the CAUV received its information or not, and thus it must consider the two possibilities. Before setting its covariance matrix P to zero and its matrix M to identity, the FAUV will store copies P_{klt} and M_{klt} of both these matrices (here, index *klt* indicates the last time step the FAUV transmitted data) and keep updating them as if no communication had taken place at all. The next time the CAUV invites the FAUV *i*, it must send along the invitation the last time *kli* it received and used information of vehicle *i* to update the SEKF. In this way, the FAUV will be able to know what copies of the couple of matrices P and M it must send to the CAUV (the ones that represent the accumulated uncertainty and vehicle dynamics since *kli*), and all the other copies will be discarded. Notice that for FAUV *i* to send the time tag *kli* does not require the synchronization of clocks between vehicles. This time tag can be thought of (or replaced by) a serial number, which is particular to each vehicle. There is no need for each vehicle to keep track of the time tag (or serial number) of the messages of the others. We must also modify the CEKOLM algorithm slightly: instead of sending in its current estimated position $\hat{X}_{Fi}(k)$, the FAUV will send in its displacement since *kli*, that is $\hat{X}_d = \hat{X}_{Fi}(k) - \hat{X}_{Fi}(kli)$.

If there is a failure in the third step, the CAUV received and used the information sent in by FAUV*i* to update the SEKF and memorized the current time as *kli* (last time information from FAUV*i* was used). The FAUV will not receive any response from the CAUV and, not knowing whether there was a failure in either the second or the third communication step, it will take the same precautions as in the case of a failure in the second step. The next time it receives an invitation from the CAUV, it will also receive the time instant *kli*, it will choose the appropriate matrices P and M and compute the displacement $\hat{X}_d$ to be sent in to the CAUV. Notice that the FAUV will have its pose and uncertainty updated only after having received the information from the CAUV. However, once a response is received, it will have the current best possible pose estimate.

By incorporating the procedures presented above, the CEKOLM algorithm is able to handle communication failures without compromising its internal consistency.

23.4.3 Decentralized CEKOLM

The key idea to decentralize the CEKOLM algorithm is to keep the cross-correlation information of the whole system in each vehicle, without a great increase in the communication burden.

The decentralized version of CEKOLM, from now on denoted by DCEKOLM, requires that the vehicles are able to broadcast a message. All the vehicles will have a SEKF running on them, as also the usual REKF. At the beginning of the operation, all the vehicles must broadcast their initial position and uncertainty so that each vehicle will have the same initial state of the SEKF.

We can assume that the vehicles are numbered from 1 to n. First, vehicle 1 will play the role of the CAUV and choose (either randomly or according to some given rule) another AUV v to observe. It sends in an invitation to vehicle v, and if accepted, then both vehicles will broadcast their pose X_1 and X_v, their accumulated error P_1 and P_v since their last observation given by the REKF running on each one of them along with their matrices M_1 and M_v and the observation itself (measured distance between them). The amount of information transmitted is approximately double of that for the centralized CEKOLM. As in CEKOLM, both vehicles involved in the observation will set to zero their accumulated error P_1 and P_v after the observation.

After receiving the broadcast information, all the vehicles of the system (including vehicles 1 and v) will update their SEKF using this information only (vehicles other than 1 and v will disregard all local information about their estimation since their last observation). In this way, the SEKF in each one of the vehicles will be exactly the same. Once this is complete, it is time for vehicle 2 to play the role of the CAUV and choose another vehicle to observe, and the process goes on cycling through all vehicles.

One difficulty that might arise is if, due to some communication failure, a certain vehicle does not receive the broadcasted information. In this case, this vehicle would not be able to update its SEKF. To solve this difficulty, a reception acknowledgment is requested from each vehicle and, in the case the information had not been received by one of them, resend it (this could be done by another AUV closer to the one that did not receive). However, this might make the system too slow, and thus another way to deal with communication failures should be considered.

Here, we outline briefly how to address this challenge. The key idea is to include a serial number of the observation in every message broadcast by each vehicle. This is a natural number which is incremented after each observation. If a vehicle does not receive a broadcast, its serial number will be smaller than the expected one when it receives the next broadcast. When its turn to play the role of CAUV arrives and it chooses another vehicle to observe, it can ask this vehicle to resend the lost information (which has to be kept by all the vehicles). With the information that has been kept and the missing information just received, the vehicle is able to reconstruct the current state and have a perfect copy of the SEKF running in the other vehicles that successfully received all the broadcast information.

23.5 Information Driven Motion Coordination

Until now, we have considered the case of the motion of each vehicle per se without any concern about his position relative to the others. But there are also cases in which it is of interest to control the relative positioning of the vehicles. For example, when the group is moving from one region to another, or when they are searching a given area, there is the possibility to position each vehicle relative to each other according to some pattern. We also exploit the extra degree of freedom of the relative positioning in order to improve the cooperative localization algorithm. We know that an observation between two vehicles in a certain relative position will give us more information than in others. So, the idea is to determine the spatial distribution of the vehicles that maximizes the information given by the observations.

In Chitre (2010) and Tan and Chitre (2011), the authors consider that the planned paths of what they call Survey AUV (similar to our FAUV) are known, and based on this information they compute the optimal trajectory of what they call Beacon AUV (AUV equipped with better navigational sensors) so that the estimated position error across all the Survey AUVs are minimized over the mission. In our case, we will consider that the vehicles have no *a priori* knowledge of each other's planned path, so we will compute the optimal relative position of the vehicles at a given time t with the current information available.

It has been shown by Whaite and Ferrie (1995) that a locally optimal strategy for observations maximizes $Det(L) = |L|$, the determinant of the prediction variance in the EKF equations, where

$$L(k) = D_X h P_p(k) D_X h^T + R(k). \tag{23.17}$$

The matrix L given by 23.17 is a function of $R(k)$, $D_X h$ (which in turn is a function of the predicted state estimate $\hat{X}_p$), and of the predicted covariance matrix P_p. So the optimal spatial configuration will be achieved determining the position of the vehicles in the system that maximizes $|D_X h P_p(k) D_X h^T + R(k)|$, and then commanding the vehicles to this position.

23.5.1 Optimal Observation with Range-Only Measurements

23.5.1.1 Single Vehicle

First, let us consider the case of a single vehicle observing a static and known feature on the environment, located at the origin, that is, $X_f = (0, 0)$. Let us compute the pose $X_v = (x_v, y_v, \theta_v)$ that the vehicle should be to maximize $|L|$.

The observation, which is range only, is given by the distance between the vehicle and the feature:

$$h = \sqrt{(x_v - x_f)^2 + (y_v - y_f)^2}.$$

The Jacobian of h w.r.t x, $D_X h$ is then given by

$$D_X h = \left[\frac{x_v - x_f}{h} \quad \frac{y_v - y_f}{h} \quad 0 \right].$$

Since the feature is at the origin, then

$$h = \sqrt{(x_v)^2 + (y_v)^2},$$

and

$$D_x h = \left[\frac{x_v}{h} \quad \frac{y_v}{h} \quad 0 \right].$$

The covariance matrix P of the vehicle is given by

$$P = \begin{bmatrix} P_{xx} & P_{xy} & P_{x\theta} \\ P_{yx} & P_{yy} & P_{y\theta} \\ P_{\theta x} & P_{\theta y} & P_{\theta\theta} \end{bmatrix}.$$

It is most likely that the measurement noise covariance matrix $R(k)$ depends on the pose (x, y, θ) of the vehicle. For a range sensor, typically, the shorter the distance between the vehicle and the feature, the smaller the error is. This would mean that the ideal position of the vehicle is as close as possible to the origin (feature position). Thus, a lower bound on the distance d of the vehicle to the feature is required.

Lemma 23.1

The value of $|L|$ does not depend on the vehicle's orientation θ, and the optimal value of $|L|$ is achieved when the vehicles are positioned so that the largest axis of its error ellipsoid (ellipsoid defined by the covariance matrix P that shows the region of confidence of the estimated state) is co-linear with the feature position.

Proof The optimization problem can be stated as follows:

$$(P) \text{ Maximize } |L(x,y)| \text{ subject to } h(x,y)=d,$$

where $h(x, y) = (x^2 + y^2)^{(1/2)}$, and

$$|L(x,y)| = \frac{1}{h^2(x,y)} \left(x^2 P_{xx} + xy(P_{xy} + P_{yx}) + y^2 P_{yy} \right) + r(x,y),$$

where $r(x, y)$ is the term originated by the covariance of the range sensor (which we assumed depends on the position of the vehicle). A necessary condition for (x^*, y^*) to be a solution to (P) is the existence of a multiplier $\lambda \neq 0$ such that $\nabla \mathcal{L}(x^*, y^*, \lambda) = 0$, where

$$\mathcal{L}(x,y,\lambda) = |L(x,y)| + \lambda(h(x,y) - d).$$

In what follows, L^*, h^*, and r^* denote the corresponding function evaluated at (x^*, y^*), and λ is a scalar multiplier.

A straightforward computation gives

$$0 = \frac{\partial \mathcal{L}^*}{\partial x} = \frac{\partial r^*}{\partial x} + \frac{x}{h^*} \lambda$$
$$+ \frac{1}{h^{*2}} \left(2x(P_{xx} - |L^*| + r^*) + y(P_{xy} + P_{yx}) \right).$$

Similarly, for y,

$$0 = \frac{\partial \mathcal{L}^*}{\partial y} = \frac{\partial r^*}{\partial y} + \frac{y}{h^*} \lambda$$
$$+ \frac{1}{h^{*2}} \left(2y(P_{yy} - |L^*| + r^*) + x(P_{xy} + P_{yx}) \right).$$

The general solution is obtained is given by $A \, col(x^*, y^*) = -\nabla r^*$ where the matrix A is defined by:

$$A = \begin{bmatrix} \dfrac{2(P_{xx} - |L^*| + r^*)}{h^{*2}} + \dfrac{\lambda}{h^*} & \dfrac{P_{xy} + P_{yx}}{h^{*2}} \\ \dfrac{P_{xy} + P_{yx}}{h^{*2}} & \dfrac{2(P_{yy} - |L^*| + r^*)}{h^{*2}} + \dfrac{\lambda}{h^*} \end{bmatrix}.$$

If r does not depend on (x, y) at the optimal, then the solution is a nonzero vector in a non-trivial null space of A. The non-triviality of the null space of A determines the value of the multiplier λ.

If we also have that $P_{xy} = P_{yx} = 0$, then the value of the multiplier λ may be chosen to obtain either $x^* = 0$ or $y^* = 0$ depending on whether the relation between P_{xx} and P_{yy}.

Suppose now that $R(k)$ is a constant function of (x, y). Then, it can be left out of the optimization problem. According to Bahr et al. (2009), this assumption is realistic, since many experiments have shown that the error in the range measurement is only weakly range-dependent and can be modeled as a Gaussian with zero mean and a fixed variance σ^2. If we write $P_{xx} = cP_{yy}, c \in \mathbb{R}$ and $P_{yy} \neq 0$, then according to Equation 23.17, and recalling that $P_{xy} = P_{yx}$, we will have

$$|L| = \begin{bmatrix} \dfrac{x}{h} & \dfrac{y}{h} & 0 \end{bmatrix} \begin{bmatrix} cP_{yy}, & P_{xy} & P_{x\theta} \\ P_{yx} & P_{yy} & P_{y\theta} \\ P_{\theta x} & P_{\theta y} & P_{\theta\theta} \end{bmatrix} \begin{bmatrix} \dfrac{x}{h} \\ \dfrac{y}{h} \\ 0 \end{bmatrix},$$
$$= \frac{P_{yy}(cx^2 + y^2) + P_{xy}(2xy)}{x^2 + y^2}.$$

Notice that the orientation θ of the vehicle does not influence $|L|$. Now, let us consider the case in which the vehicle's x and y variables are not correlated, that is, $P_{xy} = P_{yx} = 0$. If

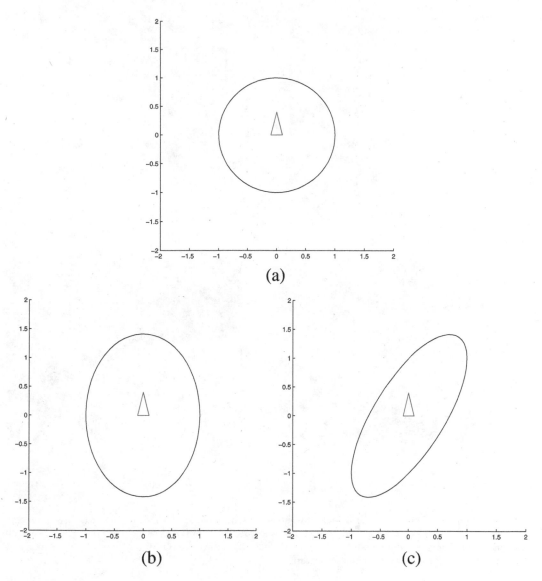

FIGURE 23.3
Uncertainty in vehicle's position represented as an ellipsoid. The axes of the ellipse are the standard deviation of the vehicle's uncertainty. (a) Round uncertainty. (b) Elliptical uncertainty without any correlation. (c) Elliptical uncertainty with some correlation.

we consider that in the present time $c = 1$, which implies that the vehicle has a uniform directional uncertainty, as show in Figure 23.3a, we conclude that

$$|L| = \frac{P_{yy}(x^2 + y^2)}{x^2 + y^2} = P_{yy},$$

which means, as expected, that the determinant of the prediction variance is constant, independently of the vehicle's position. That is, an observation from a certain position is

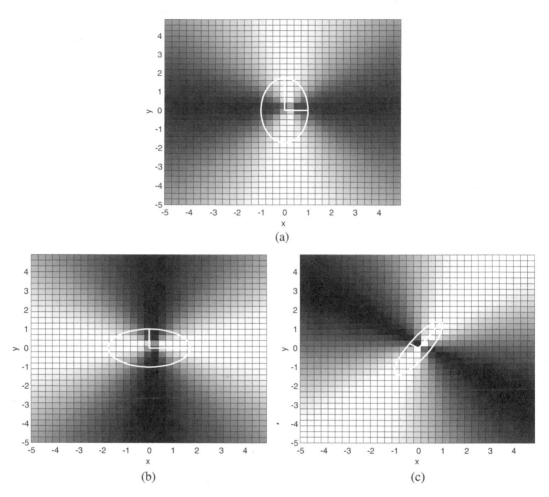

FIGURE 23.4

Computation of the determinant of L (lighter areas represents greater values) for three cases: (a) Vehicle's uncertainty without any correlation, and greater uncertainty in the y coordinate. (b) Vehicle's uncertainty without any correlation, and greater uncertainty in the x coordinate. (c) Vehicle's uncertainty with some correlation and different values for the axes (ellipsoidal uncertainty).

as good as from any other. Now, if we consider the case where $c \neq 1$ we shall have two different possibilities:

- If $c < 1$, then the vehicle has more uncertainty about its y coordinate than about his x coordinate (both in the world reference frame). In this case, $|L|$ is maximum when $x = 0$ (optimal position when vehicle is on the y axes). So, as expected, the optimal strategy is to position the vehicle in order to decrease the its uncertainty in the y axes. The graphic of $|L|$ as a function of (x, y) for this case is shown in Figure 23.4a.

- The case for which $c > 1$ is dealt with in the same way. The conclusions are as in the previous case but with the roles of x and y switched. The graphic of $|L|$ as a function of (x, y) for this case is shown in Figure 23.4b.

Suppose now that the terms $P_{xy} = P_{yx} \neq 0$. We know that the correlation between the x and y variables represents a rotation in the error ellipsoid of the vehicle, as shown in Figure 23.3c. In this case, the optimal strategy is to position the vehicle so that the greater axis of its error ellipsoid is co-linear with the origin (feature position), as show in Figure 23.4c.

23.5.1.2 Multiple Vehicles

In the case of n vehicles, $|L| = f(x_{v1}, y_{v1}, x_{v2}, y_{v2}, \ldots, x_{vn}, y_{vn})$, that is, $|L|$ is a function of (x_{vi}, y_{vi}), $1 \leq i \leq n$, the position of each vehicle. Even if we consider a limited search region for the optimal position for each vehicle, the complexity of the algorithm to determine the maximum of $|L|$ by searching all the values within the considered region will be $O(c^{2n})$, where c is the number of possible positions considered in each axis (in 2D). Clearly, for several vehicles, the complexity of the algorithm is such that its real-time execution is very difficult. However, we can see from the single vehicle scenario that—under the assumption that the sensor covariance is position invariant—the actual position of the vehicle is not relevant, but only its bearing relative to the feature—that is, the angle formed between the vehicle, the origin (feature position), and the x axis, actually matters. Does this relationship also hold true for multiple vehicles? Next, we will show that this is case.

Lemma 23.2

The computation of $|L|$ depends only on the relative bearing of the vehicles (angle formed between the vehicles, the origin and the x axis).

Proof First remark that there is no loss of generality if one of the vehicles is considered the origin of the relative position reference frame. Since the extension to n vehicles is straightforward, for the sake of simplicity, we consider the case of three vehicles. Now, the state vector, and the observation matrix, are, respectively, $X = [x_{v1} \ y_{v1} \ \theta_{v1} \ x_{v2} \ y_{v2} \ \theta_{v2} \ x_{v3} \ y_{v3} \ \theta_{v3}]^T$, and

$$
h = \begin{bmatrix} h_1 \\ h_2 \end{bmatrix} = \begin{bmatrix} \sqrt{(x_{v1} - x_{v2})^2 + (y_{v1} - y_{v2})^2} \\ \sqrt{(x_{v1} - x_{v3})^2 + (y_{v1} - y_{v3})^2} \end{bmatrix},
$$

and thus,

$$
D_X h = \begin{vmatrix} \dfrac{x_{v1} - x_{v2}}{h_1} & \dfrac{y_{v1} - y_{v2}}{h_1} & 0 & -\dfrac{x_{v1} - x_{v2}}{h_1} & -\dfrac{y_{v1} - y_{v2}}{h_1} & 0 & 0 & 0 & 0 \\ \dfrac{x_{v1} - x_{v3}}{h_2} & \dfrac{y_{v1} - y_{v3}}{h_2} & 0 & 0 & 0 & 0 & -\dfrac{x_{v1} - x_{v3}}{h_2} & -\dfrac{y_{v1} - y_{v3}}{h_2} & 0 \end{vmatrix}.
$$

Without any loss of generality, we consider one of them fixed, and with relatively to this one, we compute the position of the other vehicles. By considering that the first vehicle is at the origin $(x_{v1}, y_{v1}) = (0, 0)$, then

$$h = \begin{bmatrix} h_1 \\ h_2 \end{bmatrix} = \begin{bmatrix} \sqrt{x_{v2}^2 + y_{v2}^2} \\ \sqrt{x_{v3}^2 + y_{v3}^2} \end{bmatrix},$$

$$D_X h = \begin{bmatrix} \dfrac{-x_{v2}}{h_1} & \dfrac{-y_{v2}}{h_1} & 0 & \dfrac{x_{v2}}{h_1} & \dfrac{y_{v2}}{h_1} & 0 & 0 & 0 & 0 \\ \dfrac{-x_{v3}}{h_2} & \dfrac{-y_{v3}}{h_2} & 0 & 0 & 0 & 0 & \dfrac{x_{v3}}{h_2} & \dfrac{y_{v3}}{h_2} & 0 \end{bmatrix}.$$

Now, for each vehicle i, suppose that $y_{vi} = a_i x_{vi}$, $a_i \in \mathbb{R}$. We prove now that $|L|$ is a function of a_i (and not of both y_{vi} and x_{vi}). Now, we have

$$D_X h = \begin{bmatrix} \dfrac{x_{v2}}{h_1} & 0 \\ 0 & \dfrac{x_{v3}}{h_2} \end{bmatrix} \begin{bmatrix} -1 & -a_2 & 0 & 1 & a_2 & 0 & 0 & 0 & 0 \\ -1 & -a_3 & 0 & 0 & 0 & 0 & 1 & a_3 & 0 \end{bmatrix} = M_1 M_2.$$

Similarly, $D_X h^T = M_2^T M_1^T$, and, thus

$$L = M_1 M_2 P M_2^T M_1^T + R,$$

The matrix $M_2 P M_2^T$ is a 2×2 matrix which is a function of a_2, a_3 and P. Let us denote

$$M_2 P M_2^T = \begin{bmatrix} A & B \\ C & D \end{bmatrix}. \tag{23.18}$$

Then,

$$L = M_1 \begin{bmatrix} A & B \\ C & D \end{bmatrix} M_1^T + R,$$

$$= \begin{bmatrix} \left(\dfrac{x_{v2}}{h_1} \right)^2 A & \dfrac{x_{v2}}{h_1} \dfrac{x_{v3}}{h_2} B \\ \dfrac{x_{v2}}{h_1} \dfrac{x_{v3}}{h_2} C & \left(\dfrac{x_{v3}}{h_2} \right)^2 D \end{bmatrix} + \begin{bmatrix} r_1 & 0 \\ 0 & r_2 \end{bmatrix}, \tag{23.19}$$

$$|L| = \left(\left(\dfrac{x_{v2}}{h_1} \right)^2 A + r_1 \right) \left(\left(\dfrac{x_{v3}}{h_2} \right)^2 D + r2 \right)$$

$$- \left(\dfrac{x_{v2} x_{v3}}{h_1 h_2} \right)^2 BC$$

$$= \dfrac{x_{v2}^2 x_{v3}^2}{(x_{v2}^2 + a_2^2 x_{v2}^2)(x_{v3}^2 + a_3^2 x_{v3}^2)} (AD - BC)$$

$$+ \dfrac{x_{v2}^2 A r_2}{(x_{v2}^2 + a_2^2 x_{v2}^2)} + \dfrac{x_{v3}^2 D r_1}{(x_{v3}^2 + a_3^2 x_{v3}^2)} + r_1 r_2, \tag{23.20}$$

$$\dfrac{(AD - BC)}{(1 + a_2^2)(1 + a_3^2)} + \dfrac{A r_2}{(1 + a_2^2)} + \dfrac{D r_1}{(1 + a_3^2)} + r_1 r_2.$$

So we can conclude that $|L|$ is a function of a_i, R and P only.

This means that $|L|$ depends only on the angle of the vehicles relative to each other. □

With this simplification, the algorithm complexity is now $O(c^n)$, where c is the number of possible angles considered and n is the number of vehicles. However, the algorithm is still not practical to search the entire space for a large number of vehicles, so we must use some numerical method to find the solution.

In our implementation, the algorithm to determine the optimal spatial distribution of the vehicles is computed by the CAUV. The CAUV considers itself to be at the origin and computes the optimal position (actually the optimal angle θ_{max}) relative to itself that the CCFAUV should be. Once θ_{max} is found, it is sent to the CCFAUV.

For the CCFAUV to be able to compute the real position it should drive to, it must know the estimated position of the CAUV. The CAUV must also transmit its current heading and velocity, so that the CCFAUV will estimate the future positions it might occupy while it is not communicating with the CAUV. Hence, in between observations, each FAUV assumes that the CAUV moves in a straight line with constant longitudinal velocity. If the CAUV does not follow a trajectory close to a straight line—the "straightness" of the trajectory is in a local sense, that is, related to the frequency of CAUV's observations of the FAUVs—the method will not position the FAUVs in the optimal position, and thus the advantage will be small, as shown in Section 23.6.

One disadvantage of commanding the FAUVs to the optimal positions relative to the CAUV is that their trajectories are longer than those if they were just keeping the relative position to the CAUV all the time. This additional motion will not only increase the uncertainty of the system, but also will increase power consumption, which is a critical issue for AUVs. Therefore, in order to save power, we need to compute minimum power consumption trajectories. A similar effect was observed by Robert Sim (2005), where he presents a study of a bearings-only localization system. Because of this, it is important that the control laws applied to drive the vehicles to the respective target positions are such that the traveled distances are minimized. Next, we present a modified control law to do so.

23.5.2 Control Law

In this section, we present a method to synthesize a control law that minimizes the distance traveled by the vehicle to reach the specified target position. This optimization has the advantages of requiring less power consumption and producing trajectories with minimal process error injected into the localization system, and thus yielding minimal uncertainty on the final vehicle's pose estimate.

Aicardi et al. (1995) proposed some very simple and effective closed-loop control laws for unicycle vehicles by using the Lyapunov control theory. Consider that the position of the vehicle $X_v = (x_v, y_v, \theta_v)$ relative to the target position $X_{ref} = (x_r, y_r)$ represented in polar coordinates (as depicted in Figure 23.5) given by

$$R = \sqrt{(x_v - x_r)^2 + (y_v - y_r)^2},$$
$$\alpha = atan((y_r - y_v)/(x_r - x_v)) - \theta_v.$$

The control signals v (linear velocity along θ_v) and w (angular velocity) are given by

$$v = K_1 R \cos(\alpha),$$
$$w = K_2 \alpha + K_1 \frac{\sin(\alpha)\cos(\alpha)}{\alpha}(\alpha + K_3 \theta_v), \quad (23.21)$$

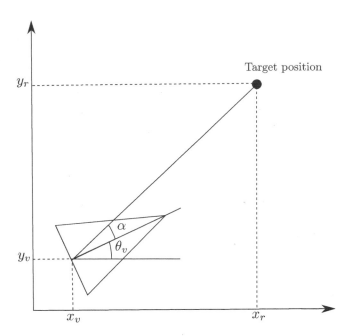

FIGURE 23.5
Vehicles pose (x_v, y_v, θ_v) and target position (x_r, y_r) represented in polar coordinates (R, α).

where K_1, K_2 and K_3 are positive gains. Notice that, due to *atan* function, this control law becomes singular when the vehicle is very close to the reference. Thus, the controller must be turned off in those situations.

This control law assumes that the vehicle can have non-positive linear velocity (Aicardi et al. 1995). But, since this is usually not possible for AUVs, if the computed velocity is negative, then we just set $v = e$, a small positive value. This value e is determined by the dynamics of the vehicle and should be such that the AUV remains robustly controllable.

In order to minimize the distance traveled by the vehicle, we propose a new controller by making use of a "virtual reference." The idea is to avoid the extra motion of the vehicle when the reference frame origin is approaching its position, as defined below. In this case, the vehicle will move as little as possible and wait for the reference to pass by. Mathematically, the virtual reference is computed whenever the projection

$$P_{X_v}^y = -\sin(\theta_v)(x_v - x_r) + \cos(\theta_v)(y_v - y_r)$$

of the position of the vehicle X_v on the y axis of the reference frame, y', has a positive value. In this case, the AUV will target a "virtual reference" located along y^r with some value w greater than $P_{X_v}^y$. When the reference passes by $\left(P_{X_v}^y < 0\right)$, the vehicle switches and targets the real reference. Figure 23.6 illustrates a case for which a virtual reference is created.

It is also necessary to decrease the linear velocity gain K_1 when the vehicle is targeting the virtual reference. K_1 is decreased proportionally to the inverse of the distance between the vehicle and the virtual reference. In this way, the vehicle will move slower as it approaches the virtual reference, and will wait for the real reference to pass by. However, the velocity should be above a minimum value e in order to preserve controllability. In order to avoid the singularity introduced when the vehicle is too close to the reference, the vehicle switches

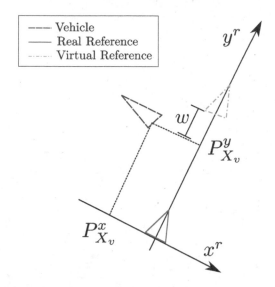

FIGURE 23.6
When the vehicle (dashed) is above the real reference (solid) along the y^r axes, that is, the projection of the position of the vehicle X_v on the y axis of the reference frame has a positive value, a virtual reference (dash-dotted) is created and targeted by the vehicle.

between virtual and real reference only after the real reference has passed a certain distance d from the vehicle $\left(P_{X_v}^y < -d\right)$.

The full control law is shown in Algorithm 23.2. In Figure 23.7, we can see a comparison between the execution of the control algorithm using the virtual reference scheme and that when not using it. Notice that with the virtual reference, the distance traveled by the vehicle is much smaller, thus injecting less noise in the vehicle's pose while saving power.

Algorithm 23.2 Control Algorithm with virtual reference

---

1. Initialize State Vector, Error Covariance Matrix and references for each FAUV
2. While true
   - (a) For $i = 1...n$, where n is the number of FAUVs in the system
     - i. Compute the reference position $X_{ref}^i(t)$ of the ith FAUV based on the reference of the last step $X_{ref}^i(t-1)$, and the FAUV current information about the heading and velocity of the CAUV
     - ii. Compute the projection $P_{X_v}^y$ of the ith FAUV estimated position on the y^r axis of the frame fixed on the reference
     - iii. If $P_{X_v}^y < -d$
       - A. Compute the virtual reference $X_{vref}^i(t)$ and set it as the target for the ith FAUV
       - B. Decrease the gain k_1 proportionally to the inverse of the distance between the ith FAUV and $X_{vref}^i(t)$
     - iv. Endif
     - v. Compute the control signal u^i by using the control laws (Equation 23.21) based on the ith FAUV current position and the reference
     - vi. Move the ith FAUVs based on the control signal u^i
   - (b) Endfor

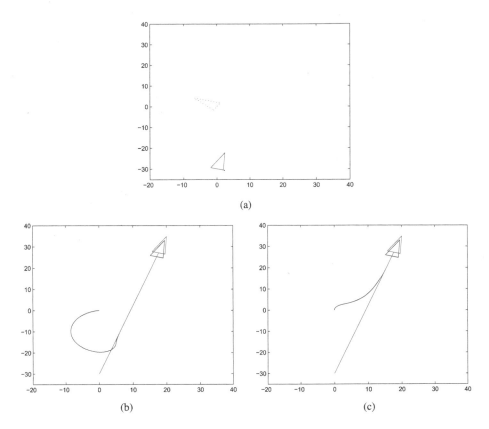

FIGURE 23.7
Comparison of control laws: (a) initial position of the vehicle (at the origin) and reference, which is moving forward in a straight line, (b) basic control law, and (c) control law with the virtual reference.

    (c) Move the CAUV based on his own internal state and current mission.
    (d) If there is an observation
        i. Update the CCFAUV information about the heading and velocity of the CAUV
        ii. Based on the covariance matrix P_{SEKF} (the one computed by the CAUV with all the vehicles), compute the angle θ_{max} of the CCFAUV relative to the CAUV that maximizes $|L|$
        iii. Compute the target position $X_{ref}(t+1)$ of the CCFAUV based on θ_{max} and some radius r (the desired distance between the FAUVS and the CAUV)
    (e) Endif
    (f) $t = t + 1$
  3. Endwhile

During an execution of the algorithm, as the CAUV communicates with the FAUVs, the optimal angle θ_{max} for each FAUV changes over time. If the optimal angle of a FAUV varies too rapidly, then this implies fast changes in the FAUV's position, and more noise is introduced in its pose estimate. Moreover, these fast changes may prevent the vehicle from reaching the optimal position. One strategy to prevent this situation is, before reaching its current target, the FAUV communicates once more with the CAUV in order to receive a new target. Therefore, we use a filter to slow down the optimal angle changes. This filter takes the weighted mean of the past n values of θ_{max} (θ_{max} $(k-1)$, θ_{max} $(k-2)$, ..., θ_{max} $(k-n)$) and the current target one θ_{max} (k).

23.6 Simulations and Results

The vehicle model used for all AUVs in the simulations was a 2D unicycle model presented in Equation 23.7. The dead reckoning information available is the control signal $u(k)$, which can be measured, for example, by an inertial measurement unit (IMU). The process noise matrix Q is a diagonal matrix with 0.02 in the first two entries and 0.35 in the third one. The external sensor, a range detector, has an error variance $R=[0.5]$. All the noises are considered to be white with zero mean. In all the simulations the time step is 0.1 seconds, the rate at which the KF runs the given proprioceptive data.

All the experiments are repeated 400 times to get the mean error value from all the executions. The error in each execution, *err*, is the mean of the Euclidean error between the true trajectory, in x and y axes, described by the CAUV (CAUV$_{true}$), and the filtered one (CAUV$_{filt}$) for each time step of the algorithm, as shown in Equation 23.22.

$$err = \frac{1}{nsteps} \sum_{k=0}^{nsteps} \| CAUV_{true}(k) - CAUV_{filt}(k) \| \tag{23.22}$$

Figure 23.8 shows a simulation of CEKOLM with three FAUVs, observation period of 5 seconds, that is, the CAUV communicates with a FAUV every 5 seconds, and no FAUV

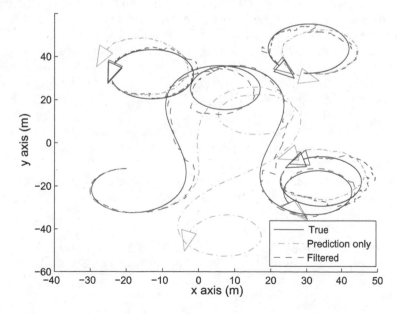

FIGURE 23.8
2D simulation of CEKOLM during 320 seconds with three FAUVs plotted in the (x, y) plane. The FAUVs describe circles at a longitudinal velocity of 2 m/s, while the CAUV performs a longer trajectory at a longitudinal speed of 1 m/s. The observation period is 5 seconds, and there is no position update by GPS by any vehicle.

has GPS update during the simulation. The figure shows the true trajectory described by the AUVs (FAUVs describes circles), the filtered trajectory estimated by the algorithm and a prediction—using the prediction equations from Equation 23.2—only estimation, using only proprioceptive information. Notice that, as expected, the prediction-only trajectory drifts away from the true trajectory for all AUVs as time increases, especially for the CAUV, which is moving faster in this simulation. On the other hand, the filtered estimation stays close to the true trajectories and drifts away more slowly.

In Table 23.1, we show the mean error of the proposed algorithm (CEKOLM), the Cooperative Navigation Algorithm (CONA) method proposed by Bahr and Leonard (2006), and the prediction-only (no filtering). In this simulation, the vehicles started at random poses in the working area and described curved trajectories. We can see that the CEKOLM algorithm can reduce the mean error of the prediction-only method by 60%, and of the method proposed by Bahr and Leonard (2006) by 70%. Notice that the variance is also greatly reduced. This poor performance by the CONA method is due to the absence of CNAs in the system, as discussed in Nogueira et al. (2010). Since CONA is based on circle intersections to estimate the CAUV position, if the noise (process and sensor) is high, the calculated circle positions have some error, and this can cause a high error on the estimated CAUV position depending on the position of the FAUVs involved, as discussed in Bahr and Leonard (2007).

To assess the handling of communication failures, we simulated failures in the three steps mentioned in Section 23.4.2. Since the amount of data transmitted in each step is different, we assigned different probabilities of communication failures for each one: 10%, 50%, and 25% to steps 1, 2, and 3, respectively. The simulations showed that, even with failures, the algorithm could perform properly and maintain its internal consistency. by running a sequence of simulations of the algorithm in two scenarios—(1) perfect communications and (2) simulated communication failures—we observed that the results were similar. The mean Euclidean error was 4.58 and 4.50 for scenarios (1) and (2), respectively. We recall that there is the need to change the observation time in scenario (1) so that the number of communication successes is the same as that in scenario (2).

We also used real experimental data from a vehicle (SeaCon, which was developed in Faculdade de Engenharia da Universidade do Porto (FEUP) and is shown in Figure 23.9) to assess our method. The vehicle used was equipped with the following sensors: a motor rpm counter (sampling frequency of 1 Hz), an IMU (50 Hz), a compass (50 Hz), a pressure sensor (15 Hz), and a GPS (1 Hz).

The vehicle is commanded for three missions of about 8 minutes each, on the surface (to use GPS as a ground truth), and the sensor data collected for each mission. The trajectory described by the vehicle can be seen in Figure 23.10. This data is then loaded on the simulator and three vehicles operating simultaneously from the data are simulated of each mission. We also simulated the communications and range measurements between the vehicles.

TABLE 23.1

Mean Euclidean Error and Variance of the Algorithms for CAUV during 300 s, with an Observation Time of 5 s and no GPS Update

Algorithm	Mean Error (m)	Variance
Prediction	8.2	20.5
CONA	11.6	21
CEKOLM	3.0	1.6

Note: The total distance traveled by the AUVs was 300 m.

FIGURE 23.9
SeaCon vehicle. (Courtesy of LSTS, FEUP.)

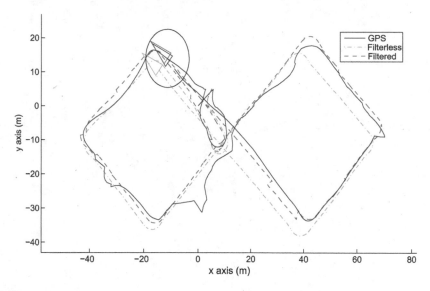

FIGURE 23.10
Result of CEKOLM using real data. Each vehicle traveled about 360 m during 450 s. The simulation considered three vehicles operating simultaneously. The simulated observation time is 10 s. The ellipse depicts the uncertainty region of the estimated vehicle position at the end of the experiment.

The range measurements were simulated with a Gaussian error of variance $R = [1]$, and the observation time was 10 seconds. In the simulations, only the 2D position of the vehicle (latitude, longitude, and heading) are considered.

Firstly, we simulated the situation where the vehicles are not equipped with a compass (ignoring information of this sensor). In this case, the filterless trajectory (prediction only) diverges very quickly from the true one, and the range measurements can greatly reduce the error of the estimated state. In this scenario, the filterless mean Euclidean error of the CAUV was 49.22 m, while the filtered trajectory shows a mean error of 15.91 m, corresponding to an error reduction of 67%.

If we use all the sensor data available, including the compass, the mean Euclidean error obtained by the filterless trajectory (of the CAUV), when compared with the GPS, is 4.89 m, while the error of the filtered trajectory using CEKOLM is 2.66 m. The computed trajectories

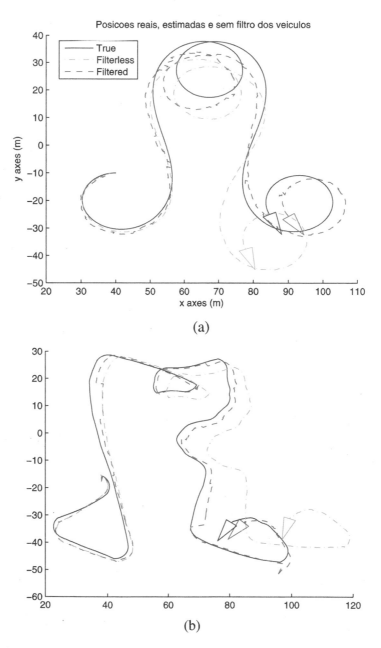

FIGURE 23.11
Simulation results for CAUV curved trajectory: (a) CAUV, and (b) one of the FAUVs. Observation time of 2 s and radius of 20 m.

described by the CAUV are shown in Figure 23.10. Although the performance of the system is not as high as in the case in which there is no compass, still the algorithm was able to reduce the error in 45%.

To compute the best position of the FAUV relative to the CAUV, since it depends only on the angle formed between the vehicle, the origin (feature position) and the x axis, as shown in Equation 23.20, we must choose a desired distance R_d, that the FAUVs should keep from the CAUV. We notice also that in a real application, the CAUV should check the target position assigned to each FAUV and change them for the detection of a collision situation. However, since it has only an estimate of the vehicle's pose, each vehicle must also have some collision avoidance system.

In Figure 23.11, we show a simulation from the CEKOLM algorithm in which the FAUVs are commanded to the best possible position to maximize the information from the observations. In this simulation, the CAUV is commanded to take a curved trajectory (solid line trajectory on Figure 23.11a). The observation time is 2 s and the desired distance R_d between CAUV and FAUVs is 20 m. The total distance traveled by the CAUV is 320 m.

Table 23.2 shows the mean error and variance of the CAUV in a system with four AUVs, for 400 executions with a curved trajectory described by the CAUV for five cases:

1. The algorithm computes the optimal angle θ_{max} for each FAUV, and commands them to the optimal position.

2. The algorithm chooses a fixed random angle at the beginning of the simulation for each FAUV, and commands them to keep this angle throughout the simulation.

3. The algorithm computes the worst possible angle $\theta_{min} = (\theta_{max} + 90°)$ for each FAUV, and then commands them to these worst possible positions.

4. All AUVs starting at random positions, and orientations and describing independent and similar predetermined trajectories (the CAUV is not followed).

5. Prediction only.

Notice that the errors of the algorithms in which the FAUVs follow the CAUV are higher than those of the case in which the AUVs describe independent predetermined trajectories. This happens because now the FAUVs have to follow the CAUV with intermittent observations, which causes them to travel greater distances and this entails in higher process errors. But, if the vehicles do not know *a priori* their target assigned by the CAUV, they

TABLE 23.2

Comparison between CEKOLM, in a Curved Trajectory, Commanding the Vehicles to the Optimal Position, to a Random Position (Based on a Random Angle), to the Opposite Position from the Optimal ($\theta_{max} + 90°$), to no Position (Predetermined Random Trajectories) and no use of the CEKOLM Algorithm (Prediction Only)

Algorithm	Mean Error (m)	Variance
CEKOLM with optimal angle	4.0	5.8
CEKOLM with fixed random angle	4.0	3.9
CEKOLM with worst possible angle	3.9	3.9
CEKOLM with random trajectories	3.0	1.6
Prediction only	8.2	20.5

Note: The observation time is 5 s, and there are four AUVs. For this kind of trajectory, all the algorithms exhibit similar performances, and reduce the error relatively to the prediction-only algorithm by approximately 50%.

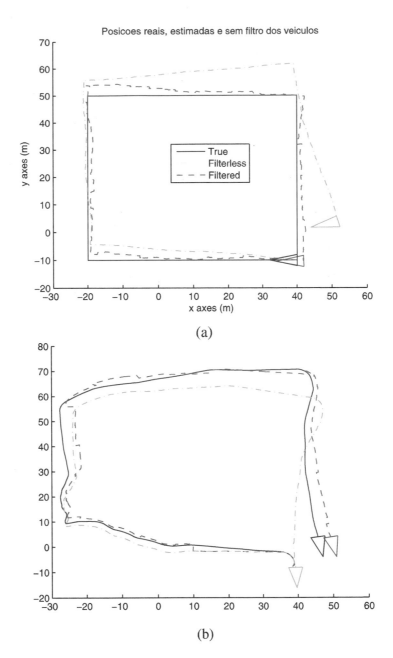

FIGURE 23.12
Results for square trajectory (a) CAUV, and (b) one of the FAUVs. Observation time of 2 s and radius of 20 m.

cannot have independent trajectories. Even if they have a common known target, albeit with independent trajectories, by making small changes in their trajectories they benefit if the changes are such that the information from the observations among them is maximized.

When compared to the prediction-only case, the error is reduced by approximately 50%. Notice however, that the three cases behave in an almost similar way. This happens because since the trajectory is highly curved and the observation time is high, the FAUVs cannot keep the optimal position relative to the CAUV.

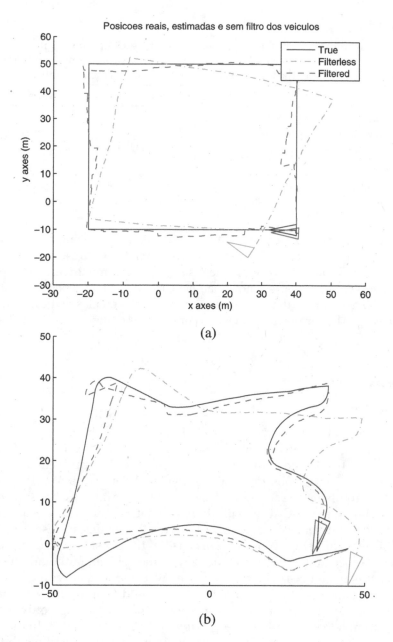

FIGURE 23.13
Results for square trajectory (a) CAUV, and (b) one of the FAUVs. Observation time of 5 s and radius of 20 m.

In order to fully use the advantages brought by the algorithm, the trajectory described by the CAUV must consist of line segments. Figures 23.12 and 23.13 show results of two simulations with this kind of trajectory. In Figure 23.12, the observation time is of 2 s, and one can observe that the trajectories described by the FAUVs are closer to the ones of the CAUV. In Figure 23.13, the observation time is 5 s, so the trajectories of the FAUVs are not so close to the one of the CAUV but are still quite similar. In both simulations, the total distance traveled by the CAUV is 320 m.

TABLE 23.3

Comparison between CEKOLM, in a Square Trajectory, Commanding the Vehicles to the Optimal Position, to a Random Position (Based on a Random Angle), to the Opposite Position from the Optimal ($\theta_{max} + 90°$) and for Prediction Only

Algorithm	Mean Error (m)	Variance
CEKOLM with optimal angle	3.8	3.4
CEKOLM with fixed random angle	4.8	9.4
CEKOLM with worst possible angle	4.8	9.3
Prediction only	8.2	17.5

Note: The observation time is 5 s and there are four AUVs. For this kind of trajectory, the algorithm with optimal angle performs better than the others.

In Table 23.3, we show the mean error and variance for 400 executions where the CAUV describes a squared trajectory similar to the one shown in the figures, where the observation time was 5 s. As in Table 23.2, we show the results for the same three cases: optimal, fixed random, and worst case. We can see now that the optimal algorithm performs a lot better. As discussed earlier, we can see that, when commanding the vehicles to the optimal positions, the algorithm yields better results than when commanding them to fixed-relative positions.

23.7 Conclusions

Cooperative localization for AUVs is a challenging area because of the challenges posed by the underwater environment, especially the communication constraints. We proposed a centralized algorithm to solve the localization problem with a group of AUVs, all with similar and possibly low-cost navigation systems. We devised a way for the FAUVs to share their trajectories and associated uncertainties transmitting the least possible information to the CAUV. With this information, the CAUV is able to maintain a consistent cross-covariance matrix involving all the vehicles present in the system, which is critical for the SLAM problem. This enables the construction of an accurate map relative to a reference frame that moves along with the vehicles (although the uncertainty of the pose of this reference frame relative to the world increases with time). We studied the relative formation that the group of vehicles should keep in order to maximize the information given the observations. It was shown that for a range-only measurement, the best formation depends only on the bearing of the FAUVs relative to the CAUV. Pure simulation and simulation incorporating experimental data results show the advantages of the proposed formation.

The proposed method works only when the vehicles execute proprioceptive navigation (using only prediction equations of the EKF) between range observations. Future work includes investigation on how exteroceptive information in between range observations can be incorporated to improve the vehicles navigation.

It was mentioned that if a large number of vehicles is required by the application, then several subgroups of vehicles, each one using the CEKOLM algorithm, can be considered. Although this benefits each group, there will be no accurate relative position between them. Thus, another line of research consists in hierarchical scaling the CEKOLM algorithm.

References

Aicardi, M., Casalino, G., Bicchi, A. and Balestrino, A. 1995. Closed loop steering of unicycle like vehicles via Lyapunov techniques, *IEEE Robotics Automation Magazine* 2(1): 27–35.

Bahr, A. 2009. *Cooperative Localization for Autonomous Underwater Vehicles*, PhD thesis, Massachusetts Institute of Technology, Cambridge, MA.

Bahr, A. and Leonard, J. J. 2006. Cooperative localization for autonomous underwater vehicles. *ISER*, pp. 387–395.

Bahr, A. and Leonard, J. J. 2007. Minimizing trilateration errors in the presence of uncertain landmark positions. *Proceedings of the 3rd European Conference on Mobile Robots (ECMR)*, Freiburg, Germany, pp. 48–53.

Bahr, A., Walter, M. and Leonard, J. 2009. Consistent cooperative localization, robotics and automation, 2009. *IEEE International Conference on ICRA '09*, pp. 3415–3422. URL: 10.1109/ROBOT.2009.5152859

Bailey, T. and Durrant-Whyte, H. 2006. Simultaneous localization and mapping (SLAM): Part II, *IEEE Robotics & Automation Magazine* 13(3): 108–117.

Chen, Q., You, K. and Song, S. 2017. Cooperative localization for autonomous underwater vehicles using parallel projection. *2017 13th IEEE International Conference on Control Automation (ICCA)*, pp. 788–793.

Chitre, M. 2010. Path planning for cooperative underwater range-only navigation using a single beacon. *2010 International Conference on Autonomous and Intelligent Systems AIS, IEEE*, pp. 1–6.

Dajun, S., Yinan, Z., Yunfeng, H. and Jucheng, Z. 2017. Preliminary study on cooperative localization for AUVs. *2017 IEEE International Conference on Signal Processing, Communications and Computing (ICSPCC)*, pp. 1–4.

Dissanayake, M., Newman, P., Clark, S., Durrant-Whyte, H. and Csorba, M. 2001. A solution to the simultaneous localization and map building (SLAM) problem, *IEEE Transactions on Robotics and Automation* 17(3): 229–241.

Durrant-Whyte, H. 1987. Uncertain geometry in robotics. *1987 IEEE International Conference on Robotics and Automation. Proceedings*, vol. 4, pp. 851–856.

Durrant-Whyte, H. and Bailey, T. 2006. Simultaneous localization and mapping: Part I, *IEEE Robotics & Automation Magazine* 13(2): 99–110.

Fox, D., Thrun, S., Burgard, W. and Dellaert, F. 2001. Particle filters for mobile robot localization. In: Doucet A., de Freitas N., Gordon N. (eds), *Sequential Monte Carlo Methods in Practice*. Statistics for Engineering and Information Science. Springer, New York, NY.

Grime, S. and Durrant-Whyte, H. 1994. Data fusion in decentralized sensor networks, *Control Engineering Practice* 2(5): 849–863. URL: http://www.sciencedirect.com/science/article/B6V2H-47X8MP0-1FW/2/ec85f9ddde044133055543e4cedc0e6b

Huang, Y., Zhang, Y., Xu, B., Wu, Z. and Chambers, J. A. 2018. A new adaptive extended Kalman filter for cooperative localization, *IEEE Transactions on Aerospace and Electronic Systems* 54(1): 353–368.

Kehagias, A., Djugash, J. and Singh, S. 2006. Range-only slam with interpolated range data, *Technical Report CMU-RI-TR-06-26*, Carnegie Mellon University, Pittsburgh, PA.

Kurazume, R., Hirose, S., Nagata, S. and Sashida, N. 1996. Study on cooperative positioning system (basic principle and measurement experiment). *1996 IEEE International Conference on Robotics and Automation, 1996. Proceedings*, vol. 2, pp. 1421–1426 vol.2.

Li, J. and Zhang, J. 2016. Research on the algorithm of multi-autonomous underwater vehicles navigation and localization based on the extended Kalman filter. *2016 IEEE International Conference on Mechatronics and Automation*, pp. 2455–2460.

Lichuan, Z., Mingyong, L., Demin, X. and Weisheng, Y. 2009. Cooperative localization for underwater vehicles. *4th IEEE Conference on Industrial Electronics and Applications, 2009. ICIEA 2009*, pp. 2524–2527. URL: 10.1109/ICIEA.2009.5138662

Matsuda, T., Maki, T., Sato, Y. and Sakamaki, T. 2014. Cooperative navigation method of multiple autonomous underwater vehicles for wide seafloor survey - sea experiment with two AUVs. *OCEANS 2014 -TAIPEI*, pp. 1–9.

Maybeck, P. S. 1979. *Stochastic Models, Estimation and Control*. Volume I. PS Maybeck and Academic Press.

Mourikis, A. and Roumeliotis, S. 2006. Performance analysis of multirobot cooperative localization, *Robotics, IEEE Transactions on* 22(4): 666–681. URL: 10.1109/TRO.2006.878957

Nettleton, E. and Durrant-Whyte, H. F. 2001. Delayed and asequent data in decentralized sensing networks, *Proc. SPIE 4571, Sensor Fusion and Decentralized Control in Robotic Systems IV*.

Newman, P. M. 2004. *EKF based navigation and SLAM*. SLAM Summer School, 2004, Toulouse.

Newman, P. M. and Leonard, J. 2003. Pure range-only sub-sea SLAM. *IEEE International Conference on Robotics and Automation, 2003. Proceedings. ICRA '03*, vol. 2, pp. 1921–1926.

Nogueira, M. B., Sousa, J. B. and Pereira, F. L. 2010. Cooperative autonomous underwater vehicle localization. *Proceedings of Oceans'10 IEEE Sydney*, Sydney, Australia.

Olson, E., Leonard, J. J. and Teller, S. 2006. Robust range-only beacon localization, *IEEE Journal of Oceanic Engineering*, 31(4): 949–958.

Paull, L., Huang, G., Seto, M. and Leonard, J. J. 2015. Communication-constrained multi-AUV cooperative SLAM. *2015 IEEE International Conference on Robotics and Automation (ICRA)*, pp. 509–516.

Roumeliotis, S. I. and Bekey, G. 2002. Distributed multirobot localization, *IEEE Transactions on Robotics and Automation* 18(5): 781–795.

Roumeliotis, S. I. and Rekleitis, I. M. 2004. Propagation of uncertainty in cooperative multi-robot localization: Analysis and experimental results, *Autonomous Robots* 17: 41–54. 10.1023/B:A URO.0000032937.98087.91. URL: http://dx.doi.org/10.1023/B:AURO.0000032937.98087.91

Sim, R. 2005. Stabilizing information-driven exploration for bearings-only SLAM using range gating. *2005 IEEE/RSJ International Conference on Intelligent Robots and Systems, 2005. (IROS 2005)*, pp. 3396–3401.

Smith, R., Self, M. and Cheeseman, P. 1990. Estimating uncertain spatial relationships in robotics. In Cox, I.J. and Wilfon, G.T. (eds) *Autonomous Robot Vehicles*, pp. 167–193, Springer-Verlag.

Tan, Y. T. and Chitre, M. 2011. Single beacon cooperative path planning using cross-entropy method. *IEEE/MTS Oceans'11 Conference*, Hawaii, USA.

Vaganay, J., Leonard, J., Curcio, J. and Willcox, J. 2004. Experimental validation of the moving long base-line navigation concept. *Autonomous Underwater Vehicles, 2004 IEEE/OES*, pp. 59–65.

Webster, S., Whitcomb, L. and Eustice, R. 2010. Preliminary results in decentralized estimation for single-beacon acoustic underwater navigation. *Proceedings of Robotics: Science and Systems*, Zaragoza, Spain.

Whaite, P. and Ferrie, F. P. 1995. Autonomous exploration: Driven by uncertainty, *IEEE Transactions on Pattern Analysis and Machine Intelligence* 19: 339–346. URL: http://citeseerx.ist.psu.edu/viewdoc/summary?doi=10.1.1.21.1276

Yao, Y., Xu, D. and Yan, W. 2009. Cooperative localization with communication delays for MAUVs. *IEEE International Conference on Intelligent Computing and Intelligent Systems, 2009. ICIS 2009*, vol. 1, pp. 244–249. URL: 10.1109/ICICISYS.2009.5357852

Yao, Y., Xu, D., Yan, W. and Gao, B. 2009. Optimal decision making for cooperative localization of MAUVs. *International Conference on Mechatronics and Automation, 2009. ICMA 2009*, pp. 1134–1139. URL: 10.1109/ICMA.2009.5246031

Zhang, L., Wang, J., Tonghao, W., Liu, M. and Gao, J. 2016. Optimal formation of multiple AUVs cooperative localization based on virtual structure, *OCEANS 2016 MTS/IEEE*, Monterey, CA, pp. 1–6.

Section IV

Applicable Platforms and Directions for Cooperative Navigation

Issues and Challenges

24

Cooperative Localization: Challenges and Future Directions

Anusna Chakraborty, Sohum Misra, Rajnikant Sharma,
Kevin M. Brink, and Clark N. Taylor

CONTENTS

24.1 Introduction

In today's world, unmanned vehicles (UVs) are steadily finding applications in both civilian and military fields, especially in dull, dirty, and dangerous mission scenarios, thus reducing potential risks to human life and increasing efficiency. Multiple UV applications are gaining precedence over single-vehicle missions since they reduce the risk of failure (single-point failure risk) and increase robustness while providing increased sensor coverage [1]. These UVs can share their sensor information with the group to aid in objective completion. One such objective is vehicle localization [10]. For a vehicle to be able to localize itself, position information is necessary, which is provided by the global positioning system (GPS) [24]. However, GPS signals may be jammed or spoofed [15] and even though technology exists to overcome such obstacles [44,46], these signals are unreliable in urban areas like canyons and near high rises. In such scenarios, for a connected sensor network, vehicles cooperate to share their odometry and relative position information to collectively estimate the pose of all vehicles in the group.

Cooperative localization [19,20,18,26,32,35,36,45] has been an active area of research over the last few decades as it presents several advantages over traditional single-vehicle systems, including robustness, increased efficiency, information exchange, and increase in virtual sensing capability. Owing to these pros, cooperative localization finds a large variety of applications such as navigation of ground [40], underwater [3], and aerial vehicles [42,43,7], patrolling [30,31], search-and-rescue missions [16], target tracking [38], and time-synchronized path following [41], to name a few.

Several different estimation approaches have been employed to solve cooperative localization, such as the Extended Kalman Filter (EKF) [40], maximum likelihood estimator [13], factor graphs using non-linear least squares [27], maximum *a priori* (MAP) [29], and so on. These estimators can be present on a single master node or distributed. When a single master node is present, it fuses the sensor information from all vehicles to provide pose estimate for the individual vehicles [47,28]. Distributed estimation techniques have been explored to reduce the computation overhead. Using such algorithms, the vehicles only exchange information with their immediate neighbors to generate their own estimates [11,5]. Regardless of the estimation technique used, the quality of the estimation or navigation results will be highly dependent on the observability of the system [14]. Simply defined, observability is the ability to distinguish between two states at different time instants given certain sensor information. Observability analysis for cooperative localization has been performed for both continuous time [39] and discrete time systems [6]. Maintaining or improving observability using controllers [37] for applications using cooperative localization is extremely important to ensure that the mission objectives are achieved in the absence of any global position reference.

Although this is widely researched area, there are many challenges faced in the practical implementation of cooperative localization, where sensor data packet loss, communication delays, vehicle failures, and random biases in sensors are common along with external conditions. These provide interesting areas of research and provide a good platform for focused future work, which can be directed toward better system modeling or more robust noise handling capabilities and have discussed in depth in [4]. The objective of this chapter is to introduce the problem of cooperative localization in Section 24.3 followed by discussion of some of the major techniques and algorithms in Section 24.4. Mathematical analysis of cooperative localization is presented in Section 24.5.1 followed by a brief introduction to controllers that aid in cooperative localization in Section 24.5.2. Some applications are presented in Section 24.6 with algorithms and results to show the effectiveness of cooperative localization in GPS-denied environments. The major outline of the chapter is as follows:

1. Problem statement

2. Centralized and distributed approaches

3. Observability analysis of cooperative localization problem

4. Observability-based control

5. Challenges and future work

24.2 Problem Statement for Cooperative Localization

As mentioned briefly in Section 24.1, the problem of cooperative localization is used to facilitate navigation in GPS-denied environments using the vehicle's motion information from its internal sensors and the information about its surroundings from its external sensors. This exchange of information between the vehicles in the group in the presence of certain features whose position information is known *a priori* helps in generating vehicle state estimates for all vehicles in the group, which enables completion of various tasks in GPS-denied or jammed or spoofed environments, or even in areas with high uncertainty in GPS signals such as canyons, urban areas, and so on. The problem of cooperative

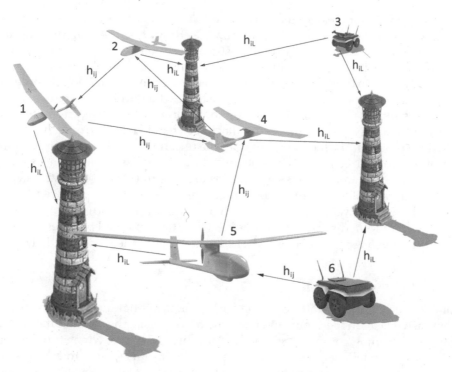

FIGURE 24.1
Sample snapshot—relative position measurement graph [6].

localization can be explained using a relative position measurement graph (RPMG), where the vehicles form the nodes and the communication between the vehicles form the edges of the graph. It can be defined as

Definition 24.1

An RPMG for n robots performing cooperative localization with l different known landmarks is a directed graph $G_n^l \triangleq V_{n,l}, \varepsilon_{n,l}$, where $V_{n,l} = 1,\ldots, n, n+1,\ldots, n+l$ is the node set consisting of n robot nodes and l landmark nodes and $\varepsilon_{n,l}(t) \subset V_{n,0} \times V_{n,l} = \eta_{ij}, i\varepsilon 1,\ldots, n, j\varepsilon 1,\ldots, n, n+1,\ldots, n+l$ is the edge set representing the availability of a relative bearing measurement. The RPMG in Figure 24.1 can be written as G_5^3 [40].

In Definition 24.1, the measurement edges in the graph are bearing measurements between the vehicles and between vehicles and landmarks. Using this RPMG, we define the problem statement for cooperative localization. The goal of cooperative localization is to estimate state X given external measurements Z and local sensor history (V and ω). There are several techniques employed to achieve such objectives and they are discussed in fairly good detail in Section 24.3.

24.3 Techniques for Estimation in Cooperative Localization

As mentioned in Sections 24.1 and 24.3, the problem of cooperative localization can be solved using a multitude of techniques with varying estimation algorithms, type of information fusion, or type of sensor measurement. Broadly, the estimation technique can be classified as centralized and distributed. In a centralized setting, a single parent node uses the observations from the different sensors to estimate the collective state of the group. This method provides the best estimates but is prone to suffer from complexity of calculations as the number of agents in the system increases and may run into node failures. In order to avoid an overdependence on a single master node, distributed algorithms were explored where each vehicle estimates its own state based on the information received from its neighbors. However, these data from the neighbors may have been handed down from other cooperating agents in a "daisy-chain" fashion, which varies the uncertainty. The agents are dealing with a subset of the centralized equivalent data and it may be delayed, and some data packets may get dropped in communications which adds to the complexities of a distributed system. Several different algorithms with various modifications have been employed to approach the problem of cooperative localization in both the centralized and distributed manner. For brevity, we refrain from discussing all possible algorithms in detail but present a couple of common techniques in Sections 24.3.1 and 24.3.2 to acclimatize the readers with the basic idea.

24.3.1 Centralized Estimation

As mentioned earlier, in a centralized estimation technique, a single node gathers all the sensor data from the vehicles in the group and fuses them to estimate the states of the vehicles. For this technique, the most common estimation algorithm is the EKF, which provides accurate state estimates for nonlinear systems and is preferred, as it is simple to execute, provides seamless information fusion from different sensors, is robust, and converges quickly when the initial state "guesses" are close to the ground truth. EKF is the most widely implemented algorithm in the literature. An EKF in an information filter form is was used in [40] to perform cooperative localization using relative bearing measurements between robots and between robots and landmarks for n ground robots. For easy understanding, we provide an example of centralized cooperative localization using an EKF for a 2D vehicle case. The vehicles are considered homogeneous with the same motion model and are equipped with cameras which are capable of providing bearing measurements. The state vector for the ith robot is $X_i = [x_i, y_i, \psi_i]^\top$ where, x_i and y_i are the x and y positions of the robot and ψ_i is the heading of the ith vehicle and the motion model is defined in Equation 24.1.

$$\dot{X}_i = f(x_i, u_i) \tag{24.1}$$

$$= \begin{bmatrix} v_i cos\psi_i \\ v_i sin\psi_i \\ \omega_i \end{bmatrix}, \tag{24.2}$$

where v_i is the linear velocity and ω_i is the angular velocity of the ith robot and $u_i = [v_i, \omega_i]^\top$ is the control input. The objective for centralized cooperative localization is to estimate the full state vector of the system is $X = [X_1, X_2, \dots, X_n]^\top$. Although the motion model defined in Equation 24.1 is for a simple 2D vehicle, it is appropriate for a wide variety of real systems and serves to demonstrate the concepts even for a 6 degree of freedom (DoF) system. For an EKF, we need to compute the motion jacobian which is defined in Equations 24.3 and 23.4.

$$A_i = \frac{\partial f_i}{\partial x} \tag{24.3}$$

$$= \begin{vmatrix} 0 & 0 & -v_i sin\psi_i \\ 0 & 0 & v_i cos\psi_i \\ 0 & 0 & 0 \end{vmatrix} \qquad (24.4)$$

The vehicles are capable of relative bearing measurements, and the measurement Equation is defined in Equation 24.5.

$$z_{\eta_{ij}} = tan^{-1}\left(\frac{y_j - y_i}{x_j - x_i}\right) - \psi_i, \qquad (24.5)$$

where, x_i and y_i are the positions of the ith vehicle and $j \in [1,...n,1,...nl]$ where n is the number of vehicles in the group and nl i the number of landmarks and $i \neq j$. The measurement jacobian is between the ith vehicle and jth vehicle or landmark is defined in Equation 24.6.

$$H_{ij} = \frac{\partial z_{\eta_{ij}}}{\partial x} \qquad (24.6)$$

$$= \left[\mathbf{0}_{3(i-1)} \cdots \frac{\Delta y_{ij}}{\rho_{ij}^2} - \frac{\Delta x_{ij}}{\rho_{ij}^2} - 1\, \mathbf{0}_{3(j-1)+3} - \frac{\Delta y_{ij}}{\rho_{ij}^2}\frac{\Delta x_{ij}}{\rho_{ij}^2} 0 \ldots \mathbf{0}_{3(n-j)} \right] \qquad (24.7)$$

where $\Delta y_{ij} = y_j - y_i$, $\Delta x_{ij} = x_j - x_i$ and $\rho_{ij} = \sqrt{\Delta y_{ij}^2 + \Delta x_{ij}^2}$. Using Equations 24.1 through 24.7, the centralized EKF algorithm is outlined in Algorithm 24.1.

Algorithm 24.1 Centralized Cooperative Localization

Initialize $\hat{X} = X_0$
for i in $1 - N$ **do**

$$\hat{X} = \hat{X} + \left(\frac{T_s}{N}\right)f(\hat{X},u)$$

$$A = \frac{\partial f}{\partial X}f(\hat{X},u)$$

$$P = P + \left(\frac{T_s}{N}\right)(AP + PA^T + Q)$$

end for
if η_{iL} received **then** (relative bearing measurement between vehicle and landmark)
  for L in $1 - n L$ **do**

$$H_{lm_{iL}} = \frac{\partial h(\hat{X},u)}{\partial X}$$

$$L_{lm_{iL}} = PH_{lm_{iL}}^T\left(R + H_{lm_{iL}}PH_{lm_{iL}}^T\right)^{-1}$$

$$P = \left(I - L_{lm_{iL}}H_{lm_{iL}}\right)P$$

$$\hat{X} = \hat{X} + L_{lm_{iL}}\left(y_{lm_{iL}} - Z(\hat{X},u)\right)$$

end for
end if
if η_{iL} received **then** (relative bearing measurement between vehicles)
    for j in $1 - n$ **do**
      if $i \neq j$ **then**

$$26 : H_{ij} = \frac{\partial h\left(\hat{X}, u\right)}{\partial X}$$
$$27 : L_{ij} = PH_{ij}^T \left(R + H_{ij}PH_{ij}^T\right)^{-1}$$
$$28 : P = (I - L_{ij}H_{ij})P$$
$$29 : \hat{X} = \hat{X} + L_{ij}\left(y_{lm_{ij}} - Z(\hat{X}, u)\right)$$

      end if
    end for
end if

Algorithm 24.1 is robust as it can fuse information from multiple vehicle sources at the same time and generate a centralized estimate for all n vehicles in the group. Using a similar EKF approach, Chakraborty et al. in [7] developed a fixed-wing cooperative localization estimator. Different sensor configurations like range-only, range and bearing, range rate, bearing rate, and bearing-only were explored and a comparison of the root-mean-square (RMS) errors for the different models were performed to depict which sensors worked better for the different given trajectories.

A centralized system provides several advantages, such as every vehicle has access to the sensor measurements and local information of every other vehicle in the group. However, as n becomes higher the amount of data to be processed by a single node becomes large, which increases the time for computation and the risk for single-point node failure. Even if extremely high-ended processors are used to speed up the data processing phase, several challenges exist in implementing these algorithms on hardware platforms with real-sensing capability like data packet drop, data delays, and sensor biases, which might create biases in estimation or provide inconsistent estimates. In order to avoid such problems, focus shifted to developing a distributed cooperative scheme which will be capable of providing estimates that may not be as good as the centralized version but enough to generate estimates that aid in mission completion. Some widely used distributed algorithms are discussed in Section 24.3.2 so as to provide the reader with both techniques which may be implemented as seen fit.

24.3.2 Distributed Estimation

Owing to the problems of the centralized estimation as discussed in Section 24.3.1, several authors have worked on distributing the cooperative localization problem so that vehicles can locally estimate their states without needing access to the information from all other vehicles in the group. Roumeliotis and Bekey attempted to create a distributed system in [33] where a central Kalman filter is decomposed into n filters and each runs on an individual robot. The paper, however, assumes that one-to-one robot communication exists between all pairs of robots, which makes this solution infeasible for larger systems. In this section, we provide an example based on the distributed cooperative localization algorithm by Nerurkar et al. in [5] based on the Covariance Intersection technique developed in [17]. A team of n 2D cooperative robots are assumed to be present, which may be fitted with either a camera or laser range finders which provide relative pose measurements from neighboring vehicles or landmarks. The motion model is a discrete-time version of Equation 24.1 and is defined as

$$X_{k+1}^i = f\left(X_k^i, u_{m,k}^i\right), \quad i = 1,\ldots,n \tag{24.8}$$

where $X_k^i = [x_k^i, y_k^i, \psi_k^i]$ is the state vector and $u_k^i = [v_k^i, \omega_k^i]^{\mathsf{T}}$ is the input vector which when added with a zero-mean white Gaussian noise with covariance Q_k^i provides $u_{m,k}^i$ The measurement model can either be range or bearing and is denoted as

$$z_{k+1}^{i,j} = h^{i,j}(X_{k+1}^i, X_{k+1}^j) + n_{k+1}^{i,j} \tag{24.9}$$

where $h^{i,j}$ is the true relative pose measurement corrupted by a zero-mean white noise $n_{k+1}^{i,j}$ to emulate a the behavior of a sensor. Using these equations, a distributed CL algorithm based on Covariance Intersection is presented in Algorithm 24.2.

In the algorithm, weight $w\varepsilon[0, 1]$ and is chosen such that the trace of $P_{k+1|k+1}^j$ is minimized. This algorithm has a complexity of $O(n)$ per robot per time step, which is much lower than that of an EKF which has a complexity of $O(n^4)$. Using the distributed algorithm as given by Algorithm 24.2, the complexity of processing and communication is reduced considerably from the centralized case. Several other estimation techniques have been developed to distribute the cooperative localization process. Nerukar et al. developed a MAP estimator based on the weighted nonlinear least-squares implementation using factor graphs to solve the distributed problem in [29]. Using the Georgia Tech Smoothing and Mapping (GTSAM) toolbox [8], a distributed solution to the cooperative localization problem is achieved, but it is complex in the amount of bookkeeping to keep track of the various vehicle nodes and the amount and time of communication. All of these can be included in cost functions in future work. The algorithm manages to reduce the space complexity in data storage from the centralized version.

An efficient distributed algorithm for navigation of underwater vehicles was developed by Bahr et al. in [2]. The paper proposed a mobile network capable of performing acoustic ranging and data exchange under water for longer durations. The major motivation of this work was to reduce the number of times the vehicles have to surface for GPS signals. The state estimate and the associated covariance is predicted using range measurements and pressure sensors on 2D planes. Using these range measurements, the vehicles build a distance matrix to keep track of the distance traveled between receiving transmissions. Based on these distance matrices, a cost function is generated based on the inverse likelihood of the vehicle traveling from x_A to x_B where the distance metric is provided by the Kullback-Leibler divergence. The likeliest position which has the smallest transition cost is found based on the intersection of the current time step m and j steps in the past. This algorithm works well when the trajectories are favorable, and it was validated on different hardware platforms to prove its efficiency. Although a distributed system uses considerably less computational power, it is heavily reliant on communication with neighbors and hence delays or loss of data can degrade the estimation quality of a distributed system significantly. Effect of communication delays on the filters are being widely researched to determine their effect on such cooperating systems.

Algorithm 24.2 Decentralized Cooperative Localization using Covariance Intersection

Initial vehicle pose $X_k^i = X_0^i$.

Propagation of state $X_{k+1|k}^{\hat{i}} = f\left(X_{k|k}^{\hat{i}}, u_{m,k}^i\right)$.

Motion jacobian $A_k^i = \dfrac{\partial f(X_k^i)}{\partial X_k^i}$.

Input jacobian $B_k^i = \dfrac{\partial f(X_k^i)}{\partial u_k^i}$

Propagation of covariance $P_{k+1|k}^i = A_k^i P_{k|k}^i (A_k^i)^\top + B_k^i Q_{k|k}^i (B_k^i)^\top$

for i in $1 - n$ **do**

  if $i \neq j$ **then**

    $X_{k+1}^{\hat{j}*} = X_{k+1|k}^{\hat{i}} + \tau_{X_{k+1|k}^i} z_{k+1}^{i,j}$ (Estimate for R_j by R_i when R_i receives a relative

    measurement from R_j

    Measurement jacobian $H_{k+1}^{\tilde{i},j} = \begin{bmatrix} I_2 & J\left(X_{k+1}^{\hat{j}*} - p_{k+1|k}^{\hat{i}}\right) \\ 0_{1\times 2} & 1 \end{bmatrix}$

    $J = \begin{bmatrix} 0 & -1 \\ 1 & 0 \end{bmatrix}$

    Estimate error $X_{k+1}^{\tilde{j}*} = H_{k+1}^{\tilde{i},j} X_{k+1|k}^{\tilde{i}} - \tau_{X_{k+1|k}^i} n_{k+1}^{i,j}$.

    Covariance update $P_{k+1}^{\tilde{j}*} = H_{k+1}^{\tilde{i},j} P_{k+1|k}^i H_{k+1}^{\tilde{i},j} + \tau_{X_{k+1|k}^i} R_{k+1}^i \tau^\tau{}_{X_{k+1|k}^i}$

    Communicate to R_j it's state estimate $X_{k+1}^{\hat{j}*}$ and corresponding co-variance P_{k+1}^{j*}

    Maintain estimation consistency $P_{k+1|k+1}^j = \left[w\left(P_{k+1|k}^j\right)^{-1} + (1-w)\left(P_{k+1|k}^{j*}\right)^{-1} \right]^\top$.

    Update state estimate $X_{k+1|k+1}^{\hat{j}} = P_{k+1|k+1}^j \left[w\left(P_{k+1|k}^j\right)^{-1} X_{k+1|k}^{\hat{j}} + (1-w)\left(P_{k+1|k}^{j*}\right)^{-1} X_{k+1}^{\hat{j}*} \right]$

  end if

end for

One of the major driving factors of these estimation algorithms is whether the quality of estimates is acceptable based on 3σ bounds. This is directly linked to the observability of a system, which is discussed in detail in Section 24.4.

24.4 Observability

As discussed in Section 24.3, the problem of cooperative localization has been solved using a number of estimation techniques in either centralized or distributed manner. Since the crux of cooperative localization is to solve an estimation problem to get an approximate idea of the system's state, these estimation errors have to be bounded or consistent to be of actual use. This phenomenon of being able to estimate a state based on the measurements received is termed as observability. A system is defined to be observable if all of its states can be estimated using the measurements from its sensors [12]. Analysis of the system to determine conditions under which the system maintain observability are discussed in Section 24.4.1, and controllers designed to ensure system observability are discussed in Section 24.4.2.

24.4.1 Analysis of the Cooperative Localization Problem

Initial results using linear observability analysis were performed by Roumeliotis and Bekey [34]. A centralized estimation algorithm was deployed to perform the system observability characteristics under a variety of conditions when none of the robots have access to GPS or when one of the robots has ground truth or when one of the robots is stationary. Their initial results indicated that the collective states are observable if any one of the robots had GPS signals or knowledge of the absolute position. Since the system was linearized, it may be argued that many of the inherent properties of the system were lost in the approximation which made the system unobservable. The paper also assumed that the absolute heading information was available and subsequently not included in the system state. Nonlinear observability analysis was performed by Martinelli and Siegwart in [21]. In this paper, a 3-state 2D vehicle model with a differential drive was considered where the states are $x = [x_i, y_i, \psi_i]$ and $i \varepsilon [1, 2]$. The motion model is defined in Equation 24.1. This paper investigates observability conditions for a pair of robots for different measurements like range and bearing, and it is determined that the relative bearing provides the best measurement. Without the presence of any landmarks, the maximum rank of the system is 3, which makes the system unobservable. Sharma et al. extended this analysis to n-vehicle case with bearing measurements [39]. A simple snapshot of the RPMG of a cooperative system is presented in Figure 24.1. The motion model remains the same as Equation 24.1 and the measurement model is defined in Equation 24.5. Nonlinear observability analysis is performed using Lie derivatives, where $L_f^0(h) = h$ is the zeroth order Lie derivative and h is the measurement model. The higher order Lie derivatives are all a function of L_f^0. The second order Lie derivative $L_f^2 = \dfrac{\partial [L_f^1(h)]}{\partial x} f$, where f is the motion model, is dependent on the first order, and so on. The observability $\mathbf{O}$ is defined as the gradient of the Lie derivatives.

Using Lie derivatives and RPMG, the observability matrix is constructed based on

1. Edge between two robots.
2. Edge between a robot and a landmark.

The observability matrix for each of the cases are determined from the Lie derivatives and using the row reduced echelon form (RREF) of the matrix, we directly obtain the maximum number of linearly independent rows in the matrix. Based on this analysis, the readers are provided with Lemmas 24.1 and 24.2 which provide conditions to maintain the maximum rank of the observability matrix.

Lemma 24.1

The rank of O_{ij} due to an edge between 2 robots is 3 if

1. $v_i > 0$.
2. $v_j > 0$.
3. The ith robot that measures the bearing does not move along the line joining the robots.
4. The jth robot does not move perpendicular to the line that joins the two robots.

where, v_i is the linear velocity of the ith robot.

Lemma 24.2

The rank of O_{ik} due to an edge between the ith robot and kth landmark is 2 if

1. $v_i > 0$.
2. The robot does not move along the line that joins the robot and landmark.

Using these lemmas and the conditions of the observability matrix in the presence and absence of known landmarks, Theorem 24.1 provides the conditions to maintain observability for a team of vehicles performing cooperative localization.

Theorem 24.1

Given a proper RPMG G_n^l, if for each robot there exists a path to at least two landmarks and the robot and the two landmarks are not on the same line ($\eta_{i1} \neq \eta_{i2} \forall i = 1, \ldots, n$), then the system is completely observable, that is, the rank of the observability matrix is $3n$.

The complete proofs for the conditions derived are present in [39]. Extensive simulations were performed to verify the derived conditions. Here, we provide a case with six vehicles moving in circular trajectories in an environment with two landmarks.

Figure 24.2 represents a snapshot of the simulation scenario. From the figure, it is clear that not all vehicles have direct paths to two landmarks. But since the RPMG is connected at every time instant, then there exists a path between every vehicle and to landmarks. This helps satisfy the observability conditions derived in Theorem 24.1. Figures 24.3 and 24.4 show the error plots with 3σ bounds. As is clear, the errors are well within the bounds, which means the filter is consistent. The 3σ bounds decrease over time, showing that the quality of estimates improves over time which means that the system is observable. These results prove the validity of the conditions in Theorem 24.1.

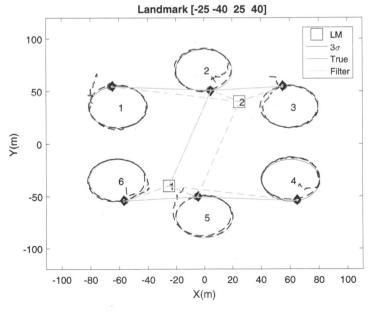

FIGURE 24.2
RPMG snapshot at kth time instant.

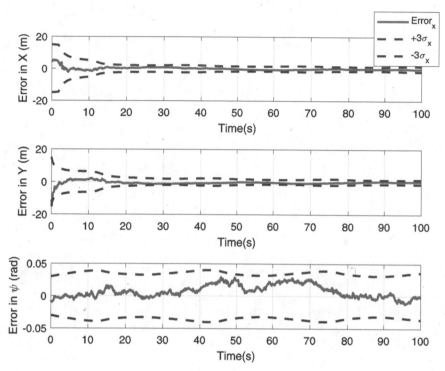

FIGURE 24.3
Error plot for Vehicle 1—continuous time observability case.

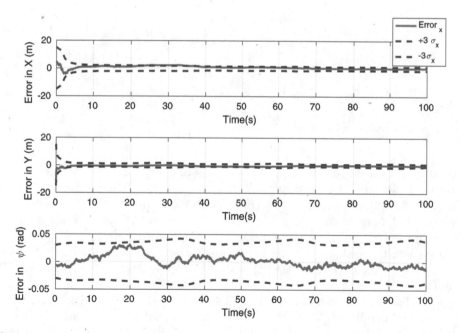

FIGURE 24.4
Error plot for Vehicle 6—continuous time observability case.

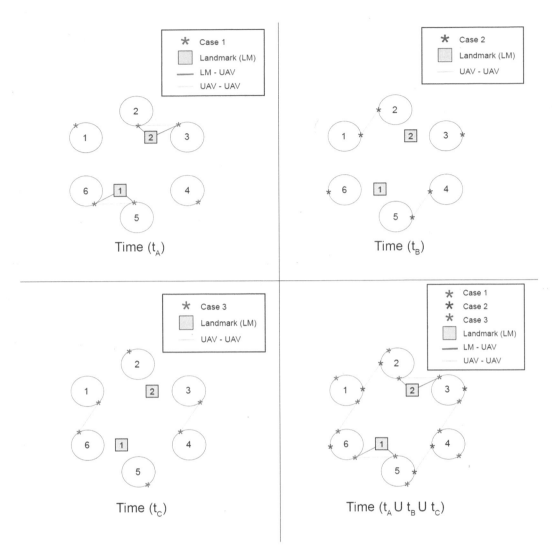

FIGURE 24.5
Discrete time-relative position measurement graph [6].

The conditions derived in Theorem 24.1 state that at every time instant, there must exist a path to two landmarks for the vehicle to remain observable. The condition may prove to be difficult to comply with in enemy terrains or in environments riddled with obstacles. Chakraborty et al. derived the same observability conditions using discrete time nonlinear analysis in [6] where it was shown that instead of maintaining path to two landmarks at all times the system, the system remains observable if the robots can maintain paths to two landmarks over a time period. Instead of an RPMG in Figure 24.1, the modified RPMG is shown in Figure 24.5.

In Figure 24.5, it is clear that at every time instant (t_A, t_B or t_C), the vehicles do not satisfy the observability conditions derived in Theorem 24.1. But when the snapshots from the time instants are combined, that is, at the union of the time instants ($t_A \cup t_B \cup t_C$), all the vehicles satisfy the condition from Theorem 24.1. In order to arrive at these conditions, discrete-time observability analysis was performed. The equation of motion is modified as given in Equation 24.12

$$X_i(k+1) = f_i(X_i(k), u_i(k))$$
$$= X_i(k) + T_s \begin{bmatrix} v_i(k)\cos\psi_i(k) \\ v_i(k)\sin\psi_i(k) \\ \omega_i(k) \end{bmatrix}, \tag{24.10}$$

where $X_i(k) = [x_i(k) \ y_i(k) \ \psi_i(k)]^\top$ is state vector of ith vehicle with 2D position $[x_i(k), y_i(k)]$ and heading angle $\psi_i(k)$, T_s is sampling time, $u_i(k) = [v_i(k) \ \omega_i(k)]^\top$ is the input vector of the ith vehicle with velocity $v_i(k)$, and turn rate $\omega_i(k)$ at the kth time instant. Each vehicle is equipped with a velocity and a turn rate sensor. Also, each vehicle has a sensor to measure bearing angle from the vehicles and landmarks that are in its sensing range. The bearing angle measured from the ith vehicle to the jth vehicle or landmark is given as

$$\eta_{ij}(k) = h_{ij}\left(X_i(k), X_j(k)\right) = \tan^{-1}\left(\frac{y_j(k) - y_i(k)}{x_j(k) - x_i(k)}\right) - \psi_i(k) \tag{24.11}$$
$$i = [1,\dots,n], \quad j = [1,\dots,n,n+1,\dots,n+l],$$

where n is the total number of vehicles and l is the total number of landmarks in the system. The topology of the RPMG is switching and not fixed. A discrete-time system may be represented as

$$X(k+1) = f(X(k), u(k)), \tag{24.12}$$

$$Z(k) = h(X(k)), \tag{24.13}$$

where $X(k+1)$ is the motion model and $Z(k)$ is the measurement model. The observability matrix over a time period can be written as

$$O(k:k+\tau) = \begin{bmatrix} \dfrac{\partial h}{\partial X}(X(k), u(k)) \\ \dfrac{\partial h}{\partial X}(X(k+1), u(k+1))\Phi(k,k+1) \\ \vdots \\ \dfrac{\partial h}{\partial X}(X(k+\tau), u(k+\tau))\Phi(k,k+\tau) \end{bmatrix}, \tag{24.14}$$

where

$$\Phi(k,k+1) = \frac{\partial f}{\partial X}(X(k), u(k)),$$
$$\Phi(k,k+\tau) = \Phi(k+\tau-1, k+\tau),\dots,\Phi(k,k+1),$$

are the state transition matrices. The observability matrix at different time steps are related as follows

$$O(k,k+\tau) = \begin{bmatrix} O(k) \\ O(k+1)\Phi(k,k+1) \\ \vdots \\ O(k+\tau)\Phi(k,k+\tau) \end{bmatrix}, \tag{24.15}$$

where, $O(k) = \dfrac{\partial h}{\partial X}(X(k))$. The system is completely observable when $rank(O(k, k + \tau)) = n$. From Equation 24.13, we find

$$H_{ij}(k) = \frac{\partial h_{ij}(X_i(k), X_j(k))}{\partial X} \tag{24.16}$$
$$= \left[\mathbf{0}_{3(i-1)} \vdots H_{ij}^i(k) \vdots \mathbf{0}_{3(j-1)-3i} \vdots H_{ij}^j(k) \vdots \mathbf{0}_{3(n-j)} \right],$$

where,

$$H_{ij}^i(k) = \begin{bmatrix} \dfrac{-\Delta y_{ij}(k)}{R_{ij}^2} & \dfrac{\Delta x_{ij}(k)}{R_{ij}^2} & -1 \end{bmatrix},$$

and

$$H_{ij}^j(k) = \begin{bmatrix} \dfrac{\Delta y_{ij}(k)}{R_{ij}^2} & \dfrac{-\Delta x_{ij}(k)}{R_{ij}^2} & 0 \end{bmatrix} \quad \forall j \in [1,n].$$

The system jacobian for cooperative localization problem is given as

$$F(k) = \frac{\partial f}{\partial X}(X(k), u(k)) = \begin{bmatrix} F_1(k) & 0 & 0 \\ 0 & \ddots & 0 \\ 0 & 0 & F_n(k) \end{bmatrix},$$

$$F_i(k) = \frac{\partial f_i}{\partial X_i}(X_i(k), u_i(k)) = \begin{bmatrix} 1 & 0 & -a_i(k) \\ 0 & 1 & -b_i(k) \\ 0 & 0 & 1 \end{bmatrix},$$

where,

$$a_i(k) = T_s v_i(k) \sin \psi_i(k),$$
$$b_i(k) = T_s v_i(k) \cos \psi_i(k).$$

The observability matrix for cooperative localization can be written as

$$O(k:k+\tau) = \begin{bmatrix} H_{ij}(k) \\ H_{ij}(k+1)F(k) \\ \vdots \\ H_{ij}(k+\tau)F(k+\tau-1)\cdots F(k) \end{bmatrix}. \tag{24.17}$$

From the above equation, we can say that the overall observability can be constructed by stacking observability matrix of all the edges η_{ij} over $k : k + \tau$ where each measurement is referred back to time k. Based on the conditions for maximum number of linearly independent rows that contribute to the rank of the observability matrix, we state Lemmas 24.3 and 24.4.

Lemma 24.3

The maximum rank of the observability matrix $O_{ij}(k : k + \tau)$ between two vehicles is three if at least three bearing measurements are taken at three different time instants between $k, k + \tau$, and if

1. $v_i > 0$ and $v_j > 0$.
2. The vehicles do not move along the straight line joining them.
3. $\psi_i(k) - \psi_j(k) \neq 0$.

Lemma 24.4

The maximum rank of the observability matrix $O_{ij}(k : k + \tau)$ between a vehicle and a landmark is two if at least two bearing measurements are taken at two different time instants between $k, k + \tau$, and if

1. $v_i > 0$
2. The vehicle is not moving along the line joining the vehicle and the landmark.

The above two lemmas show that the number of linearly independent rows contributed by two types of edges are same as obtained in continuous-time case [39]. In fact, they span the same observable space with same basis. Using this, we can state the following theorem for the overall observability of discrete-time nonlinear cooperative localization problem. The proof of the following theorem can easily be drawn from [39].

Theorem 24.2

The overall system given by Equation 24.10 is completely observable (i.e., rank($O(k, k + \tau)) = 3n$), if the union of relative position measurement graphs at different time instants as shown in Figure 24.5 is connected and graph has at least two known landmarks and each edge between two vehicle satisfies conditions in Lemma 24.3 and edge between a vehicle and a landmark satisfies condition from Lemma 24.4.

The validity of Theorem 24.2 has been proved through extensive simulation and hardware testing. A sample simulation scenario is given in Figure 24.6. Simulation videos can be found at https://www.youtube.com/watch?v=NI9TklqUwEo. Figure 24.6 shows that the vehicles have paths to two landmarks over a time-period. This periodic communication at regular intervals has been achieved using a synchronization algorithm developed in [9]. Broadly, the algorithm states that if $\psi_j = \pi + \psi_i \forall i, j\varepsilon[1, \ldots , n]$ and $i \neq j$, then the vehicles meet at fixed intervals if they are traveling in circular trajectories in the same direction. Figure 24.7 shows a simple case for maintaining synchronous communication.

Results from the simulation scenario presented in Figure 24.6 are presented in Figures 24.8 and 24.9. As seen in the figures, the errors are still within 3σ bounds even when they do

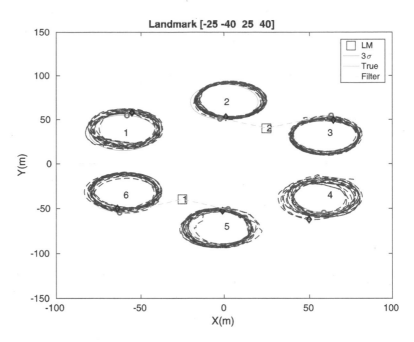

FIGURE 24.6
Discrete time-relative position measurement graph simulation scenario [6].

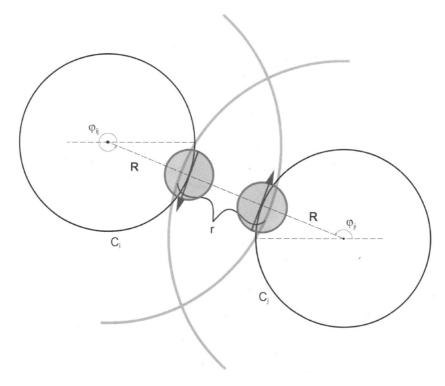

FIGURE 24.7
Synchronous communication scenario [6].

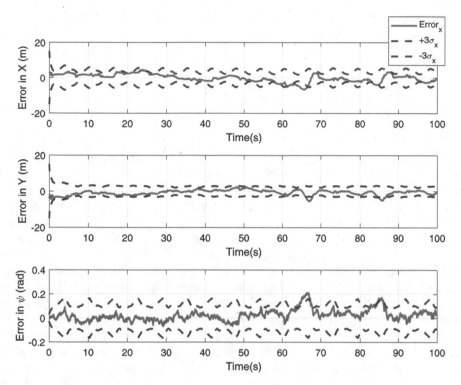

FIGURE 24.8
Error plot for Vehicle 1—discrete time observability case [6].

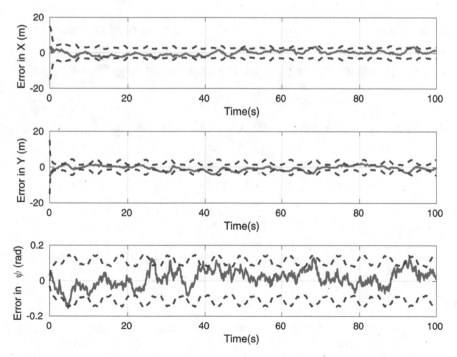

FIGURE 24.9
Error plot for Vehicle 6—discrete time observability case [6].

not have paths to two landmarks at every time instant. These results prove the validity of the conditions derived in Theorem 24.2 and provide a much more flexible condition that can be used to constrain path-planning algorithms to maintain observability.

Using the conditions from Theorems 24.1 and 24.2, different controllers to maintain a path that satisfies the derived conditions have been developed, and the broad spectrum of these algorithms is discussed next in Section 24.4.2.

24.4.2 Controllers to Maintain Observability

Since maintaining system observability is crucial to ensuring filter consistency and in turn reliable estimates, quite a lot of attention has been devoted by several authors to develop controllers that can increase the information available to the system or maintain paths that will improve system observability. As observability depends on the measurements available to the system, different factors like the RPMG orientation and UV paths play a significant role in the analysis. Several dynamic collision avoidance techniques are based on the path planning between robots, which is dependent on controllers developed to improve system observability and in turn localization accuracy. Yu et al. developed a controller in [48] that increased the observability of the obstacles in the field of view of the vehicle, which led to more robust obstacle avoidance. Zhang et al. used bearing-only measurements to determine paths for vehicles in cooperative localization in [49]. However, even though the approach was novel, it was done for a fixed sensing topology without any global information input in the form of landmarks. This technique was improved upon by Sharma et al. in [37], where cooperative localization was used to localize a target using bearing-only measurements with landmarks present in the environment. The controller designed in this paper had to satisfy three main control objectives listed as

1. Overall observability is maintained which satisfies the nonlinear conditions.
2. Localization accuracy between vehicles and target is improved.
3. Minimum distance between vehicles, landmarks, and target is maintained.

Mathematically, these conditions can be represented as

1. Graph connectivity is directly linked to the second eigenvalue of the graph Laplacian.
2. In order to improve the localization accuracy, the information available from sensors has to be improved where the information matrix is given as

$$I(t) = \sum_{i,j} w_{ij} H_{ij}^T(t) R^{-1} H_{ij}(t), \tag{24.18}$$

where, H_{ij} is the measurement jacobian and w_{ij} is the probability of detection of jth vehicle in ith vehicle's field of view.

3. $\rho_{ij} > \rho_{min} \forall i, j \varepsilon [1, n]$.

Controller design to improve observability is still an area of high interest owing to the variety of factors that can be optimized in the cost function. These controllers are generally designed to aid in some applications of cooperative localization which are discussed in detail in Section 24.5.

24.5 Applications of Cooperative Localization

Cooperative localization as a standalone technique to estimate system states is extremely robust and accurate. Over the years, several authors have used this potential and applied this technique to solve various problems of interest. Merino et al. used cooperative localization to detect fire using UAVs in [22]. A fleet of heterogeneous UAVs are used which are equipped with infrared cameras or normal cameras. The goal of this paper is to reduce the number of false alarms i.e., to decrease the uncertainty in fire detection and increase localization accuracy by cooperation and sharing information between multiple vehicles across multiple platforms. The detection algorithm itself relies heavily on image processing techniques like extracting common features from binary images using IR cameras and RGB cameras. Temperature-based thresholding is avoided, as it varies based on emissivity indices of materials that may be difficult to assess in an unstructured environment. The authors use a histogram mode for thresholding. Using cooperative localization, the detection and subsequent localization of fire has five stages—prediction, data association, update, insertion of new alarms, and not-detected alarms. The algorithm was validated experimentally using different hardware platforms.

Cooperative localization has been employed by authors to solve complex problems such as landing UAVs on moving platforms such as cars or ships. For car landing, the motion information of the vehicle is important for precise landing. In the case of landing on ship decks, the problem becomes tricky since the ship's motion is not only affected by its velocity but also by the condition of the sea. Turbulent conditions can lead to a lot of changes in the roll and pitch of the ship's deck, which has to be mapped by the UAV accurately to attempt to land. Vision-based landing in such cases may prove to be ineffective, especially under turbulent conditions when visibility is reduced. Mishra et al. attempted to solve this problem using LIDAR or other sensors that provide range information in [23] where multiple cooperating UAVs can be launched from the ship, which may aid the other vehicle to land by providing it with more information about the surroundings. Figure 24.10 shows a sample scenario where a single UAV is attempting to land on the ship deck while two other

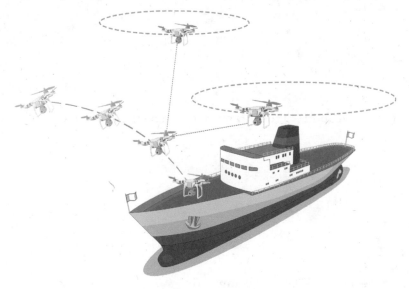

FIGURE 24.10
UAV landing on ship deck using cooperative localization.

UAVs assist in landing by providing additional measurements. The two vehicles that aid in landing are also unaware of their absolute pose information. So all the measurements are relative to the ship, which forms a moving origin to the map.

Algorithm 24.3 outlines the flow of the estimation and control problem for cooperative ship landing. In this algorithm, instead of using absolute vehicle poses, relative poses are estimated where the ship is considered as the origin. Figure 24.11 shows the true and estimated relative 3D pose of the vehicle landing on the ship, and Figure 24.12 represents the same result in 2D. In both cases, it is clear that the relative pose is estimated quite accurately using range-only measurements, and it is seen in Figure 24.11 that the estimates with cooperative localization are better than the ones without.

Algorithm 24.3 EKF-based Centralized Cooperative Localization
using Estimated Relative Pose and Orientation

Initialize $\hat{X} = X_0 + \mathcal{N}(\mu, \sigma^2)$.

for i in $1 - N$ **do** (N - number of quadcopters)

    for m in $1 - NN$ **do** (NN - number of predictions before a measurement update)

$$\hat{x}^+ = \hat{x}^- + \left(\frac{T_s}{NN}\right) f(x, u)$$

$$A = \frac{\partial f}{\partial x} f(x, u)$$

$$P^+ = P^- + \left(\frac{T_s}{NN}\right)(AP + PA^T + Q)$$

    end for

    if ρ_{i0} received **then** (range measurement between ship and ith vehicle)

$$H_{i0} = \frac{\partial h_{i0}}{\partial x}$$

$$L_{i0} = P^+ H_{i0}^T \left(\sigma_\rho + H_{i0} P^+ H_{i0}^T\right)^{-1}$$

$$\hat{x}^{++} = \hat{x}^+ + L_{i0}(\rho_{i0} - \hat{\rho}_{i0})$$

$$P^{++} = (I - L_{i0} H_{i0}) P^+$$

    end if

    for j in $1 - N$ **do**

        if ρ_{ij} received and $i \neq j$ **then** (range measurement between ith and jth vehicle)

$$H_{ij} = \frac{\partial h_{ij}}{\partial x}$$

$$L_{ij} = P^+ H_{ij}^T (\sigma_\rho + H_{ij} P^+ H_{ij}^T)^{-1}$$

$$\hat{x}^{++} = \hat{x}^+ + L(\rho_{ij} - \rho_{\hat{ij}})$$

$$P^{++} = (I - L_{ij} H_{ij}) P^+$$

        end if

    end for

    Using $\hat{x}$, generate PID control commands

end for

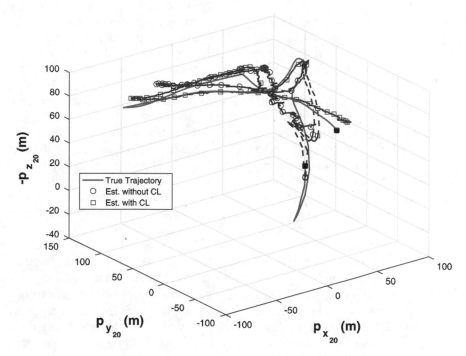

FIGURE 24.11
UAV landing on ship deck using cooperative localization—3D true and estimated trajectory.

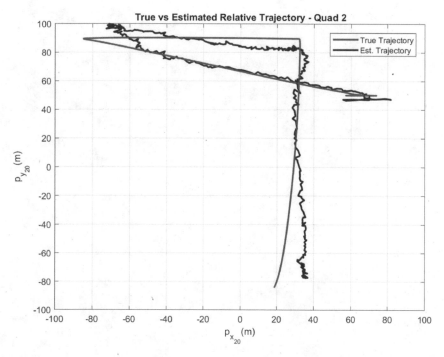

FIGURE 24.12
UAV landing on ship deck using cooperative localization—2D true and estimated trajectory.

Future work in this application includes adding more range hot-spots on the ship deck to provide more information to the UAVs to help estimate its state precisely. Also, path-planning for the other UAVs that assist in landing such that they always form a connected graph over small time intervals is an interesting aspect of this problem that can be further investigated.

Sharma et al. in [41] used cooperative localization to solve a simultaneous arrival problem to converge on a target where not all the UAVs have line-of-sight (LOS) with the target. The objective for the problem was to find a way such that multiple vehicles arrived at a common target at the same time with no access to GPS. A pictorial representation of the problem is presented in Figure 24.13. The algorithm to solve the simultaneous arrival problem without GPS is similar to Algorithm 24.3. However, the equations of motion and measurement model are different, and they are described in detail in [41]. The authors showed that if one vehicle had the target in its LOS and all the robots had a connected topology (i.e., every robot had a direct or indirect path to every other robot in the system), then the system was observable and the target states could be estimated, and all the UAVs managed to arrive at the target at the same time. This work was done for a stationary target which was extended for a moving target by Misra et al. in [25]. The authors provided a detailed observability analysis in this case demonstrating the effectiveness of the cooperative localization algorithm. Some of the results from the case where the target is moving and maneuvering are highlighted in Figures 24.14 through 24.16.

Figure 24.14 shows the true and estimated trajectory of all munitions. Using cooperative localization, only one vehicle needs to stay in LOS with the target and maintain graph connectivity for the vehicles to arrive at the same time. The consistency of the estimates and its quality is seen in Figures 24.15 and 24.16. The errors are well within the 3σ bounds

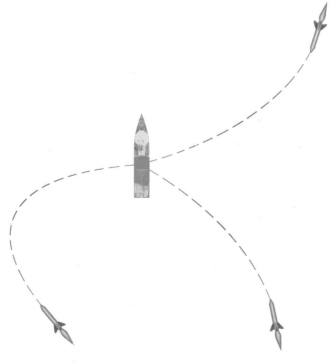

FIGURE 24.13
Simultaneous arrival of UAVs on a target.

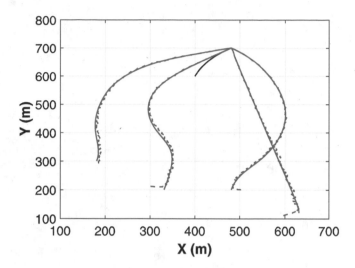

FIGURE 24.14
Simultaneous arrival of UAVs on a target—trajectory plot (true trajectory in solid lines and estimated trajectory in dotted lines).

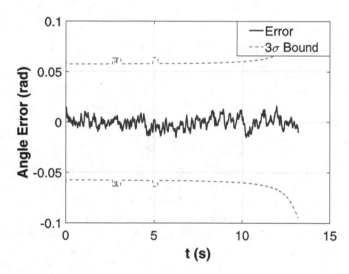

FIGURE 24.15
Simultaneous arrival of UAVs on a target—bearing error.

for both the range and bearing measurements, which show that the filter is consistent and observable. Further work in this area can be continued by placing static and dynamic avoidance algorithms in sync with the timing algorithm along with the estimation criteria. The results show the case for centralized, and it will be interesting to see the effect of a distributed estimation on such systems.

There are several other applications of cooperative localization, and different new estimation techniques are used to perform more efficient sensor fusion to create a truly distributed system which has the capability to estimate states as accurate as the centralized case. Several challenges still need to be addressed by researchers before such a system is viable, and some of these are discussed in Section 24.6.

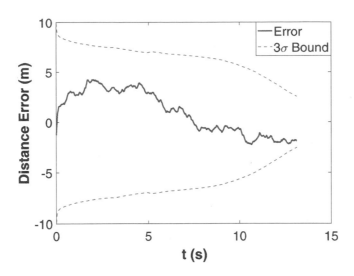

FIGURE 24.16
Simultaneous arrival of UAVs on a target—range error.

24.6 Conclusion and Challenges Faced in Cooperative Localization

This chapter has provided a brief introduction to the cooperative localization problem, and some of the widely used techniques were discussed along with the mathematical analysis for convergence. Some applications were also presented to the reader to help understand the impact of cooperative localization in aiding navigation in GPS-denied environments. Although the problem of cooperative localization has been solved using various innovative techniques over the past few decades, several of these algorithms were developed keeping in mind very specific scenarios, which renders them minimally impactful for a broad spectrum of applications. Brink discusses some of the desired attributes in systems capable of multi-vehicle navigation in GPS-denied environments in [4]. Scalability is an important characteristic for multi-vehicle hardware platforms, especially in the communication aspect where addition of more vehicles or increase in the amount of information exchanged between the existing number of vehicles should not overwhelm the onboard processing capabilities of the vehicles. Most of the algorithms in the literature assume perfect communications or delays of a negligible order. However, in hardware implementations, data delays and data loss are extremely common. Algorithms that are robust to such issues are currently being researched without making them so complex that it creates a lot of overhead that makes it unfeasible. Various optimization parameters for these algorithms are being explored like frequency of communication, data size to be communicated, need for book-keeping, and so on.

The algorithms used to solve cooperative localization generally tend to assume good *a priori* information for initialization. Some authors have stated that GPS may be used for initialization purposes. However, if the algorithms need to be re-initialized or reset altogether due to different factors like sensor bias or vehicle loss, then the information available to reset may not be accurate. This might cause the algorithms to fail. Some authors have tried to address this area using a combination of particle filters to initialize and EKF to estimate (as it is computationally more efficient). The approach is relatively new and has not been tested across platforms to check for consistency and feasibility.

Many authors have focused on estimating biases which are static (not changing over a time period). Although this may be the case for many sensors, especially the expensive ones, in cheap sensors biases can develop quickly and if not identified may provide consistent false data to the estimation algorithm, resulting in biased state estimates which are inconsistent.

The problems listed are currently being researched, and once some solutions are developed that are general enough to be applied to most platforms, a fully autonomous intelligent distributed system capable of performing intricate missions or providing support to humans can be developed.

References

1. R. Aragues, L. Carlone, G. Calafiore, and C. Sagues. Multi-agent localization from noisy relative pose measurements. In *2011 IEEE International Conference on Robotics and Automation (ICRA)*, pp. 364–369, IEEE, 2011.
2. A. Bahr, J.J. Leonard, and M.F. Fallon. Cooperative localization for autonomous underwater vehicles. *The International Journal of Robotics Research*, 28(6):714–728, 2009.
3. A. Bahr, M.R. Walter, and J.J. Leonard. Consistent cooperative localization. In *IEEE International Conference on Robotics and Automation, 2009. ICRA'09*, pp. 3415–3422, IEEE, 2009.
4. K.M. Brink. Multi-agent relative pose estimation: Approaches and applications. In *Open Architecture/Open Business Model Net-Centric Systems and Defense Transformation 2018*, volume 10651, International Society for Optics and Photonics, p. 106510D, 2018.
5. L.C. Carrillo-Arce, E.D. Nerurkar, J.L. Gordillo, and S.I. Roumeliotis. Decentralized multi-robot cooperative localization using co-variance intersection. In *2013 IEEE/RSJ International Conference on Intelligent Robots and Systems (IROS)*, pp. 1412–1417, IEEE, 2013.
6. A. Chakraborty, S. Misra, R. Sharma, and C.N. Taylor. Observability conditions for switching sensing topology for cooperative localization. *Unmanned Systems*, 5(03):141–157, 2017.
7. A. Chakraborty, C.N. Taylor, R. Sharma, and K.M. Brink. Cooperative localization for fixed wing unmanned aerial vehicles. In *2016 IEEE/ION Position, Location and Navigation Symposium (PLANS)*, pp. 106–117, IEEE, 2016.
8. F. Dellaert. *Factor graphs and GTSAM: A hands-on introduction*. Technical report, Georgia Institute of Technology, 2012.
9. J.M. Díaz-Báñez, E. Caraballo, M.A Lopez, S. Bereg, I. Maza, and A. Ollero. The synchronization problem for information exchange between aerial robots under communication constraints. In *2015 IEEE International Conference on Robotics and Automation (ICRA)*, pp. 4650–4655, IEEE, 2015.
10. M.E. El Najjar and P. Bonnifait. A road-matching method for precise vehicle localization using belief theory and Kalman filtering. *Autonomous Robots*, 19(2):173–191, 2005.
11. M.F. Fallon, G. Papadopoulos, and J.J. Leonard. A measurement distribution framework for cooperative navigation using multiple AUVs. In *2010 IEEE International Conference on Robotics and Automation (ICRA)*, pp. 4256–4263, IEEE, 2010.
12. R. Hermann and A. Krener. Nonlinear controllability and observability. *IEEE Transactions on Automatic Control*, 22(5):728–740, 1977.
13. A. Howard, M.J. Matark, and G.S. Sukhatme. Localization for mobile robot teams using maximum likelihood estimation. In *IEEE/RSJ International Conference on Intelligent Robots and Systems, 2002*, volume 1, pp. 434–439, IEEE, 2002.
14. G. Huang, N. Trawny, A.I. Mourikis, and S.I. Roumeliotis. On the consistency of multi-robot cooperative localization. In *Robotics: Science and Systems*, Seattle, WA, pp. 65–72, 2009.
15. B. Iyidir and Y. Ozkazanc. Jamming of GPS receivers. In *Proceedings of the IEEE 12th Signal Processing and Communications Applications Conference, 2004*, pp. 747–750, IEEE, 2004.

16. J.S. Jennings, G. Whelan, and W.F. Evans. Cooperative search and rescue with a team of mobile robots. In *8th International Conference on Advanced Robotics, 1997. ICAR'97 Proceedings*, pp. 193–200, IEEE, 1997.

17. S.J. Julier and J.K. Uhlmann. A non-divergent estimation algorithm in the presence of unknown correlations. In *Proceedings of the 1997 American Control Conference, 1997*, volume 4, pp. 2369–2373, IEEE, 1997.

18. R. Kurazume and S. Hirose. Study on cooperative positioning system: Optimum moving strategies for CPS-III. In *Proceedings of the IEEE International Conference on Robotics and Automation*, volume 4, pp. 2896–2903, 16–20 May 1998.

19. R. Kurazume, S. Nagata, and S. Hirose. Cooperative positioning with multiple robots. In *Proceedings of the IEEE International Conference on Robotics and Automation*, pp. 1250–1257, 8–13 May 1994.

20. R. Kurazume, S. Hirose, S. Nagata, and N. Sashida. Study on cooperative positioning system (basic principle and measurement experiment). In *Proceedings of the IEEE International Conference on Robotics and Automation, 1996*, volume 2, pp. 1421–1426, IEEE, 1996.

21. A. Martinelli and R. Siegwart. Observability analysis for mobile robot localization. In *2005 IEEE/RSJ International Conference on Intelligent Robots and Systems, 2005. (IROS 2005)*, pp. 1471–1476, IEEE, 2005.

22. L. Merino, F. Caballero, J.R. Martinez-de Dios, and A. Ollero. Cooperative fire detection using unmanned aerial vehicles. In *Proceedings of the 2005 IEEE International Conference on Robotics and Automation, 2005. ICRA 2005*, pp. 1884–1889, IEEE, 2005.

23. N. Mishra, A. Chakraborty, R. Sharma, and K.M. Brink. Cooperative relative pose estimation to aid landing of a UAV on a moving platform. In *2019. Proceedings of the 2019 Indian Control Conference*, IEEE, 2019.

24. P. Misra and P. Enge. *Global positioning system: Signals, measurements and performance*, second edition. Massachusetts: Ganga-Jamuna Press, 2006.

25. S. Misra, A. Chakraborty, R. Sharma, and K.M. Brink. Cooperative simultaneous arrival of unmanned vehicles onto a moving target in GPS-denied environment. In *2018 57th IEEE Conference on Decision and Control (CDC)*, pp. 5409–5414, IEEE, 2018.

26. A.I. Mourikis and S.I. Roumeliotis. Performance analysis of multirobot cooperative localization. *IEEE Transaction for Robotics and Autonomus Systems*, 22(4):666–681, 2006.

27. E.D. Nerurkar and S.I. Roumeliotis. *Distributed map estimation algorithm for cooperative localization*. Department of Computer Science & Engineering, University of Minnesota, Technical Report, 2008.

28. E.D. Nerurkar and S.I. Roumeliotis. Asynchronous multicentralized cooperative localization. In *2010 IEEE/RSJ International Conference on Intelligent Robots and Systems (IROS)*, pp. 4352–4359, IEEE, 2010.

29. E.D. Nerurkar, S.I. Roumeliotis, and A. Martinelli. Distributed maximum a posteriori estimation for multi-robot cooperative localization. In *ICRA'09. IEEE International Conference on Robotics and Automation, 2009*, pp. 1402–1409, IEEE, 2009.

30. F. Pasqualetti, A. Franchi, and F. Bullo. On optimal cooperative patrolling. In *2010 49th IEEE Conference on Decision and Control (CDC)*, pp. 7153–7158, IEEE, 2010.

31. F. Pasqualetti, A. Franchi, and F. Bullo. On cooperative patrolling: Optimal trajectories, complexity analysis, and approximation algorithms. *IEEE Transactions on Robotics*, 28(3):592–606, 2012.

32. I.M. Rekleitis, G. Dudek, and E.E. Milios. Multi-robot cooperative localization: A study of trade-offs between efficiency and accuracy. In *2002. IEEE/RSJ International Conference on Intelligent Robots and Systems*, volume 3, pp. 2690–2695, IEEE, 2002.

33. S.I. Roumeliotis and G.A. Bekey. Collective localization: A distributed Kalman filter approach to localization of groups of mobile robots. In *ICRA'00. IEEE International Conference on Robotics and Automation, 2000. Proceedings*, volume 3, pp. 2958–2965, IEEE, 2000.

34. S.I. Roumeliotis and G.A. Bekey. Distributed multi-robot localization. In *Distributed Autonomous Robotic Systems 4*, pp. 179–188, Springer, 2000.

35. S.I. Roumeliotis and G.A. Bekey. Distributed multirobot localization. *IEEE Transactions on Robotics and Automation*, 18(5):781–795, 2002.
36. A.C. Sanderson. Multirobot navigation using cooperative teams. In *Distributed Autonomous Robotic Systems 3*, pp. 163–172, Springer, 1998.
37. R. Sharma. Observability based control for cooperative localization. In *2014 International Conference on Unmanned Aircraft Systems (ICUAS)*, pp. 134–139, IEEE, 2014.
38. R. Sharma, R.W. Beard, C.N. Taylor, and D. Pack. Bearing-only cooperative geo-localization using unmanned aerial vehicles. In *American Control Conference (ACC), 2012*, pp. 3883–3888, IEEE, 2012.
39. R. Sharma, R.W. Beard, C.N. Taylor, and S. Quebe. Graph-based observability analysis of bearing-only cooperative localization. *IEEE Transactions on Robotics*, 28(2):522–529, 2012.
40. R. Sharma, S. Quebe, R.W. Beard, and C.N. Taylor. Bearing-only cooperative localization. *Journal of Intelligent & Robotic Systems*, 72(3–4):429–440, 2013.
41. R. Sharma and S. Rathinam. Cooperative-timing attack with smart munitions using cooperative localization in contested environments. In *AIAA Infotech@ Aerospace*, p. 1712, 2016.
42. R. Sharma and C. Taylor. Cooperative navigation of MAVs in GPS denied areas. In *IEEE International Conference on Multisensor Fusion and Integration for Intelligent Systems, 2008. MFI 2008*, pp. 481–486, IEEE, 2008.
43. R. Sharma and C.N. Taylor. Vision based distributed cooperative navigation for MAVs in GPS denied areas. In *Proceedings of the AIAA Infotech@ Aerospace Conference*, 2009.
44. D.P. Shepard, T.E. Humphreys, and A.A. Fansler. Evaluation of the vulnerability of phasor measurement units to GPS spoofing attacks. *International Journal of Critical Infrastructure Protection*, 5(3):146–153, 2012.
45. K.-T. Song, C.-Y. Tsai, and C.-H. Chiu Huang. Multi-robot cooperative sensing and localization. In *IEEE International Conference on Automation and Logistics, 2008. ICAL 2008*, pp. 431–436, IEEE, 2008.
46. H. Wen, P.Y.-R. Huang, J. Dyer, A. Archinal, and J. Fagan. Countermeasures for GPS signal spoofing. In *ION GNSS*, pp. 13–16, 2005.
47. H. Wymeersch, J. Lien, and M.Z. Win. Cooperative localization in wireless networks. *Proceedings of the IEEE*, 97(2):427–450, 2009.
48. H. Yu, R. Sharma, R.W. Beard, and C.N. Taylor. Observability-based local path planning and collision avoidance for micro air vehicles using bearing-only measurements. In *American Control Conference (ACC), 2011*, pp. 4649–4654, IEEE, 2011.
49. F. Zhang, B. Grocholsky, V. Kumar, and M. Mintz. Cooperative control for localization of mobile sensor networks. In *Cooperative Control*, pp. 241–255, Springer, 2005.

25

Cooperative Localization for Autonomous Vehicles Sharing GNSS Measurements

Khaoula Lassoued and Philippe Bonnifait

CONTENTS

25.1 Introduction

In this chapter, we propose a new formulation of multi-vehicle cooperation based on sharing global navigation satellite systems (GNSS) corrections, as a way to extend the principle of differential GPS (DGPS) to dynamic reference base stations. We present a new cooperative algorithm for vehicles positioning based on set inversion method with constraint satisfaction problem (CSP) techniques on intervals. When using V2X communications, vehicles cooperate and exchange information such that each vehicle can compute the positions of the partners with a reliable domain. The ego-motion of every vehicle can also be used to improve the cooperation. The proposed method relies on the exchange of positions estimates, pseudorange estimated errors and dead-reckoning (DR) data (see Figure 25.1). A first contribution of this work is to show that it is possible to improve the ego positioning and mutual localization between vehicles by sharing GNSS-biased pseudoranges using a model of the correlation of the pseudorange errors and through the knowledge of the local motions of the vehicles gained by DR or tracking. In the following, we present an overview of the most recent developments or work in cooperative localization over the last decade. In particular, we introduce four variants in the architecture of cooperative localization. Section 25.3 introduces a system modeling and an observability study of the cooperative system. Then, in Section 25.4, the proposed distributed estimation method is presented. It allows the data fusion of the estimated biases in a distributed fashion with no central fusion node and no base station. An experiment with two vehicles is performed in Section 25.5 to evaluate the performance in real conditions using a ground truth system. The performance of the cooperation is highlighted. In Section 25.6, a classical sequential Bayesian method is also implemented on the same data set and is compared in terms of accuracy and confidence.

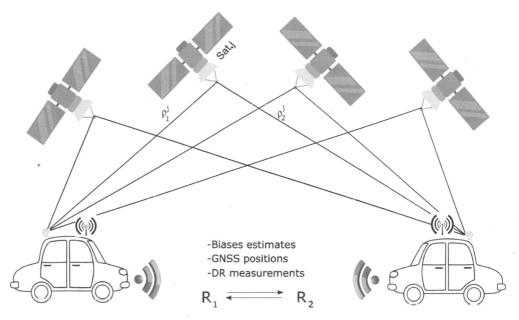

FIGURE 25.1
Experimental vehicles and shared data.

25.2 State of the Art of Cooperative Localization for Multi-Robot Systems

25.2.1 Centralized Cooperative Estimation

There are two kinds of architecture for cooperative systems: centralized and decentralized. Many variants of cooperative positioning systems are essentially based on centralized systems. A centralized system performs state estimation using a central fixed unit. This center communicates with all robots of the group. The principle of centralized estimation includes the state estimation of each robot in a unique state vector of the group. The update of the state of each robot is performed through treatment of sensor data collected from each robot. This data can be egocentric (i.e., proprioceptive) or observational (i.e., exteroceptive). Egocentric data includes robots' motions or their absolute pose. Observational data can be inter-robot or environmental or external data, such as the relative pose between robots or the range between robots and obstacles.

Hereafter, the state of the art in cooperative localization through centralized architecture is given.

The most significant developments in cooperative positioning have been achieved by Roumeliotis and Bekey [1], who presented a cooperative system for positioning of a fleet of mobile robots using a centralized approach. In this approach, they used an Extended Kalman Filter (EKF) to update the state of the fleet. Several cooperative approaches have extended Roumeliotis' and Bekeys' research to study, for example, the impacts of using relative data such as orientation measurements and relative distances in the fleets' state update process [2] or furthermore, the propagation of uncertainties [3]. Data fusion methods are not limited to the Kalman filter; other tools are proposed, such as the Posterior Maximum Estimator [4] and the Maximum Likelihood Estimation [5]. Some applications use other types of sensors; for example other research proposes to update the robots' states using acoustic or visual sensors. A camera that measures relative orientation was used in a cooperative positioning method in [6]. Acoustic waves were used to measure the relative pose between robots in [7].

Centralized approaches have advantages, such as the simplicity of the centralized merger of the data collected from each robot, allowing for an optimal update of the state of the group. The interdependencies are handled in a natural way as the data are only merged one time in the CPU, so there is no risk of the reuse of identical information in the merge process. However, in practice, centralized approaches have disadvantages related to the communication constraints with the central unit. The robots have to maintain permanent contact with the infrastructure. They must be close because the range of the communication systems is often short, as it depends on deployment provided by the phone provider, which limits the robots' area of evolution. The more remote the network access point, the greater the loss in energy level of data transmission. The rate of transmission also decreases with more data users in the same area. In order to overcome these disadvantages, the centralized architecture has evolved into a multi-centralized architecture. As a way to expand the robots' area of evolution, mobile agents gathering information are used as intermediaries between the mobile robots and the infrastructure. In the multi-centralized localization context, Worrall and Nebot have used multi-hop communications to transmit real-time vehicle state information to other vehicles [8]. However, an isolated agent problem occurs when the environment is large and sparsely populated. One possible solution to this problem is to use additional mobile agents that navigate near the isolated agent to collect updates of state information and bring them to the main network [9]. However, in these proposed solutions, the centralized system always depends on the infrastructure, and therefore a failure of the infrastructure impacts the whole system. Given the technological evolution and the availability of embedded computers in several kinds of robots, researchers have started to examine how to get rid of the central system and to distribute the estimated pose of the group within its members. This type of architecture is called distributed or decentralized architecture.

25.2.2 Decentralized Cooperative Estimation

In a distributed fusion system, each vehicle processes its measurements and communicates the results with the other vehicles. The distributed architecture does not have a central server sharing information with the members of the group. The basic concept of distributed cooperative localization is the same as the centralized approaches.

The vehicles in the group are considered as a system of systems that uses data from each member. However, in this case each agent has a version of the state of the group which it updates with its own information and with the shared data from its partners. This kind of architecture was introduced by Roumeliotis et al. [10]. The authors propose a cooperative localization system based on an EKF. This approach is based on the modification of the Kalman filter equations to distribute them to the different members of the group. This method enables the production of the same result as a centralized estimator. Other works have been inspired by this approach and propose solutions based on different sensors and mathematical tools for fully distributed information fusion. The distributed approaches can be divided into two broad categories according to the type of shared data between vehicles:

- Approaches based on the exchange of fused states or parts of them.
- Approaches based on the exchange of sensor measurements.

The concept of sensor-based exchange approaches is that each vehicle has to send its measurements to all its partners in order to have an updated version of the status in each vehicle.

In the following, each type of approach is described by providing some examples from the literature.

Hery et al. [11] proposed a cooperative localization method based on state exchange to improve the along-track position of vehicles. The vehicles are using relative measurements from perception sensors such as Lidars. They introduced a one-dimensional formulation of the localization problem under the hypothesis that lateral distances and headings of vehicles are well known. In [12], because of lateral measurements provided by cameras, for instance, cross-track errors are estimated in each vehicle using a bounded-error method and a distributed least-squares fusion elaborates north-east corrections that are applied in every vehicle. Madhavan et al. presented a distributed system of cooperative localization of heterogeneous mobile robots in [13]. The proposed approach has been applied to a fleet of two mobile robots. The cooperation is carried out by shared information about the relative pose that contains the distances between both robots, the relative orientation, and the absolute position, which contains GPS and proprioceptive measurements. Because of the exchange of data, each robot has a state version of the group. In this case, the need for permanent availability of communication between the robots is essential. Moreover, the amount of data to be shared increases according to the number of robots used in the group and the sensor rates. The estimates produced by each robot are suboptimal compared to a centralized estimator. Fox et al. [14] have adopted a similar approach to [13] in the sense that each robot has to maintain an estimate of its own pose. They used a particle filter method instead of an EKF, and a density tree approach for the fusion of the relative shared data. This approach temporarily combines the probability density functions (i.e., particles) from both robots to perform a sampling of importance.

In fused state exchange approaches, each robot only uses its own sensors to update the state of the group and only shares its final result. This type of approach limits the amount of exchanged data. However, these approaches suffer from the problem of data incest. This problem is due in particular to the re-use of the same information in the fusion process. The estimates obtained after the fusion in this case are biased and overconfident. Indeed, the problem of data consanguinity generates a problem of overconvergence, which comes from the ignorance of the correlations between the estimates of robots. Several solutions to the data incest problem have been proposed. For example, Howard et al. have used an approach based on the exchange of measured pose distributions, where each robot estimates the positions of the other robots in the group regarding its own position using the measurements of a camera and a laser ranging [15]. The fusion of shared information is performed using a particle filter. After the data exchange, each robot updates the distribution of its pose with the one received from the detected robot. In order to avoid the problem of circular updates, the author keeps up to date a dependency tree in order to only update the descendants with data that has not been merged before. Other solutions are proposed by Roumeliotis et al. and Karam et al. [16,3]. In [3], robots only share a part of their state and, during the fusion process, only a part of the collected information is considered. In this case, the obtained fused state contains suboptimal estimates since we do not use all the shared information. In [16], the overall state of the system combines all local states collected by all members of the group, and the global state is only used by local robots but never communicated again to other members.

However, this work does not address the consistency/integrity of the estimates. The covariance intersection method represents a solution of the data incest problem since it manages the dependencies between estimates, and at the same time offers reliable confidence domains. The above-mentioned methods are all based on Bayesian approaches (EKF) or particle filter. Other fusion approaches can avoid the problem of overconvergence while guaranteeing results with reliable confidence domains, such as set-membership approaches [17–20]. This will be the following approach in this study, in particular the set-membership method based on interval analysis.

25.2.3 Common Methods for Estimation

The set-membership approach is based on the assumption that errors of models and measurements are bounded. It has been successfully applied for the estimation of model parameters [21] and the estimation of robot positions when reliable confidence domains are required [22]. Meizel et al. [23] developed a set inversion method by using interval analysis techniques (set inversion via interval analysis [SIVIA]) based on bounded error observers for the localization of one robot. In [24], an efficient technique has been proposed to accurately compute a point estimate from a subpaving. However, SIVIA is not suitable for real-time applications when the initial area of research is too large, which leads to an increase in the number of bisections in the computation process. One solution is to use SIVIA while simultaneously solving a CSP to limit the calculation time of bisections [25]. Regarding real-time cooperative localization, several recent studies based on set inversion with the CSP techniques were studied. Drevelle et al. [26] operated a group of autonomous underwater vehicles (UAV) to explore a large area. In their application, they used ranging sensors to measure the inter-distance between robots.

In addition, Bethencourt [26] used distributed set-membership algorithms to accomplish a cooperative mission of a group of UAVs using intertemporal measurements. Kyoung-Hwan and Jihong also studied a cooperative localization method for several ground robots based on constraint propagation techniques [27]. In this work, the fusion of proprioceptive and exteroceptive sensors data was performed.

In summary, considerable work has been done in simulation of cooperative systems, but very few results using real data have been reported. Finally, little work has been done to evaluate the performance of the integrity of estimates from low-cost sensors.

25.2.4 Technologies for Positioning

Localization is used in many fields, such as by sailors who need to know their absolute position regularly. Today, many consumer applications utilize localization. The list of applications is very wide, and new uses have appeared regularly in recent years. In the field of robotics, localization can guide robots in difficult places that are inaccessible or contaminated to perform different tasks autonomously. In intelligent transportation systems (ITS), localization is an essential task for the navigation of intelligent vehicles. Whether localization is employed to inform the user of his position in order to guide him, or for autonomous control, the vehicle must be able to localize itself in its environment.

There are different technologies for positioning mobile robots in a free space. They are based on communicating systems that use transmitters (e.g., ground infrastructure such as beacons) and receivers (e.g., GPS, ultra-wide band [UWB], WiFi, Bluetooth, radiofrequency identification [RFID], etc.), or on embedded sensors for navigation (e.g., MEMS, camera, etc.) without relying on infrastructure or the combination of communicating systems with autonomous sensors. Among the most widespread technologies based on ground infrastructure, there are systems using radio positioning techniques based on phone networks or wireless networks (e.g., UWB, WiFi, etc.), where the mobile object is localized using the signals it transmits to the network (e.g., 2G, 3G, WiFi, RFID etc.) or receives from base stations or access points in that network. The main disadvantage of these methods is that their coverage area is limited. Geopositioning by satellite navigation systems (e.g., GNSS) offers the benefit of global coverage with a constellation of satellites in orbit (although some systems, based on geostationary satellites, have only regional coverage). Since the first satellite system, Global TRANSIT, commissioned in 1967, other systems have been developed. The NAVSTAR GPS, GALILEO, GLONASS, and BEIDOU systems are thus currently operational or being deployed. These systems provide absolute positioning with an accuracy of 10 meters, for a reasonable cost. However, satellite techniques are not effective in urban environments and closed environments, with errors of several dozen meters. An alternative method is the hybridization of GNSS positioning with other sources of information such as inertial measurements (e.g., speed and orientation measurements), relative location based on the detection of beacons (e.g., UWB, radar) or geometric elements of the route (e.g., a map). DR sensors and inertial sensors are frequently used as alternative technologies for determining position, speed, and attitude. They are very common, and measurements can easily be collected, for example from a controller area network (CAN) bus embedded on a car.

We are interested in a hybrid method that combines ground and satellite radio navigation systems with DR sensors. The technologies considered in this chapter for the positioning of vehicles are the fusion of DR measurements with radio navigation measurements such

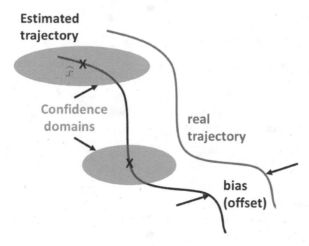

FIGURE 25.2
Illustration of the effects of biased measurements. Confidence domains are zones in which the vehicle is highly located.

as GNSS pseudorange measurements. In the following, mutual cooperation through information exchange is used to enhance positioning accuracy of a group of autonomous vehicles. We aim also to reduce uncertainty arising from the use of low-cost sensors. For example, use of mono-frequency GNSS receivers in complex environments usually leads to offsets between real and observed positions (Figure 25.2). These systematic offsets (i.e., biases) are often due to inaccurate satellite positions and atmospheric and tropospheric errors. The impact of these biases on vehicle localization should not be neglected. Cooperation and exchange of bias estimates between vehicles can significantly reduce these systematic errors [28]. However, distributed cooperative localization based on sharing estimates is subject to data incest problems. When position information is used in a safety-critical context (e.g., autonomous vehicle navigation in proximity), one should guaranty the consistency of the localization estimates. In this context, we mainly aim to improve the absolute and relative performances of vehicle localization through cooperation. Moreover, we focus on characterizing reliable confidence domains (see Figure 25.2) that contain vehicle positions with high reliability.

25.3 System Modeling and Observability Analysis for Cooperative Vehicle Systems

Here we present the mathematical models used to determine the positions of the cooperative systems. We first present basic models which are the most common in robotic applications. Then, we present a dynamic observability study based on Lie derivatives. Observability studies are essential studies for designing localization. One should check if the vehicles' positions and measurements biases can be determined before the development of any observer.

25.3.1 Modeling of Systems

25.3.1.1 Sensor Model or Distances Measurements

The GNSS raw measurements considered here are L1 pseudoranges. The pseudorange ρ_i^j [29] is the measure of each vehicle R_i ($i = \{1, \dots, n_r\}$) located at coordinates $p_i = [x_i, y_i, z_i]$ to each satellite j ($j = 1, \dots, n_s$) at position $p^j = [x^j, y^j, z^j]$. We model the common GNSS errors b^j as additive errors on the pseudoranges. The model of ρ_i^j is expressed in Equation 25.1. Please refer to [30] for further details.

$$\rho_i^j = \sqrt{(x_i - x_j)^2 + (y_i - y_j)^2 + (z_i - z_j)^2} + b^j + d_i + \beta^j \tag{25.1}$$

where d_i represents the receiver clock offset and β^j the measurement noise. Let $^i y \in \mathbb{R}^{n_s}$ be the vector of n_s pseudoranges measurements and $^i x \in \mathbb{R}^n$ be the state vector of vehicle i. The observation model at discrete time k is defined as follows

$$^i y_k = g\left(^i x_k\right) \tag{25.2}$$

Suppose that each vehicle is equipped with a GNSS receiver and DR sensors that provide speed and heading in an input vector $^i u(t) = [v_i \; \psi_i]^T$.

25.3.1.2 Vehicles

A 2D unicycle evolution model for the pose components, a linear model for the receiver clock offset, and an autoregressive (AR) model for the pseudoranges errors are concatenated and described by a continuous function f in a local east-north-up (ENU) frame (time t is omitted for clarity):

$$f(^i x, ^i u) = \begin{cases} \dot{x}_i = v_i \cos(\psi_i); & \dot{y}_i = v_i \sin(\psi_i); & \dot{z}_i = 0 \\ \dot{d}_i = dr_i; & \dot{dr}_i = 0; & \dot{b}^j = ab^j \end{cases} \tag{25.3}$$

where d_i and dr_i are respectively the receiver clock offset and its drift to be estimated. The AR parameter is $a = e^{-Te/\tau}$, T_e being the sampling time and τ the time constant of the model bias b^j ($a = 0.9995$, $\tau = 6.2$ min). The indexes of the vehicles and common satellites are denoted by $i \in \{1 \dots n_r\}$ and $j \in \{1 \dots n_s\}$, respectively.

25.3.2 Observability Study

25.3.2.1 Definitions and Theorems

Errors on pseudorange measurements are spatially correlated and similar for nearby users [29] but not directly observed with no base station. Therefore, it is essential to study the observability to evaluate if the problem is solvable. Rife and Xiao [31] have shown that it is not possible to estimate biases simply by sharing GNSS pseudorange measurements between vehicles communicating via a vehicle-to-vehicle (V2V) network in a snapshot way (Epoch by Epoch). They highlighted the limitation of distributing only GNSS data and proposed to add georeferenced information by using camera-based lane boundary sensor. A natural question that arises is whether GNSS biases are observable when adding vehicle motion information and errors evolution model.

In this section, we investigate the observability of the cooperative localization problem of vehicles sharing biases estimates when they are moving. The cooperative system described in Section 25.3.1 given DR and pseudorange measurements is nonlinear. Therefore, we use the observability rank criterion based on Lie derivatives [32] to determine the conditions under which the system is locally weakly observable. Note that Martinelli and Siegwart [33] have employed this criterion to investigate the observability of 2D cooperative localization of mobile robots. Zhou et al. [34] have used the Lie derivatives to determine the conditions for the observability of 2D relative pose of pairs of mobile robots using range measurements. In the sequel, a test of Lie derivatives is considered for vehicles sharing GNSS errors. This study is inspired by the work of Zhou et al. [34].

Definition 25.1

(Observability Rank Condition): The observability rank condition is satisfied when the observability matrix is full rank.

Theorem 25.1

(Observability Sufficient Condition): If the system satisfies the observability rank condition at a given state x^0 (at some time), then the system is locally weakly observable at x^0 [32].

25.3.2.2 Dynamic Case Study

Let consider n_r vehicles and n_s common visible satellites. Let $\hat{x}$ be the estimated state of the cooperative system (S) as follows

$$\hat{x} = [\hat{p}_1, \hat{d}_1, \ldots, \hat{p}_{n_r}, \hat{d}_{n_r}, \hat{b}^1, \ldots, \hat{b}^{n_s}]^T \tag{25.4}$$

with $\dim(\hat{x}) = 4n_r + n_s$. $\hat{p}_{1\ldots n_r}$ are the 3D vehicles positions, $\hat{d}_{1\ldots n_r}$ represent the receivers' clock offsets. $(\hat{b}^{1\ldots n_s})$ denote the biases on n_s common pseudorange measurements between vehicles.

The considered evolution model in this study consists in the first four DR equations of the system 25.3 and the last equation for the evolution of biases. Let consider $u = [v_1 \ldots v_{n_r}]^T$ the input of the system (S) in Equation 25.4. The nonlinear DR model of (S) can be written as follows

$$\dot{x} = \underbrace{\begin{bmatrix} 0 \\ 0 \\ 0 \\ dr_1 \\ \vdots \\ 0 \\ 0 \\ 0 \\ dr_{n_r} \\ ab^1 \\ \vdots \\ ab^{n_s} \end{bmatrix}}_{f_0} + \underbrace{\begin{bmatrix} \cos(\psi_1) \\ \sin(\psi_1) \\ 0 \\ 0 \\ \vdots \\ 0 \\ 0 \\ 0 \\ 0 \\ 0 \\ \vdots \\ 0 \end{bmatrix}}_{f_1} v_1 + \cdots + \underbrace{\begin{bmatrix} 0 \\ 0 \\ 0 \\ 0 \\ \vdots \\ \cos(\psi_{n_r}) \\ \sin(\psi_{n_r}) \\ 0 \\ 0 \\ 0 \\ \vdots \\ 0 \end{bmatrix}}_{f_{n_r}} v_{n_r} \tag{25.5}$$

The nonlinear observation equations are given by

$$
\mathbf{y} = n_s \times n_r \left\{
\begin{bmatrix}
\|\mathbf{p}_1 - \mathbf{p}^1\| + d_1 + b^1 \\
\vdots \\
\|\mathbf{p}_1 - \mathbf{p}^{n_s}\| + d_1 + b^{n_s} \\
\vdots \\
\|\mathbf{p}_{n_r} - \mathbf{p}^1\| + d_{n_r} + b^1 \\
\vdots \\
\|\mathbf{p}_{n_r} - \mathbf{p}^{n_s}\| + d_{n_r} + b^{n_s}
\end{bmatrix}
\right.
\tag{25.6}
$$

We compute hereafter the necessary Lie derivatives of $\mathbf{y}$ and their gradients: *Zeroth-order Lie derivatives* ($\mathcal{L}^0 \mathbf{y}$)

$$
\mathcal{L}^0 \mathbf{y} = \mathbf{y}
$$

with gradient:

$$
\nabla \mathcal{L}^0 \mathbf{y} = jacobian\ (\mathbf{y})
$$

$$
= \mathcal{G} = n_r \times n_s \left\{
\overbrace{
\begin{bmatrix}
G & 0 & \cdots & 0 & I \\
0 & G & \cdots & 0 & I \\
\vdots & \vdots & \ddots & \vdots & \vdots \\
0 & 0 & \cdots & G & I
\end{bmatrix}
}^{4n_r + n_s}
\right.
\tag{25.7}
$$

where I is the identity matrix with $\dim(I) = n_s \times n_s$ and $\mathcal{G}$ is the geometry matrix described in [31] where G is defined as follows:

$$
G = \begin{bmatrix}
(u^1)^T & 1 \\
(u^2)^T & 1 \\
\vdots & \vdots \\
(u^{n_s})^T & 1
\end{bmatrix}
\tag{25.8}
$$

with $\dim(G_i) = n_s \times 4$, the unit vector u^i in G is the estimated line of sight from the satellite j to each user receiver i. This pointing vector is the same for all users when they are assumed to be in close proximity (i.e., distance between vehicles ≤ 10 km):

$$
u^j = (\mathbf{p}_i - \mathbf{p}^j) / \| \mathbf{p}_i - \mathbf{p}^j \|.
\tag{25.9}
$$

First-order Lie derivatives $(\mathcal{L}^1_{f_0}\mathbf{y})$

$$\mathcal{L}^1_{f_0}\mathbf{y} = \nabla \mathcal{L}^0\mathbf{y} \cdot f_0 = n_s \times n_r \left\{ \begin{bmatrix} \begin{bmatrix} ab^1 + dr_1 \\ \vdots \\ ab^{n_s} + dr_1 \end{bmatrix} \\ \vdots \\ \begin{bmatrix} ab^1 + dr_{n_r} \\ \vdots \\ ab^{n_s} + dr_{n_r} \end{bmatrix} \end{bmatrix} \right.$$

with gradient:

$$\nabla \mathcal{L}^1_{f_0}\mathbf{y} = n_s \times n_r \left\{ \begin{bmatrix} \begin{bmatrix} \overset{4n_r}{\overbrace{0}} & \overset{n_s}{\overbrace{a\mathbf{I}}} \end{bmatrix} \\ \vdots & \vdots \\ \begin{bmatrix} \overset{4n_r}{\overbrace{0}} & \overset{n_s}{\overbrace{a\mathbf{I}}} \end{bmatrix} \end{bmatrix} \right.$$

The observability matrix is now:

$$\mathcal{O} = \begin{bmatrix} \nabla \mathcal{L}^0\mathbf{y} \\ \nabla \mathcal{L}^1_{f_0}\mathbf{y} \end{bmatrix}. \tag{25.10}$$

The role of the matrix in Equation 25.10 in the observability analysis of a nonlinear system is given in [32] and recalled in Definition 25.1 and Theorem 25.1.

Below, we compute the rank of the observability matrix 25.10 and determine the necessary conditions under which the system (S) can be locally weakly observable. Here, we have:

$$rank(\mathcal{O}) = rank(\nabla \mathcal{L}^0\mathbf{y}) + rank(\nabla \mathcal{L}^1_{f_0}\mathbf{y}). \tag{25.11}$$

The rank of $\nabla \mathcal{L}^0\mathbf{y}$ has been studied in [31]. They proved that $rank(\nabla \mathcal{L}^0\mathbf{y}) = 4(n_r - 1) + n_s$. It is straightforward to determine the rank of $\nabla \mathcal{L}^1_{f_0}\mathbf{y}$. Since the number of linearly independent equations in $\nabla \mathcal{L}^1_{f_0}\mathbf{y}$. appears to be n_s (i.e., rank $\left(\nabla \mathcal{L}^1_{f_0}\mathbf{y}\right) = n_s$) if we have $a \neq 0$ (i.e., AR model of the biases). So, according to 25.11 we get:

$$rank(\mathcal{O}) = 4(n_r - 1) + 2n_s \tag{25.12}$$

In order to get a full rank of $\mathcal{O}$, one must discuss the least required number of n_s common satellites between users. It is obvious that rank($\mathcal{O}$) cannot exceed the n_l unknown states of (S) which is equal to $4n_r + n_s$ (i.e., $rank(\mathcal{O}) \leq 4n_r + n_s$), so to get a full rank of $\mathcal{O}$ one must determine n_s such that

$$rank(\mathcal{O}) \geq 4n_r + n_s. \tag{25.13}$$

By replacing 25.12 in Equation 25.13 we get: $n_s \geq 4$. The observability rank condition (Definition 25.1) is obtained when this condition is satisfied. According to Theorem 25.1, it can be concluded that the system is locally weakly observable regardless of the number of users (n_r) if the biases have an AR behavior and at least four common satellites between the vehicles. Please note also that, as we have only used the Lie derivative with respect to f_0, the system is observable even if the vehicles are motionless.

25.4 Methodology for Cooperative State Estimation

This section introduces how to extend non-cooperative methods of localization to cooperative ones. Essentially, there are two methods: the probabilistic (or Bayesian) method and the bounded error (or set-membership) one.

25.4.1 Computing Reliable Confidence Domains with a Set Membership Method

25.4.1.1 Set Inversion with Constraints Propagation

To perform a state estimation in a bounded error framework with intervals, one needs solving a set inversion problem. The objective is to determine the unknown state $\mathbb{X} \subset \mathbb{R}^n$ such as $\mathbf{f}(\mathbb{X}) \subset \mathbb{X}$, where $\mathbb{Y}$ is the known set of measurements. The objective is to compute the reciprocal image $\mathbb{X} = \mathbf{f}^{-1}(\mathbb{Y})$. A guaranteed approximation of the solution set $\mathbb{X}$ can be done using 2 subpavings which bracket the solution set as follows: $\underline{\mathbb{X}} \subset \mathbb{X} \subset \overline{\mathbb{X}}$ (Figure 25.3).

A box $[\mathbf{x}]$ of $\mathbb{R}^n$ is feasible if it is inside $\mathbb{X}$ and unfeasible if it is outside $\mathbb{X}$, otherwise $[\mathbf{x}]$ is indeterminate. By using an inclusion function $[\mathbf{f}]$ of function f, one can identify the feasibility of the boxes using the following tests:

- If $[\mathbf{f}]([\mathbf{x}]) \subset \mathbb{Y}$ then $[\mathbf{x}]$ is feasible
- If $[\mathbf{f}]([\mathbf{x}]) \cap \mathbb{Y} = \varnothing$ then $[\mathbf{x}]$ is unfeasible
- Else $[\mathbf{x}]$ is indeterminate.

SIVIA solves the set inversion problem by testing recursively the feasibility of candidate boxes, starting from an arbitrarily large initial box $[\mathbf{x}_0]$ [21]. If a box is feasible, it is stored

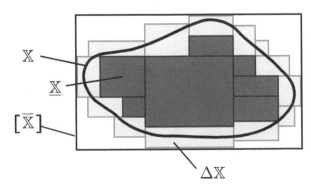

FIGURE 25.3
Bracketing of the solution set $\mathbb{X}$ between two subpavings $\underline{\mathbb{X}}$ and $\overline{\mathbb{X}} = \underline{\mathbb{X}} \cup \Delta\mathbb{X}$. $[\overline{\mathbb{X}}]$ is the hull box of $\overline{\mathbb{X}}$.

in the inner solution set $\mathbb{X}$. If a box is unfeasible, it is discarded, since the resulting box ([**f**] ([**x**])) is outside the measurement domain $\mathbb{Y}$. Finally, an indeterminate box is bisected into two sub-boxes and stored in a list $\mathfrak{L}$ waiting to be treated.

In high dimension, SIVIA is not suitable for a real-time implementation due to the computation time of the bisections. One solution is to use "SIVIAP," a SIVIA with constraints propagation (CP) on intervals. CP is very efficient in terms of computation and reduces the number of needed bisections by contracting boxes in order to speed up the processing.

SIVIAP involves the formalization of a CSP. A CSP is denoted by $\mathcal{H}$ (Equation 25.14) and provides the box [x] that satisfies all the constraints F.

$$CSP : \{\mathcal{H} : (F(x) = \mathbf{0} \mid x \in [x])\} \tag{25.14}$$

Contracting $\mathcal{H}$ means replacing [**x**] by a smaller domain [**x**′] such that the solution set remains unchanged. The contractors used in this work are forward-backward propagation and Waltz algorithm [35] intersection. More details can be found in [36].

25.4.2 Cooperative Localization Problem Statement

Vehicles share their estimated GNSS errors, their DR measurements, and their positions. The objective is to get estimates with reliable confidence domains that contain, with high probability, the true positions of the vehicles with little pessimism.

Let us describe the variables, the domains, and the constraints of the considered CSP.

25.4.2.1 Variables

In each agent R_i, there is an ego state to be estimated and a tracked position of every known other vehicle R_o in the group, where $o \in \{1, \ldots, n_r-1\}$ and $o \neq 0$.

Ego state: let $\nu \{= x_i, y_i, z_i, b^1, \ldots, b^{n_s}, d_i, dr_i\}$ be the variables of the ego state $^i x_{ego}$ of R_i of dimension $n = 5 + n_s$. Only these variables are considered in the CSP. Let denote the absolute position by $^i q = \{x_i, y_i, z_i\}$, the biases of all n_s satellites in view by $^i b = \{b^1, \ldots, b^{n_s}\}$ and the inner variables of R_i by $^i \xi = \{d_i, dr_i\}$ which contains respectively the clock offset d_i and its drift dr_i.

Tracked position: let $^i q_o = \{^i q_1, \ldots, ^i q_{n_r} -1\}$ be the positions of the other vehicles estimated by agent R_i.

The ego input of R_i is denoted by $^i u_{ego} = \{v_i, \psi_i\}$, where v_i and ψ_i are respectively the linear speed and the heading angle. $^i u_o = \{^i u_1, \ldots, ^i u_{n_r-1}\}$ represents the input of R_o composed of $^i u_o = \{v_o, \psi_o\}$, This information is received from the others.

25.4.2.2 Domains

The domains of the variables are sets which enclose the true value of the variables and are represented by boxes, that is, vectors of intervals of $\mathbb{R}^n$ as follows $[^i x_{ego}] = [[^i q]^T [^i b]^T [^i \xi]^T]^T$. Each interval contains the unknown variable $[v] = [v_{true} - \delta_v, v_{true} + \delta_v]$, where v_{true} represents the true value of v and δ_v is the bound of the error.

25.4.2.3 Constraints

The constraints that link the variables at each time k are the evolution and observation models:

$$(i) \, ^i x_{ego,k} = f_k(^i x_{ego,k-1}, \, ^i u_{ego})$$

$$(ii)\ {}^i\boldsymbol{y}_k = \boldsymbol{g}({}^i\boldsymbol{x}_{ego,k})$$

The constraint (i) corresponds to the dynamic equation of the model, while the constraint (ii) defines the CSP which is used in SIVIAP: ${}^i\boldsymbol{y}_k$ are the pseudoranges set to be inverted and ${}^i\boldsymbol{x}_{ego,k}$ is a prior feasible box coming from a prediction stage.

25.4.2.4 Solver

SIVIAP approximates the state vector ${}^i\boldsymbol{x}_{ego,k}$ such that $\boldsymbol{g}({}^i\boldsymbol{x}_{ego,k}) \subseteq [{}^i\boldsymbol{y}_k]$ using a forward-backward contractor. The set to be characterized by SIVIAP is:

$$\begin{aligned}
CSP &= \{{}^i\boldsymbol{x}_{ego,k} \subseteq [{}^i\boldsymbol{x}_{ego,k}] \setminus \boldsymbol{g}({}^i\boldsymbol{x}_{ego,k}) \subseteq [{}^i\boldsymbol{y}_k]\} \\
&= \boldsymbol{g}^{-1}([{}^i\boldsymbol{y}_k]) \cap [{}^i\boldsymbol{x}_{ego,k}]
\end{aligned} \tag{25.15}$$

25.4.3 Distributed Algorithm

The same algorithm (1) runs in every vehicle R_i. Agent R_i predicts its ego state ix_{ego} using the evolution model and its DR inputs (v_i, ψ_i) measured at high frequency (line 1). Moreover, it tracks the other vehicles (line 2) using their last received DR inputs.

Lines (4…14) of the algorithm consist in updating the predicted state ${}^i\boldsymbol{x}_{ego}$ with respect to the GNSS measurements which are available every 0.2s. In order to reduce the outliers at each time k when the GNSS measurements are available, a validation process on the measurements of every satellite is performed. For every pseudorange measurement, we check if the *SNR* (signal-to-noise ratio) of the satellite is high enough (e.g., 35 dB/Hz) and we perform an innovation test based on a punctual estimate with the center of the boxes. Afterward, we apply the SIVIAP algorithm presented in [37] with the following modifications.

Algorithm 25.1   An iteration stage of the method in R_i

Cooperation (in $: [{}^i\boldsymbol{x}_{ego}], [{}^i\boldsymbol{u}_{ego}], \left[\rho_1^1, \ldots, \rho_1^{n_s}\right], [{}^o\boldsymbol{q}], [{}^o\boldsymbol{u}_{ego}], [{}^o\boldsymbol{b}], \boldsymbol{g}$; out $: [{}^i\boldsymbol{x}_{ego}], [{}^i\boldsymbol{q}_o]$

1: $[{}^i\boldsymbol{u}_{ego}] = [v_i, \psi_i]^T = \textbf{Get}$(DR measurements)

2: **Prediction** (in: $[{}^i\boldsymbol{u}_{ego}]$; in out:$[{}^i\boldsymbol{x}_{ego}]$)

3: **Track** (in: $[{}^i\boldsymbol{q}_o], [{}^i\boldsymbol{u}_o]$; out: $[{}^i\boldsymbol{qo}]$)

4: **if** New GNSS data is available **then**

5:      $n_s =$ number of visible satellites

6:      $[\rho_1^1, \ldots, \rho_1^{n_s}] = \textbf{Get}$(GNSS measurements)

7:      Good_Pr $= \varnothing$

8:      **for** j $= 1, \ldots, n_s$ **do**

9:          **if** (ρ^j is good) **then**

10:             Add(ρ^j) to the Good_Pr list ($[{}^i\boldsymbol{y}_{good}]$)

11:         **end if**

12:     **end for**

13:     **SiviaP** (in: $\mathcal{Cs}_p, [{}^i\boldsymbol{y}_{good}], \varepsilon, \boldsymbol{g}$; in out: $[{}^i\boldsymbol{x}_{ego}]$)

14: **end if**

15: **Communication**(in: dataS; out: dataR)

16: **Track_update**(in: $[{}^o\boldsymbol{q}]$; out: $[{}^i\boldsymbol{q}_o]$)

17: **Fusion**(in out: $[{}^i\boldsymbol{b}],[{}^o\boldsymbol{b}]$)

The considered solution is the hull box of $\mathbb{X}$ which is the union of the inner $\underline{\mathbb{X}}$ and the indeterminate $\Delta \mathbb{X}$ subpavings as it is shown in Figure 25.3. In order to stop the bisections, we limit the computational time at 0.1 s for each epoch. In this case, the tolerable time communication delay is about 100 milliseconds.

In our problem, the vehicles have well-synchronized clocks using PPS pulses of the GNSS receivers. The communication delays (line 15) are neglected.

The sent (dataS) and received (dataR) data at time instant k by each vehicle have a unique identifier id in the group. The amount of the transmitted information on the communication network is low, since vehicles only exchange the lower and upper bounds of the boxes.

In line 16 of the algorithm, each vehicle i updates the tracked position of the R_o by the received estimated position $[^iq_o] := [^oq_{ego}]$. Finally, R_i merges its estimated biases with the received ones from the other vehicles R_o as follows $[^ib] = [^ib] \cap [^ob]$.

25.5 Experimental Results

The results of two scenarios, standalone (S) and cooperative (C), are reported to quantify the performance gain due to the cooperation using the proposed SIVIAP distributed method. We also compare the SIVIAP estimates with the ones of a rather conventional Bayesian procedure implementing an EKF and involving the same processes: prediction/tracking, update, communication, and fusion. Please refer to [30] for more details on the proposed cooperative EKF. The GNSS measurement errors are time correlated and, when doing sequential distributed estimation, the data fusion process that estimates the biases of these measurements incorporates loops. This induces a data incest problem. In the Bayesian framework, a usual method to address this issue is to do the fusion of the biases by using the covariance intersection (CI) operator instead of the simple convex combination (SCC), which is valid only when the errors are uncorrelated. CI is known to provide consistent estimates even when facing an unknown degree of inter-estimate correlation [38,39]. In the sequel, we denote by C-SIVIAP and C-EKF-CI the cooperative set-membership and Bayesian methods based on the CI fusion, respectively.

25.5.1 Experimental Setup

The different approaches have been tested with two experimental vehicles (Figure 25.1) and with the same data-set which was used in a post-processed way. A low-cost U-blox $4T$ GPS receiver providing raw pseudorange measurements at 5 Hz was used in each vehicle. The extraction of broadcast satellite navigation data has been done as follows. Conversion of U-blox navigation data into RINEX files and generation of satellite raw pseudoranges with ionosphere, troposphere, satellite clock offset, and time relativity classical corrections. As the localization problem is studied in a local ENU frame, the satellite positions, at their emission time, have be transformed into the ENU frame.

In every vehicle, a PolarX Septentrio receiver was used in RTK mode to provide ground truth data with heading ψ information. Indeed, when the receiver is in motion, a GNSS receiver can calculate an accurate track angle, which is the measured angle from true north in clockwise direction. When ground vehicles drive with low speed, one can assume that track is equal to heading, since slippage can be neglected. A CAN-bus gateway was used to get the linear speed v at 100 Hz rate. The inputs used by the cooperative system are $u = [v \ \psi]^T$.

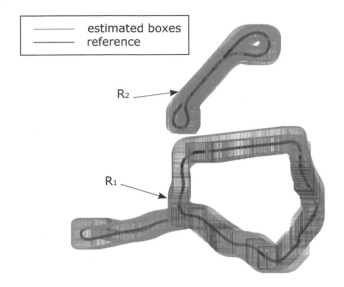

FIGURE 25.4
Trajectories of both vehicles when using C-SIVIAP. Reference and position boxes are displayed. The mean speed of vehicles was 30 Kmph. Every vehicle did several loops of its trajectory.

Ten satellites were in view during the test and five of them were at least in common, which satisfies the necessary condition of the observability study discussed in Section 25.3.2. When four common satellites between vehicles are not available, one should switch to standalone method. The GPS satellite visibility was sometimes very constrained due to buildings and trees near the test area. Vehicle R_2 has more satellite in view than vehicle R_1 during the experiment due to its favorable GNSS environments. The reported test was about 4 minutes long. Figure 25.4 shows a top view of the trajectories of both vehicles using the (C) set-membership method.

The methods are compared with the same standard deviation parameters presented in Table 25.1.

25.5.2 Set-Membership Method (SIVIAP) Performance Analysis

The distributed algorithm has been implemented in C++ using the interval library IBEX [40] and with homemade functions. Figure 25.4 shows the estimated position boxes of

TABLE 25.1

Noises Parameters Used for the Estimation Methods

Variables		Std. Deviation
$\rho(m)$	R1	$\sqrt{90000.10^{-SNR/10}}$
	R2	9
$v(m\cdot s^{-1})$	R1	$1e-3$
	R2	$2e-3$
$\psi(rad\cdot s^{-1})$	R1	$2.5e-3$
	R2	$5e-3$

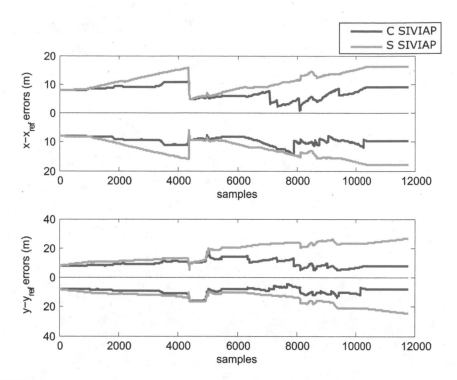

FIGURE 25.5
Bounds of position errors of R_1 centered on the reference.

both vehicles. The displayed solution is the hull box $[\overline{\mathbb{X}}]$ of the union of the inner and indeterminate subpavings: $\underline{\mathbb{X}} \cup \Delta \mathbb{X}$.

Figures 25.5 and 25.6 show the bounds of the position errors of the vehicles respectively for the x and y dimensions using the (C) and (S) methods. At first glance, one can check the consistency of both observers due to the fact that bounds always contain the zero value. It means that the RTK reference position is always included in the estimated boxes, which indicates a good tuning of the observers.

Table 25.2 gives some performance metrics for both methods in terms of absolute horizontal positioning error (HPE) and relative distance. The confidence domain size (CDS) of the resulting box is also studied; it is computed with the box volume. The CDS is evaluated via the cumulative distribution function (CDF) of vehicles' 2D boxes volumes throughout the trajectory. A net improvement is obtained for vehicle R_1 in terms of accuracy and confidence. For instance, the median of HPE is reduced from 1.43 to 0.89 m and the CDS is 66.4% condensed due to the cooperation, since the 95th percentile of the CDS is less than 475.3 m² compared to 1652 m² when using the S method. Concerning vehicle R_2, the improvement of HPE is not as significant as in R_1 since contraction of boxes can move the center away from the reference. The cumulative CDS is 52.7% reduced, since 95th percentile of the CDS is less than 256 m² compared to 541.4 m² when using the S method, which is a substantial improvement.

Regarding the estimation of the relative distance, the method improves the accuracy again through the fusion of the bias estimates. In particular, if we look at the median and standard deviation errors, they are reduced by 42% and 20%, respectively, which is a significant improvement.

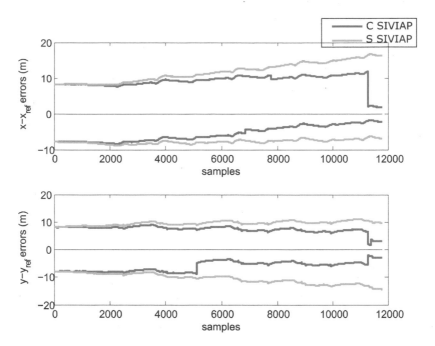

FIGURE 25.6
Bounds of position errors of R_2 centered on the reference.

TABLE 25.2

HPE, Relative Distance and CDS Statistics

		HPE R1	HPE R2	Estimated Relative Distance
Std. dev. (m)	S	2.28	1.58	2.87
	C	1.71	1.53	2.58
Max (m)	S	5.77	5.55	8.27
	C	6.21	4.30	8.87
Median (m)	S	1.43	1.83	4.93
	C	0.89	2.31	3.23
		CDS R1	CDS R2	
95th percentile (m²)	S	1652	541.4	
	C	475.3	256	

The bias on every pseudorange has been initialized with the interval $[-30, 30]$ (in meters) giving no prior knowledge. For each subplot of Figure 25.7, the x axis expresses the number of samples, the y axis displays every estimate (the center of the box) of the bias $[b^j]$ with its bounds in meters and $j = 1, \ldots, n_s$. Note that all subplots are truncated in order to observe the convergence illustrated by the horizontal final asymptotes. This convergence confirms the observability analysis. The obtained final values of the biases are very common for low-cost GNSS receivers [41].

Other results illustrating the behavior of the proposed CSIVIAP method in more detail can be found in [42].

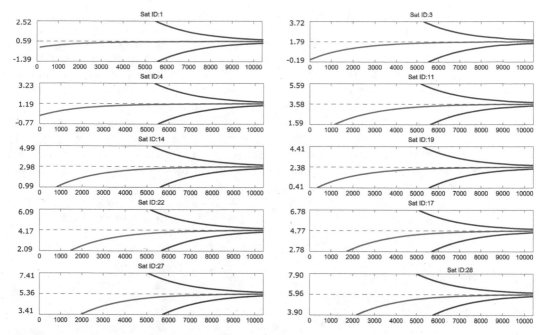

FIGURE 25.7
Estimates and bounds of pseudoranges biases (units are samples and meters).

25.6 Comparison with a Distributed Extended Kalman Filtering Implementing a Covariance Intersection

25.6.1 Backgrounds for Comparison of Both Methods

The cooperative estimation of the biases has been implemented with the CI data fusion operator [43]. Equation 25.16 shows how the covariance matrix and the state estimate are computed.

$$
\begin{cases}
P^{-1} = \omega P_1^{-1} + (1-\omega)P_2^{-1} \\
\hat{x} = P \cdot (\omega P_1^{-1} \cdot x_1 + (1-\omega)P_2^{-1} \cdot x_2)
\end{cases}
\tag{25.16}
$$

The weighting factor $\omega \in [0, 1]$ has been chosen here to minimize the determinant of the covariance matrix in order to get the smallest uncertainty. The CI fusion provides a conservative and robust result when the correlation between two estimates, $\hat{x}_1$ and $\hat{x}_2$, is unknown, as is often the case in distributed systems. However, the data fusion algorithm is not optimal [44].

When using a cooperative Bayesian method based on Kalman filtering, vehicles have to share also their estimated error covariance matrix. The algorithm used of the CI fusion of bias estimates is presented in [45]. The weight is the minimum of the covariance matrix determinant of the merged bias errors.

The C-EKF-CI has been implemented with exactly the same data as the C-SIVIAP method.

25.6.2 Comparison Criteria

The choice of good performance metrics is a key issue for assessing a system and evaluating its application. For vehicle localization systems, accuracy is an important metric which usually refers to statistical figures of merit of the position error. These errors are built with respect to ground truth. For instance, the 95th percentile of the horizontal positioning error distribution can be chosen for accessing the horizontal positioning accuracy. We also propose to evaluate the system reliability by examining the consistencies of the filters. This can be achieved by checking whether the estimated uncertainties correspond to the physical reality of the errors. The confidence bounds of estimated domains also act as decision variables and so are linked to the pessimism of a localization system. If the confidence bounds are small while keeping the estimates consistent, the localization system is considered to be not too pessimistic. So, a good localization system is a system that provides adequate confidence information and good HPE accuracy. In the following, we use two criteria.

25.6.2.1 Horizontal Protection Errors

The HPE of both methods are defined below:

$$HPE = \sqrt{e_x^2 + e_y^2} \tag{25.17}$$

where $e_x = \hat{x} - x_{ref}$, $e_y = \hat{y} - y_{ref}$. $(\hat{x}, \hat{y})$ and (x_{ref}, y_{ref}) represent respectively the 2D estimated position and the RTK reference.

For the C-SIVIAP method, the center of the estimated hull box (x_{mid}, y_{mid}) is used as a punctual estimate.

25.6.2.2 Confidence Domain Size

The CDS needs to be assessed to check if the uncertainty is well handled. To gauge this issue, the 2D Cartesian evaluation is transformed in 1D problem by using a statistical distance computation denoted $k\sigma_{HPE}$, where k is the chosen consistency risk according to a χ^2 distribution (for a 10^{-2} risk, $k = 3.035$—this is a common choice, i.e., done here). The σ_{HPE} of a Bayesian method is given by Equation 25.18 [46].

$$\sigma_{HPE} = \sqrt{\frac{1}{u_e^T P_{HPE}^{-1} u_e}}, \text{ with } u_e = \begin{pmatrix} e_x \\ e_y \end{pmatrix} / \sqrt{e_x^2 + e_y^2} \tag{25.18}$$

where u_e is the unit vector supporting the HPE and P_{HPE} is the estimated matrix of the error covariance when using C-EKF-CI (see Figure 25.8 for an illustration). For the set-membership C-SIVIAP method, the same consistency 1% risk k has been considered when setting the bounds on the pseudoranges:

$$\left[\rho_i^j\right] = \left[\rho_i^j - k\sigma_\rho, \rho_i^j + k\sigma_\rho\right] \tag{25.19}$$

where j represents the satellite index and σ_ρ represents the standard deviation of the pseudorange measurement presented in Table 25.1.

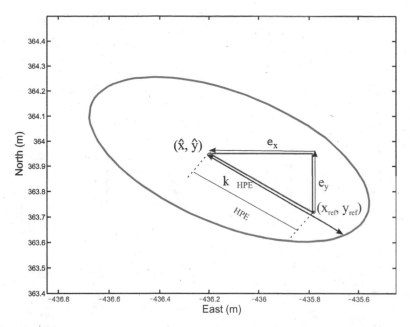

FIGURE 25.8
HPE and CDS illustration for C-EKF-CI.

In way similar to the Bayesian method, the CDS of C-SIVIAP is defined by a $k\sigma_{HPE}$ segment in the direction of the HPE vector, as shown by Figure 25.9.

25.6.3 Comparison Results

25.6.3.1 Accuracy Analysis

Figure 25.10 depicts the cumulative distribution of HPEs of both aforementioned methods. Ninety-five percent of the HPE errors for vehicle R_1 are less than 10.46 and 5.32 m, respectively, for C-EKF-CI and C-SIVIAP. For vehicle R_2 these figures are less than 7.64 and 4.41 m, respectively. The accuracy gain of C-SIVIAP compared to C-EKF-CI is 49.13% for R_1 and 42.27% for R_2. The set-membership approach is clearly more accurate than the Bayesian one.

25.6.3.2 Consistency Analysis

Let us consider now the filter consistency, which is checked if the CDS bounds actually the HPE, that is, $HPE < k\sigma_{HPE}$. Consistency tests using HPEs are fundamental to provide suitable horizontal protection levels (HPL) for cooperative vehicles [47].

Figure 25.11 presents 2D histograms in order to evaluate the consistency of the two methods for vehicles R_1 and R_2. The horizontal and vertical axes represent respectively the HPE and the CDS (i.e., $k\sigma_{HPE}$) computed for each navigation solution. Each pixel tabulates the total number of occurrences of a specific (HPE, CDS) pair. Note that the color scale is logarithmic. These histograms can be considered as simplified Stanford diagrams since we are only interested in regions where we have CDS > HPE and HPE > CDS. Points where the CDS is under the HPE error indicate a failure of integrity. In this way, the gray area corresponds to overconfident outcomes of the filters.

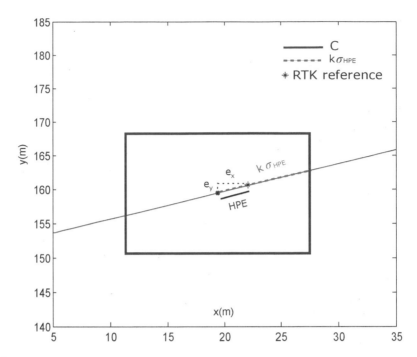

FIGURE 25.9
HPE and k σ_{HPE} illustration for C-SIVIAP.

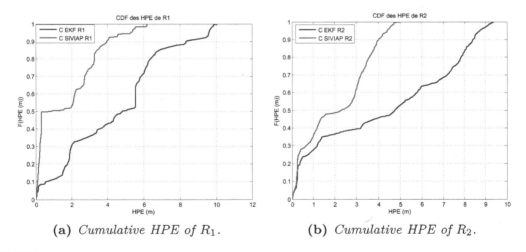

(a) *Cumulative HPE of R_1.*   **(b)** *Cumulative HPE of R_2.*

FIGURE 25.10
Plots of the cumulative distribution function (CDF) of the HPE produced by the two methods in the two vehicles (C EKF: cooperative EKF based on CI fusion, C SIVIA: cooperative set-membership method).

It appears from these results that the two methods are 100% consistent since there is no point in the gray area. Therefore, C-EKF-CI and C-SIVIAP methods are both reliable in the sense that the ground truth is always included in the estimated confidence domain.

As confidence is in practice compared to a threshold to indicate "use" or "don't use" to the client application, it is important, in terms of availability of the positioning information, to provide as small as possible confidence domains. Let us look especially at the $k\sigma_{HPE}$ of both methods for each vehicle in Figure 25.11.

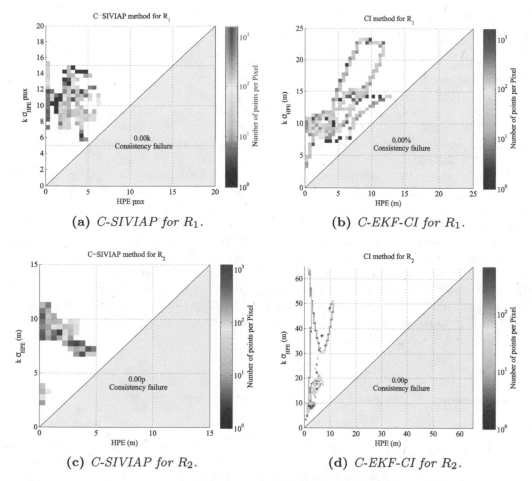

FIGURE 25.11

Simplified Stanford Diagram: Filters consistency when using cooperative set-membership and Bayesian methods for R_1 and R_2. (a) C-SIVIAP for R_1, (b) C-EKF based on CI fusion for R_1, (c) C-SIVIAP for R_2, (d) C-EKF based on CI fusion for R_2.

It can be observed that the confidence domains produced by the C-SIVIAP method are significantly tighter than the C-EKF-CI ones, in particular for vehicle R_2.

This indicates that the bounded-error method significantly reduces uncertainties compared to a Bayesian method based on covariance intersection fusion.

To resume the comparison analysis of the methods, the consistencies of set-membership and Bayesian filters are achieved. Both methods provide reliable confidence domains that contain the true positions of vehicles. Reliability is quite important for navigation missions in approach in order to avoid a collision problem. However, the cooperative set-membership method performs better than the Bayesian one in terms of accuracy and uncertainty, as it gives a significant improvement of positioning accuracy and a good decrease of the confidence domains.

If we look at performance comparison between the two vehicles when using the set-membership method, it can be observed that the best performance is obtained for vehicle R_2 which has more accurate positions (i.e., lower HPE) and less uncertainties (i.e., smaller CDS). This is due to the fact that R_1 has less satellites in view and its DR sensors are of less good quality compared to vehicle R_2.

25.7 Conclusion

This work has presented two cooperative localization methods for autonomous vehicles sharing common GNSS errors. It has been found that at least four satellites and an AR model of the biases are needed to keep the states observable, which means that the problem is solvable even if there is no fixed base station well located. The proposed cooperative bounded error approach based on set inversion method with constraint propagation provides a significant enhancement in terms of accuracy and confidence domains compared to usual standalone methods.

The reuse of identical information (which are here the estimated biases) in the fusion process is also naturally managed by a set-membership approach. This kind of method therefore deals correctly with the data incest issue. Moreover, it handles rigorously the non-linearities of the equations. We have also reported a performance comparison of the bounded-error method with a cooperative sequential Bayesian approach based on Kalman filtering and on covariance intersection fusion of the bias estimates. The experimental results indicate that both methods give reliable confidence domains of vehicle positions. However, the set-membership approach has the advantage of providing more accurate positions with smaller confidence domains. The key information deduced from this comparison is that set-membership methods are very suitable for applications requiring high integrity/accuracy in cooperative navigation contexts.

References

1. S.I. Roumeliotis and G.A. Bekey. Collective localization: A distributed Kalman filter approach to localization of groups of mobile robots. In *IEEE International Conference on Robotics and Automation (ICRA)*, volume 3, pages 2958–2965 vol.3, 2000.
2. A. Martinelli, F. Pont, and R. Siegwart. Multi-robot localization using relative observations. In *IEEE International Conference on Robotics and Automation (ICRA)*, pages 2797–2802, April 2005.
3. S.I. Roumeliotis and I.M. Rekleitis. Analysis of multirobot localization uncertainty propagation. In *IEEE/RSJ International Conference on Intelligent Robots and Systems (IROS)*, volume 2, pages 1763–1770 vol.2, Oct 2003.
4. E.D. Nerurkar, S.I. Roumeliotis, and A. Martinelli. Distributed maximum a posteriori estimation for multi-robot cooperative localization. In *IEEE International Conference on Robotics and Automation (ICRA)*, pages 1402–1409, May 2009.
5. A. Howard, M.J. Matark, and G.S. Sukhatme. Localization for mobile robot teams using maximum likelihood estimation. In *IEEE/RSJ International Conference on Intelligent Robots and Systems (IROS)*, volume 1, pages 434–439 vol.1, 2002.
6. L. Montesano, J. Gaspar, J. Santos-Victor, and L. Montano. Cooperative localization by fusing vision-based bearing measurements and motion. In *IEEE/RSJ International Conference on Intelligent Robots and Systems (IROS)*, pages 2333–2338, Aug 2005.
7. Y. Lin, P. Vernaza, J. Ham, and D.D. Lee. Cooperative relative robot localization with audible acoustic sensing. In *IEEE/RSJ International Conference on Intelligent Robots and Systems (IROS)*, pages 3764–3769, Aug 2005.
8. S. Worrall and E. Nebot. Using non-parametric filters and sparse observations to localise a fleet of mining vehicles. In *IEEE International Conference on Robotics and Automation (ICRA)*, pages 509–516, April 2007.

9. H. Mu, T. Bailey, P. Thompson, and H. Durrant-Whyte. Decentralised solutions to the cooperative multi-platform navigation problem. *IEEE Transactions on Aerospace and Electronic Systems*, 47(2):1433–1449, April 2011.
10. S.I. Roumeliotis and G.A. Bekey. Distributed multirobot localization. *IEEE Transactions on Robotics and Automation*, 18(5):781–795, Oct 2002.
11. E. Héry, P. Xu, and P. Bonnifait. Along-track localization for cooperative autonomous vehicles. In *IEEE Intelligent Vehicles Symposium (IV 2017)*, pages 511–516, Redondo Beach, CA, Jun 2017.
12. L.C. Bento, P. Bonnifait, and U. Nunes. Cooperative GNSS positioning aided by road-features measurements. *Transportation Research. Part C, Emerging Technologies*, 79:42–57, Jun 2017.
13. R. Madhavan, K. Fregene, and L.E. Parker. Distributed heterogeneous outdoor multi-robot localization. In *IEEE International Conference on Robotics and Automation*, volume 1, pages 374–381 vol.1, 2002.
14. H. Kruppa, D. Fox, W. Burgard, and S. Thrun. A probabilistic approach to collaborative multi-robot localization. *Autonomous Robots*, 8(3):325–344, 2000.
15. A. Howard, M.J. Mataric, and G.S. Sukhatme. Putting the 'i' in 'team': An ego-centric approach to cooperative localization. In *IEEE International Conference on Robotics and Automation (ICRA)*, volume 1, pages 868–874 vol.1, Sept 2003.
16. N. Karam, F. Chausse, R. Aufrere, and R. Chapuis. Localization of a group of communicating vehicles by state exchange. In *IEEE International Conference on Intelligent Robots and Systems*, pages 519–524, Oct 2006.
17. M. Kieffer. Estimation ensembliste par analyse par intervalle: Application à la localisation de véhicules. PhD thesis, Université de Paris Sud, 1999.
18. M. Kieffer and E. Walter. Interval analysis for guaranteed nonlinear parameter and state estimation. *Mathematical and Computer Modelling of Dynamical Systems*, 11(2):171–181, 2005.
19. L. Jaulin. A nonlinear set membership approach for the localization and map building of underwater robots. *IEEE Transactions on Robotics*, 25(1):88–98, Feb 2009.
20. L. Jaulin. Range-only SLAM with occupancy maps: A set-membership approach. *IEEE Transactions on Robotics*, 27(5):1004–1010, 2011.
21. Jaulin L. and Walter E. Set inversion via interval analysis for nonlinear bounded-error estimation. *Automatica*, 29(4):1053–1064, 1993.
22. V. Drevelle and Ph. Bonnifait. Localization confidence domains via set inversion on short-term trajectory. *IEEE Transactions on Robotics*, 29(5):1244–1256, 2013.
23. D. Meizel, O. Leveque, L. Jaulin, and E. Walter. Initial localization by set inversion. *IEEE Transactions on Robotics and Automation*, 18(6):966–971, 2002.
24. L.C. Bento, P. Bonnifait, and U.J. Nunes. Set-membership position estimation with GNSS pseudorange error mitigation using lane-boundary measurements. *IEEE Transactions on Intelligent Transportation Systems*, 20(1):185–194, 2018.
25. L. Jaulin. Interval constraint propagation with application to bounded-error estimation. *Automatica*, 36(10):1547–1552, 2000.
26. A. Bethencourt. Interval Analysis for swarm localization. Application to underwater robotics. PhD thesis, University of Bretagne, Sept 2014.
27. K-H. Jo and J. Lee. Cooperative localization of multiple robots with constraint propagation technique. In *IEEE International Conference on Intelligent Robots and Systems*, pages 3477–3482, Sept 2008.
28. K. Lassoued, P. Bonnifait, and I. Fantoni. Cooperative localization with reliable confidence domains between vehicles sharing GNSS pseudo-ranges errors with no base station. *IEEE Intelligent Transportation Systems Magazine*, 9(1):2234, 2017.
29. E.D. Kaplan and C. Hegarty. *Understanding GPS: Principles and Applications*. 2nd edn, Artech House, 2005.
30. K. Lassoued, I. Fantoni, and P. Bonnifait. Mutual localization and positioning of vehicles sharing GNSS pseudoranges: Sequential Bayesian approach and experiments. In *IEEE International Conference on Intelligent Transportation Systems*, Spain, Sept 2015.

31. X. Xiao and J. Rife. Estimation of spatially correlated errors in vehicular collaborative navigation with shared GNSS and road-boundary measurements. In *International Technical Meeting of The Satellite Division of the Institute of Navigation*, pages 1667–1677, Sept 2010.

32. R. Hermann and A.J. Krener. Nonlinear controllability and observability. *IEEE Transactions on Automatic Control*, 22(5):728–740, Oct 1977.

33. A. Martinelli and R. Siegwart. Observability analysis for mobile robot localization. In *IEEE International Conference on Intelligent Robots and Systems*, pages 1471–1476, Aug 2005.

34. X.S. Zhou and S.I. Roumeliotis. Robot-to-robot relative pose estimation from range measurements. *IEEE Transactions on Robotics*, 24(6):1379–1393, Dec 2008.

35. D.L. Waltz. Generating semantic descriptions from drawings of scenes with shadows. PhD thesis, MIT Artificial Intelligence Laboratory, 1972.

36. K. Lassoued, O. Stanoi, Ph. Bonnifait, and I. Fantoni. Mobile robots cooperation with biased exteroceptive measurements. In *International Conference on Control Automation Robotics Vision*, pages 1835–1840, Singapore, Dec 2014.

37. L. Jaulin, M. Kieffer, O. Didrit, and E. Walter. *Applied Interval Analysis with Examples in Parameter and State Estimation, Robust Control and Robotics*. Springer, 2001.

38. H. Li and F. Nashashibi. Cooperative multi-vehicle localization using split covariance intersection filter. *IEEE Intelligent Transportation Systems Magazine*, 5(2):33–44, Summer 2013.

39. C.Y. Chong and S. Mori. Convex combination and covariance intersection algorithms in distributed fusion. In *4th International Conference on Information Fusion*, Canada, 2001.

40. G. Chabert. IBEX (http://www.ibex-lib.org), 2007.

41. O. Le Marchand, Ph. Bonnifait, J. Ibanez-Guzman, D. Betaille, and F. Peyret. Characterization of GPS multipath for passenger vehicles across urban environments. *ATTI dell'Istituto Italiano di Navigazione*, 189(07):77–88, 2009.

42. K. Lassoued, P. Bonnifait, and I. Fantoni. Cooperative localization of vehicles sharing GNSS pseudoranges corrections with no base station using set inversion. In *IEEE Intelligent Vehicles Symposium (IV)*, Gothenburg, Sweden, June 19–22, 2016.

43. N.R. Ahmed, S.J. Julier, J.R. Schoenberg, and M.E. Campbell. Multisensor Data Fusion from Algorithms and Architectural Design to Applications, chapter Decentralized Bayesian Fusion in Networks with Non-Gaussian Uncertainties, pages 383–408. CRC Press, 2015.

44. U.W. Utete. Network management in decentralized sensing systems. PhD thesis, Department of Engineering Science, University of Oxford, 1995.

45. K. Lassoued. Localisation de robots mobiles en coopération mutuelle par observation d'état distribuée. PhD thesis, Université de Technologie de Compiègne, Sorbonne universités, 2016.

46. V. Drevelle and Ph. Bonnifait. Localization confidence domains via set inversion on short-term trajectory. *IEEE Transactions on Robotics*, 29(5):1244–1256, 2013.

47. M. Worner, F. Schuster, F. Dolitzscher, C.G. Keller, M. Haueis, and K. Dietmayer. Integrity for autonomous driving: A survey. In *IEEE/ION Position, Location and Navigation Symposium (PLANS)*, pages 666–671, April 2016.

26

Cooperative Localization and Navigation for Multiple Autonomous Underwater Vehicles

LiChuan Zhang

CONTENTS

26.1 Introduction

26.1.1 Basic Concepts

An autonomous underwater vehicle (AUV) is a robot that travels underwater without requiring input from an operator. The development of AUVs began in earnest in the 1970s. Since then, improvements in the efficiency, size, and memory capacity of computers have enhanced abilities of AUVs. Many tasks that were originally achieved with towed arrays or manned vehicles are being completely automated nowadays. In the process, accurate localization and navigation are essential to ensure the success of these tasks. The absence of global positioning system (GPS) signals underwater makes navigation and localization for AUVs a difficult challenge [2]. Therefore, the concept of cooperative localization (CL) and cooperative navigation (CN) for multiple AUVs was proposed, which will increase the aforementioned accuracy.

"Synergy" was first proposed by Henri Mazel in 1896. Synergism exists in every corner of the world, from physics to chemistry, which produces a result "$1 + 1 > 2$." In the 1990s, Kurazume Ryo, a Japanese scholar, took the lead in putting forward the concept of "cooperative positioning" in multiple-agent position areas [1]. Figure 26.1 shows cooperative processing; in CN, communication of messages between agents is vital. From the figure, the agents cannot only get the messages, such as relative distance and relative measurement angle, etc., from the adjacent agents, but also can transport their information to other agents. Meanwhile, agents use the previous location information, such as equipping inertial navigation equipment, to assist in the completion of the current location. By CN, the estimating position can be closer to real position than the position estimates by itself.

CL and CN can be divided into following types. The first is parallel form. Every AUV which takes part in the CL and CN has same structure and function. They are equal and can get the information of any other one. This kind of navigation requires every AUV to have the same high-accuracy sensor, which will cause very high costs. The second type is the leader-follower form. Few leader AUVs equip with high-level sensors, and many follower AUVs equip with low-cost sensors. The follower gets the information from the leader and then updates its own position. This form reduces the cost and reaches the request. Therefore it has been the most important approach to solve the underwater navigation of multiple AUVs. This kind of navigation balances cost and performance, and should be further studied.

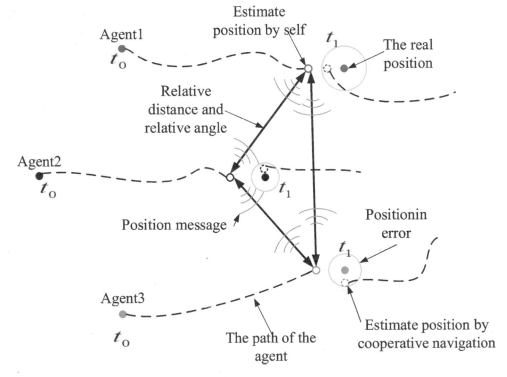

FIGURE 26.1
Cooperative navigation diagram.

26.1.2 Recent Development

With the increasing complexity of marine engineering tasks, the collaborative work of networking AUVs has become a hot issue in the current marine engineering and military marine fields. For a wide range of complex ocean observation monitoring tasks. When searching for the missing airplane Malaysia Airlines flight 370, the Bluefin-21 unmanned underwater vehicle was used to search the black box signal for the wrecked airliner; a single AUV takes a lot of time and is inefficient. Using multi-AUVs, however, cannot only save the search time, but also improve the search ability.

AUV navigation and localization in underwater environments is particularly challenging due to the rapid attenuation of GPS and radio-frequency signals [2]. Therefore, the available navigation methods underwater included terrain-matching navigation, dead reckoning (DR), image navigation, acoustic navigation, and CN.

Chen [3] reviewed the terrain-matching navigation methods and the advantages of the multibean bathymetric system in underwater topographic survey. Wang [4] proposed a comprehensive evaluation method for terrain navigation information and for constructing an underwater navigation information analysis model, which is associated with topographic features to validate the correlation of a variety of terrain parameters as terrain navigation information metrics. A terrain-aided navigation correction system has been developed for AUV, Autosub 6000. Morice [5] proposed a discussion of the biases in typical AUV navigational sensors and their influence on navigation on a map that is created in-mission by a submersible. The advantages of terrain-matching navigation are long-term, subtle, all-weather, and high accuracy; however, using terrain-matching navigation systems requires obtaining the geographic information of the target location first, and because the current marine exploration is not yet perfected, it is not suitable to use this method in wide range. To improve underwater vehicle DR, Xiao [6] proposed a developed strong tracking adaptive Kalman filter to improve AUV DR, Filaretov [7] considered an approach to creation of a reliable DR system for AUV with accommodation to incorrect sensors information, and Kapaldo [8] examined calibration methods suitable for identifying calibration coefficients in low-cost micro-electromechanical system (MEMS) gyros. Because of the error accumulation cannot be eliminated by DR and equipped with a large number of high accuracy inertial navigation equipment will dramatically increase the cost. Franco [9] reviewed the underwater simultaneous localization and mapping (SLAM) techniques—Extended Kalman Filter (EKF) SLAM, Fast SLAM, Graph SLAM—and their application in underwater environments. Ribas [10] introduced SLAM with imaging sonar to be used in underwater navigation. Due to the complexity of the underwater environment, SLAM technology needs further development to be applied to underwater navigation. Rodionov [11] analyzed the accuracy of measuring the distance between underwater vehicles using underwater acoustic modems in natural conditions and aimed at increasing the performance of autonomous underwater onboard navigation systems. Jinwoo [12] presented a method of acoustic source localization for the autonomous navigation of underwater vehicles; based on the time delay of acoustic signals received by two hydrophones, the method can provide a reliable estimation of the direction and location of the acoustic source, even under for a noisy acoustic signal.

Alexander Bahr [13] applied Kalman filter to the leader-follower multi-AUV CL to estimate the location of the AUV by minimizing the loss function, and a real aviation test was carried out; Nogueira [14] proposed a collaborative navigation algorithm process in order to achieve multi-AUV network location. Huang [15] proposed a new adaptive EKF to solve the problem of unknown noise covariance matrices inherent in the underwater CN. In [16], the probability hypothesis density filter was used in the CN to improve the position accuracy of the AUVs.

TABLE 26.1

Comparison of Underwater Navigation Methods

Navigation Methods	Advantages	Disadvantages	Adaptation
Inertial Navigation	High navigation accuracy	Navigation errors accumulate over time, equipment is expensive, and bulky	Any water area
Dead Reckoning	Simple and inexpensive equipment	Accumulated error, low navigation accuracy	Any water area
Acoustic Navigation	High navigation accuracy	Baseline layout is complex and limited in scope	Frequent work area
Optical Navigation	High navigation accuracy without cumulative error	Pre-arrange underwater targets, short viewing distance	Near distance
Terrain Matching Navigation	Passive, autonomous, no cumulative error	Pre-acquisition of geographic information database	Any water area
Cooperative Navigation	High efficiency, strong robustness, high accuracy	Reliance on underwater acoustic communications	Any water area

Zhang [17] studied the two leader AUV CN systems to solve the problem of formation collapse in the condition of position error of leader AUV increases gradually. The author of [18] studied the CN system with unmanned surface vehicle (USV) and AUV, and the results showed that considerable estimation error reduction is possible even in very deep water operations. The author of [19] proposed the divided difference filter algorithm to reduce the influence of system model truncation error to the filtering estimation and fusing the measurement information adequately by iterative use in weak observability conditions. The European Union Marine robotic systems of self-organizing logically linked physical nodes (MORPH) project [20] studied the requirements and constraints that the environment imposes on the CN and control systems that enable the concerted operation of the vehicle formation. The advantages and disadvantages of several navigation methods are shown in Table 26.1.

26.1.3 Applications of Cooperative Localization and Navigation

AUV CL and navigation is a relatively new research direction. Most of the research is still in the stage of theoretical exploration and simulation experiments, and there are not many practical systems in practice. The most representative of several multi-AUV collaborative systems are The European GREX Project with the theme "Coordination and Control of Cooperating Unmanned Systems" [21,22], the United States' Autonomous Ocean Sampling Network (AOSN) [23,24] and the New Jersey Shelf Observing System (NJSOS) [25], the MIT-developed Cooperative Autonomy for Distributed Reconnaissance and Exploration (CADRE) system [26], and the Cognitive Autonomous Diving Buddy launched by the EU's Seventh Science and Technology Framework Program [27].

26.2 Math Models of Multiple AUVs

26.2.1 Coordinate System

Before establishing a collaborative navigation mathematical model, it is necessary to determine the coordinate system so that the motion characteristics of AUV can be described in a holistic framework. The coordinate system used by underwater vehicles can be divided into two kinds: static coordinate system and dynamic coordinate system [28,29].

The static coordinate system is a coordinate system that does not change with the motion state of the vessel in the navigation system, including the geocentric inertial coordinate system (i system, $O_iX_iY_iZ_i$), the earth coordinate system (e system, $O_eX_eY_eZ_e$), and the geographic coordinate system (t system, $O_tX_tY_tZ_t$), as is shown in Figure 26.2. The origin of the i system is at the center of the earth, O_iX_i points to the vernal equinox, O_iZ_i coincides with the rotation axis of the earth, O_iY_i is determined according to the right-hand rule, and the origin of the e system is also located at the center of the earth. O_eX_e points to the intersection of the prime meridian and the equator of Greenwich. The O_eZ_e and the earth's rotation axis are merged and pointing to the north, and O_eY_e determines the direction according to the right-hand rule. Because the e system is a coordinate system built on the earth, it is stationary relative to the earth; the origin of the t-series is located at the floating center of the vessel, where O_tX_t points to the geographic north, O_tY_t points to the geographic east, O_tZ_t points to the center of the earth, and the t system is also named as the local geographic coordinate system. The relationship of the t system with respect to the earth coordinate system is the geographical length λ and the latitude α of the vessel [30].

The dynamic coordinate system, also known as the carrier coordinate system (b system, $O_bX_bY_bZ_b$), is fixed to the vehicle, as shown in Figure 26.3. In general, the origin of the b system is located at the center of the vehicle's center of gravity, O_bX_b is aligned with the axis of symmetry of the craft, pointing to the head of the craft, O_bY_b is parallel to the baseline, positive to the starboard, O_bZ_b is determined by the right-hand rule, and Point to the underside of the vehicle [31].

In a navigation task, when it is necessary to transfer the state of a point in one coordinate system to another, coordinate transformation is needed to realize. Suppose a vehicle observes that the position of a target in the carrier coordinate system is $X_b = (x_b, y_b, z_b)$, and it needs

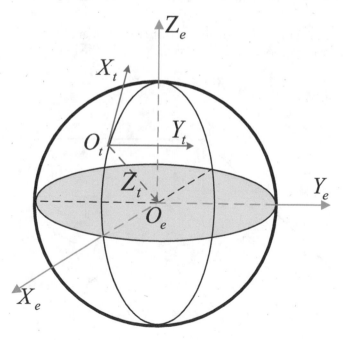

FIGURE 26.2
The earth coordinate system and the geographic coordinate system.

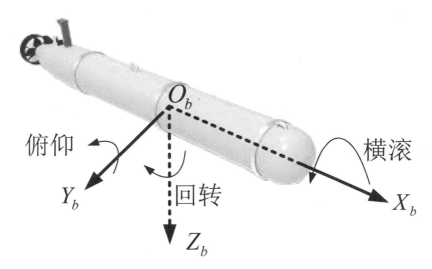

FIGURE 26.3
The carrier coordinate system.

to know its position $\mathbf{X}_t$ in the geographic coordinate system. As shown in Figure 26.4, the yaw, pitch, and roll angles of the vehicle are φ, θ, and γ, respectively. According to Euler's theorem, the coordinate transfer matrix transferred from the t system to the b system is:

$$C_b^t = \begin{bmatrix} \cos\varphi\cos\theta & \sin\varphi\cos\theta & -\sin\theta \\ -\sin\varphi\cos\gamma + \cos\varphi\sin\theta\sin\gamma & \cos\varphi\cos\gamma + \sin\gamma\sin\theta\sin\varphi & \cos\theta\sin\gamma \\ \sin\varphi\sin\gamma + \cos\varphi\cos\gamma\sin\theta & -\cos\varphi\sin\gamma + \sin\theta\sin\varphi\cos\gamma & \cos\theta\cos\gamma \end{bmatrix} \quad (26.1)$$

The coordinate transfer matrix transferred from the b system to the t system is $C_t^b = (C_b^t)^{\mathrm{T}}$. At this time, the position coordinate of the vehicle under the t system is $\mathbf{X}_{t,0}$, then $\mathbf{X}_t$ is:

$$\mathbf{X}_t = C_t^b \times \mathbf{X}_b + \mathbf{X}_{t,0} \quad (26.2)$$

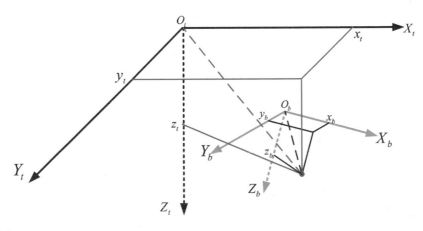

FIGURE 26.4
Attitude angle of vehicle.

26.2.2 Math Model

The core element of collaborative navigation is the information exchange between nodes [1]. Multi-AUV underwater collaborative navigation uses underwater acoustic equipment for information exchange. The navigation process is shown in Figure 26.5. Taking the form of single-leader AUV and single-follower AUV as an example, firstly, a water-acoustic communication is performed between the AUVs every Δt duration, and the leading AUV transmits the position information by underwater sound. The follower AUV receives leader AUV's position information and measures distance and orientation information from the leader AUV by underwater sound. Then, the follower estimates its own position information of the current time to correct the accumulated position error caused by the DR.

As shown in the Figure 26.5, the navigation device carried by the leader AUV has higher positioning accuracy, and the one carried by the follower AUV has lower positioning accuracy, so the position of the leader AUV is taken as a relative reference position. The relative angle of the distances of the two AUVs is determined by the underwater acoustic measuring device, the relative accurate position of the follower AUV can be estimated by the navigation algorithm, and the estimated position is closer to the real position than the position obtained by the DR estimation. Since the rate of underwater acoustic communication is slow and the amount of information transmission is not as large as that of electromagnetic waves, underwater acoustic communication measurements are not suitable for high-frequency communication and measurement. Between the two underwater acoustic communication time points, the follower AUV uses its own inertial device and Doppler velocity log (DVL) to calculate its position information.

By understanding the multi-AUV collaborative navigation process, it is first necessary to establish a kinematic model of a single AUV. In the literature [16,32] and [33], the model of coordinated navigation is established in a two-dimensional plane, ignoring factors such as interference from ocean currents. In order to facilitate the analysis without losing the generality, it is considered that the depth of the AUV is unchanged in the collaborative navigation. Moreover, the AUV has a small pitch angle during steady operation, and it can be considered that $\cos\theta \approx 0°$, and the motion state of AUV is as shown in Figure 26.6. In this figure, $O_t X_t Y_t$ is the navigation coordinate system, $O_b X_b Y_b$ is the body coordinate system of the vehicle, $\mathbf{X}_k$ is the position of the vehicle at time k, φ_k is the true heading angle of the

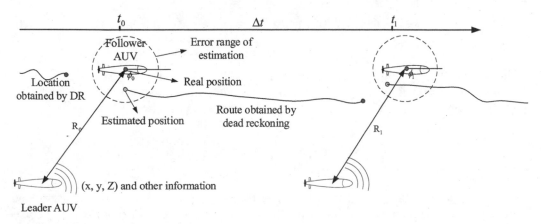

FIGURE 26.5
Process of cooperative location and navigation for multiple AUV.

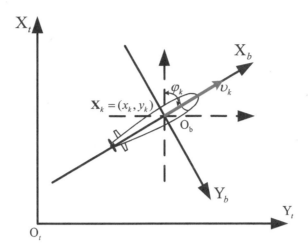

FIGURE 26.6
Motion state of AUV.

vehicle at time k, and v_k is the speed of the vehicle in the heading direction at time k. Based on the kinematics of the craft, the following kinematic equations are established:

$$\begin{cases} x_{k+1} = x_k + \Delta t \hat{v}_k \cos \hat{\varphi}_k \\ y_{k+1} = y_k + \Delta t \hat{v}_k \sin \hat{\varphi}_k \\ \hat{\varphi}_{k+1} = \varphi_{k+1} + w_\varphi \\ \hat{v}_{k+1} = v_{k+1} + w_v \end{cases} \tag{26.3}$$

In the formula (26.3), x_{k+1} and y_{k+1} are the positional components on the X-axis and Y-axis of the navigation coordinate system at time $k+1$ of the vehicle; Δt is the sampling period; $\hat{\varphi}_{k+1}$ is the heading angle measured by the gyroscope at time $k+1$; $\hat{v}_{k+1}$ is the travel speed measured by DVL at time $k+1$; w_φ and w_v are the measurement noise of the gyroscope and the measurement noise of DVL respectively, which are Gaussian white noise with a mean of 0. The formula (26.3) can be written as:

$$\mathbf{X}(k+1) = \Phi(k+1,k)\mathbf{X}(k) + \Gamma(\mathbf{u}(k) + \mathbf{w}(k)) = f[\mathbf{X}(k), \mathbf{u}(k), \mathbf{w}(k)] \tag{26.4}$$

In the formula (26.4) given above, $\mathbf{X}(k+1) = (x_k, y_k, \varphi_k)^T$ is the state vector of AUV at time k+1; $\Phi(k+1,k)$ is the state transition matrix of the system; $\Gamma(\mathbf{u}(k) + \mathbf{w}(k))$ is the system nonlinear term, and $\mathbf{u}(k) = (v_k, \varphi_k)^T$ and $\mathbf{w}(k) = (w_v, w_\varphi)^T$ are system process noise. Then the state transition matrix and the measurement covariance matrix are:

$$Q(k+1) = E\begin{bmatrix} \mathbf{w}_{k+1} & \mathbf{w}_{k+1}^T \end{bmatrix} = \begin{bmatrix} \sigma_{vk}^2 & 0 \\ 0 & \sigma_{\varphi k}^2 \end{bmatrix} \tag{26.5}$$

Because the navigation and positioning errors of the inertial combined system will accumulate over time, a hydroacoustic positioning system is needed to continuously correct the positioning error of the vehicle. According to the principle of underwater acoustic positioning, the observation equation can be written as:

$$Z_{ij}(k+1) = \left\| \mathbf{X}_i(k+1) - \mathbf{X}_j(k+1) \right\| + w_\rho(k+1)$$

$$= \sqrt{(x_i(k+1) - x_j(k+1))^2 + (y_i(k+1) - y_j(k+1))^2} + w_\rho(k+1) \qquad (26.6)$$

$$= h[\mathbf{X}_i(k+1), \mathbf{X}_j(k+1)] + w_\rho(k+1)$$

In the formula (26.6), $Z_{ij}(k+1)$ represents the relative distance from AUV_i measured by AUV_j at time $k+1$; $\mathbf{X}_i(k+1) = \begin{bmatrix} x_i(k+1) & y_i(k+1) \end{bmatrix}$ and $\mathbf{x}_j(k+1) = \begin{bmatrix} x_j(k+1) & y_j(k+1) \end{bmatrix}$ represent the position of AUV_i and AUV_j at time $k+1$, respectively; $w_\rho(k+1)$ is the measurement noise of Gaussian distribution; $h[\mathbf{X}_i(k+1), \mathbf{X}_j(k+1)]$ is about nonlinear functions of $\mathbf{X}_i(k+1)$ and $\mathbf{X}_j(k+1)$. The observed noise covariance matrix of the system can be written as:

$$R_\rho(k+1) = E\begin{bmatrix} w_\rho & w_\rho^T \end{bmatrix} = \sigma_\rho^2 \qquad (26.7)$$

Underwater acoustic measurement and underwater acoustic communication can effectively reduce the cumulative positioning error growth rate during AUV navigation. When using a fixed beacon for CN, there is no cumulative error in the positioning error of the AUV, but at this time, the navigation range is affected by the underwater acoustic measurement and the communication distance, which reduces the mobility of the system. The use of multi-AUV collaborative navigation can make the effects of mobility assort with navigation range to a certain extent.

26.3 Observability Analysis

26.3.1 The Significant of Observability Analysis

The definition of observability is introduced in document [34]. The observability of the navigation system refers to whether the AUV's estimation of the system state quantity at each sampling instant can be solved by the relative observation between the AUV and the beacon. If the navigation system is not observable, the CN algorithm will not be able to solve the AUV position, so the observability of the navigation system is one of the key issues in the study of AUV CN. In multi-AUV collaborative navigation, the estimated position of the follower AUV cannot be directly measured by the sensor. The information that the underwater acoustic device can obtain is only the relative distance and the relative angle, and further calculation is needed to obtain the estimated position of the follower AUV. Because of the complexity of the underwater environment, the CN model has serious nonlinear problems, which will have a relatively large impact on the observability of the system. It can be known from formulas (26.4) and (26.6) that the system is observable when the system can determine the state $\mathbf{X}$ of the system from a vector $\mathbf{Z} = [Z_1 \ Z_2 \ \dots \ Z_k]^T$ consisting of a set of observations.

26.3.2 Model and Method of Proving Observability

Ignoring the effects of noise, the state matrix of the system is:

$$\begin{cases} \mathbf{X}(k+1) = \Phi(k+1,k)\mathbf{X}(k) \\ Z(k+1) = H(k+1)\mathbf{X}(k+1) \end{cases} \qquad (26.8)$$

Plug $\mathbf{X}(k+1)$ into the measurement equation according to Equation 26.8 and it gives

$$Z(k+1) = H(k+1)\Phi(k+1,k)\mathbf{X}(k) \tag{26.9}$$

Then the matrix Z of the k observations of the system is

$$\begin{cases} \mathbf{X}(k+1) = \Phi(k+1,k)\mathbf{X}(k) \\ Z(k+1) = H(k+1)\mathbf{X}(k+1) \end{cases} \tag{26.10}$$

Bringing the state quantity $\mathbf{X}(k+1)$ into the measurement equation gives:

$$Z(k+1) = H(k+1)\Phi(k+1,k)\mathbf{X}(k) \tag{26.11}$$

Then the matrix Z of the k observations of the system is:

$$\begin{bmatrix} Z(1) \\ Z(2) \\ \vdots \\ Z(k) \end{bmatrix} = \begin{bmatrix} H(1)\Phi(1,k) \\ H(2)\Phi(2,k) \\ \vdots \\ H(k)\Phi(k,k) \end{bmatrix} \mathbf{X}(k) \tag{26.12}$$

Because the underwater CN system is a time-varying system, the Gram matrix is used here to analyze the observability of the system. First, both sides of the Equation 26.12 are multiplied by the vector $[\Phi(1,k)H(1)\ \Phi(2,k)H(2)\ldots\Phi(k,k)H(k)]$, we can get:

$$M(0,k)\mathbf{X}(k) = \sum_{i=1}^{k} \Phi^T(i,k)H^T(k)Z(k) \tag{26.13}$$

where $M(0,k)$ is the Gram discriminant matrix:

$$M(0,k) = \sum_{i=1}^{k} \Phi^T(i,k)H^T(i)H(i)\Phi(i,k) \tag{26.14}$$

According to the research in [35], when the $k>0$ is present such that $M(0,k)$ is a non-singular matrix, the system is observable. The necessary and sufficient condition for $M(0,k)$ to be a nonsingular matrix is:

$$\text{rank}T(k-N+1,k) = n \tag{26.15}$$

where $T(k-N+1,k)$ is the discriminant matrix of the rank of the $M(0,k)$ matrix, and n is the dimension of the state vector of the system.

$$T(k-N+1,k) - \begin{bmatrix} H(k) \\ H(k-1)\Phi(k-1,k) \\ \vdots \\ H(k-N+1)\Phi(k-N+1,k) \end{bmatrix} \tag{26.16}$$

Since the observation $Z_{ij}(k+1)$ is a one-dimensional distance measurement and the estimation of the position of the vehicle is a two-dimensional estimator, at least two observations are required to determine the position of the vehicle. Assume that during the two adjacent hydroacoustic measurements, the positions of the main AUV are (x_{k-1}^M, y_{k-1}^M) and (x_k^M, y_k^M), respectively. At this time, $N=2$, then:

$$T(k-1,k) = \begin{bmatrix} \dfrac{x_{k-1} - x_{k-1}^M}{\rho_{k-1}} & \dfrac{y_{k-1} - y_{k-1}^M}{\rho_{k-1}} \\ \dfrac{x_k - x_k^M}{\rho_k} & \dfrac{y_k - y_k^M}{\rho_k} \end{bmatrix} \tag{26.17}$$

where ρ_k is the relative distance of the master-slave AUV at time k. According to the formula (2.14), when the rank of the $T(k-1,k)$ matrix is equal to 2, there are:

$$\det T(k-1,k) = \frac{x_{k-1} - x_{k-1}^M}{\rho_{k-1}} \cdot \frac{y_k - y_k^M}{\rho_k} - \frac{y_{k-1} - y_{k-1}^M}{\rho_{k-1}} \cdot \frac{x_k - x_k^M}{\rho_k} \neq 0 \tag{26.18}$$

Regardless of the possibility of collision of the underwater vehicle during operation, let $\rho_k \neq 0$, we can know that the system is observable if and only if the following conditions are true:

$$\frac{x_{k-1} - x_{k-1}^M}{y_{k-1} - y_{k-1}^M} \neq \frac{x_k - x_k^M}{y_k - y_k^M} \tag{26.19}$$

As shown in Figure 26.7, in other words, when the measurement vectors $\mathbf{R}_k \neq \mathbf{R}_{k+1}$ during the two adjacent measurements, the system is observable; if $\mathbf{R}_k = \mathbf{R}_{k+1}$, the system is unobservable.

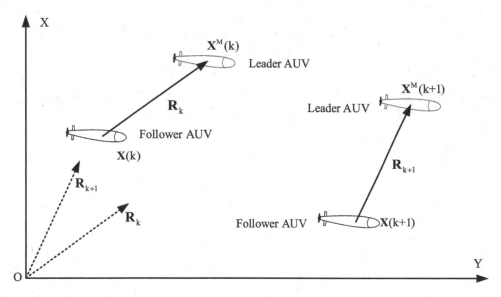

FIGURE 26.7
Schematic diagram of AUV cooperative navigation observability analysis.

26.4 GM-PHD Filter

26.4.1 What the GM-PHD Filter Is

The probability hypothesis density (PHD) filter is a novel method to solve the multiple target tracking problem in a cluttered environment. Based on the PHD filter, it is possible to solve the disturbance caused by the clutter. In the PHD filter, all states and measures are modeled as set-based values in the format of a random finite set (RFS), which makes the PHD filter a promising approach to solving these issues. The PHD filter has been used in CL for vehicles [36,37], and the experiment implied its effectiveness in clutter environments. The PHD filter was also introduced to address the visual tracking issue [38,39], aiming to solve the number variation and noise corruption of the camera. In this paper, we apply the PHD filter to solve the AUV CL problem. Meanwhile, the information entropy theory is a proper tool for determining the quality of the estimations. It has been used in optimization [40] and industrial areas to value how useful one message is [41,42] and [43]. Therefore, in this paper, underwater clutter is considered as a disturbance and the PHD filter incorporated with information entropy theory is simultaneously employed.

26.4.2 The Structure of GM-PHD Filter

Observations from sonar also contain clutter due to the complicated underwater environment. Therefore, the standard Bayesian filter cannot obtain satisfactory estimations without data association. Compared to the standard Bayesian filters, finite set statistics analysis is a convenient approach for achieving the task without considering data association [44,45]. Hence, the RFS-based Gaussian mixture-probability hypothesis density (GM-PHD) filter is utilized.

Suppose that there are N_{k-1} targets at instant t_{k-1} with states $X_{k-1} = \{x_1(k-1), x_2(k-1), \ldots, x_{N_{k-1}}(k-1)\}$, and there are N_k targets $X_k = \{x_1(k), x_2(k), \ldots, x_{N_k}(k)\}$ including newborn and dead ones. In the same manner, both Z_{k-1} and Z_k are defined. In an RFS manner, we have:

$$\begin{cases} X_k = \{x_1(k), x_2(k), \ldots, x_{N_k}(k)\} \in F(X) \\ Z_k = \{z_1(k), z_2(k), \ldots, z_{N_k}(k)\} \in F(Z) \end{cases} \tag{26.20}$$

Where $F(X)$ and $F(Z)$ denote the sets containing all states and observations, respectively. Notice that Z contains both the real measure of the target and the clutter. Then, Equation 26.20 can be simplified as

$$X_k = \left(\bigcup_{\zeta \in X(k-1)} S_{k|k-1}(x) \right) \bigcup \Gamma_k \tag{26.21}$$

$$Z_k = \left(\bigcup_{x \in X(k)} \Theta_k(x) \right) \bigcup K_k \tag{26.22}$$

where $S_{k|k-1}$ is the RFS of remaining targets at t_{k-1}, Γ_k is the set of newborn ones at t_k, $\Theta_k(x)$ is the measure set of the targets, and K_k is the clutters RFS.

Given the survival and detection rate as $p_{s,k}$ and $p_{d,k}$, the GM-PHD filter can be expressed in a Bayesian prediction-and-update form as

Prediction:

$$v_{k|k-1}(x) = v_{s,k|k-1}(x) + \gamma_k(x) \tag{26.23}$$

$$v_{k|k-1}(x) = p_{s,k} \sum_{j=1}^{J_{k-1}} \omega_{k-1}^i N(x; m_{s,k|k-1}^j, P_{s,k|k-1}^j) \tag{26.24}$$

$$m_{s,k|k-1}^j = F_{k-1} m_{k-1}^j \tag{26.25}$$

$$p_{s,k|k-1}^j = Q_{k-1} + F_{k-1} P_{k-1}^j F_{k-1}^T \tag{26.26}$$

Update:

$$\nu_k(x) = (1 - p_{d,k}) n_{k|k-1}(x) + \sum_{z \in Z_k} \nu_{d,k}(x; z) \tag{26.27}$$

$$\nu_{k|k-1}(x) = \sum_{i=1}^{J_{k|k-1}} \omega_{k-1}^i N(x; m_{k|k-1}^i, P_{k|k-1}^i) \tag{26.28}$$

$$\nu_{d,k}(x; z) = \sum_{j=1}^{J_{k|k-1}} \omega_{k-1}^j N(x; m_{k|k}^j, P_{k|k}^j) \tag{26.29}$$

$$\omega_{k-1}^i = \frac{p_{d,k} \omega_{k-1}^i q_k^i(z)}{\kappa_k(z) + p_{d,k} \sum_{l=1}^{J_{k|k-1}} \omega_{k|k-1}^l q_k^l(z)} \tag{26.30}$$

$$q_k^i(z) = N\left(z; H_k m_{k|k}^i, R_k + H_k P_{k|k-1}^i H_k^T\right) \tag{26.31}$$

$$m_{k|k}^i(z) = m_{k|k-1}^i + K_k^i(z - H_k m_{k|k-1}^i) \tag{26.32}$$

$$P_{k|k}^i = \left[I - K_k^i H_k\right] P_{k|k-1}^i \tag{26.33}$$

$$K_k^i = P_{k|k-1}^i H_k^T \left(H_k P_{k|k-1}^i H_k^T + R_k\right)^{-1} \tag{26.34}$$

where F_k and H_k are the transition matrix and the measurement matrix, respectively. Note that if the transition function f_k and the measurement function h_k are nonlinearly presented, we can take their Jacobian matrices as F_k and $H_k \cdot \gamma_k$ denotes the Gaussian mixture and is given by

$$\gamma_k(x) = \sum \omega_{\Gamma,k}^i N\left(x; m_{\Gamma,k}^i, P_{\Gamma,k}^i\right) \tag{26.35}$$

Although the PHD filter estimates both the statements and its number, it does not have all of the track information. Thus, identification of a specific state in its life cycle is not available. Hence, we have the information entropy theory to address this issue, which is introduced in the next section.

26.5 Method of Cooperative Localization and Navigation of Multiple AUVs

26.5.1 Information Entropy Cooperation Navigation

26.5.1.1 Information Entropy

Information entropy is a measure of the uncertainty of a random variable [46] and can characterize the randomness of a variable. The size of the information entropy reflects the uncertainty of the random variable. The higher the uncertainty of the random variable, the more information we can get from the experiment, the larger the information entropy. Information entropy is the self-information mathematical expectation, the average self-information of random variables, and the average uncertainty of variables.

The information entropy of a continuous source is:

$$H(X) = E\left[\log \frac{1}{p(x)}\right] = -\int p(x)\log p(x)dx \tag{26.36}$$

The information entropy of discrete sources is:

$$H(X) = E\left[\log \frac{1}{p(x_i)}\right] = -\sum p(x_i)\log p(x_i) \tag{26.37}$$

Conditional entropy is a measure of the uncertainty of another random variable in the case where a random variable is known. In general, the conditional entropy is always smaller than the unconditional entropy. This is because things are universally connected, so the two random variables X and Y are also somewhat interconnected. Then the conditional entropy under the condition that the value of a random variable is known is always not greater than the unconditional entropy of another random variable. Conditional entropy represents the uncertainty of the X "residual" after the known Y, and is the probability-weighted average of the conditional self-information.

The conditional entropy of a continuous source is:

$$E(X \mid Y) = \iint p(xy)\log p(x \mid y)dxdy = \iint p(y)p(x \mid y)\log p(x \mid y)dxdy \tag{26.38}$$

The conditional entropy of a discrete source is:

$$H(X \mid Y) = -\sum\sum p(x_i, y_i)\log p(x_i \mid y_i) = -\sum\sum p(y_i)p(x_i \mid y_i)\log p(x_i \mid y_i) \tag{26.39}$$

Two random variables X and Y. After understanding Y, the uncertainty of X is reduced by $H(X) - H(X \mid Y)$. This difference is actually the information about X provided after the value of Y is known. This difference is mutual information:

$$I(X;Y) = H(X) - H(X \mid Y) \tag{26.40}$$

Let the probability density distribution of the one-dimensional random variable X be a normal distribution, namely:

$$p(x) = \frac{1}{\sqrt{2\pi\sigma^2}} \exp\left(-\frac{(x-\mu)^2}{2\sigma^2}\right) \tag{26.41}$$

where μ is the mean of X and σ^2 is the variance of X. Thus, the entropy of the continuous source is:

$$
\begin{aligned}
h(x) \\
&= -\int_{-\infty}^{\infty} p(x) \log p(x) dx \\
&= -\int_{-\infty}^{\infty} p(x) \log\left[\frac{1}{\sqrt{2\pi\sigma^2}} \exp\left(-\frac{(x-\mu)^2}{2\sigma^2}\right)\right] dx \\
&= -\int_{-\infty}^{\infty} p(x)(-\log\sqrt{2\pi\sigma^2}) dx + \int_{-\infty}^{\infty} p(x)\left[\frac{(x-\mu)^2}{2\sigma^2}\right] dx \cdot \log e \\
&= \log\sqrt{2\pi\sigma^2} + \frac{1}{2}\log e \\
&= \frac{1}{2}\log 2\pi e\sigma^2
\end{aligned}
\tag{26.42}
$$

Among them, $\int_{-\infty}^{\infty} p(x)dx = 1$, $\int_{-\infty}^{\infty} (x-\mu)^2 p(x)dx = \sigma^2$.

It can be seen that the entropy of a continuous source of normal distribution is independent of mathematical expectation and is only related to its variance. The calculation of the entropy of multidimensional random variables are analyzed below.

The N-dimensional continuous random variable $X = (X_1 X_2 \ldots X_N)$ is a normal distribution, and this source is an N-dimensional Gaussian source. Let the mean of each variable X_i be m_i, and the joint second-order center-to-center distance between the variables is $\mu_{ij} = E[(X_i - m_i)(X_j - m_j)](i, j = 1, 2, \cdots, N)$, then the $N \times N$-order covariance matrix R:

$$
R = \begin{vmatrix}
\mu_{11} & \mu_{12} & \cdots & \mu_{1N} \\
\mu_{21} & \mu_{22} & \cdots & \mu_{2N} \\
\vdots & \vdots & \vdots & \vdots \\
\mu_{N1} & \mu_{N2} & \cdots & \mu_{NN}
\end{vmatrix}
\tag{26.43}
$$

R represents the determinant of the matrix, $|R|_{ij}$ represents the algebraic remainder, and the probability density of the random vector X is:

Thus, the information entropy of the N-dimensional Gaussian random sequence source is:

$$
\begin{aligned}
h(X) \\
&= -\int_{-\infty}^{\infty} p(x)\log p(x)dx \\
&= -\int_{-\infty}^{\infty} p(x)\log\left\{\frac{1}{(2\pi)^{1/2}\,|R|^{1/2}}\exp\left[-\frac{1}{2\,|R|}\sum_{i=1}^{N}\sum_{j=1}^{N}|R|_{ij}\,(x_i-m_i)(x_j-m_j)\right]\right\}dx_1\ldots dx_N \\
&= \log(2\pi)^{N/2}\,|R|^{1/2}+\frac{N}{2}\log e \\
&= \log\left[(2\pi e)^{N/2}\cdot|R|^{1/2}\right] \\
&= \frac{1}{2}\log(2\pi e)^{N}\,|R|
\end{aligned}
\tag{26.44}
$$

26.5.1.2 AUV Collaborative Navigation with Information Entropy

In practical CN engineering applications, the effect of improving the accuracy of CN only by improving the filtering algorithm is limited, and the selection of observations should be combined to improve the navigation accuracy. According to the information theory, the information entropy indicates the uncertainty of the state $X(k)$ before the new measurement, and the mutual information indicates the amount of information brought by the new measurement. The greater the mutual information corresponding to a certain measurement, the greater amount of information brought by the measurement and the better the measurement. The smaller the uncertainty of the state after the measurement and fusion, the higher the precision of the CN. That is, the measurement is the optimal measurement in the sense of maximum information amount.

The information entropy is used to reflect the quality of the surrounding observation data available to the AUV in the network. By calculating the information entropy, mutual information for CN observation data is obtained. The greater the mutual information measured, the better the measurement, that is, the higher the accuracy of AUV positioning. Furthermore, by establishing a performance function to measure the value of a measurement, the value of the entire measurement of the networked AUV system can be calculated. Finally, a networking AUV collaborative navigation algorithm is obtained.

The description of random variables in the Kalman filter algorithm uses a probability density function, and information entropy is the integral calculation of the probability density function of random variables. Co-navigation is the need to estimate the state quantity according to the observation. The coordinated navigation of multiple AUVs is to maximize the uncertainty of the state quantity estimation (information entropy), that is, the mutual information reaches the maximum. The acquired observations can be screened by establishing a certain performance function, and a certain number of high-performance measurements are selected for collaborative navigation solution.

In the network composed of n AUVs, it can be known from the definition of information entropy that for any $AUV_i, i = 1 \sim n$, before the arrival of the new information at time k, the information entropy is:

$$
H(X_{k|k-1},Z_{k-1}) = -\int p(X_{k|k-1}\,|\,Z_{k-1})\ln p(X_{k|k-1}\,|\,Z_{k-1})dx
\tag{26.45}
$$

After the arrival of the new interest, its information entropy is:

$$H(X_{k|k-1}, Z_k) = -\int p(X_{k|k-1} \mid Z_k) \ln p(X_{k|k-1} \mid Z_k) dx \tag{26.46}$$

The mutual information is:

$$I(X_k, Z_k) = H(X_{k|k-1}, Z_{k-1}) - H(X_{k|k-1}, Z_k) \tag{26.47}$$

It can be seen that the information entropy of the N-dimensional Gaussian random sequence source is:

$$h(X) = -\int_{-\infty}^{\infty} p(x) \log p(x) dx = \frac{1}{2} \log(2\pi e)^N \mid R \mid \tag{26.48}$$

Assuming that the state variable of the AUV obeys the Gaussian distribution, the mutual information of the AUV is:

$$
\begin{aligned}
I(X_k, Z_k) &= H(X_{k|k-1}, Z_{k-1}) - H(X_{k|k-1}, Z_k) \\
&= \frac{1}{2} \log(2\pi e)^N \mid P_{k|k-1} \mid - \frac{1}{2} \log(2\pi e)^N \mid P_k \mid \\
&= \frac{1}{2} \log \frac{\mid P_{k|k-1} \mid}{\mid P_k \mid} \\
&= \frac{1}{2} \log(2\pi e)^N \mid P_{k|k-1} \mid - \frac{1}{2} \log(2\pi e)^N \mid P_k \mid \\
&= \frac{1}{2} \log \frac{\mid P_{k|k-1} \mid}{\mid P_k \mid}
\end{aligned}
\tag{26.49}
$$

According to the meaning of mutual information, it can be used to realize the choice of observation measurement. Considering that the estimated covariance in the filtering indicates the quality of the estimated to a certain extent, the smaller the estimated covariance, the higher the quality of the estimation, and the lower the quality of the estimation. Therefore, it is necessary to consider the size of the estimated covariance when establishing the performance function. Multi-AUV information interaction inevitably requires underwater acoustic communication and detection. As the distance increases, the energy consumed by the node will also increase. In the underwater acoustic communication project, communication delay is inevitable, and the position of multi-AUV CN is updated in real time, so the influence of communication delay on coordinated navigation should be minimized. Communication delay mainly includes communication transmission delay and communication processing delay. The communication processing delay is limited by the conditions of the underwater acoustic equipment. The communication transmission delay is proportional to the relative distance of the AUV. The farther the relative distance is, the more energy the node consumes. The greater the communication transmission delay, the less the node consumes less the communication transmission delay. Therefore, the distance must be considered when establishing the performance function. When the distance of an AUV is greater than a certain threshold, the selection of the AUV measurement will be abandoned.

Taking AUV_4 as an example, it is assumed that the mutual information of AUV_i observed by AUV_4 at time k is I_i, the mutual distance is ρ_i, and the estimated covariance is P_i, of which $i = 1 \sim 9, i \neq 4$. Then AUV_4 evaluates the performance of AUV_i as:

$$\Psi^{(i)} = \omega_I^{(i)} \cdot \omega_P^{(i)} \cdot \omega_\rho^{(i)} \tag{26.50}$$

In the formula (26.50), we have:

$$\omega_I^{(i)} = I_i \bigg/ \sum_{i=1\sim9,i\neq4}^{9} I_i \tag{26.51}$$

$$\omega_P^{(i)} = \frac{1}{|P_i|} \bigg/ \sum_{i=1\sim9,i\neq4}^{9} \frac{1}{|P_i|} \tag{26.52}$$

$$\omega_P^{(i)} = \frac{1}{|P_i|} \bigg/ \sum_{i=1\sim9,i\neq4}^{9} \frac{1}{|P_i|} \tag{26.53}$$

26.5.1.3 Cooperative Navigation Algorithm Based on Information Entropy

1. Cooperative navigation algorithm flow of information entropy in EKF

Initialization

Set initial state, set sampling time, simulation step size;

Set the initial state of the EKF filter estimation $EKF(0)$;

Set the state covariance matrix of the EKF filter estimation $P(0)$;

Set the process noise variance and the observed noise variance;

Loop begin

　State prediction based on system kinematics model;

　System observation and prediction based on state prediction values;

　State equation nonlinear function Jacobian matrix update, observation equation nonlinear function Jacobian matrix update;

　Covariance matrix prediction;

　Information entropy of prior probability models;

　Kalman filter gain update;

　Covariance matrix update;

　Information entropy of posterior probability model;

　Mutual information calculation, sorted by performance function, select the optimal measurement;

　System status update;

Loop end

2. Cooperative navigation algorithm flow of information entropy in UKF

Initialization

Set initial state of AUV$_i$, set sampling time, simulation step size;

Set the initial state $UKF(0)$ of the UKF filter estimate;

Set the state covariance matrix $P(0)$ estimated by the UKF filter;

Set the process noise variance and the observed noise variance;

Loop begin

  Obtaining a set of sampling points and their corresponding weights;

  Calculating a one-step state prediction of a set of sampling points;

  Calculating the predicted value of the system state and the state prediction covariance;

  Generating a new set of sampling points using prediction state and state prediction covariance UT transformation;

  One-step prediction of observations based on a new set of sample points;

  Calculate observation prediction values, observation prediction covariance, and state observation prediction covariance;

  Information entropy of prior probability models;

  Kalman filter gain update;

  Covariance matrix update;

  Information entropy of posterior probability model;

  Mutual information calculation, sorted by performance function, select the optimal measurement;

  System status update;

Loop end

3. Cooperative navigation algorithm flow of information entropy in UPF

Initialization

Set initial state $X(0)$ of AUV$_i$, set sampling time, simulation step size;

Set the initial state UKF(0) of the UPF filter estimation;

Set the state covariance matrix $P(0)$ estimated by the UKF filter;

Set the process noise variance and the observed noise variance;

Generating an initial particle set randomly according to a prior probability density function;

Loop for

  Calculating a one-step state prediction of a particle set;

  Observing the predicted value of the system based on the one-step state prediction set;

  Calculate the weight of the particles based on the observed predicted values and normalize them;

  Calculating the information entropy of the prior probability model

According to different resampling methods, the particles with larger weights are copied, and the particles with smaller weights are discarded, and a new particle set and corresponding weights are generated;

Calculating the information entropy of the posterior probability model

Mutual information calculation, sort by performance function, select the optimal measurement

System status update

Re-sampling

Loop end

26.5.2 Cross Entropy Cooperation Navigation

26.5.2.1 Cross Entropy

Cross entropy (CE) is an important concept in the Shannon Information Theory. The amount of information can be measured by information entropy, and the CE can be used to measure the degree of difference between two sets of information.

In the "Problem of Cluster Job AUVs," the master AUV needs to move to next node to minimize the observation error, which is a combinatorial optimization problem (COP). Boer [47] gives a detailed introduction to the application of the CE method in COP. COP is usually a small probability, which needs to be accurately estimated; CE provides method a simple, efficient, and general method for solving the COP.

As Figure 26.8 shows that CE is applicable to the estimation of the hidden amount *y* bases on the input amount *x* and the output *F(x)* when the model is not clear. The CE method involves an iterative procedure where each iteration can be broken down into two phases:

1. Generate a random data sample (trajectories, vectors, etc.) according to a specified mechanism.

2. Update the parameters of the random mechanism based on the data to produce a "better" sample in the next iteration.

When the COP is complex, it is not only difficult to establish the model, but also sometimes it cannot calculate the optimal result by solving the equations. However, the CE establish the problem under the probability model, through the analysis and selection of the probability to estimate optimal solution.

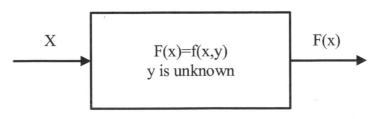

FIGURE 26.8
A black box for decoding vectory.

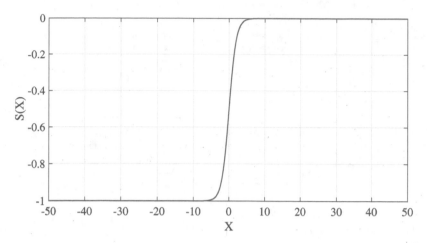

FIGURE 26.9
The change curve of the function sigmoid in the x space.

We assume that $f(X)$ is the real probability density of $\mathbf{X}$, and $h(X)$ is the estimated probability density of $\mathbf{X}$. The CE value between $f(X)$ and $h(X)$ can be written as:

$$D(f,g) = \mathbf{E}_h \ln \frac{h(\mathbf{X})}{f(\mathbf{X})} = \int h(\mathbf{X}) \ln f(\mathbf{X}) d\mathbf{X} - \int f(\mathbf{X}) \ln h(\mathbf{X}) d\mathbf{X} \qquad (26.54)$$

where $D(f,g)$ is also called the Kullback-Leibler distance [48], which is used to measure the distance between h and f.

In the regression model, the mean square deviation is sometimes used as the loss function. However, when the activation function is sigmoid [49], as following:

$$S(x) = \frac{1}{1+e^{-x}} \qquad (26.55)$$

As the training result is close to the true value, the gradient operator is minimal, which makes the convergence speed of the model become very slow, as shown in Figure 26.9. And because the CE loss function is a logarithmic function, when it comes to the upper border, which can still keep in a high gradient state. So the rate of convergence of the model will not be affected.

To show the CE training process, defining the event $\mathbf{X} = \{a, b, c, d, e\}$, which real probability distribution is $\mathbf{P}_{real} = \{p_a, p_b, p_c, p_d, p_e\}$. The learning process is shown in Figure 26.10.

26.5.2.2 *Markov Decision Process*

The Markov decision process (MDP) [50–52] is an optimal decision process of a stochastic dynamic system based on the Markov Process Theory. The MDP refers to decision-makers periodically or continuously observing a stochastic dynamic system with Markovian characteristics and making decisions in a sequential manner. That is, according to the observed state at each moment, an action is selected from the available action set to make a decision. The next (future) state of the system is random, and its state transition probability is Markovian. Decision-makers make a new decision based on the newly observed status, and then repeat this process. Markov's property refers to the nature of the probability that the future development

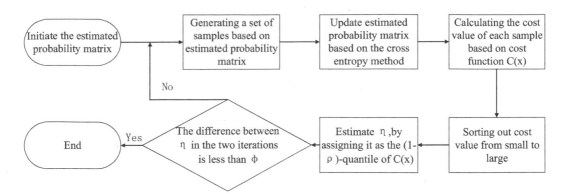

FIGURE 26.10
The learning process by CE.

of the stochastic process is independent of the history before observation. Markov's property can be simply described as the non-post-effect of state transition probability.

MDP consists of four elements. $\mathbf{M} = (\mathbf{S}, \mathbf{A}, \mathbf{P}_{sa}, \mathbf{R})$, $\mathbf{S}$ indicates the states, $\mathbf{S} = \{s_1, s_2, s_3, ..., s_n\}$; $\mathbf{A}$ indicates the actions, $\mathbf{A} = \{a_1, a_2, a_3, ..., a_m\}$; $\mathbf{P}_{sa}$ is the state transition matrix, which indicates in the current state $s \in \mathbf{S}$, the probability distribution of other states that will be transferred after action $a \in \mathbf{A}$. $\mathbf{R}$ is the reward function, which indicates that the reward value after performing the action a in the state of s can be marked as $r(s, a)$. The dynamic MDP is shown in Figure 26.11, which shows that system from the state S_x to choose an action a_i, and getting a reward r_i, then transformed to next state.

26.5.2.3 Navigation Model

In this section, we introduce the multi AUVs navigation model in the frame of MDP and CE. First, a master AUV and a slave AUV are set to CN, because increasing the number of the master AUV and slave AUV does not change the basic navigation model. Consider the MDP; we set the state $\mathbf{S}$ in the navigation system as:

$$S_t = \left\{ \varphi_t^B, \varphi_{t+1}^S, \mathbf{R}_t^{BS}, \gamma_t \right\} \tag{26.56}$$

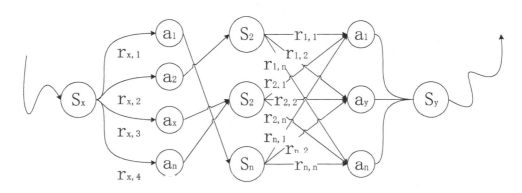

FIGURE 26.11
The model of MDP.

where S_t indicates the system state at time t; φ_t^B is the heading of master AUV at time t, φ_{t+1}^S is the heading of slave AUV at time $t+1$, R_t^{BS} is the distance between master AUV and slave AUV at time t, and γ_t is the relative measurement angle at time t. The elements in the state $\mathbf{S}$ are used to recognize the state of the system; if the number of elements is large, that will make the system model more accurate; however, it also increases the amount of calculation. It is important to find a set of appropriate elements to reduce the amount of calculation and maintain the stability of the system, $\varphi_t^B, \varphi_{t+1}^S, R_t^{BS}, \gamma_t$ can describe the system states and also reduce the amount of calculation.

Then, we define $a_t \in \mathbf{A}$ is the angle velocity of the master AUV. We set the AUVs' velocity is a constant, and the angle velocity of the master AUV is $[a_{\min}, a_{\max}]$. Master AUV adjusts the angle velocity to minimum the observation error. And the $P(S_t, a_t)$ is the probability in the state S_t to choose the action a_t, and the reward function R is defined as:

$$C(S_t, a_t) = (\varepsilon_{t+1})^2 + (\bar{\varepsilon}_{t+1})^2 \tag{26.57}$$

where the $C(S_t, a_t)$ indicates system state in the S_t to implement the action a_t, the cost is produced in this process.

In the reward function, considering to keep the communication distance in an appropriate range, when using the CE to train the navigation model, adding the distance reward to the cost function.

$$C(S_t, a_t, r_t) = C(S_t, a_t) + f_C(r_t) \tag{26.58}$$

where the $f_C(r_t)$ is the cost of the system in the distance r_t.

In summary, the final result is to get the $\mathbf{P}_{sa}$, which is trained by the CE. In the multi-AUVs CN, master AUV is based on the current state, combining with the $\mathbf{P}_{sa}$ to choose an action at to minimum the observation error. The training process is shown as follows:

1. Initial the state transition matrix $\mathbf{P}_{sa}$, cutoff cost value $\eta_0 = 0$, and $n = 0$, etc.
2. Generate the target path of the slave AUV, including N communication nodes.
3. Initial the start point of the master AUV, and then calculate the current state S_t, $t = 1, 2, 3, ..., N$. Select an action according to the $\mathbf{P}_{sa}$, and calculate the cost value $C(S_t, a_t, r_t)$ and the next state S_{t+1}. Repeat the above process until the final communication node. Generate a total of K master AUV paths.
4. Arrange the total cost value $C_i, i = 1, 2, ..., K$ of each master AUV path from small to large and set η_n as the ρ-percentile of the C_i.
5. Update the $\mathbf{P}_{sa}$ and:

$$p_{sa} = \frac{\sum_{i=1}^{k} I_{\{C(\mathbf{X}_i) \leq \eta\}} I_{\{\mathbf{X}_i \in \chi_{sa}\}}}{\sum_{i=1}^{k} I_{\{C(\mathbf{X}_i) \leq \eta\}} I_{\{\mathbf{X}_i \in \chi_s\}}} \tag{26.59}$$

where $I_{\{C(\mathbf{X}_i) \leq \eta\}} I_{\{\mathbf{X}_i \in \chi_s\}}$ is the number of the state $\mathbf{X}_i \in \chi_s$ and $C(\mathbf{X}_i) \leq \eta$ in the K paths of master AUV; $I_{\{C(\mathbf{X}_i) \leq \eta\}} I_{\{\mathbf{X}_i \in \chi_{sa}\}}$ is the number of the action $\mathbf{X}_i \in \chi_{sa}$ and $C(\mathbf{X}_i) \leq \eta$. To prevent P_{sa} from falling into local optimization, we set the $\hat{p}_{sa,n}$ as:

$$\hat{p}_{sa,n} = (1 - \mu)\, \hat{p}_{sa,n} + \mu \hat{p}_{sa,n-1} \tag{26.60}$$

This will make the process be slow, but can reduce the degree of the local optimization.

6. Repeat the process (3), until $|\eta_n - \eta_{n-1}| \leq \psi$.

References

1. T. Shiwei, D. Weiheng, L. Guangxia et al. On the research of cooperative positioning: A review. *The 4th Annual Academic Conference on Chinese Satellite Navigation*, 15, May, 2013.
2. L. Paull, S. Saeedi, M. Seto, H. Li, AUV navigation and localization: A review. *IEEE Journal of Oceanic Engineering*, 2014, 39(1); pp.131–149.
3. P. Chen, Y. Li, Y. Su, X. Chen et al. Review of AUV underwater terrain-matching navigation. *The Journal of Navigation*, 2015, 68(6); pp.1155–1172.
4. L.H. Wang, L. Yu, Y.K. Zhu, Construction method of the topographical features model for underwater terrain navigation. *Journal Citation Reports*, 2015, 22(1); pp.121–125.
5. M. Colin, V. Sandor, M. Stephen et al. Abstract. *Terrain Referencing for Autonomous Navigation of Underwater Vehicles; IEEE*; IEEE, New York, NY, 2009; pp.1–2.
6. X. Kun, F. Shao-ji, P. Yong-jie, Strong tracking adaptive Kalman filters for underwater vehicle dead reckoning. *Journal of Marine Science and Application*, 2017, 6(2); pp.19–24.
7. V.F. Filaretov, A.N. Zhirabok, A.V. Zyev et al. Design and investigation of dead reckoning system with accommodation to sensors errors for autonomous underwater vehicle. *Oceans'15 MTS/IEEE*, Washington, 19–22 Oct. 2015; pp.1–4.
8. A.J. Kapaldo. Abstract. *Gyroscope Calibration and Dead Reckoning for an Autonomous Underwater Vehicle*; Virginia Tech, 2005; pp.1–2.
9. F. Hidalgo, T. Bräunl, Perth. Review of underwater SLAM techniques. *Proceedings of the 6th International Conference on Automation, Robotics and Applications*, Feb 17–19, 2015; pp.306–311.
10. R. David, R. Pere, N. José, Abstract. *Underwater SLAM for Structured Environments Using an Imaging Sonar*; Springer-Verlag, Berlin Heidelberg, 2010; pp.1–2.
11. Yu. Rodionov, F.S. Dubrovin, P.P. Unru, S.Yu. Kulik, Experimental research of distance estimation accuracy using underwater acoustic modems to provide navigation of underwater objects. *Integrated Navigation Systems (ICINS), 2017 24th Saint Petersburg International Conference on*, 29–31 May 2017; pp.1–4.
12. J. Choi, H-T. Choi, Real-time acoustic source localization for autonomous navigation of underwater vehicles. *Oceans - St. John's*, 2014, 14–19 Sept. 2014; pp.1–5.
13. A. Bahr, J.J. Leonard, A. Martinoli, Dynamic positioning of beacon vehicles for cooperative underwater navigation. *Intelligent Robots and Systems (IROS), 2012 IEEE/RSJ International Conference on*, 7–12 Oct. 2012; pp.3760–3767.
14. M. Nogueira, J. Souza, F. Pereira, An underwater cooperative navigation scheme. *Oceans - Bergen, 2013 MTS/IEEE*, 10–14 June 2013; pp.1–7.
15. H. Yulong, Z. Yonggang, X. Bo et al. A new adaptive extended Kalman Filter for cooperative localization. *IEEE Transactions on Aerospace and Electronic Systems*, 2018, 54(1); pp.353–368.
16. L. Zhang, T. Wang, F. Zhang et al. Cooperative localization for multi-AUVs based on GM-PHD filters and information entropy theory. *Sensors*, 2017, 17(10); pp.1–16.
17. Z. Fubin, M. Peng, L. Shuqiang, A cooperative navigation algorithm for two leader AUVs based on range measurements. *Systems Engineering Theory and Practice*, 2016, 36(7); pp.1898–1904.
18. G. Salavasidis, C.A. Harris, E. Rogers, A.B. Phillips, Co-operative use of marine autonomous systems to enhance navigational accuracy of autonomous underwater vehicles. *17th Annual Conference on Towards Autonomous Robotic Systems*, Jun 26–Jul 01, 2016; pp.275–281.

19. G. Wei, L. Yalong, X. Bo, T. Lijun, Research on multiple AUV cooperative navigation algorithm based on IDDF. *Journal of Huazhong University of Science and Technology. Nature Science*, 2015, 43(6); pp.88–93.

20. A. P. Caldeira, B. Mohammadreza, B. Joao et al. Cooperative navigation and control: The EU MORPH project. *Oceans 2015 - MTS/IEEE Washington*, Oct 19–22, 2015; pp.1–10.

21. L. Brignone, J. Alves, J. Opderbecke, GREX sea trials: First experiences in multiple underwater vehicle coordination based on acoustic communication. *Oceans 2009*, Europe, 2009; pp.1–6.

22. J. Kalwa, Final results of the European Project GREX: Coordination and control of cooperating marine robots. *IFAC Proceedings Volumes*, 2010, 43(16); pp.181–186.

23. D. Fratantoni, S. Haddock, Introduction to the autonomous ocean sampling network (AOSN-II) program. *Deep Sea Research Part II: Topical Studies in Oceanography*, 2009, 56(3–5); p.61.

24. S. Ramp, R. Davis, N. Leonard et al. Preparing to predict: The second autonomous ocean sampling network (AOSN-II) experiment in the Monterey Bay. *Deep Sea Research Part II: Topical Studies in Oceanography*, 2009, 56(3–5); pp.68–86.

25. S. Glenn, O. Schofield, The New Jersey Shelf Observing System. *Oceans'02 MTS/IEEE*, 2018; pp.1680–1687.

26. S. Willcox, D. Goldberg, J. Vaganay, J. Curcio, Multi-vehicle cooperative navigation and autonomy with the bluefin cadre system. *Proceedings of IFAC (International Federation of Automatic Control) Conference*, 2006; pp.20–22.

27. P. Abreu, M. Bayat, J. Botelho, P. Gois, A. Pascoal, J. Ribeiro, M. Ribeiro, M. Rufino, L. Sebastiao, H. Silva, Cooperative control and navigation in the scope of the EC CADDY project. *Oceans 2015*, Genova, 2015; pp.1–5.

28. S. Zhao, Advanced control of autonomous underwater vehicles, PhD, University of Hawai'i, 2004.

29. J. Li, P. Lee, A neural network adaptive controller design for free-pitch-angle diving behavior of an autonomous underwater vehicle. *Robotics and Autonomous Systems*, 2005, 52(2–3); pp.132–147.

30. M. Liu, *Shui xia hang xing qi xie tong dao hang ji shu*. Guo fang gong ye chu ban she: Beijing, 2014; pp.37–38.

31. D. Zhu, Z. Hu, *Shui xia ji qi ren gu zhang zhen duan yu rong cuo kong zhi ji shu*. Guo fang gong ye chu ban she: Beijing, 2012; p.37.

32. W. Gao, Y. Liu and B. Xu, Observability analysis of cooperative navigation system for multiple AUV based on two-leaders. *Systems Engineering and Electronics*, 2013, 35(11); pp.2370–2375.

33. L. Zhang, D. Xu, M. Liu, Cooperative navigation for multiple autonomous underwater vehicles based on two hydrophones. *Systems Engineering and Electronics*, 2011, 33(7); pp.1603–1606.

34. F. Zhou, J. Zhou, J. Guo, *Xian dai kong zhi li lun ji chu*. Xi bei gong ye da xue chu ban she: Xian, 2011.

35. W. Gao, Y. Liu, B. Xu, L. Tang, Multiple-AUV cooperative navigation based on two-leader alternated navigation. *Journal of Harbin Engineering University*, 2014, 6; pp.735–740.

36. F. Zhang, C. Buckl, A. Knoll, Multiple vehicle cooperative localization with spatial registration based on a probability hypothesis density filter. *Sensors*, 2014, 14; pp.995–1009.

37. A.S. Goli, H.B. Far, O.A. Fapojuwo, Cooperative multi-sensor multi-vehicle localization in Vehicular Adhoc networks. *In Proceedings of the 2015 IEEE International Conference on Information Reuse and Integration*, San Francisco, CA, 13–15 August 2015; pp.142–149.

38. P. Feng, W. Wang, S. Dlay, S.M. Naqvi, J. Chambers, Social force model-based MCMC-OCSVM particle PHD filter for multiple human tracking. *IEEE Transactions on Multimedia*, 2017, 19; pp.725–739.

39. X. Zhou, Y. Li, B. He, T. Bai, GM-PHD-based multi-target visual tracking using entropy distribution and game theory. *IEEE Transaction on Industrial Informatics*, 2014, 10; pp.1064–1076.

40. X. Zhou, Y. Zhang, S. Hao, S. Li, A new approach for noise data detection based on cluster and information entropy. *In Proceedings of the 2015 IEEE International Conference on Cyber Technology in Automation, Control, and Intelligent Systems (CYBER)*, Shenyang, China, 8–12 June 2015; pp.1416–1419.

41. G. Zou, L. Ma, L. Zhang, L. Mo, An indoor positioning algorithm using joint information entropy based on WLAN fingerprint. *In Proceedings of the 2014 International Conference on Computing, Communication and Networking Technologies (ICCCNT)*, Hefei, China, 11–13 July 2014; pp.1–6.

42. J. Mocnej, T. Lojka, T. Zolotov, Using information entropy in smart sensors for decentralized data acquisition architecture. *In Proceedings of the 2016 IEEE 14th International Symposium on Applied Machine Intelligence and Informatics (SAMI)*, Herlany, Slovakia, 21–23 January 2016; pp.47–50.

43. X. Guo, X. Liu, G. Chen, L. Chang, X. Li, Missile weapon system-of-systems optimization method based on information entropy. *In Proceedings of the 2016 International Conference on Computer, Information and Telecommunication Systems (CITS)*, Kunming, China, 6–8 July 2016; pp.1–5.

44. R.P.S. Mahler, "Statistics 101" for multisensor, multitarget data fusion. *IEEE Aerospace and Electronic Systems Magazine*, 2004, 19; pp.53–64.

45. R. Mahler, "Statistics 102" for multisource-multitarget detection and tracking. *IEEE Journal of Selected Topics in Signal Process*, 2013, 7; pp.376–389.

46. X. Zhao, *Information Theory Foundation and Application*. China Machine Press, 2015; p.320.

47. P.T. De Boer, D.P. Kroese, S. Mannor et al. A tutorial on the cross-entropy method. *Springer Link*, 2005, 134(1); pp.19–67.

48. H. Liang, R.C. Anderson-Sprecher, R.F. Kubichek, G. Talwar, A novel approach to approximate Kullback-Leibler distance rate for hidden Markov models. *Signals, Systems and Computers, 2005. Conference Record of the Thirty-Ninth Asilomar Conference on*, 30 Oct.–2 Nov. 2005; pp.869–873.

49. V. Eleni, G. Petros, New concerns on fuzzy cognitive maps equation and sigmoid function. *Control and Automation (MED), 2017 25th Mediterranean Conference on*, 3rd–6th, July, 2017; pp.1113–1118.

50. G. D'Angelo, M. Tipaldi, L. Glielmo, S. Rampone, Spacecraft autonomy modeled via Markov decision process and associative rule-based machine learning. *Metrology for AeroSpace (MetroAeroSpace), 2017 IEEE International Workshop on*, 21st–23rd, June, 2017; pp.324–329.

51. S.K. Jayaweera, Abstract. *In Markov Decision Processes*; Wiley Telecom; pp.1–2, ISBN:9781118824818.

52. L. Liu, G. S. Sukhatme, A solution to time-varying Markov decision processes. *IEEE Robotics and Automation Letters*, July 2018, 3(3); pp.1631–1638.

27

Collaborative Self-Localization and Target Tracking under Sparse Communication

Yang Lyu, Jinwen Hu, Zhao Xu, Chunhui Zhao, and Quan Pan

CONTENTS

The problem of the collaborative self-localization and target tracking (CLAT) method under challenge environment is studied in this chapter. Specifically, the scenario with general nonlinear process and sensing model as well as sparse communication is considered by combining the distributed tracking (DT) and collaborative localization (CL) techniques. To better characterize the statistics after nonlinear transformations, the unscented transformation (UT) approach is adopted. Simulations are extensively studied to show that the proposed methods have better performance on both self-localization and target tracking than the solo CL or DT method. Further, the proposed CLAT method is validated using hardware platforms. The contribution of our method is that we propose a CLAT framework combining the correlation approximation-based CL method and covariance intersection (CI)-based DT method using UT, which is considered to have better performance in a nonlinear scenario. The proposed method does not need to store the measurement, and the measurement to both the uncooperative target and cooperative vehicles can be fused asynchronously. The remainder of this chapter is organized as follows. Section 27.1 introduces the related works, such as CL and DT. Section 27.2 formulates the CLAT problem under the system setup. Section 27.3 describes our UT-based CLAT and Section 27.4 gives the simulation results. Section 27.5 provides experimental validation based on the proposed method, and Section 27.6 concludes this chapter.

27.1 Introduction

The multi-vehicle system (MVS) has raised tremendous research interests in recent years [19]. Compared to a single-vehicle system, the MVS usually has higher efficiency and operational capability in accomplishing complex tasks such as transportation [2], search and rescue [15], and mapping [12]. Among these applications, obtaining reliable localization information for the cooperative vehicles as well as the interest targets are key signal processing tasks. In CL, each vehicle measures the relative quantities with regard to neighboring vehicles. By cooperating with other vehicles, each vehicle is able to refine the positioning information of itself. In DT, each vehicle performs measurement related to the uncooperative targets to be tracked, and then the interested states of the target can be calculated cooperatively through interaction with other vehicles. Although the problem of CL and DT are considered separately, they are correlated in a practical scenario. In the DT process, the target tracking results are based on the localization of vehicles as well as the relative measurements. To improve the target tracking results, a combined CLAT framework is considered in this chapter. Specifically, a recursive decentralized CLAT method based on sparse communication and asynchronous relative measurement is proposed.

The problem of multi-vehicle CL has drawn a lot attention in recent years. One fundamental challenge is to invest proper data fusion strategy to deal with the correlation between vehicles. To avoid double-counting, which may lead to overly confident estimates, different methods have been proposed. One approach is that the vehicle treats the neighboring vehicle fully-confident, and the uncertainty is neglected, which will lead to zero correlation, such as [18]. However, these impractical assumptions of neighboring positions may lead to overconfident estimates. A more practical method to fuse the relative measurement when the correlation is unknown is to implement a conservative method such as the CI [6] or split covariance intersection (SCI) [13]. A CL approach using the CI is proposed in [5] which is provably consistent and can handle asynchronous communication and measurement. The SCI-based approach, such as [13] and [14], further separates the covariance into a correlated part and an uncorrelated part, and the correlated part is fused using the CI method. In spite of the fact that the CI-based method is able to preserve the consistency of the estimates, they often have overly conservative results. Another popular method to deal with the CL problem is based on the factor graphs and is formulated based on the whole trajectories, such as [7]. The correlation can be explicitly tracked in the factor graph-based method, nevertheless it requires storing the whole horizon measurement so as to optimize the estimates over the full trajectories. Besides the method mentioned above, a recursive decentralized localization approach is proposed based on the extended Kalman filter (EKF) in [16], which is able to approximate the correlation and is suitable for more general scenario, such as the communication limitation, general nonlinearity, and so on.

The DT problem has also been extensively studied. Early-stage algorithms proposed to solve this problem can be roughly split into two categories, namely, the consensus-based algorithms [20] and the diffusion-based algorithms [10]. The former category in general requires multiple communication iterations during each sampling time interval and hence could bring a heavy communication burden. The latter category does not have such drawbacks but might require local joint detectabilities at every single agent, which might not be satisfied in general multi-robot target tracking scenario. A more practical DT approach, called distributed hybrid information fusion (DHIF) [23], is able to guarantee the stability and asymptotically unbiased with very mild sufficient conditions. The work is further extended to an unknown model and nonlinear scenario, respectively, in [24] and [22]. Also, the distributed particle filter has been adopted for the DT problem with regard to nonlinear, non-Gaussian systems [8]. In addition, the random finite sets (RFS) method, such as described in [4], is adopted to solve the unknown number of objects and uncertainty association problem during the DT.

The CLAT framework is beginning to arouse attention. A least-squares minimization (LSM) method is proposed in [1], where the problem is formulated using a graph and solved using the sparse optimization method. A more general factor graph-based method has been proposed combining particle-based belief propagation and gossip scheme [17]. However, the method requires all measurements to be stored within the optimization horizon so as to outperform the recursive method, such as EKF.

27.2 Problem Formulation

27.2.1 Models

Consider N cooperative vehicles $i \in \mathcal{V}$ perform the localization task and track the target t collaboratively. The dynamics of the cooperative vehicles and target are expressed respectively with following nonlinear process model

$$\mathbf{x}_{i,k+1} = f_\nu(\mathbf{x}_{i,k}, \mathbf{u}_{i,k}), \tag{27.1}$$

$$\mathbf{x}_{t,k+1} = f_t(\mathbf{x}_{t,k}, \mathbf{u}_{t,k}), \tag{27.2}$$

where $\mathbf{x}_{i,k} \in \mathbb{R}^v$ and $\mathbf{x}_{t,k} \in \mathbb{R}^t$ are, respectively, the state of vehicle i and target t. $\mathbf{u}_{i,k} \in \mathbb{R}^u$ is the control input of vehicle i, which is subject to Gaussian distribution $\mathbf{u}_i \sim \mathcal{N}(\bar{\mathbf{u}}_i, \mathbf{Q}_i)$. $\bar{\mathbf{u}}_i$ denotes the control command and $\mathbf{Q}_i$ is the control input covariance. $\mathbf{w}_{t,k} \in \mathbb{R}^w$ is the process noise of the target and is assumed to be drawn respectively from zero mean Gaussian distribution $\mathbf{w}_{i,k} \sim \mathcal{N}(\mathbf{0}, \mathbf{Q}_t)$. It is assumed that all vehicles $i \in \mathcal{V}$ share the same nonlinear transformation $f_v : \mathbb{R}^v \times \mathbb{R}^u$, and the target nonlinear transformation $f_t : \mathbb{R}^t \times \mathbb{R}^w$ is known to all vehicles.

In the cooperative localization and target tracking scenario, each vehicle i is able to measure three types of information, which are the absolute state of local vehicle and the relative measurement to the cooperative vehicles and uncooperative targets. The measurement at time instance k are denoted respectively as $\mathbf{z}_{i,k}^a \in \mathbb{R}^{za}$, $\mathbf{z}_{ij,k}^c \in \mathbb{R}^{zc}$, $j \in \{1, \cdots, N\} \setminus i$ and $\mathbf{z}_{it,k}^t \in \mathbb{R}^{zt}$. The related measurement functions for above measurement are as follows:

$$\mathbf{z}_{i,k}^a = g^a\left(\mathbf{x}_{i,k}, \mathbf{v}_{i,k}^a\right), \tag{27.3}$$

$$\mathbf{z}_{ij,k}^{c} = g^{c}\left(\mathbf{x}_{i,k}, \mathbf{x}_{t,k}, \mathbf{v}_{i,k}^{c}\right), \tag{27.4}$$

$$\mathbf{z}_{it,k}^{t} = g^{t}\left(\mathbf{x}_{i,k}, \mathbf{x}_{t,k}, \mathbf{v}_{i,k}^{t}\right), \tag{27.5}$$

where $\mathbf{v}_{i,k}^{a}, \mathbf{v}_{ij,k}^{c}$, and $\mathbf{v}_{it,k}^{t}$ are the measurement noise and assumed to be drawn from zero mean Gaussian distribution, that is, $\mathbf{v}_{i,k}^{a} \sim \mathcal{N}(\mathbf{0}, \mathbf{R}_{i}^{a}), \mathbf{v}_{i,k}^{r} \sim \mathcal{N}(\mathbf{0}, \mathbf{R}_{i}^{u}), \mathbf{v}_{i,k}^{r} \sim \mathcal{N}(\mathbf{0}, \mathbf{R}_{i}^{c})$. Furthermore, each vehicle is able to communicate with the nearby vehicles involved in the cooperative pairwise measurement. Note that the availability of measurements is dependent on the sensing range of the respective sensors.

27.2.2 Motivation and Objective

In this chapter we focus on solving the CL and DT simultaneously where DT accuracy is highly dependent on the CL accuracy. Although the existing works of CL and DT as well as CLAT are extensively studied under different system setups, we are interested in solving the problem under more practical limitations, such as both communication and storage, and more challenging system models, such as the nonlinearity in both process and measurement model. First. we implement the UT-based approach to better character the model nonlinearity, then a CLAT approach by combining the CL work [16], and CI-based distributed target tracking is proposed that can realize CLAT at the same time guarantees the consistency.

27.3 UT-Based CLAT

Suppose that at time k, each vehicle has an estimated state and an approximated error covariance of itself at previous time instance, denoted as $\hat{\mathbf{x}}_{i,k-1}$ and $\mathbf{P}_{i,k-1}$. In addition, each vehicle also holds an estimation of the target t, denoted as $\hat{\mathbf{x}}_{t,k-1}$ and $\mathbf{P}_{t,k-1}$. According to [21], when a vehicle detect a neighbors j, a correlation term $\mathbf{P}_{ij,l}$ is generated and can be arbitrarily decomposed and stored separately in vehicle i and j as below

$$\mathbf{P}_{ij,k} = \sigma_{ij,k}(\sigma_{ji,k})^{\top} \tag{27.6}$$

The required memory for each vehicle is proportional to the number of vehicles and targets.

27.3.1 Propagation

The propagation process involves the local vehicle as well as the target. According to the dynamics (27.1) and (27.2), vehicle and target can propagate separately to reduce the computational state.

Let the augmented state vector as well as the corresponding augmented covariance matrix for each vehicle's local state be denoted, respectively, as $\hat{\mathbf{x}}_{i,k-1}^{a} \in \mathbb{R}^{n_a}$ and $\mathbf{P}_{i,k-1}^{a} \in \mathbb{R}^{n_a \times n_a}$, where $n_a = n_{x_i} + n_{w_i}$,

$$\hat{\mathbf{x}}_{i,k-1}^a \triangleq \begin{bmatrix} \hat{\mathbf{x}}_{i,k-1} \\ \bar{\mathbf{u}}_{i,k-1} \end{bmatrix}, \quad \text{and} \quad \mathbf{P}_{i,k-1}^a \triangleq \begin{bmatrix} \mathbf{P}_{i,k-1} & 0 \\ 0 & \mathbf{Q}_i \end{bmatrix}.$$

A set of $2n_a + 1$ sigma points, denoted as χ^a, are selected as follows:

$$\mathcal{X}_{i,k-1}^{a,0} = \hat{\mathbf{x}}_{i,k-1}^a,$$

$$\mathcal{X}_{i,k-1}^{a,r} = \hat{\mathbf{x}}_{i,k-1}^a + \left\{ \sqrt{(n_a + \gamma)\left(\mathbf{P}_{i,k-1}^a\right)} \right\}_{(:,r)}, \quad \text{if } r \in \{1,\ldots,n_a\},$$

$$\mathcal{X}_{i,k-1}^{a,r} = \hat{\mathbf{x}}_{i,k-1}^a - \left\{ \sqrt{(n_a + \gamma)\left(\mathbf{P}_{i,k-1}^a\right)} \right\}_{(:,r-n_a)}, \quad \text{otherwise.}$$

Here $\gamma = \alpha^2(n_a + \kappa) - n_a$ is a scaling parameter, with $0 < \alpha \le 1$ and $\kappa \in \mathbb{R}$ being tuning parameters to control the spread of the sigma points. The weights for propagating the mean and covariances, denoted respectively, as $W_{i,r}^m$ and $W_{i,r}^c$, are computed as

$$W_m^0 = \gamma / (n_a + \gamma),$$

$$W_c^0 = \gamma / (n_a + \gamma) + (1 - \alpha^2 + \beta),$$

$$W_m^r = W_c^r = 1 / 2(n_a + \gamma), \quad r = 1,\ldots,2n_a,$$

where β is used to incorporate extra higher-order effects. Note that the definition of the sigma points directly implies that

$$\sum_{r=0}^{2n_a} (W_m^r) \, \mathcal{X}_{i,k-1}^{a,r} = \mathcal{X}_{i,k-1}^{a,0} = \hat{\mathbf{x}}_{i,k-1}^a,$$

or equivalently,

$$\sum_{r=0}^{2n_a} W_m^r \, \mathcal{X}_{i,k-1}^r = \hat{\mathbf{x}}_{i,k-1}, \qquad \sum_{r=0}^{2n_a} W_m^r \, \mathcal{U}_{k-1}^r = \mathbf{0},$$

where $\mathcal{X}_{i,k-1}^r$ and $\mathcal{U}_{i,k-1}^r$, collect the components of $\mathcal{X}_{i,k-1}^{a,r}$ corresponding to, respectively, $\mathbf{x}_{i,k-1}$ and $\mathbf{u}_{i,k-1}$.

The above UT is summarized as below:

$$\mathcal{X}_{i,k-1}^{a,r} = UT(\hat{\mathbf{x}}_{i,k-1}^a, \mathbf{P}_{i,k-1}^a), \qquad r = 0,\ldots,2n_a \quad . \tag{27.7}$$

Similarly, the UT of target t's estimation within vehicle i, denoted by the subscribe t_i, can be obtained as

$$\mathcal{X}_{t_i,k-1}^{a,r} = UT(\hat{\mathbf{x}}_{t_i,k-1}^a, \mathbf{P}_{t_i,k-1}^a), \qquad r = 0,\ldots,2n_a \tag{27.8}$$

Then the prior local estimates and corresponding error covariance of the current state and target are computed respectively as

$$\hat{\mathbf{x}}_{i,k} = \sum_{r=0}^{2n_a} W_m^r \, \mathcal{X}_{i,k}^r, \tag{27.9}$$

$$\mathbf{P}_{i,k} = \sum_{r=0}^{2n_a} W_c^r \left(\mathcal{X}_{i,k}^r - \hat{\mathbf{x}}_{i,k} \right) \left(\mathcal{X}_{i,k}^r - \hat{\mathbf{x}}_{i,k} \right)^\top, \tag{27.10}$$

and

$$\hat{\mathbf{x}}_{t_i,k} = \sum_{r=0}^{2n_a} W_m^r \, \mathcal{X}_{t_i,k}^r, \tag{27.11}$$

$$\mathbf{P}_{t_i,k} = \sum_{r=0}^{2n_a} W_c^r \left(\mathcal{X}_{t_i,k}^r - \hat{\mathbf{x}}_{t_i,k} \right) \left(\mathcal{X}_{t_i,k}^r - \hat{\mathbf{x}}_{t_i,k} \right)^\top, \tag{27.12}$$

where

$$\mathcal{X}_{i,k}^r = f_v \left(\mathcal{X}_{i,k-1}^r, \mathcal{U}_{i,k-1}^r \right), \quad r = 0,\dots,2n_a \tag{27.13}$$

and

$$\mathcal{X}_{t_i,k}^r = f_t \left(\mathcal{X}_{i,k-1}^r, \mathcal{W}_{i,k-1}^r \right), \quad r = 0,\dots,2n_a. \tag{27.14}$$

Furthermore, there are still correlation terms that need to be taken care of. The propagation of correlation term $\mathbf{P}_{ij}$ involves the pose and control input of both i and j. To reduce the communication, the local correlation term σ_{ij} is instead propagated with the inferred Jacobian matrices $\mathcal{F}_v$

$$\sigma_{ij,k+1} = \mathcal{F}_v \sigma_{ij,k}, \tag{27.15}$$

$\mathcal{F}_v$ is the inferred Jacobian matrix [11], defined as

$$\mathcal{F}_v = \overline{\mathbf{P}}_{i,k-1} \left(\mathbf{P}_{i,k-1} \right)^{-1}$$

where $\overline{\mathbf{P}}_{i,k-1} \approx \sum_{r=0}^{2n_a} (\mathcal{X}_{i,k}^r - \hat{x}_{i,k})(\mathcal{X}_{i,k-1}^r - \hat{x}_{i,k-1})^\top$.

27.3.2 Update

In the update stage, three types of measurement (27.3) through (27.5) are considered. When a private measurement or a target measurement is taken by vehicle i, the information is updated locally to avoid communication. When a neighbor vehicle j is within the communication/detection range, a relative measurement that involves i and j is taken, and information is exchanged.

27.3.2.1 Private Update

During the private update process, the local vehicle receives a measurement to its local pose $\mathbf{z}_{i,k}$, for example, the GPS signal, to refine its local estimation; therefore only the local pose participates in the private update process.

First, the inferred Jacobian $\mathcal{H}_{i,k}$ is obtained as

$$\mathcal{H}_{i,k} = \mathbf{P}_{i,k}^{xz} \left(\mathbf{P}_{i,k}^- \right)^{-1},$$

where

$$\mathbf{P}_{i,k}^{xz} = \sum_{r=0}^{2n_a} W_c^r \left(\mathcal{X}_{i,k}^r - \mathbf{x}_{i,k}^- \right) \left(h_i^a(\mathcal{X}_{i,k}^r) - \mathbf{z}_{i,k} \right)$$

Then the state and covariance can be updated using the typical Kalman filter formulation as

$$\mathbf{S}_{i,k} = (\mathcal{H}_{i,k})^\top \mathbf{P}_{i,k} \mathcal{H}_{i,k} + \mathbf{R}_i \tag{27.16}$$

$$\mathbf{K}_{i,k} = \mathbf{P}_{i,k} \mathcal{H}_{i,k}^\top \mathbf{S}_{i,k}^{-1} \tag{27.17}$$

$$\mathbf{x}_{i,k} = \mathbf{x}_{i,k}^- + \mathbf{K}_{i,k} \left(\mathbf{z}_{i,k} - h_i^a(\mathbf{x}_{i,k}^-) \right) \tag{27.18}$$

$$\mathbf{P}_{i,k} = (\mathbf{I} - \mathbf{K}_{i,k} \mathcal{H}_{i,k}) \mathbf{P}_{i,k} \tag{27.19}$$

The correlation term is updated as

$$\sigma_{ij,k} = (\mathbf{I} - \mathbf{K}_{i,k} \mathcal{H}_{i,k}) \sigma_{ij,k} \tag{27.20}$$

27.3.2.2 Target Measurement Update

When a target is detected by a vehicle i, a relative measurement related to the pose of both vehicle i and target t, denoted as $\mathbf{z}_{it}$, is obtained. Define the augmented state as

$$\mathbf{x}_{i,k}^b = \begin{bmatrix} \mathbf{x}_{i,k} \\ \mathbf{x}_{t_i,k} \\ 0 \end{bmatrix} \quad \text{and} \quad \mathbf{P}_{i,k}^b = \begin{bmatrix} \mathbf{P}_i & 0 & 0 \\ 0 & \mathbf{P}_t & 0 \\ 0 & 0 & \mathbf{R}_i^u \end{bmatrix}.$$

Then the augmented sigma points is obtained as

$$\mathcal{X}_{i,k}^b = UT \left(\hat{\mathbf{x}}_{i,k}^b, \mathbf{P}_{i,k}^b \right), \quad r = 0, \dots, 2n_a. \tag{27.21}$$

The inferred measurement Jacobian is

$$\begin{bmatrix} \mathcal{H}_{i,k} & \mathcal{H}_{t_i,k} & \mathcal{H}_v^k \end{bmatrix} = \mathbf{P}^{xz} \left(\mathbf{P}_{i,k}^b \right)^{-1} \tag{27.22}$$

The state and covariance are updated as

$$\mathbf{S}_{i,k} = [\, \mathcal{H}_{i,k} \quad \mathcal{H}_{t_i,k}] \begin{bmatrix} \mathbf{P}_{i,k}^- & 0 \\ 0 & \mathbf{P}_{t,k}^- \end{bmatrix} [\mathcal{H}_{i,k} \quad \mathcal{H}_{t_i,k}]^\top + \mathbf{R}_i^u \tag{27.23}$$

$$\mathbf{K}_{i,k} = \mathbf{P}_{i,k} \mathcal{H}_{i,k} \mathbf{S}_{i,k}^{-1} \tag{27.24}$$

$$\mathbf{K}_{i,k} = \mathbf{P}_{t,k} \mathcal{H}_{t_i,k} \mathbf{S}_{i,k}^{-1} \tag{27.25}$$

$$\mathbf{x}_{i,k} = \mathbf{x}_{i,k}^{-} + \mathbf{K}_{i,k}\left(\mathbf{z}_{it,k} - h_i^u\left(\mathbf{x}_{i,k}^b\right)\right) \qquad (27.26)$$

$$\mathbf{x}_{t_i,k} = \mathbf{x}_{t,k}^{-} + \mathbf{K}_{t,k}\left(\mathbf{z}_{it,k} - h_i^u\left(\mathbf{x}_{i,k}^b\right)\right) \qquad (27.27)$$

$$\mathbf{P}_{i,k} = (\mathbf{I} - \mathbf{K}_{i,k}\mathcal{H}_{i,k})\mathbf{P}_{i,k}^{-} \qquad (27.28)$$

$$\mathbf{P}_{t_i,k} = (\mathbf{I} - \mathbf{K}_{i,k}\mathcal{H}_{i,k})\mathbf{P}_{t_i,k}^{-} \qquad (27.29)$$

The correlation term is updated as

$$\mathbf{P}_{ij,k} = (\mathbf{I} - \mathbf{K}_{i,k}\mathcal{H}_{i,k})\mathbf{P}_{ij,k}^{-}. \qquad (27.30)$$

To avoid communication, the correlation term σ_{ij} is calculated as below,

$$\sigma_{ij,k} = (\mathbf{I} - \mathbf{K}_i\mathcal{H}_i)\sigma_{ij,k}^{-} \qquad (27.31)$$

In the target update process, the existence of the correlation term $\mathbf{P}_{ij} \neq \mathbf{0}$ may cause loopy information transmission, which may lead to overconfident estimate results when the interaction between different vehicles occurs. In our method, the correlation term is simply neglected.

27.3.3 Neighbor Measurement Update and Target Information Fusion

When two vehicles i and j are within a given range, a relative measurement is taken, denoted as $\mathbf{z}_{ij,k}$, and a communication link between the two vehicles is established. The target update process is as follows. First, the covariance between two vehicles, $\mathbf{P}_{ij}$ can be obtained according to Equation 27.6 and the target measurement is updated as follows. Similar to the target measurement update process, we define the augmented state as

$$\mathbf{x}_k^c = \begin{bmatrix} \mathbf{x}_{i,k} \\ \mathbf{x}_{j,k} \\ \mathbf{0} \end{bmatrix} \quad \text{and} \quad \mathbf{P}_k^c = \begin{bmatrix} \mathbf{P}_i & \mathbf{P}_{ij} & \mathbf{0} \\ \mathbf{P}_{ij}^{\top} & \mathbf{P}_j & \mathbf{0} \\ \mathbf{0} & \mathbf{0} & \mathbf{R}^c \end{bmatrix}.$$

Then the augmented sigma points is obtained as

$$\mathcal{X}_k^c = UT\left(\hat{\mathbf{x}}_k^c, \mathbf{P}_{i,k}^c\right), \quad r = 0, \ldots, 2n_a.$$

The inferred measurement Jacobian is

$$[\mathcal{H}_{i,k} \quad \mathcal{H}_{j,k} \quad \mathcal{H}_{v,k}] = \mathbf{P}^{xz}\left(\mathbf{P}_{i,k}^c\right)^{-1}$$

where

$$\mathbf{P}_{i,k}^{xz} = \sum_{r=0}^{2n_a} W_c^r\left(\mathcal{X}_{i,k}^c - \mathbf{x}_{i,k}^c\right)\left(h^c\left(\mathcal{X}_{i,k}^r\right) - \mathbf{z}_{ij,k}\right)$$

Consequently, the update process is list as below:

$$x_{i,k} = x_{i,k}^- + K_{i,k}\left(z_{ij,k} - h_i^c\left(x_{i,k}^c\right)\right),$$ (27.32)

$$x_{j,k} = x_{t,k}^- + K_{t,k}\left(z_{it,k} - h_i^c\left(x_{i,k}^c\right)\right),$$ (27.33)

$$P_{i,k} = (I - K_{i,k}\mathcal{H}_{i,k})P_{i,k} - K_{i,k}\mathcal{H}_{j,k}P_{ji},$$ (27.34)

$$P_{j,k} = (I - K_{j,k}\mathcal{H}_{j,k})P_{j,k} - K_{j,k}\mathcal{H}_{i,k}P_{ij},$$ (27.35)

$$P_{ij,k} = (I - K_{i,k}\mathcal{H}_{i,k})P_{ij,k} - K_{i,k}\mathcal{H}_{j,k}P_j,$$ (27.36)

$$P_{il,k} = (I - K_{i,k}\mathcal{H}_{i,k})P_{il,k} - K_{i,k}\mathcal{H}_{j,k}P_{jl},$$ (27.37)

where

$$S_{i,k} = \begin{bmatrix} \mathcal{H}_{i,k} & \mathcal{H}_{j,k} \end{bmatrix} \begin{bmatrix} P_{i,k} & P_{ij,k} \\ P_{ij,k}^\top & P_{j,k} \end{bmatrix} [\mathcal{H}_{i,k} \quad \mathcal{H}_{j,k}]^\top + R^c$$ (27.38)

$$K_{i,k} = (P_{i,k}\mathcal{H}_{i,k} + P_{ij,k}\mathcal{H}_{j,k})S_k^{-1}$$ (27.39)

$$K_{j,k} = (P_{ji,k}\mathcal{H}_{i,k} + P_{j,k}\mathcal{H}_{j,k})S_k^{-1}$$ (27.40)

Note that the correlation term update (27.37) involves a correlation term P_{jl} which is not available at this measurement update step, one approximation technique is as ([16]):

$$P_{il,k} = P_{i,k}\left(P_{i,k}^-\right)^{-1}P_{il,k}$$ (27.41)

After the calculation of $P_{ij,k}$, a decomposition is carried out to update the correlation term $\sigma_{ij,k}$ and $\sigma_{ji,k}$.

Additionally, the target belief $\{x_{t_i}, P_{t_i}\}$ and $\{x_{t_j}, P_{t_j}\}$ can also be fused as one. As a matter of fact, the correlation between vehicle $i \in \mathcal{V}$ and target t is not stored, a conservative CI algorithm can be used as below:

$$P_{t,k} = \left(\omega\left(P_{t_i,k}\right)^{-1} + (1-\omega)\left(P_{t_j,k}\right)^{-1}\right)^{-1},$$ (27.42)

$$x_{t,k} = P_{t,k}\left(\omega P_{t_i,k}^{-1}\left(x_{t_i,k}\right) + (1-\omega)P_{t_j,k}^{-1}\left(x_{t_j,k}\right)\right).$$ (27.43)

The weight matrix can be determined according to [23].

27.4 Simulation

In this section, the proposed method is validated using synthetic data. Specifically, four vehicles operate cooperatively to track an uncooperative target. We assume that both the vehicles and target are subject to similar nonlinear unicycle model as below:

$$\mathbf{x}_{(k+1)} \triangleq \begin{bmatrix} x_{k+1} \\ y_{k+1} \\ \theta_{k+1} \end{bmatrix} = \begin{bmatrix} x_k + \Delta_T \left(v_c + w_k^v \right) \cos(\theta^k) \\ y_k + \Delta_T \left(v_c + w_k^v \right) \sin(\theta^k) \\ \theta_k + \Delta_T \left(d_c + w_k^\omega \right) \end{bmatrix},$$

Specifically, a subscribe i or t is used for vehicles or target. The state vector $\mathbf{x}_k$ to be estimated contains three entries, namely, x_k, y_k, and θ_k, which represent, respectively, the x-position, the y-position, and the orientation of the vehicle with respect to the global frame. It is assumed that at the initial time instant, the vehicles are randomly placed on a circle centered respectively at [−5, 5], [5, 5], [−5, −5], [5, −5]. The same control command is applied on each vehicle as $\mathbf{u} = [v_c, d_c]^\top = [0.3, 0.0375]^\top$ to form four approximated circles with radius 8. The velocity and angular velocity noise is subject to Gaussian distribution with covariance $\mathbf{Q}_i = \mathrm{diag}([0.1^2, (0.5\pi/180)^2])$. The target is initialized at $[−15, −15]^\top$ in global frame and the control input is set as $\mathbf{u}_t = [0.1, 0]^\top$.

In our simulation scenario, we assume that vehicle 1 has access to its global position and orientation in global frame with following measurement model:

$$\mathbf{z}_{i,k} = \mathbf{x}_{i,k} + \mathbf{v}_{i,k},$$

where $\mathbf{v}_{i,k} \sim \mathcal{N}(\mathbf{0}, \mathrm{diag}\,[(0.1^2, 0.1^2, 0.1\pi/180)^2])$ is the control noise.

Both the cooperative vehicles and uncooperative measurement are subject to following measurement model:

$$z_{il} = \| x_l - x_i \| + v_{il}$$

l could either be an uncooperative target or a cooperative vehicle that is within the sensing range. The sensing range to target is set as $r_t = 20$ and the sensing range as well as the communication range is set as $r_c = 10$. For the cooperative measurement, the measurement noise is $v_{il} \sim \mathcal{N}(0, 0.05^2)$ and the uncooperative measurement, $v_{il} \sim \mathcal{N}(0, 0.05^2)$.

27.4.1 Scenario 1

In this section, one trial of the simulation described above is carried out. In this scenario, the target is joint observed by the four vehicles intermittently. The observation measurement availability is shown in Figure 27.1. Although the target is not observed by all vehicles, each vehicle could have an estimation of the target through communication with neighbors. The estimated trajectories of both vehicles and target are plotted in Figure 27.2. Each vehicle's self-localization result and its local target tracking results are shown in the same color with solid line and dash line, respectively. As observed

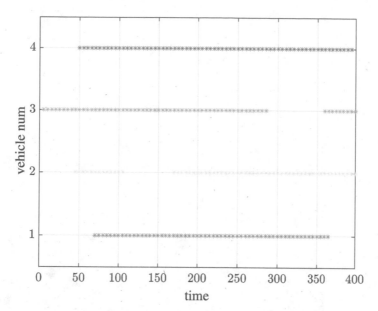

FIGURE 27.1
The measurement availability to target from the four vehicles.

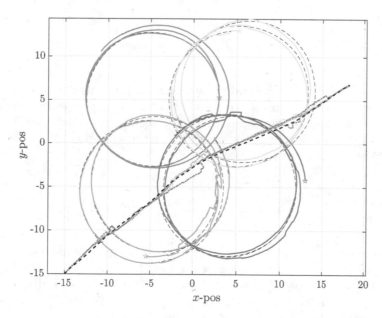

FIGURE 27.2
The estimated trajectories of vehicles and target in different colors (with black dash line as ground truth).

in Figure 27.2, the estimated trajectories by local vehicle is able to localize the true position of itself as well as track the true trajectories of the target. In Figure 27.3, the local posterior estimate errors (solid lines in colors) and the corresponding approximated three envelopes (dashed lines in the same color) of four vehicles are plotted. As observed in Figures 27.3 and 27.4, the estimate errors by each vehicle are bounded by the $\pm 3\sigma$ envelopes at the steady-state.

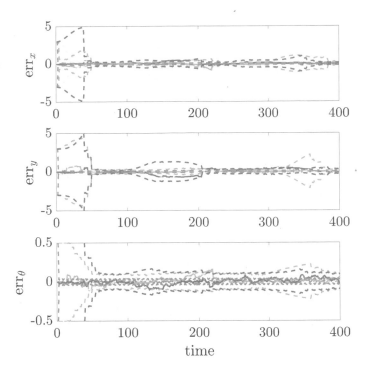

FIGURE 27.3
The self-localization errors and covariance of vehicle 1–4.

27.4.2 Scenario 2

In this part, the proposed collaborative localization and target tracking method are compared to the CL result proposed in [16] and distributed target tracking results proposed in [3] based on 1000 Monto Carlo simulations. Specifically, the simulation in Scenario 1 is repeated for 1000 times. For comparison purposes, an error index is defined as the averaged position root mean square error (PRMSE)

$$e_{i,k} = \frac{1}{1000} \sum_{m=1}^{1000} \sqrt{(x_{i,k}^m - \bar{x}_{i,k}^m)^2 + (y_{i,k}^m - \bar{y}_{i,k}^m)^2}$$

Accordingly, we have the PRMSE of the proposed CLAT method vs. the DT method plotted in Figure 27.5. Apparently, the proposed method has lower PRMSE than the DT method on the target tracking. Besides, we have the PRMSE of self-localization using both our proposed CLAT method vs. the CL method plotted in Figure 27.6. As observed, our method in general has better localization over the CL method.

27.5 CLAT for Quad-Rotors

In this section, we implement the proposed method on a multi-UAV-based target localization scenario. The system configuration is as follows (Figure 27.7). Specifically, the

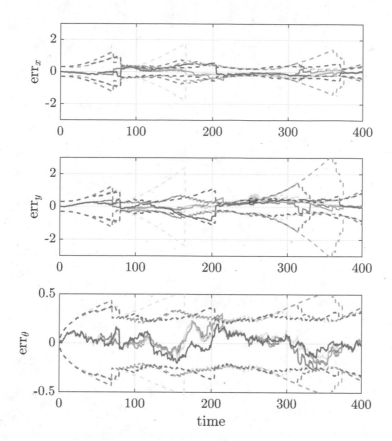

FIGURE 27.4
The target tracking errors and covariance of vehicle 1–4.

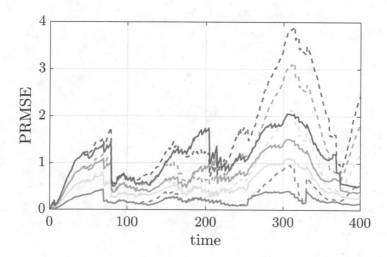

FIGURE 27.5
The different vehicle's target tracking PRMSE using the proposed CLAT method (solid line) vs. DT method (dash line).

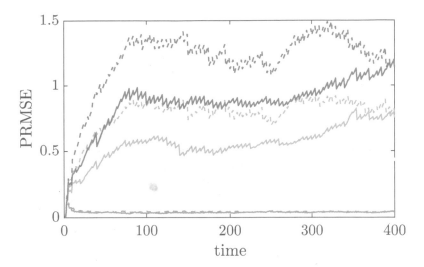

FIGURE 27.6
The different vehicle's self-localization PRMSE using the proposed CLAT method (solid line) vs. DT method (dash line).

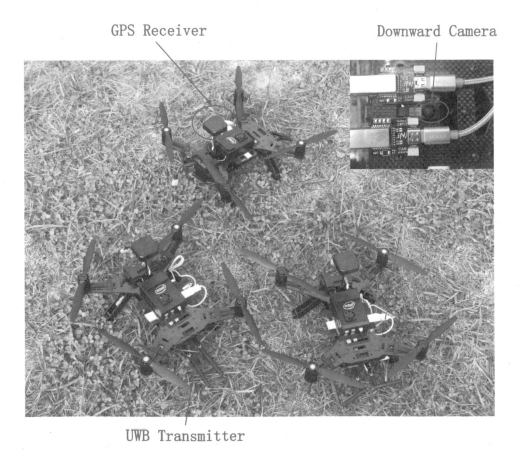

FIGURE 27.7
The snapshot of the CLAT experiment setup (Three quad-rotors and the corresponding captured image).

system consists of several Intel Aero quad-rotors equipped with the UWB sensors and a downward monochrome camera. The UWB sensors are to measure the relative distance and transmit information when two quad-rotors are within the functional range of the UWB sensors. Also, a camera is rigidly connected to the body frame of each quad-rotors. A target is detected based on the camera when the target is within the field of view (FOV) of the camera. In addition to the above onboard devices, a pack of UWB ground tags are used to record the ground truth.

27.5.1 Vehicle and Target Dynamics Model

For target-monitoring purposes, each quad-rotor i is controlled to follow a plenary circular trajectory with different centers. The state $\mathbf{x}_i$ includes the position in 3D space and the heading angle, namely $\mathbf{x}_t = [x_t, y_t, z_t, \theta_t]^\top$. $u_i \triangleq [v_i, \omega_i]^\top$ denotes the control input command, namely the linear velocity and angular velocity. The actual velocity and angular velocity are contaminated by the zero mean Gaussian noise, $w_{i,k}^v \sim \mathcal{N}(0, Q_v), w_{i,k}^\omega \sim \mathcal{N}(0, Q_\omega)$. An extra altitude noise $w_{i,k}^z \sim \mathcal{N}(0, Q_z)$ is also added to the process noise. The overall process noise covariance is denoted as $\mathbf{Q}_i = \text{blkdiag}([Q_v, Q_\omega, Q_z])$. Based on above definition, the process model for each vehicle f_i is defined as follows:

$$\mathbf{x}_{i,k+1} \triangleq \begin{bmatrix} x_{i,k+1} \\ y_{i,k+1} \\ z_{i,k+1} \\ \theta_{i,k+1} \end{bmatrix} = \begin{bmatrix} x_{i,k} + \Delta_T (v_i + w_{i,k}) \cos(\theta_{i,k}) \\ y_{i,k} + \Delta_T (v_i + w_{i,k}) \sin(\theta_{i,k}) \\ z_c + w_{i,k}^z \\ \theta_{i,k} + \left(\omega_i + w_{i,k}^\omega \right) \Delta_T \end{bmatrix} \tag{27.44}$$

Further, we model the target f_t with a similar unicycle model as Equation 27.44. The state, control input, and process noise are denoted as $\mathbf{x}_t$, $\mathbf{u}_t$, and $\mathbf{Q}_t$, respectively.

27.5.2 Measurement Model

Three types of sensing mechanisms are utilized to realize the CLAT purpose in this system setup, namely the private absolute measurement from the GPS receiver, the cooperative relative range measurement from the UWB sensors, and the relative angle measurement to target from the downward cameras. The private absolute measurement, that is, the noisy GPS signal, is modeled as below:

$$\mathbf{z}_{i,k}^a = H^a \mathbf{x}_{i,k} + \mathbf{v}_{i,k}^a \tag{27.45}$$

We assume that the GPS noise is subject to zero mean Gaussian distribution, and $\mathbf{v}_{i,k}^a \sim \mathcal{N}(\mathbf{0}, \mathbf{R}^a)$. Obviously, the absolute measurement model is a linear model; therefore, we can substitute the inferred Jacobian matrix defined in 27.3.2 with (H_i). We assume that only a subset of all vehicles is able to access to the GPS signal.

The cooperative relative measurement between vehicle i and j from UWB is a scaled distance modeled as following nonlinear model:

$$z_{ij}^u = \| \mathbf{x}_{i,k} - \mathbf{x}_{j,k} \|_D + v_{ij,k}^u, \tag{27.46}$$

where $D = \mathrm{diag}(1, 1, 1, 0)$ is to extract the position part of the state vector. The measurement noise $v_{ij,k}^u$ is assumed to be subject to zero mean Gaussian noise, $v_{ij,k}^u \sim \mathcal{N}(0, R^u)$.

The target detection measurement is defined as a vector from the local vehicle to the target with a rotation angle θ as below:

$$z_i^t = \mathcal{R}(\theta_{i,k}) \frac{\mathbf{x}_{t,k} - \mathbf{x}_{i,k}}{\| \mathbf{x}_{t,k} - \mathbf{x}_{i,k} \|} + v_{it,k}, \tag{27.47}$$

where the noise is assumed as $v_{it,k} \sim \mathcal{N}(\mathbf{0}, \mathbf{R}^t)$. The rotation matrix

$$\mathcal{R}(\theta_{i,k}) = \begin{bmatrix} \cos(\theta) & \sin(\theta) & 0 \\ -\sin(\theta) & \cos(\theta) & 0 \\ 0 & 0 & 1 \end{bmatrix}$$

In this chapter, the target detection is done using the KCF tracker [9] on the image plan, then the measurement z_i^t is then obtained based on the camera parameters; see [22] for details.

27.5.3 Experiment Results and Analysis

According to the above model description, an experiment with three quad-rotors tracking a pedestrian is carried out. The circling centers for the three quad-rotors, denoted respectively as quad1, quad2, and quad3, are set respectively as $c_1 = [-5, -5]$m, $c2 = [5, -5]$m, and $c_3 = [-5, 5]$m, with the same radius 8 m. To avoid possible collision between vehicles, the desired height for the three quad-rotors are set respectively as $z_{c,1} = 15$ m, $z_{c,2} = 16.5$ m, and $z_{c,3} = 18$ m. The control input for the circling motion is as $v_i = 0.3$ m/s and $\omega_i = 0.0375$ rad/s. The control noises are set as $Q_v = 0.05^2$, $Q_\omega = (\pi/180)^2$, and $Q_z = 0.1^2$. The initialization for the quad-rotors is as follows: $\mathbf{x}_i(0) \sim \mathcal{N}(\bar{x}i, 0, \mathbf{P}_{i,0})$ for all $i = 1, 2, 3$, where $\mathbf{P}_{i,0} = \mathrm{diag}(1^2, 1^2, 0.5^2, (10\pi/180)^2)$. The target is initialized according $\mathbf{x}_t(0) \sim \mathcal{N}(\bar{x}t, 0, \mathbf{P}_{t,0})$, where $\mathbf{P}_{t,0} = \mathrm{diag}(0.1^2, 0.1^2, 0.5^2, (1\pi/180)^2)$ and $\mathbf{x}_{t,0} = [-15, -15, 0]$. The velocity of pedestrian is around $v_t = 0.2$ m/s and $w_t = 0$ rad/s. The three types of measurement noise are set respectively as $\mathbf{R}^a = \mathrm{diag}(2^2, 2^2, 1^2)$, $\mathbf{R}^u = 0.1^2$ and $\mathbf{R}^t = \mathrm{diag}(0.01^2, 0.01^2, 0.01^2)$. In the experiment setup, only quad1 is able to get access to the GPS measurement. The CLAT estimator is running at 10 Hz.

One experiment snapshot of three quad-rotors with the corresponding captured images is shown in Figure 27.8, where the pedestrian is within the camera's FOV of all three quad-rotors. According to the CLAT algorithm, the trajectories of all three quad-rotors' self-localization results and the target tracking results are plotted in Figure 27.9. As observed in Figure 27.9, the three quad-rotors are able to localize themselves and at the same time stably track the target. It is obvious that the self-localization result from quad1 is better than the other two quad-rotors, as it can obtain the absolute GPS signal. The errors of both self-localization and target tracking are plotted in Figures 27.10 and 27.11 correspondingly. The local estimate errors (solid line) and the corresponding approximated $\pm 3\sigma$ envelops (dashed lines in the same color) of the aforementioned three quad-rotors are plotted. As observed in Figures 27.10 and 27.11, the estimate errors by each quad-rotor is bounded by the $\pm 3\sigma$ envelopes at the steady state.

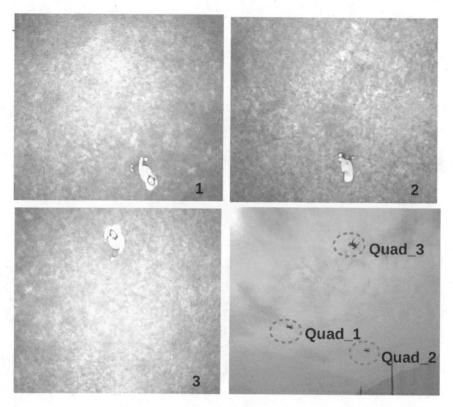

FIGURE 27.8
The trajectories of three quad-rotors self-localization results and the target tracking results with different colors, the ground-truth is provided with dash line.

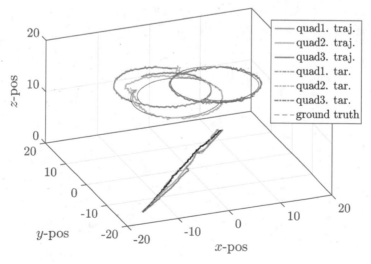

FIGURE 27.9
The self-localization errors and covariance of 3 quad-rotors.

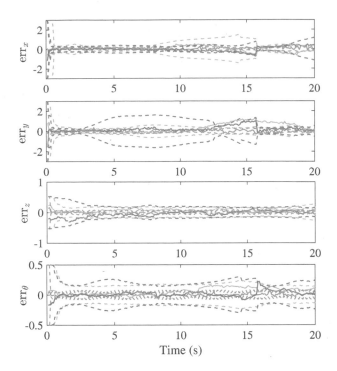

FIGURE 27.10
The object tracking results from the three quad-rotors.

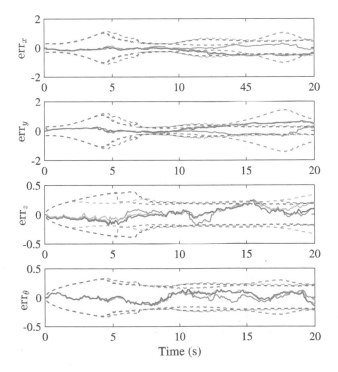

FIGURE 27.11
The self-localization results of the three quad-rotors.

27.6 Conclusion

In this chapter we consider the CL and target tracking problem in the scenario with general nonlinear process and sensing model as well as sparse communication by combining the DT and the CL techniques. To better characterize the statistics after nonlinear transformations, the UT approach is adopted. Simulations are extensively studied to show that the proposed methods have better performance in both self-localization and target tracking than the solo CL or DT method. Further, the proposed method will be implemented on multi-UAV-based target localization scenario with on-board UWB sensor and cameras.

References

1. Ahmad, A., G. D. Tipaldi, P. Lima, and W. Burgard, 2013. Cooperative robot localization and target tracking based on least squares minimization. In *Robotics and Automation (ICRA), 2013 IEEE International Conference on*, pp. 5696–5701. IEEE.
2. Alonso-Mora, J., S. Baker, and D. Rus, 2017. Multi-robot formation control and object transport in dynamic environments via constrained optimization. *The International Journal of Robotics Research* 36(9), 1000–1021.
3. Battistelli, G. and L. Chisci, 2016. Stability of consensus extended Kalman filter for distributed state estimation. *Automatica* 68, 169–178.
4. Battistelli, G., L. Chisci, C. Fantacci, A. Farina, and A. Graziano, 2013. Consensus CPHD filter for distributed multitarget tracking. *IEEE Journal of Selected Topics in Signal Processing* 7(3), 508–520.
5. Carrillo-Arce, L. C., E. D. Nerurkar, J. L. Gordillo, and S. I. Roumeliotis, 2013. Decentralized multi-robot cooperative localization using covariance intersection. In *Intelligent Robots and Systems (IROS), 2013 IEEE/RSJ International Conference on*, pp. 1412–1417. IEEE.
6. Chen, L., P. O. Arambel, and R. K. Mehra, 2002. Fusion under unknown correlation-covariance intersection as a special case. In *Information Fusion, 2002. Proceedings of the Fifth International Conference on*, Volume 2, pp. 905–912. IEEE.
7. Cunningham, A., M. Paluri, and F. Dellaert, 2010. DDF-SAM: Fully distributed SLAM using constrained factor graphs. In *Intelligent Robots and Systems (IROS), 2010 IEEE/RSJ International Conference on*, pp. 3025–3030. IEEE.
8. Farahmand, S., S. I. Roumeliotis, and G. B. Giannakis, 2011. Set-membership constrained particle filter: Distributed adaptation for sensor networks. *IEEE Transactions on Signal Processing* 59 (9), 4122–4138.
9. Henriques, J. F., R. Caseiro, P. Martins, and J. Batista, 2015. High-speed tracking with kernelized correlation filters. *IEEE Transactions on Pattern Analysis and Machine Intelligence* 37 (3), 583–596.
10. Hu, J., L. Xie, and C. Zhang, 2012. Diffusion Kalman filtering based on covariance intersection. *IEEE Transactions on Signal Processing* 60 (2), 891–902.
11. Huang, G. P. and S. I. Roumeliotis, 2008. *An observability constrained UKF for improving SLAM consistency.* University of Minnesota, Minneapolis, MN, Tech. Rep.
12. La, H. M. and W. Sheng, 2013. Distributed sensor fusion for scalar field mapping using mobile sensor networks. *IEEE Transactions on Cybernetics* 43(2), 766–778.
13. Li, H. and F. Nashashibi, 2013. Cooperative multi-vehicle localization using split covariance intersection filter. *IEEE Intelligent Transportation Systems Magazine* 5(2), 33–44.
14. Li, H., F. Nashashibi, and M. Yang, 2013. Split covariance intersection filter: Theory and its application to vehicle localization. *IEEE Transactions on Intelligent Transportation Systems* 14(4), 1860–1871.

15. Liu, Y. and G. Nejat, 2016. Multirobot cooperative learning for semi-autonomous control in urban search and rescue applications. *Journal of Field Robotics* 33(4), 512–536.

16. Luft, L. T., Schubert, S. I. Roumeliotis, and W. Burgard, 2018. Recursive decentralized localization for multi-robot systems with asynchronous pairwise communication. *International Journal of Robotics Research* 37(10), 1152–1167.

17. Meyer, F., O. Hlinka, H. Wymeersch, E. Riegler, and F. Hlawatsch, 2016. Distributed localization and tracking of mobile networks including noncooperative objects. *IEEE Transactions on Signal and Information Processing Over Networks* 2(1), 57–71.

18. Panzieri, S., F. Pascucci, and R. Setola, 2006. Multirobot localisation using interlaced extended Kalman filter. In *Intelligent Robots and Systems, 2006 IEEE/RSJ International Conference on*, pp. 2816–2821. IEEE.

19. Ren, W., R. W. Beard, and E. M. Atkins, 2007, April. Information consensus in multivehicle cooperative control. *IEEE Control Systems* 27(2), 71–82.

20. Ren, W., R. W. Beard, and D. B. Kingston, 2005. Multi-agent Kalman consensus with relative uncertainty. In *American Control Conference, 2005. Proceedings of the 2005*, pp. 1865–1870. IEEE.

21. Roumeliotis, S. I. and G. A. Bekey, 2002. Distributed multirobot localization. *IEEE Transactions on Robotics and Automation* 18(5), 781–795.

22. Wang, S., Y. Lyu, and W. Ren, 2018. Unscented-transformation-based distributed nonlinear state estimation: Algorithm, analysis, and experiments. *IEEE Transactions on Control Systems Technology* (99), 1–14.

23. Wang, S. and W. Ren, 2018. On the convergence conditions of distributed dynamic state estimation using sensor networks: A unified framework. *IEEE Transactions on Control Systems Technology* 26(4), 1300–1316.

24. Wang, S., W. Ren, and J. Chen, 2016. Fully distributed state estimation with multiple model approach. In *Decision and Control (CDC), 2016 IEEE 55th Conference on*, pp. 2920–2925. IEEE.

28

Application of Ray-Tracing Techniques to Mobile Localization

Francisco Saez de Adana, Oscar Gutiérrez Blanco, Josefa Gómez Pérez, Abdelhamid Tayebi, José Manuel Gómez Pulido, and Antonio del Corte Valiente

CONTENTS

28.1 Introduction

The interest in indoor positioning systems is becoming increasingly important nowadays. Wireless technology is used in multiple applications and has been widely investigated during the last decades. It is a fact that indoor positioning is less accurate under conditions of non-line of sight (NLoS) and multipath propagation. Thus, more effort is needed to improve the accuracy in real scenarios, where the multipath propagation changes constantly due to the presence of people, furniture, etc. Ray-tracing is a good alternative to find the position of a mobile terminal because it takes into account the multiple rays coming from different directions.

Ray-tracing analysis is used for several localization methods in order to obtain information about propagation in for indoor environments [1]. There are many works in the literature that focus on ray-tracing [2–8]. For example, [2] proposes a localization technique that makes an accurate measurement of DoA, ToA, and the Doppler shift by using ray-tracing. In [3], a 2D ray-tracing procedure for the localization of EM field sources in urban environments is explained. The proposed method in [4] combines the spatial characteristics estimated from data measurements and a ray-tracing analysis. The use of a ray-tracing analysis presented herein enables site-specific location using only a single base station. Most of these above systems rely on additional sensory hardware installations and thus are not cost and time efficient.

A fingerprinting technique is also included in the proposed localization method. The information obtained by ray-tracing is stored in a dataset during the first stage of the fingerprinting technique. During the second stage, the position of a mobile station can be calculated by comparing the measured parameters (direction of arrival [DoA] and received signal strength [RSS]) with the values stored in the dataset. The processed signals are available from the wireless devices that comprise the Wi-Fi (IEEE 802.11) and Wi-Max (IEEE 802.16) standards. In this paper, WLAN technology is used because this technique can obtain the required radio map grid spacing to reduce the location error. Moreover, the received signal strength (radio frequency) is available in all WLAN interface equipment.

28.2 The Ray-Tracing Model

Highly accurate ray-tracing analyses using three-dimensional terrain data have been carried out. In these studies, the terminals are accurately positioned using site-specific information for the measurement area. The ray-tracing model that can be obtained is a fully three-dimensional uniform theory of diffraction (UTD) model which includes ray-tracing acceleration techniques for improving the efficiency of the approach [9]. Being a ray-tracing model, the electric field levels can be obtained using the direct, reflected, transmitted, and diffracted fields. Thus, taking into account this variety of effects, the code provides good predictions [9]. These results can be used to examine the effect of varying certain sensing parameters on the precision of the system. The examined parameters include the number of antennas, the position of the antennas, and the number of tracks.

The model uses a representation of the geometry by means of plane facets to describe the environment, and the rays traveling from the transmitter antenna to the observation point are obtained including all the interactions with the geometry. All of the elements of an indoor environment are modeled by facets, and the influence of the electromagnetic properties of the material of each facet is also included. However, the main problem associated with ray-tracing approaches is that they need a great amount of time to obtain the multipath propagation. To avoid this inconvenience, the model uses the angular Z-Buffer algorithm to accelerate the computation process. In this way, a great number of case studies can be simulated in a reasonable amount of time, as shown in [9].

An advantage of using the ray-tracing technique is that, besides obtaining the power level of a series of points, information can also be obtained about multipath effects. This information can be used in a fingerprint method in order to improve the efficiency of the location system. To obtain the fingerprint database, the model employs a 3D UTD propagation analysis and provides the location of the fingerprints in a reference grid.

28.3 The Fingerprinting Technique

Radio wave propagation in indoor environments can be obstructed by physical objects located between the transmitter and the receiver. Thus, some transmitted signals can be degraded because of reflections, refractions, and diffractions from walls, floors, furniture, and people, for example. For this reason, inaccurate results follow from the assumption of LoS conditions between two wireless devices when developing localization methods. This problem is known as NLoS and is very common in indoor scenarios. The easiest way to avoid the effect of NLoS is to put more reference points in the indoor environment to reduce the likelihood that at some point there is no direct line of sight (LoS). The inclusion of more reference points leads to the concept of using a set of several nodes defined over a grid along the scenario under analysis. This technique is known as fingerprinting [10], where each node represents a fingerprint of a known position. Therefore, the unknown positions of the mobile stations are found considering the position of the "nearest" fingerprint. The cost function to determine what fingerprint is the nearest depends on the Euclidean distance considering the information of fingerprints and mobile stations. That is to say, once the database of the fingerprints is set, the localization algorithm computes as many Euclidean distances as there are fingerprints. Finally, when all the nodes are analyzed, the node with the minimum distance is given as the position of the mobile station.

Normally, fingerprints may store information related to the received signals. In a ray-tracing technique it is possible to store the time of arrival (ToA), DoA, and RSS measurements. The key advantage of using this simple technique is that each node can store not only the direct ray but also the reflected and diffracted rays. The additional storage capability becomes a very useful feature to mitigate the effects of NLoS because every propagated ray between transmitter and receiver is considered. Obviously, the accuracy when locating the mobile station is higher in the absence of NLoS than in the presence, but that is an ideal situation and rarely ever occurs in closed areas.

Depending on the kind of information that is used to locate the mobile device, the approaches are classified as follows:

a. *Localization based on time difference of arrival*: Time differences of arrival, as well as the electric field, are obtained in each fingerprint at any point by means of ray tracing by using the direct, reflected, transmitted or diffracted ray, or any combination of these effects. The information provided by the ray-tracing simulation tool is stored in two vectors: R_m corresponds to the relative time delay at every mobile and R_f corresponds to the relative time delay at every fingerprint. The algorithm computes the Euclidean distance between one mobile and all the fingerprints for all the access points. However, only mobiles and fingerprints with the same mapping effects are processed. Finally, the coordinates of the fingerprint that provides the smallest Euclidean distance are returned as the position of the mobile.

b. *Localization based on directions of arrival*: In this case, the data provided by the ray-tracing tool are stored in four vectors. The vectors θ_f and ϕ_f correspond to information on the DoA at every fingerprint (they contain the theta and phi components of the DoA from the M access points at the fingerprint f). The other two vectors, θ_m and ϕ_m, contain the same information but related to the mobile stations. These vectors are calculated at the beginning of the process and stored in the fingerprinting database. In the second stage of the localization process, the Euclidian distance between each mobile station and every fingerprint is calculated. The location of a mobile station is evaluated using the cost function corresponding to this Euclidean distance.

c. *Localization based on the received signal strength*: If the localization method uses the RSS, then the data provided by the ray-tracing tool are stored in two vectors: RSS_f and RSS_m, which contain information about the RSS in each fingerprint as well as information about the RSS of each mobile station. These vectors are also obtained at the beginning of the process and stored in the fingerprinting database. The cost function applied in this case is the Euclidean distance between those two values of the RSS.

28.4 The Combination of Direction of Arrival and Received Signal Strength

An experimental study [11] has proven that for the case of the use of ray-tracing to obtain the fingerprinting information, using the DoA or the ToA are equivalent and produce similar measurement errors. Therefore, in this chapter, the possibility of combining one of them (in this case, the DoA) with the RSS is studied.

In this case, as mentioned earlier, the fingerprinting technique is carried out in two stages. In the first stage, the radio map or RF fingerprinting database for the target environment is obtained. The location fingerprints are obtained by analyzing the relative ray-tracing delay and signal strength from multiple access points over a defined grid. In the second stage, the accuracy obtained in the localization process is analyzed. For this purpose, the developed technique uses a number of mobile stations in the area covered by the radio map and obtains the vector of received power and the directions of ray travel from different access points. In our studies, many mobile stations randomly distributed over the indoor space are considered. The position estimation is made by an algorithm that computes the Euclidean distance between the obtained vectors and each fingerprint in the radio map. The X, Y, and Z coordinates associated with the fingerprint that result in the smallest Euclidean distance are returned as the position estimation of the mobile station using a simple algorithm.

The information to apply the fingerprinting is provided by the ray-tracing tool mentioned in the previous section and is stored in six vectors. Two of them, θ_f and ϕ_f, correspond to information on the angle of arrival at every fingerprint. The first vector contains the theta component of the direction of arrival from the M access points at the fingerprint h, and the second vector contains the phi component of the direction of arrival at the same point. The other two vectors, θ_m and ϕ_m, contain the same information as the mobile stations m. Finally, the last two (RSS_f and RSS_m) vectors contain information about the RSS in each fingerprint as well as information about the RSS of each mobile station. These vectors are calculated at the beginning of the process and stored in the fingerprinting database. In the second stage of the localization process, the Euclidian distance between each mobile station and every fingerprint is calculated by considering the two parameters: the DoA and RSS.

The location of a mobile station is evaluated using the following cost function, which uses the received signal strength and the estimated directions of arrival from the received signals:

$$D(X, Y, Z) = \sum_{n=1}^{N} \eta (DoA_n^{(E)} - DoA_n^{(RT)})^2 + (1 - \eta)(RSS_n^{(E)} - RSS_n^{(RT)})^2 \tag{28.1}$$

where

$$DoA_n^{(E)} - DoA_n^{(RT)} = \sum_{i=1}^{M} \sqrt{\left| (k_x^m(i) - k_x^f(i))^2 + (k_y^m(i) - k_y^f(i))^2 + (k_z^m(i) - k_z^f(i))^2 \right|} \tag{28.2}$$

$$RSS_n^{(E)} - RSS_n^{(RT)} = \sum_{i=1}^{M} \sqrt{\left| (E_x^m(i) - E_x^f(i))^2 + (E_y^m(i) - E_y^f(i))^2 + (E_z^m(i) - E_z^f(i))^2 \right|} \tag{28.3}$$

M is the number of access points, E is the received electric field value, N is the number of rays, and η is a weighting factor that indicates the ratio between the correlation of DoAs and RSSs and $k_x = \cos\phi\sin\theta$, $k_y = \sin\phi\sin\theta$, $k_z = \cos\theta$.

In the first part of expression (28.1), the DoA of every received ray in fingerprints and every received ray in mobile stations is compared. This comparison is made according to the number of access points; that is, if a fingerprint receives just one signal from access point number 1 and a mobile station receives just one signal from access point number 3, this information is not taken into account. This restriction is applied to all cases in this study. Note that the directions of arrival between fingerprints and the mobile station are compared as long as they receive the same number of rays from the same access points. Thus, a cost function is only applied if this condition is satisfied. The process is the same for all the mobile stations; therefore, the estimated position of each mobile station can be calculated.

In the second part of expression (28.1), the method compares the electric field values received from the signals from all the access points. These levels are compared with the electric field values previously stored in the fingerprint database. In this case, the comparison is only made if the number of rays received in the fingerprint is the same as the number of received rays in the mobile station. In other words, the location algorithm computes the Euclidean distance between the DoA and RSS samples received in the unknown position and each fingerprint in the database or radio map obtained using the ray-tracing model.

Changing (X, Y, Z) inside the testing area, which is the point that minimizes the cost function, is used to estimate the position of the mobile station. Therefore, the estimated position $(\hat{X}, \hat{Y}, \hat{Z})$ is obtained as follows:

$$(\hat{X}, \hat{Y}, \hat{Z}) = \min D(X, Y, Z). \tag{28.4}$$

According to this expression (28.4), the position of the mobile station corresponds with the fingerprint whose Euclidian distance is the smallest. In order to measure the efficiency of this method, the physical distance between the estimated position $(\hat{X}, \hat{Y}, \hat{Z})$ of every mobile station and its real position (X, Y, Z) over the grid is calculated.

28.5 Experimental Results

In order to evaluate the localization performance of the proposed approach, an indoor environment (the Politecnica building in Madrid) has been analyzed with different grid densities. The geometrical model of the scene is shown in Figure 28.1. The tests were performed for LoS and NLoS situations.

The experiment considers one grid consisting of 72 × 72 fingerprints at a frequency of 2.4 GHz. The simulations also use nine access points and 99 mobile stations randomly distributed over the grids. The grid is formed by evenly distributed points, representing the 72 × 72 fingerprints; the nine triangles represent the nine access points and the circles represent the 99 mobile stations. Note that X and Y coordinates are measured in meters. The Z coordinate is always set to 1.5 meters and corresponds with the height of the mobile station from the floor. Despite this constant Z value, the method is fully 3D because the rays can reach any height (reflection on the floor, ceiling, walls, diffraction, etc.). The nearest fingerprint to each mobile station is found by using this method.

The information available in each fingerprint and mobile station are parameters that affect the precision and reliability of the location process. For this reason, a study taking into account the number of received rays in fingerprints and in mobile stations has been

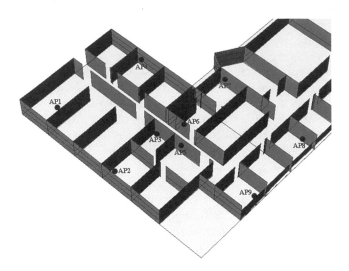

FIGURE 28.1
Geometrical model of the Politecnica building.

performed. This information generated using the 72 × 72 grid is analyzed; the distance
between each fingerprint is 0.4 meters.

Figures 28.2 and 28.3 show the results for various numbers of rays for the Politecnica
building, using a 72 × 72 grid. The mean error is shown in Figure 28.2 and variance of
the mean error is shown in Figure 28.3. The mean error is calculated as the Euclidian
distance between the real position of the mobile station and its estimated position. The
experimental results give a mean cost of about 0.3 m, as shown in Figure 28.2. Results

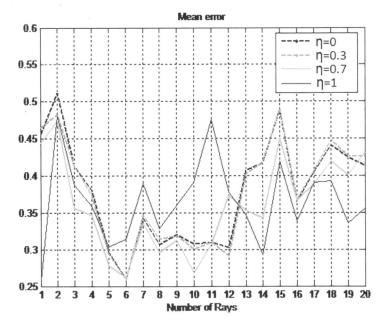

FIGURE 28.2
Mean error (in meters) as a function of the number of rays for different values of η.

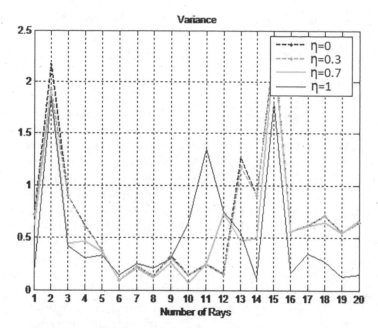

FIGURE 28.3
Variance as a function of the number of rays for different values of η.

are reasonable, since separation between each fingerprint is 0.4 m. The best results are obtained by considering 6 or 10 rays. In both cases, the mean error value is 0.261 m and the variance value is 0.09 m when η is different than 1. In other words, if the method only considers the DoA ($\eta = 1$), the results are not as accurate as other cases; other values of η do not affect the performance of the method. It is notable that the results are very similar for the rest of the η values.

28.6 Effect of Body Shadowing

Although there are several works that analyze the influence of the human body on the radio propagation channel, there are not enough contributions dealing with the effects of human body shadowing on localization methods. Due to the fact that most of the papers related to positioning methods in the literature do not consider the influence of the presence of people in indoor environments, the method presented above has been modified including this important and fundamental aspect. Therefore, it can be stated that, including this effect, the proposed method is more realistic because the presence of moving people is very common in indoor environments such as offices, airports, hotels, etc. It is almost impossible for those places to be empty, without people or furniture.

Positioning accuracy is the main goal for the location approaches. Erroneous locations are due to parameter estimation errors, oversimplified assumptions about the propagation channel, multipath effects, and NLoS conditions. Usually, the mobile station may not be visible from a determinate access point. Furthermore, in indoor scenarios characterized by dense multipath and mixed LoS and NLoS conditions, these errors become more critical. For these reasons, a positioning method that considers these unwanted effects is compulsory.

It is well known that the movement of people within indoor scenarios can cause temporal channel variations. That is to say, the presence of people in the indoor scenario can influence the RSS distribution in different ways: it may cause the RSS to decrease, increase, or remain constant. The explanation of these changes is the scattering and shadowing effects of human bodies in indoor environments. The user acts as an obstacle that obstructs the propagation path between the access point and the mobile station and therefore causes changes in the RSS value. Moreover, the WLAN signal is absorbed when the body obstructs the signal path, causing extra attenuation leading to a lower RSS value.

It is worth noting that the influence of body shadowing is included independently of the ray-tracing technique. It is considered through a statistical model that is applied to every received ray in mobile stations and fingerprints. The RSS value is attenuated by a certain level that represents the effects of body shadowing. The statistical model applied is based on [12]. In that paper, the effects of random human traffic on path loss in the communications channel at 1.8 GHz have been experimentally measured. Such effects account for an increase in path loss up to 5 dB when the transmitting and receiving antennas are not in very close proximity to the human body.

The parameters of the Gamma probability density function have been extracted from a measurement campaign. According to [12], before each measurement session, samples were taken in the absence of human traffic to know the attenuation in these conditions and then calculate the attenuation in excess. The attenuation in excess over the average without traffic yielded the following parameter statistics:

Mean attenuation in excess: 2.6 dB, standard deviation: 2.09 dB, maximum attenuation in excess: 9.18 dB, minimum attenuation in excess: −1.02 dB. The parameters of the Gamma distribution are obtained as follows:

$$E[X] = k \cdot \theta = 2.6 \tag{28.5}$$

$$V(X) = k \cdot \theta^2 = 2.09^2 \tag{28.6}$$

The resulting values are $k = 1.54$ and $\theta = 1.68$.

As mentioned earlier, [12] demonstrates that the Gamma distribution models the attenuation in excess due to body shadowing quite well. It is worthwhile to point out that the results are very similar under LoS or NLoS conditions. Therefore, the attenuation due to the presence of people in the area of propagation is independent of the direct vision conditions between the access points and the mobile stations.

RSS measurements, as noted previously, are subjected to random errors due to channel nonidealities such as multipath and shadowing. In the absence of such nonidealities, RSS measurements accurately represent the distances between the unknown node and the reference nodes. For this reason, a statistical attenuation model must be applied over the RSS values. Because the cost function in Equation 28.1 guesses an ideal channel, a realistic cost function that considers nonidealities must become a new expression.

Note that Equation 28.1 compares the non-attenuated RSS values, that is to say, the RSS is computed by considering the electric field values received in fingerprints and mobile stations from the access points without taking into account the presence of moving people. In order to include these realistic effects, RSS values are attenuated applying the conclusions obtained in [12]. The statistical behavior of the attenuation in excess introduced by the presence of people in an indoor environment can be modeled as a Gamma function.

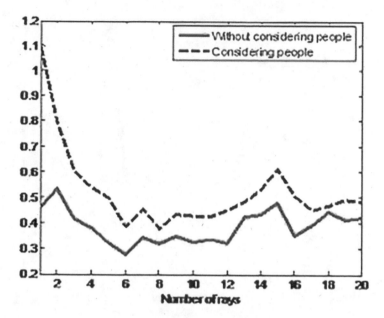

FIGURE 28.4
Comparison of the mean error (in meters) with and without the effect of moving people.

Therefore, only a portion of the total power generated at the access points is received at each fingerprint or mobile station when considering body shadowing. The percentage of attenuation is obtained by using the Gamma function as follows: first, a random number between 0 and 10 is calculated; second, the Gamma probability density function is evaluated considering the random number. As shown in [12], the resulting value is between 0 and 0.28; thus the worst scenario for the RSS level is to suffer an attenuation of 28%. However, this attenuation may happen only when the random number is near 1. If the random number is higher than 5, the attenuation percentage will be practically zero. Once these previous considerations have been performed, the part of the cost function associated to the RSS is:

$$D(X, Y, Z) = \sum_{n=1}^{N} ((1 - \Gamma(k, \theta))RSS_n - (1 - \Gamma(k, \theta))RSS_n^{(RT)})^2 \qquad (28.7)$$

Figures 28.4 and 28.5 show the mean error and the variance of the Politecnica case, including the effect of the human shadowing.

28.7 Conclusions

Under NLoS conditions and multipath contributions, indoor localization increases in error and it is necessary to find innovative solutions allowing more accurate measurements. Moreover, indoor multipath characteristics can vary considerably—they depend on building dimensions, the transmitting/receiving range, and the presence or absence of

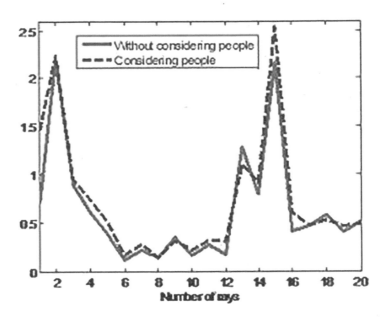

FIGURE 28.5
Comparison of the variance with and without the effect of moving people.

furniture. Due to propagation in complex indoor environments, multiple signals from different directions can be used to determine the location of the mobile terminal.

This chapter has presented an approach using the multipath behavior by combining ray-tracing measurements (DoA and RSS) and a fingerprinting technique to determine the location of a mobile station. The location algorithm computes the Euclidean distance between the DoA and RSS samples received at an unknown position and each fingerprint is stored in the database. The results of the detection method are compared for several numbers of rays.

The proposed approach is quite simple and decreases errors due to the absence of LoS and multipath reflections. Also, the approach provides accurate results by using the widely available RF-based wireless LAN infrastructure. Its advantages are that it is very useful for practical applications and there is enhanced accuracy with respect to similar approaches, especially for mixed LoS/NLoS conditions.

A fingerprinting technique using the direction of arrival and the received signal strength is relatively easy to implement and provides good results for the indoor location process. Also, the analysis presented above concludes that the method does not provide better results if the number of rays is increased and different values for η are used; it is notable that the results are very similar for all values of η. For the case where η is equal to 1, the method provides less accurate results.

On the other hand, while a great deal of time has been spent developing new localization methods, the effects of human body interactions on the position estimation remain unexplored. There are several factors that affect the propagation channel in an indoor environment such as path loss, shadowing, and multipath fading. These factors directly affect the quality of the received signal. This chapter has shown how to consider the effects of human body shadowing to provide a realistic and accurate estimation of the mobile station position in indoor scenarios.

References

1. N. Yarkoni and N. Blaunstein, "Prediction of Propagation Characteristics in Indoor Radio Communication Environments," *Progress in Electromagnetics Research, PIER* 59, 151–174, 2006.
2. N.J. Thomas, D.G.M. Cruickshank, and D.I. Laurenson, "Calculation of Mobile Location Using Scatterer Information," *Electronics Letters*, Vol. 37, No. 19, September 13, 2001.
3. S. Coco, A. Laudani, and L. Mazzurco, "A Novel 2-D Ray Tracing Procedure for the Localization of EM Field Sources in Urban Environment," *IEEE Transactions on Magnetics*, Vol. 40, No. 2, March 2004.
4. S. Kikuchi, A. Sano, and H. Tsuji, "Blind Mobile Positing in Urban Environment Based on Ray-Tracing Analysis," *EURASIP Journal on Applied Signal Processing*, Article ID38989, 1–12, 2006.
5. N. Sedaghat Alvar, A. Ghorbani, and H. Amindavar, "A Novel Hybrid Approach to Ray-Tracing Acceleration Based on Pre-Processing & Bounding Volumes," *Progress In Electromagnetics Research, PIER* 82, 19–32, 2008.
6. Y. Cocheril and R. Vauzelle, "A New Ray-Tracing Based Wave Propagation Model Including Rough Surfaces Scattering," *Progress In Electromagnetics Research, PIER* 75, 357–381, 2007.
7. C.H. Teh, F. Kung, and H.T. Chuah, "A Path-Corrected Wall Model for Ray-Tracing Propagation Modeling," *Journal of Electromagnetic Waves and Applications*, Vol. 20, No. 2, 207–214, 2006.
8. K.-S. Jin, "Fast Ray Tracing Using a Space-Division Algorithm for RCS Prediction," *Journal of Electromagnetic Waves and Applications*, Vol. 20, No. 1, 119–126, 2006.
9. F. Saez de Adana, O. Gutiérrez, I. González, J. Pérez, and M.F. Cátedra, "Propagation Model Based on Ray-Tracing for the Design of Personal Communication Systems in Indoor Environments," *IEEE Transactions on Vehicle Technology*, Vol. 49, No. 6, November 2000.
10. S.-H. Fang, T.-N. Lin, and K.-C. Lee, "A Novel Algorithm for Multipath Fingerprinting in Indoor WLAN Environments," *IEEE Transactions on Wireless Communications*, Vol. 7, No. 9, September 2008.
11. J. Gómez, A. Tayebi, A. del Corte, F. Saez de Adana, and O. Gutierrez. "A Comparative Study of Localization Methods in Indoor Environments," *Wireless Personal Communications*, Vol. 72, no. 4, October 2013.
12. C. Perez, M.L. Mediavilla, and M. C. Diez, "Efectos del Tráfico de Personas sobre la Atenuación en el Canal de Propagación en Interiores," *Simposium de la Union Internacional de Radio (URSI)*, 1997 [in Spanish].

29

Cooperative Localization in Wireless Sensor Networks via Kernel Isomap

Jyoti Kashniyal, Shekhar Verma, and Krishna Pratap Singh

CONTENTS

29.1 Introduction

Availability of low-cost sensors has made possible the use of wireless sensor networks for a wide range of applications. Most of these applications include tracking and monitoring tasks such as in military for target tracking; biomedical for patient monitoring; environment for disaster relief, precision agriculture; and industry for warehouse automation, etc. In a wireless sensor network, sensors cooperate with each other to fulfill these tasks (Mahalik, 2007). Location awareness is an essential requirement in these applications. The data sensed by a node must be accompanied with its location information. The knowledge of sensor positions is also relevant to geographical routing, cluster formation, and sensing coverage problems (Manisekaran and Venkatesan, 2014). Global positioning systems (GPS) or manual sensor placement are the most accurate positioning techniques. However, the use of expensive GPS or manual deployment for a large number of sensors is limited to unobstructed or safe environments respectively. Thus, there is a demand for an accurate and efficient localization technique through which the sensors are self-capable of localizing themselves using neighborhood information. The intersensor information can be in the form of distance, angle, hop count, etc. In traditional localization techniques like trilateration and multilateration, an unknown node localizes itself using distance information from three or more anchor nodes. Anchors represent a small fraction of nodes in the network which are aware of their position information *a priori* either through GPS or manual deployment. Due to limited communication range, as most of the unknown nodes do not have sufficient anchor information; they remain unlocalized. In cooperative or collaborative localization techniques, all sensor pairs of anchors as well as non-anchors collaborate with each other to localize themselves (Shaoo and Hwang, 2011). They ease the compulsion of an unknown node to be directly connected to three or more anchor nodes. In collaborative multilateration (Savvides et al., 2002), an unknown node having insufficient anchor neighbors localizes itself using information from more than one-hop distance anchors. Manifold learning techniques like Isomap (Tenenbaum and Silva, 2000), locally linear embedding (Roweis and Saul, 2000), Laplacian eigenmaps (Belkin and Niyogi, 2002), etc. can also be used for cooperative localization in wireless sensor networks (Patwari and Hero, 2004). They solve the sensor localization problem by considering it as a dimensionality reduction problem.

Isomap is a dimensionality reduction technique based on classical multidimensional scaling. It performs collaborative localization using the concept of geodesic distances for limited communication range sensors. However, unlike distributed algorithms, it operates in a batch mode and is inefficient in localizing a newly arrived node in the network (Li et al., 2013). Kernel Isomap (Choi and Choi, 2004) possess the generalization property using the similar spirit as in kernel principal component analysis (KPCA) (Scholkopf, 1997). It has a data-driven, positive semidefinite kernel matrix obtained after adding a constant to the geodesic distance kernel matrix. Similar to kernel Isomap, an out-of-sample extension of Isomap (Bengio et al., 2004), and Landmark Isomap (Silva and Tenenbaum, 2003) have the capability of positioning newly arrived nodes. However, in the latter two, the kernel matrix is not guaranteed to be positive definite. Other kernel-based methods like kernel principal component analysis (PCA) and kernel locality preserving projection (KLPP) (Wang et al., 2009) use an externally defined kernel like Gaussian, polynomial, etc., whose selection is based on network parameters.

In this chapter, we discuss the sensor localization problem using kernel Isomap. It is capable of localizing the new nodes without recomputing the positions for already localized nodes. Section 29.2 describes problem statement. In Section 29.3, localization using Isomap and its limitations are discussed; Section 29.4 explains kernel Isomap with several constant shifting methods and its generalization property. In Section 29.5, performance evaluation of kernel Isomap is done using simulation results. Finally, Section 29.6 concludes the chapter.

29.2 Problem Statement

Consider N sensor nodes in a two-dimensional region having communication range r_c. The first N_a nodes are anchor nodes ($N_a \ll N$). Distance d between the neighboring nodes is measured using received signal strength indicator (RSSI) which does not require any additional hardware. However, RSSI measurements are noisy, as they are significantly affected by channel environment (Shen et al., 2005). The problem of localization is to find a mapping that assigns coordinates to $(N - N_a)$ nodes using available distance and anchor node information.

Isomap, a nonlinear manifold learning technique, performs cooperative localization. Distance information between all pairs of nodes is gathered and then a classical multidimensional scaling (Cox and Cox, 2000) is applied to obtain their coordinates. However, if a new node arrives in the network, Isomap needs to recalculate the positions of all the preceding nodes in order to localize it. Reiterating the centralized operation each time a new node arrives in the network is computationally demanding. Kernel-based localization algorithms possess the capability to localize these new arrivals efficiently, but the selection of a suitable kernel which satisfies the network conditions well is another challenge.

29.3 Localization Using Isomap

Considering the localization problem as a graph $G(N, d)$ realization problem, Isomap determines the distance between all pairs of nodes in the graph. It employs the concept of geodesic distances to estimate the distances between non-neighboring nodes. Geodesic distance between two nodes is the shortest path between them in the adjoining sensor network graph obtained by Floyd's or Dijkstra's procedure. For nodes within communication range, it is represented by the estimated RSSI distance, whereas for non-neighboring nodes it is the sum of the weights of the shortest paths in the r_c-neighborhood. Figure 29.1 shows geodesic and Euclidean distances between nodes. Nodes which are within range r_c, their intersensor distance is shown by Euclidean distance (AB, BC), whereas for non-neighboring nodes (A,C), geodesic distance (AB + BC) is used, since Euclidean distance (the dotted line) between them cannot be estimated.

Once the distance between all pair nodes has been gathered, classical multidimensional scaling is applied to obtain their relative coordinates. The relative coordinates of the nodes can be transformed into respective global coordinates by using the true positions of three or more non-collinear anchor nodes (Gower, 1975). The steps of Isomap can be described as follows:

1. Create a neighborhood graph where edge length between neighbor nodes is set as their measured distance d_{ij}.

2. Compute geodesic distance matrix D as the sum of edge weights along the shortest path between all pairs of nodes.

3. Construct a matrix $K(D^2) = -1/2(HD^2H)$, where H is the centering matrix, given by $H = I - (11')/N$ and I is the N-dimensional identity matrix and 1 is a $NX1$ column vector.

4. Compute the top two eigenvalues and corresponding eigenvectors of K, to determine two-dimensional relative coordinate matrix as $Y = \lambda^{1/2}V^T$.

5. Transform relative coordinate matrix Y to respective absolute coordinates with the help of three or more anchor position information.

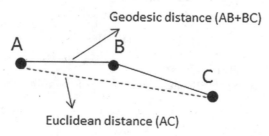

FIGURE 29.1
Geodesic and Euclidean distance.

Isomap performs a centralized localization. It means the distance information between all the pairs of nodes is needed to perform localization. Whenever a new node arrives in the network, in order to localize it Isomap repeats the energy-consuming centralized computation on the new as well as previous nodes.

29.4 Localization Using Kernel Isomap

Whenever a new node arrives in the network, running a batch version to determine its location is computationally demanding since it discards the previously computed results. Kernel Isomap (Choi and Choi, 2004), an extension of Isomap, possesses the generalization property. It has the capability to localize new nodes without discarding former computations using the kernel trick as in kernel PCA (Williams, 2002). The geodesic distance kernel matrix in Isomap is not always positive semidefinite. Kernel Isomap employs a constant adding method (step 4) to ensure its positive semidefiniteness (Cailliez, 1983). Steps of kernel Isomap are as follows:

1. Create a neighborhood graph where edge length between neighbor nodes is set as their measured distance d_{ij}.

2. Compute geodesic distance matrix as the sum of edge weights along the shortest path between all pairs of nodes.

3. Construct a matrix $K(D^2) = -1/2(HD^2H)$, where H is the centering matrix, given by $H = I - (11')/N$ and I is the N-dimensional identity matrix and 1 is a $NX1$ column vector.

4. Add a constant c to the geodesic distances d_{ij} as $d'_{ij} = |d_{ij} + c(1 - \delta_{ij})|$ such that $K' = K(D'^2)$ is guaranteed to be positive semidefinite. δ_{ij} is Kronecker delta and c is obtained by constant shifting method.

5. Compute top two eigenvalues and corresponding eigenvectors of K', and construct the two-dimensional coordinate matrix as $Y = \lambda^{1/2}V^T$.

6. Transform relative coordinate matrix Y to absolute coordinates with the help of three or more anchor position information.

29.4.1 Constant Shifting Methods

Constant shifting methods add a constant to the geodesic distances d_{ij} such that the geodesic kernel matrix is guaranteed to be positive semidefinite. One way to find a constant is to compute the largest eigenvalue, c^*, of the matrix

$$\begin{bmatrix} 0 & 2K(D^2) \\ -I & -4K(D) \end{bmatrix}$$

s.t. $d'_{ij} = d_{ij} + c$ where $c \geq c^*$ and $i \neq j$.

The other way is to add the smallest eigenvalue, c', of the matrix

$$\begin{bmatrix} 0 & 2K(D^2) \\ -I & -4K(D) \end{bmatrix}$$

s.t. $d'_{ij} = |d_{ij} + c|$ where $c < c'$ and $i \neq j$.

The constant shifted geodesic distances d'_{ij} has a Euclidean representation in the feature space.

29.4.2 Kernel Trick

The kernel trick states that the (i, j)th element in a mercer kernel matrix K can be represented as

$$K_{ij} = k(x_i, x_j) = \phi^T(x_i)\phi(x_j)$$

where $\phi(x_j)$ represents feature space coordinates of x_j obtained by non-linear mapping $\phi(\cdot)$.

29.4.3 Generalization Property

Using the kernel trick, the position of a new point in kernel PCA is computed as the projection of the centered data matrix onto the normalized eigenvectors of the respective covariance matrix. Further, if the kernel function is isotropic, kernel PCA can be interpreted as a form of metric multidimensional scaling (MDS) (Williams, 2002). Thus, kernel Isomap, having a positive semidefinite kernel matrix, inherits the generalization property of kernel PCA. It can be used to efficiently localize newly arrived nodes in a wireless sensor network. The ith coordinate of position Y_l of a new node t_l can be computed as follows:

$$[Y_l]_i = \frac{1}{\sqrt{\lambda_i}} \sum_{j=1}^{N} [v_i]_j k'(t_l, x_j)$$

where λ_i is the ith eigenvalue of K, v_i is the ith eigenvector of K, $[\cdot]_j$ represents the jth element of eigenvector and $k'(t_l, x_j)$ represents the geodesic kernel entry for the new node t_l.

$$k'(t_l, x_j) = \phi^T(t_l)\,\phi(x_j) = -1/2\left[D_{lj}'^2 - K'_{jj} - \frac{1}{N}\sum_{i=1}^{N}\left(D_{li}'^2 - K'_{ii}\right)\right]$$

where d'_{ij} is the constant-shifted geodesic distance between the new node t_l and previously localized node x_j.

29.5 Simulation Results

The performance of the kernel Isomap for localization in wireless sensor networks is evaluated under various network configurations. Nodes having communication range $= 10$ m are randomly and uniformly deployed over a $100*100$ m^2 region. For all scenarios, the number of anchors is limited to 50. The estimated sensor distance $\tilde{d}_{ij}$ between neighboring nodes is obtained by blurring the true distance d_{ij} by Gaussian noise $\mathcal{N}(0, \sigma)$ with zero mean and standard deviation σ:

$$\tilde{d}_{ij} = d_{ij} + \mathcal{N}(0, \sigma)$$

The reported results are the average of over 100 simulations.

Figure 29.2 shows the connectivity graph for 300 nodes. The neighboring nodes are connected by an edge. Figure 29.3 shows final position estimation by Isomap and kernel Isomap, respectively. Both Isomap and kernel Isomap show similar performances.

Figure 29.4 shows mean localization error of kernel Isomap as a function of node density. Distance measurement noise is fixed to 10% of communication range. As the node

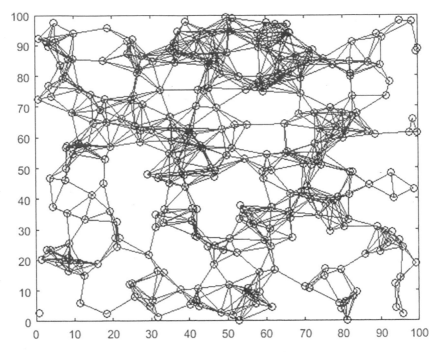

FIGURE 29.2
Connectivity diagram for 300 nodes with communication range = 10 m.

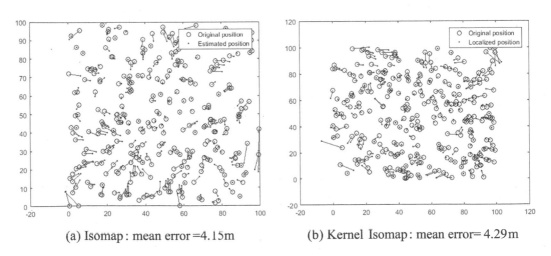

(a) Isomap: mean error = 4.15m   (b) Kernel Isomap: mean error = 4.29m

FIGURE 29.3
Location estimation for random network using distances between neighbors with 10% range error.

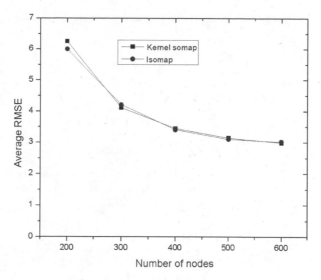

FIGURE 29.4
Localization error with different number of nodes.

density increases, localization error decreases because of the rise in the available distance information.

Figure 29.5 shows mean localization error of kernel Isomap as a function of error in distance estimation. The number of nodes in the network is fixed to 300. The higher the error in distance measurement, the lower is the accuracy of the algorithm.

Figure 29.6 shows runtime comparison for localizing varying new node density in the network. The simulation is performed on a machine with an Intel® Core™ i7-3770 CPU @ 3.40 GHz and 12 GB of RAM. Initially, there are 100 nodes in the network. Both Isomap and kernel Isomap use the centralized approach to localize them, and take similar amount of time. But as the new nodes arrive in the network, Isomap consumes more time than kernel

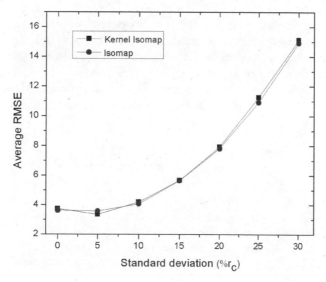

FIGURE 29.5
Localization error as a function of standard deviation.

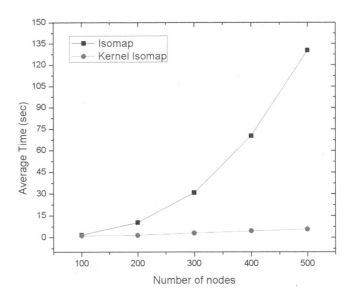

FIGURE 29.6
Comparison of running time for positioning new nodes in the network.

Isomap to localize them. This is because of the projection property of kernel Isomap to localize newly arrived nodes without discarding previous computations, whereas Isomap relocalizes the previous nodes along with newly arrived nodes.

29.6 Conclusion

The chapter explains cooperative localization using kernel Isomap. Kernel Isomap, an extension of Isomap, possesses generalization property. A comparison between the two indicates that the positive definite matrix in kernel Isomap does not improve its accuracy, but its run-time efficacy to position newly arrived nodes in the network increases remarkably. The generalization property of kernel Isomap utilizes the previously computed results to localize the new nodes, whereas Isomap repeats the whole centralized procedure whenever a new node arrives in the network, rejecting previous computations.

References

Belkin, M. and Niyogi, P. Laplacian eigenmaps and spectral techniques for embedding and clustering. In *Advances in Neural Information Processing Systems*, 2002, pp. 585–591.

Bengio, Y., Paiement, J.F., Vincent, P., Delalleau, O., Roux, N.L. and Ouimet, M. Out-of-sample extensions for LLE, Isomap, MDS, eigenmaps, and spectral clustering. In *Advances in Neural Information Processing Systems*, 2004, pp. 177–184.

Cailliez, F. The analytical solution of the additive constant problem. *Psychometrika*, 48(2), 1983: 305–308.

Choi, H. and Choi, S. Kernel Isomap. *Electronics Letters*, 40(25), 2004: 1612–1613.

Cox, T.F. and Cox, M.A. *Multidimensional Scaling.* Chapman and Hall/CRC, 2000.

Gower, J.C. Generalized Procrustes analysis. *Psychometrika*, 40(1), 1975: 33–51.

Li, B., He, Y., Guo, F. and Zuo, L. A novel localization algorithm based on Isomap and partial least squares for wireless sensor networks. *IEEE Transactions on Instrumentation and Measurement*, 62(2), 2013: 304–314.

Mahalik, N.P. *Sensor Networks and Configuration.* Springer-Verlag, Berlin Heidelberg, 2007.

Manisekaran, S.V. and Venkatesan, R. Cluster-based architecture for range-free localization in wireless sensor networks. *International Journal of Distributed Sensor Networks*, 10(4), 2014: 963473.

Patwari, N. and Hero, A.O. Manifold learning algorithms for localization in wireless sensor networks. In *Proceedings of International Conference on Acoustics, Speech, and Signal Processing (ICASSP'04)*, 2004, Vol. 3, pp. iii-857.

Roweis, S.T. and Saul, L.K. Nonlinear dimensionality reduction by locally linear embedding. *Science*, 290(5500), 2000: 2323–2326.

Sahoo, P.K. and Hwang, I. Collaborative localization algorithms for wireless sensor networks with reduced localization error. *Sensors*, 11(10), 2011: 9989–10009.

Savvides, A., Park, H. and Srivastava, M.B. The bits and flops of the n-hop multilateration primitive for node localization problems. In *Proceedings of the 1st ACM International Workshop on Wireless Sensor Networks and Applications*, 2002, pp. 112–121.

Schölkopf, B., Smola, A. and Müller, K.R. Kernel principal component analysis. In *International Conference on Artificial Neural Networks*, October 1997, pp. 583–588.

Shen, X., Wang, Z., Jiang, P., Lin, R. and Sun, Y. Connectivity and RSSI based localization scheme for wireless sensor networks. In *International Conference on Intelligent Computing*. 2005, pp. 578–587.

Silva, V.D. and Tenenbaum, J.B. Global versus local methods in nonlinear dimensionality reduction. In *Advances in Neural Information Processing Systems*, 2003, pp. 721–728.

Tenenbaum, J.B., De Silva, V. and Langford, J.C. A global geometric framework for nonlinear dimensionality reduction. *Science*, 290(5500), 2000: 2319–2323.

Wang, C., Chen, J., Sun, Y. and Shen, X. A graph embedding method for wireless sensor networks localization. In *IEEE Conference on Global Telecommunications*, 2009, pp. 1–6.

Williams, C.K. On a connection between kernel PCA and metric multidimensional scaling. *Machine Learning*, 46(1–3), 2002: 11–19.

Index